AF449373

Convex Optimization in Signal Processing and Communications

Over the past two decades there have been significant advances in the field of optimization. In particular, convex optimization has emerged as a powerful signal-processing tool, and the range of applications continues to grow rapidly. This book, written by a team of leading experts, sets out the theoretical underpinnings of the subject and provides tutorials on a wide range of convex-optimization applications. Emphasis throughout is placed on cutting-edge research and on formulating problems in convex form, making this an ideal textbook for advanced graduate courses and a useful self-study guide.

Topics covered:

- automatic code generation
- graphical models
- gradient-based algorithms for signal recovery
- semidefinite programming (SDP) relaxation
- radar waveform design via SDP
- blind source separation for image processing
- modern sampling theory
- robust broadband beamforming
- distributed multiagent optimization for networked systems
- cognitive radio systems via game theory
- the variational-inequality approach for Nash-equilibrium solutions

Daniel P. Palomar is an Assistant Professor in the Department of Electronic and Computer Engineering at the Hong Kong University of Science and Technology (HKUST). He received his Ph.D. from the Technical University of Catalonia (UPC), Spain, in 2003; was a Fulbright Scholar at Princeton University during 2004–2006; and has since received numerous awards including the 2004 Young Author Best Paper Award by the IEEE Signal Processing Society.

Yonina C. Eldar is a Professor in the Department of Electrical Engineering at the Technion, Israel Institute of Technology, and is also a Research Affiliate with the Research Laboratory of Electronics at MIT. She received her Ph.D. from the Massachusetts Institute of Technology (MIT) in 2001. She has received many awards for her research and teaching, including, the Wolf Foundation Krill Prize for Excellence in Scientific Research, the Hershel Rich Innovation Award and the Muriel & David Jacknow Award for Excellence in Teaching.

Convex Optimization in Signal Processing and Communications

Edited by

DANIEL P. PALOMAR
Hong Kong University of Science and Technology

and

YONINA C. ELDAR
Technion – Israel Institute of Technology

CAMBRIDGE UNIVERSITY PRESS
Cambridge, New York, Melbourne, Madrid, Cape Town, Singapore,
São Paulo, Delhi, Dubai, Tokyo

Cambridge University Press
The Edinburgh Building, Cambridge CB2 8RU, UK

Published in the United States of America by Cambridge University Press, New York

www.cambridge.org
Information on this title: www.cambridge.org/9780521762229

© Cambridge University Press 2010

First published 2010

Printed in the United Kingdom at the University Press, Cambridge

A catalog record for this publication is available from the British Library

ISBN 978-0-521-76222-9 Hardback

Contents

Contributors

Sergio Barbarossa
University of Rome – La Sapienza
Italy

Amir Beck
Technion – Israel Institute
   of Technology
Haifa
Israel

Stephen Boyd
Stanford University
California
USA

Tsung-Han Chan
National Tsing Hua University
Hsinchu
Taiwan

Tsung-Hui Chang
National Tsing Hua University
Hsinchu
Taiwan

Chong-Yung Chi
National Tsing Hua University
Hsinchu
Taiwan

Joachim Dahl
Anybody Technology A/S
Denmark

Yonina C. Eldar
Technion – Israel Institute of Technology
Haifa
Israel

Amr El-Keyi
Alexandia University
Egypt

Francisco Facchinei
University of Rome – La Sapienza
Rome
Italy

Alex B. Gershman
Darmstadt University of Technology
Darmstadt
Germany

Yongwei Huang
Hong Kong University of Science
   and Technology
Hong Kong

Thia Kirubarajan
McMaster University
Hamilton, Ontario
Canada

Zhi-Quan Luo
University of Minnesota
Minneapolis
USA

Wing-Kin Ma
Chinese University of Hong Kong
Hong Kong

Antonio De Maio
Università degli Studi di Napoli –
    Federico II
Naples
Italy

Jacob Mattingley
Stanford University
California
USA

Tomer Michaeli
Technion – Israel Institute
    of Technology
Haifa
Israel

Angelia Nedić
University of Illinois at
    Urbana-Champaign
Illinois
USA

Asuman Ozdaglar
Massachusetts Institute of Technology
Boston, Massachusetts
USA

Daniel P. Palomar
Hong Kong University of
    Science and Technology
Hong Kong

Jong-Shi Pang
University of Illinois
    at Urbana-Champaign
Illinois
USA

Michael Rübsamen
Darmstadt University
    of Technology
Darmstadt
Germany

Gesualdo Scutari
Hong Kong University of Science
    and Technology
Hong Kong

Anthony Man-Cho So
Chinese University of Hong Kong
Hong Kong

Jitkomut Songsiri
University of California
Los Angeles, California
USA

Marc Teboulle
Tel-Aviv University
Tel-Aviv
Israel

Lieven Vandenberghe
University of California
Los Angeles, California
USA

Yue Wang
Virginia Polytechnic Institute
    and State University
Arlington
USA

Yinyu Ye
Stanford University
California
USA

Shuzhong Zhang
Chinese University of Hong Kong
Hong Kong

Preface

The past two decades have witnessed the onset of a surge of research in optimization. This includes theoretical aspects, as well as algorithmic developments such as generalizations of interior-point methods to a rich class of convex-optimization problems. The development of general-purpose software tools together with insight generated by the underlying theory have substantially enlarged the set of engineering-design problems that can be reliably solved in an efficient manner. The engineering community has greatly benefited from these recent advances to the point where convex optimization has now emerged as a major signal-processing technique. On the other hand, innovative applications of convex optimization in signal processing combined with the need for robust and efficient methods that can operate in real time have motivated the optimization community to develop additional needed results and methods. The combined efforts in both the optimization and signal-processing communities have led to technical breakthroughs in a wide variety of topics due to the use of convex optimization. This includes solutions to numerous problems previously considered intractable; recognizing and solving convex-optimization problems that arise in applications of interest; utilizing the theory of convex optimization to characterize and gain insight into the optimal-solution structure and to derive performance bounds; formulating convex relaxations of difficult problems; and developing general purpose or application-driven specific algorithms, including those that enable large-scale optimization by exploiting the problem structure.

This book aims at providing the reader with a series of tutorials on a wide variety of convex-optimization applications in signal processing and communications, written by worldwide leading experts, and contributing to the diffusion of these new developments within the signal-processing community. The goal is to introduce convex optimization to a broad signal-processing community, provide insights into how convex optimization can be used in a variety of different contexts, and showcase some notable successes. The topics included are automatic code generation for real-time solvers, graphical models for autoregressive processes, gradient-based algorithms for signal-recovery applications, semidefinite programming (SDP) relaxation with worst-case approximation performance, radar waveform design via SDP, blind non-negative source separation for image processing, modern sampling theory, robust broadband beamforming techniques, distributed multiagent optimization for networked systems, cognitive radio systems via game theory, and the variational-inequality approach for Nash-equilibrium solutions.

There are excellent textbooks that introduce nonlinear and convex optimization, providing the reader with all the basics on convex analysis, reformulation of optimization problems, algorithms, and a number of insightful engineering applications. This book is targeted at advanced graduate students, or advanced researchers that are already familiar with the basics of convex optimization. It can be used as a textbook for an advanced graduate course emphasizing applications, or as a complement to an introductory textbook that provides up-to-date applications in engineering. It can also be used for self-study to become acquainted with the state of-the-art in a wide variety of engineering topics.

This book contains 12 diverse chapters written by recognized leading experts worldwide, covering a large variety of topics. Due to the diverse nature of the book chapters, it is not possible to organize the book into thematic areas and each chapter should be treated independently of the others. A brief account of each chapter is given next.

In Chapter 1, Mattingley and Boyd elaborate on the concept of convex optimization in real-time embedded systems and automatic code generation. As opposed to generic solvers that work for general classes of problems, in real-time embedded optimization the same optimization problem is solved many times, with different data, often with a hard real-time deadline. Within this setup, the authors propose an automatic code-generation system that can then be compiled to yield an extremely efficient custom solver for the problem family.

In Chapter 2, Beck and Teboulle provide a unified view of gradient-based algorithms for possibly nonconvex and non-differentiable problems, with applications to signal recovery. They start by rederiving the gradient method from several different perspectives and suggest a modification that overcomes the slow convergence of the algorithm. They then apply the developed framework to different image-processing problems such as ℓ_1-based regularization, TV-based denoising, and TV-based deblurring, as well as communication applications like source localization.

In Chapter 3, Songsiri, Dahl, and Vandenberghe consider graphical models for autoregressive processes. They take a parametric approach for maximum-likelihood and maximum-entropy estimation of autoregressive models with conditional independence constraints, which translates into a sparsity pattern on the inverse of the spectral-density matrix. These constraints turn out to be nonconvex. To treat them, the authors propose a relaxation which in some cases is an exact reformulation of the original problem. The proposed methodology allows the selection of graphical models by fitting autoregressive processes to different topologies and is illustrated in different applications.

The following three chapters deal with optimization problems closely related to SDP and relaxation techniques.

In Chapter 4, Luo and Chang consider the SDP relaxation for several classes of quadratic-optimization problems such as separable quadratically constrained quadratic programs (QCQPs) and fractional QCQPs, with applications in communications and signal processing. They identify cases for which the relaxation is tight as well as classes of quadratic-optimization problems whose relaxation provides a guaranteed, finite worst-case approximation performance. Numerical simulations are carried out to assess the efficacy of the SDP-relaxation approach.

In Chapter 5, So and Ye perform a probabilistic analysis of SDP relaxations. They consider the problem of maximum-likelihood detection for multiple-input–multiple-output systems via SDP relaxation plus a randomization rounding procedure and study its loss in performance. In particular, the authors derive an approximation guarantee based on SDP weak-duality and concentration inequalities for the largest singular value of the channel matrix. For example, for MPSK constellations, the relaxed SDP detector is shown to yield a constant factor approximation to the ML detector in the low signal-to-noise ratio region.

In Chapter 6, Huang, De Maio, and Zhang treat the problem of radar design based on convex optimization. The design problem is formulated as a nonconvex QCQP. Using matrix rank-1 decompositions they show that nonetheless strong duality holds for the nonconvex QCQP radar code-design problem. Therefore, it can be solved in polynomial time by SDP relaxation. This allows the design of optimal coded waveforms in the presence of colored Gaussian disturbance that maximize the detection performance under a control both on the region of achievable values for the Doppler-estimation accuracy and on the similarity with a given radar code.

The next three chapters consider very different problems, namely, blind source separation, modern sampling theory, and robust broadband beamforming.

In Chapter 7, Ma, Chan, Chi, and Wang consider blind non-negative source separation with applications in imaging. They approach the problem from a convex-analysis perspective using convex-geometry concepts. It turns out that solving the blind separation problem boils down to finding the extreme points of a polyhedral set, which can be efficiently solved by a series of linear programs. The method is based on a deterministic property of the sources called local dominance which is satisfied in many applications with sparse or high-contrast images. A robust method is then developed to relax the assumption. A number of numerical simulations show the effectiveness of the method in practice.

In Chapter 8, Michaeli and Eldar provide a modern perspective on sampling theory from an optimization point of view. Traditionally, sampling theories have addressed the problem of perfect reconstruction of a given class of signals from their samples. During the last two decades, it has been recognized that these theories can be viewed in a broader sense of projections onto appropriate subspaces. The authors introduce a complementary viewpoint on sampling based on optimization theory. They provide extensions and generalizations of known sampling algorithms by constructing optimization problems that take into account the goodness of fit of the recovery to the samples as well as any other prior information on the signal. A variety of formulations are considered including aspects such as noiseless/noisy samples, different signal priors, and different least-squares/minimax objectives.

In Chapter 9, Rübsamen, El-Keyi, Gershman, and Kirubarajan develop several worst-case broadband beamforming techniques with improved robustness against array manifold errors. The methods show a robustness matched to the presumed amount of uncertainty, each of them offering a different trade-off in terms of interference suppression capability, robustness against signal self-nulling, and computational complexity.

The authors obtain convex second-order cone programming and SDP reformulations of the proposed beamformer designs which lead to efficient implementation.

The last three chapters deal with optimization of systems with multiple nodes. Chapter 10 takes an optimization approach with cooperative agents, whereas Chapters 11 and 12 follow a game-theoretic perspective with noncooperative nodes.

In Chapter 10, Nedic and Ozdaglar study the problem of distributed optimization and control of multiagent networked systems. Within this setup, a network of agents has to cooperatively optimize in a distributed way a global-objective function, which is a combination of local-objective functions, subject to local and possibly global constraints. The authors present both classical results as well as recent advances on design and analysis of distributed-optimization algorithms, with recent applications. Two main approaches are considered depending on whether the global objective is separable or not; in the former case, the classical Lagrange dual decompositions can be employed, whereas in the latter case consensus algorithms are the fundamental building block. Practical issues associated with the implementation of the optimization algorithms over networked systems are also considered such as delays, asynchronism, and quantization effects in the network implementation.

In Chapter 11, Scutari, Palomar, and Barbarossa apply the framework of game theory to different communication systems, namely, ad-hoc networks and cognitive radio systems. Game theory describes and analyzes scenarios with interactive decisions among different players, with possibly conflicting goals, and is very suitable for multiuser systems where users compete for the resources. For some problem formulations, however, game theory may fall short, and it is then necessary to use the more general framework of variational-inequality (VI) theory. The authors show how many resource-allocation problems in ad-hoc networks and in the emerging field of cognitive radio networks fit naturally either in the game-theoretical paradigm or in the more general theory of VI (further elaborated in the following chapter). This allows the study of existence/uniqueness of Nash-equilibrium points as well as the design of practical algorithms with provable converge to an equilibrium.

In Chapter 12, Facchinei and Pang present a comprehensive mathematical treatment of the Nash-equilibrium problem based on the variational-inequality and complementarity approach. They develop new results on existence of equilibria based on degree theory, global uniqueness, local-sensitivity analysis to data variation, and iterative algorithms with convergence conditions. The results are then illustrated with an application in communication systems.

1 Automatic code generation for real-time convex optimization

Jacob Mattingley and Stephen Boyd

This chapter concerns the use of convex optimization in real-time embedded systems, in areas such as signal processing, automatic control, real-time estimation, real-time resource allocation and decision making, and fast automated trading. By "embedded" we mean that the optimization algorithm is part of a larger, fully automated system, that executes automatically with newly arriving data or changing conditions, and without any human intervention or action. By "real-time" we mean that the optimization algorithm executes much faster than a typical or generic method with a human in the loop, in times measured in milliseconds or microseconds for small and medium size problems, and (a few) seconds for larger problems. In real-time embedded convex optimization the same optimization problem is solved many times, with different data, often with a hard real-time deadline. In this chapter we propose an automatic code generation system for real-time embedded convex optimization. Such a system scans a description of the problem family, and performs much of the analysis and optimization of the algorithm, such as choosing variable orderings used with sparse factorizations and determining storage structures, at code generation time. Compiling the generated source code yields an extremely efficient custom solver for the problem family. We describe a preliminary implementation, built on the Python-based modeling framework CVXMOD, and give some timing results for several examples.

1.1 Introduction

1.1.1 Advisory optimization

Mathematical optimization is traditionally thought of as an aid to human decision making. For example, a tool for portfolio optimization *suggests* a portfolio to a human decision maker, who possibly carries out the proposed trades. Optimization is also used in many aspects of engineering design; in most cases, an engineer is in the decision loop, continually reviewing the proposed designs and changing parameters in the problem specification, if needed.

When optimization is used in an advisory role, the solution algorithms do not need to be especially fast; an acceptable time might be a few seconds (for example, when analyzing scenarios with a spreadsheet), or even tens of minutes or hours for very large

Convex Optimization in Signal Processing and Communication, eds. Daniel P. Palomar and Yonina C. Eldar. Published by Cambridge University Press © Cambridge University Press, 2010.

problems (e.g., engineering design synthesis, or scheduling). Some unreliability in the solution methods can be tolerated, since the human decision maker will review the proposed solutions, and hopefully catch problems.

Much effort has gone into the development of optimization algorithms for these settings. For adequate performance, they must detect and exploit a generic problem structure not known (to the algorithm) until the particular problem instance is solved. A good generic *linear programming* (LP) solver, for example, can solve, on human-based time scales, large problems in digital circuit design, supply chain management, filter design, or automatic control. Such solvers are often coupled with optimization modeling languages, which allow the user to efficiently describe optimization problems in a high level format. This permits the user to rapidly see the effect of new terms or constraints.

This is all based on the conceptual model of a human in the loop, with most previous and current solver development effort focusing on scaling to *large* problem instances. Not much effort, by contrast, goes into developing algorithms that solve small- or medium-sized problems on fast (millisecond or microsecond) time scales, and with great reliability.

1.1.2   Embedded optimization

In this chapter we focus on embedded optimization, where solving optimization problems is part of a wider, automated algorithm. Here the optimization is deeply embedded in the application, and no human is in the loop. In the introduction to the book *Convex Optimization* [1], Boyd and Vandenberghe state:

A relatively recent phenomenon opens the possibility of many other applications for mathematical optimization. With the proliferation of computers embedded in products, we have seen a rapid growth in *embedded optimization*. In these embedded applications, optimization is used to automatically make real-time choices, and even carry out the associated actions, with no (or little) human intervention or oversight. In some application areas, this blending of traditional automatic control systems and embedded optimization is well under way; in others, it is just starting. Embedded real-time optimization raises some new challenges: in particular, it requires solution methods that are extremely reliable, and solve problems in a predictable amount of time (and memory).

In real-time embedded optimization, different instances of the same small- or medium-size problem must be solved extremely quickly, for example, on millisecond or microsecond time scales; in many cases the result must be obtained before a strict real-time deadline. This is in direct contrast to generic algorithms, which take a variable amount of time, and exit only when a certain precision has been achieved.

An early example of this kind of embedded optimization, though not on the time scales that we envision, is *model predictive control* (MPC), a form of feedback control system. Traditional (but still widely used) control schemes have relatively simple control policies, requiring only a few basic operations like matrix-vector multiplies and lookup table searches at each time step [2, 3]. This allows traditional control policies to be executed rapidly, with strict time constraints and high reliability. While the control policies themselves are simple, great effort is expended in developing and tuning (i.e., choosing parameters in) them. By contrast, with MPC, at each step the control action is determined by solving an optimization problem, typically a (convex) *quadratic program* (QP). It was first deployed in the late 1980s in the chemical process industry, where the hard real-time deadlines were in the order of 15 minutes to an hour per optimization problem [4]. Since then, we have seen huge computer processing power increases, as well as substantial advances in algorithms, which allow MPC to be carried out on the same fast time scales as many conventional control methods [5, 6]. Still, MPC is generally not considered by most control engineers, even though there is much evidence that MPC provides better control performance than conventional algorithms, especially when the control inputs are constrained.

Another example of embedded optimization is program or algorithmic trading, in which computers initiate stock trades without human intervention. While it is hard to find out what is used in practice due to trade secrets, we can assume that at least some of these algorithms involve the repeated solution of linear or quadratic programs, on short, if not sub-second, time scales. The trading algorithms that run on faster time scales are presumably just like those used in automatic control; in other words, simple and quickly executable. As with traditional automatic control, huge design effort is expended to develop and tune the algorithms.

In signal processing, an algorithm is used to extract some desired signal or information from a received noisy or corrupted signal. In *off-line signal processing*, the entire noisy signal is available, and while faster processing is better, there are no hard real-time deadlines. This is the case, for example, in the restoration of audio from wax cylinder recordings, image enhancement, or geophysics inversion problems, where optimization is already widely used. In *on-line* or *real-time signal processing*, the data signal samples arrive continuously, typically at regular time intervals, and the results must be computed within some fixed time (typically, a fixed number of samples). In these applications, the algorithms in use, like those in traditional control, are still relatively simple [7].

Another relevant field is communications. Here a noise-corrupted signal is received, and a decision as to which bit string was transmitted (i.e., the decoding) must be made within some fixed (and often small) period of time. Typical algorithms are simple, and hence fast. Recent theoretical studies suggest that decoding methods based on convex optimization can deliver improved performance [8–11], but the standard methods for these problems are too slow for most practical applications. One approach has been the development of custom solvers for communications decoding, which can execute far faster than generic methods [12].

We also envisage real-time optimization being used in statistics and machine learning. At the moment, most statistical analysis has a human in the loop. But we are starting to see some real-time applications, e.g., spam filtering, web search, and automatic fault detection. Optimization techniques, such as support vector machines (SVMs), are heavily used in such applications, but much like in traditional control design, the optimization problems are solved on long time scales to produce a set of model parameters or weights. These parameters are then used in the real-time algorithm, which typically involves not much more than computing a weighted sum of features, and so can be done quickly. We can imagine applications where the weights are updated rapidly, using some real-time, optimization-based method. Another setting in which an optimization problem might be solved on a fast time scale is real-time statistical inference, in which estimates of the probabilities of unknown variables are formed soon after new information (in the form of some known variables) arrives.

Finally, we note that the ideas behind real-time embedded optimization could also be useful in more conventional situations with no real-time deadlines. The ability to extremely rapidly solve problem instances from a specific problem family gives us the ability to solve large numbers of similar problem instances quickly. Some example uses of this are listed below.

- *Trade-off analysis.* An engineer formulating a design problem as an optimization problem solves a large number of instances of the problem, while varying the constraints, to obtain a sampling of the optimal trade-off surface. This provides useful design guidelines.
- *Global optimization.* A combinatorial optimization problem is solved using branch-and-bound or a similar global optimization method. Such methods require the solution of a large number of problem instances from a (typically convex, often LP) problem family. Being able to solve each instance very quickly makes it possible to solve the overall problem much faster.
- *Monte Carlo performance analysis.* With Monte Carlo simulation, we can find the distribution of minimum cost of an optimization problem that depends on some random parameters. These parameters (e.g., prices of some resources or demands for products) are random with some given distribution, but will be known before the optimization is carried out. To find the distribution of optimized costs, we use Monte Carlo: we generate a large number of samples of the price vector (say), and for each one we carry out optimization to find the minimal cost. Here, too, we end up solving a large number of instances of a given problem family.

1.1.3   Convex optimization

Convex optimization has many advantages over general nonlinear optimization, such as the existence of efficient algorithms that can reliably find a globally optimal solution. A less appreciated advantage is that algorithms for specific convex optimization problem families can be highly robust and reliable; unlike many general purpose optimization algorithms, they do not have parameters that must be manually tuned for particular problem instances. Convex optimization problems are, therefore, ideally suited to real-time embedded applications, because they can be reliably solved.

A large number of problems arising in application areas like signal processing, control, finance, statistics and machine learning, and network operation can be cast (exactly, or with reasonable approximations) as convex problems. In many other problems, convex optimization can provide a good heuristic for approximate solution of the problem [13, 14].

In any case, much of what we say in this chapter carries over to local optimization methods for nonconvex problems, although without the global optimality guarantee, and with some loss in reliability. Even simple methods of extending the methods of convex optimization can work very well in pratice. For example, we can use a basic interior-point method as if the problem were convex, replacing nonconvex portions with appropriate convex approximations at each iteration.

1.1.4 Outline

In Section 1.2, we describe problem families and the specification languages used to formally model them, and two general approaches to solving problem instances described this way: via a parser-solver, and via code generation. We list some specific example applications of real-time convex optimization in Section 1.3. In Section 1.4 we describe, in general terms, some requirements on solvers used in real-time optimization applications, along with some of the attributes of real-time optimization problems that we can exploit. We give a more detailed description of how a code generator can be constructed in Section 1.5, briefly describe a preliminary implementation of a code generator in Section 1.6, and report some numerical results in Section 1.7. We give a summary and conclusions in Section 1.8.

1.1.5 Previous and related work

Here we list some representative references that focus on various aspects of real-time embedded optimization or closely related areas.

Control

Plenty of work focuses on traditional real-time control [15–17], or basic model predictive control [18–23]. Several recent papers describe methods for solving various associated QPs quickly. One approach is *explicit MPC*, pioneered by Bemporad and Morari [24], who exploit the fact that the solution of the QP is a piecewise linear function of the problem data, which can be determined analytically ahead of time. Solving instances of the QP then reduces to evaluating a piecewise linear function. Interior-point methods [25], including fast custom interior-point methods [6], can also be used to provide rapid solutions. For fast solution of the QPs arising in evaluation of control-Lyapunov policies (a special case of MPC), see [26]. Several authors consider fast solution of nonlinear control problems using an MPC framework [27–29]. Others discuss various real-time applications [30, 31], especially those in robotics [32–34].

Signal processing, communications, and networking

Work on convex optimization in signal processing includes ℓ_1-norm minimization for sparse signal recovery, recovery of noisy signals, or statistical estimation [35, 36], or linear programming for error correction [37]. Goldfarb and Yin discuss interior-point algorithms for solving total variation image restoration problems [38]. Some combinatorial optimization problems in signal processing that are approximately, and very quickly, solved using convex relaxations and local search are static fault detection [14], dynamic fault detection [39], query model estimation [40] and sensor selection [13]. In communications, convex optimization is used in DSL [41], radar [42], and CDMA [43], to list just a few examples.

Since the publication of the paper by Kelly *et al.* [44], which poses the optimal network flow control as a convex optimization problem, many authors have looked at optimization-based network flow methods [45–48], or optimization of power and bandwidth [49, 50].

Code generation

The idea of automatic generation of source code is quite old. Parser-generators such as Yacc [51], or more recent tools like GNU Bison [52], are commonly used to simplify the writing of compilers. For engineering problems, in particular, there are a range of code generators: one widely used commercial tool is Simulink [53], while the open-source Ptolemy project [54] provides a modeling environment for embedded systems. Domain-specific code generators are found in many different fields [55–58].

Generating source code for optimization solvers is nothing new either; in 1988 Oohori and Ohuchi [59] explored code generation for LPs, and generated explicit Cholesky factorization code ahead of time. Various researchers have focused on code generation for convex optimization. McGovern, in his PhD thesis [60], gives a computational complexity analysis of real-time convex optimization. Hazan considers algorithms for on-line convex optimization [61], and Das and Fuller [62] hold a patent on an active-set method for real-time QP.

1.2   Solvers and specification languages

It will be important for us to carefully distinguish between an instance of an optimization problem, and a parameterized family of optimization problems, since one of the key features of real-time embedded optimization applications is that each of the specific problems to be solved comes from a single family.

1.2.1   Problem families and instances

We consider continuously parameterized families of optimization problems, of the form

$$
\begin{aligned}
\text{minimize} \quad & F_0(x, a) \\
\text{subject to} \quad & F_i(x, a) \leq 0, \quad i = 1, \ldots, m \\
& H_i(x, a) = 0, \quad i = 1, \ldots, p,
\end{aligned}
\tag{1.1}
$$

where $x \in \mathbf{R}^n$ is the (vector) optimization variable, and $a \in \mathcal{A} \subset \mathbf{R}^\ell$ is a parameter or data vector that specifies the problem instance. To specify the problem family (1.1), we need descriptions of the functions $F_0, \ldots, F_m, H_1, \ldots, H_p$, and the parameter set $\mathcal{A}$. When we fix the value of the parameters, by fixing the value of a, we obtain a problem *instance*.

As a simple example, consider the QP

$$
\begin{aligned}
\text{minimize} \quad & (1/2)x^T P x + q^T x \\
\text{subject to} \quad & Gx \leq h, \quad Ax = b,
\end{aligned}
\tag{1.2}
$$

with variable $x \in \mathbf{R}^n$, where the inequality between vectors means componentwise. Let us assume that in all instances we care about, the equality constraints are the same, that is, A and b are fixed. The matrices and vectors P, q, G, and h can vary, although P must be symmetric positive semidefinite. For this problem family we have

$$a = (P, q, G, h) \in \mathcal{A} = \mathbf{S}_+^n \times \mathbf{R}^n \times \mathbf{R}^{m \times n} \times \mathbf{R}^m,$$

where $\mathbf{S}_+^n$ denotes the set of symmetric $n \times n$ positive semidefinite matrices. We can identify a with an element of $\mathbf{R}^\ell$, with total dimension

$$\ell = \underbrace{n(n+1)/2}_{P} + \underbrace{n}_{q} + \underbrace{mn}_{G} + \underbrace{m}_{h}.$$

In this example, we have

$$F_0(x, a) = (1/2)x^T P x + q^T x,$$

$$F_i(x, a) = g_i^T x - h_i, \quad i = 1, \ldots, m,$$

$$H_i(x, a) = \tilde{a}_i^T x - b_i, \quad i = 1, \ldots, p,$$

where g_i^T is the ith row of G, and $\tilde{a}_i^T$ is the ith row of A. Note that the equality constraint functions H_i do not depend on the parameter vector a; the matrix A and vector b are constants in the problem family (1.2).

Here we assume that the data matrices have no structure, such as sparsity. But in many cases, problem families do have structure. For example, suppose that we are interested in the problem family in which P is tridiagonal, and the matrix G has some specific sparsity pattern, with N (possibly) nonzero entries. Then $\mathcal{A}$ changes, as does the total parameter dimension, which becomes

$$\ell = \underbrace{2n - 1}_{P} + \underbrace{n}_{q} + \underbrace{N}_{G} + \underbrace{m}_{h}.$$

In a more general treatment, we could also consider the dimensions and sparsity patterns as (discrete) parameters that one specifies when fixing a particular problem instance. Certainly when we refer to QP generally, we refer to families of QPs with any dimensions, and not just a family of QPs with some specific set of dimensions and sparsity patterns. In this chapter, however, we restrict our attention to continuously parameterized problem families, as described above; in particular, the dimensions n, m, and p are fixed, as are the sparsity patterns in the data.

The idea of a parameterized problem family is a central concept in optimization (although in most cases, a family is considered to have variable dimensions). For example, the idea of a solution algorithm for a problem family is sensible, but the idea of a solution algorithm for a problem instance is not. (The best solution algorithm for a problem instance is, of course, to output a pre-computed solution.)

Nesterov and Nemirovsky refer to families of convex optimization problems, with constant structure and parameterized by finite dimensional parameter vectors as *well structured problem (families)* [63].

1.2.2   Solvers

A *solver* or *solution method* for a problem family is an algorithm that, given the parameter value $a \in \mathcal{A}$, finds an optimal point $x^{\star}(a)$ for the problem instance, or determines that the problem instance is infeasible or unbounded.

Traditional solvers [1,64,65] can handle problem families with a range of dimensions (e.g., QPs with the form (1.2), any values for m, n, and p, and any sparsity patterns in the data matrices). With traditional solvers, the dimensions, sparsity patterns and all other problem data a are specified only at solve time, that is, when the solver is invoked. This is extremely useful, since a single solver can handle a very wide range of problems, and exploit (for efficiency) a wide variety of sparsity patterns. The disadvantage is that analysis and utilization of problem structure can only be carried out as each problem instance is solved, which is then included in the per-instance solve time. This also limits the reasonable scope of efficiency gains: there is no point in spending longer looking for an efficient method than it would take to solve the problem with a simpler method.

This traditional approach is far from ideal for real-time embedded applications, in which a very large number of problems, from the same continuously parameterized family, will be solved, hopefully very quickly. For such problems, the dimensions and sparsity patterns are known ahead of time, so much of the problem and efficiency analysis can be done ahead of time (and in relative leisure).

It is possible to develop a custom solver for a specific, continuously parameterized problem family. This is typically done by hand, in which case the development effort can be substantial. On the other hand, the problem structure and other attributes of the particular problem family can be exploited, so the resulting solver can be far more efficient than a generic solver [6,66].

1.2.3   Specification languages

A *specification language* allows a user to describe a problem instance or problem family to a computer, in a convenient, high-level algebraic form. All specification languages have the ability to declare optimization variables; some also have the ability to declare parameters. Expressions involving variables, parameters, constants, supported operators, and functions from a library can be formed; these can be used to specify objectives and constraints. When the specification language supports the declaration of parameters, it can also be used to describe $\mathcal{A}$, the set of valid parameters. (The domains of functions used in the specification may also implicitly impose constraints on the parameters.)

Some specification languages impose few restrictions on the expressions that can be formed, and the objective and constraints that can be specified. Others impose strong restrictions to ensure that specified problems have some useful property such as convexity, or are transformable to some standard form such as an LP or a *semidefinite program* (SDP).

1.2.4 Parser-solvers

A *parser-solver* is a system that scans a specification language description of a problem *instance*, checks its validity, carries out problem transformations, calls an appropriate solver, and transforms the solution back to the original form. Parser-solvers accept directives that specify which solver to use, or that override algorithm parameter defaults, such as required accuracy.

Parser-solvers are widely used. Early (and still widely used) parser-solvers include AMPL [67] and GAMS [68], which are general purpose. Parser-solvers that handle more restricted problem types include SDPSOL [69], LMILAB [70], and LMITOOL [71] for SDPs and *linear matrix inequalities* (LMIs), and GGPLAB [72] for generalized geometric programs. More recent examples, which focus on convex optimization, include YALMIP [73], CVX [74], CVXMOD [75], and Pyomo [76]. Some tools [77–79] are used as post-processors, and attempt to detect convexity of a problem expressed in a general purpose modeling language.

As an example, an *instance* of the QP problem (1.2) can be specified in CVXMOD as

```
P = matrix(...); q = matrix(...); A = matrix(...)
b = matrix(...); G = matrix(...); h = matrix(...)
x = optvar('x', n)
qpinst = problem(minimize(0.5*quadform(x, P) + tp(q)*x),
                 [G*x <= h, A*x == b])
```

The first two (only partially shown) lines assign names to specific numeric values, with appropriate dimensions and values. The third line declares x to be an optimization variable of dimension n, which we presume has a fixed numeric value. The last line generates the problem instance itself (but does not solve it), and assigns it the name qpinst. This problem instance can then be solved with

```
qpinst.solve()
```

which returns either `'optimal'` or `'infeasible'`, and, if optimal, sets `x.value` to an optimal value $x^\star$.

For specification languages that support parameter declaration, numeric values must be attached to the parameters before the solver is called. For example, the QP problem *family* (1.2) is specified in CVXMOD as

```
A = matrix(...); b = matrix(...)
P = param('P', n, n, psd=True); q = param('q', n)
G = param('G', m, n); h = param('h', m)
x = optvar('x', n)
qpfam = problem(minimize(0.5*quadform(x, P) + tp(q)*x),
                [G*x <= h, A*x == b])
```

In this code segment, as in the example above, m and n are fixed integers. In the first line, A and b are still assigned fixed values, but in the second and third lines, P, q, G, and h are declared instead as parameters with appropriate dimensions. Additionally, P is specified as symmetric positive semidefinite. As before, x is declared to be an optimization variable. In the final line, the QP problem family is constructed (with identical syntax), and assigned the name qpfam.

If we called qpfam.solve() right away, it would fail, since the parameters have no numeric values. However (with an overloading of semantics), if values are attached to each parameter first, qpfam.solve() will create a problem instance and solve that:

```
P.value = matrix(...); q.value = matrix(...)
G.value = matrix(...); h.value = matrix(...)
qpfam.solve()   # Instantiates, then solves.
```

This works since the solve method will solve the particular instance of a problem family specified by the numeric values in the value attributes of the parameters.

1.2.5   Code generators

A *code generator* takes a description of a problem family, scans it and checks its validity, carries out various problem transformations, and then generates source code that compiles into a (hopefully very efficient) solver for that problem family. Figures 1.1 and 1.2 show the difference between code generators and parser-solvers.

A code generator will have options configuring the type of code it generates, including, for example, the target language and libraries, the solution algorithm (and algorithm parameters) to use, and the handling of infeasible problem instances. In addition to source code for solving the optimization problem family, the output might also include:

- Auxiliary functions for checking parameter validity, setting up problem instances, preparing a workspace in memory, and cleaning up after problem solution.

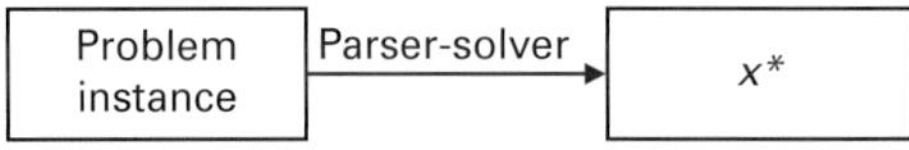

Figure 1.1   A parser-solver processes and solves a single problem instance.

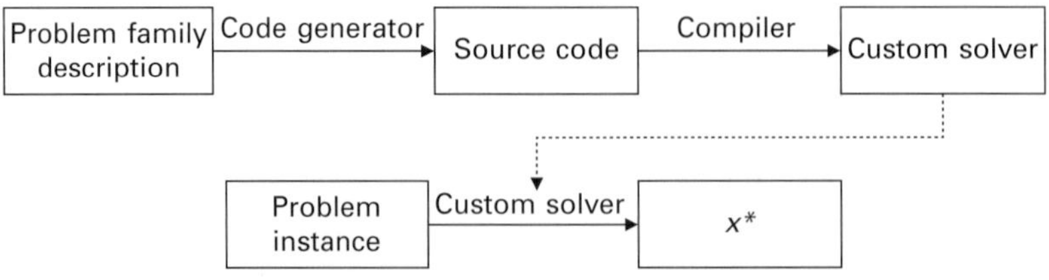

Figure 1.2   A code generator processes a problem family, generating a fast, custom solver, which is used to rapidly solve problem instances.

- Documentation describing the problem family and how to use the code.
- Documentation describing any problem transformations.
- An automated test framework.
- Custom functions for converting problem data to or from a range of formats or environments.
- A system for automatically building and testing the code (such as a `Makefile`).

1.2.6   Example from CVXMOD

In this section we focus on the preliminary code generator in CVXMOD, which generates solver code for the C programming language. To generate code for the problem family described in `qpfam`, we use

```
qpfam.codegen('qpfam/')
```

This tells CVXMOD to generate code and auxiliary files and place them in the `qpfam/` directory. Upon completion, the directory will contain the following files:

- `solver.c`, which includes the actual solver function `solve`, and three initialization functions (`initparams`, `initvars`, and `initwork`).
- `template.c`, a simple file illustrating basic usage of the solver and initialization functions.
- README, which contains code generation information, including a list of the generated files and information about the targeted problem family.
- `doc.tex`, which provides LaTeX source for a document that describes the problem family, transformations performed to generate the internal standard form, and reference information about how to use the solver.
- `Makefile`, which has rules for compiling and testing the solver.

The file `template.c` contains the following:

```c
#include solver.h
int main(int argc, char **argv) {
    Params params = initparams();
    Vars vars = initvars();
    Workspace work = initwork(vars);
    for (;;) { // Real-time loop.
        // Get new parameter values here.
        status = solve(params, vars, work);
        // Test status, export variables, etc here.
    }
}
```

The main real-time loop (here represented, crudely, as an asynchronous infinite loop) repeatedly carries out the following:

1. Get a new problem instance, that is, get new parameter values.

2. Solve this instance of the problem, set new values for the variables, and return an indication of solver success (optimality, infeasibility, failure).
3. Test the status and respond appropriately. If optimization succeeded, export the variables to the particular application.

Some complexity is hidden from the user. For example, the allocated optimization variables include not just the variables specified by the user in the problem specification, but also other, automatically generated intermediate variables, such as slack variables. Similarly, the workspace variables stored within work need not concern someone wanting to just get the algorithm working – they are only relevant when the user wants to adjust the various configurable algorithm parameters.

1.3   Examples

In this section we describe several examples of real-time optimization applications. Some we describe in a general setting (e.g., model predictive control); others we decribe in a more specific setting (e.g., optimal order execution). We first list some broad categories of applications, which are not meant to be exclusive or exhaustive.

Real-time adaptation

Real-time optimization is used to optimally allocate multiple resources, as the amounts of resources available, the system requirements or objective, or the system model dynamically change. Here, real-time optimization is used to adapt the system to the changes, to maintain optimal performance. In simple adaptation, we ignore any effect the current choice has on future resource availability or requirements. In this case we are simply solving a sequence of independent optimization problem instances, with different data. If the changes in data are modest, warm-start can be used. To be effective, real-time optimization has to be carried out at a rate fast enough to track the changes. Real-time adaptation can be either event-driven (taking place, say, whenever the parameters have shifted significantly) or synchronous, with re-optimization occurring at regular time intervals.

Real-time trajectory planning

In trajectory planning we choose a sequence of inputs to a dynamical system that optimizes some objective, while observing some constraints. (This is also called input generation or shaping, or open-loop control.) Typically this is done asynchronously: a higher level task planner occasionally issues a command such as "sell this number of shares of this asset over this time period" or "move the robot end effector to this position at this time". An optimization problem is then solved, with parameters that depend on the current state of the system, the particular command issued, and other relevant data; the result is a sequence of inputs to the system that will (optimally) carry out the high level command.

Feedback control

In feedback control, real-time optimization is used to determine actions to be taken, based on periodic measurements of some dynamic system, in which current actions *do* affect the future. This task is sometimes divided into two conceptual parts: optimally sensing or estimating the system state, given the measurements, and choosing an optimal action, based on the estimated system state. (Each of these can be carried out by real-time optimization.) To be effective, the feedback control updates should occur on a time scale at least as fast as the underlying dynamics of the system being controlled. Feedback control is typically synchronous.

Real-time sensing, estimation, or detection

Real-time optimization is used to estimate quantities, or detect events, based on sensor measurements or other periodically arriving information. In a static system, the quantities to be estimated at each step are independent, so we simply solve an independent problem instance with each new set of measurements. In a dynamic system, the quantities to be estimated are related by some underlying dynamics. In a dynamic system we can have a delay (or look-ahead): we form an estimate of the quantities at time period $t - d$ (where d is the delay), based on measurements up to time period t, or the measurements in some sliding time window.

Real-time system identification

Real-time optimization is used to estimate the parameters in a dynamical model of a system, based on recent measurements of the system outputs (and, possibly, inputs). Here the optimization is used to track changes in the dynamic system; the resulting time-varying dynamic model can in turn be used for prediction, control, or dynamic optimization.

1.3.1 Adaptive filtering and equalization

In adaptive filtering or equalization, a high rate signal is processed in real-time by some (typically linear) operation, parameterized by some coefficients, weights, or gains, that can change with time. The simplest example is a static linear combining filter,

$$y_t = w_t^T u_t,$$

where $u_t \in \mathbf{R}^n$ and $y_t \in \mathbf{R}$ are the vector input and (filtered or equalized) scalar output signals, and $w_t \in \mathbf{R}^n$ is the filter parameter vector, at time $t \in \mathbf{Z}$. The filter parameter w_t is found by solving an (often convex) optimization problem that depends on changing data, such as estimates of noise covariances or channel gains. The filter parameter can be updated (i.e., re-optimized) at every step, synchronously every K steps, or asynchronously in an event-driven scheme.

When the problem is sufficiently simple, for example, unconstrained quadratic minimization, the weight updates can be carried out by an analytical method [7, 80, 81]. Subgradient-type or stochastic gradient methods, in which the parameters are updated (usually, slightly) in each step, can also be used [82, 83]. These methods have low update complexity, but only find the optimal weight in the limit of (many) iterations, by which time the data that determined the weight design have already changed. The weight updates can instead be carried out by real-time convex optimization.

To give a specific example, suppose that w_t is chosen to solve the problem

$$
\begin{array}{ll}
\text{maximize} & w_t^T f_t \\
\text{subject to} & |w_t^T g_t^{(i)}| \le 1, \quad i = 1, \dots, m,
\end{array}
$$

with data $f_t, g_t^{(1)}, \dots, g_t^{(m)}$. Here f_t is a direction associated with the desired signal, while $g_t^{(i)}$ are directions associated with interference or noise signals. This convex problem can be solved every K steps, say, based on the most recent data available.

1.3.2   Optimal order execution

A sell or buy order, for some number of shares of some asset, is to be executed over a (usually short) time interval, which we divide into T discrete time periods. We have a statistical model of the price in each period, which includes a random component, as well as the effect on the prices due to the amounts sold in the current and previous periods. We may also add constraints, such as a limit on the amount sold per period. The goal is to maximize the expected total revenue from the sale. We can also maximize a variance-adjusted revenue.

In the open-loop version of this problem, we commit to the sales in all periods beforehand. In the closed-loop version, we have recourse: in each period we are told the price (without the current sales impact), and can then adjust the amount we sell. While some forms of this problem have analytical solutions [84, 85], we consider here a more general form.

To give a specific example, suppose that the prices $p = (p_1, \dots, p_T)$ are modeled as

$$
p = p_0 - As,
$$

where $s = (s_1, \dots, s_T)$ are sales, the matrix A (which is lower triangular with non-negative elements) describes the effect of sales on current and future prices, and $p_0 \sim \mathcal{N}(\bar{p}, \Sigma)$ is a random price component. The total achieved sales revenue is

$$
R = p^T s \sim \mathcal{N}(\bar{p}^T s - s^T As, s^T \Sigma s).
$$

We will choose how to sell $\mathbf{1}^T s = S$ shares, subject to per-period sales limits $0 \le s \le S^{\max}$, to maximize the risk-adjusted total revenue,

$$
\mathbf{E}\, R - \gamma\, \mathbf{var}\, R = \bar{p}^T s - s^T Q s,
$$

where $\gamma > 0$ is a risk-aversion parameter, and

$$
Q = \gamma \Sigma + (1/2)(A + A^T).
$$

(We can assume that $Q \succeq 0$, i.e., Q is positive semidefinite.)

In the open-loop setting, this results in the (convex) QP

$$\begin{array}{ll} \text{maximize} & \bar{p}^T s - s^T Q s \\ \text{subject to} & 0 \le s \le S^{\max}, \quad \mathbf{1}^T s = S, \end{array}$$

with variable $s \in \mathbf{R}^T$. The parameters are $\bar{p}$, Q (which depends on the original problem data Σ, A, and γ), $S^{\max}$, and S. An obvious initialization is $s = (S/T)\mathbf{1}$, that is, constant sales over the time interval.

Real-time optimization for this problem might work as follows. When an order is placed, the problem parameters are determined, and the above QP is solved to find the sales schedule. At least some of these parameters will depend (in part) on the most recently available data; for example, $\bar{p}$, which is a prediction of the mean prices over the next T periods, if no sales occurred.

The basic technique in MPC can be used as a very good heuristic for the closed-loop problem. At each time step t, we solve the problem again, using the most recent values of the parameters, and fixing the values of the previous sales $s_1, \ldots, s_{t-1}$ to their (already chosen) values. We then sell the amount s_t from the solution. At the last step no optimization is needed: we simply sell $s_T = S - \sum_{t=1}^{T-1}$, that is, the remaining unsold shares.

1.3.3 Sliding-window smoothing

We are given a noise-corrupted scalar signal y_t, $t \in \mathbf{Z}$, and want to form an estimate of the underlying signal, which we denote x_t, $t \in \mathbf{Z}$. We form our estimate $\hat{x}_t$ by examining a window of the corrupted signal, $(y_{t-p}, \ldots, y_{t+q})$, and solving the problem

$$\begin{array}{ll} \text{minimize} & \sum_{\tau=t-p}^{t+q}(y_\tau - \tilde{x}_\tau)^2 + \lambda\phi(\tilde{x}_{t-p}, \ldots, \tilde{x}_{t+q}) \\ \text{subject to} & (\tilde{x}_{t-p}, \ldots, \tilde{x}_{t+q}) \in \mathcal{C}, \end{array}$$

with variables $(\tilde{x}_{t-p}, \ldots, \tilde{x}_{t+q}) \in \mathbf{R}^{p+1+1}$. Here $\phi : \mathbf{R}^{p+q+1} \to \mathbf{R}$ is a (typically convex) function that measures the implausibility of $(\tilde{x}_{t-p}, \ldots, \tilde{x}_{t+q})$, and $\mathcal{C} \subset \mathbf{R}^{p+q+1}$ is a (typically convex) constraint set representing prior information about the signal. The parameter $\lambda > 0$ is used to trade-off fit and implausibility. The integer $p \ge 0$ is the look-behind length, that is, how far back in time we look at the corrupted signal in forming our estimate; $q \ge 0$ is the look-ahead length, that is, how far forward in time we look at the corrupted signal. Our estimate of x_t is $\hat{x}_t = \tilde{x}_t^\star$, where $\tilde{x}^\star$ is a solution of the problem above.

The implausibility function ϕ is often chosen to penalize rapidly varying signals, in which case the estimated signal $\hat{x}$ can be interpreted as a smoothed version of y. One interesting case is $\phi(z) = \sum_{t=1}^{p+q} |z_{t+1} - z_t|$, the total variation of z [86]. Another interesting case is $\phi(z) = \sum_{t=2}^{p+q} |z_{t+1} - 2z_t + z_{t-1}|$, the ℓ_1 norm of the second-order difference (or Laplacian); the resulting filter is called an ℓ_1-*trend filter* [87].

One simple initialization for the problem above is $\tilde{x}_\tau = y_\tau$, $\tau = t - p, \ldots, t + q$; another is to shift the previous solution in time.

1.3.4   Sliding-window estimation

Sliding-window estimation, also known as *moving-horizon estimation* (MHE) uses optimization to form an estimate of the state of a dynamical system [21,88,89].

A linear dynamical system is modeled as

$$x_{t+1} = Ax_t + w_t,$$

where $x_t \in \mathcal{X} \subset \mathbf{R}^n$ is the state and w_t is a process noise at time period $t \in \mathbf{Z}$. We have linear noise-corrupted measurements of the state,

$$y_t = Cx_t + v_t,$$

where $y_t \in \mathbf{R}^p$ is the measured signal and v_t is measurement noise. The goal is to estimate x_t, based on prior information, that is, A, C, $\mathcal{X}$, and the last T measurements, that is, $y_{t-T+1}, \ldots, y_t$, along with our estimate of x_{t-T}.

A sliding-window estimator chooses the estimate of x_t, which we denote as $\hat{x}_t$, as follows. We first solve the problem

$$
\begin{array}{ll}
\text{minimize} & \sum_{\tau=t-T+1}^{t} (\phi_w(\tilde{x}_\tau - A\tilde{x}_{\tau-1}) + \phi_v(y_\tau - C\tilde{x}_\tau)) \\
\text{subject to} & \tilde{x}_{t-T} = \hat{x}_{t-T}, \quad \tilde{x}_\tau \in \mathcal{X}, \quad \tau = t - T + 1, \ldots, t,
\end{array}
$$

with variables $\tilde{x}_{t-T}, \ldots, \tilde{x}_t$. Our estimate is then $\hat{x}_t = \tilde{x}_t^\star$, where $\tilde{x}^\star$ is a solution of the problem above. When $\mathcal{X}$, ϕ_w, and ϕ_v are convex, the problem above is convex.

Several variations of this problem are also used. We can add a cost term associated with $\tilde{x}$, meant to express prior information we have about the state. We can replace the equality constraint $\tilde{x}_{t-T} = \hat{x}_{t-T}$ (which corresponds to the assumption that our estimate of x_{t-T} is perfect) with a cost function term that penalizes deviation of $\tilde{x}_{t-T}$ from $\hat{x}_{t-T}$.

We interpret the cost function term $\phi_w(w)$ as measuring the implausibility of the process noise taking on the value w. Similarly, $\phi_v(v)$ measures the implausibility of the measurement noise taking on the value v. One common choice for these functions is the negative logarithm of the densities of w_t and v_t, respectively, in which case the sliding-window estimate is the maximum-likelihood estimate of x_t (assuming the estimate of x_{t-T} was perfect, and the noises w_t are *independent and identically distributed* (IID), and v_t are IID).

One particular example is $\phi_w(w) = (1/2)\|w\|_2^2$, $\phi_v(v) = (1/2\sigma^2)\|v\|_2^2$, which corresponds to the statistical assumptions $w_t \sim \mathcal{N}(0, I)$, $v_t \sim \mathcal{N}(0, \sigma^2 I)$. We can also use cost functions that give robust estimates, that is, estimates of x_t that are not greatly affected by occasional large values of w_t and v_t. (These correspond to sudden unexpected changes in the state trajectory, or outliers in the measurements, respectively.) For example, using the (vector) Huber measurement cost function

$$
\phi_v(v) = \begin{cases} (1/2)\|v\|_2^2 & \|v\|_2 \leq 1 \\ \|v\|_1 - 1/2 & \|v\|_2 \geq 1 \end{cases}
$$

yields state estimates that are surprisingly immune to occasional large values of the measurement noise v_t [1, Section 6.1.2].

We can initialize the problem above with the previously computed state trajectory, shifted in time, or with one obtained by a linear-estimation method, such as Kalman filtering, that ignores the state constraints and, if needed, approximates the cost functions as quadratic.

1.3.5 Real-time input design

We consider a linear dynamical system

$$x_{t+1} = Ax_t + Bu_t,$$

where $x_t \in \mathbf{R}^n$ is the state, and $u_t \in \mathbf{R}^m$ is the control input at time period $t \in \mathbf{Z}$. We are interested in choosing $u_t, \ldots, u_{t+T-1}$, given x_t (the current state) and some convex constraints and objective on $u_t, \ldots, u_{t+T-1}$ and $x_{t+1}, \ldots, x_T$.

As a specific example, we consider minimum norm state transfer to a desired state x^{des}, with input and state bounds. This can be formulated as the QP

$$
\begin{array}{ll}
\text{minimize} & \sum_{\tau=t}^{T-1} \|u_\tau\|_2^2 \\
\text{subject to} & x_{\tau+1} = Ax_\tau + Bu_\tau, \quad \tau = t, \ldots, t+T-1 \\
& u^{\mathrm{min}} \leq u_\tau \leq u^{\mathrm{max}}, \quad \tau = t, \ldots, t+T-1 \\
& x^{\mathrm{min}} \leq x_\tau \leq x^{\mathrm{max}}, \quad \tau = t, \ldots, t+T, \\
& x_T = x^{\mathrm{des}},
\end{array}
$$

with variables $u_t, \ldots, u_{t+T-1}, x_t, \ldots, x_{t+T}$. (The inequalities on u_τ and x_τ are componentwise.)

1.3.6 Model predictive control

We consider a linear dynamical system

$$x_{t+1} = Ax_t + Bu_t + w_t, \quad t = 1, 2, \ldots,$$

where $x_t \in \mathbf{R}^n$ is the state, $u_t \in \mathcal{U} \subset \mathbf{R}^m$ is the control input, and $w_t \in \mathbf{R}^n$ is a zero mean random process noise, at time period $t \in \mathbf{Z}_+$. The set $\mathcal{U}$, which is called the input constraint set, is defined by a set of linear inequalities; a typical case is a box,

$$\mathcal{U} = \{v \mid \|v\|_\infty \leq U^{\mathrm{max}}\}.$$

We use a state feedback function (control policy) $\varphi : \mathbf{R}^n \to \mathcal{U}$, with $u(t) = \varphi(x_t)$, so the "closed-loop" system dynamics are

$$x_{t+1} = Ax_t + B\varphi(x_t) + w_t, \quad t = 1, 2, \ldots.$$

The goal is to choose the control policy φ to minimize the average stage cost, defined as

$$J = \lim_{T \to \infty} \frac{1}{T} \sum_{t=1}^{T} \mathbf{E} \left(x_t^T Q x_t + u_t^T R u_t \right),$$

where $Q \succeq 0$ and $R \succeq 0$. The expectation here is over the process noise.

Model predictive control is a general method for finding a good (if not optimal) control policy. To find $u_t = \varphi^{\mathrm{mpc}}(x_t)$, we first solve the optimization problem

$$
\begin{array}{ll}
\text{minimize} & \frac{1}{T} \sum_{t=1}^{T} \left(z_t^T Q z_t + v_t^T R v_t \right) + z_{T+1}^T Q_f z_{T+1} \\
\text{subject to} & z_{t+1} = A z_t + B v_t, \quad t = 1, \ldots, T \\
& v_t \in \mathcal{U}, \quad t = 1, \ldots, T \\
& z_1 = x_t,
\end{array}
\tag{1.3}
$$

with variables $v_1, \ldots, v_T \in \mathbf{R}^m$ and $z_1, \ldots, z_{T+1} \in \mathbf{R}^n$. Here T is called the MPC horizon, and $Q_f \succeq 0$ defines the final state cost. We can interpret the solution to this problem as a plan for the next T time steps, starting from the current state, and ignoring the disturbance. Our control policy is

$$u_t = \varphi^{\mathrm{mpc}}(x_t) = v_1^\star,$$

where $v^\star$ is a solution of the problem (1.3). Roughly speaking, in MPC we compute a *plan of action* for the next T steps, but then execute only the *first control input* from the plan.

The difference between real-time trajectory planning and MPC is *recourse* (or feedback). In real-time trajectory planning an input sequence is chosen, and then executed. In MPC, a trajectory plan is carried out at each step, based on the most current information. In trajectory planning, the system model is deterministic, so no recourse is needed.

One important special case of MPC is when the MPC horizon is $T = 1$, in which case the control policy is

$$u_t = \underset{v \in \mathcal{U}}{\mathrm{argmin}} \left(v^T R v + (A x_t + B v)^T Q_f (A x_t + B v) \right). \tag{1.4}$$

In this case the control policy is referred to as a control-Lyapunov policy [90, 91].

To evaluate $\varphi(x_t)$, we must solve instances of the QP (1.3) or (1.4). The only parameter in these problem families is x_t; the other problem data $(A, B, \mathcal{U}, Q, R, Q_f, T)$ are fixed and known.

There are several useful initializations for the QP (1.3) [6]. One option is to use a linear state feedback gain for an associated, unconstrained control problem. Another is to propagate a solution from the previous time step forward.

1.3.7   Optimal network flow rates

This is an example of a resource allocation or resource sharing problem, where the resource to be allocated is the bandwidth over each of a set of links [92, 93],

[94, Section 8]. We consider a network with m edges or links, labeled $1, \ldots, m$, and n flows, labeled $1, \ldots, n$. Each flow has an associated non-negative flow rate f_j; each edge or link has an associated positive capacity c_i. Each flow passes over a fixed set of links (its route); the total traffic t_i on link i is the sum of the flow rates over all flows that pass through link i. The flow routes are described by a routing matrix $R \in \{0, 1\}^{m \times n}$, defined as

$$R_{ij} = \begin{cases} 1 & \text{flow } j \text{ passes through link } i \\ 0 & \text{otherwise.} \end{cases}$$

Thus, the vector of link traffic, $t \in \mathbf{R}^m$, is given by $t = Rf$. The link capacity constraints can be expressed as $Rf \le c$.

With a given flow vector f, we associate a total utility

$$U(f) = U_1(f_1) + \cdots + U_n(f_n),$$

where U_i is the utility for flow i, which we assume is concave and nondecreasing. We will choose flow rates that maximize total utility, in other words, that are solutions of

$$\begin{array}{ll} \text{maximize} & U(f) \\ \text{subject to} & Rf \le c, \quad f \ge 0, \end{array}$$

with variable f. This is called the *network utility maximization* (NUM) problem.

Typical utility functions include linear, with $U_i(f_i) = w_i f_i$, where w_i is a positive constant; logarithmic, with $U_i(f_i) = w_i \log f_i$, and saturated linear, with $U_i(f_i) = w_i \min\{f_i, s_i\}$, where w_i is a positive weight and s_i is a positive satiation level. With saturated linear utilities, there is no reason for any flow to exceed its satiation level, so the NUM problem can be cast as

$$\begin{array}{ll} \text{maximize} & w^T f \\ \text{subject to} & Rf \le c, \quad 0 \le f \le s, \end{array} \tag{1.5}$$

with variable f.

In a real-time setting, we can imagine that R, and the form of each utility function, are fixed; the link capacities and flow utility weights or satiation flow rates change with time. We solve the NUM problem repeatedly, to adapt the flow rates to changes in link capacities or in the utility functions.

Several initializations for (1.5) can be used. One simple one is $f = \alpha \mathbf{1}$, with $\alpha = \min_i c_i / k_i$, where k_i is the number of flows that pass over link i.

1.3.8   Optimal power generation and distribution

This is an example of a single commodity, network flow optimization problem. We consider a single commodity network, such as an electrical power network, with n nodes, labeled $1, \ldots, n$, and m directed edges, labeled $1, \ldots, m$. Sources (generators) are connected to a subset $\mathcal{G}$ of the nodes, and sinks (loads) are connected to a subset $\mathcal{L}$ of the nodes. Power can flow along the edges (lines), with a loss that depends on the flow.

We let p_j^{in} denote the (non-negative) power that enters the tail of edge j; p_j^{out} will denote the (non-negative) power that emerges from the head of edge j. These are related by

$$p_j^{\text{in}} = p_j^{\text{out}} + \ell_j(p_j^{\text{in}}), \quad j = 1, \dots, m, \tag{1.6}$$

where $\ell_j(p_j^{\text{in}})$ is the loss on edge j. We assume that ℓ_j is a non-negative, increasing, and convex function. Each line also has a maximum allowed input power: $p_j^{\text{in}} \le P_j^{\text{max}}$, $j = 1, \dots, m$.

At each node the total incoming power, from lines entering the node and a generator, if one is attached to the node, is converted and routed to the outgoing nodes, and to any attached loads. We assume the conversion has an efficiency $\eta_i \in (0, 1)$. Thus we have

$$l_i + \sum_{j \in \mathcal{I}(i)} p_j^{\text{out}} = \eta_i \left(g_i + \sum_{j \in \mathcal{O}(j)} p_j^{\text{in}} \right), \quad i = 1, \dots, n, \tag{1.7}$$

where l_i is the load power at node i, g_i is the generator input power at node i, $\mathcal{I}(i)$ is the set of incoming edges to node i, and $\mathcal{O}(i)$ is the set of outgoing edges from node i. We take $l_i = 0$ if $i \notin \mathcal{L}$, and $g_i = 0$ if $i \notin \mathcal{G}$.

In the problem of optimal generation and distribution, the node loads l_i are given; the goal is to find generator powers $g_i \le G_i^{\text{max}}$, and line power flows p_i^{in} and p_j^{out}, that minimize the total generating cost, which we take to be a linear function of the powers, $c^T g$. Here c_i is the (positive) cost per watt for generator i. The problem is thus

$$\begin{array}{ll} \text{minimize} & c^T g \\ \text{subject to} & (1.6), \ (1.7) \\ & 0 \le g \le G^{\text{max}} \\ & 0 \le p^{\text{in}} \le P^{\text{max}}, \quad 0 \le p^{\text{out}} \end{array}$$

with variables g_i, for $i \in \mathcal{G}$; $p^{\text{in}} \in \mathbf{R}^m$, and $p^{\text{out}} \in \mathbf{R}^m$. (We take $g_i = 0$ for $i \notin \mathcal{G}$.)

Relaxing the line equations (1.6) to the inequalities

$$p_j^{\text{in}} \ge p_j^{\text{out}} + \ell_j(p_j^{\text{in}}), \quad j = 1, \dots, m,$$

we obtain a convex optimization problem. (It can be shown that every solution of the relaxed problem satisfies the line loss equations [1.6].)

The problem described above is the basic static version of the problem. There are several interesting dynamic versions of the problem. In the simplest, the problem data (e.g., the loads and generation costs) vary with time; in each time period, the optimal generation and power flows are to be determined by solving the static problem. We can add constraints that couple the variables across time periods; for example, we can add a constraint that limits the increase or decrease of each generator power in each time period. We can also add energy storage elements at some nodes, with various inefficiencies, costs, and limits; the resulting problem could be handled by (say) model predictive control.

1.3.9 Processor-speed scheduling

We first describe the deterministic finite-horizon version of the problem. We must choose the speed of a processor in each of T time periods, which we denote $s_1, \ldots, s_T$. These must lie between given minimum and maximum values, $s^{\min}$ and $s^{\max}$. The energy consumed by the processor in period t is given by $\phi(s_t)$, where $\phi : \mathbf{R} \to \mathbf{R}$ is increasing and convex. (A very common model, based on simultaneously adjusting the processor voltage with its speed, is quadratic: $\phi(s_t) = \alpha s_t^2$.) The total energy consumed over all the periods is $E = \sum_{t=1}^{T} \phi(s_t)$.

Over the T time periods, the processor must handle a set of n jobs. Each job has an availability time $A_i \in \{1, \ldots, T\}$, and a deadline $D_i \in \{1, \ldots, T\}$, with $D_i \geq A_i$. The processor cannot start work on job i until period $t = A_i$, and must complete the job by the end of period D_i. Each job i involves a (non-negative) total work W_i.

In period t, the processor allocates its total speed s_t across the n jobs as

$$s_t = S_{t1} + \cdots + S_{tn},$$

where $S_{ti} \geq 0$ is the effective speed the processor devotes to job i during period t. To complete the jobs we must have

$$\sum_{t=A_i}^{D_i} S_{ti} \geq W_i, \quad i = 1, \ldots, n. \tag{1.8}$$

(The optimal allocation will automatically respect the availability and deadline constraints, i.e., satisfy $S_{ti} = 0$ for $t < A_i$ or $t > D_i$.)

We will choose the processor speeds, and job allocations, to minimize the total energy consumed:

$$
\begin{array}{ll}
\text{minimize} & E = \sum_{t=1}^{T} \phi(s_t) \\
\text{subject to} & s^{\min} \leq s \leq s^{\max}, \quad s = S\mathbf{1}, \quad S \geq 0 \\
& (1.8),
\end{array}
$$

with variables $s \in \mathbf{R}^T$ and $S \in \mathbf{R}^{T \times n}$. (The inequalities here are all elementwise.)

In the simplest embedded real-time setting, the speeds and allocations are found for consecutive blocks of time, each T periods long, with no jobs spanning two blocks of periods. The speed-allocation problem is solved for each block separately; these optimization problems have differing job data (availability time, deadline, and total work).

We can also schedule the speed over a rolling horizon, that extends T periods into the future. At time period t, we schedule processor speed and allocation for the periods $t, t+1, \ldots, t+T$. We interpret n as the maximum number of jobs that can be simultaneously active over such a horizon. Jobs are dynamically added and deleted from the list of active jobs. When a job is finished, it is removed; if a job has already been allocated speed in previous periods, we simply set its availability time to t, and change its required work to be the remaining work to be done. For jobs with deadlines beyond our horizon, we set the deadline to be $t + T$ (the end of our rolling horizon), and linearly interpolate the required work. This gives us a model predictive control method, where we solve the resulting (changing) processor speed and allocation problems in each period, and use the processor speed and allocation corresponding to the current time period. Such a method can dynamically adapt to changing job workloads, new jobs, canceled jobs, or changes in availability and deadlines. This scheme requires the solution of a scheduling problem in each period.

1.4   Algorithm considerations

1.4.1   Requirements

The requirements and desirable features of algorithms for real-time embedded optimization applications differ from those for traditional applications. We first list some important requirements for algorithms used in real-time applications.

Stability and reliability

The algorithm should work well on all, or almost all, $a \in \mathcal{A}$. In contrast, a small failure rate is expected and tolerated in traditional generic algorithms, as a price paid for the ability to efficiently solve a wide range of problems.

Graceful handling of infeasibility

When the particular problem instance is infeasible, or near the feasible–infeasible boundary, a point that is closest to feasible, in some sense, is typically needed. Such points can be found with a traditional Phase I method [1, Section 11.4], which minimizes the maximum constraint violation, or a sum of constraint violations. In industrial implementations of MPC controllers, for example, the state bound constraints are replaced with what are called soft constraints, in other words, penalties for violating the state constraints that are added to the objective function [21, Section 3.4]. Another option is to use an infeasible Newton-based method [1, Section 10.3], in which all iterates satisfy the inequality constraints, but not necessarily the equality constraints, and simply terminate this after a fixed number of steps, whether or not the equality constraints are satisfied [6].

Guaranteed-run time bounds

Algorithms used in a real-time setting must be fast, with execution time that is *predictable* and *bounded*. Any algorithm in a real-time loop must have a finite maximum execution time, so results become available in time for the rest of the real-time loop to proceed. Most traditional optimization algorithms have variable run times, since they exit only when certain residuals are small enough.

Another option, that can be useful in synchronous or asynchronous real-time optimization applications, is to employ an *any-time* algorithm, that is, an algorithm which can be interrupted at any time (after some minimum), and shortly thereafter returns a reasonable approximation of the solution [95, 96].

1.4.2　Exploitable features

On the other hand, real-time applications present us with several features that can work to our advantage, compared to traditional generic applications.

Known (and often modest) accuracy requirements

Most general-purpose solvers provide high levels of accuracy, commonly providing optimal values accurate to six or more significant figures. In a real-time setting, such high accuracy is usually unnecessary. For any specific real-time application, the required accuracy is usually known, and is typically far less than six figures. There are several reasons that high accuracy is often not needed in real-time applications. The variables might represent actions that can be only carried out with some finite fixed resolution (as in a control actuator), so accuracy beyond this resolution is meaningless. As another example, the problem data might be (or come from) physical measurements, which themselves have relatively low accuracy; solving the optimization problem to high accuracy, when the data itself has low accuracy, is unnecessary. And finally, the model (such as a linear dynamical system model or a statistical model) used to form the real-time optimization problem might not hold to high accuracy; so once again solving the problem to high accuracy is unnecessary.

In many real-time applications, the optimization problem can be solved to low or even very low accuracy, without substantial deterioration in the performance of the overall system. This is especially the case in real-time feedback control, or systems that have recourse, where feedback helps to correct errors from solving previous problem instances inaccurately. For example, Wang and Boyd recently found that, even when the QPs arising in MPC are solved very crudely, high quality control is still achieved [6].

Good initializations are often available

In real-time optimization applications, we often have access to a good initial guess for $x^\star$. In some problems, this comes from a heuristic or approximation specific to the application. For example, in MPC we can initialize the trajectory with one found (quickly) from a classical control method, with a simple projection to ensure the inequality constraints are satisfied. In other real-time optimization applications, the successive problem instances to be solved are near each other, so the optimal point from the last solved problem instance is a good starting point for the current problem instance. MPC provides a good example here, as noted earlier: the most recently computed trajectory can be shifted by one time step, with the boundaries suitably adjusted.

Using a previous solution, or any other good guess, as an initialization for a new problem instance is called *warm starting* [97], and in some cases can dramatically reduce the time required to compute the new solution.

Variable ranges can be determined

A generic solver must work well for data (and solutions) that vary over large ranges of values, that are not typically specified ahead of time. In any particular real-time embedded application, however, we can obtain rather good data about the range of values of variables and parameters. This can be done through simulation of the system, with historical data, or with randomly generated data from an appropriate distribution. The knowledge that a variable lies between 0 and 10, for example, can be used to impose (inactive) bounds on it, even when no bounds are present in the original problem statement. Adding bounds like this, which are meant to be inactive, can considerably improve the reliability of, for example, interior-point methods. Other solution methods can use these bounds to tune algorithm parameters.

1.4.3 Interior-point methods

Many methods can be used to solve optimization problems in a real-time setting. For example, Diehl *et al.* [28, 29, 98] have used active set methods for real-time nonlinear MPC. First order methods, such as classical projected-gradient methods [99], or the more recently developed mirror-descent methods [100], can also be attractive, especially when warm-started, since the accuracy requirements for embedded applications can sometimes be low. The authors have had several successful experiences with interior-point methods. These methods typically require several tens of steps, each of which involves solving a set of equations associated with Newton's method.

Simple primal-barrier methods solve a sequence of smooth, equality-constrained problems using Newton's method, with a barrier parameter κ that controls the accuracy or duality gap [1, Section 11], [101]. For some real-time embedded applications, we can fix the accuracy parameter κ at some suitable value, and limit the number of Newton steps taken. With proper choice of κ, and warm-start initialization, good application performance can be obtained with just a few Newton steps. This approach is used in [32] to compute optimal robot-grasping forces, and in [6] for MPC.

More sophisticated interior-point methods, such as primal-dual methods [64, Section 19, 65] are also very good candidates for real-time embedded applications. These methods can reliably solve problem instances to high accuracy in several tens of steps, but we have found that in many cases, accuracy that is more than adequate for real-time embedded applications is obtained in just a few steps.

1.4.4 Solving systems with KKT-like structure

The dominant effort required in each iteration of an interior-point method is typically the calculation of the search direction, which is found by solving one or two sets of linear equations with KKT (Karush–Kuhn–Tucker) structure:

$$\begin{bmatrix} H & A^T \\ A & 0 \end{bmatrix} \begin{bmatrix} \Delta x \\ \Delta y \end{bmatrix} = \begin{bmatrix} r_1 \\ r_2 \end{bmatrix}. \tag{1.9}$$

Here H is positive semidefinite, A is full rank and fat (i.e., has fewer rows than columns), and Δx and Δy are (or are used to find) the search direction for the primal and dual variables. The data in the KKT system change in each iteration, but the sparsity patterns in H and A are often the same for all iterations, and all problem instances to be solved. This common sparsity pattern derives from the original problem family.

The KKT equations (1.9) can be solved by several general methods. An iterative solver can provide good performance for very large problems, or when extreme speed is needed, but can require substantial tuning of the parameters [102, 103]. For small- and medium-size problems, though, we can employ a direct method, such as LDL^T ("signed Cholesky") factorization, possibly using block elimination [104]. We find a permutation P (also called an elimination ordering or pivot sequence), and a lower triangular matrix L and a diagonal matrix D (both invertible) such that

$$P \begin{bmatrix} H & A^T \\ A & 0 \end{bmatrix} P^T = LDL^T. \tag{1.10}$$

Once the LDL^T factorization has been found, we use backward and forward elimination to solve the KKT system (1.9) [1, Section C.4.2], [105]. The overall effort required depends on the sparsity pattern of L; more specifically, on the number of nonzero entries. This number is always at least as large as the number of nonzero entries in the lower triangular part of the KKT matrix; additional nonzero entries in L, that are not in the KKT matrix, are called "fill-in entries", and the number of fill-in entries is referred to as the "fill-in". The smaller the fill-in, the more efficiently the KKT system can be solved.

The permutation P is chosen to reduce fill-in, while avoiding numerical instability (such as dividing by a very small number) in the factorization, in other words, the computation of L and D. In an extreme case, we can encounter a divide-by-zero in our attempt to compute the LDL^T factorization (1.10), which means that such a factorization does not exist for that particular KKT matrix instance, and that choice of permutation. (The factorization exists if, and only if, every leading principal submatrix of the permuted KKT matrix is nonsingular.)

Static pivoting or *symbolic permutation* refers to the case when P is chosen based only on the sparsity pattern of the KKT matrix. In contrast, *"dynamic pivoting"* refers to the case when P is chosen, in part, based on the numeric values in the partially factorized matrix. Most general-purpose, sparse-equation solvers use dynamic pivoting; static pivoting is used in some special cases, such as when H is positive definite and A is absent. For real-time embedded applications, static pivoting has several advantages. It results in a simple algorithm with no conditionals, which allows us to bound the run-time (and memory requirements) reliably, and allows much compiler optimization (since the algorithm is branch free). So we proceed assuming that static permutation will be employed. In other words, we will choose one permutation P and use it to solve the KKT system arising in each interior-point iteration in each problem instance to be solved.

Methods for choosing P, based on the sparsity pattern in the KKT matrix, generally use a heuristic for minimizing fill-in, while guaranteeing that the LDL^T factorization exists. KKT matrices have special structure, which may be exploited when selecting the permutation [106–108]. A recent example is the KKTDirect package, developed by Bridson [109], which chooses a permutation that guarantees existence of the LDL^T factorization, provided H is positive definite and A is full rank, and tends to achieve low fill-in. Other methods include approximate minimum-degree ordering [110], or METIS [111], which may be applied to the positive definite portion of the KKT matrix, and again after a block reduction. While existence of the factorization does not guarantee numerical stability, it has been observed in practice. (Additional methods, described below, can be used to guard against numerical instability.)

One pathology that can occur is when H is singular (but still positive semidefinite). One solution is to solve the (equivalent) linear equations

$$\begin{bmatrix} H + A^T QA & A^T \\ A & 0 \end{bmatrix} \begin{bmatrix} \Delta x \\ \Delta y \end{bmatrix} = \begin{bmatrix} r_1 + A^T Qr_2 \\ r_2 \end{bmatrix},$$

where Q is any positive semidefinite matrix for which $H + A^T QA$ is positive definite [1, Section 10.4.2]. The key here is to choose Q (if possible), so that the number of nonzero entries in $H + A^T QA$ is not too much more than the number of nonzero entries in A.

Some standard tricks used in optimization computations can also be used in the context of real-time embedded applications. One is to dynamically add diagonal elements to the KKT matrix, during factorization, to avoid division by small numbers (including zero) [112]. In this case we end up with the factorization of a matrix that is close to, but not the same as, the KKT matrix; the search direction computed using this approximate factorization is only an approximate solution of the original KKT system. One option is to simply use the resulting approximate search direction in the interior-point method, as if it were the exact search direction [65, Section 11]. Another option is to use a few steps of iterative refinement, using the exact KKT matrix and the approximate factorization [113].

1.5   Code generation

1.5.1   Custom KKT solver

To generate a custom solver based on an interior-point method, we start by generating code to carry out the LDL^T factorization. The most important point here is that the sparsity pattern of L can be determined symbolically, from the sparsity patterns of H and A (which, in turn, derive from the structure of the original problem family), and the permutation matrix P. In particular, the sparsity pattern of L, as well as the exact list of operations to be carried out in the factorization, are known at code generation time. Thus, at code generation time, we can carry out the following tasks.

1. *Choose the permutation P.* This is done to (approximately) minimize fill-in while ensuring existence of the factorization. Considerable effort can be expended in this task, since it is done at code generation time.
2. *Determine storage schemes.* Once the permutation is fixed, we can choose a storage scheme for the permuted KKT matrix (if we in fact form it explicitly), and its factor L.
3. *Generate code.* We can now generate code to perform the following tasks.

   - Fill the entries of the permuted KKT matrix, from the parameter a and the current primal and dual variables.

- Factor the permuted KKT matrix, that is, compute the values of L and D.
- Solve the permuted KKT system, by backward and forward substitution.

Thus, we generate custom code that quickly solves the KKT system (1.9). Note that once the code is generated, we know the exact number of floating-point operations required to solve the KKT system.

1.5.2 Additional optimizations

The biggest reduction in solution time comes from careful choice of algorithm, problem transformations, and permutations used in computing the search directions. Together, these fix the number of floating-point operations ("flops") that a solver requires. Floating-point arithmetic is typically computationally expensive, even when dedicated hardware is available.

A secondary goal, then, is to maximize utilization of a computer's floating-point unit or units. This is a standard code optimization task, some parts of which a good code generator can perform automatically. Here are some techniques that may be worth exploring. For general information about compilers, see [114].

- *Memory optimization.* We want to store parameter and working problem data as efficiently as possible in memory, to access it quickly, maximize locality of reference, and minimize total memory used. The code generator should choose the most appropriate scheme ahead of time.
  Traditional methods for working with sparse matrices, for example, in packages like UMFPACK [115] and CHOLMOD [116], require indices to be stored along with problem data. At code generation time we already know all locations of nonzero entries, so we have the choice of removing explicit indices, designing a specific storage structure, and then referring to the nonzero entries directly.
- *Unrolling code.* Similar to unrolling loops, we can "unroll" factorizations and multiplies. For example, when multiplying two sparse matrices, one option is to write each (scalar) operation explicitly in code. This eliminates loops and branches to make fast, linear code, but also makes the code more verbose.
- *Caching arithmetic results.* If certain intermediate arithmetic results are used multiple times, it may be worth trading additional storage for reduced floating-point computations.
- *Re-ordering arithmetic.* As long as the algorithm remains mathematically correct, it may be helpful to re-order arithmetic instructions to better use caches. It may also be worth replacing costly divides (say) with additional multiplies.
- *Targeting specific hardware.* Targeting specific hardware features may allow performance gains. Some processors may include particular instruction sets (such as SSE [117], which allows faster floating-point operations). This typically requires carefully arranged memory access patterns, which the code generator may be able to provide. Using more exotic hardware is possible too; graphics processors allow high-speed parallel operations [118], and some recent work has investigated using FPGAs for MPC [119, 120].

- *Parallelism.* Interior-point algorithms offer many opportunities for parallelism, especially because all instructions can be scheduled at code generation time. A powerful code generator may be able to generate parallel code to efficiently use multiple cores or other parallel hardware.
- *Empirical optimization.* At the expense of extra code generation time, a code generator may have the opportunity to use empirical optimization to find the best of multiple options.
- *Compiler selection.* A good compiler with carefully chosen optimization flags can significantly improve the speed of program code.

We emphasize that all of this optimization can take place at code generation time, and thus, can be done in relative leisure. In a real-time optimization setting, longer code generation and compile times can be tolerated, especially when the benefit is solver code that runs very fast.

1.6 CVXMOD: a preliminary implementation

We have implemented a code generator, within the CVXMOD framework, to test some of these ideas. It is a work in progress; we report here only a preliminary implementation. It can handle any problem family that is representable via disciplined convex programming [121–123] as a QP (including, in particular, LP). Problems are expressed naturally in CVXMOD, using QP-representable functions such as `min`, `max`, `norm1`, and `norminf`.

A wide variety of algorithms can be used to solve QPs. We briefly describe here the particular primal-dual method we use in our current implementation. While it has excellent performance (as we will see from the experimental results), we do not claim that it is any better than other, similar methods.

1.6.1 Algorithm

CVXMOD begins by transforming the given problem into the standard QP form (1.2). The optimization variable therein includes the optimization variables in the original problem, and possibly other, automatically introduced variables. Code is then prepared for a Mehrotra predictor-corrector, primal-dual, interior-point method [124].

We start by introducing a slack variable $s \in \mathbf{R}^m$, which results in the problem

$$
\begin{array}{ll}
\text{minimize} & (1/2)x^T P x + q^T x \\
\text{subject to} & Gx + s = h, \quad Ax = b, \quad s \geq 0,
\end{array}
$$

with (primal) variables x and s. Introducing dual variables $y \in \mathbf{R}^n$ and $z \in \mathbf{R}^m$, and defining $X = \mathbf{diag}(x)$ and $S = \mathbf{diag}(s)$, the KKT optimality conditions for this problem are

$$Px + q + G^T z + A^T y = 0$$
$$Gx + s = h, \quad Ax = b$$
$$s \geq 0, \quad z \geq 0$$
$$ZS = 0.$$

The first and second lines (which are linear equations) correspond to dual and primal feasibility, respectively. The third gives the inequality constraints, and the last line (a set of nonlinear equations) is the complementary slackness condition.

In a primal-dual, interior-point method, the optimality conditions are solved by a modified Newton method, maintaining strictly positive s and z (including by appropriate choice of step length), and linearizing the complementary slackness condition at each step. The linearized equations are

$$\begin{bmatrix} P & 0 & G^T & A^T \\ 0 & Z & S & 0 \\ G & I & 0 & 0 \\ A & 0 & 0 & 0 \end{bmatrix} \begin{bmatrix} \Delta x^{(i)} \\ \Delta s^{(i)} \\ \Delta z^{(i)} \\ \Delta y^{(i)} \end{bmatrix} = \begin{bmatrix} r_1^{(i)} \\ r_2^{(i)} \\ r_3^{(i)} \\ r_4^{(i)} \end{bmatrix},$$

which, in a Mehrotra predictor-corrector scheme, we need to solve with two different right-hand sides [124]. This system of linear equations is nonsymmetric, but can be put in standard KKT form (1.9) by a simple scaling of variables:

$$\begin{bmatrix} P & 0 & G^T & A^T \\ 0 & S^{-1}Z & I & 0 \\ G & I & 0 & 0 \\ A & 0 & 0 & 0 \end{bmatrix} \begin{bmatrix} \Delta x^{(i)} \\ \Delta s^{(i)} \\ \Delta z^{(i)} \\ \Delta y^{(i)} \end{bmatrix} = \begin{bmatrix} r_1^{(i)} \\ S^{-1} r_2^{(i)} \\ r_3^{(i)} \\ r_4^{(i)} \end{bmatrix}.$$

The (1,1) block here is positive semidefinite.

The current implementation of CVXMOD performs two steps of block elimination on this block 4×4 system of equations, which results in a reduced block 2×2 system, also of KKT form. We determine a permutation P for the reduced system using KKTDirect [109].

The remainder of the implementation of the primal-dual algorithm is straightforward, and is described elsewhere [64, 65, 124]. Significant performance improvements are achieved by using many of the additional optimizations described in Section 1.5.2.

1.7   Numerical examples

To give a rough idea of the performance achieved by our preliminary code generation implementation, we conducted numerical experiments. These were performed on an unloaded Intel Core Duo 1.7 GHz, with 2 GB of RAM and Debian GNU Linux 2.6. The results for several different examples are given below.

1.7.1   Model predictive control

We consider a model predictive control problem as described in Section 1.3.6, with state dimension $n = 10$, $m = 3$ actuators, and horizon $T = 10$. We generate A and B with $\mathcal{N}(0, 1)$ entries, and then scale A so that its spectral radius is one (which makes the control challenging, and therefore interesting). The input constraint set $\mathcal{U}$ is a box, with $U^{\max} = 0.15$. (This causes about half of the control inputs to saturate.) Our objective is defined by $Q = I$, $R = I$. We take $Q_f = 0$, instead adding the constraint that $x(10) = 0$. All of these data are constants in the problem family (1.3); the only parameter is $x(1)$, the initial state. We generate $10\,000$ instances of the problem family by choosing the components of $x(1)$ from a (uniform) $\mathcal{U}[-1, 1]$ distribution.

The resulting QP (1.3), both before and after transformation to CVXMOD's standard form, has 140 variables, 120 equality, and 70 inequality constraints. The lower triangular portion of the KKT matrix has 1740 nonzeros; after ordering, the factor L has 3140 nonzeros. (This represents a fill-in factor of just under two.)

We set the maximum number of iterations to be four, terminating early if sufficient accuracy is attained in fewer steps. (The level of accuracy obtained is more than adequate to provide excellent control performance [6].) The performance results are summarized in Table 1.1.

The resulting QP may be solved at well over 1000 times per second, meaning that MPC can run at over 1 kHz. (Kilohertz rates can be obtained using explicit MPC methods, but this problem is too large for that approach.)

1.7.2   Optimal order execution

We consider the open-loop, optimal execution problem described in Section 1.3.2. We use a simple affine model for the mean price trajectory. We fix the mean starting price $\bar{p}_1$ and set

$$\bar{p}_i = \bar{p}_1 + (d\bar{p}_1/(T - 1))(i - 1), \quad i = 2, \ldots, T,$$

where d is a parameter (the price drift). The final mean price $\bar{p}_T$ is a factor $1 + d$ times the mean starting price.

We model price variation as a random walk, parameterized by the single-step variance σ^2/T. This corresponds to the covariance matrix with

$$\Sigma_{ij} = (\sigma^2/T) \min(i, j), \quad i, j = 1, \ldots, T.$$

The standard deviation of the final price is σ.

Table 1.1. Model predictive control. Performance results for 10 000 solves.

Original problem		Transformed problem		Performance (per solve)	
Variables	140	n	140	Step limit	4
Parameters	140	p	120	Steps (avg)	3.3
Equalities	120	m	70	Final gap (avg)	0.9%
Inequalities	60	$\mathbf{nnz}(KKT)$	1740	Time (avg)	425 μs
		$\mathbf{nnz}(L)$	3140	Time (max)	515 μs

We model the effect of sales on prices with

$$
A_{ij} = \begin{cases} (\alpha \bar{p}_1 / S_{\max}) e^{(j-i)/\beta} & i \geq j \\ 0 & i < j, \end{cases}
$$

where α and β are parameters. The parameter α gives the immediate decrease in price, relative to the initial mean price, when we sell the maximum number of shares, $S_{\max}$. The parameter β, which has units of periods, determines (roughly) the number of periods over which a sale impacts prices: after around β periods, the price impact is around $1/e \approx 38\%$ of the initial impact.

To test the performance of the generated code, we generate 1000 problem instances, for a problem family with $T = 20$ periods. We fix the starting price as $\bar{p}_1 = 10$, the risk-aversion parameter as $\gamma = 0.5$, the final price standard deviation as $\sigma = 4$, and the maximum shares sold per period as $S_{\max} = 10\,000$. For each problem instance we randomly generate parameters as follows. We take $d \sim \mathcal{U}[-0.3, 0.3]$ (representing a mean price movement between 30% decrease and 30% increase) and choose $S \sim \mathcal{U}[0, 10\,000]$. We take $\alpha \sim \mathcal{U}[0.05, 0.2]$ (meaning an immediate price impact for the maximum sale ranges between 5% and 20%), and $\beta \sim \mathcal{U}[1, 10]$ (meaning the price increase effect persists between 1 and 10 periods).

CVXMOD transforms the original problem, which has 20 variables, one equality constraint, and 20 inequality constraints, into a standard form problem with 20 variables, 40 inequality constraints, and one equality constraint. The lower triangular portion of the KKT matrix has a total of 231 nonzero entries. After ordering, L also has 231 entries, so there is no fill-in.

We fix the maximum number of iterations at four, terminating earlier if the duality gap is less than 500. (The average objective value is around 250 000, so this represents a required accuracy around 0.2%.) The performance results are summarized in Table 1.2. We can see that well over 100 000 problem instances can be solved in one second.

We should mention that the solve speed obtained is far faster than what is needed in any real-time implementation. One practical use of the very fast solve time is for Monte Carlo simulation, which might be used to test the optimal execution algorithm, tune various model parameters (e.g., A), and so on.

Table 1.2. Optimal-execution problem. Performance results for 1000 solves.

Original problem		Transformed problem		Performance (per solve)	
Variables	20	n	20	Step limit	4
Parameters	232	p	1	Steps (avg)	3.0
Equalities	1	m	40	Final gap (avg)	0.05%
Inequalities	40	**nnz**(KKT)	231	Time (avg)	49 μs
		nnz(L)	231	Time (max)	65 μs

1.7.3   Optimal network flow rates

We consider the NUM problem with satiation (1.5). We choose (and fix) a random routing matrix $R \in \{0, 1\}^{50 \times 50}$ by setting three randomly chosen entries in each column to 1. This corresponds to a network with 50 flows and 50 links, with each flow passing over 3 links.

CVXMOD transforms the original problem, which has 50 scalar variables and 150 inequality constraints, into a problem in standard form, which also has 50 scalar variables and 150 inequality constraints. The lower triangular portion of the KKT matrix has a total of 244 nonzero entries. After ordering, L has 796 nonzero entries. This corresponds to a fill-in factor of a little over three.

To test the performance of the generated code, we generate 100 000 problem instances, with random parameters, chosen as follows. The weights are uniform on $[1, 5]$, the satiation levels are uniform on $[5000, 15\,000]$, and the link capacities are uniform on $[10\,000, 30\,000]$.

We fix a step limit of six, terminating earlier if sufficient accuracy is attained (in this case, if the duality gap passes below 2000, which typically represents about 1.5% accuracy). The performance results are summarized in Table 1.3.

We can see that this optimal flow problem can be solved several thousand times per second, making it capable of responding to changing link capacities, satiation levels, or flow utilities on the millisecond level. As far as we know, flow control on networks is not currently done by explicitly solving optimization problems; instead, it is carried out using protocols that adjust rates based on a simple feedback mechanism, using the number of lost packets, or round-trip delay times, for each flow. Our high solver speed suggests that, in some cases, flow control could be done by explicit solution of an optimization problem.

1.7.4   Real-time actuator optimization

A rigid body has n actuators, each of which applies a force (or torque) at a particular position, in a particular direction. These forces and torques must lie in given intervals. For example, a thruster, or a force transferred by tension in a cable, can only apply a non-negative force, which corresponds to a lower limit of zero. The bounds can also represent limits on the magnitude of the force or torque for each actuator. The forces and torques yield a net force ($\in \mathbf{R}^3$) and net moment ($\in \mathbf{R}^3$) that are linear functions of

Table 1.3. NUM problem. Performance results for 100 000 solves.

Original problem		Transformed problem		Performance (per solve)	
Variables	50	n	50	Step limit	6
Parameters	300	p	0	Steps (avg)	5.7
Equalities	0	m	150	Final gap (avg)	0.8%
Inequalities	150	**nnz**(KKT)	244	Time (avg)	230 μs
		nnz(L)	496	Time (max)	245 μs

the actuator values. The goal is to achieve a given desired net force and moment with minimum actuator cost.

Taking the lower limit to be zero (i.e., all forces must be non-negative), and a linear cost function, we have the problem

$$\begin{array}{ll} \text{minimize} & c^T f \\ \text{subject to} & F^{\text{des}} = Af, \quad \Omega^{\text{des}} = Bf, \\ & 0 \le f \le F^{\text{max}}, \end{array}$$

with variable $f \in \mathbf{R}^n$. Here $c \in \mathbf{R}^n$ defines the cost, and $A \in \mathbf{R}^{3 \times n}$ ($B \in \mathbf{R}^{3 \times n}$) relates the applied forces (moments) to the net force (moment). (These matrices depend on the location and direction of the actuator forces.) The problem data are A, B, F^{des}, and Ω^{des}.

This actuator-optimization problem can be used to determine the necessary actuator signals in real-time. A high-level control algorithm determines the desired net force and moment to be applied at each sampling interval, and the problem above is solved to determine how best the actuators should achieve the required net values.

To test the performance of the generated code, we solve 100 000 problem instances. At each instance we choose an $A, B \in \mathbf{R}^{3 \times 100}$ with entries uniform on $[-2.5, 2.5]$, set the entries of c uniform on $[0.1, 1]$, the entries of F^{net} and Ω^{net} uniform on $[-5, 5]$ and $F^{\text{max}} = 1$.

We fix a step limit of seven, terminating earlier if sufficient accuracy is attained (in this case, if the duality gap passes below 0.007, which typically represents about 1% accuracy). The performance results are summarized in Table 1.4.

We see that the problem instances can be solved at a rate of several thousand per second, which means that this actuator optimization method can be embedded in a system running at several kHz.

1.8   Summary, conclusions, and implications

In real-time embedded optimization we must solve many instances of an optimization problem from a given family, often at regular time intervals. A generic solver can be used if the time intervals are long enough, and the problems small enough. But for many interesting applications, particularly those in which problem instances must be solved

Table 1.4. Actuator problem. Performance results for 100 000 solves.

Original problem		Transformed problem		Performance (per solve)	
Variables	50	n	50	Step limit	7
Parameters	300	p	6	Steps (avg)	6.4
Equalities	6	m	100	Final gap (avg)	0.4%
Inequalities	100	**nnz**(KKT)	356	Time (avg)	170 μs
		nnz(L)	1317	Time (max)	190 μs

in milliseconds (or faster), a generic solver is not fast enough. In these cases a custom solver can always be developed "by hand", but this requires much time and expertise.

We propose that code generation should be used to rapidly generate source code for custom solvers for specific problem families. Much optimization of the algorithm (for example the ordering of variable elimination), can be carried out automatically during code generation. While code generation and subsequent compilation can be slow, the resulting custom solver is very fast, and has a well-defined maximum run-time, which makes it suitable for real-time applications.

We have implemented a basic code generator for convex-optimization problems that can be transformed to QPs, and demonstrated that extremely fast (worst-case) execution times can be reliably achieved. On a 1.7 GHz processor, the generated code solves problem instances with tens, or even more than a hundred variables, in times measured in microseconds. (We believe that these times can be further reduced by various improvements in the code generation.) These are execution times several orders of magnitude faster than those achieved by generic solvers.

There are several other uses for very fast custom solvers, even in cases where the raw speed is not needed in the real-time application. One obvious example is simulation, where many instances of the problem must be solved. Suppose, for example, that a real-time trading system requires solution of a QP instance each second, which (assuming the problem size is modest) is easily achieved with a generic QP solver. If a custom solver can solve this QP in 100 μs, we can carry out simulations of the trading system 10 000 times faster than real-time. This makes it easier to judge performance and tune algorithm parameters.

Our main interest is in real-time embedded applications, in which problem instances are solved repeatedly, and very quickly. An immediate application is MPC, a well-developed, real-time control method that relies on the solution of a QP in each time step. Until now, however, general MPC was mostly considered practical only for "slow" systems, in which the time steps are long, say, seconds or minutes. (One obvious exception is the recent development of explicit MPC, which is limited to systems with a small number of states, actuators, and constraints, and a short horizon.) We believe that MPC should be much more widely used than it is currently, especially in applications with update times measured in milliseconds.

We can think of many other potential applications of real-time embedded optimization, in areas such as robotics, signal processing, and machine learning, to name just a few. Many of the real-time methods in widespread use today are simple, requiring only a few matrix–vector (or even vector–vector) operations at run-time. The parameters or weights, however, are calculated with considerable effort, often by optimization, and off-line. We suspect that methods that solve optimization problems on-line can out perform methods that use only simple calculations on-line. (This is certainly the case in control, which suggests that it should be the case in other application areas as well.)

There might seem to be a clear dividing line between traditional real-time control- and signal processing methods, which rely on simple calculations in each step, and "computational" methods, that carry out what appear to be more complex calculations in each step, such as solving a QP. The classical example here is the distinction between a classical feedback control law, which requires a handful of matrix–matrix and matrix–vector operations in each step, and MPC, which requires the solution of a QP in each time step. We argue that no such clean dividing line exists: we can often solve a QP (well enough to obtain good performance) in a time comparable to that required to compute a classical control law, and, in any case, fast enough for a wide variety of applications. We think that a smearing of the boundaries between real-time optimization and control will occur, with many applications benefitting from solution of optimization problems in real time.

We should make a comment concerning the particular choices we made in the implementation of our prototype code generator. These should be interpreted merely as incidental selections, and not as an assertion that these are the best choices for all applications. In particular, we do not claim that primal-dual, interior-point methods are better than active-set or first-order methods; we do not claim that dynamic pivoting should always be avoided; and we do not claim that using an LDL^T factorization of the reduced KKT system is the best way to solve for the search direction. We do claim that these choices appear to be good enough to result in very fast, and very reliable solvers, suitable for use in embedded real-time optimization applications.

Finally, we mention that our preliminary code generator, and numerical examples, focus on relatively small problems, with tens of variables, and very fast solve times (measured in microseconds). But many of the same ideas apply to much larger problems, say with thousands of variables. Custom code automatically generated for such a problem family would be much faster than a generic solver.

Acknowledgments

We are grateful to several people for very helpful discussions, including Arkadi Nemirovsky, Eric Feron, Yang Wang, and Behcet Acikmese.

The research reported here was supported in part by NSF award 0529426, AFOSR award FA9550-06-1-0514, and by DARPA contract N66001-08-1-2066. Jacob Mattingley was supported in part by a Lucent Technologies Stanford Graduate Fellowship.

References

[1] S. Boyd and L. Vandenberghe, *Convex Optimization*. New York: Cambridge University Press, 2004.

[2] G. F. Franklin, J. D. Powell, and A. Emami-Naeini, *Feedback Control of Dynamic Systems*. Boston, MA: Prentice Hall, 1991.

[3] G. F. Franklin, M. L. Workman, and D. Powell, *Digital Control of Dynamic Systems*. Boston, MA: Addison-Wesley, 1997.

[4] S. J. Qin and T. A. Badgwell, "A survey of industrial model predictive control technology," *Control Engineering Practice*, vol. 11, no. 7, pp. 733–64, 2003.

[5] A. Bemporad and C. Filippi, "Suboptimal explicit receding horizon control via approximate multiparametric quadratic programming," *Journal of Optimization Theory and Applications*, vol. 117, no. 1, pp. 9–38, 2004.

[6] Y. Wang and S. Boyd, "Fast model predictive control using online optimization," in *Proceedings IFAC World Congress*, Jul. 2008, pp. 6974–97.

[7] A. H. Sayed, *Fundamentals of Adaptive Filtering*. IEEE Press, Hoboken, NJ: 2003.

[8] E. J. Candès and T. Tao, "Decoding by linear programming," *IEEE Transactions on Information Theory*, vol. 51, no. 12, pp. 4203–15, 2005.

[9] J. Feldman, D. R. Karger, and M. J. Wainwright, "LP decoding," in *Proceedings, Annual Allerton Conference on Communication Control and Computing*, vol. 41, no. 2, pp. 951–60, 2003.

[10] J. Feldman, "Decoding error-correcting codes via linear programming," PhD dissertation, Massachusetts Institute of Technology, 2003.

[11] J. Jalden, C. Martin, and B. Ottersten, "Semidefinite programming for detection in linear systems—optimality conditions and space-time decoding," *IEEE International Conference on Acoustics, Speech, and Signal Processing*, vol. 4, 2003.

[12] M. Kisialiou and Z.-Q. Luo, "Performance analysis of quasi-maximum-likelihood detector based on semi-definite programming," *IEEE International Conference on Acoustics, Speech, and Signal Processing*, vol. 3, pp. 433–36, 2005.

[13] S. Joshi and S. Boyd, "Sensor selection via convex optimization," *IEEE Transactions on Signal Processing*, to be published.

[14] A. Zymnis, S. Boyd, and D. Gorinevsky. (2008, May). Relaxed maximum a posteriori fault identification. Available: www.stanford.edu/ boyd/papers/fault_det.html

[15] I. M. Ross and F. Fahroo, "Issues in the real-time computation of optimal control," *Mathematical and Computer Modelling*, vol. 43, no. 9-10, pp. 1172–88, 2006.

[16] G. C. Goodwin, M. Seron, and J. D. Dona, *Constrained Control and Estimation: An Optimisation Approach*. London: Springer, 2005.

[17] B. Bäuml and G. Hirzinger, "When hard realtime matters: software for complex mechatronic systems," *Journal of Robotics and Autonomous Systems*, vol. 56, pp. 5–13, 2008.

[18] E. F. Camacho and C. Bordons, *Model Predictive Control*. London: Springer, 2004.

[19] D. E. Seborg, T. F. Edgar, and D. A. Mellichamp, *Process dynamics and control*. New York: Wiley, 1989.

[20] F. Allgöwer and A. Zheng, *Nonlinear Model Predictive Control*. Basel, Germany: Birkhauser, 2000.

[21] J. M. Maciejowski, *Predictive Control with Constraints*. Harlow, UK: Prentice Hall, 2002.

[22] I. Dzafic, S. Tesnjak, and M. Glavic, "Automatic object-oriented code generation to power system on-line optimization and analysis," in *21st IASTED International Conference on Modeling, Identification and Control*, 2002.

[23] R. Soeterboek, *Predictive Control: A Unified Approach*. Upper Saddle River, NJ: Prentice Hall, 1992.

[24] A. Bemporad, M. Morari, V. Dua, and E. N. Pistikopoulos, "The explicit linear quadratic regulator for constrained systems," *Automatica*, vol. 38, no. 1, pp. 3–20, 2002.

[25] C. V. Rao, S. J. Wright, and J. B. Rawlings, "Application of interior-point methods to model predictive control," *Journal of Optimization Theory and Applications*, vol. 99, no. 3, pp. 723–57, 1998.

[26] Y. Wang and S. Boyd, "Performance bounds for linear stochastic control," working manuscript, 2008.

[27] V. M. Zavala and L. T. Biegler, "The advanced-step NMPC controller: optimality, stability and robustness," *Automatica*, 2007, submitted for publication.

[28] M. Diehl, H. G. Bock, and J. P. Schlöder, "A real-time iteration scheme for nonlinear optimization in optimal feedback control," *SIAM Journal on Control and Optimization*, vol. 43, no. 5, pp. 1714–36, 2005.

[29] M. Diehl, R. Findeisen, F. Allgöwer, H. G. Bock, and J. P. Schlöder, "Nominal stability of the real-time iteration scheme for nonlinear model predictive control," *Proceedings Control Theory Applications*, vol. 152, no. 3, pp. 296–308, 2005.

[30] E. Frazzoli, Z. H. Mao, J. H. Oh, and E. Feron, "Aircraft conflict resolution via semidefinite programming," *AIAA Journal of Guidance, Control, and Dynamics*, vol. 24, no. 1, pp. 79–86, 2001.

[31] T. Ohtsuka and H. A. Fujii, "Nonlinear receding-horizon state estimation by real-time optimization technique," *Journal of Guidance, Control, and Dynamics*, vol. 19, no. 4, pp. 863–70, 1996.

[32] S. P. Boyd and B. Wegbreit, "Fast computation of optimal contact forces," *IEEE Transactions on Robotics*, vol. 23, no. 6, pp. 1117–32, 2007.

[33] D. Verscheure, B. Demeulenaere, J. Swevers, J. De Schutter, and M. Diehl, "Practical time-optimal trajectory planning for robots: a convex optimization approach," *IEEE Transactions on Automatic Control*, 2008, submitted for publication.

[34] J. Zhao, M. Diehl, R. W. Longman, H. G. Bock, and J. P. Schloder, "Nonlinear model predictive control of robots using real-time optimization," *Advances in the Astronautical Sciences*, vol. 120, pp. 1023–42, 2004.

[35] E. J. Candes, M. B. Wakin, and S. Boyd, "Enhancing sparsity by reweighted l1 minimization," *Journal of Fourier Analysis and Applications*, vol. 14, no. 5, pp. 877–905, 2008.

[36] J. A. Tropp, "Just relax: convex programming methods for identifying sparse signals in noise," *IEEE Transactions on Information Theory*, vol. 52, no. 3, pp. 1030–51, 2006.

[37] E. Candès, M. Rudelson, T. Tao, and R. Vershynin, "Error correction via linear programming," *Annual symposium on foundations of computer science*, vol. 46, p. 295, 2005.

[38] D. Goldfarb and W. Yin, "Second-order cone programming methods for total variation-based image restoration," *SIAM Journal on Scientific Computing*, vol. 27, no. 2, pp. 622–45, 2005.

[39] A. Zymnis, S. Boyd, and D. Gorinevsky, "Mixed state estimation for a linear Gaussian Markov model," *IEEE Conference on Decision and Control*, pp. 3219–26, 2008.

[40] K. Collins-Thompson, "Estimating robust query models with convex optimization," *Advances in Neural Information Processing Systems*, 2008, submitted for publication.

[41] G. Ginis and J. M. Cioffi, "Vectored transmission for digital subscriber line systems," *IEEE Journal on Selected Areas in Communications*, vol. 20, no. 5, pp. 1085–104, 2002.

[42] P. Stoica, J. Li, and Y. Xie, "On probing signal design for MIMO radar," *IEEE Transactions on Signal Processing*, vol. 55, no. 8, pp. 4151–61, 2007.

[43] W.-K. Ma, T. N. Davidson, K. M. Wong, Z.-Q. Luo, and P.-C. Ching, "Quasi-maximum-likelihood multiuser detection using semi-definite relaxation with application to synchronous CDMA," *IEEE Transactions on Signal Processing*, vol. 50, pp. 912–22, 2002.

[44] F. P. Kelly, A. K. Maulloo, and D. K. H. Tan, "Rate control for communication networks: shadow prices, proportional fairness and stability," *Journal of Operational Research Society*, vol. 49, no. 3, pp. 237–52, 1998.

[45] D. X. Wei, C. Jin, S. H. Low, and S. Hegde, "FAST TCP: motivation, architecture, algorithms, performance," *IEEE/ACM Transactions on Networking*, vol. 14, no. 6, pp. 1246–59, 2006.

[46] M. Chiang, S. H. Low, A. R. Calderbank, and J. C. Doyle, "Layering as optimization decomposition: a mathematical theory of network architectures," *Proceedings of the IEEE*, vol. 95, no. 1, pp. 255–312, 2007.

[47] S. Shakkottai and R. Srikant, *Network Optimization and Control*. Boston, MA: Now Publishers, 2008.

[48] S. Meyn, "Stability and optimization of queueing networks and their fluid models," *Mathematics of Stochastic Manufacturing Systems*, vol. 33, pp. 175–200, 1997.

[49] M. Chiang, C. W. Tan, D. P. Palomar, D. O'Neill, and D. Julian, "Power control by geometric programming," *IEEE Transactions on Wireless Communications*, vol. 6, no. 7, pp. 2640–51, 2007.

[50] D. O'Neill, A. Goldsmith, and S. Boyd, "Optimizing adaptive modulation in wireless networks via utility maximization," *in Proceedings of the IEEE International Conference on Communications*, pp. 3372–77, 2008.

[51] S. C. Johnson, "Yacc: yet another compiler-compiler," *Computing Science Technical Report*, vol. 32, 1975.

[52] C. Donnely and R. Stallman, *Bison version 2.3*, 2006.

[53] The Mathworks, Inc. (2008, Oct.). *Simulink: Simulation and model-based design*. Available: www.mathworks.com/products/simulink

[54] J. Eker, J. W. Janneck, E. A. Lee, J. Liu, X. Liu, J. Ludvig, S. Neuendorffer, S. Sachs, and Y. Xiong, "Taming heterogeneity—the Ptolemy approach," *Proceedings of the IEEE*, vol. 91, no. 1, pp. 127–44, 2003.

[55] E. Kant, "Synthesis of mathematical-modeling software," *IEEE Software*, vol. 10, no. 3, pp. 30–41, 1993.

[56] R. Bacher, "Automatic generation of optimization code based on symbolic non-linear domain formulation," *Proceedings International Symposium on Symbolic and Algebraic Computation*, pp. 283–91, 1996.

[57] ——, "Combining symbolic and numeric tools for power system network optimization," *Maple Technical Newsletter*, vol. 4, no. 2, pp. 41–51, 1997.

[58] C. Shi and R. W. Brodersen, "Automated fixed-point data-type optimization tool for signal processing and communication systems," *ACM IEEE Design Automation Conference*, pp. 478–83, 2004.

[59] T. Oohori and A. Ohuchi, "An efficient implementation of Karmarkar's algorithm for large sparse linear programs," *Proceedings IEEE Conference on Systems, Man, and Cybernetics*, vol. 2, 1988, pp. 1389–92.

[60] L. K. McGovern, "Computational analysis of real-time convex optimization for control systems," PhD dissertation, Massachusetts Institute of Technology, 2000.

[61] E. Hazan, "Efficient algorithms for online convex optimization and their applications," PhD dissertation, Department of Computer Science, Princeton University, Sep. 2006.

[62] I. Das and J. W. Fuller, "Real-time quadratic programming for control of dynamical systems," US Patent 7328074, 2008.

[63] Y. Nesterov and A. Nemirovskii, *Interior Point Polynomial Algorithms in Convex Programming*. Philadelphia, PA: SIAM, 1994.

[64] J. Nocedal and S. J. Wright, *Numerical Optimization*. New York: Springer, 1999.

[65] S. J. Wright, *Primal-Dual Interior-Point Methods*. Philadelphia, PA: SIAM, 1997.

[66] S. Boyd and B. Wegbreit, "Fast computation of optimal contact forces," *IEEE Transactions on Robotics*, vol. 23, no. 6, pp. 1117–32, 2006.

[67] R. Fourer, D. Gay, and B. Kernighan, *AMPL: A Modeling Language for Mathematical Programming*. Pacific Grove, CA: Duxbury Press, 1999.

[68] A. Brooke, D. Kendrick, A. Meeraus, and R. Raman, *GAMS: A User's Guide*. San Francisco, CA: The Scientific Press, 1998.

[69] S.-P. Wu and S. Boyd, "SDPSOL: a parser/solver for semidefinite programs with matrix structure," in *Recent Advances in LMI Methods for Control*, L. El Ghaoui and S.-I. Niculescu, eds. Philadelphia, PA: SIAM, 2000, pp. 79–91.

[70] I. The Mathworks. (2002) *LMI control toolbox 1.0.8* (software package). Available: www.mathworks.com/products/lmi.

[71] L. El Ghaoui, J.-L. Commeau, F. Delebecque, and R. Nikoukhah. (1999). *LMITOOL 2.1 (software package)*. Available: http://robotics.eecs.berkeley.edu/~elghaoui/lmitool/lmitool.html

[72] A. Mutapcic, K. Koh, S.-J. Kim, L. Vandenberghe, and S. Boyd (2005). *GGPLAB: A simple matlab toolbox for geometric programming*. Available: www.stanford.edu/boyd/ggplab/

[73] J. Löfberg, "YALMIP: A toolbox for modeling and optimization in MATLAB," in *Proceedings of the CACSD Conference*, Taipei, Taiwan, 2004. Available: http://control.ee.ethz.ch/~joloef/yalmip.php

[74] M. Grant and S. Boyd. (2008, July). *CVX: Matlab software for disciplined convex programming*. [Web page and software]. Available: www.stanford.edu/~boyd/cvx/

[75] J. Mattingley and S. Boyd. (2008). *CVXMOD: Convex optimization software in Python*. [Web page and software]. Available: http://cvxmod.net/.

[76] W. E. Hart, "Python optimization modeling objects (Pyomo)," in *Proceedings, INFORMS Computing Society Conference*, 2009, to be published.

[77] D. Orban and B. Fourer, "Dr. Ampl: A meta solver for optimization," presented at the CORS/INFORMS Joint International Meeting, 2004.

[78] I. P. Nenov, D. H. Fylstra, and L. V. Kolev, "Convexity determination in the Microsoft Excel solver using automatic differentiation techniques," *International Workshop on Automatic Differentiation*, 2004.

[79] Y. Lucet, H. H. Bauschke, and M. Trienis, "The piecewise linear-quadratic model for computational convex analysis," *Computational Optimization and Applications*, pp. 1–24, 2007.

[80] B. Widrow, J. M. McCool, M. G. Larimore, and C. R. Johnson, Jr., "Stationary and non-stationary learning characteristics of the LMS adaptive filter," *Proceedings of the IEEE*, vol. 64, no. 8, pp. 1151–62, 1976.

[81] S. Haykin, *Adaptive Filter Theory*. Upper Saddle River, NJ: Prentice Hall, 1996.

[82] A. T. Erdogan and C. Kizilkale, "Fast and low complexity blind equalization via subgradient projections," *IEEE Transactions on Signal Processing*, vol. 53, no. 7, pp. 2513–24, 2005.

[83] R. L. G. Cavalcante and I. Yamada, "Multiaccess interference suppression in orthogonal space-time block coded mimo systems by adaptive projected subgradient method," *IEEE Transactions on Signal Processing*, vol. 56, no. 3, pp. 1028–42, 2008.

[84] D. Bertsimas and A. W. Lo, "Optimal control of execution costs," *Journal of Financial Markets*, vol. 1, no. 1, pp. 1–50, 1998.

[85] R. Almgren and N. Chriss, "Optimal execution of portfolio transactions," *Journal of Risk*, vol. 3, no. 2, pp. 5–39, 2000.

[86] L. Rudin, S. Osher, and E. Fatemi, "Nonlinear total variation based noise removal algorithms," *Physica D*, vol. 60, no. 1-4, pp. 259–68, 1992.

[87] S.-J. Kim, K. Koh, M. Lustig, S. Boyd, and D. Gorinevsky, "A method for large-scale l1-regularized least squares," *IEEE Journal on Selected Topics in Signal Processing*, vol. 3, pp. 117–20, 2007.

[88] A. Bemporad, D. Mignone, and M. Morari, "Moving horizon estimation for hybrid systems and fault detection," in *Proceedings of the American Control Conference*, vol. 4, 1999.

[89] B. Kouvaritakis and M. Cannon, *Nonlinear Predictive Control: Theory and Practice*. London: IET, 2001.

[90] M. Corless and G. Leitmann, "Controller design for uncertain system via Lyapunov functions," in *Proceedings of the American Control Conference*, vol. 3, 1988, pp. 2019–25.

[91] E. D. Sontag, "A Lyapunov-like characterization of asymptotic controllability," *SIAM Journal on Control and Optimization*, vol. 21, no. 3, pp. 462–71, 1983.

[92] A. Zymnis, N. Trichakis, S. Boyd, and D. O. Neill, "An interior-point method for large scale network utility maximization," *Proceedings of the Allerton Conference on Communication, Control, and Computing*, 2007.

[93] R. Srikant, *The Mathematics of Internet Congestion Control*. Boston, MA: Birkhäuser, 2004.

[94] D. Bertsekas, *Network Optimization: Continuous and Discrete Models*. Nashna, NH: Athena Scientific, 1998.

[95] M. Boddy and T. L. Dean, "Solving time-dependent planning problems," *Proceedings of the Eleventh International Joint Conference on Artificial Intelligence*, vol. 2, pp. 979–84, 1989.

[96] E. A. Hansen and S. Zilberstein, "Monitoring and control of anytime algorithms: a dynamic programming approach," *Artificial Intelligence*, vol. 126, pp. 139–7, 2001.

[97] E. A. Yildirim and S. J. Wright, "Warm-start strategies in interior-point methods for linear programming," *SIAM Journal on Optimization*, vol. 12, no. 3, pp. 782–810, 2002.

[98] M. Diehl, "Real-time optimization for large scale nonlinear processes," PhD dissertation, University of Heidelberg, 2001.

[99] D. Bertsekas, *Nonlinear Programming*, 2nd ed. Nashna, NH: Athena Scientific, 1999.

[100] A. Nemirovski and D. Yudin, *Problem Complexity and Method Efficiency in Optimization*. Hoboken, NJ: Wiley, 1983.

[101] A. V. Fiacco and G. P. McCormick, "Nonlinear programming: sequential unconstrained minimization techniques," DTIC Research Report AD0679036, 1968.

[102] A. Aggarwal and T. H. Meng, "A convex interior-point method for optimal OFDM PAR reduction," *IEEE International Conference on Communications*, vol. 3, 2005.

[103] Y. Y. Saad and H. A. van der Vorst, "Iterative solution of linear systems in the 20th century," *Journal of Computational and Applied Mathematics*, vol. 123, no. 1-2, pp. 1–33, 2000.

[104] M. A. Saunders and J. A. Tomlin, "Stable reduction to KKT systems in barrier methods for linear and quadratic programming," *International Symposium on Optimization and Computation*, 2000.

[105] G. Golub and C. F. V. Loan, *Matrix Computations*, 2nd ed. Baltimore, MD: Johns Hopkins University Press, 1989.

[106] M. Tuma, "A note on the LDL^T decomposition of matrices from saddle-point problems," *SIAM Journal on Matrix Analysis and Applications*, vol. 23, no. 4, pp. 903–15, 2002.

[107] R. J. Vanderbei, "Symmetric quasi-definite matrices," *SIAM Journal on Optimization*, vol. 5, no. 1, pp. 100–13, 1995.

[108] R. J. Vanderbei and T. J. Carpenter, "Symmetric indefinite systems for interior point methods," *Mathematical Programming*, vol. 58, no. 1, pp. 1–32, 1993.

[109] R. Bridson, "An ordering method for the direct solution of saddle-point matrices," Preprint, 2007 (available at www.cs.ubc.ca/-rbridson/kktdirect/).

[110] P. R. Amestoy, T. A. Davis, and I. S. Duff, "Algorithm 837: AMD, an approximate minimum degree ordering algorithm," *ACM Transactions on Mathematical Software (TOMS)*, vol. 30, no. 3, pp. 381–8, 2004.

[111] G. Karypis and V. Kumar, "A Fast and High Quality Multilevel Scheme for Partitioning Irregular Graphs," *SIAM Journal on Scientific Computing*, vol. 20, pp. 359–2, 1999.

[112] S. J. Wright, "Modified Cholesky factorizations in interior-point algorithms for linear programming," *SIAM Journal on Optimization*, vol. 9, pp. 1159–91, 1999.

[113] P. E. Gill, W. Murray, D. B. Ponceleon, and M. A. Saunders, "Solving reduced KKT systems in barrier methods for linear and quadratic programming," Technical Report SOL 91-7, 1991.

[114] A. Aho, M. Lam, R. Sethi, and J. Ullman, *Compilers: Principles, Techniques, and Tools*, Pearson Education, Upper Saddle River, NJ: 2007.

[115] T. A. Davis (2003). UMFPACK User Guide. Available: www.cise.ufl.edu/ research/sparse/ umfpack.

[116] ——. (2006). "CHOLMOD User Guide". Available: www.cise.ufl.edu/ research/sparse/ cholmod/.

[117] D. Aberdeen and J. Baxter, "Emmerald: a fast matrix-matrix multiply using Intel's SSE instructions," *Concurrency and Computation: Practice and Experience*, vol. 13, no. 2, pp. 103–19, 2001.

[118] E. S. Larsen and D. McAllister, "Fast matrix multiplies using graphics hardware," in *Proceedings of the ACM/IEEE Conference on Supercomputing*, 2001, pp. 55–60.

[119] K. V. Ling, B. F. Wu, and J. M. Maciejowski, "Embedded model predictive control (MPC) using a FPGA," in *Proceedings IFAC World Congress*, 2008, pp. 15 250–5.

[120] M. S. K. Lau, S. P. Yue, K. V. Ling, and J. M. Maciejowski, "A comparison of interior point and active set methods for FPGA implementation of model predictive control," *Proceedings, ECC*, 2009, submitted for publication.

[121] M. Grant, "Disciplined convex programming," PhD dissertation, Department of Electrical Engineering, Stanford University, 2004.

[122] M. Grant, S. Boyd, and Y. Ye, "Disciplined convex programming," in *Global Optimization: from Theory to Implementation*, ser. Nonconvex Optimization and Its Applications, L. Liberti and N. Maculan, eds. New York: Springer Science & Business Media, Inc., 2006, pp. 155–210.

[123] M. Grant and S. Boyd, "Graph implementations for nonsmooth convex programs," in *Recent Advances in Learning and Control (a Tribute to M. Vidyasagar)*, V. Blondel, S. Boyd, and H. Kimura, eds. Berlin, Germany: Springer, 2008, pp. 95–110.

[124] S. Mehrotra, "On the implementation of a primal-dual interior point method," *SIAM Journal on Optimization*, vol. 2, p. 575, 1992.

2   Gradient-based algorithms with applications to signal-recovery problems

Amir Beck and Marc Teboulle

This chapter presents, in a self-contained manner, recent advances in the design and analysis of gradient-based schemes for specially structured, smooth and nonsmooth minimization problems. We focus on the mathematical elements and ideas for building fast gradient-based methods and derive their complexity bounds. Throughout the chapter, the resulting schemes and results are illustrated and applied on a variety of problems arising in several specific key applications such as sparse approximation of signals, total variation-based image-processing problems, and sensor-location problems.

2.1   Introduction

The gradient method is probably one of the oldest optimization algorithms going back as early as 1847 with the initial work of Cauchy. Nowadays, gradient-based methods[1] have attracted a revived and intensive interest among researches both in theoretical optimization, and in scientific applications. Indeed, the very large-scale nature of problems arising in many scientific applications, combined with an increase in the power of computer technology have motivated a "return" to the "old and simple" methods that can overcome the curse of dimensionality; a task which is usually out of reach for the current more sophisticated algorithms.

One of the main drawbacks of gradient-based methods is their speed of convergence, which is known to be slow. However, with proper modeling of the problem at hand, combined with some key ideas, it turns out that it is possible to build fast gradient schemes for various classes of problems arising in applications and, in particular, signal-recovery problems.

The purpose of this chapter is to present, in a self-contained manner, such recent advances. We focus on the essential tools needed to build and analyze fast gradient schemes, and present successful applications to some key scientific problems. To achieve these goals our emphasis will focus on:

- Optimization models/formulations.
- Building approximation models for gradient schemes.

---

[1] We also use the term "gradient" instead of "subgradient" in case of nonsmooth functions.

Convex Optimization in Signal Processing and Communication, eds. Daniel P. Palomar and Yonina C. Eldar. Published by Cambridge University Press ©Cambridge University Press, 2010.

- Fundamental mathematical tools for convergence and complexity analysis.
- Fast gradient schemes with better complexity.

On the application front, we review some recent and challenging problems that can benefit from the above theoretical and algorithmic framework, and we include gradient-based methods applied to:

- Sparse approximation of signals.
- Total variation-based image-processing problems.
- Sensor-location problems.

The contents and organization of the chapter is well summarized by the two lists of items above. We will strive to provide a broad picture of the current research in this area, as well as to motivate further research within the gradient-based framework.

To avoid cutting the flow of the chapter, we refrain from citing references within the text. Rather, the last section of this chapter includes bibliographical notes. While we did not attempt to give a complete bibliography on the covered topics (which is very large), we did try to include earlier works and influential papers, to cite all the sources for the results we used in this chapter, and to indicate some pointers on very recent developments that hopefully will motivate further research in the field. We apologize in advance for any possible omission.

2.2 The general optimization model

2.2.1 Generic problem formulation

Consider the following generic optimization model:

$$(\text{M}) \quad \min \{ F(\mathbf{x}) = f(\mathbf{x}) + g(\mathbf{x}) : x \in \mathbb{E} \},$$

where

- $\mathbb{E}$ is a finite dimensional Euclidean space with inner product $\langle \cdot, \cdot \rangle$ and norm $\| \cdot \| = \langle \cdot, \cdot \rangle^{1/2}$.
- $g : \mathbb{E} \to (-\infty, \infty]$ is a proper, closed and convex function which is assumed subdifferentiable over dom g.[2]
- $f : \mathbb{E} \to (-\infty, \infty)$ is a continuously differentiable function over $\mathbb{E}$.

The model (M) is rich enough to recover generic classes of smooth/nonsmooth convex minimization problems, as well as smooth nonconvex problems. This is illustrated in the following examples.

---

[2] Throughout this paper all necessary notations/definitions/results from convex analysis not explicitly given are standard and can be found in the classic monograph [50].

Example 2.1a Convex-minimization problems   Pick $f \equiv 0$ and $g = h_0 + \delta_C$ where $h_0 : \mathbb{E} \to (-\infty, \infty)$ is a convex function (possibly nonsmooth) and δ_C is the indicator function defined by

$$\delta_C(\mathbf{x}) = \begin{cases} 0 & \mathbf{x} \in C, \\ \infty & \mathbf{x} \notin C, \end{cases}$$

where $C \subseteq \mathbb{E}$ is a closed and convex set. The model (M) reduces to the generic convex-optimization problem

$$\min \{h_0(\mathbf{x}) : \mathbf{x} \in C\}.$$

In particular, if C is described by convex inequality constraints, that is, with

$$C = \{\mathbf{x} \in \mathbb{E} : h_i(\mathbf{x}) \leq 0, \ i = 1, \ldots, m\},$$

where h_i are some given proper, closed convex functions on $\mathbb{E}$, we recover the following general form of convex programs:

$$\min \{h_0(\mathbf{x}) : h_i(\mathbf{x}) \leq 0, \ i = 1, \ldots, m\}.$$

Example 2.1b Smooth constrained minimization   Set $g = \delta_C$ with $C \subseteq \mathbb{E}$ being a closed convex set. Then (M) reduces to the problem of minimizing a smooth (possibly nonconvex) function over C, that is,

$$\min \{f(\mathbf{x}) : \mathbf{x} \in C\}.$$

A more specific example that can be modeled by (M) is from the field of signal recovery and is now described.

2.2.2   Signal recovery via nonsmooth regularization

A basic linear inverse problem is to estimate an unknown signal $\mathbf{x}$ satisfying the relation

$$\mathbf{Ax} = \mathbf{b} + \mathbf{w},$$

where $\mathbf{A} \in \mathbb{R}^{m \times n}$ and $\mathbf{b} \in \mathbb{R}^m$ are known, and $\mathbf{w}$ is an unknown noise vector. The basic problem is then to recover the signal $\mathbf{x}$ from the noisy measurements $\mathbf{b}$. A common approach for this estimation problem is to solve the "regularized least-squares" (RLS) minimization problem

$$\text{(RLS)} \quad \min_{\mathbf{x}} \left\{ \|\mathbf{Ax} - \mathbf{b}\|^2 + \lambda R(\mathbf{x}) \right\}, \tag{2.1}$$

where $\|\mathbf{A}\mathbf{x} - \mathbf{b}\|^2$ is a least-squares term that measures the distance between $\mathbf{b}$ and $\mathbf{A}\mathbf{x}$ in an l_2 norm sense,[3] $R(\cdot)$ is a convex regularizer used to stabilize the solution, and $\lambda > 0$ is a regularization parameter providing the trade-off between fidelity to measurements and noise sensitivity. Model (RLS) is of course a special case of model (M) by setting $f(\mathbf{x}) \equiv \|\mathbf{A}\mathbf{x} - \mathbf{b}\|^2$ and $g(\mathbf{x}) \equiv \lambda R(\mathbf{x})$.

Popular choices for $R(\cdot)$ are dictated from the application in mind and include, for example, the following model:

$$R(\mathbf{x}) = \sum_{i=1}^{s} \|\mathbf{L}_i\mathbf{x}\|_p^p, \tag{2.2}$$

where $s \geq 1$ is an integer number, $p \geq 1$ and $\mathbf{L}_i : \mathbb{R}^n \to \mathbb{R}^{d_i}$ ($d_1, \ldots, d_s$ being positive integers) are linear maps. Of particular interest for signal-processing applications are the following cases:

1. *Tikhonov regularization.* By setting $s = 1, \mathbf{L}_i = \mathbf{L}, p = 2$, we obtain the standard Tikhonov regularization problem:

$$\min_{\mathbf{x}} \|\mathbf{A}\mathbf{x} - \mathbf{b}\|^2 + \lambda \|\mathbf{L}\mathbf{x}\|^2.$$

2. *l_1 regularization.* By setting $s = 1, \mathbf{L}_i = \mathbf{I}, p = 1$ we obtain the l_1 regularization problem:

$$\min_{\mathbf{x}} \|\mathbf{A}\mathbf{x} - \mathbf{b}\|^2 + \lambda \|\mathbf{x}\|_1.$$

Other closely related problems include for example,

$$\min\{\|\mathbf{x}\|_1 : \|\mathbf{A}\mathbf{x} - \mathbf{b}\|^2 \leq \epsilon\} \quad \text{and} \quad \min\{\|\mathbf{A}\mathbf{x} - \mathbf{b}\|^2 : \|\mathbf{x}\|_1 \leq \epsilon\}.$$

The above are typical formulations in statistic regression (LASSO, basis pursuit), as well as in the emerging technology of compressive sensing.

3. *Wavelet-based regularization.* By choosing $p = 1, s = 1, \mathbf{L}_i = \mathbf{W}$ where $\mathbf{W}$ is a wavelet transform matrix, we recover the wavelet-based regularization problem

$$\min_{\mathbf{x}} \|\mathbf{A}\mathbf{x} - \mathbf{b}\|^2 + \lambda \|\mathbf{W}\mathbf{x}\|_1.$$

4. *TV-based regularization.* When the set $\mathbb{E}$ is the set of $m \times n$ real-valued matrices representing the set of all $m \times n$ images, it is often the case that one chooses the regularizer to be a total variation function which has the form

$$R(\mathbf{x}) \equiv \text{TV}(\mathbf{x}) = \sum_{i=1}^{m} \sum_{j=1}^{n} \|(\nabla\mathbf{x})_{i,j}\|.$$

A more precise definition of R will be given in Section 2.7.

---

[3] Throughout the chapter, unless otherwise stated, the norm $\| \cdot \|$ stands for the Euclidean norm associated with $\mathbb{E}$.

Note that the last three examples deal with nonsmooth regularizers. The reason for using such seemingly difficult regularization functions and not the more standard smooth quadratic Tikhonov regularization will be explained in Sections 2.6 and 2.7.

In the forthcoming sections, many of these problems will be described in more detail, and gradient-based methods will be the focus of relevant schemes for their solution.

2.3   Building gradient-based schemes

In this section we describe the elements needed to generate a gradient-based method for solving problems of the form (M). These rely mainly on building "good approximation models" and on applying fixed-point methods on corresponding optimality conditions. In Sections 2.3.1 and 2.3.2 we will present two different derivations – corresponding to the two building techniques – of a method called the 'proximal gradient' algorithm. We will then show in Section 2.3.3 the connection to the so-called majorization–minimization approach. Finally, in Section 2.3.4 we will explain the connection of the devised techniques to Weiszfeld's method for the Fermat–Weber location problem.

2.3.1   The quadratic-approximation model for (M)

Let us begin with the simplest, unconstrained minimization problem of a continuously differentiable function f on $\mathbb{E}$ (i.e., we set $g \equiv 0$ in (M)):

$$(U) \quad \min\{f(\mathbf{x}) : \mathbf{x} \in \mathbb{E}\}.$$

The well-known basic gradient method generates a sequence $\{\mathbf{x}_k\}$ via

$$\mathbf{x}_0 \in \mathbb{E}, \quad \mathbf{x}_k = \mathbf{x}_{k-1} - t_k \nabla f(\mathbf{x}_{k-1}) \quad (k \geq 1), \tag{2.3}$$

where $t_k > 0$ is a suitable step size. The gradient method thus takes, at each iteration, a step along the negative gradient direction, which is the direction of "steepest descent". This interpretation of the method, although straightforward and natural, cannot be extended to the more general model (M). Another simple way to interpret the above scheme is via an approximation model that would replace the original problem (U) with a "reasonable" approximation of the objective function. The simplest idea is to consider the quadratic model

$$q_t(\mathbf{x}, \mathbf{y}) := f(\mathbf{y}) + \langle \mathbf{x} - \mathbf{y}, \nabla f(\mathbf{y}) \rangle + \frac{1}{2t} \|\mathbf{x} - \mathbf{y}\|^2, \tag{2.4}$$

namely, the linearized part of f at some given point $\mathbf{y}$, regularized by a quadratic proximal term that would measure the "local error" in the approximation, and also results in a well-defined, that is, a strongly convex approximate-minimization problem for (U):

$$(\hat{U}_t) \quad \min\{q_t(\mathbf{x}, \mathbf{y}) : \mathbf{x} \in \mathbb{E}\}.$$

For a fixed given point $\mathbf{y} := \mathbf{x}_{k-1} \in \mathbb{E}$, the unique minimizer $\mathbf{x}_k$ solving $(\hat{U}_{t_k})$ is

$$\mathbf{x}_k = \operatorname{argmin}\left\{ q_{t_k}(\mathbf{x}, \mathbf{x}_{k-1}) : \mathbf{x} \in \mathbb{E} \right\},$$

which yields the same gradient scheme (2.3).

Simple algebra also shows that (2.4) can be written as,

$$q_t(\mathbf{x}, \mathbf{y}) = \frac{1}{2t}\|\mathbf{x} - (\mathbf{y} - t\nabla f(\mathbf{y}))\|^2 - \frac{t}{2}\|\nabla f(\mathbf{y})\|^2 + f(\mathbf{y}). \tag{2.5}$$

Using the above identity also allows us to easily pass from the unconstrained minimization problem (U) to an approximation model for the constrained model

$$(P) \quad \min\{f(\mathbf{x}) : \mathbf{x} \in C\},$$

where $C \subseteq \mathbb{E}$ is a given closed convex set. Ignoring the constant terms in (2.5) leads us to solve (P) via the scheme

$$\mathbf{x}_k = \operatorname*{argmin}_{\mathbf{x} \in C} \frac{1}{2}\left\| \mathbf{x} - (\mathbf{x}_{k-1} - t_k \nabla f(\mathbf{x}_{k-1})) \right\|^2, \tag{2.6}$$

which is the so-called gradient projection method (GPM):

$$\mathbf{x}_k = \Pi_C(\mathbf{x}_{k-1} - t_k \nabla f(\mathbf{x}_{k-1})).$$

Here Π_C denotes the orthogonal projection operator defined by

$$\Pi_C(\mathbf{x}) = \operatorname*{argmin}_{\mathbf{z} \in C} \|\mathbf{z} - \mathbf{x}\|^2.$$

Turning back to our general model (M), one could naturally suggest to consider the following approximation in place of $f(\mathbf{x}) + g(\mathbf{x})$:

$$q(\mathbf{x}, \mathbf{y}) = f(\mathbf{y}) + \langle \mathbf{x} - \mathbf{y}, \nabla f(\mathbf{y}) \rangle + \frac{1}{2t}\|\mathbf{x} - \mathbf{y}\|^2 + g(\mathbf{x}).$$

That is, we leave the nonsmooth part $g(\cdot)$ untouched.

Indeed, in accordance with the previous framework, the corresponding scheme would then read:

$$\mathbf{x}_k = \operatorname*{argmin}_{\mathbf{x} \in \mathbb{E}}\left\{ g(\mathbf{x}) + \frac{1}{2t_k}\|\mathbf{x} - (\mathbf{x}_{k-1} - t_k \nabla f(\mathbf{x}_{k-1}))\|^2 \right\}. \tag{2.7}$$

In fact, the latter leads to another interesting way to write the above scheme via the fundamental proximal operator. For any scalar $t > 0$, the proximal map associated with g is defined by

$$\operatorname{prox}_t(g)(\mathbf{z}) = \operatorname*{argmin}_{\mathbf{u} \in \mathbb{E}}\left\{ g(\mathbf{u}) + \frac{1}{2t}\|\mathbf{u} - \mathbf{z}\|^2 \right\}. \tag{2.8}$$

With this notation, the scheme (2.7), which consists of a proximal step at a resulting gradient point, will be called the "proximal gradient" method, and reads as:

$$\mathbf{x}_k = \mathrm{prox}_{t_k}(g)(\mathbf{x}_{k-1} - t_k \nabla f(\mathbf{x}_{k-1})). \tag{2.9}$$

An alternative and useful derivation of the proximal gradient method is via the "fixed point approach" developed next.

2.3.2   The fixed-point approach

Consider the nonconvex and nonsmooth optimization model (M). If $\mathbf{x}^* \in \mathbb{E}$ is a local minimum of (M), then it is a stationary point of (M), that is, one has

$$\mathbf{0} \in \nabla f(\mathbf{x}^*) + \partial g(\mathbf{x}^*), \tag{2.10}$$

where $\partial g(\cdot)$ is the subdifferential of g. Note that whenever f is also convex, the latter condition is necessary and sufficient for $\mathbf{x}^*$ to be a global minimum of (M).

Now, fix any $t > 0$, then (2.10) holds if and only if the following equivalent statements hold:

$$\mathbf{0} \in t\nabla f(\mathbf{x}^*) + t\partial g(\mathbf{x}^*),$$

$$\mathbf{0} \in t\nabla f(\mathbf{x}^*) - \mathbf{x}^* + \mathbf{x}^* + t\partial g(\mathbf{x}^*),$$

$$(I + t\partial g)(\mathbf{x}^*) \in (I - t\nabla f)(\mathbf{x}^*),$$

$$\mathbf{x}^* = (I + t\partial g)^{-1}(I - t\nabla f)(\mathbf{x}^*),$$

where the last relation is an equality (and not an inclusion) thanks to the properties of the proximal map (cf. the first part of Lemma 2.2 below). The last equation naturally calls for the "fixed-point scheme" that generates a sequence $\{\mathbf{x}_k\}$ via:

$$\mathbf{x}_0 \in \mathbb{E}, \quad \mathbf{x}_k = (I + t_k \partial g)^{-1}(I - t_k \nabla f)(\mathbf{x}_{k-1}) \quad (t_k > 0). \tag{2.11}$$

Using the identity $(I + t_k \partial g)^{-1} = \mathrm{prox}_{t_k}(g)$ (cf., first part of Lemma 2.2), it follows that the scheme (2.11) is nothing else but the proximal gradient method devised in Section 2.3.1. Note that the scheme (2.11) is in fact a special case of the so-called "proximal backward–forward" scheme, which was originally devised for finding a zero of the more general inclusion problem:

$$\mathbf{0} \in T_1(\mathbf{x}^*) + T_2(\mathbf{x}^*),$$

where T_1, T_2 are maximal monotone set valued maps (encompassing (2.10) with f, g convex and $T_1 := \nabla f, T_2 := \partial g$).

2.3.3 Majorization–minimization technique

The idea

A popular technique to devise gradient-based methods in the statistical and engineering literature is the MM approach where the first M stands for majorization and the second M for minimization[4] (maximization problems are similarly handled with minorization replacing majorization).

The MM technique follows in fact from the same idea of approximation models described in Section 2.3.1, except that the approximation model in the MM technique does not have to be quadratic.

The basic idea of MM relies on finding a *relevant* approximation to the objective function F of model (M) that satisfies:

(i) $M(\mathbf{x}, \mathbf{x}) = F(\mathbf{x})$ for every $\mathbf{x} \in \mathbb{E}$.
(ii) $M(\mathbf{x}, \mathbf{y}) \geq F(\mathbf{x})$ for every $\mathbf{x}, \mathbf{y} \in \mathbb{E}$.

Geometrically, this means that $\mathbf{x} \mapsto M(\mathbf{x}, \mathbf{y})$ lies above $F(\mathbf{x})$ and is tangent at $\mathbf{x} = \mathbf{y}$. From the above definition of $M(\cdot, \cdot)$, a natural and simple minimization scheme consists of solving

$$\mathbf{x}_k \in \operatorname*{argmin}_{\mathbf{x} \in \mathbb{E}} M(\mathbf{x}, \mathbf{x}_{k-1}).$$

This scheme immediately implies that

$$M(\mathbf{x}_k, \mathbf{x}_{k-1}) \leq M(\mathbf{x}, \mathbf{x}_{k-1}) \text{ for every } \mathbf{x} \in \mathbb{E}, \tag{2.12}$$

and hence from (i) and (ii) it follows that

$$F(\mathbf{x}_k) \overset{(ii)}{\leq} M(\mathbf{x}_k, \mathbf{x}_{k-1}) \overset{(2.12)}{\leq} M(\mathbf{x}_{k-1}, \mathbf{x}_{k-1}) \overset{(i)}{=} F(\mathbf{x}_{k-1}) \text{ for every } k \geq 1,$$

thus naturally producing a descent scheme for minimizing problem (M).

Clearly, the key question is then how to generate a "good" upper bounding function $M(\cdot, \cdot)$ satisfying (i) and (ii). A universal rule to determine the function M does not exist, and most often the structure of the problem at hand provides helpful hints to achieve this task. This will be illustrated below.

The MM method for the RLS problem

An interesting example of the usage of MM methods is in the class of RLS problems described in Section 2.2.2. This example will also demonstrate the intimate relations between the MM approach and the basic approximation model.

Consider the RLS problem from Section 2.2.2, Problem (2.1), that is, the general model (M) with $f(\mathbf{x}) = \|\mathbf{A}\mathbf{x} - \mathbf{b}\|^2$ and $g(\mathbf{x}) = \lambda R(\mathbf{x})$. Since f is a quadratic function,

---

[4] MM algorithms also appear under different terminology such as surrogate/transfer function approach and bound-optimization algorithms.

easy algebra shows that for any $\mathbf{x}, \mathbf{y}$:

$$f(\mathbf{x}) = f(\mathbf{y}) + 2\langle \mathbf{A}(\mathbf{x} - \mathbf{y}), \mathbf{A}\mathbf{y} - \mathbf{b} \rangle + \langle \mathbf{A}^T \mathbf{A}(\mathbf{x} - \mathbf{y}), \mathbf{x} - \mathbf{y} \rangle.$$

Let $\mathbf{D}$ be any matrix satisfying $\mathbf{D} \succeq \mathbf{A}^T \mathbf{A}$. Then

$$f(\mathbf{x}) \leq f(\mathbf{y}) + 2\langle \mathbf{A}(\mathbf{x} - \mathbf{y}), \mathbf{A}\mathbf{y} - \mathbf{b} \rangle + \langle \mathbf{D}(\mathbf{x} - \mathbf{y}), \mathbf{x} - \mathbf{y} \rangle,$$

and hence with

$$M(\mathbf{x}, \mathbf{y}) := g(\mathbf{x}) + f(\mathbf{y}) + 2\langle \mathbf{A}(\mathbf{x} - \mathbf{y}), \mathbf{A}\mathbf{y} - \mathbf{b} \rangle + \langle \mathbf{D}(\mathbf{x} - \mathbf{y}), \mathbf{x} - \mathbf{y} \rangle,$$

we have

$$M(\mathbf{x}, \mathbf{x}) = F(\mathbf{x}) \text{ for every } \mathbf{x} \in \mathbb{E},$$

$$M(\mathbf{x}, \mathbf{y}) \geq F(\mathbf{x}) \text{ for every } \mathbf{x}, \mathbf{y} \in \mathbb{E}.$$

In particular, with $\mathbf{D} = \mathbf{I}$ ($\mathbf{I}$ being the identity matrix), the stated assumption on $\mathbf{D}$ reduces to $\lambda_{\max}(\mathbf{A}^T \mathbf{A}) \leq 1$, and a little algebra shows that M reduces in that case to

$$M(\mathbf{x}, \mathbf{y}) = g(\mathbf{x}) + \|\mathbf{A}\mathbf{x} - \mathbf{b}\|^2 - \|\mathbf{A}\mathbf{x} - \mathbf{A}\mathbf{y}\|^2 + \|\mathbf{x} - \mathbf{y}\|^2.$$

The resulting iterative scheme is given by

$$\mathbf{x}_k = \underset{\mathbf{x}}{\operatorname{argmin}} \, M(\mathbf{x}, \mathbf{x}_{k-1}). \tag{2.13}$$

This scheme has been used extensively in the signal-processing literature (see bibliographic notes) to devise convergent schemes for solving the RLS problem. A close inspection of the explanation above indicates that the MM approach for building iterative schemes to solve (RLS) is in fact equivalent to the basic approximation model discussed in Section 2.3.1. Indeed, opening the squares in $M(\cdot, \cdot)$ and collecting terms we obtain:

$$M(\mathbf{x}, \mathbf{y}) = g(\mathbf{x}) + \|\mathbf{x} - \{\mathbf{y} - \mathbf{A}^T(\mathbf{A}\mathbf{y} - \mathbf{b})\}\|^2 + C(\mathbf{b}, \mathbf{y}),$$

where $C(\mathbf{b}, \mathbf{y})$ is constant with respect to $\mathbf{x}$. Since

$$\nabla f(\mathbf{y}) = 2\mathbf{A}^T(\mathbf{A}\mathbf{y} - \mathbf{b}),$$

it follows that (2.13) is just the scheme devised in (2.70) with constant step size $t_k \equiv \frac{1}{2}$. Further ways to derive MM-based schemes that do not involve quadratic functions exploit tools and properties such as convexity of the objective; standard inequalities, for example, Cauchy–Schwartz; topological properties of f, for example, Lipschitz gradient; see the bibliographic notes.

2.3.4 Fermat–Weber location problem

In the early 17th century the French mathematician Pierre de Fermat challenged the mathematicians at the time (relax, this is not the "big" one!) with the following problem:

Fermat's problem: given three points on the plane, find another point so that the sum of the distances to the existing points is minimum.

At the beginning of the 20th century, the German economist Weber studied an extension of this problem: given n points on the plane, find another point such that the weighted sum of the Euclidean distances to these n points is minimal.

In mathematical terms, given m points $\mathbf{a}_1, \ldots, \mathbf{a}_m \in \mathbb{R}^n$, we wish to find the location of $\mathbf{x} \in \mathbb{R}^n$ solving

$$\min_{\mathbf{x} \in \mathbb{R}^n} \left\{ f(\mathbf{x}) \equiv \sum_{i=1}^m \omega_i \|\mathbf{x} - \mathbf{a}_i\| \right\}.$$

In 1937, Weiszfeld proposed an algorithm for solving the Fermat–Weber problem. This algorithm, although not identical to the proximal gradient method, demonstrates well the two principles alluded to in the previous sections for constructing gradient-based methods. On one hand, the algorithm can be viewed as a fixed-point method employed on the optimality condition of the problem, and on the other hand, each iteration can be equivalently constructed via minimization of a quadratic approximation of the problem at the previous iteration.

Let us begin with the first derivation of Weiszfeld's method. The optimality condition of the problem is $\nabla f(\mathbf{x}^*) = \mathbf{0}$. Of course we may encounter problems if $\mathbf{x}^*$ happens to be one of the points $\mathbf{a}_i$, because $f(\mathbf{x})$ is not differentiable at these points, but for the moment, let us assume that this is not the case (see Section 2.8 for how to properly handle nonsmoothness). The gradient of the problem is given by

$$\nabla f(\mathbf{x}) = \sum_{i=1}^m \omega_i \frac{\mathbf{x} - \mathbf{a}_i}{\|\mathbf{x} - \mathbf{a}_i\|},$$

and thus after rearranging the terms, the optimality condition can be written as

$$\mathbf{x}^* \sum_{i=1}^m \omega_i \frac{1}{\|\mathbf{x}^* - \mathbf{a}_i\|} = \sum_{i=1}^m \omega_i \frac{\mathbf{a}_i}{\|\mathbf{x}^* - \mathbf{a}_i\|},$$

or equivalently as

$$\mathbf{x}^* = \frac{\sum_{i=1}^m \omega_i \dfrac{\mathbf{a}_i}{\|\mathbf{x}^* - \mathbf{a}_i\|}}{\sum_{i=1}^m \dfrac{\omega_i}{\|\mathbf{x}^* - \mathbf{a}_i\|}}.$$

Weiszfeld's method is nothing else but the fixed-point iterations associated with the latter equation:

$$
\mathbf{x}_k = \frac{\sum_{i=1}^{m} \omega_i \dfrac{\mathbf{a}_i}{\|\mathbf{x}_{k-1} - \mathbf{a}_i\|}}{\sum_{i=1}^{m} \dfrac{\omega_i}{\|\mathbf{x}_{k-1} - \mathbf{a}_i\|}}
\tag{2.14}
$$

with $\mathbf{x}_0$ a given arbitrary point.

The second derivation of Weiszfeld's method relies on the simple observation that the general step (2.14) can be equivalently written as

$$
\mathbf{x}_k = \underset{\mathbf{x}}{\operatorname{argmin}} \sum_{i=1}^{m} \omega_i \frac{\|\mathbf{x} - \mathbf{a}_i\|^2}{\|\mathbf{x}_{k-1} - \mathbf{a}_i\|}.
\tag{2.15}
$$

Therefore, the scheme (2.14) has the representation

$$
\mathbf{x}_k = \underset{\mathbf{x}}{\operatorname{argmin}} \, h(\mathbf{x}, \mathbf{x}_{k-1}),
\tag{2.16}
$$

where the auxiliary function $h(\cdot, \cdot)$ is defined by

$$
h(\mathbf{x}, \mathbf{y}) := \sum_{i=1}^{m} \omega_i \frac{\|\mathbf{x} - \mathbf{a}_i\|^2}{\|\mathbf{y} - \mathbf{a}_i\|}.
$$

This approximation is completely different from the quadratic approximation described in Section 2.3.1 and it also cannot be considered as an MM method since the auxiliary function $h(\mathbf{x}, \mathbf{y})$ is not an upper bound of the objective function $f(\mathbf{x})$.

Despite the above, Weiszfeld's method, like MM schemes, is a descent method. This is due to a nice property of the function h which is stated below.

LEMMA 2.1   *For every* $\mathbf{x}, \mathbf{y} \in \mathbb{R}^n$ *such that* $\mathbf{y} \notin \{\mathbf{a}_1, \ldots, \mathbf{a}_m\}$

$$
h(\mathbf{x}, \mathbf{x}) = f(\mathbf{x}),
\tag{2.17}
$$

$$
h(\mathbf{x}, \mathbf{y}) \geq 2f(\mathbf{x}) - f(\mathbf{y}).
\tag{2.18}
$$

Proof   The first property follows by substitution. To prove the second property (2.18), note that for every two real numbers $a \in \mathbb{R}, b > 0$, the inequality

$$
\frac{a^2}{b} \geq 2a - b,
$$

holds true. Therefore, for every $i = 1, \ldots, m$

$$
\frac{\|\mathbf{x} - \mathbf{a}_i\|^2}{\|\mathbf{y} - \mathbf{a}_i\|} \geq 2\|\mathbf{x} - \mathbf{a}_i\| - \|\mathbf{y} - \mathbf{a}_i\|.
$$

Multiplying the latter inequality by ω_i and summing over $i = 1,\ldots,m$, (2.18) follows. $\blacksquare$

Recall that in order to prove the descent property of an MM method we used the fact that the auxiliary function is an upper bound of the objective function. For the Fermat–Weber problem this is not the case, however the new property (2.18) is sufficient to prove the monotonicity. Indeed,

$$f(\mathbf{x}_{k-1}) \overset{(2.17)}{=} h(\mathbf{x}_{k-1},\mathbf{x}_{k-1}) \overset{(2.16)}{\geq} h(\mathbf{x}_k,\mathbf{x}_{k-1}) \overset{(2.18)}{\geq} 2f(\mathbf{x}_k) - f(\mathbf{x}_{k-1}).$$

Therefore, $f(\mathbf{x}_{k-1}) \geq 2f(\mathbf{x}_k) - f(\mathbf{x}_{k-1})$, implying the descent property $f(\mathbf{x}_k) \leq f(\mathbf{x}_{k-1})$.

As a final note, we mention the fact that Weiszfeld's method is, in fact, a gradient method

$$\mathbf{x}_k = \mathbf{x}_{k-1} - t_k \nabla f(\mathbf{x}_{k-1})$$

with a special choice of the stepsize t_k given by

$$t_k = \left(\sum_{i=1}^{m} \frac{\omega_i}{\|\mathbf{x}_{k-1} - \mathbf{a}_i\|} \right)^{-1}.$$

To conclude, Weiszfeld's method for the Fermat–Weber problem is one example of a gradient-based method that can be constructed by either fixed-point ideas, or by approximation models. The derivation of the method is different from what was described in previous sections, thus emphasizing the fact that the specific structure of the problem can and should be exploited. Another interesting example related to location and communication will be presented in Section 2.8.

In the forthcoming sections of this chapter we will focus on gradient-based methods emerging from the fixed-point approach, and relying on the quadratic approximation. A special emphasis will be given to the proximal gradient method and its accelerations.

2.4   Convergence results for the proximal-gradient method

In this section we make the setting more precise and introduce the main computational objects, study their properties, and establish some key generic inequalities that serve as the principal vehicle to establish convergence, and rate of convergence results of the proximal-gradient method and its extensions. The rate of convergence of the proximal-gradient method will be established in this section, while the analysis of extensions and/or accelerations of the method will be studied in the following sections. In the sequel, we make the standing assumption that there exists an optimal solution $\mathbf{x}^*$ to problem (M) and we set $F_* = F(\mathbf{x}^*)$.

2.4.1   The prox-grad map

Following Section 2.3.1, we adopt the following approximation model for F. For any $L > 0$, and any $\mathbf{x}, \mathbf{y} \in \mathbb{E}$, define

$$Q_L(\mathbf{x}, \mathbf{y}) := f(\mathbf{y}) + \langle \mathbf{x} - \mathbf{y}, \nabla f(\mathbf{y}) \rangle + \frac{L}{2}\|\mathbf{x} - \mathbf{y}\|^2 + g(\mathbf{x}),$$

and

$$p_L^{f,g}(\mathbf{y}) := \operatorname{argmin}\{Q_L(\mathbf{x}, \mathbf{y}) : \mathbf{x} \in \mathbb{E}\}.$$

Ignoring the constant terms in $\mathbf{y}$, this reduces to (see also (2.7)):

$$p_L^{f,g}(\mathbf{y}) = \operatorname*{argmin}_{\mathbf{x} \in \mathbb{E}} \left\{ g(\mathbf{x}) + \frac{L}{2}\left\|\mathbf{x} - \left(\mathbf{y} - \frac{1}{L}\nabla f(\mathbf{y})\right)\right\|^2 \right\}$$
$$= \operatorname{prox}_{\frac{1}{L}}(g)\left(\mathbf{y} - \frac{1}{L}\nabla f(\mathbf{y})\right). \tag{2.19}$$

We call this composition of the proximal map with a gradient step of f the "prox-grad map" associated with f and g. The prox-grad map $p_L^{f,g}(\cdot)$ is well defined by the underlying assumptions on f and g. To simplify notation, we will omit the superscripts f and g and simply write p_L instead of $p_L^{f,g}$ whenever no confusion arises. First, we recall basic properties of Moreau's proximal map.

LEMMA 2.2   *Let $g : \mathbb{E} \to (-\infty, \infty]$ be a closed, proper convex function and for any $t > 0$, let*

$$g_t(\mathbf{z}) = \min_{\mathbf{u}} \left\{ g(\mathbf{u}) + \frac{1}{2t}\|\mathbf{u} - \mathbf{z}\|^2 \right\}. \tag{2.20}$$

Then,

1. *The minimum in (2.20) is attained at the unique point $\operatorname{prox}_t(g)(\mathbf{z})$. As a consequence, the map $(I + t\partial g)^{-1}$ is single valued from $\mathbb{E}$ into itself and*

$$\operatorname{prox}_t(g)(\mathbf{z}) = (I + t\partial g)^{-1}(\mathbf{z}) \text{ for every } \mathbf{z} \in \mathbb{E}.$$

2. *The function $g_t(\cdot)$ is continuously differentiable on $\mathbb{E}$ with a $\frac{1}{t}$-Lipschitz gradient given by*

$$\nabla g_t(\mathbf{z}) = \frac{1}{t}(I - \operatorname{prox}_t(g)(\mathbf{z})) \text{ for every } \mathbf{z} \in \mathbb{E}.$$

In particular, if $g \equiv \delta_C$, with $C \subseteq \mathbb{E}$ closed and convex, then $\operatorname{prox}_t(g) = (I + t\partial g)^{-1} = \Pi_C$, the orthogonal projection on C and we have

$$g_t(\mathbf{z}) = \frac{1}{2t}\|\mathbf{z} - \Pi_C(\mathbf{z})\|^2.$$

2.4.2 Fundamental inequalities

We develop here some key inequalities that play a central role in the analysis of the proximal-gradient method and, in fact, for any gradient-based method. Throughout the rest of this chapter we assume that ∇f is Lipschitz on $\mathbb{E}$, namely, there exists $L(f) > 0$ such that

$$\|\nabla f(\mathbf{x}) - \nabla f(\mathbf{y})\| \leq L(f)\|\mathbf{x} - \mathbf{y}\| \text{ for every } \mathbf{x}, \mathbf{y} \in \mathbb{E}.$$

For convenience we denote this class by $C_{L(f)}^{1,1}$. The first result is a well-known, important property of smooth functions.

LEMMA 2.3 Descent lemma *Let $f : \mathbb{E} \to (-\infty, \infty)$ be $C_{L(f)}^{1,1}$. Then for any $L \geq L(f)$,*

$$f(\mathbf{x}) \leq f(\mathbf{y}) + \langle \mathbf{x} - \mathbf{y}, \nabla f(\mathbf{y}) \rangle + \frac{L}{2}\|\mathbf{x} - \mathbf{y}\|^2 \text{ for every } \mathbf{x}, \mathbf{y} \in \mathbb{E}.$$

The next result gives a useful inequality for the prox-grad map which, in turn, can be used in the characterization of $p_L(\cdot)$. For a function f we define

$$l_f(\mathbf{x}, \mathbf{y}) := f(\mathbf{x}) - f(\mathbf{y}) - \langle \mathbf{x} - \mathbf{y}, \nabla f(\mathbf{y}) \rangle.$$

LEMMA 2.4 *Let $\boldsymbol{\xi} = prox_t(g)(\mathbf{z})$ for some $\mathbf{z} \in \mathbb{E}$ and let $t > 0$. Then*

$$2t(g(\boldsymbol{\xi}) - g(\mathbf{u})) \leq \|\mathbf{u} - \mathbf{z}\|^2 - \|\mathbf{u} - \boldsymbol{\xi}\|^2 - \|\boldsymbol{\xi} - \mathbf{z}\|^2 \text{ for every } \mathbf{u} \in dom\, g.$$

Proof By definition of $\boldsymbol{\xi}$ we have

$$\boldsymbol{\xi} = \underset{\mathbf{u}}{\operatorname{argmin}} \left\{ g(\mathbf{u}) + \frac{1}{2t}\|\mathbf{u} - \mathbf{z}\|^2 \right\}.$$

Writing the optimality condition for the above minimization problem yields

$$\langle \mathbf{u} - \boldsymbol{\xi}, \boldsymbol{\xi} - \mathbf{z} + t\boldsymbol{\gamma} \rangle \geq 0 \text{ for every } \mathbf{u} \in dom\, g, \tag{2.21}$$

where $\boldsymbol{\gamma} \in \partial g(\boldsymbol{\xi})$. Since g is convex with $\boldsymbol{\gamma} \in \partial g(\boldsymbol{\xi})$, we also have

$$g(\boldsymbol{\xi}) - g(\mathbf{u}) \leq \langle \boldsymbol{\xi} - \mathbf{u}, \boldsymbol{\gamma} \rangle,$$

which, combined with (2.21), and the fact that $t > 0$ yields

$$2t(g(\boldsymbol{\xi}) - g(\mathbf{u}) \leq 2\langle \mathbf{u} - \boldsymbol{\xi}, \boldsymbol{\xi} - \mathbf{z} \rangle,$$

then the desired result follows from the identity

$$2\langle \mathbf{u} - \boldsymbol{\xi}, \boldsymbol{\xi} - \mathbf{z} \rangle = \|\mathbf{u} - \mathbf{z}\|^2 - \|\mathbf{u} - \boldsymbol{\xi}\|^2 - \|\boldsymbol{\xi} - \mathbf{z}\|^2. \tag{2.22}$$

$\blacksquare$

Since $p_L(\mathbf{y}) = \text{prox}_{1/L}(g)\left(\mathbf{y} - \frac{1}{L}\nabla f(\mathbf{y})\right)$, invoking Lemma 2.4, we now obtain a useful characterization of p_L. For further reference we denote for any $\mathbf{y} \in \mathbb{E}$:

$$\xi_L(\mathbf{y}) := \mathbf{y} - \frac{1}{L}\nabla f(\mathbf{y}). \tag{2.23}$$

LEMMA 2.5   *For any* $\mathbf{x} \in \text{dom } g, \mathbf{y} \in \mathbb{E}$, *the prox-grad map* p_L *satisfies*

$$\frac{2}{L}\left[g(p_L(\mathbf{y})) - g(\mathbf{x})\right] \le \|\mathbf{x} - \xi_L(\mathbf{y})\|^2 - \|\mathbf{x} - p_L(\mathbf{y})\|^2 - \|p_L(\mathbf{y}) - \xi_L(\mathbf{y})\|^2, \tag{2.24}$$

where $\xi_L(\mathbf{y})$ *is given in (2.23).*

Proof   Follows from Lemma 2.4 with $t = \frac{1}{L}, \boldsymbol{\xi} = p_L(\mathbf{y})$ and $\mathbf{z} = \xi_L(\mathbf{y}) = \mathbf{y} - \frac{1}{L}\nabla f(\mathbf{y})$. ∎

Our last result combines all the above to produce the main pillar of the analysis.

LEMMA 2.6   *Let* $\mathbf{x} \in \text{dom } g, \mathbf{y} \in \mathbb{E}$ *and let* $L > 0$ *be such that the inequality*

$$F(p_L(\mathbf{y})) \le Q(p_L(\mathbf{y}), \mathbf{y}) \tag{2.25}$$

is satisfied. Then

$$\frac{2}{L}(F(\mathbf{x}) - F(p_L(\mathbf{y}))) \ge \frac{2}{L}l_f(\mathbf{x}, \mathbf{y}) + \|\mathbf{x} - p_L(\mathbf{y})\|^2 - \|\mathbf{x} - \mathbf{y}\|^2.$$

Furthermore, if f *is also convex then*

$$\frac{2}{L}(F(\mathbf{x}) - F(p_L(\mathbf{y}))) \ge \|\mathbf{x} - p_L(\mathbf{y})\|^2 - \|\mathbf{x} - \mathbf{y}\|^2.$$

Proof   Recalling that

$$p_L(\mathbf{y}) = \underset{\mathbf{x}}{\text{argmin}}\, Q_L(\mathbf{x}, \mathbf{y}),$$

and using the definition of $Q_L(\cdot, \cdot)$ we have:

$$Q(p_L(\mathbf{y}), \mathbf{y}) = f(\mathbf{y}) + \langle p_L(\mathbf{y}) - \mathbf{y}, \nabla f(\mathbf{y})\rangle + \frac{L}{2}\|p_L(\mathbf{y}) - \mathbf{y}\|^2 + g(p_L(\mathbf{y})).$$

Therefore, using (2.25) it follows that

$$\begin{aligned}
F(\mathbf{x}) - F(p_L(\mathbf{y})) &\ge F(\mathbf{x}) - Q_L(p_L(\mathbf{y}), \mathbf{y}) \\
&= f(\mathbf{x}) - f(\mathbf{y}) - \langle p_L(\mathbf{y}) - \mathbf{y}, \nabla f(\mathbf{y})\rangle \\
&\quad - \frac{L}{2}\|p_L(\mathbf{y}) - \mathbf{y}\|^2 + g(\mathbf{x}) - g(p_L(\mathbf{y})) \\
&= l_f(\mathbf{x}, \mathbf{y}) + \langle \mathbf{x} - p_L(\mathbf{y}), \nabla f(\mathbf{y})\rangle \\
&\quad - \frac{L}{2}\|p_L(\mathbf{y}) - \mathbf{y}\|^2 + g(\mathbf{x}) - g(p_L(\mathbf{y})).
\end{aligned}$$

Now, invoking Lemma 2.5 and (2.22) we obtain,

$$\frac{2}{L}(g(\mathbf{x}) - g(p_L(\mathbf{y}))) \geq 2\langle \mathbf{x} - p_L(\mathbf{y}), \xi_L(\mathbf{y}) - p_L(\mathbf{y})\rangle,$$

which, substituted in the above inequality, and recalling that $\xi_L(\mathbf{y}) = \mathbf{y} - \frac{1}{L}\nabla f(\mathbf{y})$, yields

$$\frac{2}{L}(F(\mathbf{x}) - F(p_L(\mathbf{y}))) \geq \frac{2}{L}l_f(\mathbf{x}, \mathbf{y}) + 2\langle \mathbf{x} - p_L(\mathbf{y}), \mathbf{y} - p_L(\mathbf{y})\rangle - \|p_L(\mathbf{y}) - \mathbf{y}\|^2$$

$$= \frac{2}{L}l_f(\mathbf{x}, \mathbf{y}) + \|p_L(\mathbf{y}) - \mathbf{y}\|^2 + 2\langle \mathbf{y} - \mathbf{x}, p_L(\mathbf{y}) - \mathbf{y}\rangle$$

$$= \frac{2}{L}l_f(\mathbf{x}, \mathbf{y}) + \|\mathbf{x} - p_L(\mathbf{y})\|^2 - \|\mathbf{x} - \mathbf{y}\|^2,$$

proving the first inequality. When f is convex we have $l_f(\mathbf{x}, \mathbf{y}) \geq 0$, and hence the second inequality follows. ∎

Note that condition (2.25) of Lemma 2.6 is always satisfied for $p_L(\mathbf{y})$ with $L \geq L(f)$, thanks to the descent lemma (Lemma 2.3).

2.4.3 Convergence of the proximal-gradient method: the convex case

We consider the proximal-gradient method scheme for solving the model (M) when f is assumed convex. Since g is also assumed convex, the general model (M) is in this case convex. When $L(f) > 0$ is known, we can define the proximal-gradient method with a constant stepsize rule.

Proximal-gradient method with constant stepsize

Input: $L = L(f)$ - A Lipschitz constant of ∇f.

Step 0. Take $\mathbf{x}_0 \in \mathbb{E}$.

Step k. $(k \geq 1)$ Compute

$$\mathbf{x}_k = p_L(\mathbf{x}_{k-1}).$$

An evident possible drawback of the above scheme is that the Lipschitz constant $L(f)$ is not always known, or not easily computable (e.g., in the example of Section 2.6 where one needs to know $\lambda_{\max}(\mathbf{A}^T\mathbf{A})$). To overcome this potential difficulty, we also suggest and analyze the proximal-gradient method with an easy backtracking stepsize rule. This is the next algorithm described below.

> **Proximal-gradient method with backtracking**
> **Step 0.** Take $L_0 > 0$, some $\eta > 1$ and $\mathbf{x}_0 \in \mathbb{E}$.
> **Step k.** ($k \geq 1$) Find the smallest non-negative integer i_k such that with, $\bar{L} = \eta^{i_k} L_{k-1}$:
>
> $$F(p_{\bar{L}}(\mathbf{x}_{k-1})) \leq Q_{\bar{L}}(p_{\bar{L}}(\mathbf{x}_{k-1}), \mathbf{x}_{k-1})). \tag{2.26}$$
>
> Set $L_k = \eta^{i_k} L_{k-1}$ and compute
>
> $$\mathbf{x}_k = p_{L_k}(\mathbf{x}_{k-1}). \tag{2.27}$$

REMARK 2.1 *The sequence of function values $\{F(\mathbf{x}_k)\}$ produced by the proximal-gradient method with either constant, or backtracking stepsize rules is nonincreasing. Indeed, for every $k \geq 1$:*

$$F(\mathbf{x}_k) \leq Q_{L_k}(\mathbf{x}_k, \mathbf{x}_{k-1}) = Q_{L_k}(\mathbf{x}_{k-1}, \mathbf{x}_{k-1}) = F(\mathbf{x}_{k-1}),$$

where L_k is chosen by the backtracking rule, or $L_k \equiv L(f)$, whenever the Lipschitz constant of ∇f is known.

REMARK 2.2 *Since (2.26) holds for $\bar{L} \geq L(f)$, then for the proximal-gradient method with backtracking, it holds that $L_k \leq \eta L(f)$ for every $k \geq 1$, so that overall*

$$\beta L(f) \leq L_k \leq \alpha L(f), \tag{2.28}$$

where $\alpha = \beta = 1$ for the constant stepsize setting, and $\alpha = \eta, \beta = \frac{L_0}{L(f)}$ for the backtracking case.

The next result shows that the proximal-gradient method (with either of the two constant stepsize rules) converge at a sublinear rate in function values. Recall that since for $g \equiv 0$ and $g = \delta_C$, our model (M) recovers the basic gradient and gradient projection methods respectively, the result demonstrates that the presence of the nonsmooth term g in the model (M) does not deteriorate with the same rate of convergence which is known to be valid for smooth problems.

THEOREM 2.1 (Sublinear rate of convergence of the proximal-gradient method.) *Let $\{\mathbf{x}_k\}$ be the sequence generated by the proximal-gradient method with either a constant or a backtracking stepsize rule. Then for every $k \geq 1$:*

$$F(\mathbf{x}_k) - F(\mathbf{x}^*) \leq \frac{\alpha L(f)\|\mathbf{x}_0 - \mathbf{x}^*\|^2}{2k}$$

for every optimal solution $\mathbf{x}^*$.

Proof Invoking Lemma 2.6 with $\mathbf{x} = \mathbf{x}^*, \mathbf{y} = \mathbf{x}_n$ and $L = L_{n+1}$, we obtain

$$\frac{2}{L_{n+1}}\left(F(\mathbf{x}^*) - F(\mathbf{x}_{n+1})\right) \geq \|\mathbf{x}^* - \mathbf{x}_{n+1}\|^2 - \|\mathbf{x}^* - \mathbf{x}_n\|^2,$$

which combined with (2.28), and the fact that $F(\mathbf{x}^*) - F(\mathbf{x}_{n+1}) \leq 0$, yields

$$\frac{2}{\alpha L(f)}\left(F(\mathbf{x}^*) - F(\mathbf{x}_{n+1})\right) \geq \|\mathbf{x}^* - \mathbf{x}_{n+1}\|^2 - \|\mathbf{x}^* - \mathbf{x}_n\|^2. \tag{2.29}$$

Summing this inequality over $n = 0, \ldots, k-1$ gives

$$\frac{2}{\alpha L(f)}\left(kF(\mathbf{x}^*) - \sum_{n=0}^{k-1} F(\mathbf{x}_{n+1})\right) \geq \|\mathbf{x}^* - \mathbf{x}_k\|^2 - \|\mathbf{x}^* - \mathbf{x}_0\|^2. \tag{2.30}$$

Invoking Lemma 2.6 one more time with $\mathbf{x} = \mathbf{y} = \mathbf{x}_n, L = L_{n+1}$, yields

$$\frac{2}{L_{n+1}}(F(\mathbf{x}_n) - F(\mathbf{x}_{n+1})) \geq \|\mathbf{x}_n - \mathbf{x}_{n+1}\|^2.$$

Since we have $L_{n+1} \geq \beta L(f)$ (see (2.28)) and $F(\mathbf{x}_n) - F(\mathbf{x}_{n+1}) \geq 0$, it follows that

$$\frac{2}{\beta L(f)}(F(\mathbf{x}_n) - F(\mathbf{x}_{n+1})) \geq \|\mathbf{x}_n - \mathbf{x}_{n+1}\|^2.$$

Multiplying the last inequality by n and summing over $n = 0, \ldots k-1$, we obtain,

$$\frac{2}{\beta L(f)}\sum_{n=0}^{k-1}(nF(\mathbf{x}_n) - (n+1)F(\mathbf{x}_{n+1}) + F(\mathbf{x}_{n+1})) \geq \sum_{n=0}^{k-1} n\|\mathbf{x}_n - \mathbf{x}_{n+1}\|^2,$$

which simplifies to:

$$\frac{2}{\beta L(f)}\left(-kF(\mathbf{x}_k) + \sum_{n=0}^{k-1} F(\mathbf{x}_{n+1})\right) \geq \sum_{n=0}^{k-1} n\|\mathbf{x}_n - \mathbf{x}_{n+1}\|^2. \tag{2.31}$$

Adding (2.30) and (2.31) times β/α, we get

$$\frac{2k}{\alpha L(f)}\left(F(\mathbf{x}^*) - F(\mathbf{x}_k)\right) \geq \|\mathbf{x}^* - \mathbf{x}_k\|^2 + \frac{\beta}{\alpha}\sum_{n=0}^{k-1} n\|\mathbf{x}_n - \mathbf{x}_{n+1}\|^2 - \|\mathbf{x}^* - \mathbf{x}_0\|^2,$$

and hence it follows that

$$F(\mathbf{x}_k) - F(\mathbf{x}^*) \leq \frac{\alpha L(f)\|\mathbf{x} - \mathbf{x}_0\|^2}{2k}.$$

∎

This result demonstrates that in order to obtain an ϵ-optimal solution of (M), that is, a point $\hat{\mathbf{x}}$ such that $F(\hat{\mathbf{x}}) - F(\mathbf{x}^*) \leq \epsilon$, one requires at most $\left\lceil \frac{C}{\epsilon} \right\rceil$ iterations, where $C = \frac{\alpha L(f)\|\mathbf{x}_0 - \mathbf{x}^*\|^2}{2}$. Thus, even for low accuracy requirements, the proximal-gradient method can be very slow and inadequate for most applications. Later on, in Section 2.5 we will present an acceleration of the proximal gradient method that is equally simple, but possesses a significantly improved complexity rate.

It is also possible to prove the convergence of the sequence generated by the proximal gradient method, and not only the convergence of function values. This result relies on the Fejer monotonicity property of the generated sequence.

THEOREM 2.2 (Convergence of the sequence generated by the proximal gradient method.)   *Let $\{\mathbf{x}_k\}$ be the sequence generated by the proximal-gradient method with either a constant or a backtracking stepsize rule. Then*

1. **Fejer monotonicity.** *For every optimal solution $\mathbf{x}^*$ of the model (M) and any $k \geq 1$:*

$$\|\mathbf{x}_k - \mathbf{x}^*\| \leq \|\mathbf{x}_{k-1} - \mathbf{x}^*\|. \tag{2.32}$$

2. *The sequence $\{\mathbf{x}_k\}$ converges to an optimal solution of problem (M).*

Proof   1. Invoking Lemma 2.6 with $\mathbf{x} = \mathbf{x}^*, \mathbf{y} = \mathbf{x}_{k-1}$, and $L = L_k$ (for the constant stepsize rule), $L_k \equiv L(f)$, we obtain

$$\frac{2}{L_k}(F(\mathbf{x}^*) - F(\mathbf{x}_k)) \geq \|\mathbf{x}^* - \mathbf{x}_k\|^2 - \|\mathbf{x}^* - \mathbf{x}_{k-1}\|^2.$$

Since $F(\mathbf{x}^*) - F(\mathbf{x}_k) \leq 0$, property (2.32) follows.

2. By Fejer monotonicity it follows that for a given optimal solution $\mathbf{x}^*$

$$\|\mathbf{x}_k - \mathbf{x}^*\| \leq \|\mathbf{x}_0 - \mathbf{x}^*\|.$$

Therefore, the sequence $\{\mathbf{x}_k\}$ is bounded. To prove the convergence of $\{\mathbf{x}_k\}$, it only remains to show that all converging subsequences have the same limit. Suppose in contradiction that two subsequences $\{\mathbf{x}_{k_j}\}, \{\mathbf{x}_{n_j}\}$ converge to different limits $\mathbf{x}^\infty, \mathbf{y}^\infty$, respectively ($\mathbf{x}^\infty \neq \mathbf{y}^\infty$). Since $F(\mathbf{x}_{k_j}), F(\mathbf{x}_{n_j}) \to F_*$ (recalling that F_* is the optimal function value), it follows that $\mathbf{x}^\infty$ and $\mathbf{y}^\infty$ are optimal solutions of (M). Now, by Fejer monotonicity of the sequence $\{\mathbf{x}_k\}$, it follows that the sequence $\{\|\mathbf{x}_k - \mathbf{x}^\infty\|\}$ is bounded and nonincreasing, and thus has a limit $\lim_{k \to \infty} \|\mathbf{x}_k - \mathbf{x}^\infty\| = l_1$. However, we also have $\lim_{k \to \infty} \|\mathbf{x}_k - \mathbf{x}^\infty\| = \lim_{j \to \infty} \|\mathbf{x}_{k_j} - \mathbf{x}^\infty\| = 0$, and $\lim_{k \to \infty} \|\mathbf{x}_k - \mathbf{x}^\infty\| = \lim_{j \to \infty} \|\mathbf{x}_{n_j} - \mathbf{x}^\infty\| = \|\mathbf{y}^\infty - \mathbf{x}^\infty\|$, so that $l_1 = 0 = \|\mathbf{x}^\infty - \mathbf{y}^\infty\|$; which is obviously a contradiction.   ∎

2.4.4   The nonconvex case

When f is nonconvex, the convergence result is of course weaker. Convergence to a global minimum is out of reach. Recall that for a fixed $L > 0$, the condition $\mathbf{x}^* = p_L(\mathbf{x}^*)$ is a necessary condition for $\mathbf{x}^*$ to be an optimal solution of (M). Therefore, the convergence of the sequence to a stationary point can be measured by the quantity $\|\mathbf{x} - p_L(\mathbf{x})\|$. This is done in the next result.

THEOREM 2.3 (Convergence of the proximal-gradient method in the nonconvex case.) *Let $\{\mathbf{x}_k\}$ be the sequence generated by the proximal-gradient method with either a constant or a backtracking stepsize rule. Then for every $n \geq 1$ we have*

$$\gamma_n \leq \frac{1}{\sqrt{n}} \left(\frac{2(F(\mathbf{x}_0) - F_*)}{\beta L(f)} \right)^{1/2},$$

where

$$\gamma_n := \min_{1 \leq k \leq n} \|\mathbf{x}_{k-1} - p_{L_k}(\mathbf{x}_{k-1})\|.$$

Moreover, $\|\mathbf{x}_{k-1} - p_{L_k}(\mathbf{x}_{k-1})\| \to 0$ as $k \to \infty$.

Proof　Invoking Lemma 2.6 with $\mathbf{x} = \mathbf{y} = \mathbf{x}_{k-1}, L = L_k$ and using the relation $\mathbf{x}_k = p_{L_k}(\mathbf{x}_{k-1})$, it follows that

$$\frac{2}{L_k}(F(\mathbf{x}_{k-1}) - F(\mathbf{x}_k)) \geq \|\mathbf{x}_{k-1} - \mathbf{x}_k\|^2, \tag{2.33}$$

where we also used the fact that $l_f(\mathbf{x}, \mathbf{x}) = 0$. By (2.28), $L_k \geq \beta L(f)$, which combined with (2.33) results with the inequality

$$\frac{\beta L(f)}{2} \|\mathbf{x}_{k-1} - \mathbf{x}_k\|^2 \leq F(\mathbf{x}_{k-1}) - F(\mathbf{x}_k).$$

Summing over $k = 1, \ldots, n$ we obtain

$$\frac{\beta L(f)}{2} \sum_{k=1}^{n} \|\mathbf{x}_{k-1} - \mathbf{x}_k\|^2 \leq F(\mathbf{x}_0) - F(\mathbf{x}_n),$$

which readily implies that $\|\mathbf{x}_{k-1} - p_{L_k}(\mathbf{x}_{k-1})\| \to 0$ and that

$$\min_{1 \leq k \leq n} \|\mathbf{x}_{k-1} - p_{L_k}(\mathbf{x}_{k-1})\|^2 \leq \frac{2(F(\mathbf{x}_0) - F_*)}{\beta L(f) n}.$$

$\blacksquare$

REMARK 2.3　*If $g \equiv 0$, the proximal-gradient method reduces to the gradient method for the unconstrained nonconvex problem*

$$\min_{\mathbf{x} \in \mathbb{E}} f(\mathbf{x}).$$

In this case

$$\mathbf{x}_{k-1} - p_{L_k}(\mathbf{x}_{k-1}) = \mathbf{x}_{k-1} - \left(\mathbf{x}_{k-1} - \frac{1}{L_k} \nabla f(\mathbf{x}_{k-1}) \right) = \frac{1}{L_k} \nabla f(\mathbf{x}_{k-1}),$$

and Theorem 2.3 reduces to

$$\min_{1 \le k \le n} \|\nabla f(\mathbf{x}_{k-1})\| \le \frac{1}{\sqrt{n}} \left(\frac{2\alpha^2 L(f)(F(\mathbf{x}_0) - F_*)}{\beta} \right)^{1/2},$$

recovering the classical rate of convergence of the gradient method, in other words, $\nabla f(\mathbf{x}_k) \to 0$ *at a rate of* $O(1/\sqrt{k})$.

2.5   A fast proximal-gradient method

2.5.1   Idea of the method

In this section we return to the convex scenario, that is, we assume that f is convex. The basic gradient method relies on using information on the previous iterate only. On the other hand, the so-called conjugate gradient method does use "memory", that is, it generates steps which exploit the *two* previous iterates, and has been known to often improve the performance of basic gradient methods. Similar ideas have been followed to handle *nonsmooth* problems, in particular, the so-called R-algorithm of Shor (see bibliography notes).

However, such methods have not been proven to exhibit a better complexity rate than $O(1/k)$, furthermore, they also often involve some matrix operations that can be problematic in large-scale applications.

Therefore, here the objective is double, namely to build a gradient-based method that

1. Keeps the simplicity of the proximal gradient method to solve model (M).
2. Is proven to be significantly faster, both theoretically and practically.

Both tasks will be achieved by considering again the basic model (M) of Section 2.2 in the convex case. Specifically, we will build a method that is very similar to the proximal-gradient method and is of the form

$$\mathbf{x}_k = p_L(\mathbf{y}_k),$$

where the new point $\mathbf{y}_k$ will be smartly chosen in terms of the two previous iterates $\{\mathbf{x}_{k-1}, \mathbf{x}_{k-2}\}$ and is very easy to compute. Thus, here also we follow the idea of building a scheme with memory, but which is much simpler than the methods alluded to above, and as shown below, will be proven to exhibit a faster rate of convergence.

2.5.2   A fast proximal-gradient method using two past iterations

We begin by presenting the algorithm with a constant stepsize.

> **Fast proximal gradient method with constant stepsize**
>
> **Input:** $L = L(f)$ - A Lipschitz constant of ∇f.
> **Step 0.** Take $\mathbf{y}_1 = \mathbf{x}_0 \in \mathbb{E}$, $t_1 = 1$.
> **Step k.** $(k \geq 1)$ Compute
>
> $$\mathbf{x}_k = p_L(\mathbf{y}_k), \tag{2.34}$$
>
> $$t_{k+1} = \frac{1 + \sqrt{1 + 4t_k^2}}{2}, \tag{2.35}$$
>
> $$\mathbf{y}_{k+1} = \mathbf{x}_k + \left(\frac{t_k - 1}{t_{k+1}}\right)(\mathbf{x}_k - \mathbf{x}_{k-1}). \tag{2.36}$$

The main difference between the above algorithm and the proximal gradient method is that the prox-grad operation $p_L(\cdot)$ is not employed on the previous point $\mathbf{x}_{k-1}$, but rather at the point $\mathbf{y}_k$ – which uses a very specific linear combination of the previous two points $\{\mathbf{x}_{k-1}, \mathbf{x}_{k-2}\}$. Obviously, the main computational effort in both the basic and fast versions of the proximal-gradient method remains the same, namely in the operator p_L. The requested additional computation for the fast proximal-gradient method in the steps (2.35) and (2.36) is clearly marginal. The specific formula for (2.35) emerges from the recursive relation that will be established below in Lemma 2.7.

For the same reasons already explained in Section 2.4.3, we will also analyze the fast proximal-gradient method with a backtracking stepsize rule, which we now explicitly state.

> **Fast proximal-gradient method with backtracking**
>
> **Step 0.** Take $L_0 > 0$, some $\eta > 1$ and $\mathbf{x}_0 \in \mathbb{E}$. Set $\mathbf{y}_1 = \mathbf{x}_0$, $t_1 = 1$.
> **Step k.** $(k \geq 1)$ Find the smallest non-negative integer i_k such that with $\bar{L} = \eta^{i_k} L_{k-1}$:
>
> $$F(p_{\bar{L}}(\mathbf{y}_k)) \leq Q_{\bar{L}}(p_{\bar{L}}(\mathbf{y}_k), \mathbf{y}_k).$$
>
> Set $L_k = \eta^{i_k} L_{k-1}$ and compute
>
> $$\mathbf{x}_k = p_{L_k}(\mathbf{y}_k),$$
>
> $$t_{k+1} = \frac{1 + \sqrt{1 + 4t_k^2}}{2},$$
>
> $$\mathbf{y}_{k+1} = \mathbf{x}_k + \left(\frac{t_k - 1}{t_{k+1}}\right)(\mathbf{x}_k - \mathbf{x}_{k-1}).$$

Note that the upper and lower bounds on L_k given in Remark 2.2 still hold true for the fast proximal-gradient method, namely

$$\beta L(f) \leq L_k \leq \alpha L(f).$$

The next result provides the key recursive relation for the sequence $\{F(\mathbf{x}_k) - F(\mathbf{x}^*)\}$ that will imply the better complexity rate $O(1/k^2)$. As we shall see, Lemma 2.6 of Section 2.4.2 plays a central role in the proofs.

LEMMA 2.7   *The sequences $\{\mathbf{x}_k, \mathbf{y}_k\}$ generated via the fast proximal-gradient method with either a constant or backtracking stepsize rule satisfy for every $k \geq 1$*

$$\frac{2}{L_k} t_k^2 v_k - \frac{2}{L_{k+1}} t_{k+1}^2 v_{k+1} \geq \|\mathbf{u}_{k+1}\|^2 - \|\mathbf{u}_k\|^2,$$

where

$$v_k := F(\mathbf{x}_k) - F(\mathbf{x}^*), \tag{2.37}$$

$$\mathbf{u}_k := t_k \mathbf{x}_k - (t_k - 1)\mathbf{x}_{k-1} - \mathbf{x}^*. \tag{2.38}$$

Proof   Invoking Lemma 2.6 with $\mathbf{x} = t_{k+1}^{-1}\mathbf{x}^* + (1 - t_{k+1}^{-1})\mathbf{x}_k, \mathbf{y} = \mathbf{y}_{k+1}$ and $L = L_{k+1}$ we have

$$\frac{2}{L_{k+1}}(F(t_{k+1}^{-1}\mathbf{x}^* + (1 - t_{k+1}^{-1})\mathbf{x}_k) - F(\mathbf{x}_{k+1})) \tag{2.39}$$

$$\geq \frac{1}{t_{k+1}^2}\{\|t_{k+1}\mathbf{x}_{k+1} - (\mathbf{x}^* + (t_{k+1} - 1)\mathbf{x}_k)\|^2 - \|t_{k+1}\mathbf{y}_{k+1} - (\mathbf{x}^* + (t_{k+1} - 1)\mathbf{x}_k)\|^2\}.$$

By the convexity of F we also have

$$F(t_{k+1}^{-1}\mathbf{x}^* + (1 - t_{k+1}^{-1})\mathbf{x}_k) \leq t_{k+1}^{-1}F(\mathbf{x}^*) + (1 - t_{k+1}^{-1})F(\mathbf{x}_k),$$

which, combined with (2.39), yields

$$\frac{2}{L_{k+1}}((1 - t_{k+1}^{-1})v_k - v_{k+1}) \geq \frac{1}{t_{k+1}^2}\{\|t_{k+1}\mathbf{x}_{k+1} - (\mathbf{x}^* + (t_{k+1} - 1)\mathbf{x}_k)\|^2$$

$$- \|t_{k+1}\mathbf{y}_{k+1} - (\mathbf{x}^* + (t_{k+1} - 1)\mathbf{x}_k)\|^2\}.$$

Using the relation $t_k^2 = t_{k+1}^2 - t_{k+1}$, the latter is equivalent to

$$\frac{2}{L_{k+1}}(t_k^2 v_k - t_{k+1}^2 v_{k+1}) \geq \|\mathbf{u}_{k+1}\|^2 - \|\mathbf{u}_k\|^2,$$

where we used the definition of $\mathbf{u}_k$ (2.38), and the definition of $\mathbf{y}_{k+1}$ (2.36) to simplify the right-hand side. Since $L_{k+1} \geq L_k$, the desired result follows.   ∎

We also need the following trivial facts.

LEMMA 2.8 *Let $\{a_k, b_k\}$ be positive sequences of reals satisfying*

$$a_k - a_{k+1} \geq b_{k+1} - b_k, \forall k \geq 1, \ \text{with } a_1 + b_1 \leq c, \ c > 0.$$

Then, $a_k \leq c$ for every $k \geq 1$.

LEMMA 2.9 *The positive sequence $\{t_k\}$ generated by the fast proximal-gradient method via (2.35) with $t_1 = 1$ satisfies $t_k \geq (k+1)/2$ for all $k \geq 1$.*

We are now ready to prove the promised improved complexity result for the fast proximal-gradient method.

THEOREM 2.4 *Let $\{\mathbf{x}_k\}$, $\{\mathbf{y}_k\}$ be generated by the fast proximal-gradient method with either a constant, or a backtracking stepsize rule. Then for any $k \geq 1$*

$$F(\mathbf{x}_k) - F(\mathbf{x}^*) \leq \frac{2\alpha L(f) \|\mathbf{x}_0 - \mathbf{x}^*\|^2}{(k+1)^2}, \quad \forall \mathbf{x}^* \in X_*, \tag{2.40}$$

where $\alpha = 1$ for the constant stepsize setting and $\alpha = \eta$ for the backtracking stepsize setting.

Proof Let us define the following quantities:

$$a_k := \frac{2}{L_k} t_k^2 v_k, \quad b_k := \|\mathbf{u}_k\|^2, \quad c := \|\mathbf{y}_1 - \mathbf{x}^*\|^2 = \|\mathbf{x}_0 - \mathbf{x}^*\|^2,$$

and recall (cf. Lemma 2.7) that $v_k := F(\mathbf{x}_k) - F(\mathbf{x}^*)$. Then, by Lemma 2.7 we have, for every $k \geq 1$,

$$a_k - a_{k+1} \geq b_{k+1} - b_k,$$

and hence assuming that $a_1 + b_1 \leq c$ holds true, invoking Lemma 2.8, we obtain that

$$\frac{2}{L_k} t_k^2 v_k \leq \|\mathbf{x}_0 - \mathbf{x}^*\|^2,$$

which, combined with $t_k \geq (k+1)/2$ (by Lemma 2.9), yields

$$v_k \leq \frac{2 L_k \|\mathbf{x}_0 - \mathbf{x}^*\|^2}{(k+1)^2}.$$

Utilizing the upper bound on L_k given in (2.28), the desired result (2.40) follows. Thus, all that remains is to prove the validity of the relation $a_1 + b_1 \leq c$. Since $t_1 = 1$, and using the definition of $\mathbf{u}_k$ (2.38), we have here:

$$a_1 = \frac{2}{L_1} t_1 v_1 = \frac{2}{L_1} v_1, \quad b_1 = \|\mathbf{u}_1\|^2 = \|\mathbf{x}_1 - \mathbf{x}^*\|^2.$$

Applying Lemma 2.6 to the points $\mathbf{x} := \mathbf{x}^*, \mathbf{y} := \mathbf{y}_1$ with $L = L_1$, we get

$$\frac{2}{L_1}(F(\mathbf{x}^*) - F(\mathbf{x}_1)) \geq \|\mathbf{x}_1 - \mathbf{x}^*\|^2 - \|\mathbf{y}_1 - \mathbf{x}^*\|^2, \tag{2.41}$$

namely

$$\frac{2}{L_1} v_1 \leq \|\mathbf{y}_1 - \mathbf{x}^*\|^2 - \|\mathbf{x}_1 - \mathbf{x}^*\|^2,$$

that is, $a_1 + b_1 \leq c$ holds true.   ∎

The number of iterations of the proximal-gradient method required to obtain an ε-optimal solution, that is an $\tilde{\mathbf{x}}$ such that $F(\tilde{\mathbf{x}}) - F_* \leq \varepsilon$, is at most $\lceil C/\sqrt{\varepsilon} - 1 \rceil$, where $C = \sqrt{2\alpha L(f)} \|\mathbf{x}_0 - \mathbf{x}^*\|^2$, and which clearly improves the convergence of the basic proximal-gradient method. In Section 2.6 we illustrate the practical value of this theoretical global convergence rate estimate on the l_1-based regularization problem and demonstrate its applicability in wavelet-based image deblurring.

2.5.3   Monotone versus nonmonotone

The fast proximal-gradient method, as opposed to the standard proximal-gradient one, is *not* a monotone algorithm, that is, the function values are not guaranteed to be nonincreasing. Monotonicity seems to be a desirable property of minimization algorithms, but it is not required in the proof of convergence of the fast proximal-gradient method. Moreover, numerical simulations show that, in fact, the algorithm is "almost monotone", that is, except for very few iterations, the algorithm exhibits a monotonicity property.

However, for some applications the prox operation cannot be computed exactly, see for example the total-variation based deblurring example in Section 2.7. In these situations, monotonicity becomes an important issue. It might happen that due to the inexact computations of the prox map, the algorithm might become extremely nonmonotone and in fact can even diverge! This is illustrated in the numerical examples of Section 2.7.3. This is one motivation to introduce a monotone version of the fast proximal-gradient method, which is now explicitly stated in the constant stepsize rule setting.

Monotone fast proximal-gradient method

Input: $L \geq L(f)$ - An upper bound on the Lipschitz constant of ∇f.

Step 0. Take $\mathbf{y}_1 = \mathbf{x}_0 \in \mathbb{E}$, $t_1 = 1$.

Step k. $(k \geq 1)$ Compute

$$\mathbf{z}_k = p_L(\mathbf{y}_k),$$

$$t_{k+1} = \frac{1 + \sqrt{1 + 4t_k^2}}{2}, \tag{2.42}$$

$$\mathbf{x}_k = \operatorname{argmin}\{F(\mathbf{x}) : \mathbf{x} = \mathbf{z}_k, \mathbf{x}_{k-1}\} \tag{2.43}$$

$$\mathbf{y}_{k+1} = \mathbf{x}_k + \left(\frac{t_k}{t_{k+1}}\right)(\mathbf{z}_k - \mathbf{x}_k) + \left(\frac{t_k - 1}{t_{k+1}}\right)(\mathbf{x}_k - \mathbf{x}_{k-1}). \tag{2.44}$$

Clearly, with this modification, we now have a monotone algorithm which is easily seen to be as simple as the fast proximal-gradient method regarding its computational steps. Moreover, it turns out that this modification does not affect the theoretical rate of convergence. Indeed, the convergence rate result for the monotone version remains the same as the convergence rate result of the nonmonotone method:

THEOREM 2.5   *Let $\{\mathbf{x}_k\}$ be generated by the monotone proximal-gradient method. Then for any $k \geq 1$ and any optimal solution $\mathbf{x}^*$:*

$$F(\mathbf{x}_k) - F(\mathbf{x}^*) \leq \frac{2L(f)\|\mathbf{x}_0 - \mathbf{x}^*\|^2}{(k+1)^2}.$$

2.6   Algorithms for l_1-based regularization problems

2.6.1   Problem formulation

In this section we return to the RLS problem discussed in Section 2.2.2. We concentrate on the l_1-based regularization problem in which one seeks to find the solution of

$$\min_{\mathbf{x}}\{F(\mathbf{x}) \equiv \|\mathbf{A}\mathbf{x} - \mathbf{b}\|^2 + \lambda\|\mathbf{x}\|_1\}, \tag{2.45}$$

which is the general model (M) with $f(\mathbf{x}) = \|\mathbf{A}\mathbf{x} - \mathbf{b}\|^2, g(\mathbf{x}) = \lambda\|\mathbf{x}\|_1$. In image deblurring applications, and in particular in wavelet-based restoration methods, $\mathbf{A}$ is often chosen as $\mathbf{A} = \mathbf{R}\mathbf{W}$, where $\mathbf{R}$ is the blurring matrix and $\mathbf{W}$ contains a wavelet basis (i.e., multiplying by $\mathbf{W}$ corresponds to performing an inverse wavelet transform). The vector $\mathbf{x}$ contains the coefficients of the unknown image. The underlying philosophy in dealing with the l_1 norm regularization criterion is that most images have a sparse representation in the wavelet domain. The presence of the l_1 term in (2.45) is used to induce sparsity of the solution. Another important advantage of the l_1-based regularization (2.45) over the l_2-based Tikhonov regularization is that, as opposed to the latter, l_1 regularization is less sensitive to outliers, which in image-processing applications correspond to sharp edges.

The convex optimization problem (2.45) can be cast as a second-order cone programming problem, and thus could be solved via interior-point methods. However, in most applications, for example, in image deblurring, the problem is not only large scale (can reach millions of decision variables), but also involves dense matrix data, which often precludes the use and potential advantage of sophisticated interior-point methods. This motivated the search for simpler gradient-based algorithms for solving (2.45), where the dominant computational effort is relatively cheap matrix–vector multiplications involving $\mathbf{A}$ and $\mathbf{A}^T$.

2.6.2 ISTA: iterative shrinkage/thresholding algorithm

One popular method to solve problem (2.45) is to employ the proximal-gradient method. The proximal map associated with $g(\mathbf{x}) = \lambda \|\mathbf{x}\|_1$ can be analytically computed:

$$\mathrm{prox}_t(g)(\mathbf{y}) = \underset{\mathbf{u}}{\mathrm{argmin}} \left\{ \frac{1}{2t} \|\mathbf{u} - \mathbf{y}\|^2 + \lambda \|\mathbf{u}\|_1 \right\} = \mathcal{T}_{\lambda t}(\mathbf{y}),$$

where $\mathcal{T}_\alpha : \mathbb{R}^n \to \mathbb{R}^n$ is the shrinkage or soft threshold operator defined by

$$\mathcal{T}_\alpha(\mathbf{x})_i = (|x_i| - \alpha)_+ \mathrm{sgn}\,(x_i). \tag{2.46}$$

The arising method is the so-called "iterative shrinkage/thresholding" algorithm (ISTA),[5] which we now explicitly write for the constant stepsize setting.

ISTA with constant stepsize
Input: $L = L(f)$ - A Lipschitz constant of ∇f.
Step 0. Take $\mathbf{x}_0 \in \mathbb{E}$.
Step k. $(k \geq 1)$ Compute

$$\mathbf{x}_k = \mathcal{T}_{\lambda/L}\left(\mathbf{x}_{k-1} - \frac{2}{L}\mathbf{A}^T(\mathbf{A}\mathbf{x}_{k-1} - \mathbf{b})\right).$$

We note that the Lipschitz constant of the gradient of $f(\mathbf{x}) = \|\mathbf{A}\mathbf{x} - \mathbf{b}\|^2$ is given by $L(f) = 2\lambda_{\max}(\mathbf{A}^T\mathbf{A})$. It is of course also possible to incorporate a backtracking stepsize rule in ISTA, as defined in Section 2.4.3.

2.6.3 FISTA: fast ISTA

The function values of the sequence generated by ISTA, which is just a special case of the proximal-gradient method, converge to the optimal function value at a rate of $O(1/k)$, k being the iteration index. An acceleration of ISTA can be achieved by invoking the fast proximal-gradient method for the l_1-based regularization problem (2.45) discussed in Section 2.5. The fast version ISTA algorithm is called FISTA and is now explicitly stated.

---

[5] Other names in the signal processing literature include, for example, the threshold Landweber method, iterative denoising, deconvolution algorithms.

FISTA with constant stepsize
Input: $L = L(f)$ - A Lipschitz constant of ∇f.
Step 0. Take $\mathbf{y}_1 = \mathbf{x}_0 \in \mathbb{E}$, $t_1 = 1$.
Step k. $(k \geq 1)$ Compute

$$\mathbf{x}_k = \mathcal{T}_{\lambda/L}\left(\mathbf{y}_k - \frac{2}{L}\mathbf{A}^T(\mathbf{A}\mathbf{y}_k - \mathbf{b})\right),$$

$$t_{k+1} = \frac{1 + \sqrt{1 + 4t_k^2}}{2},$$

$$\mathbf{y}_{k+1} = \mathbf{x}_k + \left(\frac{t_k - 1}{t_{k+1}}\right)(\mathbf{x}_k - \mathbf{x}_{k-1}).$$

Invoking Theorem 2.4, the rate of convergence of FISTA is $O(1/k^2)$ – a substantial improvement of the rate of convergence of ISTA. Next, we demonstrate through representative examples the practical value of this theoretical, global convergence-rate estimate derived for FISTA on the l_1 wavelet-based regularization problem (2.45).

2.6.4   Numerical examples

Consider the 256×256 cameraman test image whose pixels were scaled into the range between 0 and 1. The image went through a Gaussian blur of size 9×9 and standard deviation 4, followed by an additive, zero-mean, white Gaussian noise with standard deviation 10^{-3}. The original and observed images are given in Figure 2.1.

For these experiments we assume reflexive (Neumann) boundary conditions. We then tested ISTA and FISTA for solving problem (2.45), where $\mathbf{b}$ represents the (vectorized) observed image, and $\mathbf{A} = \mathbf{RW}$, where $\mathbf{R}$ is the matrix representing the blur operator, and

Original   Blurred and noisy

Figure 2.1   Blurring of the cameraman.

Figure 2.2　Iterations of ISTA and FISTA methods for deblurring of the cameraman.

$\mathbf{W}$ is the inverse of a three-stage Haar wavelet transform. The regularization parameter was chosen to be $\lambda = 2e{-}5$ and the initial image was the blurred image. The Lipschitz constant was computable in this example (and those in the sequel) since the eigenvalues of the matrix $\mathbf{A}^T\mathbf{A}$ can be easily calculated using the two-dimensional cosine transform.

Iterations 100 and 200 are described in Figure 2.2. The function value at iteration k is denoted by F_k. The images produced by FISTA are of a better quality than those created by ISTA. The function value of FISTA was consistently lower than the function value of ISTA. We also computed the function values produced after 1000 iterations for ISTA and FISTA which were, respectively, 2.45e-1 and 2.23e-1. Note that the function value of ISTA after 1000 iterations is still worse (that is, larger) than the function value of FISTA after 100 iterations.

From the previous example it seems that, practically, FISTA is able to reach accuracies that are beyond the capabilities of ISTA. To test this hypothesis we also considered an example in which the optimal solution is known. In that instance we considered a

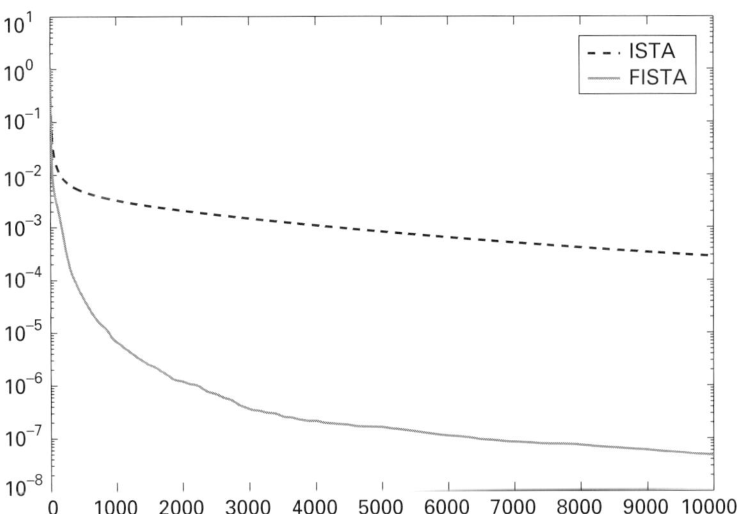

Figure 2.3   Comparison of function values errors $F(\mathbf{x}_k) - F(\mathbf{x}^*)$ of ISTA and FISTA.

64×64 image which undergoes the same blur operator as in the previous example. No noise was added and we solved the least squares problem, that is, $\lambda = 0$. The optimal solution of this problem is zero. The function values of the two methods for $10\,000$ iterations are described in Figure 2.3. The results produced by FISTA are better than those produced by ISTA by several orders of magnitude, and clearly demonstrate the effective performance of FISTA. One can see that after $10\,000$ iterations FISTA reaches accuracy of approximately 10^{-7} while ISTA reaches accuracy of only 10^{-3}. Finally, we observe that the values obtained by ISTA at iteration $10\,000$ was already obtained by FISTA at iteration 275.

2.7   TV-based restoration problems

2.7.1   Problem formulation

Consider images that are defined on rectangle domains. Let $\mathbf{b} \in \mathbb{R}^{m \times n}$ be an observed noisy image, $\mathbf{x} \in \mathbb{R}^{m \times n}$, the true (original) image to be recovered, $\mathcal{A}$ an affine map representing a blurring operator, and $\mathbf{w} \in \mathbb{R}^{m \times n}$ a corresponding additive unknown noise satisfying the relation:

$$\mathbf{b} = \mathcal{A}(\mathbf{x}) + \mathbf{w}. \tag{2.47}$$

The problem of finding an $\mathbf{x}$ from the above relation is a special case of the basic, discrete linear inverse problem discussed in Section 2.2.2. Here we are concerned with total variation (TV)-based regularization, which, given $\mathcal{A}$ and $\mathbf{b}$, seeks to recover $\mathbf{x}$ by solving the convex nonsmooth minimization problem

$$\min_{\mathbf{x}}\{\|\mathcal{A}(\mathbf{x}) - \mathbf{b}\|_F^2 + 2\lambda \text{TV}(\mathbf{x})\}, \tag{2.48}$$

where $\lambda > 0$ and $\text{TV}(\cdot)$ stands for the discrete total variation function. The underlying Euclidean space $\mathbb{E}$ comprises all $m \times n$ matrices with the usual inner product: $\langle \mathbf{a}, \mathbf{b} \rangle = \text{Tr}(\mathbf{b}^T \mathbf{a})$ and the induced Frobenius norm $\| \cdot \|_F$. The identity map will be denoted by $\mathcal{I}$, and with $\mathcal{A} \equiv \mathcal{I}$, problem (2.48) reduces to the so-called "denoising" problem.

Two popular choices for the discrete TV are the isotropic TV defined by

$$\mathbf{x} \in \mathbb{R}^{m \times n}, \quad \text{TV}_I(\mathbf{x}) = \sum_{i=1}^{m-1} \sum_{j=1}^{n-1} \sqrt{(x_{i,j} - x_{i+1,j})^2 + (x_{i,j} - x_{i,j+1})^2}$$
$$+ \sum_{i=1}^{m-1} |x_{i,n} - x_{i+1,n}| + \sum_{j=1}^{n-1} |x_{m,j} - x_{m,j+1}|$$

and the l_1-based, anisotropic TV defined by

$$\mathbf{x} \in \mathbb{R}^{m \times n}, \quad \text{TV}_{l_1}(\mathbf{x}) = \sum_{i=1}^{m-1} \sum_{j=1}^{n-1} \left\{ |x_{i,j} - x_{i+1,j}| + |x_{i,j} - x_{i,j+1}| \right\}$$
$$+ \sum_{i=1}^{m-1} |x_{i,n} - x_{i+1,n}| + \sum_{j=1}^{n-1} |x_{m,j} - x_{m,j+1}|,$$

where, in the above formulae, we assumed the (standard) reflexive boundary conditions:

$$x_{m+1,j} - x_{m,j} = 0, \ \forall j \text{ and } x_{i,n+1} - x_{i,n} = 0, \ \forall i.$$

2.7.2   TV-based denoising

As was already mentioned, when $\mathcal{A} = \mathcal{I}$, problem (2.48) reduces to the denoising problem

$$\min \|\mathbf{x} - \mathbf{b}\|_F^2 + 2\lambda \text{TV}(\mathbf{x}), \tag{2.49}$$

where the nonsmooth regularizer function TV is either the isotropic TV_I or anisotropic TV_{l_1} function. Although this problem can be viewed as a special case of the general model (M) by substituting $f(\mathbf{x}) = \|\mathbf{x} - \mathbf{b}\|^2$ and $g(\mathbf{x}) = 2\lambda \text{TV}(\mathbf{x})$, it is not possible to solve it via the proximal gradient method (or its extensions). This is due to the fact that computation of the prox map amounts to solving a denoising problem of the exact same form.

A common approach for solving the denoising problem is to formulate its dual problem and solve it via a gradient-based method. In order to define the dual problem, some notation is in order:

- $\mathcal{P}$ is the set of matrix-pairs $(\mathbf{p}, \mathbf{q})$ where $\mathbf{p} \in \mathbb{R}^{(m-1) \times n}$ and $\mathbf{q} \in \mathbb{R}^{m \times (n-1)}$ that satisfy

$$p_{i,j}^2 + q_{i,j}^2 \leq 1, \quad i = 1, \ldots, m-1, j = 1, \ldots, n-1,$$
$$|p_{i,n}| \leq 1, \quad i = 1, \ldots, m-1,$$
$$|q_{m,j}| \leq 1, \quad j = 1, \ldots, n-1.$$

- The linear operation $\mathcal{L} : \mathbb{R}^{(m-1)\times n} \times \mathbb{R}^{m \times (n-1)} \to \mathbb{R}^{m \times n}$ is defined by the formula

$$\mathcal{L}(\mathbf{p}, \mathbf{q})_{i,j} = p_{i,j} + q_{i,j} - p_{i-1,j} - q_{i,j-1}, \quad i = 1, \ldots, m, j = 1, \ldots, n,$$

where we assume that $p_{0,j} = p_{m,j} = q_{i,0} = q_{i,n} \equiv 0$ for every $i = 1, \ldots, m$ and $j = 1, \ldots, n$.

The formulation of the dual problem is now recalled.

PROPOSITION 2.1 *Let $(\mathbf{p}, \mathbf{q}) \in \mathcal{P}$ be the optimal solution of the problem*

$$\max_{(\mathbf{p},\mathbf{q})\in\mathcal{P}} -\|\mathbf{b} - \lambda\mathcal{L}(\mathbf{p}, \mathbf{q})\|_F^2. \tag{2.50}$$

Then the optimal solution of (2.49) with $TV = TV_I$ is given by

$$\mathbf{x} = \mathbf{b} - \lambda\mathcal{L}(\mathbf{p}, \mathbf{q}). \tag{2.51}$$

Proof First note that the relations

$$\sqrt{x^2 + y^2} = \max_{p_1,p_2}\{p_1 x + p_2 y : p_1^2 + p_2^2 \le 1\},$$

$$|x| = \max_p\{px : |p| \le 1\}$$

hold true. Hence, we can write

$$TV_I(\mathbf{x}) = \max_{(\mathbf{p},\mathbf{q})\in\mathcal{P}} T(\mathbf{x}, \mathbf{p}, \mathbf{q}),$$

where

$$T(\mathbf{x}, \mathbf{p}, \mathbf{q}) = \sum_{i=1}^{m-1}\sum_{j=1}^{n-1}[p_{i,j}(x_{i,j} - x_{i+1,j}) + q_{i,j}(x_{i,j} - x_{i,j+1})]$$
$$+ \sum_{i=1}^{m-1} p_{i,n}(x_{i,n} - x_{i+1,n}) + \sum_{j=1}^{n-1} q_{m,j}(x_{m,j} - x_{m,j+1}).$$

With this notation we have

$$T(\mathbf{x}, \mathbf{p}, \mathbf{q}) = \mathrm{Tr}(\mathcal{L}(\mathbf{p}, \mathbf{q})^T \mathbf{x}).$$

The problem (2.49) therefore becomes

$$\min_{\mathbf{x}} \max_{(\mathbf{p},\mathbf{q})\in\mathcal{P}} \left\{ \|\mathbf{x} - \mathbf{b}\|_F^2 + 2\lambda\mathrm{Tr}(\mathcal{L}(\mathbf{p}, \mathbf{q})^T \mathbf{x}) \right\}. \tag{2.52}$$

Since the objective function is convex in $\mathbf{x}$ and concave in $\mathbf{p}, \mathbf{q}$, we can exchange the order of the minimum and maximum and obtain the equivalent formulation

$$\max_{(\mathbf{p},\mathbf{q})\in\mathcal{P}} \min_{\mathbf{x}} \left\{ \|\mathbf{x} - \mathbf{b}\|_F^2 + 2\lambda\mathrm{Tr}(\mathcal{L}(\mathbf{p}, \mathbf{q})^T \mathbf{x}) \right\}.$$

The optimal solution of the inner minimization problem is

$$\mathbf{x} = \mathbf{b} - \lambda \mathcal{L}(\mathbf{p}, \mathbf{q}).$$

Plugging the above expression for $\mathbf{x}$ back into (2.52), and omitting constant terms, we obtain the dual problem (2.50). ∎

REMARK 2.4 *The only difference in the dual problem corresponding to the case $TV = TV_{l_1}$ (in comparison to the case $TV = TV_I$), is that the minimization in the dual problem is not done over the set $\mathcal{P}$, but over the set $\mathcal{P}_1$, which consists of all pairs of matrices $(\mathbf{p}, \mathbf{q})$ where $\mathbf{p} \in \mathbb{R}^{(m-1) \times n}$ and $\mathbf{q} \in \mathbb{R}^{m \times (n-1)}$ satisfying*

$$|p_{i,j}| \leq 1, \quad i = 1, \ldots, m-1, \quad j = 1, \ldots, n,$$

$$|q_{i,j}| \leq 1, \quad i = 1, \ldots, m, \quad j = 1, \ldots, n-1.$$

The dual problem (2.50), when formulated as a minimization problem:

$$\min_{(\mathbf{p}, \mathbf{q}) \in \mathcal{P}} \|\mathbf{b} - \lambda \mathcal{L}(\mathbf{p}, \mathbf{q})\|_F^2 \tag{2.53}$$

falls into the category of model (M) by taking f to be the objective function of (2.53) and $g \equiv \delta_{\mathcal{P}}$ – to be the indicator function of $\mathcal{P}$. The objective function, being quadratic, has a Lipschitz gradient and, as a result, we can invoke either the proximal gradient method, which coincides with the gradient projection method in this case, or the fast proximal gradient method. The exact details of computations of the Lipschitz constant and of the gradient are omitted. The slower method will be called GP (for "gradient projection") and the faster method will be called FGP (for "fast gradient projection").

To demonstrate the advantage of FGP over GP, we have taken a small 10×10 image for which we added normally distributed white noise with standard deviation 0.1. The parameter λ was chosen as 0.1. Since the problem is small, we were able to find its exact solution. Figure 2.4 shows the difference $F(\mathbf{x}_k) - F_*$ (in log scale) for $k = 1, \ldots, 100$.

Clearly, FGP reaches greater accuracies than those obtained by GP. After 100 iterations FGP reached an accuracy of 10^{-5} while GP reached an accuracy of only $10^{-2.5}$. Moreover, the function value reached by GP at iteration 100 was already obtained by GP after 25 iterations. Another interesting phenomena can be seen at iterations 81 and 82, and is marked on the figure. As opposed to the GP method, FGP is not a monotone method. This does not have an influence on the convergence of the sequence and we see that in most iterations there is a decrease in the function value. In the next section, we will see that this nonmonotonicity phenomena can have a severe impact on the convergence of a related two-steps method for the image deblurring problem.

2.7.3   TV-based deblurring

Consider now the TV-based deblurring optimization model

$$\min \|\mathcal{A}(\mathbf{x}) - \mathbf{b}\|_F^2 + 2\lambda \mathrm{TV}(\mathbf{x}), \tag{2.54}$$

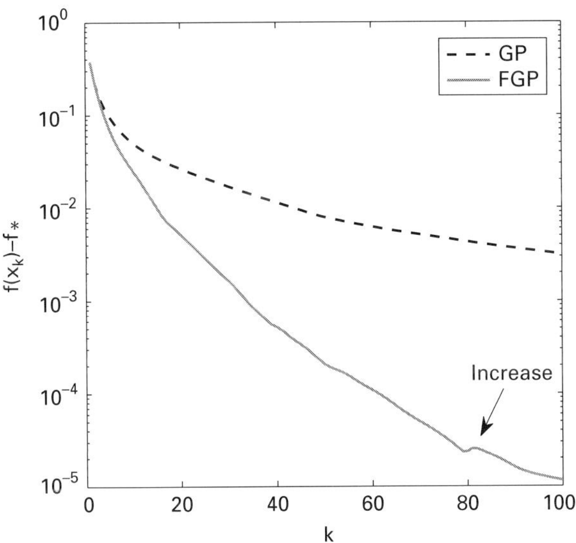

Figure 2.4   Accuracy of FGP compared with GP.

where $\mathbf{x} \in \mathbb{R}^{m \times n}$ is the original image to be restored, $\mathcal{A} : \mathbb{R}^{m \times n} \to \mathbb{R}^{m \times n}$ is a linear transformation representing some blurring operator, $\mathbf{b}$ is the noisy and blurred image, and $\lambda > 0$ is a regularization parameter. Obviously problem (2.54) is within the setting of the general model (M) with

$$f(\mathbf{x}) = \|\mathcal{A}(\mathbf{x}) - \mathbf{b}\|^2, \quad g(\mathbf{x}) = 2\lambda \mathrm{TV}(\mathbf{x}), \quad \text{and} \quad \mathbb{E} = \mathbb{R}^{m \times n}.$$

Deblurring is of course more challenging than denoising. Indeed, to construct an equivalent smooth optimization problem for (2.54) via its dual along the approach of Section 2.7.2, it is easy to realize that one would need to invert the operator $\mathcal{A}^T \mathcal{A}$, which is clearly an ill-posed problem, in other words, such an approach is not viable. This is in sharp contrast to the denoising problem, where a smooth dual problem was constructed, and was the basis of efficient solution methods. Instead, we suggest to solve the deblurring problem by the fast proximal gradient method. Each iteration of the method will require the computation of the prox map, which in this case amounts to solving a denoising problem. More precisely, if we denote the optimal solution of the constrained *denoising* problem (2.49) with observed image $\mathbf{b}$, and the regularization parameter λ by $D_C(\mathbf{b}, \lambda)$, then with this notation, the prox-grad map $p_L(\cdot)$ can be simply written as:

$$p_L(\mathbf{Y}) = D_C \left(\mathbf{Y} - \frac{2}{L} \mathcal{A}^T (\mathcal{A}(\mathbf{Y}) - \mathbf{b}), \frac{2\lambda}{L} \right).$$

Thus, each iteration involves the solution of a subproblem that should be solved using an iterative method such as GP or FGP. Note also that this is in contrast to the situation with the simpler l_1-based regularization problem where ISTA or FISTA requires only the computation of a gradient step and a shrinkage, which in that case is an *explicit* operation

(Section 2.6). The fact that the prox operation does not have an explicit expression but is rather computed via an iterative algorithm can have a profound impact on the performance of the method. This is illustrated in the following section.

2.7.4   Numerical example

Consider a 64×64 image that was cut from the cameraman test image (whose pixels are scaled to be between 0 and 1). The image goes through a Gaussian blur of size 9×9 and standard deviation 4, followed by an additive, zero-mean, white Gaussian noise with standard deviation 10^{-2}. The regularization parameter λ is chosen to be 0.01. We adopt the same terminology used for the l_1-based regularization, and use the named ISTA for the proximal gradient method, and the named FISTA for the fast proximal gradient method.

Figure 2.5 presents three graphs showing the function values of the FISTA method applied to (2.54), in which the denoising subproblems are solved using FGP with a number of FGP iterations, denoted by N, taking the values 5, 10, and 20. In the left image the denoising subproblems are solved using FGP and in the right image the denoising subproblems are solved using GP. Clearly FISTA, in combination with either GP or FGP, diverges when $N = 5$, although it seems that the combination FISTA/GP is worse than FISTA/FGP. For $N = 10$ FISTA/FGP seems to converge to a value which is slightly higher than the one obtained by the same method with $N = 20$, and FISTA/GP with $N = 10$ is still very much erratic and does not seem to converge.

From this example we can conclude that (1) FISTA can diverge when the subproblems are not solved exactly, and (2) the combination FISTA/FGP seems to be better than FISTA/GP. The latter conclusion is further numerical evidence (in addition to the results of Section 2.6) to the superiority of FGP over GP. The first conclusion motivates us to

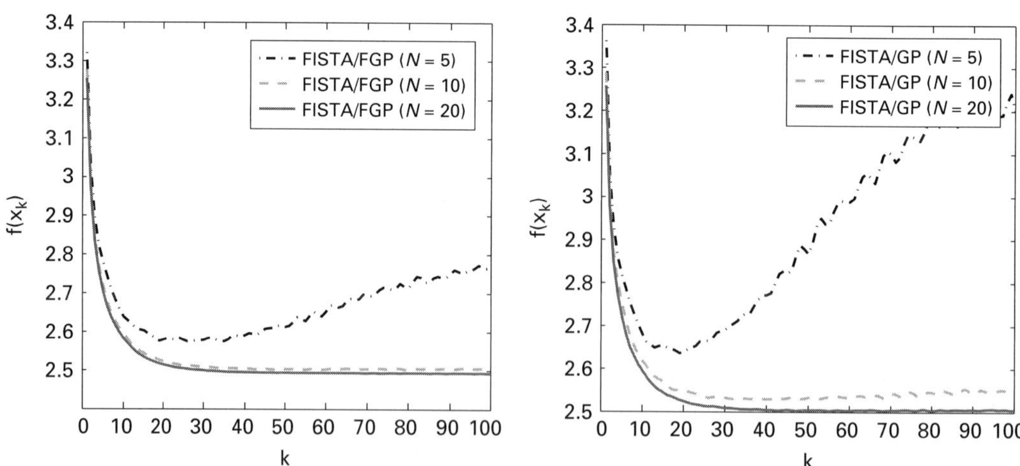

Figure 2.5   Function values of the first 100 iterations of FISTA. The denoising subproblems are solved using FGP (left image) or GP (right image) with $N = 5, 10, 20$.

use the monotone version of FISTA, which we term MFISTA, and which was introduced in Section 2.5.3 for the general model (M). We ran MFISTA on the exact same problem and the results are shown in Figure 2.5. Clearly the monotone version of FISTA seems much more robust and stable. Therefore, it seems that there is a clear advantage in using MFISTA instead of FISTA when the prox map cannot be computed exactly.

2.8 The source-localization problem

2.8.1 Problem formulation

Consider the problem of locating a single radiating source from noisy range measurements collected using a network of passive sensors. More precisely, consider an array of m sensors, and let $\mathbf{a}_j \in \mathbb{R}^n$ denote the coordinates of the jth sensor.[6] Let $\mathbf{x} \in \mathbb{R}^n$ denote the unknown source's coordinate vector, and let $d_j > 0$ be a noisy observation of the range between the source and the jth sensor:

$$d_j = \|\mathbf{x} - \mathbf{a}_j\| + \varepsilon_j, \quad j = 1,\ldots,m, \tag{2.55}$$

where $\boldsymbol{\varepsilon} = (\varepsilon_1,\ldots,\varepsilon_m)^T$ denotes the unknown noise vector. Such observations can be obtained, for example, from the time-of-arrival measurements in a constant-velocity propagation medium. The source-localization problem is the following:

The source-localization problem
Given the observed range measurements $d_j > 0$, find a "good" approximation of the source $\mathbf{x}$.

The "source-localization" (SL) problem has received significant attention in the signal-processing literature and, specifically, in the field of mobile phone localization.

There are many possible mathematical formulations for the source-localization problem. A natural and common approach is to consider a least-squares criterion in which the optimization problems seeks to minimize the squared sum of errors:

$$\text{(SL):} \quad \min_{\mathbf{x}\in\mathbb{R}^n} \left\{ f(\mathbf{x}) \equiv \sum_{j=1}^{m}(\|\mathbf{x} - \mathbf{a}_j\| - d_j)^2 \right\}. \tag{2.56}$$

The above criterion also has a statistical interpretation. When $\boldsymbol{\varepsilon}$ follows a Gaussian distribution with a covariance matrix proportional to the identity matrix, the optimal solution of (SL) is in fact the maximum-likelihood estimate.

The SL problem is a nonsmooth, nonconvex problem and, as such, is not an easy problem to solve. In the following we will show how to construct two simple methods using the concepts explained in Section 2.3. The derivation of the algorithms is inspired

---

[6] In practical applications $n = 2$ or 3.

by Weiszfeld's algorithm for the Fermat–Weber problem which was described in Section 2.3.4. Throughout this section, we denote the set of sensors by $\mathcal{A} := \{\mathbf{a}_1, \ldots, \mathbf{a}_m\}$.

2.8.2   The simple fixed-point algorithm: definition and analysis

Similarly to Weiszfeld's method, our starting point for constructing a fixed-point algorithm to solve the SL problem is to write the optimality condition and "extract" $\mathbf{x}$. Assuming that $\mathbf{x} \notin \mathcal{A}$ we have that $\mathbf{x}$ is a stationary point for problem (SL) if, and only if

$$\nabla f(\mathbf{x}) = 2 \sum_{j=1}^{m} (\|\mathbf{x} - \mathbf{a}_j\| - d_j) \frac{\mathbf{x} - \mathbf{a}_j}{\|\mathbf{x} - \mathbf{a}_j\|} = \mathbf{0}, \tag{2.57}$$

which can be written as

$$\mathbf{x} = \frac{1}{m} \left\{ \sum_{j=1}^{m} \mathbf{a}_j + \sum_{j=1}^{m} d_j \frac{\mathbf{x} - \mathbf{a}_j}{\|\mathbf{x} - \mathbf{a}_j\|} \right\}.$$

The latter relation calls for the following fixed-point algorithm which we term the "standard fixed-point (SFP) scheme":

Algorithm SFP

$$\mathbf{x}_k = \frac{1}{m} \left\{ \sum_{j=1}^{m} \mathbf{a}_j + \sum_{j=1}^{m} d_j \frac{\mathbf{x}_{k-1} - \mathbf{a}_j}{\|\mathbf{x}_{k-1} - \mathbf{a}_j\|} \right\}, \quad k \geq 1. \tag{2.58}$$

Like in Weiszfeld's algorithm, the SFP scheme is not well defined if $\mathbf{x}_k \in \mathcal{A}$ for some k. In the sequel we will state a result claiming that by carefully selecting the initial vector $\mathbf{x}_0$ we can *guarantee* that the iterates are not in the sensor set $\mathcal{A}$, therefore establishing that the method is well defined.

Before proceeding with the analysis of the SFP method, we record the fact that the SFP scheme is actually a gradient method with a fixed stepsize.

PROPOSITION 2.2   *Let $\{\mathbf{x}_k\}$ be the sequence generated by the SFP method (2.58) and suppose that $\mathbf{x}_k \notin \mathcal{A}$ for all $k \geq 0$. Then for every $k \geq 1$:*

$$\mathbf{x}_k = \mathbf{x}_{k-1} - \frac{1}{2m} \nabla f(\mathbf{x}_{k-1}). \tag{2.59}$$

Proof   Follows by a straightforward calculation, using the gradient of f computed in (2.57).   ∎

It is interesting to note that the SFP method belongs to the class of MM methods (see Section 2.3.3). That is, a function $h(\cdot,\cdot)$ exists such that $h(\mathbf{x},\mathbf{y}) \geq f(\mathbf{x})$ and $h(\mathbf{x},\mathbf{x}) = f(\mathbf{x})$, for which

$$\mathbf{x}_k = \operatorname*{argmin}_{\mathbf{x}} h(\mathbf{x},\mathbf{x}_{k-1}).$$

The only departure from the philosophy of MM methods is that special care should be given to the sensors set $\mathcal{A}$. We define the auxiliary function h as

$$h(\mathbf{x},\mathbf{y}) \equiv \sum_{j=1}^{m} \|\mathbf{x} - \mathbf{a}_j - d_j r_j(\mathbf{y})\|^2, \qquad \text{for every } \mathbf{x} \in \mathbb{R}^n, \mathbf{y} \in \mathbb{R}^n \setminus \mathcal{A}, \tag{2.60}$$

where

$$r_j(\mathbf{y}) \equiv \frac{\mathbf{y} - \mathbf{a}_j}{\|\mathbf{y} - \mathbf{a}_j\|}, \quad j = 1,\ldots,m.$$

Note that for every $\mathbf{y} \notin \mathcal{A}$, the following relations hold for every $j = 1,\ldots,m$:

$$\|r_j(\mathbf{y})\| = 1, \tag{2.61}$$

$$(\mathbf{y} - \mathbf{a}_j)^T r_j(\mathbf{y}) = \|\mathbf{y} - \mathbf{a}_j\|. \tag{2.62}$$

In Lemma 2.10 below, we prove the key properties of the auxiliary function h defined in (2.60). These properties verify the fact that this is, in fact, an MM method.

LEMMA 2.10 (a) $h(\mathbf{x},\mathbf{x}) = f(\mathbf{x})$ for every $\mathbf{x} \notin \mathcal{A}$.
(b) $h(\mathbf{x},\mathbf{y}) \geq f(\mathbf{x})$ for every $\mathbf{x} \in \mathbb{R}^n, \mathbf{y} \in \mathbb{R}^n \setminus \mathcal{A}$.
(c) If $\mathbf{y} \notin \mathcal{A}$ then

$$\mathbf{y} - \frac{1}{2m}\nabla f(\mathbf{y}) = \operatorname*{argmin}_{\mathbf{x} \in \mathbb{R}^n} h(\mathbf{x},\mathbf{y}). \tag{2.63}$$

Proof (a) For every $\mathbf{x} \notin \mathcal{A}$,

$$f(\mathbf{x}) = \sum_{j=1}^{m}(\|\mathbf{x} - \mathbf{a}_j\| - d_j)^2$$

$$= \sum_{j=1}^{m}(\|\mathbf{x} - \mathbf{a}_j\|^2 - 2d_j\|\mathbf{x} - \mathbf{a}_j\| + d_j^2)$$

$$\overset{(2.61),(2.62)}{=} \sum_{j=1}^{m}(\|\mathbf{x} - \mathbf{a}_j\|^2 - 2d_j(\mathbf{x} - \mathbf{a}_j)^T r_j(\mathbf{x}) + d_j^2\|r_j(\mathbf{x})\|^2) = h(\mathbf{x},\mathbf{x}),$$

where the last equation follows from (2.60).

(b) Using the definition of f and h given respectively in (2.56),(2.60), and the fact (2.61), a short computation shows that for every $\mathbf{x} \in \mathbb{R}^n, \mathbf{y} \in \mathbb{R}^n \setminus \mathcal{A}$,

$$h(\mathbf{x},\mathbf{y}) - f(\mathbf{x}) = 2\sum_{j=1}^{m} d_j \left(\|\mathbf{x} - \mathbf{a}_j\| - (\mathbf{x} - \mathbf{a}_j)^T r_j(\mathbf{y}) \right) \geq 0,$$

where the last inequality follows from the Cauchy–Schwartz inequality and using again (2.61).

(c) For any $\mathbf{y} \in \mathbb{R}^n \setminus \mathcal{A}$, the function $\mathbf{x} \mapsto h(\mathbf{x},\mathbf{y})$ is strictly convex on $\mathbb{R}^n$, and consequently admits a unique minimizer $\mathbf{x}^*$ satisfying

$$\nabla_{\mathbf{x}} h(\mathbf{x}^*,\mathbf{y}) = \mathbf{0}.$$

Using the definition of h given in (2.60), the latter identity can be explicitly written as

$$\sum_{j=1}^{m} (\mathbf{x}^* - \mathbf{a}_j - d_j r_j(\mathbf{y})) = \mathbf{0},$$

which, by simple algebraic manipulation, can be shown to be equivalent to $\mathbf{x}^* = \mathbf{y} - \frac{1}{2m}\nabla f(\mathbf{y})$. ∎

By the properties just established, it follows that the SFP method is an MM method and, as such, is a descent scheme. It is also possible to prove the convergence result given in Theorem 2.6 below. Not surprisingly, since the problem is nonconvex, only convergence to stationary points is established.

THEOREM 2.6 (Convergence of the SFP Method.) *Let $\{\mathbf{x}_k\}$ be generated by (2.58) such that $\mathbf{x}_0$ satisfies*

$$f(\mathbf{x}_0) < \min_{j=1,\ldots,m} f(\mathbf{a}_j). \tag{2.64}$$

Then,

(a) $\mathbf{x}_k \notin \mathcal{A}$ for every $k \geq 0$.
(b) For every $k \geq 1, f(\mathbf{x}_k) \leq f(\mathbf{x}_{k-1})$ and equality is satisfied if and only if $\mathbf{x}_k = \mathbf{x}_{k-1}$.
(c) The sequence of function values $\{f(\mathbf{x}_k)\}$ converges.
(d) The sequence $\{\mathbf{x}_k\}$ is bounded.
(e) Any limit point of $\{\mathbf{x}_k\}$ is a stationary point of f.

The condition (2.64) is very mild in the sense that it is not difficult to find an initial vector $\mathbf{x}_0$ satisfying it (see bibliographic notes for details).

Next we show how to construct a different method for solving the source-localization problem using a completely different approximating auxiliary function.

2.8.3 The SWLS algorithm

To motivate the construction of the second method, let us first go back again to Weiszfeld's scheme, and recall that Weiszfeld's method can also be written as (see also [2.15]):

$$\mathbf{x}_k = \operatorname*{argmin}_{\mathbf{x}\in\mathbb{R}^n} h(\mathbf{x}, \mathbf{x}_{k-1}),$$

where

$$h(\mathbf{x}, \mathbf{y}) \equiv \sum_{j=1}^{m} \omega_j \frac{\|\mathbf{x} - \mathbf{a}_j\|^2}{\|\mathbf{y} - \mathbf{a}_j\|} \quad \text{for every } \mathbf{x} \in \mathbb{R}^n, \mathbf{y} \in \mathbb{R}^n \setminus \mathcal{A}.$$

The auxiliary function h was essentially constructed from the objective function of the Fermat–Weber location problem, by replacing the norm terms $\|\mathbf{x} - \mathbf{a}_j\|$ with $\frac{\|\mathbf{x}-\mathbf{a}_j\|^2}{\|\mathbf{y}-\mathbf{a}_j\|}$. Mimicking this observation for the SL problem under study, we will use an auxiliary function in which each norm term $\|\mathbf{x} - \mathbf{a}_j\|$ in the objective function (2.56) is replaced with $\frac{\|\mathbf{x}-\mathbf{a}_j\|^2}{\|\mathbf{y}-\mathbf{a}_j\|}$, resulting in the following auxiliary function:

$$g(\mathbf{x}, \mathbf{y}) \equiv \sum_{i=1}^{m} \left(\frac{\|\mathbf{x} - \mathbf{a}_i\|^2}{\|\mathbf{y} - \mathbf{a}_i\|} - d_i \right)^2, \quad \mathbf{x} \in \mathbb{R}^n, \mathbf{y} \in \mathbb{R}^n \setminus \mathcal{A}. \tag{2.65}$$

The general step of the algorithm for solving problem (SL), termed "the sequential weighted least-squares" (SWLS) method, is now given by

$$\mathbf{x}_k \in \operatorname*{argmin}_{\mathbf{x}\in\mathbb{R}^n} g(\mathbf{x}, \mathbf{x}_{k-1}).$$

or more explicitly by

Algorithm SWLS

$$\mathbf{x}_k \in \operatorname*{argmin}_{\mathbf{x}\in\mathbb{R}^n} \sum_{j=1}^{m} \left(\frac{\|\mathbf{x} - \mathbf{a}_j\|^2}{\|\mathbf{x}_{k-1} - \mathbf{a}_j\|} - d_j \right)^2. \tag{2.66}$$

The name SWLS stems from the fact that at each iteration k we are required to solve the following weighted, nonlinear least-squares problem:

$$\text{(NLS):} \quad \min_{\mathbf{x}} \sum_{j=1}^{m} \omega_j^k (\|\mathbf{x} - \mathbf{c}_j\|^2 - \beta_j^k)^2, \tag{2.67}$$

with

$$\mathbf{c}_j = \mathbf{a}_j, \beta_j^k = d_j \|\mathbf{x}_{k-1} - \mathbf{a}_j\|, \omega_j^k = \frac{1}{\|\mathbf{x}_{k-1} - \mathbf{a}_j\|^2}. \tag{2.68}$$

Note that the SWLS algorithm as presented above is not defined for iterations in which $\mathbf{x}_{k-1} \in \mathcal{A}$. However, as in the SFP method, it is possible to find an initial point ensuring that the iterates are not in the sensor set $\mathcal{A}$.

The NLS problem is a nonconvex problem, but it can still be solved globally and efficiently by transforming it into a problem of minimizing a quadratic function subject to a single quadratic constraint. Indeed, for a given fixed k, we can transform (2.67) into a constrained-minimization problem (the index k is omitted):

$$\min_{\mathbf{x} \in \mathbb{R}^n, \alpha \in \mathbb{R}} \left\{ \sum_{j=1}^m \omega_j(\alpha - 2\mathbf{c}_j^T \mathbf{x} + \|\mathbf{c}_j\|^2 - \beta_j)^2 : \|\mathbf{x}\|^2 = \alpha \right\}, \tag{2.69}$$

which can also be written as (using the substitution $\mathbf{y} = (\mathbf{x}^T, \alpha)^T$)

$$\min_{\mathbf{y} \in \mathbb{R}^{n+1}} \left\{ \|\mathbf{A}\mathbf{y} - \mathbf{b}\|^2 : \mathbf{y}^T \mathbf{D} \mathbf{y} + 2\mathbf{f}^T \mathbf{y} = 0 \right\}, \tag{2.70}$$

where

$$\mathbf{A} = \begin{pmatrix} -2\sqrt{\omega_1}\mathbf{c}_1^T & \sqrt{\omega_1} \\ \vdots & \vdots \\ -2\sqrt{\omega_m}\mathbf{c}_m^T & \sqrt{\omega_m} \end{pmatrix}, \; \mathbf{b} = \begin{pmatrix} \sqrt{\omega_1}(\beta_1 - \|\mathbf{c}_1\|^2) \\ \vdots \\ \sqrt{\omega_m}(\beta_m - \|\mathbf{c}_m\|^2) \end{pmatrix}$$

and

$$\mathbf{D} = \begin{pmatrix} \mathbf{I}_n & \mathbf{0}_{n \times 1} \\ \mathbf{0}_{1 \times n} & 0 \end{pmatrix}, \; \mathbf{f} = \begin{pmatrix} 0 \\ -0.5 \end{pmatrix}.$$

Problem (2.70) belongs to the class of problems consisting of minimizing a quadratic function subject to a single quadratic constraint (without any convexity assumptions). Problems of this type are called "generalized trust-region subproblems" (GTRSs). GTRS problems possess necessary and sufficient optimality conditions from which efficient solution methods can be derived.

The analysis of the SWLS method is more complicated than the analysis of the SFP method and we only state the main results. For the theoretical convergence analysis two rather mild assumptions are required:

Assumption 1 The matrix

$$\mathbf{A} = \begin{pmatrix} 1 & \mathbf{a}_1^T \\ 1 & \mathbf{a}_2^T \\ \vdots & \vdots \\ 1 & \mathbf{a}_m^T \end{pmatrix}$$

is of full column rank.

For example, when $n = 2$, the assumption states that $\mathbf{a}_1, \ldots, \mathbf{a}_m$ are not on the same line. The second assumption states that the value of the initial vector $\mathbf{x}_0$ is "small enough".

Assumption 2 $f(\mathbf{x}_0) < \frac{\min_j \{d_j\}^2}{4}$.

A similar assumption was made for the SFP method (see condition [2.64]). Note that for the true source location $\mathbf{x}_{\text{true}}$ one has $f(\mathbf{x}_{\text{true}}) = \sum_{j=1}^m \varepsilon_j^2$. Therefore, $\mathbf{x}_{\text{true}}$ satisfies Assumption 2 if the errors ε_j are smaller in some sense from the range measurements d_j. This is a very reasonable assumption, since in real applications the errors ε_j are often smaller in an order of magnitude than d_j. Now, if the initial point $\mathbf{x}_0$ is "good enough" in the sense that it is close to the true source location, then Assumption 2 will be satisfied.

Under the above assumption it is possible to prove the following key properties of the auxiliary function g:

$$g(\mathbf{x}, \mathbf{x}) = f(\mathbf{x}), \text{ for every } \mathbf{x} \in \mathbb{R}^n,$$

$$g(\mathbf{x}, \mathbf{x}_{k-1}) \geq 2f(\mathbf{x}) - f(\mathbf{x}_{k-1}), \text{ for every } \mathbf{x} \in \mathbb{R}^n, k \geq 1. \tag{2.71}$$

Therefore, the SWLS method, as opposed to the SFP method, is not an MM method, since the auxiliary function $g(\cdot, \cdot)$ is not an upper bound on the objective function. However, similarly to Weiszfeld's method for the Fermat–Weber problem (see Section 2.15), the property (2.71) implies the descent property of the SWLS method, and it can also be used in order to prove the convergence result of the method which is given below.

THEOREM 2.7 (Convergence of the SWLS Method.) *Let $\{\mathbf{x}_k\}$ be the sequence generated by the SWLS method. Suppose that Assumptions 1 and 2 hold true. Then*

(a) $\mathbf{x}_k \notin \mathcal{A}$ *for $k \geq 0$.*
(b) *For every $k \geq 1$, $f(\mathbf{x}_k) \leq f(\mathbf{x}_{k-1})$ and equality holds, if, and only if $\mathbf{x}_k = \mathbf{x}_{k-1}$.*
(c) *The sequence of function values $\{f(\mathbf{x}^k)\}$ converges.*
(d) *The sequence $\{\mathbf{x}^k\}$ is bounded.*
(e) *Any limit point of $\{\mathbf{x}^k\}$ is a stationary point of f.*

2.9   Bibliographic notes

Section 2.2 The class of optimization problems (M) has been first studied in [4] and provides a natural vehicle to study various generic optimization models under a common framework. Linear inverse problems arise in a wide range of diverse applications, and the literature is vast [26]. A popular regularization technique is the Tikhonov smooth quadratic regularization [56] which has been extensively studied and extended [33–35]. Early works promoting the use of the convex, nonsmooth l_1 regularization appear, for example, in [18, 20, 38]. The l_1 regularization has now attracted an intensive revived interest in the signal-processing literature, in particular in compressed sensing, which has led to a large amount of literature [11, 24, 29].

Section 2.3 The gradient method is one of the very first methods for unconstrained minimization, going back to 1847 with the work of Cauchy [12]. Gradient methods and their variants have been studied by many authors. We mention, in particular, the classical works developed in the 60s and 70s by [1,22,32,41,47], and for more modern presentations with many results, including the extension to problems with constraints, and the resulting gradient-projection method given in 2.3.1, see the books of [9, 46, 49] and references therein. The quadratic-approximation model in 2.3.1 is a very well-known interpretation of the gradient method as a proximal regularization of the linearized part of a differentiable function f [49]. The proximal map was introduced by Moreau [42]. The extension to handle the nonsmooth model (M), as given in Sections 2.3.1 and 2.3.2 is a special case of the proximal forward–backward method for finding the zero of the sum of two maximal monotone operators, originally proposed by [48]. The terminology "proximal gradient" is used to emphasize the specific composite operation when applied to a minimization problem of the form (M). The majorization–minimization idea discussed in 2.3.3 has been developed by many authors, and for a recent tutorial on MM algorithms, applications and many references see [37], and for its use in signal processing see, for instance, [23,28]. The material on the Fermat–Weber location problem presented in Section 2.3.4 is classical. The original Weiszfeld algorithm can be found in [58] and is further analyzed in [40]. It has been intensively and further studied in the location theory literature [52].

Section 2.4 For simplicity of exposition, we assumed the existence of an optimal solution for the optimization model (M). For classical and more advanced techniques to handle the existence of minimizers, see [2]. The proximal map and regularization of a closed, proper convex function is due to Moreau; see [42] for the proof of Lemma 2.2. The results of Section 2.4.2 are well known and can be found in [9,49]. Lemma 2.6 is a slight modification of a recent result proven in [8]. The material of Section 2.4.3 follows [8], except for the pointwise convergence Theorem 2.2. More general convergence results of the sequence $\mathbf{x}_k$ can be found in [27], and, in particular, we refer to the comprehensive recent work [21]. The nonconvex case in Theorem 2.3 seems to be new, and naturally extends the known result [46] for the smooth unconstrained case, cf. Remark 2.3.

Section 2.5 A well-known and popular gradient method based on two steps memory is the conjugate-gradient algorithm [9, 49]. In the nonsmooth case, a similar idea was developed by Shor with the R-algorithm [54]. However, such methods do not appear to improve the complexity rate of basic gradient-like methods. This goal was achieved by Nesterov [44], who was the first to introduce a new idea and algorithm for minimizing a smooth convex function proven to be an *optimal-gradient* method in the sense of complexity analysis [43]. This algorithm was recently extended to the convex nonsmooth model (M) in [8], and all the material in this section is from [8], except for the nonmonotone case and Theorem 2.5, which were very recently developed in [7]. For recent alternative gradient-based methods, including methods based on two or more gradient steps, and that could speed-up the proximal-gradient method for solving the special case of model (M) with f being the least-squares objective, see, for instance, [10,25,29]. The speedup gained by these methods has been shown through numerical experiments, but global, nonasymptotic rate of convergence results have not been established. In the recent work of [45], a multistep fast gradient method that solves model (M) has been developed and proven to share the same complexity rate $O(1/k^2)$ as derived here. The new method of [45] is remarkably different conceptually and computationally from the fast proximal-gradient method; it uses the accumulated history of past iterates, and requires two projection-like operations per iteration [8, 45]. For a recent study on a gradient scheme based on non-Euclidean distances for solving smooth, conic convex problems, and which shares the same fast complexity rate, see the recent work [3].

Section 2.6 The quadratic l_1-based regularization model has attracted a considerable amount of attention in the signal-processing literature [13,23,30]; for the iterative shrinkage/thresholding algorithm (ISTA) and for more recent works, including new algorithms, applications, and many pointers to relevant literature, see [21,25,29,39]. The results and examples presented in Section 2.6.3 are from the recent work [8] where more details, references, and examples are given.

Section 2.7 The total variation (TV)-based model has been introduced by [51]. The literature on numerical methods for solving model (2.48) abounds [14, 17, 31, 36, 57]. This list is just given as an indicator of the intense research in the field and is far from being comprehensive. The work of Chambolle [15, 16] is of particular interest. There, he introduced and developed a globally convergent, gradient, dual-based algorithm for the denoising problem, which was shown to be faster than primal-based schemes. His works motivated our recent analysis and algorithmic developments given in [7] for the more involved constrained TV-based deblurring problem, which when combined with FISTA, produces fast gradient methods. This section has presented some results and numerical examples from [7], to which we refer the reader for further reading.

Section 2.8 The single source-localization problem has received significant attention in the field of signal processing [5, 19, 53, 55]. The algorithms and results given in this section are taken from the recent work [6], where more details, results, and proofs of the theorems can be found.

References

[1] L. Armijo, "Minimization of functions having continuous partial derivatives," *Pacific Journal of Mathematics*, vol. 16, pp. 1–3, 1966.

[2] A. Auslender and M. Teboulle, *Asymptotic Cones and Functions in Optimization and Variational Inequalities*. Springer Monographs in Mathematics. New York: Springer, 2003.

[3] A. Auslender and M. Teboulle, "Interior gradient and proximal methods for convex and conic optimization," *SIAM Journal of Optimization*, vol. 16. no. 3, pp. 697–725, 2006.

[4] A. Auslender, "Minimisation de fonctions localement Lipschitziennes: applications a la programmation mi-convexe, mi-differentiable," in *Nonlinear Programming 3*, O. L. Mangasarian, R. R. Meyer, and S. M. Robinson, eds. New York: Academic Press, 1978, pp. 429–60.

[5] A. Beck, P. Stoica, and J. Li, "Exact and approximate solutions of source localization problems," *IEEE Transactions on Signal Processing*, vol. 56. no. 5, pp. 1770–8, 2008.

[6] A. Beck, M. Teboulle, and Z. Chikishev, "Iterative minimization schemes for solving the single source localization problem," *SIAM Journal on Optimization*, vol. 19, no. 3, pp. 1397–1416, 2008.

[7] A. Beck and M. Teboulle, "Fast gradient-based algorithms for constrained total variation image denoising and deblurring problems," Accepted for publication in *IEEE Transactions on Image Processing*.

[8] A. Beck and M. Teboulle, "A fast iterative shrinkage-thresholding algorithm for linear inverse problems," *SIAM Journal on Imaging Sciences*, vol. 2, no. 1, pp. 183–202, 2009.

[9] D. P. Bertsekas, *Nonlinear Programming*, 2nd ed. Belmont, MA: Athena Scientific, 1999.

[10] J. Bioucas-Dias and M. Figueiredo, "A new TwIST: two-step iterative shrinkage/thresholding algorithms for image restoration," *IEEE Transactions on Image Processing*, vol. 16, pp. 2992–3004, 2007.

[11] E. J. Candès, J. K. Romberg, and T. Tao, "Stable signal recovery from incomplete and inaccurate measurements," *Communications on Pure & Applied Mathematics*, vol. 59, no. 8, pp. 1207–23, 2006.

[12] A-L. Cauchy, "Méthode generales pour la résolution des systèmes d'equations simultanées," *Comptes Rendues de l'Acdémie des Sciences Paris*, vol. 25, pp. 536–8, 1847.

[13] A. Chambolle, R. A. DeVore, N. Y. Lee, and B. J. Lucier, "Nonlinear wavelet image processing: Variational problems, compression, and noise removal through wavelet shrinkage," *IEEE Transactions on Image Processing*, vol. 7, pp. 319–35, 1998.

[14] A. Chambolle and P. L. Lions, "Image recovery via total variation minimization and related problmes," *Numerische Mathematik*, vol. 76, pp. 167–88, 1997.

[15] A. Chambolle, "An algorithm for total variation minimization and applications," *Journal of Mathematical Imaging & Vision*, vol. 20, nos. 1–2, pp. 89–97, 2004. Special issue on mathematics and image analysis.

[16] A. Chambolle, "Total variation minimization and a class of binary MRF models," *Lecture Notes in Computer Sciences*, vol. 3757, 2005, pp. 136–52.

[17] T. F. Chan, G. H. Golub, and P. Mulet, "A nonlinear primal-dual method for total variation-based image restoration," *SIAM Journal on Scientific Computing*, vol. 20, no. 6, pp. 1964–77, 1999.

[18] S. S. Chen, D. L. Donoho, and M. A. Saunders, "Atomic decomposition by basis pursuit", *SIAM Journal on Scientific Computing*, vol. 20, no. 1, pp. 33–61, 1998.

[19] K. W. Cheung, W. K. Ma, and H. C. So, "Accurate approximation algorithm for TOA-based maximum likelihood mobile location using semidefinite programming," *Proceedings of ICASSP*, vol. 2, 2004, pp. 145–8.

[20] J. Claerbout and F. Muir, "Robust modelling of erratic data," *Geophysics*, vol. 38, pp. 826–44, 1973.

[21] P. Combettes and V. Wajs, "Signal recovery by proximal forward-backward splitting," *Multiscale Modeling and Simulation*, vol. 4, pp. 1168–200, 2005.

[22] J. W. Daniel, *The Approximate Minimization of Functionals*. Englewood Cliffs, NJ: Prentice-Hall, 1971.

[23] I. Daubechies, M. Defrise, and C. De Mol, "An iterative thresholding algorithm for linear inverse problems with a sparsity constraint," *Communications on Pure & Applied Mathematics*, vol. 57, no. 11, pp. 1413–57, 2004.

[24] D. L. Donoho, "Compressed sensing," *IEEE Transactions on Information Theory*, vol. 52, no. 4, pp. 1289–306, 2006.

[25] M. Elad, B. Matalon, J. Shtok, and M. Zibulevsky, "A wide-angle view at iterated shrinkage algorithms," presented at *SPIE (Wavelet XII) 2007*, San-Diego, CA, August 26–29, 2007.

[26] H. W. Engl, M. Hanke, and A. Neubauer, "Regularization of inverse problems," in *Mathematics and its Applications*, vol. 375. Dordrecht: Kluwer Academic Publishers Group, 1996.

[27] F. Facchinei and J. S. Pang, *Finite-dimensional Variational Inequalities and Complementarity Problems, Vol. II*. Springer Series in Operations Research. New York: Springer-Verlag, 2003.

[28] M. A. T. Figueiredo, J. Bioucas-Dias, and R. Nowak, "Majorization-minimization algorithms for wavelet-based image restoration," *IEEE Transactions on Image Processing*, vol. 16, no. 12, pp. 2980–91, 2007.

[29] M. A. T. Figueiredo, R. D. Nowak, and S. J. Wright, "Gradient projection for sparse reconstruction: Application to compressed sensing and other inverse problems," *IEEE Journal on Selected Topics in Signal Processing*, to be published.

[30] M. A. T. Figueiredo and R. D. Nowak, "An EM algorithm for wavelet-based image restoration," *IEEE Transactions on Image Processing*, vol. 12, no. 8, pp. 906–16, 2003.

[31] D. Goldfarb and W. Yin, "Second-order cone programming methods for total variation-based image restoration," *SIAM Journal on Scientific Computing*, pp. 622–45, 2005.

[32] A. A. Goldstein, "Cauchy's method for minimization," *Numerisch Mathematik*, vol. 4, pp. 146–50, 1962.

[33] G. H. Golub, P. C. Hansen, and D. P. O'Leary, "Tikhonov regularization and total least squares," *SIAM Journal on Matrix Analysis & Applications*, vol. 21, no. 2, pp. 185–94, 1999.

[34] M. Hanke and P. C. Hansen, "Regularization methods for large-scale problems," *Surveys on Mathematics for Industry*, vol. 3, no. 4, pp. 253–315, 1993.

[35] P. C. Hansen, "The use of the L-curve in the regularization of discrete ill-posed problems," *SIAM Journal on Scientific & Statistical Computing*, vol. 14, pp. 1487–503, 1993.

[36] M. Hintermuller and G. Stadler, "An infeasible primal-dual algorithm for tv-based inf convolution-type image restoration," *SIAM Journal on Scientific Computing*, vol. 28, pp. 1–23, 2006.

[37] D. R. Hunter and K. Lange, "A tutorial on MM algorithms," *The American Statistician*, vol. 58, no. 1, pp. 30–7, 2004.

[38] S. Bank H. Taylor and J. McCoy, "Deconvolution with the l_1-norm," *Geophysics*, vol. 44, pp. 39–52, 1979.

[39] S. J. Kim, K. Koh, M. Lustig, S. Boyd, and D. Gorinevsky, "A Method for Large-Scale l1-Regularized Least Squares", *IEEE Journal on Selected Topics in Signal Processing*, vol. 1, no. 4, pp. 606–17, 2007.

[40] H. W. Kuhn, "A note on Fermat's problem," *Mathematical Programming*, vol. 4, pp. 98–107, 1973.

[41] E. S. Levitin and B. T. Polyak, "Constrained minimization methods," *USSR Computational Mathematics & Mathematical Physics*, vol. 6, pp. 787–823, 1966.

[42] J. J. Moreau, "Proximitéet dualité dans un espace hilbertien," *Bulletin de La Société Mathématique de France*, vol. 93, pp. 273–99, 1965.

[43] A. S. Nemirovsky and D. B. Yudin, *Problem Complexity and Method Efficiency in Optimization*. A Wiley-Interscience Publication. New York: John Wiley & Sons Inc., 1983. Translated from the Russian and with a preface by E. R. Dawson, Wiley-Interscience Series in Discrete Mathematics.

[44] Y. E. Nesterov, "A method for solving the convex programming problem with convergence rate $O(1/k^2)$," *Doklady Akademii Nauk ISSSR*, vol. 269, no. 3, pp. 543–7, 1983.

[45] Y. E. Nesterov, "Gradient methods for minimizing composite objective function," 2007. CORE Report. Available: www.ecore.beDPs/dp1191313936.pdf

[46] Y. Nesterov, *Introductory Lectures on Convex Optimization*. Boston, MA: Kluwer, 2004.

[47] J. M. Ortega and W. C. Rheinboldt, "Iterative solution of nonlinear equations in several variables," in *Classics in Applied Mathematics*, vol. 30. Philadelphia, PA: Society for Industrial and Applied Mathematics (SIAM), 2000. Reprint of the 1970 original.

[48] G. B. Passty, "Ergodic convergence to a zero of the sum of monotone operators in Hilbert space," *Journal of Mathematical Analysis & Applications*, vol. 72, no. 2, pp. 383–90, 1979.

[49] B. T. Polyak, *Introduction to Optimization*. Translations Series in Mathematics and Engineering. New York: Optimization Software Inc. Publications Division, 1987. Translated from the Russian, with a foreword by Dimitri P. Bertsekas.

[50] R. T. Rockafellar, *Convex Analysis*. Princeton, NJ: Princeton University Press, 1970.

[51] L. I. Rudin, S. J. Osher, and E. Fatemi, "Nonlinear total variation based noise removal algorithms," *Physica D*, vol. 60, pp. 259–68, 1992.

[52] J. G. Morris, R. F. Love, and G. O. Wesolowsky, *Facilities Location: Models and Methods*. New York: North-Holland Publishing Co., 1988.

[53] A. H. Sayed, A. Tarighat, and N. Khajehnouri, "Network-based wireless location," *IEEE Signal Processing Magazine*, vol. 22, no. 4, pp. 24–40, 2005.

[54] N. Z. Shor, *Minimization Methods for Nondifferentiable Functions*. New York: Springer-Verlag, 1985.

[55] J. O. Smith and J. S. Abel, "Closed-form least-squares source location estimation from range-difference measurements," *IEEE Transactions on Acoustics, Speech and Signal Processing*, vol. 12, pp. 1661–9, 1987.

[56] A. N. Tikhonov and V. Y. Arsenin, *Solution of Ill-Posed Problems*. Washington, DC: V.H. Winston, 1977.

[57] C. R. Vogel and M. E. Oman, "Iterative methods for total variation denoising," *SIAM Journal of Scientific Computing*, vol. 17, pp. 227–38, 1996.

[58] E. Weiszfeld, "Sur le point pour lequel la somme des distances de n points donnés est minimum," *Tohoku Mathematical Journal*, vol. 43, pp. 355–86, 1937.

3 Graphical models of autoregressive processes

Jitkomut Songsiri, Joachim Dahl, and Lieven Vandenberghe

We consider the problem of fitting a Gaussian autoregressive model to a time series, subject to conditional independence constraints. This is an extension of the classical covariance selection problem to time series. The conditional independence constraints impose a sparsity pattern on the inverse of the spectral density matrix, and result in nonconvex quadratic equality constraints in the maximum likelihood formulation of the model estimation problem. We present a semidefinite relaxation, and prove that the relaxation is exact when the sample covariance matrix is block-Toeplitz. We also give experimental results suggesting that the relaxation is often exact when the sample covariance matrix is not block-Toeplitz. In combination with model selection criteria the estimation method can be used for topology selection. Experiments with randomly generated and several real data sets are also included.

3.1 Introduction

Graphical models give a graph representation of relations between random variables. The simplest example is a *Gaussian graphical model*, in which an undirected graph with n nodes is used to describe conditional independence relations between the components of an n-dimensional random variable $x \sim N(0, \Sigma)$. The absence of an edge between two nodes of the graph indicates that the corresponding components of x are independent, conditional on the other components. Other common examples of graphical models include *contingency tables*, which describe conditional independence relations in multinomial distributions, and *Bayesian networks*, which use directed acyclic graphs to represent causal or temporal relations. Graphical models find applications in bio-informatics, speech and image processing, combinatorial optimization, coding theory, and many other fields. Graphical representations of probability distributions not only offer insight in the structure of the distribution, they can also be exploited to improve the efficiency of statistical calculations, such as the computation of conditional or marginal probabilities. For further background we refer the reader to several books and survey papers on the subject [1–7].

Estimation problems in graphical modeling can be divided in two classes, depending on whether the topology of the graph is given or not. In a Gaussian graphical model of

Convex Optimization in Signal Processing and Communication, eds. Daniel P. Palomar and Yonina C. Eldar.
Published by Cambridge University Press © Cambridge University Press, 2010.

$x \sim N(0, \Sigma)$, for example, the conditional independence relations between components of x correspond to zero entries in the inverse covariance matrix [8]. This follows from the fact that the conditional distribution of two variables x_i, x_j, given the remaining variables, is Gaussian, with covariance matrix

$$\begin{bmatrix} (\Sigma^{-1})_{ii} & (\Sigma^{-1})_{ij} \\ (\Sigma^{-1})_{ji} & (\Sigma^{-1})_{jj} \end{bmatrix}^{-1}.$$

Hence x_i and x_j are conditionally independent if, and only if

$$(\Sigma^{-1})_{ij} = 0.$$

Specifying the graph topology of a Gaussian graphical model is therefore equivalent to specifying the sparsity pattern of the inverse covariance matrix. This property allows us to formulate the *maximum-likelihood* (ML) estimation problem of a Gaussian graphical model, for a given graph topology, as

$$\begin{aligned} &\text{maximize} && -\log \det \Sigma - \mathbf{tr}(C\Sigma^{-1}) \\ &\text{subject to} && (\Sigma^{-1})_{ij} = 0, \quad (i,j) \in \mathcal{V}, \end{aligned} \tag{3.1}$$

where C is the sample covariance matrix, and $\mathcal{V}$ are the pairs of nodes (i,j) that are not connected by an edge, in other words, for which x_i and x_j are conditionally independent. (Throughout the chapter we take as the domain of the function $\log \det X$ the set of positive definite matrices.) A change of variables $X = \Sigma^{-1}$ results in a convex problem

$$\begin{aligned} &\text{maximize} && \log \det X - \mathbf{tr}(CX) \\ &\text{subject to} && X_{ij} = 0, \quad (i,j) \in \mathcal{V}. \end{aligned} \tag{3.2}$$

This is known as the *covariance selection problem* [8], [2, Section 5.2]. The corresponding dual problem is

$$\begin{aligned} &\text{minimize} && \log \det Z^{-1} \\ &\text{subject to} && Z_{ij} = C_{ij}, \quad (i,j) \notin \mathcal{V}, \end{aligned} \tag{3.3}$$

with variable $Z \in \mathbf{S}^n$ (the set of symmetric matrices of order n). It can be shown that $Z = X^{-1} = \Sigma$ at the optimum of (3.1), (3.2), and (3.3). The ML estimate of the covariance matrix in a Gaussian graphical model is the maximum determinant (or maximum entropy) completion of the sample covariance matrix [9, 10].

The problem of estimating the topology in a Gaussian graphical model is more involved. One approach is to formulate hypothesis testing problems to decide about the presence or absence of edges between two nodes [2, Section 5.3.3]. Another possibility is to enumerate different topologies, and use information-theoretic criteria (such as the Akaike or Bayes information criteria) to rank the models. A more recent development is the use of convex methods based on ℓ_1-norm regularization to estimate sparse inverse covariance matrices [11–13].

In this chapter we address the extension of estimation methods for Gaussian graphical models to *autoregressive* (AR) Gaussian processes

$$x(t) = -\sum_{k=1}^{p} A_k x(t-k) + w(t), \tag{3.4}$$

where $x(t) \in \mathbf{R}^n$, and $w(t) \sim N(0, \Sigma)$ is Gaussian white noise. It is known that conditional independence between components of a multivariate stationary Gaussian process can be characterized in terms of the inverse of the spectral density matrix $S(\omega)$: two components $x_i(t)$ and $x_j(t)$ are independent, conditional on the other components of $x(t)$, if, and only if

$$(S(\omega)^{-1})_{ij} = 0$$

for all ω [14, 15]. This connection allows us to include conditional independence constraints in AR estimation methods by placing restrictions on the sparsity pattern of the inverse spectral density matrix. As we will see in Section 3.3.1, the conditional independence constraints impose quadratic equality constraints on the AR parameters. The main contribution of the chapter is to show that under certain conditions the constrained estimation problem can be solved efficiently via a convex (semidefinite programming) relaxation. This convex formulation can be used to estimate graphical models where the AR parameters are constrained with respect to a given graph structure. In combination with model selection criteria they can also be used to identify the conditional independence structure of an AR process. In Section 3.4 we present experimental results using randomly generated and real data sets.

Graphical models of AR processes have several applications [16–22]. Most previous work on this subject is concerned with statistical tests for topology selection. Dahlhaus [15] derives a statistical test for the existence of an edge in the graph, based on the maximum of a nonparametric estimate of the normalized inverse spectrum $S(\omega)^{-1}$; see [16–22] for applications of this approach. Eichler [23] presents a more general approach by introducing a hypothesis test based on the norm of some suitable function of the spectral density matrix. Related problems have also been studied in [24, 25]. Bach and Jordan [24] consider the problem of learning the structure of the graphical model of a time series from sample estimates of the joint spectral density matrix. Eichler [25] uses Whittle's approximation of the exact-likelihood function, and imposes sparsity constraints on the inverse covariance functions via algorithms extended from covariance selection. Numerical algorithms for the estimation of graphical AR models have been explored in [22, 25, 26]. The convex framework proposed in this chapter provides an alternative and more direct approach, and readily leads to efficient estimation algorithms.

Notation

$\mathbf{R}^{m \times n}$ denotes the set of real matrices of size $m \times n$, $\mathbf{S}^n$ is the set of real symmetric matrices of order n, and $\mathbf{M}^{n,p}$ is the set of matrices

$$X = \begin{bmatrix} X_0 & X_1 & \cdots & X_p \end{bmatrix}$$

with $X_0 \in \mathbf{S}^n$ and $X_1, \ldots, X_p \in \mathbf{R}^{n \times n}$. The standard trace inner product $\mathbf{tr}(X^T Y)$ is used on each of these three vector spaces. $\mathbf{S}^n_+$ ($\mathbf{S}^n_{++}$) is the set of symmetric positive semidefinite (positive definite) matrices of order n. X^H denotes the complex conjugate transpose of X.

The linear mapping $\mathrm{T} : \mathbf{M}^{n,p} \to \mathbf{S}^{n(p+1)}$ constructs a symmetric block-Toeplitz matrix from its first block row: if $X \in \mathbf{M}^{n,p}$, then

$$\mathrm{T}(X) = \begin{bmatrix} X_0 & X_1 & \cdots & X_p \\ X_1^T & X_0 & \cdots & X_{p-1} \\ \vdots & \vdots & \ddots & \vdots \\ X_p^T & X_{p-1}^T & \cdots & X_0 \end{bmatrix}. \tag{3.5}$$

The adjoint of T is a mapping $\mathrm{D} : \mathbf{S}^{n(p+1)} \to \mathbf{M}^{n,p}$ defined as follows. If $S \in \mathbf{S}^{n(p+1)}$ is partitioned as

$$S = \begin{bmatrix} S_{00} & S_{01} & \cdots & S_{0p} \\ S_{01}^T & S_{11} & \cdots & S_{1p} \\ \vdots & \vdots & & \vdots \\ S_{0p}^T & S_{1p}^T & \cdots & S_{pp} \end{bmatrix},$$

then $\mathrm{D}(S) = \begin{bmatrix} \mathrm{D}_0(S) & \mathrm{D}_1(S) & \cdots & \mathrm{D}_p(S) \end{bmatrix}$ where

$$\mathrm{D}_0(S) = \sum_{i=0}^{p} S_{ii}, \qquad \mathrm{D}_k(S) = 2 \sum_{i=0}^{p-k} S_{i,i+k}, \quad k = 1, \ldots, p. \tag{3.6}$$

A symmetric sparsity pattern of a sparse matrix X of order n will be defined by giving the set of indices $\mathcal{V} \subseteq \{1, \ldots, n\} \times \{1, \ldots, n\}$ of its zero entries. $\mathrm{P}_{\mathcal{V}}(X)$ denotes the projection of a matrix $X \in \mathbf{S}^n$ or $X \in \mathbf{R}^{n \times n}$ on the complement of the sparsity pattern $\mathcal{V}$:

$$\mathrm{P}_{\mathcal{V}}(X)_{ij} = \begin{cases} X_{ij} & (i,j) \in \mathcal{V} \\ 0 & \text{otherwise.} \end{cases} \tag{3.7}$$

The same notation will be used for $\mathrm{P}_{\mathcal{V}}$ as a mapping from $\mathbf{R}^{n \times n} \to \mathbf{R}^{n \times n}$ and as a mapping from $\mathbf{S}^n \to \mathbf{S}^n$. In both cases, $\mathrm{P}_{\mathcal{V}}$ is self-adjoint. If X is a $p \times q$ block matrix with i,j block X_{ij}, and each block is square of order n, then $\mathrm{P}_{\mathcal{V}}(X)$ denotes the $p \times q$ block matrix with i,j block $\mathrm{P}_{\mathcal{V}}(X)_{ij} = \mathrm{P}_{\mathcal{V}}(X_{ij})$. The subscript of $\mathrm{P}_{\mathcal{V}}$ is omitted if the sparsity pattern $\mathcal{V}$ is clear from the context.

3.2   Autoregressive processes

This section provides some necessary background on AR processes and AR estimation methods. The material is standard and can be found in many textbooks [27–31].

We use the notation (3.4) for an AR model of order p. Occasionally the equivalent model

$$B_0 x(t) = -\sum_{k=1}^{p} B_k x(t-k) + v(t), \tag{3.8}$$

with $v(t) \sim N(0, I)$, will also be useful. The coefficients in the two models are related by $B_0 = \Sigma^{-1/2}$, $B_k = \Sigma^{-1/2} A_k$ for $k = 1, \ldots, p$.

The autocovariance sequence of the AR process is defined as

$$R_k = \mathbf{E}\, x(t+k) x(t)^T,$$

where $\mathbf{E}$ denotes the expected value. We have $R_{-k} = R_k^T$ since $x(t)$ is real. It is easily shown that the AR model parameters A_k, Σ, and the first $p+1$ covariance matrices R_k are related by the linear equations

$$\begin{bmatrix} R_0 & R_1 & \cdots & R_p \\ R_1^T & R_0 & \cdots & R_{p-1} \\ \vdots & \vdots & \ddots & \vdots \\ R_p^T & R_{p-1}^T & \cdots & R_0 \end{bmatrix} \begin{bmatrix} I \\ A_1^T \\ \vdots \\ A_p^T \end{bmatrix} = \begin{bmatrix} \Sigma \\ 0 \\ \vdots \\ 0 \end{bmatrix}. \tag{3.9}$$

These equations are called the *Yule–Walker equations* or *normal equations*.

The transfer function from w to x is $\mathbf{A}(z)^{-1}$ where

$$\mathbf{A}(z) = I + z^{-1} A_1 + \cdots + z^{-p} A_p.$$

The AR process is stationary if the poles of $\mathbf{A}$ are inside the unit circle. The spectral density matrix is defined as the Fourier transform of the autocovariance sequence,

$$S(\omega) = \sum_{k=-\infty}^{\infty} R_k e^{-jk\omega}$$

(where $j = \sqrt{-1}$), and can be expressed as $S(\omega) = \mathbf{A}(e^{j\omega})^{-1} \Sigma \mathbf{A}(e^{j\omega})^{-H}$. The inverse spectrum of an AR process is therefore a trigonometric matrix polynomial

$$S(\omega)^{-1} = \mathbf{A}(e^{j\omega})^H \Sigma^{-1} \mathbf{A}(e^{j\omega}) = Y_0 + \sum_{k=1}^{p} (e^{-jk\omega} Y_k + e^{jk\omega} Y_k^T) \tag{3.10}$$

where

$$Y_k = \sum_{i=0}^{p-k} A_i^T \Sigma^{-1} A_{i+k} = \sum_{i=0}^{p-k} B_i^T B_{i+k} \tag{3.11}$$

(with $A_0 = I$).

3.2.1   Least-squares linear prediction

Suppose $x(t)$ is a stationary process (not necessarily autoregressive). Consider the problem of finding an optimal linear prediction

$$\hat{x}(t) = -\sum_{k=1}^{p} A_k x(t-k),$$

of $x(t)$, based on past values $x(t-1), \ldots, x(t-p)$. This problem can also be interpreted as approximating the process $x(t)$ by the AR model with coefficients A_k. The prediction error between $x(t)$ and $\hat{x}(t)$ is

$$e(t) = x(t) - \hat{x}(t) = x(t) + \sum_{k=1}^{p} A_k x(t-k).$$

To find the coefficients $A_1, \ldots, A_p$, we can minimize the mean-squared prediction error $\mathbf{E}\,\|e(t)\|_2^2$. The mean-squared error can be expressed in terms of the coefficients A_k and the covariance function of x as $\mathbf{E}\,\|e(t)\|_2^2 = \mathbf{tr}(A\,\mathrm{T}(R)A^T)$ where

$$A = \begin{bmatrix} I & A_1 & \cdots & A_p \end{bmatrix}, \qquad R = \begin{bmatrix} R_0 & R_1 & \cdots & R_p \end{bmatrix},$$

$R_k = \mathbf{E}\,x(t+k)x(t)^T$, and $\mathrm{T}(R)$ is the block-Toeplitz matrix with R as its first block row (see the Notation section at the end of Section 3.1). Minimizing the prediction error is therefore equivalent to the quadratic optimization problem

$$\text{minimize} \quad \mathbf{tr}(A\,\mathrm{T}(R)A^T) \tag{3.12}$$

with variables $A_1, \ldots, A_p$.

In practice, the covariance matrix $\mathrm{T}(R)$ in (3.12) is replaced by an estimate C computed from samples of $x(t)$. Two common choices are as follows. Suppose samples $x(1), x(2), \ldots, x(N)$ are available.

- The *autocorrelation method* uses the *windowed estimate*

$$C = \frac{1}{N} H H^T, \tag{3.13}$$

where

$$H = \begin{bmatrix} x(1) & x(2) & \cdots & x(p+1) & \cdots & x(N) & 0 & \cdots & 0 \\ 0 & x(1) & \cdots & x(p) & \cdots & x(N-1) & x(N) & \cdots & 0 \\ \vdots & \vdots & \ddots & \vdots & & \vdots & \vdots & \ddots & \vdots \\ 0 & 0 & \cdots & x(1) & \cdots & x(N-p) & x(N-p+1) & \cdots & x(N) \end{bmatrix}. \tag{3.14}$$

Note that the matrix C is block-Toeplitz, and that it is positive definite (unless the sequence $x(1), \ldots, x(N)$ is identically zero).

- The *covariance method* uses the *non-windowed estimate*

$$C = \frac{1}{N-p} HH^T,$$ (3.15)

where

$$H = \begin{bmatrix} x(p+1) & x(p+2) & \cdots & x(N) \\ x(p) & x(p+1) & \cdots & x(N-1) \\ \vdots & \vdots & & \vdots \\ x(1) & x(2) & \cdots & x(N-p) \end{bmatrix}.$$ (3.16)

In this case the matrix C is not block-Toeplitz.

To summarize, least-squares estimation of AR models reduces to an unconstrained quadratic optimization problem

$$\text{minimize} \quad \mathbf{tr}(ACA^T).$$ (3.17)

Here, C is the exact covariance matrix, if available, or one of the two sample estimates (3.13) and (3.15). The first of these estimates is a block-Toeplitz matrix, while the second one is in general not block-Toeplitz. The covariance method is known to be slightly more accurate in practice if N is small [31, page 94]. The correlation method, on the other hand, has some important theoretical and practical properties, that are easily explained from the optimality conditions of (3.17). If we define $\hat{\Sigma} = ACA^T$ (i.e., the estimate of the prediction error $\mathbf{E}\,\|e(t)\|_2^2$ obtained by substituting C for T(R)), then the optimality conditions can be expressed as

$$\begin{bmatrix} C_{00} & C_{01} & \cdots & C_{pp} \\ C_{10} & C_{11} & \cdots & C_{1p} \\ \vdots & \vdots & & \vdots \\ C_{p0} & C_{p1} & \cdots & C_{pp} \end{bmatrix} \begin{bmatrix} I \\ A_1^T \\ \vdots \\ A_p^T \end{bmatrix} = \begin{bmatrix} \hat{\Sigma} \\ 0 \\ \vdots \\ 0 \end{bmatrix}.$$ (3.18)

If C is block-Toeplitz, these equations have the same form as the Yule–Walker equations (3.9), and can be solved more efficiently than when C is not block-Toeplitz. Another advantage is that the solution of (3.18) always provides a stable model if C is block-Toeplitz and positive definite. This can be proved as follows [32]. Suppose z is a zero of $\mathbf{A}(z)$, that is, a nonzero w, exists such that $w^H \mathbf{A}(z) = 0$. Define $u_1 = w$ and $u_k = A_{k-1}^T w + \bar{z} u_{k-1}$ for $k = 2, \ldots, p$. Then we have

$$u = A^T w + \bar{z}\tilde{u}$$

where $u = (u_1, u_2, \ldots, u_p, 0)$, $\tilde{u} = (0, u_1, u_2, \ldots, u_p)$. From this and (3.18),

$$u^H Cu = w^H \hat{\Sigma} w + |z|^2 \tilde{u}^H C\tilde{u}.$$

The first term on the right-hand side is positive because $\hat{\Sigma} \succ 0$. Also, $u^H C u = \tilde{u}^H C \tilde{u}$ since C is block-Toeplitz. Therefore $|z| < 1$.

In the following two sections we give alternative interpretations of the covariance and correlation variants of the least-squares estimation method, in terms of maximum-likelihood and maximum-entropy estimation, respectively.

3.2.2   Maximum-likelihood estimation

The exact likelihood function of an AR model (3.4), based on observations $x(1), \ldots, x(N)$, is complicated to derive and difficult to maximize [28, 33]. A standard simplification is to treat $x(1)$, $x(2)$, ..., $x(p)$ as fixed, and to define the likelihood function in terms of the conditional distribution of a sequence $x(t), x(t+1), \ldots, x(t+N-p-1)$, given $x(t-1), \ldots, x(t-p)$. This is called the *conditional* maximum-likelihood estimation method [33, Section 5.1].

The conditional likelihood function of the AR process (3.4) is

$$
\frac{1}{((2\pi)^n \det \Sigma)^{(N-p)/2}} \exp\left(-\frac{1}{2} \sum_{t=p+1}^{N} \mathbf{x}(t)^T A^T \Sigma^{-1} A \mathbf{x}(t) \right)
$$
$$
= \left(\frac{\det B_0}{(2\pi)^{n/2}} \right)^{N-p} \exp\left(-\frac{1}{2} \sum_{t=p+1}^{N} \mathbf{x}(t)^T B^T B \mathbf{x}(t) \right) \tag{3.19}
$$

where $\mathbf{x}(t)$ is the $((p+1)n)$-vector $\mathbf{x}(t) = (x(t), x(t-1), \ldots, x(t-p))$ and

$$
A = \begin{bmatrix} I & A_1 & \cdots & A_p \end{bmatrix}, \qquad B = \begin{bmatrix} B_0 & B_1 & \cdots & B_p \end{bmatrix},
$$

with $B_0 = \Sigma^{-1/2}$, $B_k = \Sigma^{-1/2} A_k$, $k = 1, \ldots, p$. Taking the logarithm of (3.19) we obtain the conditional log-likelihood function (up to constant terms and factors)

$$
L(B) = (N-p) \log \det B_0 - \frac{1}{2} \operatorname{tr}(BHH^T B^T)
$$

where H is the matrix (3.16). If we define $C = (1/(N-p))HH^T$, we can then write the conditional ML estimation problem as

$$
\text{minimize} \quad -2 \log \det B_0 + \operatorname{tr}(CB^T B) \tag{3.20}
$$

with variable $B \in \mathbf{M}^{n,p}$. This problem is easily solved by setting the gradient equal to zero: the optimal B satisfies $CB^T = (B_0^{-1}, 0, \ldots, 0)$. Written in terms of the model parameters $A_k = B_0^{-1} B_k$, $\Sigma = B_0^{-2}$, this yields

$$
C \begin{bmatrix} I \\ A_1^T \\ \vdots \\ A_p^T \end{bmatrix} = \begin{bmatrix} \Sigma \\ 0 \\ \vdots \\ 0 \end{bmatrix},
$$

in other words, the Yule–Walker equations with the block-Toeplitz coefficient matrix replaced by C. The conditional ML estimate is therefore equal to the least-squares estimate from the covariance method.

3.2.3 Maximum-entropy estimation

Consider the *maximum-entropy* (ME) problem introduced by Burg [34]:

$$\text{maximize} \quad \frac{1}{2\pi}\int_{-\pi}^{\pi}\log\det S(\omega)d\omega$$
$$\text{subject to} \quad \frac{1}{2\pi}\int_{-\pi}^{\pi}S(\omega)e^{jk\omega}d\omega = \bar{R}_k, \quad 0\le k\le p. \tag{3.21}$$

The matrices $\bar{R}_k$ are given. The variable is the spectral density $S(\omega)$ of a real stationary Gaussian process $x(t)$, that is, the Fourier transform of the covariance function $R_k = \mathbf{E}\,x(t+k)x(t)^T$:

$$S(\omega) = R_0 + \sum_{k=0}^{\infty}\left(R_k e^{-jk\omega} + R_k^T e^{jk\omega}\right), \qquad R_k = \frac{1}{2\pi}\int_{-\pi}^{\pi}S(\omega)e^{jk\omega}d\omega.$$

The constraints in (3.21) therefore fix the first $p+1$ covariance matrices to be equal to $\bar{R}_k$. The problem is to extend these covariances so that the entropy rate of the process is maximized. It is known that the solution of (3.21) is a Gaussian AR process of order p, and that the model parameters A_k, Σ follow from the Yule–Walker equations (3.9) with $\bar{R}_k$ substituted for R_k.

To relate the ME problem to the estimation methods of the preceding sections, we derive a dual problem. To simplify the notation later on, we multiply the two sides of the equality constraints $k = 1,\dots,p$ by 2. We introduce a Lagrange multiplier $Y_0 \in \mathbf{S}^n$ for the first equality constraint ($k = 0$), and multipliers $Y_k \in \mathbf{R}^{n\times n}$, $k = 1,\dots,p$, for the other p equality constraints. If we change the sign of the objective, the Lagrangian is

$$-\frac{1}{2\pi}\int_{-\pi}^{\pi}\log\det S(\omega)d\omega + \mathbf{tr}(Y_0(R_0 - \bar{R}_0)) + 2\sum_{k=1}^{p}\mathbf{tr}(Y_k^T(R_k - \bar{R}_k)).$$

Differentiating with respect to R_k gives

$$\frac{1}{2\pi}\int_{-\pi}^{\pi}S^{-1}(\omega)e^{j\omega k}d\omega = Y_k, \quad 0\le k\le p \tag{3.22}$$

and hence

$$S^{-1}(\omega) = Y_0 + \sum_{k=1}^{p}\left(Y_k e^{-jk\omega} + Y_k^T e^{jk\omega}\right) \triangleq Y(\omega).$$

Substituting this in the Lagrangian gives the dual problem

$$\text{minimize} \quad -\frac{1}{2\pi}\int_{-\pi}^{\pi} \log\det Y(\omega) + \mathbf{tr}(Y_0^T\bar{R}_0) + 2\sum_{k=1}^{p}\mathbf{tr}(Y_k^T\bar{R}_k) - n, \qquad (3.23)$$

with variables Y_k. The first term in the objective can be rewritten by using Kolmogorov's formula [35]:

$$\frac{1}{2\pi}\int_{-\pi}^{\pi}\log\det Y(\omega)d\omega = \log\det(B_0^T B_0),$$

where $Y(\omega) = \mathbf{B}(e^{j\omega})^H\mathbf{B}(e^{j\omega})$ and $\mathbf{B}(z) = \sum_{k=0}^{p} z^{-k}B_k$ is the minimum-phase spectral factor of Y. The second term in the objective of the dual problem (3.23) can also be expressed in terms of the coefficients B_k, using the relations $Y_k = \sum_{i=0}^{p-k} B_i^T B_{i+k}$ for $0 \le k \le p$. This gives

$$\mathbf{tr}(Y_0\bar{R}_0) + 2\sum_{k=1}^{p}\mathbf{tr}(Y_k^T\bar{R}_k) = \mathbf{tr}(\mathrm{T}(\bar{R})B^T B),$$

where $\bar{R} = \begin{bmatrix} \bar{R}_0 & \bar{R}_1 \cdots & \bar{R}_p \end{bmatrix}$ and $B = \begin{bmatrix} B_0 & B_1 & \cdots & B_p \end{bmatrix}$. The dual problem (3.23) thus reduces to

$$\text{minimize} \quad -2\log\det B_0 + \mathbf{tr}(CB^T B) \qquad (3.24)$$

where $C = \mathrm{T}(\bar{R})$. Without loss of generality, we can choose B_0 to be symmetric positive definite. The problem is then formally the same as the ML estimation problem (3.20), except for the definition of C. In (3.24) C is a block-Toeplitz matrix. If we choose for $\bar{R}_k$ the sample estimates

$$\bar{R}_k = \frac{1}{N}\sum_{t=1}^{N-k} x(t+k)x(t)^T,$$

then C is identical to the block-Toeplitz matrix (3.13) used in the autocorrelation variant of the least-squares method.

3.3   Autoregressive graphical models

In this section we first characterize conditional independence relations in multivariate Gaussian processes, and specialize the definition to AR processes. We then add the conditional independence constraints to the ML and ME estimation problems derived in the previous section, and investigate convex optimization techniques for solving the modified estimation problems.

3.3.1 Conditional independence in time series

Let $x(t)$ be an n-dimensional, stationary, zero-mean Gaussian process with spectrum $S(\omega)$:

$$S(\omega) = \sum_{k=-\infty}^{\infty} R_k e^{-jk\omega}, \qquad R_k = \mathbf{E}\, x(t+k) x(t)^T.$$

We assume that S is invertible for all ω. Components $x_i(t)$ and $x_j(t)$ are said to be independent, conditional on the other components of $x(t)$, if

$$(S(\omega)^{-1})_{ij} = 0$$

for all ω. This definition can be interpreted and justified as follows (see Brillinger [36, Section 8.1]). Let $u(t) = (x_i(t), x_j(t))$ and let $v(t)$ be the $(n-2)$-vector containing the remaining components of $x(t)$. Define $e(t)$ as the error

$$e(t) = u(t) - \sum_{k=-\infty}^{\infty} H_k v(t-k)$$

between $u(t)$ and the linear filter of $v(t)$ that minimizes $\mathbf{E}\,\|e(t)\|_2^2$. Then it can be shown that the spectrum of the error process $e(t)$ is

$$\begin{bmatrix} (S(\omega)^{-1})_{ii} & (S(\omega)^{-1})_{ij} \\ (S(\omega)^{-1})_{ji} & (S(\omega)^{-1})_{jj} \end{bmatrix}^{-1}. \tag{3.25}$$

This is the Schur complement of the submatrix in $S(\omega)$ indexed by $\{1,\ldots,n\} \setminus \{i,j\}$. The off-diagonal entry in the error spectrum (3.25) is called the *partial cross-spectrum* of x_i and x_j, after removing the effects of v. The partial cross-spectrum is zero if, and only if, the error covariances $\mathbf{E}\, e(t+k) e(t)^T$ are diagonal, in other words, the two components of the error process $e(t)$ are independent.

We can apply this to an AR process (3.4) using the relation between the inverse spectrum $S(\omega)$ and the AR coefficients given in (3.10) and (3.11). These expressions show that $(S(\omega)^{-1})_{ij} = 0$ if, and only if the i,j entries of Y_k are zero for $k = 0,\ldots,p$, where Y_k is given in (3.11). Using the notation defined in (3.6), we can write this as $(\mathrm{D}_k(A^T \Sigma^{-1} A))_{ij} = 0$, where $A = \begin{bmatrix} I & A_1 & \cdots & A_p \end{bmatrix}$, or as

$$\left(\mathrm{D}_k (B^T B) \right)_{ij} = 0, \qquad k = 0,\ldots,p, \tag{3.26}$$

where $B = \begin{bmatrix} B_0 & B_1 & \cdots & B_p \end{bmatrix}$.

3.3.2 Maximum-likelihood and maximum-entropy estimation

We now return to the ML and ME estimation methods for AR processes, described in Sections 3.2.2 and 3.2.3, and extend the methods to include conditional independence constraints. As we have seen, the ML and ME estimation problems can be expressed as a convex optimization problem (3.20) and (3.24), with different choices of the matrix C. The distinction will turn out to be important later, but for now we make no assumptions on C, except that it is positive definite.

As for the Gaussian graphical models mentioned in the introduction, we assume that the conditional independence constraints are specified via an index set $\mathcal{V}$, with $(i,j) \in \mathcal{V}$ if the processes $x_i(t)$ and $x_j(t)$ are conditionally independent. We write the constraints (3.26) for $(i,j) \in \mathcal{V}$ as

$$\mathrm{P}_{\mathcal{V}}\left(\mathrm{D}(B^T B)\right) = 0,$$

where $\mathrm{P}_{\mathcal{V}}$ is the projection operator defined in (3.7). We assume that $\mathcal{V}$ does not contain the diagonal entries (i,i) and that it is symmetric (if $(i,j) \in \mathcal{V}$, then $(j,i) \in \mathcal{V}$). The ML and ME estimation with conditional independence constraints can therefore be expressed as

$$
\begin{array}{ll}
\text{minimize} & -2\log\det B_0 + \mathbf{tr}(CB^T B) \\
\text{subject to} & \mathrm{P}(\mathrm{D}(B^T B)) = 0.
\end{array}
\tag{3.27}
$$

(Henceforth we drop the subscript of $\mathrm{P}_{\mathcal{V}}$.) The variable is $B = \begin{bmatrix} B_0 & B_1 & \cdots & B_p \end{bmatrix} \in \mathbf{M}^{n,p}$.

The problem (3.27) includes quadratic equality constraints and is therefore nonconvex. The quadratic terms in B suggest the convex relaxation

$$
\begin{array}{ll}
\text{minimize} & -\log\det X_{00} + \mathbf{tr}(CX) \\
\text{subject to} & \mathrm{P}(\mathrm{D}(X)) = 0 \\
& X \succeq 0
\end{array}
\tag{3.28}
$$

with variable $X \in \mathbf{S}^{n(p+1)}$ (X_{00} denotes the leading $n \times n$ subblock of X). The convex optimization problem (3.28) is a relaxation of (3.27) and only equivalent to (3.27) if the optimal solution X has rank n, so that it can be factored as $X = B^T B$. We will see later that this is the case if C is block-Toeplitz.

The proof of exactness of the relaxation under assumption of block-Toeplitz structure will follow from the dual of (3.28). We introduce a Lagrange multiplier $Z = \begin{bmatrix} Z_0 & Z_1 & \cdots & Z_p \end{bmatrix} \in \mathbf{M}^{n,p}$ for the equality constraints and a multiplier $U \in \mathbf{S}^{n(p+1)}$ for the inequality constraint. The Lagrangian is

$$
\begin{aligned}
L(X,Z,U) &= -\log\det X_{00} + \mathbf{tr}(CX) + \mathbf{tr}(Z^T \mathrm{P}(\mathrm{D}(X))) - \mathbf{tr}(UX) \\
&= -\log\det X_{00} + \mathbf{tr}\left((C + \mathrm{T}(\mathrm{P}(Z)) - U)X\right).
\end{aligned}
$$

Here we made use of the fact that the mappings T and D are adjoints, and that P is self-adjoint. The dual function is the infimum of L over all X with $X_{00} \succ 0$. Setting the gradient with respect to X equal to zero gives

$$C + T(P(Z)) - U = \begin{bmatrix} X_{00}^{-1} & 0 & \cdots & 0 \\ 0 & 0 & \cdots & 0 \\ \vdots & \vdots & & \vdots \\ 0 & 0 & \cdots & 0 \end{bmatrix}.$$

This shows that Z, U are dual feasible if $C + T(P(Z)) - U$ is zero, except for the $0, 0$ block, which must be positive definite. If U and Z satisfy these conditions, the Lagrangian is minimized by any X with $X_{00} = (C_{00} + P(Z_0) - U_{00})^{-1}$ (where C_{00} and U_{00} denote the leading $n \times n$ blocks of C and U). Hence we arrive at the dual problem

$$\begin{aligned} \text{maximize} \quad & \log \det(C_{00} + P(Z_0) - U_{00}) + n \\ \text{subject to} \quad & C_{i,i+k} + P(Z_k) - U_{i,i+k} = 0, \quad k = 1, \ldots, p, \quad i = 0, \ldots, p - k \\ & U \succeq 0. \end{aligned}$$

If we define $W = C_{00} + P(Z_0) - U_{00}$ and eliminate the slack variable U, we can write this more simply as

$$\begin{aligned} \text{maximize} \quad & \log \det W + n \\ \text{subject to} \quad & \begin{bmatrix} W & 0 \\ 0 & 0 \end{bmatrix} \preceq C + T(P(Z)). \end{aligned} \tag{3.29}$$

Note that for $p = 0$ problem (3.28) reduces to the covariance selection problem (3.2), and the dual problem reduces to the maximum-determinant completion problem

$$\text{maximize} \quad \log \det(C + P(Z)) + n,$$

which is equivalent to (3.3).

We note the following properties of the primal problem (3.28) and the dual problem (3.29).

- The primal problem is strictly feasible ($X = I$ is strictly feasible), so Slater's condition holds. This implies strong duality, and also that the dual optimum is attained if the optimal value is finite.
- We have assumed that $C \succ 0$, and this implies that the primal objective function is bounded below, and that the primal optimum is attained. This also follows from the fact that the dual is strictly feasible ($Z = 0$ is strictly feasible if we take W small enough), so Slater's condition holds for the dual.

Therefore, if $C \succ 0$, we have strong duality and the primal and dual optimal values are attained. The *Karush–Kuhn–Tucker* (KKT) conditions are therefore necessary and sufficient for optimality of X, Z, W. The KKT conditions are:

1. *Primal feasibility.*

$$X \succeq 0, \qquad X_{00} \succ 0, \qquad P(D(X)) = 0. \tag{3.30}$$

2. *Dual feasibility.*

$$W \succ 0, \qquad C + \mathrm{T}(\mathrm{P}(Z)) \succeq \begin{bmatrix} W & 0 \\ 0 & 0 \end{bmatrix}. \tag{3.31}$$

3. *Zero duality gap.*

$$X_{00}^{-1} = W, \qquad \mathbf{tr}\left(X \left(C + \mathrm{T}(\mathrm{P}(Z)) - \begin{bmatrix} W & 0 \\ 0 & 0 \end{bmatrix} \right) \right) = 0. \tag{3.32}$$

The last condition can also be written as

$$X \left(C + \mathrm{T}(\mathrm{P}(Z)) - \begin{bmatrix} W & 0 \\ 0 & 0 \end{bmatrix} \right) = 0. \tag{3.33}$$

3.3.3   Properties of block-Toeplitz sample covariances

In this section we study in more detail the solution of the primal and dual problems (3.28), and (3.29) if C is block-Toeplitz. The results can be derived from connections between spectral factorization, semidefinite programming, and orthogonal matrix polynomials discussed in [37, Section 6.1.1]. In this section, we provide alternative and self-contained proofs.

Assume $C = \mathrm{T}(R)$ for some $R \in \mathbf{M}^{n,p}$ and that C is positive definite.

Exactness of the relaxation

We first show that the relaxation (3.28) is exact when C is block-Toeplitz, that is, the optimal X^* has rank n and the optimal B can be computed by factoring X^* as $X^* = B^T B$. We prove this result from the optimality conditions (3.30)–(3.33).

Assume X^*, W^*, Z^* are optimal. Clearly $\mathbf{rank}\, X^* \geq n$, since its $0,0$ block is nonsingular. We will show that $C + \mathrm{T}(\mathrm{P}(Z^*)) \succ 0$. Therefore the rank of

$$C + \mathrm{T}(\mathrm{P}(Z^*)) - \begin{bmatrix} W^* & 0 \\ 0 & 0 \end{bmatrix}$$

is at least np, and the complementary slackness condition (3.33) implies that X^* has rank at most n, so we can conclude that

$$\mathbf{rank}\, X^* = n.$$

The positive definiteness of $C + \mathrm{T}(\mathrm{P}(Z^*))$ follows from the dual-feasibility condition (3.31) and the following basic property of block-Toeplitz matrices: if $\mathrm{T}(S)$ is a symmetric block-Toeplitz matrix, with $S \in \mathbf{M}^{n,p}$, and

$$\mathrm{T}(S) \succeq \begin{bmatrix} Q & 0 \\ 0 & 0 \end{bmatrix} \tag{3.34}$$

for some $Q \in \mathbf{S}_{++}^{n}$, then $\mathrm{T}(S) \succ 0$. We can verify this by induction on p. The property is obviously true for $p = 0$, since the inequality (3.34) then reduces to $S = S_0 \succeq Q$. Suppose the property holds for $p - 1$. Then (3.34) implies that the leading $np \times np$ submatrix of $\mathrm{T}(S)$, which is a block-Toeplitz matrix with first row $\begin{bmatrix} S_0 & \cdots & S_{p-1} \end{bmatrix}$, is positive definite. Let us denote this matrix by V. Using the Toeplitz structure, we can partition $T(S)$ as

$$\mathrm{T}(S) = \begin{bmatrix} S_0 & U^T \\ U & V \end{bmatrix},$$

where $V \succ 0$. The inequality (3.34) implies that the Schur complement of V in the matrix $\mathrm{T}(S)$ satisfies

$$S_0 - U^T V^{-1} U \succeq Q \succ 0.$$

Combined with $V \succ 0$ this shows that $\mathrm{T}(S) \succ 0$.

Stability of estimated models

It follows from (3.30)–(3.33) and the factorization $X^* = B^T B$, that

$$(C + \mathrm{T}(\mathrm{P}(Z))) \begin{bmatrix} I \\ A_1^T \\ \vdots \\ A_p^T \end{bmatrix} = \begin{bmatrix} \Sigma \\ 0 \\ \vdots \\ 0 \end{bmatrix}, \tag{3.35}$$

if we define $\Sigma = B_0^{-2}$, $A_k = B_0^{-1} B_k$. These equations are Yule–Walker equations with a positive definite block-Toeplitz coefficient matrix. As mentioned at the end of section 3.2.1, this implies that the zeros of $\mathbf{A}(z) = I + z^{-1} A_1 + \cdots + z^{-p} A_p$ are inside the unit circle. Therefore the solution to the convex problem (3.28) provides a stable AR model.

3.3.4 Summary

We have proposed convex relaxations for the problems of conditional ML and ME estimation of AR models with conditional independent constraints. The two problems have the same form with different choices for the sample covariance matrix C. For the ME problem, C is given by (3.13), while for the conditional ML problem, it is given by (3.15). In both cases, C is positive definite if the information matrix H has full rank. This is sufficient to guarantee that the relaxed problem (3.28) is bounded below.

The relaxation is exact if the matrix C is block-Toeplitz, that is, for the ME problem. The Toeplitz structure also ensures stability of the estimated AR model. In the conditional ML problem, C is, in general, not block-Toeplitz, but approaches a block-Toeplitz matrix as N goes to infinity. We conjecture that the relaxation of the ML problem is exact with high probability even for moderate values of N. This will be illustrated by the experimental results in the next section.

3.4   Numerical examples

In this section we evaluate the ML and ME estimation methods on several data sets. The convex optimization package CVX [38,39] was used to solve the ML and ME estimation problems.

3.4.1   Randomly generated data

The first set of experiments uses data randomly generated from AR models with sparse inverse spectra. The purpose is to examine the quality of the semidefinite relaxation (3.28) of the ML estimation problem for finite N. We generated 50 sets of time series from four AR models of different dimensions. We solved (3.28) for different N. Figure 3.1 shows the percentage of the 50 data sets for which the relaxation was exact (the optimal X in (3.28) had rank n). The results illustrate that the relaxation is often exact for moderate values of N, even when the matrix C is not block-Toeplitz.

The next figure shows the convergence rate of the ML and ME estimates, with and without imposed conditional independence constraints, to the true model, as a function of the number of samples. The data were generated from an AR model of dimension $n = p = 6$ with nine zeros in the inverse spectrum. Figure 3.2 shows the *Kullback–Leibler* (KL) divergence [24] between the estimated and the true spectra as a function of N, for four estimation methods: the ML and ME estimation methods without conditional independence constraints, and the ML and ME estimation methods with the correct conditional independence constraints. We notice that the KL divergences decrease at the same rate for the four estimates. However, the ML and ME estimates without the sparsity constraints give models with substantially larger values of KL divergence when

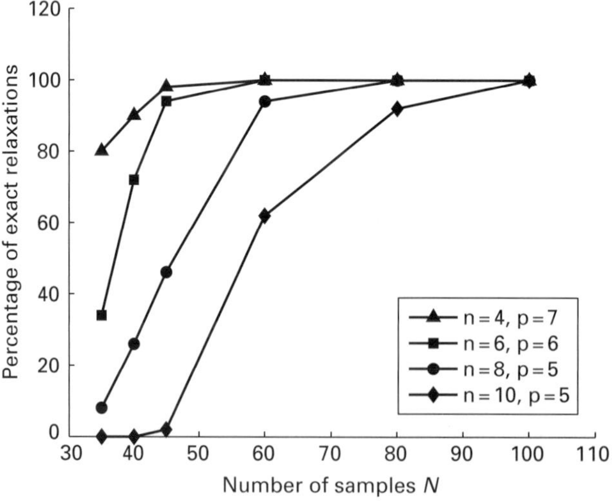

Figure 3.1   Number of cases where the convex relaxation of the ML problem is exact, versus the number of samples.

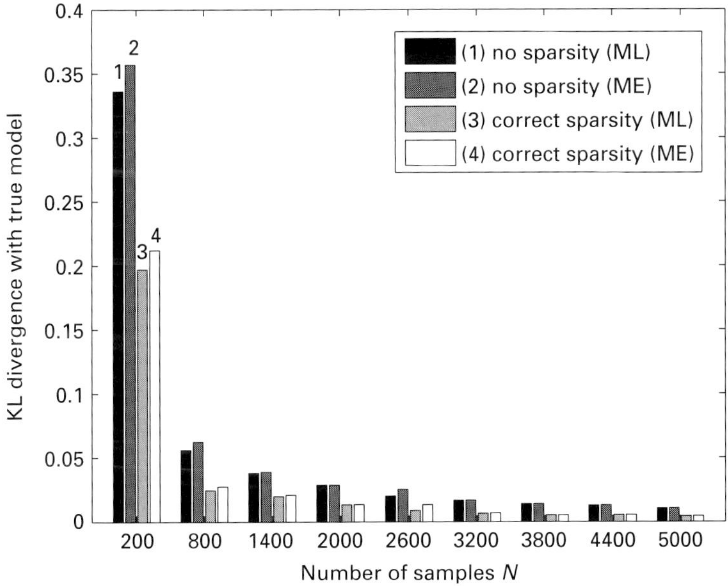

Figure 3.2   KL divergence between estimated AR models and the true model ($n = 6, p = 6$) versus the number of samples.

N is small. For sample size under 3000, the ME estimates (with and without the sparsity constraints) are also found to be less accurate than their ML counterparts. This effect is well known in spectral analysis [31, page 94]. As N increases, the difference between the ME and ML methods disappears.

3.4.2   Model selection

The next experiment is concerned with the problem of topology selection in graphical AR models.

Three popular model selection criteria are the *Akaike information criterion* (AIC), the *second-order variant of AIC* (AIC_c), and the *Bayes information criterion* (BIC) [40]. These criteria are used to make a fair comparison between models of different complexity. They assign to an estimated model a score equal to $-2L$, where L is the likelihood of the model, augmented with a term that depends on the effective number of parameters k in the model:

$$\text{AIC} = -2L + 2k, \qquad \text{AIC}_c = -2L + \frac{2kN}{N - k - 1}, \qquad \text{BIC} = -2L + k \log N.$$

The second term places a penalty on models with high complexity. When comparing different models, we rank them according to one of the criteria and select the model with the lowest score. Of these three criteria, the AIC is known to perform poorly if N is small compared to the number of parameters k. The AIC_c was developed as a correction to the AIC for small N. For large N the BIC favors simpler models than the AIC or AIC_c.

To select a suitable graphical AR model for observed samples of an n-dimensional time series, we can enumerate models of different lengths p and with different graphs. For each model, we solve the ML estimation problem, calculate the AIC, AIC_c, or BIC score, and select the model with the best (lowest) score. Obviously, an exhaustive search of all sparsity patterns is only feasible for small n (say, $n \leq 6$), since there are

$$\sum_{m=0}^{n(n-1)/2} \binom{n(n-1)/2}{m} = 2^{n(n-1)/2} \tag{3.36}$$

different graphs with n nodes.

In the experiment we generate $N = 1000$ samples from an AR model of dimension $n = 5, p = 4$, and zeros in positions $(1,2), (1,3), (1,4), (2,4), (2,5), (4,5)$ of the inverse spectrum. We show only results for the BIC. In the BIC we substitute the conditional likelihood discussed in Section 3.2.2 for the exact likelihood L. (For sufficiently large N the difference is negligible.) As an effective number of parameters we take

$$k = \frac{n(n+1)}{2} - |\mathcal{V}| + p(n^2 - 2|\mathcal{V}|)$$

where $|\mathcal{V}|$ is the number of conditional independence constraints, in other words, the number of zeros in the lower triangular part of the inverse spectrum.

Figure 3.3 shows the scores of the estimated models as a function of p. For each p the score shown is the best score among all graph topologies. The BIC selects the correct model order $p = 4$. Figure 3.4 shows the seven best models according to the BIC. The subgraphs labeled #1 to #7 show the estimated model order p, and the selected sparsity pattern. The corresponding scores are shown in the first subgraph, and the true sparsity pattern is shown in the second subgraph. The BIC identified the correct sparsity pattern.

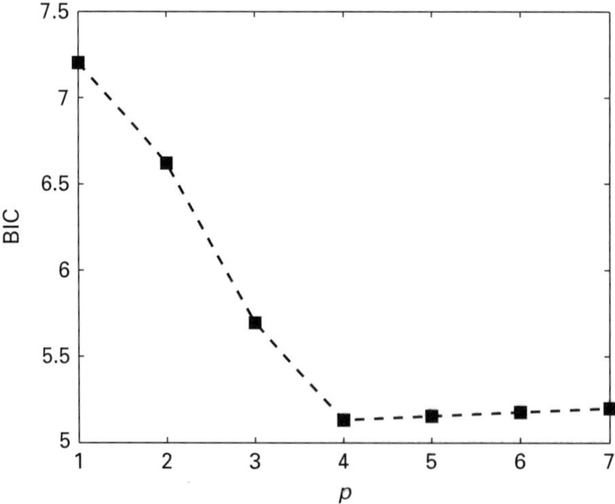

Figure 3.3   BIC score scaled by $1/N$ of AR models of order p.

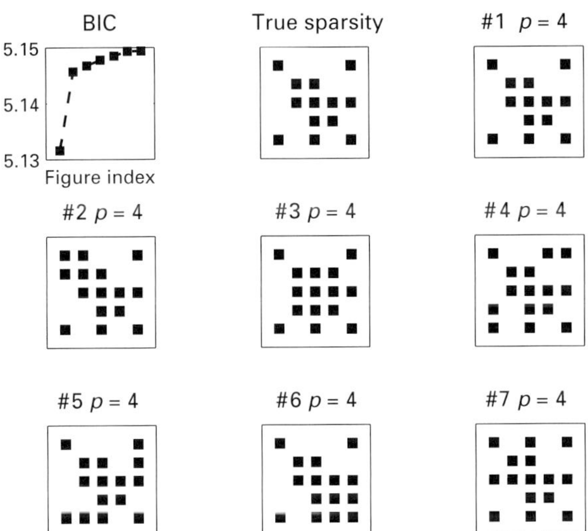

Figure 3.4 Seven best-ranked topologies according to the BIC.

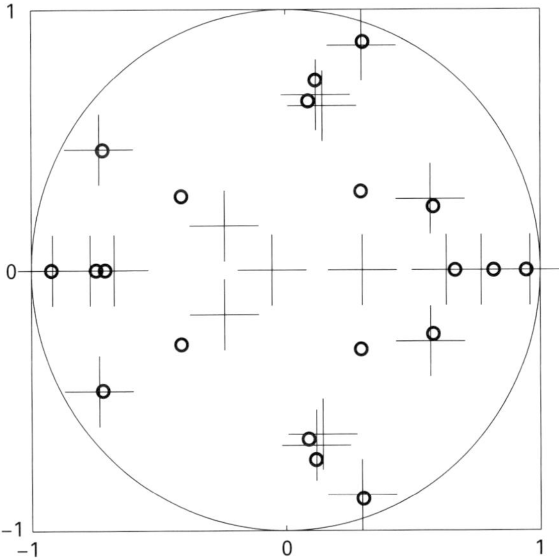

Figure 3.5 Poles of the true model (plus signs) and the estimated model (circles).

Figure 3.5 shows the location of the poles of the true AR model and the model selected by the BIC.

In Figures 3.6 and 3.7 we compare the spectrum of the model, selected by the BIC with the spectrum of the true model, and with a nonparametric estimate of the spectrum. The lower half of the figures show the *coherence spectrum*, that is, the spectrum normalized

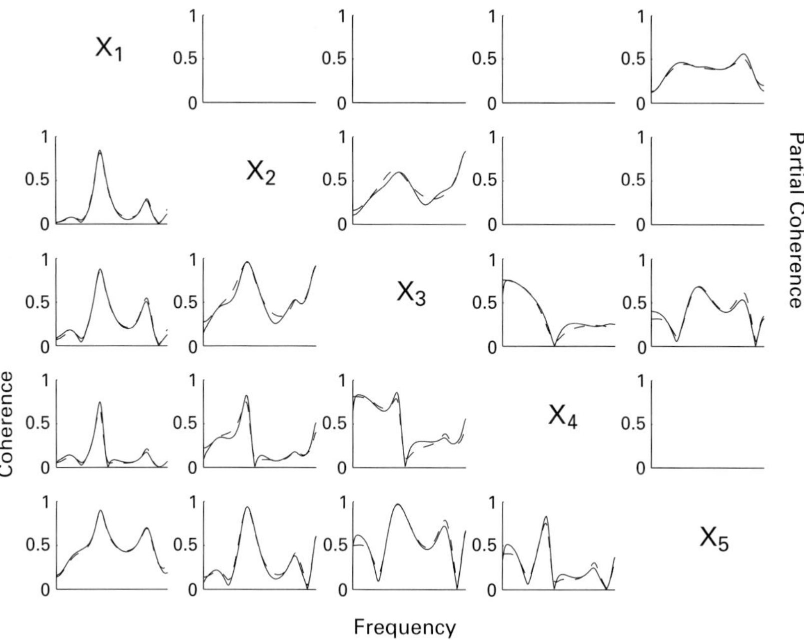

Figure 3.6   Partial coherence and coherence spectra of the AR model: true spectrum (dashed lines) and ML estimates (solid lines).

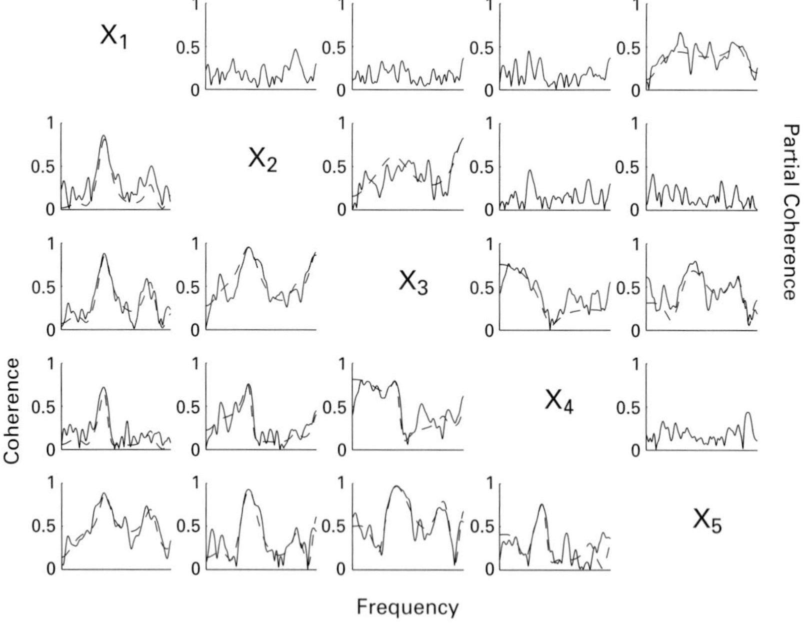

Figure 3.7   Partial coherence and coherence spectra of the AR model: true spectrum (dashed lines) and nonparametric estimates (solid lines).

to have diagonal one:

$$\mathbf{diag}(S(\omega))^{-1/2}S(\omega)\,\mathbf{diag}(S(\omega))^{-1/2},$$

where $\mathbf{diag}(S)$ is the diagonal part of S. The upper half shows the *partial coherence spectrum*, that is, the inverse spectrum normalized to have diagonal one:

$$\mathbf{diag}(S(\omega)^{-1})^{-1/2}S(\omega)^{-1}\,\mathbf{diag}(S(\omega)^{-1})^{-1/2}.$$

The i,j entry of the coherence spectrum is a measure of how dependent components i and j of the time series are. The i,j entry of the partial coherence spectrum, on the other hand, is a measure of *conditional* dependence. The dashed lines show the spectra of the true model. The solid lines in Figure 3.6 are the spectra of the ML estimates. The solid lines in Figure 3.7 are nonparametric estimates of the spectrum, obtained with Welch's method [41, Section 12.2.2] using a Hamming window of length 40 [41, page 642]. The nonparametric estimate of the partial coherence spectrum clearly gives a poor indication of the correct sparsity pattern.

3.4.3 Air pollution data

The data set used in this section consists of a time series of dimension $n = 5$. The components are four air pollutants, CO, NO, NO_2, O_3, and the solar radiation intensity R, recorded hourly during 2006 at Azusa, California. The entire data set consists of $N = 8370$ observations, and was obtained from Air Quality and Meteorological Information System (AQMIS) (`www.arb.ca.gov/aqd/aqdcd/aqdcd.htm`). The daily averages over one year are shown in Figure 3.8. A similar data set was studied previously in [15], using a nonparametric approach.

We use the BIC to compare models with orders ranging from $p = 1$ to $p = 8$. Table 3.1 lists the models with the best ten BIC scores (which differ by only 0.84%).

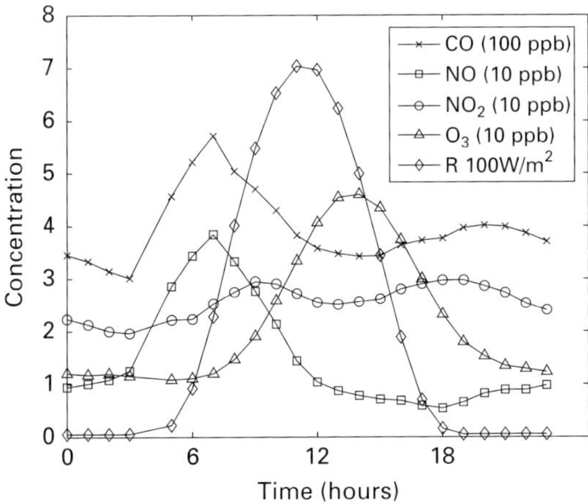

Figure 3.8 Average of daily concentration of CO, NO, NO_2, and O_3, and the solar radiation (R).

Table 3.1. Models with the lowest BIC scores for the air pollution data, determined by an exhaustive search of all models of orders $p = 1, \ldots, 8$. $\mathcal{V}$ is the set of conditionally independent pairs in the model.

Rank	p	BIC score	$\mathcal{V}$
1	4	15414	(NO, R)
2	5	15455	(NO, R)
3	4	15461	
4	4	15494	$(\text{CO}, \text{O}_3), (\text{CO}, \text{R})$
5	4	15502	(CO, R)
6	5	15509	$(\text{CO}, \text{O}_3), (\text{CO}, \text{R})$
7	5	15512	
8	4	15527	(CO, O_3)
9	6	15532	(NO, R)
10	5	15544	(CO, R)

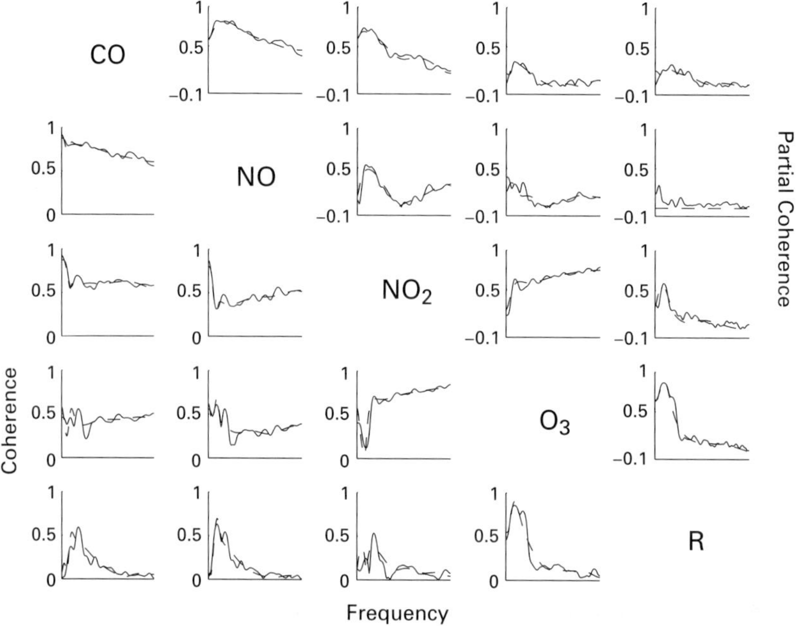

Figure 3.9   Coherence (lower half) and partial coherence spectra (upper half) for the first model in Table 3.1. Nonparametric estimates are in solid lines, and ML estimates in dashed lines.

Figure 3.9 shows the coherence and partial coherence spectra obtained from a nonparametric estimation (solid lines), and the ML model with the best BIC score (dashed lines).

From Table 3.1, the lowest BIC scores of each model of order $p = 4, 5, 6$ correspond to the missing edge between NO and the solar radiation. This agrees with the empirical partial coherence in Figure 3.9 where the pair NO–R is weakest. Table 3.1 also suggests that other weak links are (CO, O_3) and (CO, R). The partial coherence spectra of these pairs are not identically zero, but are relatively small compared to the other pairs.

The presence of the stronger components in the partial coherence spectra are consistent with the discussion in [15]. For example, the solar radiation plays a role in the photolysis of NO_2 and the generation of O_3. The concentration of CO and NO are highly correlated because both are generated by traffic.

3.4.4   International stock markets

We consider a multivariate time series of five stock market indices: the S&P 500 composite index (U.S.), Nikkei 225 share index (Japan), the Hang Seng stock composite index (Hong Kong), the FTSE 100 share index (United Kingdom), and the Frankfurt DAX 30 composite index (Germany). The data were recorded from 4 June 1997 to 15 June 1999, and were downloaded from www.globalfinancialdata.com. (The data were converted to US dollars to take the volatility of exchange rates into account. We also replaced missing data due to national holidays by the most recent values.) For each market we use as variable the return between trading day $k - 1$ and k, defined as

$$r_k = 100 \log(p_k/p_{k-1}), \tag{3.37}$$

where p_k is the closing price on day k. The resulting five-dimensional time series of length 528 is shown in Figure 3.10. This data set is a subset of the data set used in [42].

We enumerate all graphical models of orders ranging from $p = 1$ to $p = 9$. Because of the relatively small number of samples, the AIC_c criterion will be used to compare

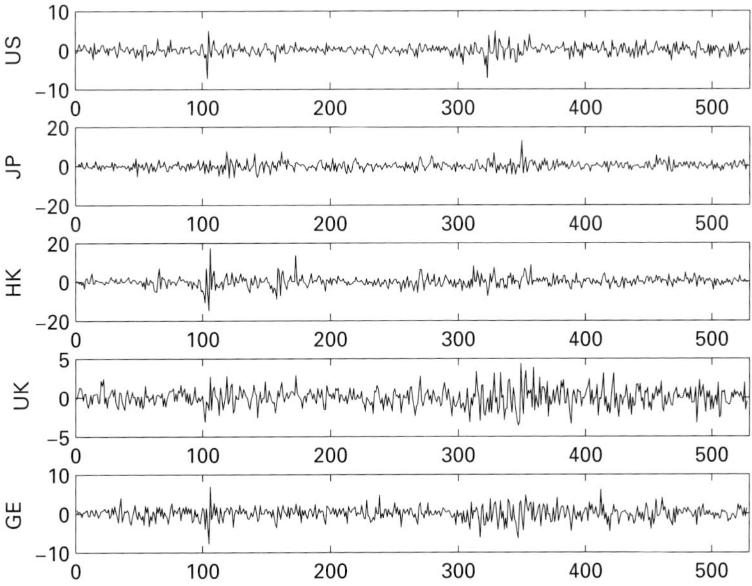

Figure 3.10   Detrended daily returns for five stock market indices between 4 June 1997 and 15 June 1999.

Table 3.2. Five best AR models, ranked according to AIC_c scores, for the international stock market data.

Rank	p	AIC_c score	$\mathcal{V}$
1	2	4645.5	(US,JP), (JP,GE)
2	2	4648.0	(US,JP)
3	1	4651.1	(US,JP), (JP,GE)
4	1	4651.6	(US,JP)
5	2	4653.1	(JP,GE)

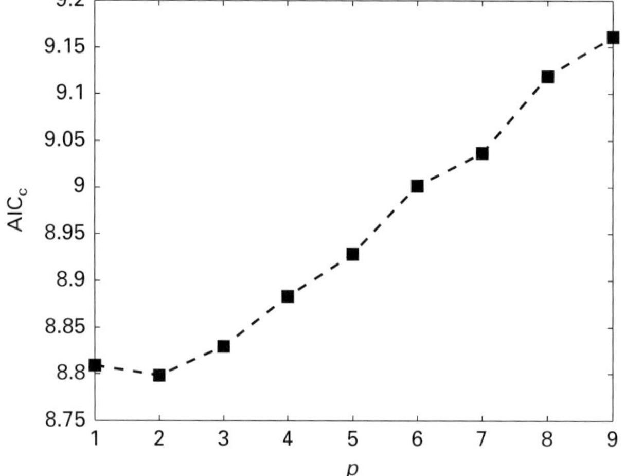

Figure 3.11   Minimized AIC_c scores (scaled by $1/N$) of pth-order models for the stock market return data.

the models. Figure 3.11 shows the optimal AIC_c (optimized over all models of a given lag p) versus p. Table 3.2 shows the model order and topology of the five models with the best AIC_c scores. The column labeled $\mathcal{V}$ shows the list of conditionally independent pairs of variables.

Figure 3.12 shows the coherence (bottom half) and partial coherence (upper half) spectra for the model selected by the AIC_c, and for a nonparametric estimate.

It is interesting to compare the results with the conclusions in [42]. For example, the authors of [42] mention a strong connection between the German and the other European stock markets, in particular, the UK. This agrees with the high value of the UK–GE component of the partial coherence spectrum in Figure 3.12. The lower strength of the connections between the Japanese and the other stock markets is also consistent with the findings in [42]. Another conclusion from [42] is that the volatility in the US stock markets transmits to the world through the German and Hong Kong markets. As far as the German market is concerned, this seems to be confirmed by the strength of the US–GE component in the partial coherence spectrum.

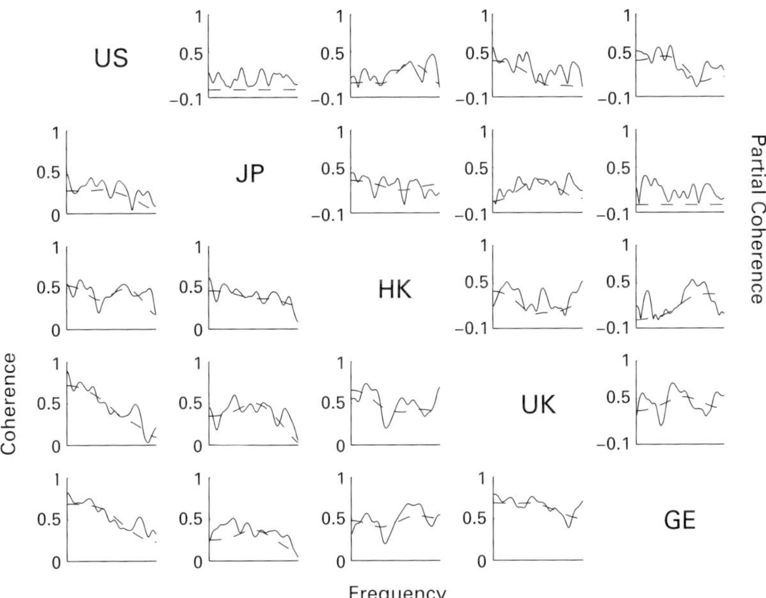

Figure 3.12 Coherence and partial coherence spectra of international stock market data, for the first model in Table 3.2. Nonparametric estimates are shown in solid lines and ML estimates are shown in dashed lines.

3.4.5 European stock markets

This data set is similar to the previous one. We consider a five-dimensional time series consisting of the following stock market indices: the FTSE 100 share index (United Kingdom), CAC 40 (France), the Frankfurt DAX 30 composite index (Germany), MIBTEL (Italy), and the Austrian Traded Index ATX (Austria). The data were stock index closing prices recorded from 1 January 1999 to 31 July 2008, and obtained from www.globalfinancialdata.com. The stock market daily returns were computed from (3.37), resulting in a five-dimensional time series of length $N = 2458$.

The BIC selects a model with lag $p = 1$, and with (UK,IT), (FR,AU), and (GE, AU) as the conditionally independent pairs. The coherence and partial coherence spectra for this model are shown in Figure 3.13. The partial coherence spectrum suggests that the French stock market is the market on the Continent most strongly connected to the UK market. The French, German, and Italian stock markets are highly inter-dependent, while the Austrian market is more weakly connected to the other markets. These results agree with conclusions from the analysis in [43].

3.5 Conclusion

We have considered a parametric approach for maximum-likelihood estimation of autoregressive models with conditional independence constraints. These constraints impose a

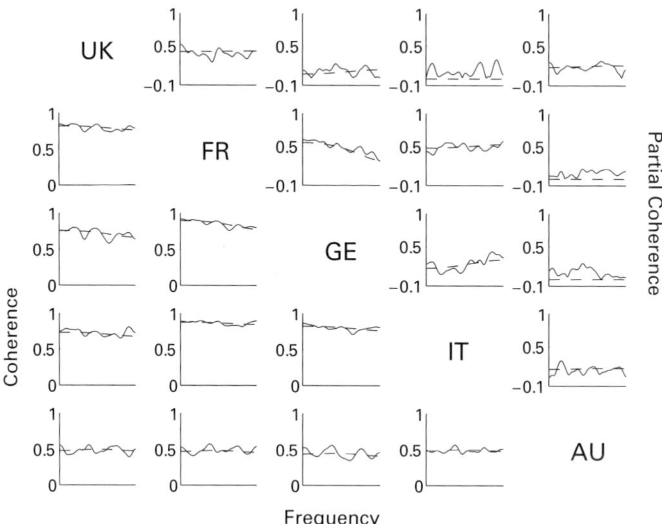

Figure 3.13   Coherence and partial coherence spectrum of the model for the European stock return data. Nonparametric estimates (solid lines) and ML estimates (dashed lines) for the best model selected by the BIC.

sparsity pattern on the inverse of the spectral density matrix, and result in nonconvex equalities in the estimation problem. We have formulated a convex relaxation of the ML estimation problem and shown that the relaxation is exact when the sample covariance matrix in the objective of the estimation problem is block-Toeplitz. We have also noted from experiments that the relaxation is often exact for covariance matrices that are not block-Toeplitz.

The convex formulation allows us to select graphical models by fitting autoregressive models to different topologies, and ranking the topologies using information-theoretic model selection criteria. The approach was illustrated with randomly generated and real data, and works well when the number of models in the comparison is small, or the number of nodes is small enough for an exhaustive search. For larger model selection problems, it will be of interest to extend recent techniques for covariance selection [12, 13] to time series.

Acknowledgments

The research was supported in part by NSF grants ECS-0524663 and ECCS-0824003, and a Royal Thai government scholarship. Part of the research by Joachim Dahl was carried out during his affiliation with Aalborg University, Denmark.

References

[1] J. Pearl, *Probabilistic Reasoning in Intelligent Systems*. San Mateo, CA: Morgan Kaufmann, 1988.

[2] S. L. Lauritzen, *Graphical Models*. New York: Oxford University Press, 1996.

[3] D. Edwards, *Introduction to Graphical Modelling*. New York: Springer, 2000.

[4] J. Whittaker, *Graphical Models in Applied Multivariate Statistics*. New York: Wiley, 1990.

[5] M. I. Jordan, Ed., *Learning in Graphical Models*. Cambridge, MA: MIT Press, 1999.

[6] J. Pearl, *Causality. Models, Reasoning, and Inference*. New York: Cambridge University Press, 2000.

[7] M. I. Jordan, "Graphical models," *Statistical Science*, vol. 19, pp. 140–55, 2004.

[8] A. P. Dempster, "Covariance selection," *Biometrics*, vol. 28, pp. 157–75, 1972.

[9] R. Grone, C. R. Johnson, E. M. Sá, and H. Wolkowicz, "Positive definite completions of partial Hermitian matrices," *Linear Algebra and Applications*, vol. 58, pp. 109–24, 1984.

[10] J. Dahl, L. Vandenberghe, and V. Roychowdhury, "Covariance selection for non-chordal graphs via chordal embedding," *Optimization Methods and Software*, vol. 23, no. 4, pp. 501–20, 2008.

[11] N. Meinshausen and P. Bühlmann, "High-dimensional graphs and variable selection with the Lasso," *Annals of Statistics*, vol. 34, no. 3, pp. 1436–62, 2006.

[12] O. Banerjee, L. El Ghaoui, and A. d'Aspremont, "Model selection through sparse maximum likelihood estimation for multivariate Gaussian or binary data," *Journal of Machine Learning Research*, vol. 9, pp. 485–516, 2008.

[13] Z. Lu, "Adaptive first-order methods for general sparse inverse covariance selection," 2008, unpublished manuscript.

[14] D. R. Brillinger, "Remarks concerning graphical models for time series and point processes," *Revista de Econometria*, vol. 16, pp. 1–23, 1996.

[15] R. Dahlhaus, "Graphical interaction models for multivariate time series," *Metrika*, vol. 51, no. 2, pp. 157–72, 2000.

[16] R. Dahlhaus, M. Eichler, and J. Sandkühler, "Identification of synaptic connections in neural ensembles by graphical models," *Journal of Neuroscience Methods*, vol. 77, no. 1, pp. 93–107, 1997.

[17] M. Eichler, R. Dahlhaus, and J. Sandkühler, "Partial correlation analysis for the identification of synaptic connections," *Biological Cybernetics*, vol. 89, no. 4, pp. 289–302, 2003.

[18] R. Salvador, J. Suckling, C. Schwarzbauer, and E. Bullmore, "Undirected graphs of frequency-dependent functional connectivity in whole brain networks," *Philosophical Transactions of the Royal Society B: Biological Sciences*, vol. 360, no. 1457, pp. 937–46, 2005.

[19] U. Gather and M. Imhoff and R. Fried, "Graphical models for multivariate time series from intensive care monitoring," *Statistics in Medicine*, vol. 21, no. 18, pp. 2685–701, 2002.

[20] J. Timmer, M. Lauk, S. Häußler, V. Radt, B. Köster, B. Hellwig, B. Guschlbauer, C. Lücking, M. Eichler, and G. Deuschl, "Cross-spectral analysis of tremor time series," *International Journal of Bifurcation and Chaos in Applied Sciences and Engineering*, vol. 10, no. 11, pp. 2595–610, 2000.

[21] S. Feiler, K. Müller, A. Müller, R. Dahlhaus, and W. Eich, "Using interaction graphs for analysing the therapy process," *Psychotherapy Psychosomatics*, vol. 74, no. 2, pp. 93–9, 2005.

[22] R. Fried and V. Didelez, "Decomposability and selection of graphical models for multivariate time series," *Biometrika*, vol. 90, no. 2, p. 251–67, 2003.

[23] M. Eichler, "Testing nonparametric and semiparametric hypotheses in vector stationary processes," *Journal of Multivariate Analysis*, vol. 99, no. 5, pp. 968–1009, 2008.

[24] F. R. Bach and M. I. Jordan, "Learning graphical models for stationary time series," *IEEE Transactions on Signal Processing*, vol. 52, no. 8, pp. 2189–99, 2004.

[25] M. Eichler, "Fitting graphical interaction models to multivariate time series," *Proceedings of the 22nd Conference on Uncertainty in Artificial Intelligence*, 2006.

[26] R. Dahlhaus and M. Eichler, "Causality and graphical models in time series analysis," in *Highly Structured Stochastic Systems*, P. Green, N. L. Hjort and S. Richardson, eds. New York: Oxford University Press, 2003, pp. 115–44.

[27] T. Söderström and P. Stoica, *System Identification*. London: Prentice Hall International, 1989.

[28] G. Box and G. Jenkins, *Time Series Analysis, Forecasting and Control*. San Francisco, CA: Holden-Day, Incorporated, 1976.

[29] S. Marple, *Digital Spectral Analysis with Applications*. Englewood Cliffs, NJ: Prentice-Hall, Inc., 1987.

[30] S. Kay, *Modern Spectral Estimation: Theory and Application*. Englewood Cliffs, NJ: Prentice Hall, 1988.

[31] P. Stoica and R. Moses, *Introduction to Spectral Analysis*. Upper Saddle River, NJ: Prentice Hall, Inc., 1997.

[32] P. Stoica and A. Nehorai, "On stability and root location of linear prediction models," *IEEE Transactions on Acoustics, Speech, and Signal Processing*, vol. 35, pp. 582–4, 1987.

[33] G. C. Reinsel, *Elements of Multivariate Time Series Analysis*, 2nd ed. New York: Springer, 1997.

[34] J. P. Burg, "Maximum entropy spectral analysis," PhD dissertation, Stanford University, 1975.

[35] E. J. Hannon, *Multiple Time Series*. New York: John Wiley and Sons, Inc., 1970.

[36] D. Brillinger, *Time Series Analysis: Data Analysis and Theory*. New York: Holt, Rinehart & Winston, Inc., 1975.

[37] Y. Hachez, "Convex optimization over non-negative polynomials: structured algorithms and applications," PhD dissertation, Université catholique de Louvain, Belgium, 2003.

[38] M. Grant and S. Boyd. (2008). *CVX: Matlab software for disciplined convex programming*. [Web page and software]. Available: http://stanford.edu/~boyd/cvx

[39] ——, "Graph implementations for nonsmooth convex programs," in *Recent Advances in Learning and Control (A Tribute to M. Vidyasagar)*, V. Blondel, S. Boyd, and H. Kimura, eds. New York: Springer, 2008.

[40] K. Burnham and D. Anderson, *Model Selection and Multimodel Inference: A Practical Information-theoretic Approach*. New York: Springer, 2002.

[41] J. Proakis, *Digital Communications*, 4th ed. Boston, MA: McGraw-Hill, 2001.

[42] D. Bessler and J. Yang, "The structure of interdependence in international stock markets," *Journal of International Money and Finance*, vol. 22, no. 2, pp. 261–87, 2003.

[43] J. Yang, I. Min, and Q. Li, "European stock market integration: Does EMU matter?" *Journal of Business Finance & Accounting*, vol. 30, no. 9-10, pp. 1253–76, 2003.

4 SDP relaxation of homogeneous quadratic optimization: approximation bounds and applications

Zhi-Quan Luo and Tsung-Hui Chang

Many important engineering problems can be cast in the form of a *quadratically constrained quadratic program* (QCQP) or a *fractional* QCQP. In general, these problems are nonconvex and NP-hard. This chapter introduces a *semidefinite programming* (SDP) relaxation procedure for this class of quadratic optimization problems which can generate a provably approximately optimal solution with a randomized polynomial time complexity. We illustrate the use of SDP relaxation in the context of downlink transmit beamforming, and show that the SDP relaxation approach can either generate the global optimum solution, or provide an approximately optimal solution with a guaranteed worst-case approximation performance. Moreover, we describe how the SDP relaxation approach can be used in magnitude filter design and in magnetic resonance imaging systems.

4.1 Introduction

In this chapter, we consider several classes of nonconvex quadratic constrained quadratic programs (QCQPs) and a class of nonconvex fractional QCQPs. The importance of these classes of problems lies in their wide-ranging applications in signal processing and communications which include:

- the Boolean least-squares (LS) problem in digital communications [1];
- the noncoherent maximum-likelihood detection problem in *multiple-input multiple-output* (MIMO) communications [2, 3];
- the MAXCUT problem in network optimization [4];
- the large-margin parameter estimation problem in automatic speech recognition [5–8];
- the optimum coded waveform design for radar detection [9];
- the image segmentation problem in pattern recognition [10];
- the magnitude filter design problem in digital signal processing [11];
- the transmit B_1 shim and specific absorption rate computation in *magnetic resonance imaging* (MRI) systems [12, 13];
- the downlink transmit beamforming problem [14–18] and the Capon beamforming for microphone array [19];
- the network beamforming [20, 21].

Convex Optimization in Signal Processing and Communication, eds. Daniel P. Palomar and Yonina C. Eldar.
Published by Cambridge University Press © Cambridge University Press, 2010.

It is known that the nonconvex QCQP problems are in general NP-hard [15]. This makes the existence of an efficient algorithm to find a global optimal solution of nonconvex QCQP unlikely. As a result, we are naturally led to consider polynomial-time approximation algorithms for these problems. In this chapter, we introduce a semidefinite programming (SDP) relaxation approach for the QCQP problems, which not only has a polynomial-time computational complexity, but also guarantees a worst-case approximation performance. From the computational standpoint, the SDP relaxation approach consists of solving a convex SDP followed by a simple randomization procedure, both of which can be performed in polynomial time. As for the approximation quality, the SDP approach can guarantee an objective value that is within a constant factor of optimality. In realistic downlink beamforming setups, this performance gap translates to at most $2 \sim 3$ dB SNR loss, which is acceptable. It is worth noting that, prior to the emergence of SDP relaxation for the QCQP problems, there has been no effective approach to deal with these difficult NP-hard problems. Most of the previously proposed algorithms for QCQP rely on either exhaustive search (and its variations), or on ad hoc relaxation methods which can not guarantee satisfactory approximation performance.

The organization of this chapter is as follows. First, we will introduce the ideas of SDP relaxation and present how it can be used for the approximation of nonconvex QCQPs. To illustrate the use of SDP relaxation techniques, we focus on the specific application in downlink transmit beamforming and develop the associated SDP relaxation algorithms. Then, we will present an analysis framework which bounds the worst-case approximation performance of the SDP relaxation method. Numerical results will be used to to illustrate the approximation performance of the SDP relaxation method. Finally, the applications of SDP relaxation to the magnitude filter design problem and the transmit B_1 shim problem in MRI will also be presented.

4.2   Nonconvex QCQPs and SDP relaxation

Mathematically, a QCQP can be written as

$$\min_{x} \ x^T P_0 x + q_0^T x + r_0 \tag{4.1a}$$

$$\text{s.t.} \ x^T P_i x + q_i^T x + r_i \le 0, \quad i = 1, \ldots, m, \tag{4.1b}$$

where x, $q_i \in \mathbb{R}^K$, $r_i \in \mathbb{R}$ and $P_i \in \mathbb{R}^{K \times K}$ are symmetric. If all the P_i are *positive semidefinite* (p.s.d.) [i.e., $P_i \succeq 0$], the QCQP in (4.1) is convex and can be efficiently solved to the global optimum. However, if at least one of the P_i is not p.s.d., the QCQP is nonconvex and, in general, is computationally difficult to solve. Two classical examples of the nonconvex QCQPs are given as follows.

Example 4.1 Boolean least-squares problem   The Boolean *least-squares* (LS) problem is a fundamental and important problem in digital communications. It can be expressed as

$$\min_{x} \ \|Ax - b\|^2 \tag{4.2a}$$

$$\text{s.t.} \ x_i^2 = 1, \quad i = 1, \ldots, K, \tag{4.2b}$$

where $x = [x_1, x_2, \ldots, x_K]^T \in \mathbb{R}^K$, $b \in \mathbb{R}^K$, $A \in \mathbb{R}^{K \times K}$. Expanding the LS objective function, (4.2) can be written as a nonconvex QCQP:

$$\min_{x} \ x^T A^T A x - 2b^T A x + b^T b \tag{4.3a}$$

$$\text{s.t.} \ x_i^2 = 1, \quad i = 1, \ldots, K. \tag{4.3b}$$

Since $x_i^2 = 1$ is equivalent to $x_i \in \{-1, 1\}$, one can solve this problem by checking all the 2^K points in $\{-1, 1\}^K$, which, however, is impractical when K is large. In fact, problem (4.2) has been shown to be NP-hard in general, and is very difficult to solve in practice.

Example 4.2 Partitioning problem   Let us consider a two-way partitioning problem that, given a set of K elements $\{1, \ldots, K\}$, one would like to partition into two sets while minimizing the total cost. Let $x = [x_1, x_2, \ldots, x_K]^T \in \{\pm 1\}^K$, which corresponds to the partition

$$\{1, \ldots, K\} = \{i \mid x_i = 1\} \cup \{i \mid x_i = -1\}.$$

Denote by $W_{i,j}$ the cost of having the elements i and j in the same set. The partitioning problem can be expressed as follows:

$$\min_{x} \ x^T W x \tag{4.4a}$$

$$\text{s.t.} \ x_i^2 = 1, \quad i = 1, \ldots, K, \tag{4.4b}$$

where $W = [W_{i,j}] \in \mathbb{S}^K$ (the space of K by K symmetric matrices) and $W_{i,i} = 0$ for $i = 1, \ldots, K$. It should be noted that if $W_{i,j} \geq 0$ for all $i \neq j$ then (4.4) is the MAXCUT, a classical problem in network optimization [4]. Problem (4.4) is also NP-hard in general.

4.2.1 SDP relaxation

Since the nonconvex QCQP is NP-hard in general, a polynomial-time approximation method is desired. In the subsection, we introduce such a method, based on relaxing the nonconvex QCQP into a semidefinite program (SDP), referred to as the SDP relaxation or the *semidefinite relaxation* (SDR). To put the SDP relaxation into context, let us consider the following homogeneous QCQP:

$$v_{qp} = \min_{x} \; x^T P_0 x + r_0 \tag{4.5a}$$

$$\text{s.t. } x^T P_i x + r_i \leq 0, \quad i = 1, \ldots, m. \tag{4.5b}$$

For nonhomogeneous quadratic functions with linear terms as in (4.1), we can "homogenize" them by introducing an additional variable x_0 and an additional constraint $x_0^2 = 1$. Then each linear term $q_i^T x$ can be written as a quadratic term $x_0 q_i^T x$.

The SDP relaxation makes use of the following fundamental observation:

$$X = xx^T \iff X \succeq 0, \; \text{rank}(X) = 1.$$

Using this observation, we can "linearize" the QCQP problem (4.5) by representing it in terms of the matrix variable X. Specifically, we note that $x^T P_i x = \text{Tr}(P_i X)$, so (4.5) can be rewritten as

$$\min_{x} \; \text{Tr}(P_0 X) + r_0 \tag{4.6a}$$

$$\text{s.t. } \text{Tr}(P_i X) + r_i \leq 0, \quad i = 1, \ldots, m, \tag{4.6b}$$

$$X \succeq 0, \tag{4.6c}$$

$$\text{rank}(X) = 1. \tag{4.6d}$$

Since in (4.6) the only nonconvex constraint is $\text{rank}(X) = 1$, one can directly relax the last constraint, in other words, dropping the nonconvex $\text{rank}(X) = 1$ and keeping only $X \succeq 0$, to obtain the following SDP:

$$v_{sdp} = \min_{x} \; \text{Tr}(P_0 X) + r_0 \tag{4.7a}$$

$$\text{s.t. } \text{Tr}(P_i X) + r_i \leq 0, \quad i = 1, \ldots, m, \tag{4.7b}$$

$$X \succeq 0. \tag{4.7c}$$

The relaxed problem (4.7) gives a lower bound on the optimal objective value, that is, $v_{sdp} \leq v_{qp}$. In fact, (4.7) gives the same lower bound as the Lagrangian dual of (4.5) because it can be shown that the SDP relaxation problem (4.7) is, in essence, the bi-dual of (4.5) [22]. In contrast to the original homogeneous QCQP (4.5), the SDP relaxation problem (4.7) is a convex optimization problem, and thus can be efficiently solved by interior-point methods [23] in polynomial time.

4.2.2   Extracting a rank-1 solution: Gaussian sampling

The optimum solution of the SDP relaxation problem (4.7) is a matrix, denoted by $X^\star$, which is not necessarily rank-1. We need to extract a rank-1 component xx^T (or equivalently a vector x) from $X^\star$ so that x is feasible for the original problem (4.5) and serves as a good approximate solution. One straightforward approach is to perform the rank-1 approximation by first taking the principal eigenvector of $X^\star$ and then projecting it into the feasible set of (4.5). Another equally simple but more effective approach is to pick a random vector x from the Gaussian distribution $N(\mathbf{0}, X^\star)$ with zero mean and covariance matrix equal to $X^\star$, and then project it to the feasible region of (4.5). This random sampling can be performed multiple times (polynomially many times in theory) and one can pick only the best approximate solution. Since

$$E[x^T P_i x + r_i] = \mathrm{Tr}(P_i X^\star) + r_i,$$

a random Gaussian vector $x \sim N(\mathbf{0}, X^\star)$ actually solves the homogeneous QCQP (4.5) "in expectation", that is,

$$\min_x \; E[x^T P_0 x + r_0] \tag{4.8a}$$

$$\text{s.t.} \; E[x^T P_i x + r_i] \le 0, \quad i = 1, \ldots, m. \tag{4.8b}$$

This suggests that a good approximate solution $\hat{x}$ can be obtained by sampling enough times from the Gaussian distribution $N(\mathbf{0}, X^\star)$.

Example 4.3 SDP relaxation of the Boolean LS problem   Let us show how the SDP relaxation can be applied to the Boolean LS problem in Example 4.1. We first homogenize the problem (4.3). Let $x = t\tilde{x}$, where $t \in \{\pm 1\}$ and $\tilde{x} \in \{\pm 1\}^K$. By substituting $x = t\tilde{x}$ into (4.3), one can rewrite (4.3) as

$$\min_{\tilde{x}, t} \; \tilde{x}^T A^T A \tilde{x} - 2b^T A(t\tilde{x}) + \|b\|^2 \tag{4.9a}$$

$$\text{s.t.} \; \tilde{x}_i^2 = 1, \quad i = 1, \ldots, K, \tag{4.9b}$$

$$t^2 = 1. \tag{4.9c}$$

Define $y = [\tilde{x}^T, t]^T \in \{\pm 1\}^{K+1}$. Then (4.9) can be formulated as a homogeneous QCQP as follows:

$$\min_y \; y^T P y \tag{4.10a}$$

$$\text{s.t.} \; y_i^2 = 1, \quad i = 1, \ldots, K + 1, \tag{4.10b}$$

where

$$P = \begin{bmatrix} A^T A & -b \\ -b^T & \|b\|^2 \end{bmatrix}.$$

Let $Y = yy^T$. Following the steps in (4.6) and (4.7), the SDP relaxation of (4.10) is obtained as

$$Y^\star = \arg\min_{Y} \ \mathrm{Tr}(PY) \tag{4.11a}$$

$$\text{s.t.} \ \ Y_{i,i} = 1, \quad i = 1,\ldots,K+1, \tag{4.11b}$$

$$Y \succeq 0. \tag{4.11c}$$

We illustrate how the Gaussian sampling idea can be used to obtain an approximate solution of (4.10). Let $L > 0$ be an integer. We generate a set of random vectors $\boldsymbol{\xi}_\ell$, $\ell = 1,\ldots,L$, from the Gaussian distribution $N(0, Y^\star)$, and quantize each of them into the binary vector $\hat{\boldsymbol{y}}_\ell = \mathrm{sign}(\boldsymbol{\xi}_\ell) \in \{\pm 1\}^{K+1}$, where $\mathrm{sign} : \mathbb{R}^{K+1} \to \{\pm 1\}^{K+1}$. An approximate solution of (4.10) can be obtained as

$$\hat{\boldsymbol{y}} = \min_{\ell=1,\ldots,L} \hat{\boldsymbol{y}}_\ell^T P \hat{\boldsymbol{y}}_\ell. \tag{4.12}$$

Let $\hat{\boldsymbol{y}} = [\hat{\boldsymbol{x}}^T, \hat{t}]^T$. Finally, the associated approximate solution to (4.9) is given by $\hat{\boldsymbol{x}} := \hat{t}\hat{\boldsymbol{x}}$. It is empirically found that $L = 50 \sim 100$ is sufficient to obtain near-optimum approximation performance. Readers are referred to [1] for the details.

In practice, after obtaining an approximate solution through the aforementioned Gaussian sampling procedure, we can employ a local optimization procedure (e.g., gradient-type or Newton-type algorithms [24]) to further optimize the nonconvex quadratic function, subject to the quadratic constraints. Sometimes such a local optimization step can provide an appreciable improvement over a Gaussian-sampled vector.

4.2.3   Approximation ratio

An important question regarding the SDP relaxation method is how good the approximate solution $\hat{\boldsymbol{x}}$ is in terms of its achieved objective value. Specifically, let $v_{\mathrm{qp}}(\hat{\boldsymbol{x}}) = \hat{\boldsymbol{x}}^T P_0 \hat{\boldsymbol{x}} + r_0$, and let v_{qp} and v_{sdp} be the optimum objective values of the nonconvex QCQP (4.5) and its SDP relaxation problem (4.7), respectively. The approximation quality of the SDP relaxation solution can be measured by the ratio of

$$\gamma = \max\left\{ \frac{v_{\mathrm{qp}}}{v_{\mathrm{sdp}}}, \frac{v_{\mathrm{qp}}(\hat{\boldsymbol{x}})}{v_{\mathrm{qp}}} \right\}. \tag{4.13}$$

Clearly, the approximation ratio is always greater than or equal to 1. If $\gamma = 1$ for some particular problem instance, then there is no relaxation gap between (4.5) and (4.7) ($v_{\mathrm{qp}} = v_{\mathrm{sdp}}$), and the corresponding nonconvex problem (4.5) is solved to global optimality. Equivalently, for this particular problem instance, the SDP relaxation admits a rank-1 optimal solution $X^\star$. If $\gamma > 1$, the SDP relaxation method generates an approximate solution $\hat{x}$ that is within $(\gamma - 1)$ fraction to the optimum minimum v_{qp}. It also indicates that the gap between the optimum value of the original problem (4.5) v_{qp} and the optimum value of the relaxation problem v_{sdp} is less than $(\gamma - 1)v_{qp}$. Our interest in the ensuing sections is to identify conditions under which the approximation ratio γ is equal to 1, or can be bounded by a constant factor independent of the problem dimension n and the data matrices P_i and r_i. If a finite (data independent) γ exists, then we say that the SDP relaxation method provides a guaranteed, finite, worst-case approximation performance for the nonconvex QCQP problem (4.5).

4.3   SDP relaxation for separable homogeneous QCQPs

Motivated by applications in the downlink transmit beamforming for wireless communications and sensor localization problem in wireless sensor networks, we consider, in this section, a class of separable homogeneous QCQPs, and develop an SDP relaxation approach to approximately solve them. Moreover, we will analyze the approximation performance of the SDP relaxation methods.

4.3.1   Separable homogeneous QCQPs

Let us consider the following separable homogeneous QCQP :

$$\min_{w_i \in \mathbb{H}^{K_i}} \quad \sum_{i=1}^{n} w_i^\dagger C_i w_i \tag{4.14a}$$

$$\text{s.t.} \quad w_1^\dagger A_{11} w_1 + w_2^\dagger A_{12} w_2 + \cdots + w_n^\dagger A_{1n} w_n \geq b_1, \tag{4.14b}$$

$$w_1^\dagger A_{21} w_1 + w_2^\dagger A_{22} w_2 + \cdots + w_n^\dagger A_{2n} w_n \geq b_1, \tag{4.14c}$$

$$\vdots$$

$$w_1^\dagger A_{m1} w_1 + w_2^\dagger A_{m2} w_2 + \cdots + w_n^\dagger A_{mn} w_n \geq b_m, \tag{4.14d}$$

where C_i, $A_{i,j}$ are Hermitian matrices, $\mathbb{H}$ is either $\mathbb{R}$ or $\mathbb{C}$, and the superscript $()^\dagger = ()^T$ when $\mathbb{H} = \mathbb{R}$, and $()^\dagger = ()^H$ when $\mathbb{H} = \mathbb{C}$. The separable homogeneous QCQP is NP-hard in general and is typically hard to solve. Similar to the problem in (4.5), we consider the following SDP relaxation of (4.14):

$$\min_{W_i \in \mathbb{H}^{K_i \times K_i}} \quad \sum_{i=1}^{n} \text{Tr}(C_i W_i) \tag{4.15a}$$

$$\text{s.t.} \quad \text{Tr}(A_{11} W_1) + \text{Tr}(A_{12} W_2) + \cdots + \text{Tr}(A_{1n} W_n) \geq b_1, \tag{4.15b}$$

$$\mathrm{Tr}(A_{21}W_1) + \mathrm{Tr}(A_{22}W_2) + \cdots + \mathrm{Tr}(A_{2n}W_n) \geq b_1, \tag{4.15c}$$

$$\vdots$$

$$\mathrm{Tr}(A_{m1}W_1) + \mathrm{Tr}(A_{m2}W_2) + \cdots + \mathrm{Tr}(A_{mn}W_n) \geq b_m, \tag{4.15d}$$

$$W_i \succeq 0, \quad i = 1, \ldots, n, \tag{4.15e}$$

and will discuss the associated approximation quality in the ensuing subsections. In general, the SDP relaxation for separable homogeneous QCQPs is not tight, as shown by a simple example below.

Example 4.4   Consider the following problem:

$$v_{\mathrm{qp}} = \min_{x,y \in \mathbb{R}} \; x^2 + y^2 \tag{4.16a}$$

$$\text{s.t.} \; y^2 \geq 1, \; x^2 - Mxy \geq 1, \; x^2 + Mxy \geq 1, \tag{4.16b}$$

where $M > 0$, and its SDP relaxation

$$v_{\mathrm{sdp}} = \min_{X} \; X_{11} + X_{22} \tag{4.17a}$$

$$\text{s.t.} \; X_{22} \geq 1, \; X_{11} - MX_{12} \geq 1, \; X_{11} + MX_{12} \geq 1, \tag{4.17b}$$

$$X = \begin{bmatrix} X_{11} & X_{12} \\ X_{12} & X_{22} \end{bmatrix} \succeq 0. \tag{4.17c}$$

Notice that the last two constraints in (4.16b) imply $x^2 \geq M|x||y| + 1$ and $x^2 \geq 1$. In light of the constraint $y^2 \geq 1$, this further implies $x^2 \geq M + 1$, and thus $v_{\mathrm{qp}} \geq M + 2$. For the relaxation problem (4.17), it is clear that $X = I$ is a feasible solution, and therefore $v_{\mathrm{sdp}} \leq 2$. Hence one can see that the ratio $v_{\mathrm{qp}}/v_{\mathrm{sdp}}$ is at least $(M+2)/2$, which can be arbitrarily large.

On the other hand, for some special cases of the separable homogeneous QCQP, the SDP relaxation can be tight, that is, $\gamma = v_{\mathrm{qp}}/v_{\mathrm{sdp}} = 1$. The following examples present two such cases.

Example 4.5 Special case of all $K_i = 1$ Assume that $K_i = 1$ for all i in the separable homogeneous QCQP (4.14). Then $\boldsymbol{w}_i = w_i$, $\boldsymbol{C}_i = c_i$, and $\boldsymbol{A}_{ij} = a_{ij}$, all of which are just scalars, and thus (4.14) reduces to

$$\min_{w_i} \ \sum_{i=1}^{n} |w_i|^2 c_i \tag{4.18a}$$

$$\text{s.t.} \ \sum_{i=1}^{n} |w_i|^2 a_{ij} \geq b_j, \quad j = 1, \ldots, m. \tag{4.18b}$$

By replacing $|w_i|^2$ with $W_i \geq 0$, we see that the SDP relaxation of (4.18) is a linear program and is equivalent to the original problem (4.18).

Example 4.6 Special case of $m = n = 1$ Consider the nonconvex QCQP (4.14) with $m = n = 1$:

$$\min_{w} \ w^{\dagger} C w \ \text{ s.t. } \ w^{\dagger} A w \geq b. \tag{4.19}$$

By Lagrangian theory, the optimum solution $w^{\star}$ must satisfy

$$C w^{\star} = \lambda A w^{\star}$$

for some $\lambda = (w^{\star})^{\dagger} C w^{\star}$. Hence, (4.19) is a generalized eigenvalue problem for matrices C and A that can be efficiently computed. As it will be shown later, the SDP relaxation of (4.19) admits a rank-1 solution, and thus the SDP relaxation is tight.

4.3.2 Downlink transmit beamforming

Our interest in this separable homogenous QCQP (4.14) is primarily motivated by the downlink transmit beamforming problem for multicast applications [15, 16, 18].

Example 4.7 Multicast transmit beamforming Consider the downlink transmit beamforming system as depicted in Figure 4.1, in which a base station equipped with K transmit antennas tries to broadcast n $(n > 1)$ independent data streams to m $(m = \sum_{i=1}^{n} m_i)$ single-antenna receivers over a common frequency band. Each of the receivers belongs to one of the n groups, $\mathcal{G}_i, \ i = 1, \ldots, n$, with receivers from the same group interested in a common data stream. Let $s_i(t)$ and $w_i \in \mathbb{C}^K$ denote the broadcasting data stream and the transmit weight vector (or beamforming vector) for the ith group, respectively. The transmitted signal at the base station is given by $\sum_{i=1}^{n} w_i s_i(t)$. Assume that $s_i(t), \ k = 1, \ldots, n$, are statistically independent and are temporally white with zero mean and unit variance. Let $h_k \in \mathbb{C}^K$ denote the channel vector between the base station

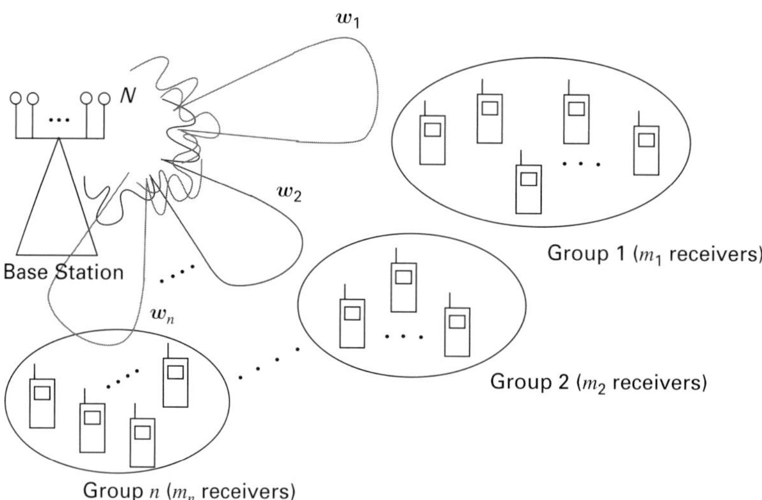

Figure 4.1   System diagram of a downlink transmit beamforming system with multiple multicast. There are K antennas equipped in the base station, n groups, and totally $m = m_1 + m_2 + \cdots + m_n$ single-antenna receivers.

and the kth receiver where $k \in \mathcal{G}_i$. The received signal at the receiver $k \in \mathcal{G}_i$ is given by

$$\underbrace{s_i(t)w_i^H h_k}_{\text{desired signal}} + \underbrace{\sum_{j\neq i} s_j(t)w_j^H h_k + n_k(t)}_{\text{interference and noise terms}}, \tag{4.20}$$

where $n_k(t)$ represents the received noise signal with power σ_k^2.

The idea of transmit beamforming is to design the beamforming weight vectors w_i, $i = 1,\ldots,n$, such that each receiver can retrieve the signal of interest with desired "quality of service" (QoS). The QoS is usually measured in terms of the "signal-to-interference-plus-noise-ratio" (SINR) which, by (4.20), is given by

$$\text{SINR}_k = \frac{|w_i^H h_k|^2}{\sum_{j\neq i} |w_j^H h_k|^2 + \sigma_k^2}, \quad \forall\, k \in \mathcal{G}_i, \ i = 1,\ldots,n. \tag{4.21}$$

For the case that the receiver is equipped with multiple antennas, say Q antennas, the SINR then is given by

$$\text{SINR}_k = \frac{w_i^H H_k w_i}{\sum_{j\neq i} w_j^H H_k w_j + \sigma_k^2}, \quad \forall\, k \in \mathcal{G}_i, \ i = 1,\ldots,n, \tag{4.22}$$

in which $H_k = \sum_{q=1}^{Q} h_{k,q} h_{k,q}^H \succeq 0$.

To achieve a guaranteed QoS, one design criterion that has been widely used in the literature is to minimize the transmit power at the base station, while making sure that the SINR in each receiver is no smaller than some specified value. Mathematically, it can be written as the following optimization problem:

$$\min_{\substack{\boldsymbol{w}_i \in \mathbb{C}^K \\ i=1,\ldots,n}} \sum_{i=1}^{n} \|\boldsymbol{w}_i\|^2 \tag{4.23a}$$

$$\text{s.t.} \quad \frac{\boldsymbol{w}_i^H \boldsymbol{H}_k \boldsymbol{w}_i}{\sum_{j \neq i} \boldsymbol{w}_j^H \boldsymbol{H}_k \boldsymbol{w}_j + \sigma_k^2} \geq \rho_k, \quad \forall\, k \in \mathcal{G}_i, \ i = 1,\ldots,n, \tag{4.23b}$$

where $\rho_k > 0$ is the target SINR value for the receiver $k \in \mathcal{G}_i$. By some simple reformulations, the above transmit beamforming problem can be recast as a separable homogeneous QCQP as follows:

$$\min_{\substack{\boldsymbol{w}_i \in \mathbb{C}^K \\ i=1,\ldots,n}} \sum_{i=1}^{n} \|\boldsymbol{w}_i\|^2 \tag{4.24a}$$

$$\text{s.t.} \quad \boldsymbol{w}_i^H \boldsymbol{H}_k \boldsymbol{w}_i - \rho_k \sum_{j \neq i} \boldsymbol{w}_j^H \boldsymbol{H}_k \boldsymbol{w}_j \geq \sigma_k^2 \rho_k, \tag{4.24b}$$

$$\forall\, k \in \mathcal{G}_i, \ i = 1,\ldots,n. \tag{4.24c}$$

As in (4.15), the SDP relaxation of the above multicast transmit beamforming problem can be easily derived. The effectiveness of the SDP relaxation to this transmit beamforming problem has been investigated through extensive computer simulations. Promising performance results have been observed in either simulated or measured VDSL channels [15–17, 25].

A general observation from these simulation results is that the SDP relaxation method works surprisingly well, and in some scenarios the SDP relaxation even yields the global optimum of the original nonconvex problem (4.23). These exciting observations have motivated research into whether a theoretically guaranteed performance for the SDP relaxation method exists. Indeed, as we will show in the upcoming subsections, if the transmit beamforming problem (4.23) degenerates into the unicast case ($m_i = 1$ for all i), or if the channel vectors $\boldsymbol{h}_k$ possess a Vandermonde structure, its SDP relaxation is tight and the corresponding relaxation gap is zero ($\gamma = 1$). Moreover, if there is only one group in the system ($n = 1$), the SDP relaxation for (4.23), though not being tight, has a guaranteed worst-case approximation performance.

4.3.3 SDP relaxation for unicast transmit beamforming

Now let us consider the classical unicast transmit beamforming problem in which each signal $s_i(k)$ is dedicated to one receiver instead of multiple receivers (Figure 4.1). Mathematically, it is equivalent to having problem (4.23) with $m_i = 1$ for all i, and thus the unicast transmit beamforming problem is given by

$$\min_{\substack{\boldsymbol{w}_i \in \mathbb{C}^K \\ i=1,\ldots,n}} \quad \sum_{i=1}^{n} \|\boldsymbol{w}_i\|^2 \tag{4.25a}$$

$$\text{s.t.} \quad \boldsymbol{w}_i^H \boldsymbol{H}_i \boldsymbol{w}_i - \rho_i \sum_{j \neq i} \boldsymbol{w}_j^H \boldsymbol{H}_i \boldsymbol{w}_j \geq \sigma_i^2 \rho_i, \quad i = 1,\ldots,n. \tag{4.25b}$$

Note that in this case, we have $m = n$. If $\boldsymbol{H}_i = \boldsymbol{h}_i \boldsymbol{h}_i^H$ for $i = 1,\ldots,n$, the above unicast problem can be shown [26] to be a convex "second-order cone program" (SOCP). To illustrate this, one can see that the constraints in (4.25b) can be rewritten as

$$|\boldsymbol{w}_i^H \boldsymbol{h}_i|^2 - \rho_i \sum_{j \neq i} |\boldsymbol{w}_j^H \boldsymbol{h}_i|^2 \geq \sigma_i^2 \rho_i, \tag{4.26}$$

when $\boldsymbol{H}_i = \boldsymbol{h}_i \boldsymbol{h}_i^H$ for $i = 1,\ldots,n$. Since the objective value in (4.25b) does not change for any phase rotation on each $\boldsymbol{w}_i$, it is without loss of generality to rewrite (4.27) as

$$\boldsymbol{w}_i^H \boldsymbol{h}_i \geq \sqrt{\sigma_i^2 \rho_i + \rho_i \sum_{j \neq i} |\boldsymbol{w}_j^H \boldsymbol{h}_i|^2}, \tag{4.27}$$

which then is a SOC constraint. Therefore, in this case, the unicast transmit beamforming problem itself is convex, and hence can be efficiently solved.

In fact, even without the condition of $\boldsymbol{H}_i = \boldsymbol{h}_i \boldsymbol{h}_i^H$ for $i = 1,\ldots,n$, the unicast problem (4.25) can be globally solved by its SDP relaxation problem. This profound result was first proved by Bengtsson and Ottersten in [14] using the Lagrangian duality and the Perron–Frobenius theorem. Here we present a more general statement to show the same result. The SDP relaxation of (4.25) is given as follows:

$$\min_{\boldsymbol{W}_i \in \mathbb{C}^{K \times K}} \quad \sum_{i=1}^{n} \text{Tr}(\boldsymbol{W}_i) \tag{4.28a}$$

$$\text{s.t.} \quad \text{Tr}(\boldsymbol{H}_i \boldsymbol{W}_i) - \rho_i \sum_{j \neq i} \text{Tr}(\boldsymbol{H}_i \boldsymbol{W}_j) \geq \sigma_i^2 \rho_i, \tag{4.28b}$$

$$\boldsymbol{W}_i \succeq \boldsymbol{0}, \quad i = 1,\ldots,n. \tag{4.28c}$$

Suppose that the above SDP has a solution $\{\boldsymbol{W}_i^\star\}_{i=1}^{n}$, and let $r(\boldsymbol{W}_i^\star)$ denote the rank of $\boldsymbol{W}_i^\star$. According to Pataki's result (Theorem 2.2 in [27]), one can find, in polynomial time, a solution $\{\boldsymbol{W}_i^\star\}_{i=1}^{n}$ that satisfies

$$\sum_{i=1}^{n} \frac{r(\boldsymbol{W}_i^\star)(r(\boldsymbol{W}_i^\star) + 1)}{2} \leq m = n. \tag{4.29}$$

On the other hand, since all feasible beamforming solutions have $\boldsymbol{W}_i \neq \boldsymbol{0}$, we have $r(\boldsymbol{W}_i^\star) \geq 1$, and therefore (4.29) implies that a solution $\{\boldsymbol{W}_i^\star\}_{i=1}^{n}$ exists with all the $\boldsymbol{W}_i^\star$

being of rank-1. Hence this demonstrates that the SDP relaxation problem (4.28) attains the global optimum of (4.25).

4.3.4　SDP relaxation for far-field, multicast transmit beamforming

The unicast scenario is not the only case of the SDP relaxation being tight. The same can also be true in the general multicast scenario (4.23) under the assumption that the channel vectors $\boldsymbol{h}_k$ possess a Vandermonde structure [17], for example,

$$\boldsymbol{h}_k = [1, e^{j\phi_k}, e^{j2\phi_k}, \ldots, e^{j(K-1)\phi_k}]^T := \boldsymbol{a}(\phi_k) \tag{4.30}$$

where $j = \sqrt{-1}$ and $\phi_k \in (0, 2\pi]$. Vandermonde channel vectors arise when a *uniform linear antenna array* (ULA) is used at the base station under far-field, line-of-sight propagation and narrowband signal conditions. These conditions are very realistic in wireless backhaul scenarios and can be found in the IEEE 802.16e standards. The SDP relaxation of (4.24) is given by

$$\min_{\substack{\boldsymbol{W}_i \in \mathbb{C}^{K \times K} \\ i=1,\ldots,n}} \sum_{i=1}^{n} \mathrm{Tr}(\boldsymbol{W}_i) \tag{4.31a}$$

$$\mathrm{s.t.}\ \ \mathrm{Tr}(\boldsymbol{H}_k \boldsymbol{W}_i) - \rho_k \sum_{j \neq i} \mathrm{Tr}(\boldsymbol{H}_k \boldsymbol{W}_j) \geq \sigma_k^2 \rho_k, \quad \forall\, k \in \mathcal{G}_i, \tag{4.31b}$$

$$\boldsymbol{W}_i \succeq \boldsymbol{0}, \quad i = 1, \ldots, n. \tag{4.31c}$$

Next we show that the SDP problem (4.31) admits a rank-1 solution when $\boldsymbol{H}_k = \boldsymbol{a}(\phi_k)\boldsymbol{a}^H(\phi_k)$, which implies that the SDP relaxation of (4.24) is tight. Let $\{\boldsymbol{W}_i^\star\}_{i=1}^n$ be one of the optimum solutions of (4.31). Suppose that $r(\boldsymbol{W}_i^\star) = r_i$, one can decompose each $\boldsymbol{W}_i^\star$ as

$$\boldsymbol{W}_i^\star = \sum_{\ell=1}^{r_i} \boldsymbol{w}_{i,\ell}^\star (\boldsymbol{w}_{i,\ell}^\star)^H \tag{4.32}$$

where $\boldsymbol{w}_{i,\ell}^\star \in \mathbb{C}^K$. Therefore, each trace term in (4.31b) can be expressed as

$$\mathrm{Tr}(\boldsymbol{H}_k \boldsymbol{W}_i^\star) = \sum_{\ell=1}^{r_i} |\boldsymbol{a}(\phi_k)^H \boldsymbol{w}_{i,\ell}^\star|^2 \geq 0, \tag{4.33}$$

which is a non-negative, real-valued, complex trigonometric polynomial. According to the Riesz–Fejer theorem [28], a vector $\bar{\boldsymbol{w}}_i^\star \in \mathbb{R} \times \mathbb{C}^{K-1}$ exists that is independent of the phases ϕ_k, $\forall\, k \in \mathcal{G}_i$ and satisfies

$$\sum_{\ell=1}^{r_i} |\boldsymbol{a}(\phi_k)^H \boldsymbol{w}_{i,\ell}^\star|^2 = |\boldsymbol{a}(\phi_k)^H \bar{\boldsymbol{w}}_i^\star|^2 \tag{4.34}$$

$$= \mathrm{Tr}(\boldsymbol{H}_k \bar{\boldsymbol{w}}_i^\star (\bar{\boldsymbol{w}}_i^\star)^H). \tag{4.35}$$

Define $\bar{W}_i^\star = \bar{w}_i^\star (\bar{w}_i^\star)^H$. Then by (4.33) and (4.35) we have

$$\mathrm{Tr}(H_k W_i^\star) = \mathrm{Tr}(H_k \bar{W}_i^\star). \qquad (4.36)$$

Moreover, by integrating out ϕ_k in (4.34) over $(0, 2\pi]$, one can obtain

$$\mathrm{Tr}(W_i^\star) = \mathrm{Tr}(\bar{W}_i^\star). \qquad (4.37)$$

Therefore we see that the rank-1 matrices $\{\bar{W}_i^\star\}_{i=1}^n$ form an optimum solution of (4.31), demonstrating that the SDP relaxation of (4.24) is tight for Vandermonde channels.

4.3.5 SDP relaxation for single-group, multicast transmit beamforming

While the SDP relaxation of the transmit beamforming problem (4.23) is tight for Vandermonde channels, the same is not true for general channel vectors. However, as we show in this subsection, for the case of $n = 1$ (single-group multicast), the SDP relaxation has a bounded worst-case approximation performance [15].

In this single-group scenario, all the m receivers are from the same group and they wish to receive a common broadcast signal. Therefore, there is no co-channel interference. As a result, the single-group, multicast beamforming problem (4.23) can be written as follows

$$v_{\mathrm{qp}} = \min_{w \in \mathbb{H}^K} \ \|w\|^2 \qquad (4.38\mathrm{a})$$

$$\text{s.t.} \ \ w^\dagger H_k w \geq 1, \quad k = 1, \ldots, m. \qquad (4.38\mathrm{b})$$

Note that we have set $\rho_k = 1$ for all k for simplicity. Besides, we have replaced $\mathbb{C}$ by $\mathbb{H}$ in order to accommodate both the real $\mathbb{R}$ and complex $\mathbb{C}$ cases in the ensuing analysis. While complex-valued signal models ($\mathbb{H} = \mathbb{C}$) are generally used in wireless communications, real-valued signals also arise in other communication systems. For example, in the pulse-based, ultra-wideband systems [29–31], both the transmitted baseband pulses and the received signals are real-valued.

The SDP relaxation of (4.38) is given by

$$v_{\mathrm{sdp}} = \min_{W \in \mathbb{H}^{K \times K}} \ \mathrm{Tr}(W) \qquad (4.39\mathrm{a})$$

$$\text{s.t.} \ \ \mathrm{Tr}(H_k W) \geq 1, \quad k = 1, \ldots, m, \qquad (4.39\mathrm{b})$$

$$W \succeq 0. \qquad (4.39\mathrm{c})$$

Though problem (4.38) has a much simplified structure as compared with (4.23), it is still an NP-hard problem in general [32], and its SDP relaxation (4.39) is generally not tight (see Example 4.4). To obtain a feasible approximate solution $\hat{w}$ for (4.38), we resort to the Gaussian sampling idea discussed in Section 4.2.2. Let $W^\star$ be the obtained optimum solution of (4.39). The following randomization procedure can be used to construct a feasible solution of (4.38):

> **Box 1.** Gaussian randomization procedure for (4.38)
>
> Let $L > 0$ be an integer.
>
> 1. For $\ell = 1, \ldots, L$, generate a random vector $\boldsymbol{\xi}_\ell \in \mathbb{H}^K$ from the Gaussian distribution $N(0, \boldsymbol{W}^*)$, and let
>
> $$w(\boldsymbol{\xi}_\ell) = \boldsymbol{\xi}_\ell / \min_{1 \le k \le m} \sqrt{\boldsymbol{\xi}_\ell^\dagger \boldsymbol{H}_k \boldsymbol{\xi}_\ell}. \qquad (4.40)$$
>
> 2. Let
>
> $$\hat{w} = \arg \min_{w(\xi_\ell), \ell=1,\ldots,L} \|w(\boldsymbol{\xi}_\ell)\|^2 \qquad (4.41)$$
>
> be the approximate solution of (4.38).

Note that when $\mathbb{H} = \mathbb{R}$, the random vectors $\boldsymbol{\xi}_\ell$ are drawn from the real-valued Gaussian $N_\mathbb{R}(0, \boldsymbol{W}^*)$, whereas when $\mathbb{H} = \mathbb{C}$, the random vectors $\boldsymbol{\xi}_\ell$ are drawn from the complex-valued Gaussian $N_\mathbb{C}(0, \boldsymbol{W}^*)$.

Let $v_{\text{qp}}(\hat{w}) = \|\hat{w}\|^2$. Next we show that the worst-case approximation ratio, defined as

$$\gamma_{\text{sdr}} := \max_{\boldsymbol{H}_k, K} \max \left\{ \frac{v_{\text{qp}}}{v_{\text{sdp}}}, \frac{v_{\text{qp}}(\hat{w})}{v_{\text{qp}}} \right\} \qquad (4.42)$$

for this single-group multicast scenario can be upper bounded by a constant factor which is proportional to m and m^2 for $\mathbb{H} = \mathbb{C}$ and $\mathbb{H} = \mathbb{R}$, respectively. It will also be shown the same order of approximation bounds hold even when one of the $\boldsymbol{H}_k$ is indefinite. Numerical results will be presented to demonstrate the empirical approximation performance of the SDP relaxation method to this single-group multicast problem (4.38).

Worst-case approximation bounds

Let us present an analysis for the approximation performance of the SDP relaxation method. The above Gaussian randomization procedure will be used in the analysis. First, it is noticed from Box 1 that the random vectors $\boldsymbol{\xi}_\ell$ generated by the Gaussian distribution $N(0, \boldsymbol{W}^\star)$ satisfy all the constraints of (4.38) in expectation, that is

$$E\{\boldsymbol{\xi}_\ell^\dagger \boldsymbol{H}_k \boldsymbol{\xi}_\ell\} = \text{Tr}(\boldsymbol{H}_k \boldsymbol{W}^\star) \ge 1, \quad \forall\, k = 1, \ldots, m,$$

$$E\{\|\boldsymbol{\xi}_\ell\|^2\} = \text{Tr}(\boldsymbol{W}^\star). \qquad (4.43)$$

This observation motivates a question whether it is possible to generate a random vector $\boldsymbol{\xi}_\ell$ such that

$$\boldsymbol{\xi}_\ell^\dagger \boldsymbol{H}_k \boldsymbol{\xi}_\ell \ge \eta, \quad \forall\, k = 1, \ldots, m, \quad \text{and} \quad \|\boldsymbol{\xi}_\ell\|^2 \le \mu \text{Tr}(\boldsymbol{W}^\star) \qquad (4.44)$$

for some $\eta, \mu > 0$. If (4.44) is true, then it implies that the associated $w(\xi_\ell)$ in (4.40) is feasible to (4.38) and satisfies

$$
v_{\mathrm{qp}} \leq \|\hat{w}\|^2 \leq \|w(\xi_\ell)\|^2 = \frac{\|\xi_\ell\|^2}{\min_k \xi_\ell^T H_k \xi_\ell} \leq \frac{\mu \operatorname{Tr}(W^\star)}{\eta} = \frac{\mu}{\eta} v_{\mathrm{sdp}}, \tag{4.45}
$$

where the last equality uses $\operatorname{Tr}(W^\star) = v_{\mathrm{sdp}}$. Therefore, we see that the relaxation ratio of $v_{\mathrm{qp}}/v_{\mathrm{sdp}}$ is bounded as

$$
\frac{v_{\mathrm{qp}}}{v_{\mathrm{sdp}}} \leq \frac{\mu}{\eta}. \tag{4.46}
$$

Indeed, it has been shown in [32] that η, μ and $c > 0$ exist such that

$$
\operatorname{Prob}\left(\min_{1 \leq k \leq m} \xi_\ell^\dagger H_k \xi_\ell \geq \eta, \ \|\xi_\ell\|^2 \leq \mu \operatorname{Tr}(W^*) \right) > c, \tag{4.47}
$$

which indicates that it is possible (with probability at least c) to have (4.44) hold true for some η and $\mu > 0$. Specifically, it has been shown [32] that when $\mathbb{H} = \mathbb{R}$, (4.47) holds for $\mu = 3$, $\eta = \pi/(9m^2)$, and $c = 0.0758$, and when $\mathbb{H} = \mathbb{C}$, (4.47) holds for $\mu = 2$, $\eta = 1/(4m)$, and $c = 1/6$. Plugging these numbers into (4.46), we see that the ratio between the objective values of the original problem (4.38) and the relaxation problem (4.39) is bounded as

$$
\frac{v_{\mathrm{qp}}}{v_{\mathrm{sdp}}} \leq \frac{27m^2}{\pi} \quad \text{and} \quad \frac{v_{\mathrm{qp}}}{v_{\mathrm{sdp}}} \leq 8m, \tag{4.48}
$$

for $\mathbb{H} = \mathbb{R}$ and $\mathbb{H} = \mathbb{C}$, respectively.

On the other hand, since there are L independent ξ_ℓ, $\ell = 1, \ldots, L$, it follows from (4.47) that we can obtain, with probability at least $1 - (1 - c)^L$, a ξ_ℓ satisfying (4.44) and thus (4.45). As a result, it is with at least probability $1 - (1 - c)^L$ that the approximate solution $\hat{w}$ in (4.41) achieves the following quality bounds:

$$
\frac{v_{\mathrm{qp}}(\hat{w})}{v_{\mathrm{qp}}} \leq \frac{27m^2}{\pi} \quad \text{and} \quad \frac{v_{\mathrm{qp}}(\hat{w})}{v_{\mathrm{qp}}} \leq 8m, \tag{4.49}
$$

for the real and complex cases, respectively. If $L = 100$, we see that (4.49) holds with probability almost equal to one (at least 0.999), demonstrating that the guaranteed performance bounds can be achieved with high probability. From (4.48) and (4.49), it can be observed that the worst-case performance of SDP relaxation deteriorates *linearly* with the number of quadratic constraints for complex case problems, in contrast to the *quadratic* rate of deterioration in the real case. Thus, the SDP relaxation can yield better performance in the complex case. This is in the same spirit as the results in [33], which showed that the quality of SDP relaxation improves by a constant factor for certain quadratic *maximization* problems when the underlying vector space is changed from $\mathbb{R}$ to $\mathbb{C}$. For a more unified treatment of the analysis discussed in this subsection, readers are referred to [34].

The worst-case performance bounds in (4.48) are tight up to a constant scalar, as demonstrated in the following example for $\mathbb{H} = \mathbb{R}$.

Example 4.8 For any $m \geq 2$ and $K \geq 2$, consider a special instance of (4.38) whereby $\boldsymbol{H}_k = \boldsymbol{h}_k \boldsymbol{h}_k^T$ and

$$
\boldsymbol{h}_k = \left(\cos\left(\frac{k\pi}{m} \right), \sin\left(\frac{k\pi}{m} \right), 0, \ldots, 0 \right)^T, \qquad k = 1, \ldots, m.
$$

Let $\boldsymbol{w}^\star = (w_1^\star, w_2^\star, \ldots, w_K^\star)^T \in \mathbb{R}^K$ be an optimal solution of (4.38) corresponding to the above choice of steering vectors $\boldsymbol{h}_k$. We can write

$$
(w_1^\star, w_2^\star) = \rho(\cos\theta, \sin\theta), \quad \text{for some } \theta \in [0, 2\pi).
$$

Since $\{ k\pi/m, \ k = 1, \ldots, m \}$ is uniformly spaced on $[0, \pi)$, there must exist an integer k such that

$$
\text{either} \quad \left| \theta - \frac{k\pi}{m} - \frac{\pi}{2} \right| \leq \frac{\pi}{2m}, \quad \text{or} \quad \left| \theta - \frac{k\pi}{m} + \frac{\pi}{2} \right| \leq \frac{\pi}{2m}.
$$

For simplicity, we assume the first case. (The second case can be treated similarly.) Since the last $(K - 2)$ entries of $\boldsymbol{h}_k$ are zero, it is readily checked that

$$
|\boldsymbol{h}_k^T \boldsymbol{w}^\star| = \rho \left| \cos\left(\theta - \frac{k\pi}{m} \right) \right| = \rho \left| \sin\left(\theta - \frac{k\pi}{m} - \frac{\pi}{2} \right) \right| \leq \rho \left| \sin\left(\frac{\pi}{2m} \right) \right| \leq \frac{\rho\pi}{2m}.
$$

Since $\boldsymbol{w}^\star$ satisfies the constraint $|\boldsymbol{h}_k^T \boldsymbol{w}^\star| \geq 1$, it follows that

$$
\|\boldsymbol{w}^\star\| \geq \rho \geq \frac{2m|\boldsymbol{h}_k^T \boldsymbol{w}^\star|}{\pi} \geq \frac{2m}{\pi},
$$

implying

$$
\upsilon_{\mathrm{qp}} = \|\boldsymbol{w}^\star\|^2 \geq \frac{4m^2}{\pi^2}.
$$

On the other hand, the positive semidefinite matrix

$$
\boldsymbol{W}^\star = \mathrm{Diag}\{1, 1, 0, \ldots, 0\}
$$

is feasible for the SDP relaxation (4.39), and it has an objective value of $\mathrm{Tr}(\boldsymbol{W}^\star) = 2$. Thus, for this problem instance, we have

$$
\upsilon_{\mathrm{qp}} \geq \frac{2m^2}{\pi^2} \upsilon_{\mathrm{sdp}}.
$$

The preceding example indicates that the SDP relaxation (4.39) can be weak if the number of quadratic constraints is large, especially when the steering vectors $\boldsymbol{h}_k$ are in a certain sense "uniformly distributed" in space. The same is true for the complex case. Specifically, by considering a problem instance of (4.38) for $m \geq 2$ and $K \geq 2$ with $\boldsymbol{H}_k = \boldsymbol{h}_k(\boldsymbol{h}_k)^H$ and

$$\boldsymbol{h}_k = \left(\cos\frac{i\pi}{M}, \sin\frac{i\pi}{M} e^{\frac{j2\ell\pi}{M}}, 0, \ldots, 0 \right)^T, \tag{4.50}$$

$k = iM - M + \ell$, $i, \ell = 1, \ldots, M$, in which $j = \sqrt{-1}$ and $M = \lceil\sqrt{m}\rceil$, one can show [32] that

$$v_{\text{qp}} \geq \frac{2m}{\pi^2(3 + \pi/2)^2} \, v_{\text{sdp}},$$

implying that the approximation bound for the complex case is also tight up to a constant.

We have assumed so far that all the homogeneous quadratic constraints in the homogeneous QP (4.38) are concave (all the matrices $\boldsymbol{H}_k$ are p.s.d.). It would be interesting to know whether the analyzed performance bounds still hold when some of the $\boldsymbol{H}_k$ are indefinite. To see this, let us recall the simple QP in Example 4.4:

$$v_{\text{qp}} = \min_{x,y} \; x^2 + y^2 \tag{4.51a}$$

$$\text{s.t.} \; y^2 \geq 1, \tag{4.51b}$$

$$[x \; y] \begin{bmatrix} 1 & -M/2 \\ -M/2 & 0 \end{bmatrix} \begin{bmatrix} x \\ y \end{bmatrix} \geq 1, \tag{4.51c}$$

$$[x \; y] \begin{bmatrix} 1 & M/2 \\ M/2 & 0 \end{bmatrix} \begin{bmatrix} x \\ y \end{bmatrix} \geq 1, \tag{4.51d}$$

in which the two matrices associated with (4.51c) and (4.51d) are indefinite. As shown in (4.4), the ratio of $v_{\text{qp}}/v_{\text{sdp}}$ can be arbitrarily large with M and thus is not bounded. Indeed, for the homogeneous QP (4.38), there is, in general, no data-independent upper bound on $v_{\text{qp}}/v_{\text{sdp}}$ when there are two or more indefinite $\boldsymbol{H}_k$. However, for (4.38) with only one indefinite $\boldsymbol{H}_k$, the same order of approximation bounds in (4.48) can be preserved. Interested readers are referred to [35] for a detailed description. The example below shows that in the case of more than two indefinite quadratic constraints present, the SDP relaxation may not admit any *finite* quality bound.

Example 4.9   Consider the following QCQP:

$$\min \; x_4^2$$
$$\text{s.t.} \; x_1 x_2 + x_3^2 + x_4^2 \geq 1,$$
$$-x_1 x_2 + x_3^2 + x_4^2 \geq 1,$$
$$\tfrac{1}{2} x_1^2 - x_3^2 \geq 1,$$
$$\tfrac{1}{2} x_2^2 - x_3^2 \geq 1,$$

where $x_1, x_2, x_3, x_4 \in \mathbb{R}$. Note that all the four quadratic constraints are indefinite. The first two constraints are equivalent to $|x_1 x_2| \leq x_3^2 + x_4^2 - 1$, meanwhile the last two constraints imply $|x_1 x_2| \geq 2(x_3^2 + 1)$. Combining these two inequalities yields

$$x_3^2 + x_4^2 - 1 \geq 2(x_3^2 + 1),$$

which further implies $x_4^2 \geq 3$ (thus $v_{\mathrm{qp}} \geq 3$). However, it is easy to check that the matrix

$$\begin{bmatrix} 4 & 0 & 0 & 0 \\ 0 & 4 & 0 & 0 \\ 0 & 0 & 1 & 0 \\ 0 & 0 & 0 & 0 \end{bmatrix}$$

is feasible for the corresponding SDP relaxation problem and attains an objective value of 0. Thus, it must be optimal and $v_{\mathrm{sdp}} = 0$. Therefore, we have $v_{\mathrm{qp}}/v_{\mathrm{sdp}} = \infty$ in this case.

Now let us summarize the primary results discussed in the subsection:

THEOREM 4.1 *For the QCQP (4.38) and its SDP relaxation (4.39), let $\hat{w}$ be the approximate solution obtained from the Gaussian randomization procedure in Box 1. Then the worse-case approximation ratio*

$$\gamma_{\mathrm{sdr}} := \max_{H_k, K} \max \left\{ \frac{v_{\mathrm{qp}}}{v_{\mathrm{sdp}}}, \frac{v_{\mathrm{qp}}(\hat{w})}{v_{\mathrm{qp}}} \right\} \tag{4.52}$$

is upper bound by $(27m^2)/\pi$ for $\mathbb{H} = \mathbb{R}$ and by $8m$ for $\mathbb{H} = \mathbb{C}$, with high probability. The same orders of bounds also hold true if one of the H_k is indefinite.

Numerical results

While theoretical worst-case analysis is very useful, empirical analysis of the ratio

$$\gamma = \max \left\{ \frac{v_{\mathrm{qp}}}{v_{\mathrm{sdp}}}, \frac{v_{\mathrm{qp}}(\hat{w})}{v_{\mathrm{qp}}} \right\} \tag{4.53}$$

through simulations with randomly generated channel vectors h_k is often equally important. Since v_{qp} is not practically available, we use the empirical ratio of $v_{\mathrm{qp}}(\hat{w})/v_{\mathrm{sdp}}$ ($\geq \gamma$) to estimate γ. The single-group multicast problem (4.38) for $\mathbb{H} = \mathbb{R}$ and $\mathbb{H} = \mathbb{C}$ was considered, with $m = 8$ and $K = 4$. The coefficients of each channel vector h_k were *independent and identically distributed* (IID) (real-valued or complex-valued) Gaussian distributed with zero mean and unit variance. The Gaussian randomized procedure in Box 1 was used with $L = 30Km$. Figure 4.2 shows the simulation results for the real-valued (4.38) and (4.39) from 300 independent trials; whereas Figure 4.3 presents the results for the complex-valued problems.

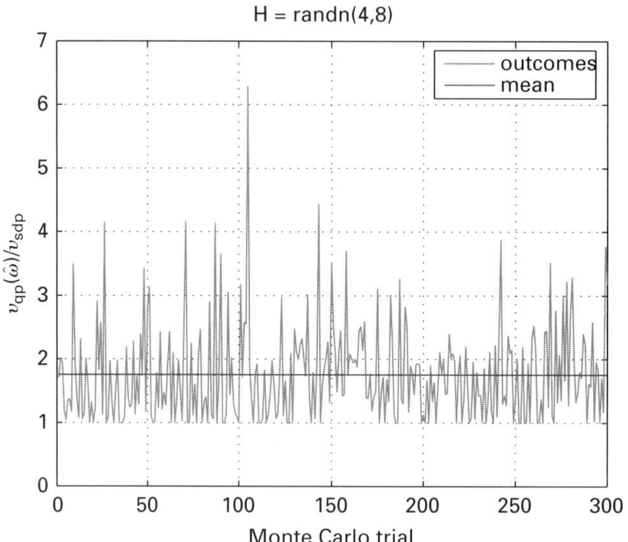

Figure 4.2 Empirical ratio of $v_{qp}(\hat{w})/v_{sdp}$ for $m = 8$ and $K = 4$, 300 realizations of real Gaussian IID channel vector entries.

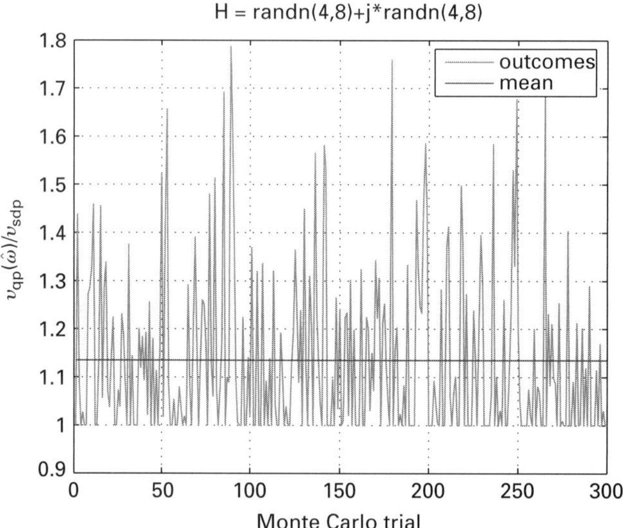

Figure 4.3 Empirical ratio of $v_{qp}(\hat{w})/v_{sdp}$ for $m = 8$ and $K = 4$, 300 realizations of complex Gaussian IID channel vector entries.

It can be observed from these two figures that, on average, the approximation ratio is less than 2 (the mean value), which demonstrates the superiority of the SDP relaxation technique in practical situations. Both the mean and the maximum of the upper bound $v_{qp}(\hat{w})/v_{sdp}$ are lower in the complex case. The simulation results indicate that SDP approximation is better in the complex case, not only in the worst case, but also on average, complementing our theoretical worst-case analysis for the class of problems considered herein.

For more simulation results, readers are referred to [15]. Two insights from the simulation results are summarized as follows:

- For moderate values of m, K (e.g., $m = 24$, $K = 8$), and IID, complex-valued, circular Gaussian (IID Rayleigh) entries of the channel vectors $\{\boldsymbol{h}_k\}$, the average value of γ is under 3 – much lower than the worst-case value predicted by our analysis.
- In experiments with measured VDSL channel data, for which the channel vectors follow a correlated log-normal distribution, $\gamma = 1$ in over 50% of instances.

Next, let us present some simulation results about the approximation performance of SDP relaxation to the homogeneous QCQP (4.38) with indefinite $\boldsymbol{H}_k$. We considered (4.38) for $\mathbb{H} = \mathbb{R}$, $K = 10$, and $m = 5, 10, 15, \ldots, 100$. For full rank $\boldsymbol{H}_k$, we set

$$\boldsymbol{H}_k = \boldsymbol{U}^H \boldsymbol{D} \boldsymbol{U} \tag{4.54}$$

where $\boldsymbol{D} = \mathrm{Diag}\{\lambda_1, \ldots, \lambda_K\}$ in which $\lambda_i > 0$ were randomly generated, and $\boldsymbol{U} \in \mathbb{C}^{K \times K}$ is a unitary matrix obtained by QR factorization of a randomly generated $K \times K$ real matrix. For rank-1 $\boldsymbol{H}_k$, all $\lambda_i = 0$ for $i \neq 1$ while $\lambda_1 > 0$ was randomly generated. We respectively considered four scenarios for each m: (a) generate 1000 random $\boldsymbol{H}_k$ such that only one of them is indefinite and all the others are positive definite; (b) generate 1000 random $\boldsymbol{H}_k$ such that only 10% of them are indefinite and all the others are positive definite; (c) generate 1000 random $\boldsymbol{H}_k$ such that only one of them is indefinite and all the others are rank-1 and positive definite; (d) generate 1000 random $\boldsymbol{H}_k$ such that only 10% of them are indefinite and all the others are rank-1 and positive definite. Figure 4.4 presents the simulation results. As seen from Figure 4.4(a) and (c) we have quality bounds of $O(m^2)$ for the worst case, which is consistent with Theorem 4.1; while for the cases of (b) and (d) there is no worst-case theoretical bound. On the other hand, one can observe that the empirical ratios are very small for case (a) and moderate for case (c), and are indeed large for cases (b) and (d).

4.4 SDP relaxation for maximization homogeneous QCQPs

In this section, we turn our attention to a class of maximization homogeneous QCQPs. Specifically, we will investigate the performance of SDP relaxation for the following class of problems:

$$v_{\mathrm{qp}} = \max_{\boldsymbol{w} \in \mathbb{H}^K} \ \|\boldsymbol{w}\|^2 \tag{4.55a}$$

$$\text{s.t.} \ \boldsymbol{w}^\dagger \boldsymbol{H}_k \boldsymbol{w} \leq 1, \quad k = 1, \ldots, m, \tag{4.55b}$$

where $\boldsymbol{H}_k \succeq \boldsymbol{0}$ for all k. Opposed to the downlink transmit beamforming problem described in Section 4.3, the maximization QP (4.55) is motivated by the (uplink) receiver

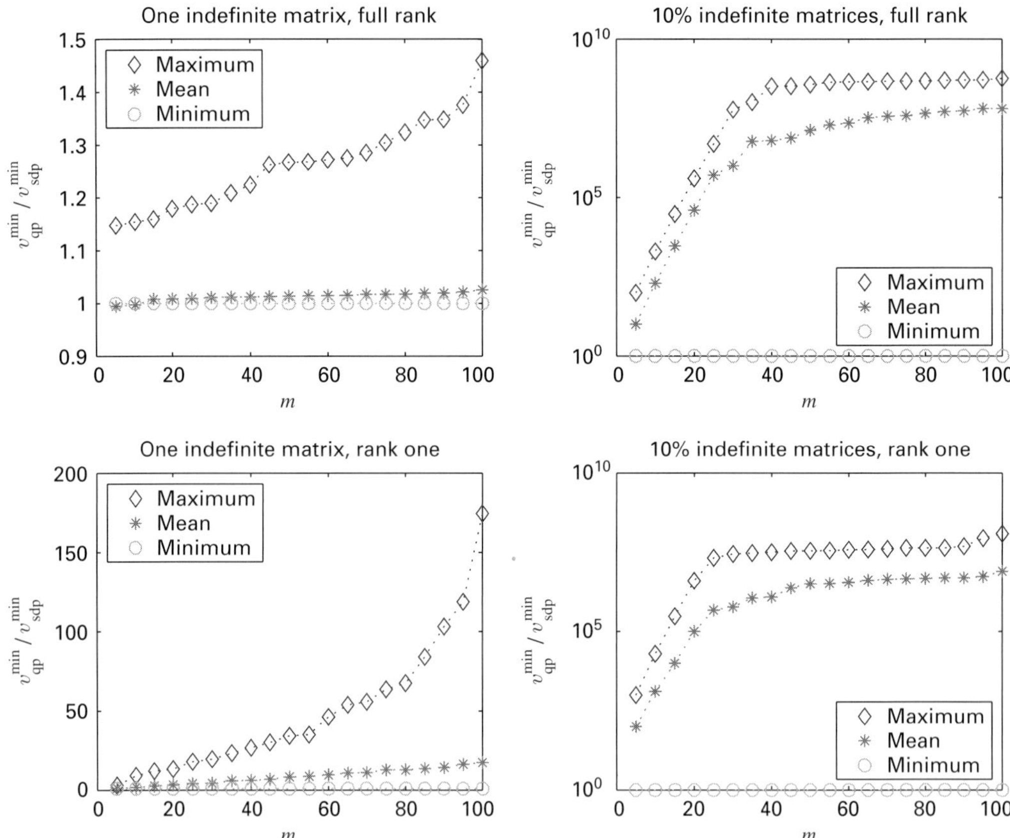

Figure 4.4   Empirical ratio of $v_{qp}(\hat{w})/v_{sdp}$ for problem (4.38) and (4.39) with $\mathbb{H} = \mathbb{R}$, $K = 10$, and $m = 5, 10, \ldots, 100$.

intercept beamforming problem in which the base station, equipped with an antenna array, is capable of suppressing signals impinging from irrelevant or hostile emitters, for example, jammers, and meanwhile achieving as high a gain as possible for desired signals. The channel vectors (spatial signatures, or "footprints") of jammers have been previously estimated, and known to the base station. The jammer suppression capability is captured in the constraints of (4.55b), and covers the case where a jammer employs more than one transmit antenna (e.g., $\boldsymbol{H}_k = \sum_{i=1}^{N} \boldsymbol{h}_{k,i}\boldsymbol{h}_{k,i}^H$ where N denotes the number of transmit antennas in each jammer and $\boldsymbol{h}_{k,i} \in \mathbb{C}^K$ represents the channel vector between the base station and the ith antenna of jammer k).

The maximization of the objective $\|\boldsymbol{w}\|^2$ can be motivated as follows. In intercept applications, the channel vector of the emitter of interest, $\boldsymbol{h}$, is a-priori unknown, and is naturally modeled as random. A natural optimization objective is then the average beamformer output power, measured by $E\{|\boldsymbol{h}^H\boldsymbol{w}|^2\}$. Under the assumption that the entries of $\boldsymbol{h}$ are uncorrelated and have equal average power, it follows that $E\{|\boldsymbol{h}^H\boldsymbol{w}|^2\}$ is proportional to $\|\boldsymbol{w}\|^2$, which is often referred to as the beamformer's "white noise gain."

Like its minimization counterpart in (4.38), problem (4.55) is NP-hard in general, and the polynomial-time SDP relaxation approximation technique can be applied as well. The SDP relaxation of (4.55) is given by

$$v_{\text{sdp}} = \max_{\boldsymbol{W} \in \mathbb{H}^{K \times K}} \ \text{Tr}(\boldsymbol{W}) \tag{4.56a}$$

$$\text{s.t.} \ \ \text{Tr}(\boldsymbol{H}_k \boldsymbol{W}) \leq 1, \quad k = 1, \dots, m, \tag{4.56b}$$

$$\boldsymbol{W} \succeq \boldsymbol{0}. \tag{4.56c}$$

Since the SDP relaxation (4.56) is not tight to (4.55) in general, the Gaussian sampling idea in Section 4.2.2 can be applied to obtain a rank-1 feasible approximate solution. In particular, the following Gaussian randomization procedure can be used:

Box 2. Gaussian randomization procedure for (4.55)

Let $L > 0$ be an integer.

1. For $\ell = 1, \dots, L$, generate a random vector $\boldsymbol{\xi}_\ell \in \mathbb{H}^K$ from the complex-valued normal distribution $N(0, \boldsymbol{W}^\star)$, and let

$$\boldsymbol{w}(\boldsymbol{\xi}_\ell) = \boldsymbol{\xi}_\ell / \max_{1 \leq k \leq m} \sqrt{\boldsymbol{\xi}_\ell^\dagger \boldsymbol{H}_k \boldsymbol{\xi}_\ell}. \tag{4.57}$$

2. Let

$$\hat{\boldsymbol{w}} = \arg \max_{\boldsymbol{w}(\boldsymbol{\xi}_\ell), \ell = 1, \dots, L} \ \|\boldsymbol{w}(\boldsymbol{\xi}_\ell)\|^2 \tag{4.58}$$

be the approximate solution of (4.55).

Because (4.55) is a maximization problem, we have $v_{\text{sdp}} \geq v_{\text{qp}}$ and $v_{\text{qp}} \geq v_{\text{qp}}(\hat{\boldsymbol{w}}) := \|\hat{\boldsymbol{w}}\|^2$. The worst-case approximation ratio of the maximization QP then is defined as

$$\gamma_{\text{sdr}} := \max_{\boldsymbol{H}_k, K} \max \left\{ \frac{v_{\text{sdp}}}{v_{\text{qp}}}, \frac{v_{\text{qp}}}{v_{\text{qp}}(\hat{\boldsymbol{w}})} \right\}. \tag{4.59}$$

Next let us show that the SDP relaxation method can provide a data-independent upper bound for γ_{sdr}.

4.4.1 Worst-case approximation bounds

The ingredients used here are very similar to those in Section 1. First, we would like to show that through the Gaussian randomization procedure, it is possible to have a random vector $\boldsymbol{\xi}_\ell$ such that

$$\max_{1 \leq k \leq m} \boldsymbol{\xi}_\ell^H \boldsymbol{H}_k \boldsymbol{\xi}_\ell \leq \eta, \quad \text{and} \quad \|\boldsymbol{\xi}_\ell\|^2 \geq \mu \text{Tr}(\boldsymbol{W}^\star) \tag{4.60}$$

for some $\eta, \mu > 0$. In that case, the associated $w(\xi_\ell)$ in (4.57) satisfies

$$\frac{\mu}{\eta}v_{\mathrm{qp}} \leq \frac{\mu}{\eta}v_{\mathrm{sdp}} = \mathrm{Tr}(W^\star)\frac{\mu}{\eta} \leq \|w(\xi_\ell)\|^2 \leq \|\hat{w}\|^2 \leq v_{\mathrm{qp}},$$

which leads to a performance upper bound of

$$\max\left\{\frac{v_{\mathrm{sdp}}}{v_{\mathrm{qp}}}, \frac{v_{\mathrm{qp}}}{\|\hat{w}\|^2}\right\} \leq \frac{\eta}{\mu}. \tag{4.61}$$

In [32], it has been shown that the equation

$$\mathrm{Prob}\left(\max_{1 \leq k \leq m} \xi_\ell^\dagger H_k \xi_\ell \leq \eta, \ \|\xi_\ell\|^2 \geq \mu\mathrm{Tr}(W^\star)\right) > c \tag{4.62}$$

holds for $\mu = 1/4$, $\eta = \ln(100M)$, and $c = 0.00898$ when $\mathbb{H} = \mathbb{C}$, and for $\mu = 0.01$, $\eta = 2\ln(50M)$, and $c = 0.0022$ when $\mathbb{H} = \mathbb{R}$, where $M := \sum_{k=1}^m \min\{\mathrm{rank}(H_k), \sqrt{m}\}$. This implies that for either a real or complex case, it is with high probability that $\gamma_{\mathrm{sdr}} \leq O(\ln M)$. It should be pointed out that the same order of approximation bounds can be established even if one of the H_k in (4.55) is indefinite [35], similar to its minimization counterpart in (4.38). However, if there is more than one indefinite H_k, no data-independent, worst-case approximation bound exists, as the following example demonstrates:

Example 4.10   Consider the following QP in the maximization form:

$$\max \ x_1^2 + \frac{1}{M}x_2^2,$$

$$\text{s.t. } Mx_1x_2 + x_2^2 \leq 1,$$

$$- Mx_1x_2 + x_2^2 \leq 1,$$

$$M(x_1^2 - x_2^2) \leq 1,$$

where $M > 0$ is an arbitrarily large positive constant. Note that the first and the third constraints are not convex. Its SDP relaxation is

$$\max \ X_{11} + \frac{1}{M}X_{22}$$

$$\text{s.t. } MX_{12} + X_{22} \leq 1, \ -MX_{12} + X_{22} \leq 1, \ M(X_{11} - X_{22}) \leq 1,$$

$$\begin{bmatrix} X_{11} & X_{12} \\ X_{21} & X_{22} \end{bmatrix} \succeq 0.$$

For this quadratic program, the first two constraints imply that $|x_1 x_2| \leq \frac{1-x_2^2}{M} \leq \frac{1}{M}$ and so $x_1^2 \leq \frac{1}{M^2 x_2^2}$. The third inequality assures that $x_1^2 \leq \frac{1}{M} + x_2^2$. Therefore, $x_1^2 \leq$

$\min \left\{ \frac{1}{M^2 x_2^2}, \frac{1}{M} + x_2^2 \right\} \leq \frac{\sqrt{5}+1}{2M} \approx \frac{1.618}{M}$. Moreover, $x_2^2 \leq 1$, and so $v_{\text{qp}} \leq \frac{2.618}{M}$. Notice that the 2×2 identity matrix is a feasible solution for the SDP relaxation. Moreover, it achieves an objective value of $1 + \frac{1}{M} > 1$. Therefore, for this example, the approximation ratio is $\frac{v_{\text{sdp}}}{v_{\text{qp}}} \geq \frac{M}{2.618} \approx 0.382M$, which can be arbitrarily large, depending on the size of M.

We summarize the results presented in this subsection in the following theorem:

THEOREM 4.2 *Consider the maximization QP (4.55) and its SDP relaxation (4.56). Let $\hat{w}$ be the approximate solution obtained from the Gaussian randomization procedure in Box 2. Then for either $\mathbb{H} = \mathbb{R}$ or $\mathbb{H} = \mathbb{C}$, it is with high probability that*

$$\gamma_{\text{sdr}} = \max_{\boldsymbol{H}_k, K} \max \left\{ \frac{v_{\text{sdp}}}{v_{\text{qp}}}, \frac{v_{\text{qp}}}{v_{\text{qp}}(\hat{\boldsymbol{w}})} \right\} \leq O(\ln M), \tag{4.63}$$

where $M = \sum_{k=1}^{m} \min\{\text{rank}(\boldsymbol{H}_k), \sqrt{m}\}$. The same orders of bounds also hold true if one of the $\boldsymbol{H}_k s$ is indefinite.

In a related work, Nemirovski *et al.* [36] proved the same order of approximation bounds when the objective function is an indefinite quadratic function, but all the constraints are convex quadratic. That is, for the following QP:

$$v_{\text{qp}} = \max_{\boldsymbol{w} \in \mathbb{H}^K} \boldsymbol{w}^\dagger \boldsymbol{C} \boldsymbol{w} \tag{4.64a}$$

$$\text{s.t.} \quad \boldsymbol{w}^\dagger \boldsymbol{H}_k \boldsymbol{w} \leq 1, \quad k = 1, \ldots, m, \tag{4.64b}$$

where $\boldsymbol{C} \in \mathbb{H}^{K \times K}$ is Hermitian and can be indefinite, while $\boldsymbol{H}_k s$ are positive semidefinite. Suppose that v_{sdp} is the optimum value of its SDP relaxation. Then

$$\frac{v_{\text{sdp}}}{v_{\text{qp}}} \leq O(\ln M). \tag{4.65}$$

An example is also provided in [36] which shows that the bounds in (4.63) and (4.65) are tight up to a constant factor.

If the matrices $\boldsymbol{C}$ and $\boldsymbol{H}_k$ possess some special structures, the associated upper bound for (4.64) can be further tightened. Readers are referred to [4, 37] for further details. Here we summarize some of the important results.

- For $\mathbb{H} = \mathbb{R}$, if $K = m$, $\boldsymbol{H}_k = \boldsymbol{e}_k \boldsymbol{e}_k^T$ for all $k = 1, \ldots, m$, where $\boldsymbol{e}_k \in \mathbb{R}^K$ is the kth unit vector, and $\boldsymbol{C}$ is a p.s.d. matrix with non-negative diagonal entries, then

$$\frac{v_{\text{sdp}}}{v_{\text{qp}}} \leq 1.13822 \ldots$$

Note that by letting $\boldsymbol{C} = -\boldsymbol{W}$ then this problem instance corresponds to the MAXCUT problem discussed in Example 4.2.

- For $\mathbb{H} = \mathbb{R}$, if $K = m$, $\boldsymbol{H}_k = \boldsymbol{e}_k\boldsymbol{e}_k^T$ for all $k = 1, \ldots, m$, and $\boldsymbol{C}$ is an arbitrary p.s.d. matrix, then

$$\frac{v_{\text{sdp}}}{v_{\text{qp}}} \leq 1.57 \ldots$$

One of the applications of this problem instance is the blind *maximum-likelihood* (ML) detection problem of orthogonal space-time block codes [2] in MIMO communications, in which the SDP relaxation method has been shown to be very effective in obtaining good approximate ML solutions.

4.4.2   Numerical results

We now present some numerical results to illustrate the empirical approximation performance of SDP relaxation to the maximization QCQP (4.55) with indefinite $\boldsymbol{H}_k$. We considered problem (4.55) and (4.56) for $\mathbb{H} = \mathbb{R}$, $K = 10$, and $m = 5, 10, 15, \ldots, 100$.

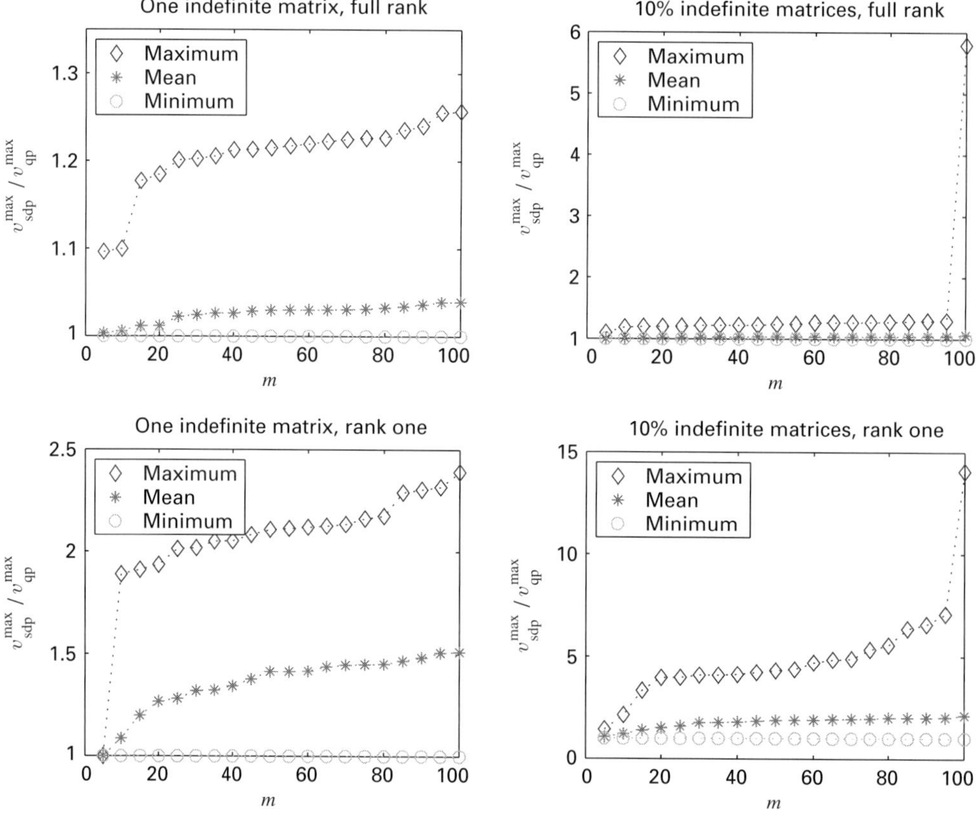

Figure 4.5   Empirical ratio of $v_{\text{sdp}}/v_{\text{qp}}(\hat{\boldsymbol{w}})$ for problems (4.55) and (4.56) with $\mathbb{H} = \mathbb{R}$, $K = 10$, and $m = 5, 10, \ldots, 100$.

The matrices $\boldsymbol{H}_k$ were generated in the same way as those in Section 4.3.5. The four scenarios: (a) generate 1000 random $\boldsymbol{H}_k$ such that only one of them is indefinite and all the others are positive definite; (b) generate 1000 random $\boldsymbol{H}_k$ such that only 10% of them are indefinite and all the others are positive definite; (c) generate 1000 random $\boldsymbol{H}_k$ such that only one of them is indefinite and all the others are rank-1 and positive definite; (d) generate 1000 random $\boldsymbol{H}_k$ such that only 10% of them are indefinite and all the others are rank-1 and positive definite, were simulated. The empirical ratio of $v_{\text{sdp}}/v_{\text{qp}}(\hat{\boldsymbol{w}})$ was computed as an estimate of the true raio of $v_{\text{sdp}}/v_{\text{qp}}$.

Figure 4.5 presents the simulation results. As seen, the worst-case approximation ratios for cases (a) and (c) are very small; while they are somewhat larger for cases (b) and (d). Comparing Figure 4.5 with 4.4, it can be observed that in general, the SDP relaxation for the maximization QPs yields a better approximation performance than for the minimization QPs.

4.5 SDP relaxation for fractional QCQPs

Motivated by the recent application in the network beamforming problem, we consider in this section a class of optimization problems in which the objective function exhibits a fractional quadratic form. We show in the first subsection that the SDP relaxation technique can still be applied in this case with a guaranteed data-independent, worst-case approximation performance. In the later subsections, we consider the more complicated problem of maximizing the minimum of a set of fractional quadratic terms within a ball. The latter problem is motivated by the downlink multicast transmit beamforming problem presented in Example 4.7 under the max–min fairness beamforming design criterion. We show that the SDP relaxation provides a data-independent, worst-case approximation performance in this case as well.

4.5.1 SDP relaxation for fractional QCQPs

Let us consider the following nonconvex fractional quadratic optimization problem

$$v_{\text{fqp}} = \max_{\boldsymbol{w}\in\mathbb{H}^K} \frac{\boldsymbol{w}^\dagger \boldsymbol{R}\boldsymbol{w}}{\boldsymbol{w}^\dagger \boldsymbol{Q}\boldsymbol{w} + 1} \tag{4.66a}$$

$$\text{s.t. } \boldsymbol{w}^\dagger \boldsymbol{G}_k \boldsymbol{w} \leq 1, \quad k = 1,\ldots,m, \tag{4.66b}$$

where $\boldsymbol{R}$, $\boldsymbol{Q}$, and $\boldsymbol{G}_k$, $k = 1,\ldots,m$, are Hermitian and positive semidefinite matrices. The problem (4.66) is NP-hard, since it reduces to the NP-hard QCQP (4.64) problem when $\boldsymbol{Q} = 0$. The following example is an application of the fractional QCQP (4.66) in the network beamforming context [20, 21].

Example 4.11 Network beamforming [20,21] Let us consider the network beamforming problem in Figure 4.6 where one source node wants to communicate with a destination

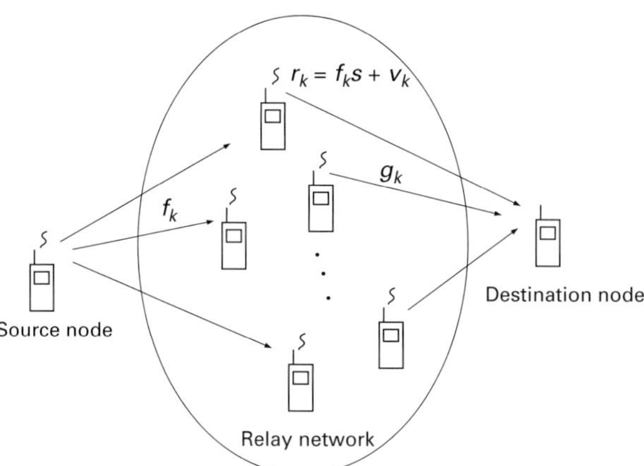

Figure 4.6   System diagram of network beamforming where the source node communicates with the destination node through the collaboration between a set of relay nodes.

node through the collaboration with a set of m relay nodes. The relay nodes collaborate with each other to form a beam at the destination for good reception performance.

As shown in Figure 4.6, suppose that the received signal at relay k is given by

$$r_k = f_k s + v_k, \tag{4.67}$$

where s is the signal transmitted by the source node, f_k is the fading channel coefficient between the source node and the kth relay, and v_k is the white noise at relay k. To form a beam at the destination node, each relay amplifies its received signal r_k by complex weight w_k and forwards it to the destination node. Thus, the received signal at the destination node is given by

$$y = \sum_{k=1}^{m} g_k w_i r_k + w_k = \underbrace{\sum_{k=1}^{m} g_k w_k f_k s}_{\text{desired signal}} + \underbrace{\sum_{k=1}^{m} g_k w_k v_k + w}_{\text{noise terms}},$$

where g_k is the fading channel coefficient between the kth relay and the destination node and w is the noise at the destination node. To calculate the received "signal-to-noise" ratio (SNR) at the destination node, we assume that f_k, g_k, v_k, w are statistically independent, and that v_k, $k = 1, \ldots, m$ are zero mean, with variance equal to σ_v^2 and are uncorrelated to each other. Then the average signal power can be calculated as

$$E\left\{\left|\sum_{k=1}^{m} g_k w_k f_k s\right|^2\right\} = E\left\{\left|\sum_{k=1}^{m} g_k w_k f_k\right|^2\right\} E\{|s|^2\} = w^H R w, \tag{4.68}$$

where we have assumed that $E\{|s|^2\} = 1$, and set $\boldsymbol{w} = [w_1, \ldots, w_m]^T$, $\boldsymbol{h} = [g_1 f_1, \ldots, g_m f_m]^T$, and $\boldsymbol{R} = E\{\boldsymbol{h}\boldsymbol{h}^H\}$. The average noise power is given by

$$E\left\{\left|\sum_{k=1}^{m} g_k w_k v_k + w\right|^2\right\} = \sigma_v^2 E\left\{\sum_{k=1}^{m} |g_k w_k|^2\right\} + \sigma_w^2$$

$$= \boldsymbol{w}^H \boldsymbol{Q} \boldsymbol{w} + \sigma_w^2, \tag{4.69}$$

where $\boldsymbol{Q} = \sigma_v^2 \mathrm{Diag}\{|g_1|^2, \ldots, |g_m|^2\}$. Hence the SNR at the destination node is given by the fractional quadratic form

$$\mathrm{SNR} = \frac{\boldsymbol{w}^H \boldsymbol{R} \boldsymbol{w}}{\boldsymbol{w}^H \boldsymbol{Q} \boldsymbol{w} + \sigma_w^2}.$$

Suppose that there is a constraint on the power of w_k at each relay such that $|w_k|^2 \leq P_k$. Then one of the possible design criterion of $\boldsymbol{w}$ is to maximize the SNR at the destination node subject to the power constraints:

$$\max_{\boldsymbol{w} \in \mathbb{C}^m} \frac{\boldsymbol{w}^H \boldsymbol{R} \boldsymbol{w}}{\boldsymbol{w}^H \boldsymbol{Q} \boldsymbol{w} + \sigma_w^2} \quad \text{s.t.} \quad |w_i|^2 \leq P_k, \quad k = 1, \ldots, m, \tag{4.70}$$

which is exactly a fractional QCQP defined in (4.66).

By applying the SDP relaxation procedure to this fractional QCQP (4.66), we obtain the following problem:

$$v_{\mathrm{sdp}} = \max_{\boldsymbol{W} \in \mathbb{H}^{K \times K}} \frac{\mathrm{Tr}(\boldsymbol{R}\boldsymbol{W})}{\mathrm{Tr}(\boldsymbol{Q}\boldsymbol{W}) + 1} \tag{4.71a}$$

$$\text{s.t.} \quad \mathrm{Tr}(\boldsymbol{G}_k \boldsymbol{W}) \leq 1, \quad k = 1, \ldots, m, \tag{4.71b}$$

$$\boldsymbol{W} \succeq \boldsymbol{0}. \tag{4.71c}$$

Note that, unlike (4.56), the objective function of the relaxed problem (4.71) is fractional linear and therefore quasi-convex. To solve this quasi-convex problem, we can employ the classical bisection method. To illustrate this, let us rewrite (4.71) in the epigraph form by introducing a new variable $t \geq 0$:

$$v_{\mathrm{sdp}} = \max_{\boldsymbol{W}, t \geq 0} t \tag{4.72a}$$

$$\text{s.t.} \quad \mathrm{Tr}(\boldsymbol{G}_k \boldsymbol{W}) \leq 1, \quad k = 1, \ldots, m, \tag{4.72b}$$

$$\mathrm{Tr}(\boldsymbol{R}\boldsymbol{W}) \geq t\,\mathrm{Tr}(\boldsymbol{Q}\boldsymbol{W}) + t, \tag{4.72c}$$

$$\boldsymbol{W} \succeq \boldsymbol{0}. \tag{4.72d}$$

Given the value of $t \geq 0$, the bisection method checks the following feasibility problem:

$$\text{Find } \boldsymbol{W} \tag{4.73a}$$

$$\text{s.t. } \text{Tr}(\boldsymbol{G}_k \boldsymbol{W}) \leq 1, \quad k = 1, \ldots, m, \tag{4.73b}$$

$$\text{Tr}(\boldsymbol{R}\boldsymbol{W}) \geq t\text{Tr}(\boldsymbol{Q}\boldsymbol{W}) + t, \tag{4.73c}$$

$$\boldsymbol{W} \succeq \boldsymbol{0}. \tag{4.73d}$$

If (4.73) is feasible, it means that $v_{\text{sdp}} \geq t$, otherwise $v_{\text{sdp}} < t$. Based on this observation, (4.71) can be solved by the following bisection procedure.

Box 3. Bisection algorithm for solving (4.71)

Given an interval $[\ell, u]$ such that $v_{\text{sdp}} \in [\ell, u]$, and a tolerance $\epsilon > 0$.

1. Set $t = \frac{\ell + u}{2}$, and solve problem (4.73).
2. If (4.73) is feasible, then set $\ell = t$; otherwise set $u = t$.
3. If $u - \ell \leq \epsilon$, then terminate; otherwise go to Step 1.

After obtaining the optimum solution of (4.71), denoted by $\boldsymbol{W}^\star$, one may obtain an approximate solution of (4.66) by a Gaussian randomization procedure as follows:

Box 4. Gaussian randomization procedure for (4.66)

Let $L > 0$ be an integer.

1. For $\ell = 1, \ldots, L$, generate a random vector $\boldsymbol{\xi}_\ell \in \mathbb{H}^K$ from the Gaussian distribution $N(0, \boldsymbol{W}^\star)$, and let

$$w(\boldsymbol{\xi}_\ell) = \boldsymbol{\xi}_\ell / \max_{1 \leq k \leq m} \sqrt{\boldsymbol{\xi}_\ell^\dagger \boldsymbol{G}_k \boldsymbol{\xi}_\ell}. \tag{4.74}$$

2. Let

$$\hat{w} = \arg \max_{w(\boldsymbol{\xi}_\ell), \ell = 1, \ldots, L} \frac{(w(\boldsymbol{\xi}_\ell))^\dagger \boldsymbol{R} w(\boldsymbol{\xi}_\ell)}{(w(\boldsymbol{\xi}_\ell))^\dagger \boldsymbol{Q} w(\boldsymbol{\xi}_\ell) + 1} \tag{4.75}$$

be the approximate solution of (4.66).

Next we will show that the SDP relaxation for this fractional QCQP can ensure a data-independent, worst-case approximation upper bound.

Worst-case approximation bounds

Here let us show that the SDP relaxation problem (4.71) can at least provide an $O((\ln m)^{-1})$ approximation to the fractional QCQP (4.66). To prove this, let us consider the following SDP

$$\max_{X \in \mathbb{H}^{K \times K}} \operatorname{Tr}(RX) - v_{\text{sdp}}(\operatorname{Tr}(QX) + 1) \tag{4.76a}$$

$$\text{s.t. } \operatorname{Tr}(G_k X) \le 1, \quad k = 1, \ldots, m, \tag{4.76b}$$

$$X \succeq 0. \tag{4.76c}$$

It is not hard to see that the optimum solution of (4.71), $W^\star$, is feasible to (4.76) and attains its optimum objective value 0. If not, then there is an $X^\star$ such that

$$\frac{\operatorname{Tr}(RX^\star)}{\operatorname{Tr}(QX^\star) + 1} > v_{\text{sdp}},$$

which contradicts the optimality of $W^\star$ to (4.71). Note that (4.76) is the SDP relaxation of the following maximization QCQP:

$$\max_{w \in \mathbb{C}^K} x^H R x - v_{\text{sdp}}(x^H Q x + 1) \tag{4.77a}$$

$$\text{s.t. } x^H G_k x \le 1, \quad k = 1, \ldots, m. \tag{4.77b}$$

According to Theorem 4.2, by drawing the random vector ξ_ℓ from the complex Gaussian distribution $N(0, W^\star)$ there exists (with a positive probability) a ξ_ℓ such that

$$\frac{1}{c}\left(\operatorname{Tr}(RW^\star) - v_{\text{sdp}}\operatorname{Tr}(QW^\star)\right) = \frac{1}{c}v_{\text{sdp}}$$

$$\le (w(\xi_\ell))^H R w(\xi_\ell) - v_{\text{sdp}}(w(\xi_\ell))^H Q w(\xi_\ell), \tag{4.78}$$

where $c = O(\ln m)$. The above inequality (4.78) can be further manipulated as follows:

$$(w(\xi_\ell))^H R w(\xi_\ell) \ge \frac{1}{c}v_{\text{sdp}} + v_{\text{sdp}}(w(\xi_\ell))^H Q w(\xi_\ell)$$

$$\ge \frac{1}{c}v_{\text{sdp}}(1 + (w(\xi_\ell))^H Q w(\xi_\ell)), \tag{4.79}$$

which implies that

$$v_{\text{fqp}} \le v_{\text{sdp}} \le c \cdot \frac{(w(\xi_\ell))^H R w(\xi_\ell)}{1 + (w(\xi_\ell))^H Q w(\xi_\ell)} \le c \cdot \frac{\hat{w}^H R \hat{w}}{1 + \hat{w}^H Q \hat{w}} \le c \cdot v_{\text{fqp}}.$$

Therefore, the worst-case approximation ratio is upper bounded as

$$\gamma_{\text{sdp}} := \max_{R, Q, G_k, K} \max\left\{\frac{v_{\text{sdp}}}{v_{\text{fqp}}}, \frac{v_{\text{fqp}}}{v_{\text{fqp}}(\hat{w})}\right\} \le O(\ln m)$$

with a positive probability.

The above analysis in fact suggests that a more general result is true: the SDP relaxation of the fractional QCQP (4.66) has a data-independent, worst-case approximation bound if that of the maximization form QCQP (4.64) does. Since the SDP relaxation of (4.64) has a worst-case approximation bound of $O(\ln m)$, even when one of the $\boldsymbol{H}_k$ is indefinite (see Section 4.4.1), it follows that the same $O(\ln m)$ bound holds true for the SDP relaxation of the fractional QCQP (4.66) when one of the $\boldsymbol{G}_k$ is indefinite.

4.5.2   SDP relaxation for generalized fractional QCQPs

In this subsection, we consider a more complex fractional QCQP which is in the form of maximizing the minimum of a set of fractional quadratic terms

$$v_{\text{fqp}} = \max_{\boldsymbol{w} \in \mathbb{H}^K} \min_{j=1,\ldots,m} \frac{\boldsymbol{w}^\dagger \boldsymbol{R}_j \boldsymbol{w}}{\boldsymbol{w}^\dagger \boldsymbol{Q}_j \boldsymbol{w} + 1} \tag{4.80a}$$

$$\text{s.t. } \boldsymbol{w}^\dagger \boldsymbol{w} \leq P, \tag{4.80b}$$

where $P > 0$, $\boldsymbol{R}_j \neq \boldsymbol{0}$, $\boldsymbol{R}_j \succeq \boldsymbol{0}$ and $\boldsymbol{Q}_j \neq \boldsymbol{0}$, $\boldsymbol{Q}_j \succeq \boldsymbol{0}$ for $j = 1, \ldots, m$. This generalized fractional QCQP (4.80) is nonconvex and NP-hard in general. To motivate the fractional QCQP problem, let us consider the max–min–fair transmit beamforming problem in the multiple-group multicast scenario discussed in Example 4.7.

Example 4.12 Max–min–fair transmit beamforming   Recall from (4.22) that the SINR at the kth receiver in group $\mathcal{G}_i$ is given by

$$\text{SINR}_k = \frac{\boldsymbol{w}_i^H \boldsymbol{H}_k \boldsymbol{w}_i}{\sum_{j \neq i} \boldsymbol{w}_j^H \boldsymbol{H}_k \boldsymbol{w}_j + \sigma_k^2} \quad \forall\, k \in \mathcal{G}_i, \quad i = 1, \ldots, n. \tag{4.81}$$

In this max–min–fair design criterion, the base station designs the beamforming weights $\boldsymbol{w}_i$ by maximizing the minimum SINR value among m receivers subject to the power constraint $P > 0$ at the base station. Mathematically, it can be formulated as

$$\max_{\substack{\boldsymbol{w}_i \in \mathbb{C}^N \\ i=1,\ldots,n}} \min_{\substack{k \in \mathcal{G}_i \\ i=1,\ldots,n}} \frac{\boldsymbol{w}_i^H \boldsymbol{H}_k \boldsymbol{w}_i}{\sum_{j \neq k} \boldsymbol{w}_i^H \boldsymbol{H}_k \boldsymbol{w}_i + \sigma_k^2} \tag{4.82a}$$

$$\text{s.t. } \sum_{i=1}^{n} \|\boldsymbol{w}_i\|^2 \leq P. \tag{4.82b}$$

The max–min–fair transmit beamforming problem (4.82) is a special instance of the generalized fractional QCQP (4.80). To see this, let us define $\boldsymbol{w} = [\boldsymbol{w}_1^T, \ldots, \boldsymbol{w}_G^T]^T \in \mathbb{C}^K$ with $K = GN$, and let $\boldsymbol{R}_j$ and $\boldsymbol{Q}_j$ be block diagonal matrices defined by channel correlation matrices $\boldsymbol{H}_k$ and noise variance σ_k^2. Then (4.82) can be recast as a problem in the form of (4.80).

For the max–min–fair problem (4.82), a SDP, relaxation-based approximation method has been proposed in [16]. In this subsection, we will analyze the approximation performance based on the generalized fractional QCQP (4.80). Analogous to (4.71), the SDP relaxation of (4.80) is given by

$$v_{\text{sdp}} = \max_{W \in \mathbb{H}^{K \times K}} \min_{j=1,\ldots,m} \frac{\text{Tr}(R_j W)}{\text{Tr}(Q_j W) + 1} \tag{4.83a}$$

$$\text{s.t. } \text{Tr}(W) \leq P, \tag{4.83b}$$

$$W \succeq 0. \tag{4.83c}$$

Problem (4.83) is a generalized quasi-convex problem and can be solved by the bisection method as for (4.71) [see Box 3]. Once the optimum solution of (4.83), denoted by $W^\star$, is obtained, the following Gaussian randomization procedure can be used to obtain an approximation of (4.80).

Box 5. Gaussian randomization procedure for (4.80)

Let $L > 0$ be an integer.

1. For $\ell = 1, \ldots, L$, generate a random vector $\xi_\ell \in \mathbb{H}^K$ from the Gaussian distribution $N(0, W^\star)$, and let

$$w(\xi_\ell) = \sqrt{P}\xi_\ell / \|\xi_\ell\|. \tag{4.84}$$

2. Let

$$\hat{w} = \arg \max_{w(\xi_\ell),\ell=1,\ldots,L} \min_{j=1,\ldots,m} \frac{(w(\xi_\ell))^\dagger R_j w(\xi_\ell)}{(w(\xi_\ell))^\dagger Q_j w(\xi_\ell) + 1} \tag{4.85}$$

be the approximate solution of (4.80).

As we will analyze next, the SDP relaxation problem (4.83), together with the above randomization procedure, has a guaranteed, worst-case approximation bound, which is proportional to m (the number of fractional quadratic terms) for $\mathbb{H} = \mathbb{C}$, and proportional to m^2 for $\mathbb{H} = \mathbb{R}$. Some numerical results will also be presented in later subsections.

Worst-case approximation bounds

The analysis idea in this subsection extends those in Section 4.3.5. The extension deals with the fractional quadratic terms. Let us first focus on the complex case $\mathbb{H} = \mathbb{C}$.

LEMMA 4.1 *There exists an optimum solution $W^\star$ of problem (4.83) whose rank is upper bounded by $\sqrt{m}$.*

Proof It is easy to show that problem (4.83) is equivalent to the following optimization problem

$$P = \min_{W \in \mathbb{C}^{K \times K}} \mathrm{Tr}(W) \tag{4.86a}$$

$$\text{s.t.} \quad \frac{\mathrm{Tr}(R_j W)}{\mathrm{Tr}(Q_j W) + 1} \geq v_{\text{sdp}}, \quad j = 1, \ldots, m, \tag{4.86b}$$

$$W \succeq 0. \tag{4.86c}$$

Since problem (4.86) is a complex-valued SDP, the result of [38] implies that there exists an optimum solution with $\mathrm{rank}(W^\star) \leq \sqrt{m}$. ∎

We also need the following key lemma to bound the probability that a random, fractional quadratic quantity falls in the small neighborhood of the origin.

LEMMA 4.2 *Let $R \in \mathbb{C}^{K \times K}$ and $Q \in \mathbb{C}^{K \times K}$ be two Hermitian positive semidefinite matrices ($R \succeq 0$ and $Q \neq 0$, $Q \succeq 0$), and $\xi \in \mathbb{C}^K$ be a random vector with complex Gaussian distribution $N_{\mathbb{C}}(0, W^\star)$. Then*

$$\mathrm{Prob}\left(\frac{\xi^H R \xi}{\xi^H Q \xi + 1} < \eta \, \frac{\mathrm{E}\{\xi^H R \xi\}}{\mathrm{E}\{\xi^H Q \xi\} + 1} \right) \leq \max\left\{ \frac{3\eta}{\alpha - 2\eta}, \left(\frac{5\eta}{\frac{1-\alpha}{\bar{r}-1} - 3\eta} \right)^2 \right\} \tag{4.87}$$

where $\bar{r} = \min\{\mathrm{rank}(R), \mathrm{rank}(W^\star)\}$, $0 \leq \eta < \min\{\frac{\alpha}{2}, \frac{1}{3}\frac{1-\alpha}{\bar{r}-1}\}$ and $0 < \alpha < 1$.

Interested readers are referred to [18] for the complete proof. With Lemmas 4.1 and 4.2, we can have the following theorem:

THEOREM 4.3 *Let $\hat{w}$ be obtained by applying the polynomial-time Gaussian randomization procedure in Box 5 to problem (4.80) and its relaxation problem (4.83) with $\mathbb{H} = \mathbb{C}$. Let*

$$v_{\text{fqp}}(\hat{w}) := \min_{j=1,\ldots,m} \frac{\hat{w}^H R_j \hat{w}}{\hat{w}^H Q_j \hat{w} + 1}.$$

Then

$$\gamma_{\text{sdr}} := \max_{R_j, Q_j, P, K} \max\left\{ \frac{v_{\text{sdp}}}{v_{\text{fqp}}}, \frac{v_{\text{fqp}}}{v_{\text{fqp}}(\hat{w})} \right\} \leq 30m, \tag{4.88}$$

holds with probability at least $1 - (0.9393)^L$.

Proof By Lemma 4.1, for $m \leq 3$ there exists a solution of problem (4.83) with $\mathrm{rank}(W^\star) = 1$. Hence for $m \leq 3$, $v_{\text{sdp}} = v_{\text{fqp}}$. This rank-1 solution, which is feasible to (4.80) and has an objective value $v_{\text{fqp}}(\hat{w})$ equal to v_{fqp}, can always be obtained via a matrix decomposition procedure [38]. Therefore, $\gamma_{\text{sdr}} = 1$ for $m \leq 3$. To obtain (4.88) for $m > 3$, we need prove that

$$\text{Prob}\left(\min_{j=1,\dots,m} \frac{\boldsymbol{\xi}^H \boldsymbol{R}_j \boldsymbol{\xi}}{\boldsymbol{\xi}^H \boldsymbol{Q}_j \boldsymbol{\xi} + 1} \geq \eta \, v_{\text{sdp}}, \ \boldsymbol{\xi}^H \boldsymbol{\xi} \leq \mu P\right) > 0 \qquad (4.89)$$

for $\eta = 1/(16m)$ and $\mu = 15/8$, where $\boldsymbol{\xi} \in \mathbb{C}^K$ is a random vector with complex Gaussian distribution $N_{\mathbb{C}}(\boldsymbol{0}, \boldsymbol{W}^\star)$. If (4.89) is true, then there exists a realization of $\boldsymbol{\xi}$ which satisfies $\boldsymbol{\xi}^H \boldsymbol{\xi} \leq (15/8)P$ and

$$\min_{j=1,\dots,m} \frac{\boldsymbol{\xi}^H \boldsymbol{R}_j \boldsymbol{\xi}}{\boldsymbol{\xi}^H \boldsymbol{Q}_j \boldsymbol{\xi} + 1} \geq \left(\frac{1}{16m}\right) v_{\text{sdp}}.$$

Let $\bar{\boldsymbol{\xi}} = \sqrt{8/15}\boldsymbol{\xi}$ which then is feasible for problem (4.80) (i.e., $\bar{\boldsymbol{\xi}}^H \bar{\boldsymbol{\xi}} \leq P$) and satisfies

$$\left(\frac{1}{30m}\right) v_{\text{sdp}} \leq \min_{j=1,\dots,m} \frac{\bar{\boldsymbol{\xi}}^H \boldsymbol{R}_j \bar{\boldsymbol{\xi}}}{\bar{\boldsymbol{\xi}}^H \boldsymbol{Q}_j \bar{\boldsymbol{\xi}} + 1} \leq v_{\text{fqp}}, \qquad (4.90)$$

which is part of (4.88).

We now prove (4.89). Note that the left-hand side (LHS) of (4.89) can be lower bounded as follows:

$$\text{Prob}\left(\min_{j=1,\dots,m} \frac{\boldsymbol{\xi}^H \boldsymbol{R}_j \boldsymbol{\xi}}{\boldsymbol{\xi}^H \boldsymbol{Q}_j \boldsymbol{\xi} + 1} \geq \eta v_{\text{sdp}}, \ \boldsymbol{\xi}^H \boldsymbol{\xi} \leq \mu P\right)$$

$$\geq 1 - \sum_{j=1}^{m} \text{Prob}\left(\frac{\boldsymbol{\xi}^H \boldsymbol{R}_j \boldsymbol{\xi}}{\boldsymbol{\xi}^H \boldsymbol{Q}_j \boldsymbol{\xi} + 1} < \eta v_{\text{sdp}}\right) - \text{Prob}\left(\boldsymbol{\xi}^H \boldsymbol{\xi} > \mu P\right)$$

$$\geq 1 - \sum_{j=1}^{m} \text{Prob}\left(\frac{\boldsymbol{\xi}^H \boldsymbol{R}_j \boldsymbol{\xi}}{\boldsymbol{\xi}^H \boldsymbol{Q}_j \boldsymbol{\xi} + 1} < \eta \frac{\text{Tr}(\boldsymbol{R}_j \boldsymbol{W}^\star)}{\text{Tr}(\boldsymbol{Q}_j \boldsymbol{W}^\star) + 1}\right) - \text{Prob}\left(\boldsymbol{\xi}^H \boldsymbol{\xi} > \mu \cdot \text{Tr}(\boldsymbol{W}^\star)\right)$$

(by (4.83)) $\qquad\qquad\qquad\qquad\qquad\qquad\qquad\qquad\qquad\qquad\qquad\qquad\qquad\qquad (4.91)$

$$= 1 - \sum_{j=1}^{m} \text{Prob}\left(\frac{\boldsymbol{\xi}^H \boldsymbol{R}_j \boldsymbol{\xi}}{\boldsymbol{\xi}^H \boldsymbol{Q}_j \boldsymbol{\xi} + 1} < \eta \frac{\text{E}\{\boldsymbol{\xi}^H \boldsymbol{R}_j \boldsymbol{\xi}\}}{\text{E}\{\boldsymbol{\xi}^H \boldsymbol{Q}_j \boldsymbol{\xi}\} + 1}\right) - \text{Prob}\left(\boldsymbol{\xi}^H \boldsymbol{\xi} > \mu \text{E}\{\boldsymbol{\xi}^H \boldsymbol{\xi}\}\right)$$

(since $\boldsymbol{\xi} \sim \mathcal{N}_{\mathbb{C}}(\boldsymbol{0}, \boldsymbol{W}^\star)$)

$$\geq 1 - \sum_{j=1}^{m} \text{Prob}\left(\frac{\boldsymbol{\xi}^H \boldsymbol{R}_j \boldsymbol{\xi}}{\boldsymbol{\xi}^H \boldsymbol{Q}_j \boldsymbol{\xi} + 1} < \eta \frac{\text{E}\{\boldsymbol{\xi}^H \boldsymbol{R}_j \boldsymbol{\xi}\}}{\text{E}\{\boldsymbol{\xi}^H \boldsymbol{Q}_j \boldsymbol{\xi}\} + 1}\right) - \frac{1}{\mu} \quad \text{(by Markov inequality)}$$

$$\geq \frac{\mu - 1}{\mu} - \sum_{j=1}^{m} \max\left\{\frac{3\eta}{\alpha - 2\eta}, \left(\frac{5\eta}{\frac{1-\alpha}{\bar{r}-1} - 3\eta}\right)^2\right\}, \qquad (4.92)$$

where the last step in (4.92) is due to Lemma 4.2. Because $\bar{r} \leq \sqrt{m}$ (Lemma 4.1), by choosing $\alpha = 0.4932$ and $\eta = 1/(16m)$, we can show that for $m > 3$,

$$\frac{3\eta}{\alpha - 2\eta} \geq \left(\frac{5\eta}{\frac{1-\alpha}{\bar{r}-1} - 3\eta} \right)^2. \tag{4.93}$$

Hence, for $m > 3$ and $\mu = 15/8$,

$$\mathrm{Prob}\left(\min_{j=1,\ldots,m} \frac{\boldsymbol{\xi}^H \boldsymbol{R}_j \boldsymbol{\xi}}{\boldsymbol{\xi}^H \boldsymbol{Q}_j \boldsymbol{\xi} + 1} \geq \eta v_{\mathrm{sdp}}, \ \boldsymbol{\xi}^H \boldsymbol{\xi} \leq \mu P \right)$$

$$\geq \frac{7}{15} - m\frac{3\eta}{\alpha - 2\eta} = \frac{7}{15} - \frac{3}{7.8912 - 2/m} > 0.0607, \tag{4.94}$$

which establishes (4.89). This further implies that (4.90) holds.

To complete the proof, recall the Gaussian randomization procedure in Box 5. Then, for each ℓ, it follows from (4.90) and (4.94) that $\boldsymbol{w}(\boldsymbol{\xi}_\ell) = \sqrt{P}\boldsymbol{\xi}_\ell/\|\boldsymbol{\xi}_\ell\|$ satisfies

$$\left(\frac{1}{30m} \right) v_{\mathrm{fqp}} \leq \left(\frac{1}{30m} \right) v_{\mathrm{sdp}} \leq v_{\mathrm{fqp}}^{(\ell)} := \min_{j=1,\ldots,m} \frac{(\boldsymbol{w}(\boldsymbol{\xi}_\ell))^H \boldsymbol{R}_j \boldsymbol{w}(\boldsymbol{\xi}_\ell)}{(\boldsymbol{w}(\boldsymbol{\xi}_\ell))^H \boldsymbol{Q}_j \boldsymbol{w}(\boldsymbol{\xi}_\ell) + 1}, \tag{4.95}$$

with probability at least 0.0607. If one generates L independent realizations of $\boldsymbol{\xi}$ from the distribution $N_{\mathbb{C}}(\boldsymbol{0}, \boldsymbol{W}^\star)$, then it is with probability at least[1] $1 - (1 - 0.0607)^L$ to obtain one $\boldsymbol{\xi}$ which can achieve the approximation quality in (4.95). Since $v_{\mathrm{fqp}}(\hat{\boldsymbol{w}}) = \max\{v_{\mathrm{fqp}}^{(1)}, \ldots, v_{\mathrm{fqp}}^{(L)}\}$, it follows that

$$\frac{v_{\mathrm{fqp}}}{30m} \leq v_{\mathrm{fqp}}(\hat{\boldsymbol{w}}) \leq v_{\mathrm{fqp}}, \tag{4.96}$$

holds with probability at least $1 - (0.9393)^L$. Theorem 4.3 is proved.   ∎

Theorem 4.3 implies that the SDP relaxation of (4.80) has a worst-case $O(1/m)$ approximation to the optimum solution of v_{fqp}. It also implies that the SDP relaxation can have at least an $O(1/m)$ approximation to the optimum solution of the max–min–fair transmit beamforming problem in (4.82). The following example demonstrates that this worst-case approximation ratio estimated in (4.88) is tight up to a constant factor.

Example 4.13   Let us consider a special instance of (4.80) that is analogous to that in (4.50). Let $\boldsymbol{R}_j = \boldsymbol{h}_j \boldsymbol{h}_j^H$ where

$$\boldsymbol{h}_j = \left(\cos\left(\frac{k\pi}{M} \right), \sin\left(\frac{k\pi}{M} \right) e^{i2\pi\ell/M}, 0, \ldots, 0 \right)^T \in \mathbb{C}^K,$$

---

[1] For $L = 50$, this probability is 0.9563.

in which $i = \sqrt{-1}$, $M = \lceil \sqrt{m} \rceil$, $m \geq 2$ (thus $M \geq 2$), $K \geq 2$, and $j = kM - M + \ell$ with $k, \ell = 1, \ldots, M$. Let $Q_j = \mathrm{Diag}\{1/P, 1/P, 0, \ldots, 0\} \in \mathbb{R}^{K \times K}$. Without loss of generality, assume that $w^\star = \left(\rho \cos \theta, \rho(\sin \theta)e^{i\phi}, \ldots \right)^T \in \mathbb{C}^K$ where $\rho > 0$ and $\theta, \phi \in [0, 2\pi)$. It has been shown in [32] that there exists an index $j \in \{1, \ldots, m\}$ such that

$$|h_j^H w^\star|^2 \leq \rho^2 \frac{\pi^2 (3K + \pi)^2}{4M^4}. \tag{4.97}$$

By (4.80) and (4.97), one can have

$$v_{\mathrm{fqp}} \leq \frac{|h_j^H w^\star|^2}{(w^\star)^H Q_j w^\star + 1} \leq |h_j^H w^\star|^2 \leq \rho^2 \frac{\pi^2 (3K + \pi)^2}{4M^4} \leq \left(\frac{P}{4} \right) \frac{\pi^2 (3M + \pi)^2}{M^4}. \tag{4.98}$$

A feasible point of (4.83) can be $\bar{W} = \mathrm{Diag}\{P/2, P/2, 0, \ldots, 0\}$ which leads to

$$\min_{j=1,\ldots,m} \frac{\mathrm{Tr}(R_j \bar{W})}{\mathrm{Tr}(Q_j \bar{W}) + 1} = \frac{P}{4} \leq v_{\mathrm{sdp}}. \tag{4.99}$$

Combining (4.98) and (4.99) gives

$$v_{\mathrm{sdp}} \geq \frac{M^2}{\pi^2 (3 + \pi/M)^2} v_{\mathrm{fqp}} \geq \frac{m}{\pi^2 (3 + \pi/2)^2} v_{\mathrm{fqp}}, \tag{4.100}$$

which indicates that the upper bound given by (4.88) can be attained within a constant factor. Equation (4.100) also serves as a theoretical lower bound for the value of γ_{sdr}.

Interestingly, for (4.80) and (4.83) with $\mathbb{H} = \mathbb{R}$, the corresponding worst-case approximation ratio γ_{sdr} deteriorates to $O(m^2)$. We briefly summarize the results here. Interested readers are referred to [18] for the detailed descriptions.

THEOREM 4.4 *Let $\hat{w}$ be obtained by applying the polynomial-time Gaussian randomization procedure in Box 5 (with ξ_ℓ drawn from the real-valued Gaussian $\mathcal{N}_{\mathbb{R}}(0, W^\star)$) to problem (4.80) and its relaxation problem (4.83) with $\mathbb{H} = \mathbb{R}$. Let*

$$v_{\mathrm{fqp}}(\hat{w}) := \min_{j=1,\ldots,m} \frac{\hat{w}^T R_j \hat{w}}{\hat{w}^T Q_j \hat{w} + 1}.$$

Then

$$\gamma_{\mathrm{sdr}} := \max_{R_j, Q_j, P, K} \max \left\{ \frac{v_{\mathrm{sdp}}}{v_{\mathrm{fqp}}}, \frac{v_{\mathrm{fqp}}}{v_{\mathrm{fqp}}(\hat{w})} \right\} \leq 80m^2, \tag{4.101}$$

holds with probability at least $1 - (0.9521)^L$.

Numerical results

Let us present some simulation results to illustrate the effectiveness of the SDP relaxation for the generalized fractional QCQP (4.80). We generate 1000 random problem instances of (4.80) for $\mathbb{H} = \mathbb{C}$ and $\mathbb{H} = \mathbb{R}$, respectively. For each problem instance, the positive semidefinite matrices R_j and Q_j were generated in a similar way as the H_ks in Section 4.3.5. We respectively tested for R_j having full rank and rank-1; while Q_j were all of full rank. The problem (4.83) was solved by the bisection algorithm [22] wherein SeDuMi [39] was employed to handle the associated semidefinite feasibility problems. The Gaussian randomization procedure in Box 5 was implemented with $L = 50$ for each problem instance. The empirical approximation ratio $v_{\text{sdp}}/v_{\text{fqp}}(\hat{w})$ was used to approximate the true ratio $\max\left\{v_{\text{sdp}}/v_{\text{fqp}}, v_{\text{fqp}}/v_{\text{fqp}}(\hat{w})\right\}$. For the real-valued (4.80) and

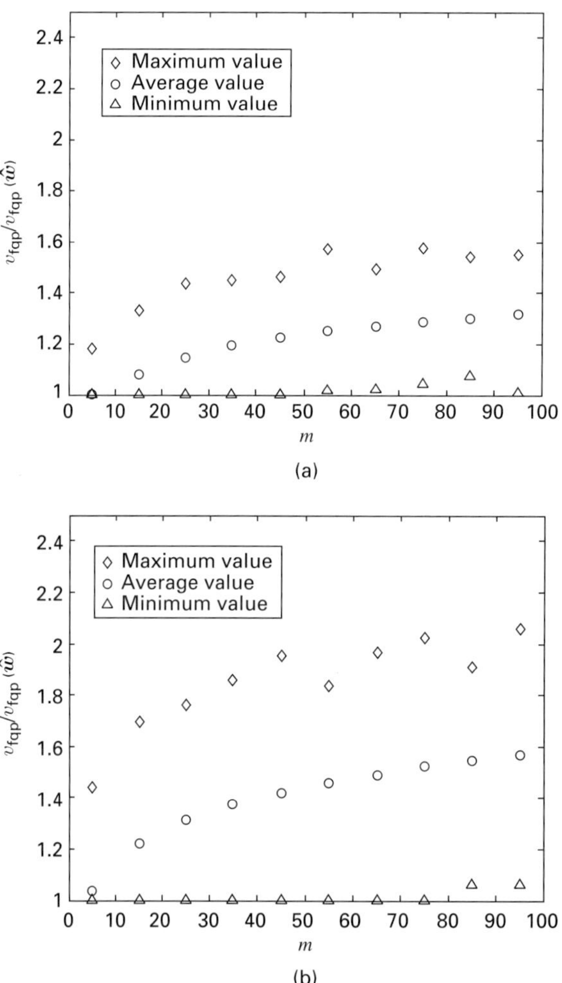

Figure 4.7   Empirical approximation ratios for $K = 10$, $P = 10$, and full rank R_j for a (a) complex-valued problem (b) real-valued problem.

(4.83), the associated Gaussian randomization procedure is the same as that in Box 5, but random vector $\boldsymbol{\xi}$s were drawn from the real Gaussian distribution $N_{\mathbb{R}}(\mathbf{0}, \boldsymbol{W}^{\star})$.

We first consider the results when matrices $\boldsymbol{R}_j$ are full rank. Figure 4.7 shows the empirical approximation ratios for $K = 10$ and $P = 10$. One can see from these figures that the empirical approximation ratios get larger when m increases, and the approximation ratios of complex-valued problems are smaller than those of real-valued problems. It can also be seen from these figures that, in the average sense, the SDP relaxation provides very good approximation qualities (max $\left\{v_{\mathrm{sdp}}/v_{\mathrm{fqp}}, v_{\mathrm{fqp}}/v_{\mathrm{fqp}}(\hat{\boldsymbol{w}})\right\} < 1.6$ for complex-valued problems).

Figure 4.8 illustrates the empirical approximation ratios for $K = 10$ and $P = 10$ when matrices $\boldsymbol{R}_j$ are rank one. By comparing Fig. 4.7 and Fig. 4.8, one

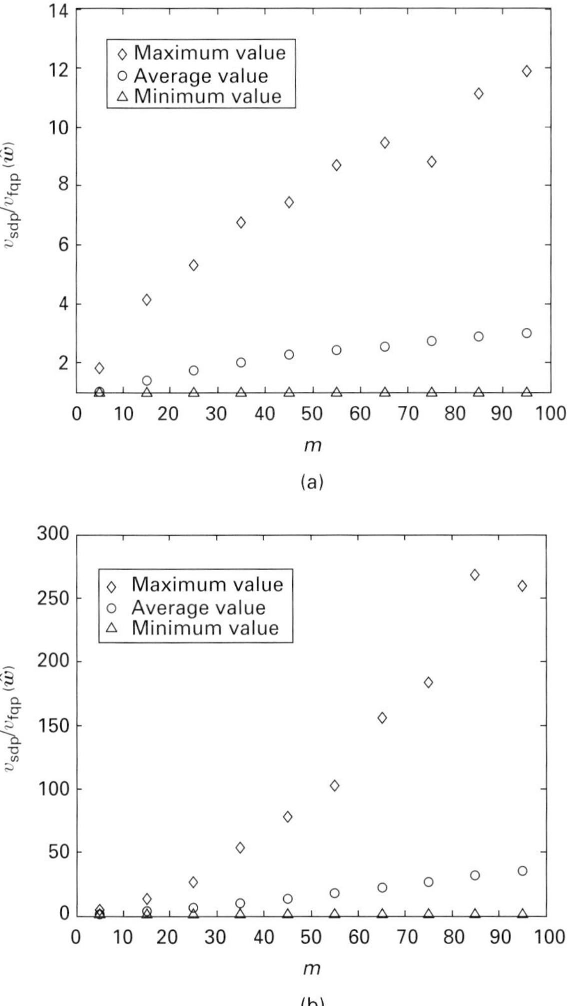

Figure 4.8 Empirical approximation ratios for $n = 10$, $P = 10$ and rank one $\boldsymbol{R}_j$ for a (a) complex-valued problem (b) real-valued problem.

can observe that the approximation ratios for full rank $\boldsymbol{R}_j$ are much smaller than those for rank one $\boldsymbol{R}_j$. One can also observe from Figure 4.8 that the maximum values of $v_{\text{sdp}}/v_{\text{fqp}}(\hat{\boldsymbol{w}})$ increase with m roughly in a linear manner for the complex-valued problem (Figure 4.8(a)), and in a quadratic manner for the real-valued problem (Figure 4.8(b)). These results coincide with our analytic results and Example 4.13 in Section 4.5.2, where the proposed approximation bound in (4.88) is tight (to the first-order of m) in a specific problem instance with all $\boldsymbol{R}_j$ being rank one.

4.6   More applications of SDP relaxation

In this section, we consider a *magnitude least-squares* (MLS) problem and a *magnitude-squared least-squares* (MSLS) problem to which the SDP relaxation can also be applied. These classes of problems are important in a wide range of applications; for example, one-dimensional (multidimensional) magnitude filter design, approximate polynomial factorization, radiation pattern synthesis (see [11] and references therein), and some problems arising from "magnetic resonance imaging" (MRI) systems [12, 13, 40].

4.6.1   SDP relaxation for the magnitude least-squares (MLS) problem

Given a non-negative vector $\boldsymbol{b} = [b_1, \ldots, b_m]^T$ and a complex matrix $\boldsymbol{A} = [\boldsymbol{a}_1, \ldots, \boldsymbol{a}_m]^H \in \mathbb{C}^{m \times K}$, the MLS problem can be formulated as follows,

$$\min_{\boldsymbol{x} \in \mathbb{C}^K} \sum_{i=1}^{m} \left(|\boldsymbol{a}_i^H \boldsymbol{x}| - b_i \right)^2 . \tag{4.102}$$

The MLS problem can be interpreted as a magnitude fitting problem. It is not hard to verify that the MLS problem (4.102) is not convex, in contrast to the popular convex LS problem

$$\min_{\boldsymbol{x} \in \mathbb{C}^K} \sum_{i=1}^{m} \left| \boldsymbol{a}_i^H \boldsymbol{x} - b_i \right|^2 = \min_{\boldsymbol{x} \in \mathbb{C}^K} \| \boldsymbol{A}\boldsymbol{x} - \boldsymbol{b} \|^2 . \tag{4.103}$$

One can observe that the nonconvexity of (4.102) arises from the absolute terms of $|\boldsymbol{a}_i^H \boldsymbol{x}|$. These absolute terms, however, can be substituted by introducing m unit-modulus variables, as demonstrated by the following theorem [11]:

THEOREM 4.5   *The MLS problem* (4.102) *is equivalent to the following quadratic problem:*

$$\min_{\boldsymbol{x} \in \mathbb{C}^K, \, \boldsymbol{z} \in \mathbb{C}^m} \sum_{i=1}^{m} \left| \boldsymbol{a}_i^H \boldsymbol{x} - b_i z_i \right|^2 \tag{4.104a}$$

$$\text{s.t. } |z_i| = 1, \quad i = 1, \ldots, m. \tag{4.104b}$$

The reformulation in (4.104) suggests a simple, alternating least-squares method to solve (4.102), provided that an initial estimate of x is given. This local method is based on variable exchange, which solves (4.104) with respect to x and $z = [z_1, \ldots, z_m]^T$, iteratively. To illustrate it, given an x, this local method solves

$$\min_{z \in \mathbb{C}^m} \ \sum_{i=1}^{m} \left| a_i^H x - b_i z_i \right|^2 \tag{4.105a}$$

$$\text{s.t. } |z_i| = 1, \quad i = 1, \ldots, m. \tag{4.105b}$$

Let $a_i^H x = |a_i^H x| e^{j\phi_i}$. It can be readily seen that the optimum solution of (4.105) is given by $z^\star = [z_1^\star, \ldots, z_m^\star]^T$ with $z_i^\star = e^{-j\phi_i}$. On the other hand, given a z, the local method solves

$$\min_{x \in \mathbb{C}^K} \ \sum_{i=1}^{m} \left| a_i^H x - b_i z_i \right|^2 = \min_{x \in \mathbb{C}^K} \ \|Ax - c\|^2, \tag{4.106}$$

where $c = [b_1 z_1, \ldots, b_m z_m]^T$. Since (4.106) is just a convex LS problem, it has a closed form solution of $x^\star = A^\dagger c$, where $A^\dagger$ denotes the matrix pseudo-inverse of A. The alternating least-squares method updates the estimates of x and z iteratively until some predefined convergence criterion is satisfied.

The reformulation in Theorem 4.5 also enables an SDP relaxation approximation method for (4.102). To illustrate this, we rewrite (4.104) as follows

$$\min_{z \in \mathbb{C}^m} \ \left\{ \min_{x \in \mathbb{C}^K} \ \|Ax - \mathrm{Diag}(b)z\|^2 \right\} \tag{4.107a}$$

$$\text{s.t. } |z_i| = 1, \quad i = 1, \ldots, m. \tag{4.107b}$$

Since the inner minimization term has a closed form solution of

$$x^\star = A^\dagger \mathrm{Diag}(b)z, \tag{4.108}$$

one can express (4.107) as

$$\min_{z \in \mathbb{C}^m} \ \|AA^\dagger \mathrm{Diag}(b)z - \mathrm{Diag}(b)z\|^2 \tag{4.109a}$$

$$\text{s.t. } |z_i| = 1, \quad i = 1, \ldots, m. \tag{4.109b}$$

By defining

$$C = \left(AA^\dagger \mathrm{Diag}(b) - \mathrm{Diag}(b) \right)^H \left(AA^\dagger \mathrm{Diag}(b) - \mathrm{Diag}(b) \right)$$

$$= \mathrm{Diag}(b) \left(I_m - AA^\dagger \right) \mathrm{Diag}(b), \tag{4.110}$$

(4.104) can be expressed as the following homogeneous nonconvex QP:

$$\min_{z \in \mathbb{C}^m} \ z^H C z \tag{4.111a}$$

$$\text{s.t. } |z_i| = 1, \quad i = 1, \ldots, m. \tag{4.111b}$$

Therefore, the idea of the SDP relaxation method can be readily applied to (4.111) to obtain an approximate solution. After solving the corresponding SDP relaxation problem of (4.111), the Gaussian randomization procedure proposed in [33] can be applied to obtain a rank-1, feasible approximate solution. In particular, one can first generate a Gaussian random vector ξ from the distribution of $\mathcal{N}_{\mathbb{C}}(0, Z^\star)$ where $Z^\star$ denotes the optimum solution of the SDP relaxation of (4.111), followed by rounding ξ into a feasible vector by letting

$$z_i = \xi_i / |\xi_i|, \quad i = 1, \ldots, m. \tag{4.112}$$

After repeating the randomization steps a number of times, a good approximate solution $\hat{z}^\star$ of (4.111) can be obtained by choosing the one with the minimum objective value. Then an approximate solution to the MLS problem (4.102) is obtained as $\hat{x}^\star = A^\dagger \mathrm{Diag}(b) \hat{z}^\star$ [see (4.108)].

It is worthwhile to note that the SDP approximate solution $\hat{x}^\star$ can also be used as an initial point of the previously-mentioned, alternating least-squares method for further performance improvement. Since the analysis results in previous sections suggest that the worst-case approximation performance generally deteriorates with the number of constraints m, this solution refinement procedure is particularly useful when m is large. For more information about the performance of SDP relaxation for the MLS problem, readers are referred to [11].

4.6.2   SDP relaxation for the magnitude-squared least-squares (MSLS) problem

Since the SDP relaxation for the MLS problem (4.102) requires the solution an SDP with problem dimension equal to m, the computational complexity may not be affordable if m is large. In this case, one can consider an alternative criterion which minimizes the sum of squared magnitude differences as follows

$$\min_{x \in \mathbb{C}^K} \sum_{i=1}^{m} \left(|a_i^H x|^2 - b_i^2 \right)^2. \tag{4.113}$$

Though the *magnitude-squared least-squared* (MSLS) problem (4.113) is similar to the MLS problem (4.102), its SDP relaxation only involves solving a convex problem with dimension equal to K. Inherent from the quadratic nature of $|a_i^H x|^2$, one can see that the

SDP relaxation of (4.113) leads to the following convex quadratic SDP:

$$\min_{X \in \mathbb{C}^{K \times K}} \sum_{i=1}^{m} \left(\mathrm{Tr}(a_i a_i^H X) - b_i^2 \right)^2 \tag{4.114a}$$

$$\text{s.t. } X \succeq 0. \tag{4.114b}$$

It can be seen that (4.114) has a problem dimension equal to K.

In some applications, minimizing the worst-case magnitude difference instead of the total sum of magnitude differences is of interest. In that case, (4.114) is replaced by the following minmax problem:

$$\min_{x \in \mathbb{C}^K} \left\{ \max_{i=1,\ldots,m} \left(|a_i^H x|^2 - b_i^2 \right)^2 \right\}. \tag{4.115}$$

It is easy to see that the SDP relaxation method can also be applied to (4.115) for obtaining an approximate solution. The following example presents an application of (4.115) to the transmit B_1 shim problem in an MRI system.

Example 4.14 Transmit B_1 Shim in MRI In a typical MRI system, a set of RF coils are required to generate a magnetic field across the load under imaging (e.g., human head or human body). Due to the complex interactions between the electrical-magnetic field and the loaded tissues, the resultant *magnetic magnitude field* (or $|B_1|$ field) exhibits strong inhomogeneity across the load. This $|B_1|$ inhomogeneity would lead to severe artifacts in final MR images, and thus the B_1 shim technique, which designs the input RF signals in order to homogenize the $|B_1|$ field, is required.

Denote by $M_k \in \mathbb{C}^{p \times q}$ the B_1 field map due to the kth RF coil where $k = 1, \ldots, K$ and K is the total number of coils. Let $x_i \in \mathbb{C}$ be the input RF signal strength (including the electrical current magnitude and phase) of the kth RF coil. The transmit B_1 shim problem designs the input signals x_k, such that the magnitude distribution of the superimposed map

$$M(x) = M_1 x_1 + M_2 x_2 + \cdots + M_K x_K$$

can be uniform and homogenous, where $x = [x_1, \ldots, x_K]^T \in \mathbb{C}^K$. The map $M(x = 1)$ (without B_1 shim) usually gives a nonuniform $|B_1|$ distribution. Define

$$A = \left[\mathrm{vec}(M_1), \ldots, \mathrm{vec}(M_K) \right] := [a_1, \ldots, a_m]^H \in \mathbb{C}^{m \times K}, \tag{4.116}$$

where $\mathrm{vec}(\cdot)$ denotes the column-by-column vectorization of matrices, $m = p \times q$ is the total number of pixels in the map, and $a_i \in \mathbb{C}^K$. Then $Ax = \mathrm{vec}(M(x))$.

One of the B_1 shim criteria is to predefine a uniform target $|B_1|$ map $b = [b, \ldots, b]^T \in \mathbb{R}^m$ where $b > 0$, and adjust the input signals x, such that the distribution of $|a_i^H x|$, $i = 1, \ldots, m$, can be close to b. Besides, we wish to ensure that each pixel is as close to b as possible in order to prevent producing a map with weak B_1 local spots. To this end, the minimax problem formulation (4.115) can be applied. However, since the number of pixels, m, is typically a very large number (hundreds to thousands), the approximation performance of SDP relaxation to (4.115) is, in general, not satisfying. Nevertheless, the approximate solution provided by SDP relaxation can be used as a good starting point in any local optimization method for (4.115) for further performance improvement. The choice of local methods can be either the simple gradient descent method, Newton method, or the iterative linearization method. As will be shown, the SDP relaxation approximate solution serves as a robust initial point for the optimization problem (4.115). In other words, the SDP relaxation can serve as a robust initialization step for the presented B_1 shim technique. In the numerical simulations, it is completely insensitive to the initial set of RF amplitudes and phases, since the SDP relaxation problem is a convex SDP. This is in strong contrast to most of the existing B_1 shim methods which typically solve a nonconvex optimization problem directly by using random starts and, therefore, their performance is quite sensitive to the initial setting of RF amplitudes and phases. For example, a commonly used B_1 shim method is to directly minimize the normalized variance of $|a_i^H x|$, $i = 1, \ldots, m$ [41], in other words,

$$\min_{x \in \mathbb{C}^K} \frac{\sqrt{\sum_{i=1}^m \left(|a_i^H x| - \frac{1}{m} \sum_{i=1}^m |a_i^H x| \right)^2}}{\frac{1}{m} \sum_{i=1}^m |a_i^H x|}. \tag{4.117}$$

Since (4.117) is a nonconvex problem, the resultant B_1 shim performance varies significantly with the chosen initial points.

We have tested the SDR initialization technique on a human brain in a 7 tesla MRI system. The B_1 maps were experimentally measured in an axial slice of the human brain at 7 tesla for each channel of a 16-element transceiver stripline array [42]. The experiment was conducted on a healthy male volunteer in the Center of Magnetic Resonance Research at the University of Minnesota. Figure 4.9(a) shows the original $|B_1|$ distribution without any B_1 shim optimization. It can be observed that the $|B_1|$ distribution is highly nonuniform. Figures 4.9(b) and 4.9(c) present the results by solving (4.117) using the quasi-Newton method with two different initial points; while Figure 4.9(d) shows the results by solving (4.115) using the iterative linearization method with the

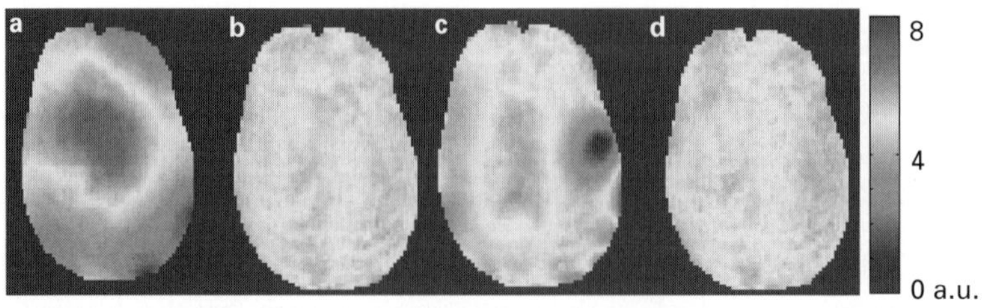

Figure 4.9   Comparison results between B_1 shim methods based on measured B_1 maps (arbitrary unit (a.u.))

SDP relaxation approximate solution as the initial point. The value b was set as the mean $|B_1|$ value of the non-optimized map in Figure 4.9(a). The SDP relaxation approximate solution was obtained by a Gaussian randomization procedure in which 500 complex Gaussian vectors were generated based on the optimum solution of the SDP relaxation problem (4.115). One can see from these figures that the performance of the B_1 shim method by directly solving (4.117) varies dramatically with initial points. In contrast, the SDR initialization can provide a reliable good B_1 shim result, independent of the initial RF amplitudes and phases.

4.7   Summary and discussion

In this chapter, we have considered the SDP relaxation for several classes of quadratic optimization problems. These include the separable homogeneous QCQPs, maximization-form homogeneous QCQPs, and fractional QCQPs. All of these quadratic optimization problems are strongly motivated by important applications in signal processing which include transmit beamforming, network beamforming, and MRI B_1 shim, to name just a few. To assess the performance of SDP relaxation, we have identified the following cases of the separable homogeneous QCQP [cf. (4.23)] for which the SDP relaxation is tight:

- the unicast case ($n = m$);
- the Vandermonde channel case arising in ULA, far-field, light-of-sight transmissions.

Moreover, we have identified several classes of quadratic optimization problems for which the SDP relaxation provides a guaranteed, finite worst-case approximation performance. These include

- single-group multicasting ($n = 1$, $\boldsymbol{H}_k \succeq \boldsymbol{0}$) [cf. (4.38)],
- receive beamforming in the presence of an intercept receiver [cf. (4.64)] ,
- max–min–fair multicast transmit beamforming [cf. (4.80)].

In each of the above cases, and under either the real field $\mathbb{R}$ or the complex field $\mathbb{C}$, we have given explicit bounds on the worst-case SDP approximation performance, which are tight up to a constant. In addition, we have given examples that suggest these bounds cannot be extended to more general cases.

A substantial amount of simulation has been performed to assess the efficacy of the SDP relaxation approach. For the aforementioned applications, we have found that SDP relaxation is a highly effective tool, yielding excellent approximately optimal solutions in a reliable and efficient manner. Even for quadratic optimization problems for which no worst-case performance bounds exist, computer simulations show that the SDP technique still provides a good initial point for any local optimization methods used to solve the original nonconvex quadratic problem.

Finally, for ease of reference, we summarize the main analytical results presented in this chapter in the following table.

Table 4.1. Summary of the worst-case approximation bounds of SDP relaxation.

	$\mathbb{H} = \mathbb{R}, d = 0, 1$	$\mathbb{H} = \mathbb{C}, d = 0, 1$	$\mathbb{R}$ or $\mathbb{C}, d \geq 2$
$\min\ w^H A_0 w$ s.t. $w^H A_i w \geq 1,\ w^H B_j w \geq 1$	$O(m^2)$	$O(m)$	∞
$\max\ w^H B_0 w$ s.t. $w^H A_i w \leq 1,\ w^H B_j w \leq 1$	$O(\log m)$	$O(\log m)$	∞
$\max\ \dfrac{w^H A_0 w}{w^H \bar{A}_0 w + \sigma^2}$ s.t. $w^H A_i w \leq 1,\ w^H B_j w \leq 1$	$O(\log m)$	$O(\log m)$	N.A.
$\max\ \min\limits_{1 \leq i \leq m} \dfrac{w^H A_i w}{w^H \bar{A}_i w + \sigma^2}$ s.t. $\|w\|^2 \leq P$	$O(m^2)$	$O(m)$	N.A.

Notation: $A_i, \bar{A}_i \succeq 0$, $i = 0, \ldots, m$, $B_j \not\succeq 0$ (indefinite), $j = 0, \ldots, d$. N.A. = not applicable

Acknowledgments

This research is supported in part by the National Science Foundation, Grant No. DMS-0312416, and in part by the direct grant of National Tsing Hua University, No. 97N2309E1.

References

[1] W.-K. Ma, T. N. Davidson, K. M. Wong, Z.-Q. Luo, and P.-C. Ching, "Quasi-maximum-likelihood multiuser detection using semidefinite relaxation with applications to synchronous CDMA," *IEEE Transactions on Signal Processing*, vol. 50, no. 4, pp. 912–22, 2002.

[2] W.-K. Ma, B.-N. Vo, T. N. Davidson, and P.-C. Ching, "Blind ML detection of orthogonal space-time block codes: efficient high-performance implementations," *IEEE Transactions on Signal Processing*, vol. 54, no. 2, pp. 738–51, 2006.

[3] T.-H. Chang, W.-K. Ma, and C.-Y. Chi, "Maximum-likelihood detection of orthogonal space-time block coded OFDM in unknown block fading channels," *IEEE Transactions on Signal Processing*, vol. 56, no. 4, pp. 1637–49, 2008.

[4] M. X. Goemans and D. P. Williamson, "Improved approximation algorithms for maximum cut and satisfiability problem using semi-definite programming," *Journal of ACM*, vol. 42, pp. 1115–45, 1995.

[5] H. Jiang, X. Li, and C. Liu, "Large margin hidden Markov models for speech recognition," *IEEE Transactions on Audio, Speech and Language Processing*, vol. 14, no. 5, pp. 1584–95, 2006.

[6] H. Jiang and X. Li, "Incorporating training errors for large margin hmms under semi-definite programming framework," in *Proceedings of the IEEE ICASSP*, vol. 4, Honolulu, HI, April 15–20, 2007, pp. 629–32.

[7] T.-H. Chang, Z.-Q. Luo, L. Deng, and C.-Y. Chi, "A convex optimization method for joint mean and variance parameter estimation of large-margin CDHMM," in *Proceedings of the IEEE ICASSP*, Las Vegas, Nevada, March 30–April 4, 2008, pp. 4053–6.

[8] F. Sha and L. K. Saul, "Large margin hidden markov models for automatic speech recognition," in *Advances in Neural Information Processing Systems*. Cambridge, MA: MIT Press, 2007.

[9] A. D. Miao, S. D. Nicola, Y. Huang, S. Zhang, and A. Farino, "Code design to optimize radar detection performance under accuracy and similarity constraints," *IEEE Transactions on Signal Processing*, vol. 56, no. 11, pp. 5618–26, 2008.

[10] J. Keuchel, M. Heiler, and C. Schnörr, "Hierarchical image segmentation based on semidefinite program," in *Proceedings of the 26th Dagm Symp. Pattern Recognition*, Tübingen, Germany, Aug. 30–Sept. 1, 2004, pp. 120–8.

[11] P. W. Kassakian, "Convex approximation and optimization with applications in magnitude filter design and radiation pattern analysis," PhD dissertation, Dept. of EECS, University of California, Berkeley, 2006.

[12] T.-H. Chang, Z.-Q. Luo, X. Wu, C. Akgun, J. Vaughan, K. Ugurbil, and P.-F. V. de Moortele, "Transmit B1 shimming at high field with sar constraints: a two stage optimization method independent of the initial set of RF phases and amplitudes," in *Proceedings of the 16th International Society for Magnetic Resonance in Medicine (ISMRM)*, Toronto, Ontario, Canada, May 3–9, 2008.

[13] X. Wu, T.-H. Chang, Z.-Q. Luo, C. Akgun, J. Vaughan, K. Ugurbil, and P.-F. V. de Moortele, "Worst case SAR scenario as a new metric for SAR analysis in B1 phase shim," in *Proceedings of the 16th International Society for Magnetic Resonance in Medicine (ISMRM)*, Toronto, Ontario, Canada, May 3–9, 2008.

[14] M. Bengtsson and B. Ottersten, "Optimal and suboptimal transmit beamforming," in *Handbook of Antennas in Wireless Communications*, L. C. Godara, ed, Boca Raton: FL: CRC Press, 2001.

[15] N. D. Sidiropoulos, T. D. Davidson, and Z.-Q. Luo, "Transmit beamforming for physical-layer multicasting," *IEEE Transactions on Signal Processing*, vol. 54, no. 6, pp. 2239–51, 2006.

[16] E. Karipidis, N. D. Sidiropoulos, and Z.-Q. Luo, "Quality of service and max-min-fair transmit beamforming to multiple co-channel multicast groups," *IEEE Transactions on Signal Processing*, vol. 56, no. 3, pp. 1268–79, 2008.

[17] ——, "Far-field multicast beamforming for uniform linear antenna arrays," *IEEE Transactions on Signal Processing*, vol. 55, no. 10, pp. 4916–27, 2007.

[18] T.-H. Chang, Z.-Q. Luo, and C.-Y. Chi, "Approximation bounds for semidefinite relaxation of max-min-fair multicast transmit beamforming problem," *IEEE Transactions on Signal Processing*, vol. 56, no. 8, pp. 3932–43, 2008.

[19] W.-K. Ma, P.-C. Ching, and B.-N. Vo, "Crosstalk resilient interference cancellation in microphone arrays using Capon beamforming," *IEEE Transactions on Signal Processing*, vol. 12, no. 5, pp. 468–77, 2004.

[20] V. Havary-Nassab, S. Shahbazpanahi, A. Grami, and Z.-Q. Luo, "Network beamforming based on second-order statistics of the channel station information," in *Proceedings of the IEEE ICASSP*, Las Vegas, Nevada, March 30–April 4, 2008, pp. 2605–8.

[21] ——, "Distributed beamforming for relay networks based on second-order statistics of the channel state information," *IEEE Transactins on Signal Processing*, vol. 56, no. 9, pp. 4306–15, 2008.

[22] S. Boyd and L. Vandenberghe, *Convex Optimization*. Cambridge: Cambridge University Press, 2004.

[23] C. Helmberg, F. Rendl, R. Vanderbei, and H. Wolkowicz, "An interior-point method for semidefinite programming," *SIAM Journal of Optimization*, vol. 6, no. 2, pp. 342–61, 1996.

[24] D. P. Bertsekas, *Nonlinear Programming: Second Edition*. Belmont, MA: Athena Scientific, 2003.

[25] E. Karipidis, N. D. Sidiropoulos, and Z.-Q. Luo, "Transmit beamforming to multiple co-channel multicast groups," in *Proceedings of the IEEE CAMSAP*, Puerto Vallarta, Mexico, Dec. 13–15, 2005, pp. 109–12.

[26] A. Wiesel, Y. C. Eldar, and S. Shamai, "Linear precoding vis conic optimization for fixed MIMO receivers," *IEEE Transactions on Signal Processing*, vol. 54, no. 1, pp. 161–76, 2006.

[27] G. Pataki, "On the rank of extreme matrices in semidefinite programs and the multiplicity of optimal eigenvalues," *Mathematics of Operations Research*, vol. 23, pp. 339–58, 1998.

[28] G. Szegö, *Orthogonal Polynomials*. San Francisco, CA: American Mathematical Society, 1939.

[29] K. Usuda, H. Zhang, and M. Nakagawa, "Pre-rake performance for pulse based UWB system in a standardized UWB short-range channel," in *Proceedings of the IEEE WCNC*, Atlanta, GA, March 22–25, 2004, pp. 920–5.

[30] H. Liu, R. C. Qiu, and Z. Tian, "Error performance of pulse-based ultra-wideband MIMO systems over indoor wireless channels," *IEEE Transactions on Wireless Communications*, vol. 4, no. 6, pp. 2939–44, 2005.

[31] A. Sibille and V. P. Tran, "Spatial multiplexing in pulse based ultrawideband communications," *European Transactions on Telecommunications*, vol. 18, pp. 627–37, May 2007.

[32] Z.-Q. Luo, N. D. Sidiropoulos, P. Tseng, and S. Zhang, "Approximation bounds for quadratic optimization with homogeneous quadratic constraints," *SIAM Journal on Optimization*, vol. 18, no. 1, pp. 1–28, 2007.

[33] S. Zhang and Y. Huang, "Complex quadratic optimization and semidefinite programming," *SIAM Journal on Optimization*, vol. 16, pp. 871–90, 2006.

[34] A. M.-C. So, Y. Ye, and J. Zhang, "A unified theorem on SDP rank reduction," *Mathematics of Operations Research*, vol. 33, no. 4, pp. 910–20, 2008.

[35] S. He, Z.-Q. Luo, J. Nie, and S. Zhang, "Semidefinite relaxation bounds for indefinite homogeneous quadratic optimization," *SIAM Journal on Optimization*, vol. 19, no. 2, pp. 503–23, 2008.

[36] A. Nemirovski, C. Roos, and T. Terlaky, "On maximization of quadratic form over intersection of ellipsoids with common center," *Mathematical Programming*, vol. 86, pp. 463–73, 1999.

[37] Y. Nesterov, "Semidefinite relaxation and non-convex quadratic optimization," *Optimization Methods and Software*, vol. 12, pp. 1–20, 1997.

[38] Y. Huang and S. Zhang, "Complex matrix decomposition and quadratic programming," *Mathematics of Operations Research*, no. 32, pp. 758–68, 2007.

[39] J. F. Sturm, "Using SeDuMi 1.02, a MATLAB toolbox for optimization over symmetric cones," *Optimization Methods and Software*, vol. 11–12, pp. 625–53, 1999. Available: http://sedumi.mcmaster.ca/

[40] K. Setsompop, L. Wald, V. Alagappan, B. Gagoski, and E. Adalsteinsson, "Magnitude least squares optimization for parallel radio frequency excitation design demonstrated at 7 Tesla with eight channels," *Magnetic Resonance in Medicine*, vol. 59, pp. 908–15, 2008.

[41] C. A. van den Berg, B. van den Bergen, J. B. vande Kamer, B. W. Raaymakers, H. Kroeze, L. W. Bartels, and J. J. Lagendijk, "Simultaneous B1+ homogenization and specific absorption rate hopspot suppression using a magnetic resonance phased array transmit coil," *Magnetic Resonance in Medicine*, vol. 57, pp. 577–86, 2007.

[42] G. Adriany, P.-F. V. D. Moortele, J. Ritter, S. Moeller, E. Auerbach, C. Akgun, C. Snyder, J.-T. Vaughan, and K. Ugurbil, "A geometrically adjustable 16-channel transmit/receive transmission line array for improved RF efficiency and parallel imaging performance at 7 tesla," *Magnetic Resonance in Medicine*, vol. 59, pp. 590–7, 2008.

5 Probabilistic analysis of semidefinite relaxation detectors for multiple-input, multiple-output systems

Anthony Man-Cho So and Yinyu Ye

Due to their computational efficiency and strong empirical performance, *semidefinite relaxation* (SDR)-based algorithms have gained much attention in *multiple-input, multiple-output* (MIMO) detection. However, the theoretical performance of those algorithms, especially when applied to constellations other than the *binary phase-shift keying* (BPSK) constellation, is still not very well-understood. In this chapter we describe a recently-developed approach for analyzing the approximation guarantees of various SDR-based algorithms in the low *signal-to-noise ratio* (SNR) region. Using such an approach, we show that in the case of *M-ary phase-shift keying* (MPSK) and *quadrature amplitude modulation* (QAM) constellations, various SDR-based algorithms will return solutions with near-optimal log-likelihood values with high probability. The results described in this chapter can be viewed as average-case analyses of certain SDP relaxations, where the input distribution is motivated by physical considerations. More importantly, they give some theoretical justification for using SDR-based algorithms for MIMO detection in the low SNR region.

5.1 Introduction

Semidefinite programming (SDP) has now become an important algorithm design tool for a wide variety of optimization problems. From a practical standpoint, SDP-based algorithms have proven to be effective in dealing with various fundamental engineering problems, such as control system design [1, 2], structural design [3], signal detection [4, 5], and network localization [6–8]. From a theoretical standpoint, SDP is playing an important role in advancing the theory of algorithms. For instance, it has been used to design approximation algorithms (with the best approximation guarantees known to date) for a host of NP-hard problems [9–14], and it forms the basis for the sums of squares approach to solving polynomial optimization problems [15]. One of the problems that has received considerable attention over the years, and to which SDP techniques have been applied, is the *multiple-input, multiple-output* (MIMO) detection problem. In that problem, one is interested in detecting a vector of information-carrying symbols that is

Convex Optimization in Signal Processing and Communication, eds. Daniel P. Palomar and Yonina C. Eldar. Published by Cambridge University Press © Cambridge University Press, 2010.

being transmitted over a MIMO communication channel. Such a problem arises in many modern communication systems [16, 17], and a good solution will certainly enhance their performance (for instance, by increasing their capacity and reliability). Before we formulate the MIMO detection problem, let us fix some notation. Let $\mathbb{F}$ be either the real, or complex scalar field. Let $\mathcal{S}$ be a finite set representing the signal constellation (e.g., $\mathcal{S} = \{-1, +1\}$), and let $x \in \mathcal{S}^n$ be a vector of transmitted symbols. We assume that the vector $y \in \mathbb{F}^m$ of received signals is given by a linear combination of the transmitted symbols, subject to an *additive white Gaussian noise* (AWGN) $v \in \mathbb{F}^m$. In particular, we model the input–output relationship of the MIMO channel as

$$y = Hx + v, \tag{5.1}$$

where $H \in \mathbb{F}^{m \times n}$ is the *channel matrix* (with $m \geq n$), $y \in \mathbb{F}^m$ is the vector of received signals, and $v \in \mathbb{F}^m$ is an additive white Gaussian noise (i.e., a Gaussian vector with IID zero mean components). The common variance σ^2 of the components of v is known as the *noise variance* and it dictates the *signal-to-noise ratio* (SNR) ρ of the channel. Roughly speaking, the SNR ρ is inversely proportional to the noise variance σ^2. Now, given the channel model (5.1), the goal of the MIMO detection problem is to recover the vector of transmitted symbols x from the vector of received signals y, assuming that we only have full knowledge of the channel matrix H. Specifically, we would like to design a *detector* $\varphi : \mathbb{F}^m \times \mathbb{F}^{m \times n} \to \mathcal{S}^n$ that takes the vector of received signals $y \in \mathbb{F}^m$ and the channel matrix $H \in \mathbb{F}^{m \times n}$ as inputs and produces an estimate $\hat{x} \in \mathcal{S}^n$ of the transmitted vector $x \in \mathcal{S}^n$ as output. It turns out that under some mild assumptions, the *maximum-likelihood* (ML) detector, which is given by:

$$\hat{x} = \underset{x \in \mathcal{S}^n}{\mathrm{argmin}} \, \|y - Hx\|_2^2, \tag{5.2}$$

minimizes the error probability $\Pr(\hat{x} \neq x)$ [16]. Unfortunately, whenever $|\mathcal{S}| > 1$, the problem of computing $\hat{x}$ via (5.2) is NP-hard in general [18]. On the other hand, in the context of communications, the channel matrix H usually follows a certain probability distribution. However, it is still not known whether there exists a provably efficient algorithm for solving such instances. Therefore, much of the recent research has focused on developing heuristics that not only are efficient, but can also achieve near-ML performance. One such heuristic is the so–called semidefinite relaxation (SDR) detector, which solves an SDP relaxation of (5.2) and produces, via some rounding procedure, an approximate solution to the detection problem in polynomial time. The SDR detector was first proposed by Tan and Rasmussen [4] and Ma *et al.* [5] to handle the case where $\mathcal{S} = \{-1, +1\}$ (known as the *binary phase-shift keying* (BPSK) constellation). They showed, via simulations, that the SDR detector can be a very effective heuristic for solving the detection problem. In an attempt to understand this phenomenon, researchers have focused on the case where $\mathbb{F} = \mathbb{R}$ and $\mathcal{S} = \{-1, +1\}$ and proposed various approaches to analyze the performance of the SDR detector. These include:

- *Approximation analysis.* Since the SDR detector is based on an SDP relaxation of the quadratic minimization problem (5.2), we can evaluate the quality of such a relaxation by its *approximation guarantee*. Specifically, we would like to know whether there exists an $\alpha \geq 1$ (called the *approximation ratio*) such that for *any* instance (H, y) of problem (5.2), the SDR detector would compute a feasible solution whose objective value is at most $\alpha \cdot \text{OPT}(H, y)$, where

$$\text{OPT}(H, y) \equiv \min_{x \in \mathcal{S}^n} \|y - Hx\|_2^2$$

denotes the optimal value of the instance (H, y). Unfortunately, if we consider the objective value of the solution returned by the SDR detector and compare it directly to the optimal value of the SDP relaxation (as is done in almost all approximation analyses of SDP relaxations), then we cannot hope to get a bounded approximation ratio in general. This is due to the fact that there exist instances (H, y) of problem (5.2) whose optimal values are positive, and yet the optimal values of the corresponding SDP relaxations are zero. However, such a difficulty can be circumvented if one considers only those instances that are generated by the so-called Rayleigh fading channel model. Specifically, under the assumptions that (a) $m = n$, in other words there is an equal number of inputs and outputs; (b) $H \in \mathbb{R}^{n \times n}$ is a real Gaussian random matrix (i.e., the entries of H are IID standard, real Gaussian random variables) and is independent of the noise vector v; and (c) the SNR ρ is sufficiently small, Kisialiou and Luo [19] were able to show that the probability of the SDR detector yielding a constant factor approximation to problem (5.2) tends to 1 as $n \to \infty$. Here, the probability is taken over all possible realizations of (H, v), as well as the randomness in the rounding procedure.

- *Error analysis.* Recall that the goal of the MIMO detection problem is to recover the vector of transmitted symbols from the vector of received signals. Hence, we can also evaluate the performance of the SDR detector by analyzing its *error probability*, that is, the probability that the symbol vector $\hat{x}$ returned by the SDR detector differs from the transmitted symbol vector x. Such an approach has recently been pursued by Jaldén and Ottersten [20], who showed that when $H \in \mathbb{R}^{m \times n}$ is a real Gaussian random matrix with $m \geq n$ fixed and is independent of the noise vector v, the error probability of the SDR detector is asymptotically (as the SNR ρ tends to infinity) on the order of $\rho^{-m/2}$. More precisely, they showed that

$$\lim_{\rho \to \infty} \frac{\log \text{Pr}(\hat{x} \neq x)}{\log \rho} = -\frac{m}{2},$$

where the probability is taken over all possible realizations of (H, v).

Although the results above offer some insights into the performance of the SDR detector, they are all asymptotic in nature. In particular, they do not fully explain the performance of the SDR detector *in practical settings*, that is, when the SNR ρ and the channel size parameters m, n are finite. Moreover, the techniques used in establishing those results depend crucially on the fact that $\mathbb{F} = \mathbb{R}$ and $\mathcal{S} = \{-1, +1\}$, and hence they do not easily generalize to cover other cases.

In this chapter we give an overview of an approach for obtaining *non-asymptotic* guarantees on the approximation ratio of the SDR detector in the low SNR region. The approach was first introduced in [21] and it incorporates ideas from Kisialiou and Luo [19], as well as results from non-asymptotic random matrix theory [22, 23]. As an illustration, we consider the case where $\mathbb{F} = \mathbb{C}$ and $\mathcal{S}$ is the *M-ary phase-shift keying* (MPSK) constellation. We show that in the low SNR region, the probability of the SDR detector yielding a constant factor approximation to problem (5.2) tends to 1 *exponentially fast* as the channel size increases. Besides yielding non-asymptotic performance guarantees, our approach is also quite general and can be applied to other variants of the SDR detector as well (see Section 5.4 and cf. [24]).

The rest of this chapter is organized as follows. In Section 5.2 we give a more precise formulation of the MIMO detection problem and introduce the SDR detector as a heuristic for solving it. We then analyze the approximation guarantee of the SDR detector for the MPSK constellations in Section 5.3. As mentioned earlier, the techniques we developed are quite general. We illustrate this by showing how our approach can be used to analyze the approximation guarantee of the SDR detector for the *quadrature amplitude modulation* (QAM) constellations in Section 5.4. Finally, we close with some concluding remarks in Section 5.5.

5.2 Problem formulation

To begin our discussion, consider the scenario where symbols from an M-ary phase-shift keying (MPSK) constellation are transmitted across a Rayleigh fading channel. Specifically, consider a channel of the form (5.1), where $H \in \mathbb{C}^{m \times n}$ is the channel matrix whose entries are IID standard, complex Gaussian random variables:

$$H_{pq} = g_{pq}^1 + j g_{pq}^2 \quad \text{for } 1 \le p \le m, \ 1 \le q \le n,$$

where $j \equiv \sqrt{-1}$, and g_{pq}^1 and g_{pq}^2 are independent, real Gaussian random variables with mean 0 and variance $1/2$ [17, Appendix A]; $v \in \mathbb{C}^m$ is an additive white Gaussian noise with variance σ^2 (i.e., v is a circular, symmetric, complex Gaussian random vector with covariance matrix $\sigma^2 I$ and is independent of H [17, Appendix A]); $x \in \mathcal{S}_M^n$ is the vector of transmitted symbols, where $\mathcal{S}_M$ ($M = 2, 3, \ldots$) is the MPSK constellation:

$$\mathcal{S}_M = \{\exp(2\pi l j / M) : l = 0, 1, \ldots, M - 1\},$$

and $y \in \mathbb{C}^m$ is the vector of received signals. We remark that the above channel model is physically motivated. For instance, it has been used to model multiple-antenna wireless MIMO channels with rich scattering [17].

As mentioned in the introduction, the ML detector attempts to recover the transmitted symbol vector $x \in \mathcal{S}_M^n$ from both the received signal vector $y \in \mathbb{C}^m$ and a realization of the channel $H \in \mathbb{C}^{m \times n}$ (which is known to the detector) by solving the following discrete least-squares problem:

$$v_{ml} = \min_{x \in \mathcal{S}_M^n} \|y - Hx\|_2^2. \tag{5.3}$$

(The value v_{ml} is sometimes known as the *optimal log-likelihood value* of the detection problem.) Now, since it is still not known whether there exists a provably efficient algorithm for solving (5.3), many heuristics have been proposed. One such heuristic is based on solving an SDP relaxation of (5.3). To derive the SDP relaxation, we first observe that problem (5.3) is equivalent to the following homogenized problem:

$$
\begin{aligned}
v_{ml} &= \min_{(x,t) \in \mathcal{S}_M^{n+1}} \|yt - Hx\|_2^2 \\
&= \min_{(x,t) \in \mathcal{S}_M^{n+1}} \left\{ |t|^2 \|y\|_2^2 - \bar{t}y^*Hx - tx^*H^*y + x^*H^*Hx \right\} \\
&= \min_{z \in \mathcal{S}_M^{n+1}} \operatorname{tr}(Qzz^*),
\end{aligned}
\tag{5.4}
$$

where H^* (resp. z^*) denotes the conjugate transpose of H (resp. z), and

$$
Q = \begin{bmatrix} H^*H & -H^*y \\ -y^*H & \|y\|_2^2 \end{bmatrix} \in \mathbb{C}^{(n+1) \times (n+1)}.
$$

Note that problem (5.4) is nonconvex, and hence is difficult to solve in general. However, observe that $zz^* \succeq \mathbf{0}$ for any $z \in \mathbb{C}^{n+1}$, and that for $z \in \mathcal{S}_M^{n+1}$, we have

$$
\operatorname{diag}(zz^*) = (z_1 z_1^*, \ldots, z_{n+1} z_{n+1}^*) = \mathbf{e},
$$

where $\mathbf{e} \in \mathbb{R}^{n+1}$ is the vector of all ones. Thus, we may relax problem (5.4) to the following *complex* SDP [25, 26]:

$$
\begin{aligned}
v_{sdp} = \quad &\inf \quad &&\operatorname{tr}(QZ) \\
&\text{subject to} \quad &&\operatorname{diag}(Z) = \mathbf{e} \\
& &&Z \succeq \mathbf{0},
\end{aligned}
\tag{5.5}
$$

where $Z \in \mathbb{C}^{(n+1) \times (n+1)}$ is a Hermitian positive semidefinite matrix. In particular, if $\hat{z} \in \mathcal{S}_M^{n+1}$ is a feasible solution to problem (5.4), then $\hat{Z} = \hat{z}\hat{z}^* \in \mathbb{C}^{(n+1) \times (n+1)}$ is a feasible solution to problem (5.5). Moreover, we have

$$
\operatorname{tr}(Q\hat{z}\hat{z}^*) = \operatorname{tr}(Q\hat{Z}). \tag{5.6}
$$

However, the converse need not be true. In other words, given a feasible solution $\hat{Z} \in \mathbb{C}^{(n+1) \times (n+1)}$ to problem (5.5), there may not be a corresponding feasible solution $\hat{z} \in \mathcal{S}_M^{n+1}$ to problem (5.4) that satisfies (5.6).

Now, since $Q \succeq \mathbf{0}$, and problem (5.5) is a relaxation of problem (5.4), we have $0 \leq v_{sdp} \leq v_{ml}$. We should emphasize that both v_{ml} and v_{sdp} depend on the particular realizations of H *and* v, since y is related to H and v via (5.1).

One of the upshots of the complex SDP formulation (5.5) is that it can be solved to any desired accuracy in polynomial time using either the ellipsoid method [27] or the more efficient interior-point method [28, 29] (see also the discussion in [30]). However, since problem (5.5) is a relaxation of problem (5.3), we still need a procedure that converts a feasible solution $\hat{Z} \in \mathbb{C}^{(n+1)\times(n+1)}$ to (5.5) into a feasible solution $\hat{x} \in \mathcal{S}_M^n$ to (5.3). Below is one such procedure:

1. Partition the matrix $\hat{Z} \in \mathbb{C}^{(n+1)\times(n+1)}$ as:

$$
\hat{Z} = \begin{bmatrix} U & u \\ u^* & 1 \end{bmatrix},
\tag{5.7}
$$

where $u \in \mathbb{C}^n$ and $U \in \mathbb{C}^{n\times n}$. Note that since $\hat{Z} \succeq \mathbf{0}$ and $\mathrm{diag}(\hat{Z}) = \mathbf{e}$, we must have $|u_k| \leq 1$ for $k = 1, \ldots, n$. Indeed, suppose that $|u_l| > 1$ for some $l = 1, \ldots, n$. Let $v \in \mathbb{C}^{n+1}$ be such that $v_l = u_l$, $v_{n+1} = -1$, and $v_i = 0$ otherwise. Then, we have

$$
v^* \hat{Z} v = |u_l|^2 + 1 - 2|u_l|^2 = 1 - |u_l|^2 < 0,
$$

which contradicts the fact that $\hat{Z} \succeq \mathbf{0}$.

2. Let $z^i = (z_1^i, \ldots, z_{n+1}^i) \in \mathbb{C}^{n+1}$, where $i = 1, \ldots, m$, be m independent random vectors, where the entries of each are independently distributed according to the following distribution:

$$
\Pr\left(z_k^i = e^{2\pi j l/M}\right) = \frac{1 + \Re(u_k e^{-2\pi j l/M})}{M} \quad \text{for } 1 \leq k \leq n,\ 0 \leq l \leq M - 1,
$$

$$
\Pr\left(z_{n+1}^i = e^{2\pi j l/M}\right) = \frac{1 + \Re(e^{-2\pi j l/M})}{M} \quad \text{for } 0 \leq l \leq M - 1.
\tag{5.8}
$$

Here, $\Re(z)$ denotes the real part of $z \in \mathbb{C}$. Note that (5.8) defines a valid probability distribution on $\mathcal{S}_M$. Indeed, since $|u_k| \leq 1$, we have

$$
\frac{1 + \Re(u_k e^{-2\pi j l/M})}{M} \geq 0 \quad \text{for } k = 1, \ldots, n,\ l = 0, 1, \ldots, M - 1.
$$

Similarly, we have

$$
\frac{1 + \Re(e^{-2\pi j l/M})}{M} \geq 0 \quad \text{for } l = 0, 1, \ldots, M - 1.
$$

Moreover, we have

$$\frac{1}{M}\sum_{l=0}^{M-1}\left(1+\Re(u_k e^{-2\pi jl/M})\right)=1+\frac{1}{M}\Re\left(u_k\sum_{l=0}^{M-1}e^{-2\pi jl/M}\right)=1$$

for $k=1,\ldots,n$, and

$$\frac{1}{M}\sum_{l=0}^{M-1}\left(1+\Re(e^{-2\pi jl/M})\right)=1.$$

Consequently, the vectors $z^1,\ldots,z^m$ are all feasible for (5.4).

3. Let

$$\bar{i}=\operatorname*{argmin}_{1\le i\le m}\left(z^i\right)^*Qz^i$$

and define $\hat{z}=z^{\bar{i}}$. Set $v_{sdr}=\hat{z}^*Q\hat{z}$ and return

$$\hat{x}=\overline{\hat{z}_{n+1}}(\hat{z}_1,\ldots,\hat{z}_n)\in\mathcal{S}_M^n$$

as our candidate solution to (5.3). Note that $v_{sdr}=\|y-H\hat{x}\|_2^2$, and hence we have $v_{ml}\le v_{sdr}$.

Naturally, we are interested in the performance of the above rounding procedure, and one measure is the approximation ratio. Specifically, we would like to establish a probabilistic upper bound on the ratio v_{sdr}/v_{ml}, where the probability is computed over all possible realizations of H and v, as well as the random vectors generated according to (5.8). Intuitively, if the ratio is close to 1, then we may conclude that the solution generated by the rounding procedure is close (in terms of the log-likelihood value) to the optimal ML solution. In the next section we show that the aforementioned SDP-based procedure will actually achieve a *constant* approximation ratio (i.e., independent of m and n) in the low SNR region with high probability. This gives a strong indication that the SDR detector is a good heuristic for solving the MIMO detection problem, at least in the low SNR region.

5.3   Analysis of the SDR detector for the MPSK constellations

5.3.1   The $M\ge 3$ case

We shall first focus on the case where $M\ge 3$. For reasons that will become clear, the case where $M=2$ requires a slightly different treatment and will be dealt with later. Our goal in this section is to prove the following theorem:

THEOREM 5.1　*Let $m \geq \max\{n, 7\}$. Consider a Rayleigh fading channel with a MPSK constellation, where $M \geq 3$. Suppose that the noise variance σ^2 satisfies $\sigma^2 > 63n$, so that*

$$\Lambda \equiv \left[\frac{n + \sigma^2}{2} - 4\sqrt{n(n + \sigma^2)}\right] > 0.$$

Then, we have

$$\Pr_{(H,v,z)}\left[v_{sdr} \leq 2\left(\frac{1}{4} + \frac{8n}{\Lambda} + \frac{3(n + \sigma^2)}{2\Lambda}\right) v_{sdp}\right] \geq 1 - 3\exp(-m/6) - 2^{-m} > 0,$$

$$(5.9)$$

where $\Pr_{(H,v,z)}(\cdot)$ means that the probability is computed over all possible realizations of (H, v) and the random vectors $z^1, \ldots, z^m$ in Step 2 of the randomized rounding procedure.

Before we prove Theorem 5.1, let us briefly comment on the result it claims. First, observe that for any fixed $n \geq 1$, the function $\alpha_n : \mathbb{R}_+ \to \mathbb{R}_+$, where

$$\alpha_n(\sigma) = 2\left(\frac{1}{4} + \frac{8n}{\Lambda} + \frac{3(n + \sigma^2)}{2\Lambda}\right),$$

is monotonically decreasing (see Figure 5.1). Now, upon setting $\sigma^2 = \theta n$ for some $\theta > 63$, we obtain

$$\alpha_n(\theta n) = \bar{\alpha}(\theta) \equiv \frac{1}{2} + \frac{32 + 6(\theta + 1)}{\theta + 1 - 8\sqrt{\theta + 1}} \qquad \text{for } n \geq 1$$

(see Figure 5.1). In particular, we see that if $\theta > 63$ is fixed, then $\alpha_n(\theta n)$ is a constant that is independent of n. Moreover, we have

$$\lim_{\sigma^2 \to \infty} \alpha_n(\sigma) = \frac{13}{2} \qquad \text{for } n \geq 1.$$

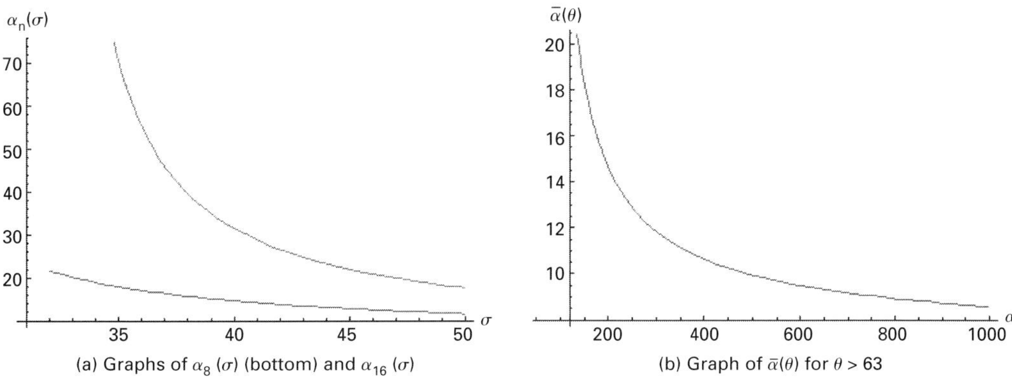

Figure 5.1　　Behavior of the approximation ratio $\alpha_n(\sigma)$.

Thus, Theorem 5.1 implies that in the low SNR region (i.e., when $\sigma^2 > 63n$), the SDR detector will produce a constant factor approximate solution to the MIMO detection problem (5.3) with exponentially high probability as the channel size increases. We remark that the constants in the bound (5.9) are chosen to simplify the exposition and have not been optimized. With a more refined analysis, those constants can certainly be improved.

The proof of Theorem 5.1 consists of two steps. The first step is to show that, conditioned on a particular realization of (H, v), the value v_{sdr} is, with high probability, at most $O(v_{sdp} + \Delta)$, where Δ depends only on (H, v). Then, in the second step we analyze what effect does the distribution of (H, v) have on the value v_{sdr}. In particular, we will show that v_{sdp} and Δ are comparable (i.e., of the same order) with high probability. This will then imply Theorem 5.1.

To begin, consider a particular realization of (H, v) (and hence of Q). Let $\hat{Z} \in \mathbb{C}^{(n+1)\times(n+1)}$ be a feasible solution to (5.5) with objective value v_{sdp}, and partition $\hat{Z}$ according to (5.7). Let $z^1, \ldots, z^m$ be the random vectors generated in Step 2 of the randomized rounding procedure. Set $\Gamma \equiv \mathbb{E}_z\left[(z^1)^* Q z^1\right]$, where $\mathbb{E}_z$ denotes the mathematical expectation with respect to the distribution defined in (5.8). Then, by Markov's inequality and the fact that the random vectors $z^1, \ldots, z^m$ are IID, we have

$$\Pr_z\left(v_{sdr} \geq 2\Gamma\right) = \left[\Pr_z\left((z^1)^* Q z^1 \geq 2\Gamma\right)\right]^m \leq 2^{-m}. \tag{5.10}$$

To get a hold on the value of Γ, we need the following proposition:

PROPOSITION 5.1   *([31, Lemma 2.1]) Let $u_1, \ldots, u_n \in \mathbb{C}$ be given by (5.7), and set $u_{n+1} = 1$. Let z^1 be generated according to the distribution defined in (5.8). Then, for $M \geq 3$ and $1 \leq k \neq k' \leq n+1$, we have*

$$\mathbb{E}_z\left[z_k^1 \overline{z_{k'}^1}\right] = \frac{1}{4} u_k \overline{u_{k'}}.$$

Now, set $\hat{u} = (u, 1) \in \mathbb{C}^{n+1}$. Armed with Proposition 5.1, let us compute Γ for the case where $M \geq 3$:

$$\Gamma = \mathbb{E}_z\left[\sum_{k=1}^{n+1}\sum_{k'=1}^{n+1} Q_{kk'} z_k^1 \overline{z_{k'}^1}\right]$$

$$= \sum_{k=1}^{n+1} Q_{kk} + 2\Re\left(\frac{1}{4}\sum_{1\leq k<k'\leq n} Q_{kk'} u_k \overline{u_{k'}}\right) + 2\Re\left(\frac{1}{4}\sum_{k=1}^{n} Q_{k,n+1} u_k\right)$$

$$= \frac{1}{4}\hat{u}^* Q \hat{u} + \sum_{k=1}^{n} Q_{kk}\left(1 - \frac{1}{4}|u_k|^2\right) + \frac{3}{4} Q_{n+1,n+1}$$

$$\leq \frac{1}{4}\hat{u}^* Q \hat{u} + \sum_{k=1}^{n} Q_{kk} + \frac{3}{4}\|y\|_2^2. \tag{5.11}$$

Since $\hat{Z} \succeq \mathbf{0}$, we have $U \succeq uu^*$ by the Schur complement. It follows that $\hat{Z} \succeq \hat{u}\hat{u}^*$, whence

$$\hat{u}^*Q\hat{u} = \operatorname{tr}(Q\hat{u}\hat{u}^*) \leq \operatorname{tr}(Q\hat{Z}) = v_{sdp}.$$

In particular, we conclude that

$$\Gamma \leq \frac{1}{4}v_{sdp} + \operatorname{tr}(H^*H) + \frac{3}{4}\|y\|_2^2 \leq \frac{1}{4}v_{sdp} + n \cdot \lambda_{max}(H^*H) + \frac{3}{4}\|y\|_2^2, \tag{5.12}$$

where $\lambda_{max}(H^*H)$ is the largest eigenvalue of H^*H.

Now, if we could show that the values v_{sdp}, $n \cdot \lambda_{max}(H^*H)$ and $\|y\|_2^2$ are all within a constant factor of each other with high probability (with respect to the realizations of (H, v)), then (5.10) and (5.12) would imply that v_{sdr} and v_{sdp} are within a constant factor of each other with high probability (with respect to the realizations of (H, v) *and* $\hat{z}$). To carry out this idea, we first need estimates on the largest eigenvalue of the matrix H^*H and the squared norm $\|y\|_2^2$. These are given below, and their proofs can be found in Sections A5.1 and A5.4:

- **Estimate on $\lambda_{max}(H^*H)$.**

$$\Pr_{H} \left(\lambda_{max}(H^*H) > 8m \right) \leq \exp(-m/2). \tag{5.13}$$

- **Estimates on $\|y\|_2^2$.**

$$\Pr_{(H,v)} \left[\|y\|_2^2 \leq \frac{1}{2}m(n + \sigma^2) \right] \leq \exp(-m/6). \tag{5.14}$$

$$\Pr_{(H,v)} \left[\|y\|_2^2 \geq 2m(n + \sigma^2) \right] \leq \exp(-m/4). \tag{5.15}$$

The above estimates imply that in order to prove Theorem 5.1, it suffices to show that v_{sdp} is large (say, on the order of mn) with high probability (with respect to the realizations of (H, v)). By the SDP weak-duality theorem [32], it suffices to consider the dual of (5.5) and exhibit a dual feasible solution with large objective value. We remark that such an idea has been used in the work of Kisialiou and Luo [19]. However, our approach differs from that of [19] in that we are able to obtain a non-asymptotic result.

To begin, let us write down the dual of (5.5):

$$\sup \quad \operatorname{tr}(W)$$

$$\text{subject to} \quad Q - W \succeq \mathbf{0}$$

$$W \in \mathbb{R}^{(n+1)\times(n+1)} \text{ diagonal.} \tag{5.16}$$

Let $\alpha > 0$ and $\beta \in \mathbb{R}$ be parameters to be chosen, and define

$$\hat{W} = \begin{bmatrix} -\alpha I & \mathbf{0} \\ \mathbf{0}^T & \beta \end{bmatrix}.$$

In order for $\hat{W}$ to be feasible for (5.16), we must have $Q - \hat{W} \succeq 0$. By the Schur complement, this is equivalent to

$$y^* \left[I - H \left(H^*H + \alpha I \right)^{-1} H^* \right] y \geq \beta. \tag{5.17}$$

(Note that $H^*H + \alpha I$ is invertible for any $\alpha > 0$.) Now, observe that

$$y^* \left[I - H \left(H^*H + \alpha I \right)^{-1} H^* \right] y \geq \left[1 - \lambda_{\max} \left(H \left(H^*H + \alpha I \right)^{-1} H^* \right) \right] \cdot \|y\|_2^2$$

$$\geq \left(1 - \frac{\lambda_{max}(H^*H)}{\alpha} \right) \cdot \|y\|_2^2,$$

where the last inequality follows from the fact that

$$\lambda_{max} \left[H \left(H^*H + \alpha I \right)^{-1} H^* \right] \leq \lambda_{max}(HH^*) \cdot \lambda_{max} \left[\left(H^*H + \alpha I \right)^{-1} \right]$$

$$\leq \frac{\lambda_{max}(H^*H)}{\alpha}.$$

Thus, by setting $\beta = \beta_0$ in (5.17), where

$$\beta_0 \equiv \frac{1}{2} \left(1 - \frac{8m}{\alpha} \right) m(n + \sigma^2),$$

we conclude from (5.13) and (5.14) that the matrix $\hat{W}$ will be feasible for (5.16) with probability at least $1 - \exp(-m/2) - \exp(-m/6)$. In that event we have

$$v_{sdp} \geq \mathrm{tr}(\hat{W}) = \beta_0 - n\alpha$$

by the SDP weak-duality theorem, and upon optimizing over $\alpha > 0$, we see that whenever $\sigma^2 > 63n$, the following inequalities hold:

$$v_{sdp} \geq \left[\frac{n + \sigma^2}{2} - 4\sqrt{n(n + \sigma^2)} \right] m > 0. \tag{5.18}$$

Now, using (5.18), we are ready to finish the proof of Theorem 5.1. First, note that by (5.12), (5.13), and (5.15), we have

$$\Gamma \leq \frac{1}{4} v_{sdp} + 8mn + \frac{3}{2} m(n + \sigma^2)$$

with probability at least $1 - \exp(-m/2) - \exp(-m/4)$. Hence, with probability at least $1 - 3\exp(-m/6)$, we have

$$\Gamma \leq \left(\frac{1}{4} + \frac{8n}{\Lambda} + \frac{3(n + \sigma^2)}{2\Lambda} \right) v_{sdp},$$

where

$$\Lambda \equiv \left[\frac{n + \sigma^2}{2} - 4\sqrt{n(n + \sigma^2)} \right] > 0.$$

This, together with (5.10), implies that

$$\Pr_{(H,v,z)} \left[v_{sdr} \le 2 \left(\frac{1}{4} + \frac{8n}{\Lambda} + \frac{3(n + \sigma^2)}{2\Lambda} \right) v_{sdp} \right] \ge 1 - 3\exp(-m/6) - 2^{-m},$$

and the proof of Theorem 5.1 is completed.

5.3.2   The $M = 2$ case

It is a bit unfortunate that the above argument does not readily extend to cover the $M = 2$ case. The main difficulty is the following. Proposition 5.1 now gives

$$\mathbb{E}_z[z_k^1 \overline{z_{k'}^1}] = \Re(u_k)\Re(u_{k'}) \quad \text{for } 1 \le k \ne k' \le n + 1,$$

which implies the bound

$$\Gamma \le \Re(\hat{u})^T Q \Re(\hat{u}) + n \cdot \lambda_{max}(H^* H),$$

where $\Re(\hat{u}) = (\Re(\hat{u}_1), \ldots, \Re(\hat{u}_{n+1})) \in \mathbb{R}^{n+1}$ (cf. (5.12)). However, there is no clear relationship between the quantities $\Re(\hat{u})^T Q \Re(\hat{u})$ and $v_{sdp} = \hat{u}^* Q \hat{u}$. To circumvent this difficulty, we may proceed as follows. Observe that the discrete least-squares problem (5.4) can be written as

$$v_{ml} = \min_{(x,t)\in\{-1,+1\}^{n+1}} \|\tilde{y}t - \tilde{H}x\|_2^2 = \min_{z\in\{-1,+1\}^{n+1}} \text{tr}(\tilde{Q}zz^T), \tag{5.19}$$

where

$$\tilde{y} = \begin{bmatrix} \Re(y) \\ \Im(y) \end{bmatrix} \in \mathbb{R}^{2m}, \quad \tilde{H} = \begin{bmatrix} \Re(H) \\ \Im(H) \end{bmatrix} \in \mathbb{R}^{2m \times n}, \quad \tilde{v} = \begin{bmatrix} \Re(v) \\ \Im(v) \end{bmatrix} \in \mathbb{R}^{2m},$$

and

$$\tilde{Q} = \begin{bmatrix} \tilde{H}^T \tilde{H} & -\tilde{H}^T \tilde{y} \\ -\tilde{y}^T \tilde{H} & \|\tilde{y}\|_2^2 \end{bmatrix} \in \mathbb{R}^{(n+1) \times (n+1)}.$$

Thus, problem (5.19) can be relaxed to a *real* SDP of the form (5.5), where Q is replaced by $\tilde{Q}$. Now, let $\hat{Z}$ be a feasible solution to the SDP with objective value v_{sdp} (note that $\hat{Z}$ is now an $(n+1) \times (n+1)$ *real* matrix). Clearly, we can still apply the randomized rounding procedure in Section 5.2 on $\hat{Z}$. Since $\Re(u_k) = u_k$ for $k = 1, \ldots, n + 1$, the candidate solution $\hat{x}$ returned by the rounding procedure will be feasible (i.e., $\hat{x} \in \{-1, +1\}^n$). Moreover, its objective value satisfies

$$\Pr_z(v_{sdr} \geq 2\mathbb{E}_z[(z^1)^T \tilde{Q} z^1]) \leq 2^{-m}. \tag{5.20}$$

Now, it can be readily verified that

$$\mathbb{E}_z[z_k^1 z_{k'}^1] = u_k u_{k'} \quad \text{for } 1 \leq k \neq k' \leq n+1,$$

whence

$$\mathbb{E}_z[(z^1)^T \tilde{Q} z^1] = \sum_{k=1}^{n+1} \tilde{Q}_{kk} + 2 \sum_{1 \leq k < k' \leq n} \tilde{Q}_{kk'} u_k u_{k'} + 2 \sum_{k=1}^{n} \tilde{Q}_{k,n+1} u_k$$

$$\leq v_{sdp} + n \cdot \lambda_{max}(\tilde{H}^T \tilde{H}) \tag{5.21}$$

(cf. (5.11)). Since $\tilde{H}$ is a random matrix whose entries are IID, real Gaussian random variables with mean 0 and variance $1/2$, we can use the machinery developed in Section A5.1 to obtain the following estimate:

- **Estimate on $\lambda_{max}(\tilde{H}^T \tilde{H})$.**

$$\Pr_H \left(\lambda_{max}(\tilde{H}^T \tilde{H}) > 6m \right) \leq \exp(-m/2).$$

(see Section A5.3 for the proof). Furthermore, by the results in Section A5.4, we have:

- **Estimates on $\|\tilde{y}\|_2^2$.**

$$\Pr_{(H,v)} \left[\|\tilde{y}\|_2^2 \leq \frac{1}{2} m(n + \sigma^2) \right] \leq \exp(-m/6).$$

Hence, upon following the argument in Section 5.3.1, we see that with probability at least $1 - \exp(-m/2) - \exp(-m/6)$, we have

$$v_{sdp} \geq \left[\frac{n + \sigma^2}{2} - \sqrt{12n(n + \sigma^2)} \right] m. \tag{5.22}$$

Moreover, whenever $\sigma^2 > 47n$, the quantity on the right-hand side of (5.22) is strictly positive. Now, using (5.21) and (5.22), we conclude that the inequality

$$\mathbb{E}_z[(z^1)^T \tilde{Q} z^1] \leq \left(1 + \frac{6n}{\Lambda} \right) v_{sdp}$$

will hold with probability at least $1 - 2\exp(-m/6)$, where

$$\Lambda \equiv \left[\frac{n + \sigma^2}{2} - \sqrt{12n(n + \sigma^2)} \right] > 0.$$

This, together with (5.20), implies that

$$\Pr_{(H,v,z)}\left[v_{sdr} \le 2\left(1 + \frac{6n}{\Lambda}\right)v_{sdp}\right] \ge 1 - 2\exp(-m/6) - 2^{-m}.$$

We now summarize our result as follows:

THEOREM 5.2   *Let $m \ge \max\{n,5\}$. Consider a Rayleigh fading channel with a BPSK constellation. Suppose that the noise variance σ^2 satisfies $\sigma^2 > 47n$, so that*

$$\Lambda \equiv \left[\frac{n + \sigma^2}{2} - \sqrt{12n(n + \sigma^2)}\right] > 0.$$

Then, we have

$$\Pr_{(H,v,z)}\left[v_{sdr} \le 2\left(1 + \frac{6n}{\Lambda}\right)v_{sdp}\right] \ge 1 - 2\exp(-m/6) - 2^{-m} > 0.$$

5.4   Extension to the QAM constellations

In the last section we saw how SDP duality theory and results in non-asymptotic random matrix theory can be used to establish approximation bounds for the SDR detector in the case of MPSK constellations. It turns out that those techniques are quite general and can be used to analyze the performance of SDR detectors for other constellations. As an illustration, let us apply those techniques to analyze the performance of a version of the SDR detector for the QAM constellations.

As with the case of MPSK constellations, the starting point is the discrete least-squares problem

$$v_{ml} = \min_{x \in \mathcal{S}_q^n} \|y - Hx\|_2^2.$$

Here, $H \in \mathbb{C}^{m \times n}$, $v \in \mathbb{C}^m$ and $y = Hx + v \in \mathbb{C}^m$ are as in Section 5.2, but $\mathcal{S}_q$ is now the 4^q-QAM constellation, where $q \ge 1$ is some fixed integer:

$$\mathcal{S}_q = \left\{s_R + js_I : s_R, s_I \in \{\pm 1, \pm 3, \pm 5, \ldots, \pm(2^q - 1)\}\right\}.$$

Upon setting

$$\tilde{y} = \begin{bmatrix}\Re(y)\\\Im(y)\end{bmatrix} \in \mathbb{R}^{2m}, \quad \tilde{H} = \begin{bmatrix}\Re(H) & -\Im(H)\\\Im(H) & \Re(H)\end{bmatrix} \in \mathbb{R}^{2m \times 2n}, \quad \tilde{v} = \begin{bmatrix}\Re(v)\\\Im(v)\end{bmatrix} \in \mathbb{R}^{2m},$$

it can be readily verified that

$$\tilde{y} = \tilde{H}\begin{bmatrix}\Re(x)\\\Im(x)\end{bmatrix} + \tilde{v}.$$

Hence, we have

$$v_{ml} = \min_{x \in \tilde{\mathcal{S}}_q^{2n}} \|\tilde{y} - \tilde{H}x\|_2^2 \quad \text{where } \tilde{\mathcal{S}}_q = \big\{ \pm 1, \pm 3, \pm 5, \ldots, \pm(2^q - 1) \big\}. \tag{5.23}$$

Now, there are many ways to relax problem (5.23) into an SDP [33–37]. For the sake of simplicity, we shall follow the approach of Mao *et al.* [35]. We remark that such a choice does not limit the applicability of our results, as the recent work of Ma *et al.* [38] allows us to transfer those results to other SDP relaxations as well.

The main observation of Mao *et al.* [35] is that given any integer $q \geq 1$, we have

$$\tilde{\mathcal{S}}_q^{2n} = \big\{x_1 + 2x_2 + \cdots + 2^{q-1}x_q : x_1, \ldots, x_q \in \{-1, 1\}^{2n}\big\}. \tag{5.24}$$

In other words, every symbol $s \in \tilde{\mathcal{S}}_q$ can be expressed as

$$s = s_1 + 2s_2 + \cdots + 2^{q-1}s_q$$

for some $s_1, \ldots, s_q \in \{-1, 1\}$. Note that the bits $s_1, \ldots, s_q$ need not correspond to the actual information bits that are mapped into the symbol s. In particular, the discrete least-squares problem

$$v_{ml} = \min_{x \in \{-1,1\}^{2qn}} \|\tilde{y} - \hat{H}x\|_2^2 \tag{5.25}$$

where $\hat{H} = \begin{bmatrix} \tilde{H} & 2\tilde{H} & 4\tilde{H} & \cdots & 2^{q-1}\tilde{H} \end{bmatrix} \in \mathbb{R}^{2m \times 2qn}$ – which is equivalent to problem (5.23) due to the representation (5.24) – does not depend on how the information bits are mapped to the symbols. Now, the upshot of (5.25) is that it can be treated using the techniques developed for the BPSK case (see Section 5.3.2). In particular, problem (5.25) can be relaxed to a real SDP of the form (5.5), and using the rounding procedure given in Section 5.2, one can extract from any feasible solution to the SDP, a feasible solution to (5.25). Moreover, upon following the argument in Section 5.3.2, it is clear that in order to determine the approximation guarantee of the SDP, it suffices to estimate the largest singular value of $\hat{H}$. Towards that end, we need the following proposition:

PROPOSITION 5.2 *Let $A \in \mathbb{R}^{m \times n}$ be an arbitrary matrix, and let $q \geq 1$ be an integer. Set*

$$\hat{A} = \begin{bmatrix} A & 2A & 4A & \cdots & 2^{q-1}A \end{bmatrix} \in \mathbb{R}^{m \times qn}.$$

Then, we have

$$\|\hat{A}\|_\infty = \sqrt{\frac{4^q - 1}{3}} \cdot \|A\|_\infty.$$

Proof Let $\mathbb{S}^{n-1} = \{x \in \mathbb{R}^n : \|x\|_2 = 1\}$. Recall that

$$\|\hat{A}\|_\infty = \sup_{u \in \mathbb{S}^{m-1}} \sup_{v \in \mathbb{S}^{qn-1}} u^T \hat{A} v.$$

Now, for any $v \in \mathbb{S}^{qn-1}$, we write $v = (v_1, \ldots, v_q)$, where $v_1, \ldots, v_q \in \mathbb{R}^n$. Then, given a fixed $u \in \mathbb{S}^{m-1}$, we have

$$\sup_{v \in \mathbb{S}^{qn-1}} u^T \hat{A} v = \sup_{v=(v_1,\ldots,v_q) \in \mathbb{S}^{qn-1}} \sum_{k=1}^{q} 2^{k-1} u^T A v_k$$

$$= \|A^T u\|_2 \cdot \left(\sup_{\|\theta\|_2 \leq 1} \sum_{k=1}^{q} 2^{k-1} \theta_k \right).$$

By the Karush–Kuhn–Tucker (KKT) theorem [39], the optimal solution to the problem

$$\sup_{\|\theta\|_2 \leq 1} \sum_{k=1}^{q} 2^{k-1} \theta_k$$

is given by

$$\theta_k^* = 2^{k-1} \cdot \sqrt{\frac{3}{4^q - 1}} \qquad \text{for } k = 1, \ldots, q,$$

and the optimal value is $\sqrt{(4^q - 1)/3}$. It follows that

$$\|\hat{A}\|_\infty = \sqrt{\frac{4^q - 1}{3}} \cdot \sup_{u \in \mathbb{S}^{m-1}} \|A^T u\|_2 = \sqrt{\frac{4^q - 1}{3}} \cdot \|A\|_\infty,$$

as desired. $\blacksquare$

Now, Propositions 5.2 and A5.3 imply the following:

- **Estimate on $\lambda_{max}(\hat{H}^T \hat{H})$.**

$$\Pr_H \left(\lambda_{max}(\hat{H}^T \hat{H}) > \frac{8(4^q - 1)}{3} m \right) \leq \exp(-m/2).$$

Hence, upon following the argument in Section 5.3.2, we obtain the following result:

THEOREM 5.3 *Let $m \geq \max\{n, 5\}$. Consider a Rayleigh fading channel with a 4^q–QAM constellation, where $q \geq 1$ is a fixed integer. Suppose that the noise variance σ^2 satisfies*

$$\sigma^2 > \frac{64q(4^q - 1) - 3}{3} n,$$

so that

$$\Lambda \equiv \left[\frac{n + \sigma^2}{2} - \sqrt{\frac{16q(4^q - 1)}{3} n(n + \sigma^2)} \right] > 0.$$

Then, we have

$$\Pr_{(H,v,z)}\left[v_{sdr} \le 2\left(1 + \frac{8(4^q - 1)n}{3\Lambda}\right)v_{sdp}\right] \ge 1 - 2\exp(-m/6) - 2^{-m} > 0.$$

We refer the reader to [24] for further details.

5.5   Concluding remarks

In this chapter we gave an overview of a general approach for obtaining non-asymptotic performance guarantees of the SDR detector. Specifically, we showed that the approximation guarantee of the SDR detector can be obtained using the SDP weak-duality theorem and concentration inequalities for the largest singular value of the channel matrix. As an illustration, we considered the case where symbols from an MPSK constellation are transmitted across a Rayleigh fading channel. We showed that in this case, the SDR detector will yield a constant factor approximation to the ML detector (in terms of the optimal log-likelihood value) in the low SNR region with probability that increases to 1 exponentially fast. To further demonstrate the power of our techniques, we also considered the case where symbols are drawn from the QAM constellations, and showed that a similar approximation result holds. We believe that the techniques introduced in this chapter will find further applications in the probabilistic analysis of SDP relaxations. Also, it will be interesting to see whether concentration inequalities for the extremal singular values of the channel matrix can be used to derive non-asymptotic bounds on the error probability of the SDR detector.

Acknowledgments

Research by the first author is supported by CUHK Direct Grant No. 2050401 and Hong Kong Research Grants Council (RGC) General Research Fund (GRF) Project No. CUHK 2150603.

A5   Appendix to Chapter 5. Some probabilistic estimates

A5.1   Bounding the largest singular value of an arbitrary matrix

Let $A \in \mathbb{R}^{m \times n}$ be an arbitrary matrix. It is well known that the largest singular value $\|A\|_\infty$ of A can be estimated by a so-called ϵ-net argument. Before we present the argument, let us begin with a definition.

DEFINITION A5.1   *Let $D \subset \mathbb{R}^n$ and $\epsilon > 0$ be fixed. We say that a subset $N \subset D$ is an ϵ-net of D if for every $p \in D$, there exists a $p' \in N$ such that $\|p - p'\|_2 \le \epsilon$.*

Now, let $\mathbb{S}^{n-1} = \{x \in \mathbb{R}^n : \|x\|_2 = 1\}$. It is well known that for any given $\epsilon > 0$ and $S \subset \mathbb{S}^{n-1}$, an ϵ-net of S of small cardinality exists [23, Proposition 2.1]. For completeness sake, we include a proof here.

PROPOSITION A5.1   *Let $\epsilon > 0$ and $S \subset \mathbb{S}^{n-1}$ be given. Then, there exists an ϵ–net N of S with*

$$|N| \leq 2n \left(1 + \frac{2}{\epsilon}\right)^{n-1}.$$

Proof   Without loss of generality, we may assume that $\epsilon < 2$, for otherwise any single point in S forms the desired ϵ-net. Now, let N be a maximal cardinality subset of S such that for any distinct $p, p' \in N$, we have $\|p - p'\|_2 > \epsilon$. By the maximal cardinality property, we see that N is an ϵ-net of S. To estimate its size, observe that $B(p, \epsilon/2) \cap B(p', \epsilon/2) = \emptyset$ for every distinct $p, p' \in N$, where $B(x, \epsilon)$ is the ball centered at $x \in \mathbb{R}^n$ with radius ϵ. Moreover, we have

$$\bigcup_{p \in N} B(p, \epsilon/2) \subset B(\mathbf{0}, 1 + \epsilon/2) \backslash B(\mathbf{0}, 1 - \epsilon/2).$$

Hence, by comparing the volumes of the balls, we have

$$|N| \cdot \mathrm{vol}(B(p, \epsilon/2)) \leq \mathrm{vol}(B(\mathbf{0}, 1 + \epsilon/2)) - \mathrm{vol}(B(\mathbf{0}, 1 - \epsilon/2)). \tag{A5.1}$$

Now, recall that

$$\mathrm{vol}(B(x, \epsilon)) = \epsilon^n \mathrm{vol}(B(\mathbf{0}, 1)) \quad \text{for any } x \in \mathbb{R}^n.$$

Hence, upon dividing both sides of (A5.1) by $\mathrm{vol}(B(\mathbf{0}, 1))$, we obtain

$$|N| \cdot \left(\frac{\epsilon}{2}\right)^n \leq \left(1 + \frac{\epsilon}{2}\right)^n - \left(1 - \frac{\epsilon}{2}\right)^n.$$

Using the inequality

$$(1 + x)^l - (1 - x)^l \leq 2lx(1 + x)^{l-1},$$

which is valid for all $x \in (0, 1)$, we conclude that

$$|N| \leq n\epsilon \left(\frac{2}{\epsilon}\right)^n \left(1 + \frac{\epsilon}{2}\right)^{n-1} = 2n \left(1 + \frac{2}{\epsilon}\right)^{n-1},$$

as desired.   ∎

The following result shows that the largest singular value of an arbitrary $m \times n$ real matrix A can be estimated using appropriate ϵ-nets:

PROPOSITION A5.2   *Let $\epsilon, \delta > 0$ be fixed. Let N be an ϵ-net of $\mathbb{S}^{n-1}$, and let M be a δ-net of $\mathbb{S}^{m-1}$. Then, for any $m \times n$ real matrix A, we have*

$$\|A\|_\infty \leq \frac{1}{(1-\epsilon)(1-\delta)} \sup_{p \in M, q \in N} \left| p^T A q \right|.$$

Proof   (cf. [23, Proposition 2.3]) Since N is an ϵ-net of $\mathbb{S}^{n-1}$, every $z \in \mathbb{S}^{n-1}$ can be decomposed as $z = q + h$, where $q \in N$ and $\|h\|_2 \leq \epsilon$. In particular, we have

$$\|A\|_\infty = \sup_{z \in \mathbb{S}^{n-1}} \|Az\|_2$$

$$\leq \sup_{q \in N} \|Aq\|_2 + \sup_{h \in \mathbb{R}^n : \|h\|_2 \leq \epsilon} \|Ah\|_2$$

$$\leq \sup_{q \in N} \|Aq\|_2 + \epsilon \|A\|_\infty,$$

or equivalently,

$$\|A\|_\infty \leq \frac{1}{1-\epsilon} \sup_{q \in N} \|Aq\|_2. \tag{A5.2}$$

Now, let $q \in N$ be fixed. By a similar argument, we have

$$\|Aq\|_2 = \sup_{y \in \mathbb{S}^{m-1}} |y^T A q|$$

$$\leq \sup_{p \in M} |p^T A q| + \sup_{h \in \mathbb{R}^m : \|h\|_2 \leq \delta} |h^T A q|$$

$$\leq \sup_{p \in M} |p^T A q| + \delta \|Aq\|_2,$$

or equivalently,

$$\|Aq\|_2 \leq \frac{1}{1-\delta} \sup_{p \in M} |p^T A q|. \tag{A5.3}$$

The desired result now follows from (A5.2) and (A5.3).   ■

A5.2   The largest singular value of a complex Gaussian random matrix

Let $H \in \mathbb{C}^{m \times n}$ be a matrix whose entries are IID, standard, complex Gaussian random variables. Note that the largest singular value $\|H\|_\infty$ of H is equal to the largest singular value $\|\tilde{H}\|_\infty$ of the $2m \times 2n$ *real* matrix $\tilde{H}$ defined by

$$\tilde{H} = \begin{bmatrix} \Re(H) & -\Im(H) \\ \Im(H) & \Re(H) \end{bmatrix} \in \mathbb{R}^{2m \times 2n}. \tag{A5.4}$$

Here, $\Re(H)$ (resp. $\Im(H)$) denotes the real (resp. imaginary) part of H. To see this, observe that for any $u, v \in \mathbb{R}^n$, we have

$$(u + jv)^* H^* H (u + jv)$$

$$= (u^T - jv^T)(\Re(H)^T - j\Im(H)^T)(\Re(H) + j\Im(H))(u + jv)$$

$$= u^T A u + u^T B v - v^T B u + v^T A v + j(u^T A v - v^T A u - u^T B u - v^T B v),$$

where

$$A = \Re(H)^T \Re(H) + \Im(H)^T \Im(H),$$

$$B = \Im(H)^T \Re(H) - \Re(H)^T \Im(H),$$

and $H^* H = A - jB$. In particular, we see that $A \in \mathbb{R}^{n \times n}$ is a symmetric matrix, and that $B \in \mathbb{R}^{n \times n}$ is a skew-symmetric matrix. This implies that $u^T A v = v^T A u$ and $u^T B u = v^T B v = 0$, and hence

$$(u + jv)^* H^* H (u + jv) = \begin{bmatrix} u^T & v^T \end{bmatrix} \tilde{H}^T \tilde{H} \begin{bmatrix} u \\ v \end{bmatrix}.$$

Now, the desired conclusion follows from the Courant–Fischer theorem (see, e.g., [40, Theorem 7.3.10]).

Using the machinery developed in the previous section, we can estimate $\|\tilde{H}\|_\infty$ as follows:

PROPOSITION A5.3 *Let $\tilde{H}$ be as in (A5.4), with $m \geq n \geq 1$. Then, we have*

$$\Pr_H \left(\|\tilde{H}\|_\infty > \sqrt{8m} \right) = \Pr_H \left(\lambda_{max}(H^* H) > 8m \right) \leq \exp(-m/2).$$

Proof Let N be a $(1/2)$-net of $\mathbb{S}^{2n-1}$, and let M be a $(1/2)$-net of $\mathbb{S}^{2m-1}$. By Proposition A5.1, these nets can be chosen with $|M| \leq 6^{2m}$ and $|N| \leq 6^{2n}$. Now, let $p = (p^1, p^2) \in M$ and $q = (q^1, q^2) \in N$, where $p^1, p^2 \in \mathbb{R}^m$ and $q^1, q^2 \in \mathbb{R}^n$. We claim that $p^T \tilde{H} q$ is a Gaussian random variable with mean 0 and variance $1/2$. Indeed, we compute

$$p^T \tilde{H} q = (p^1)^T \Re(H) q^1 + (p^2)^T \Im(H) q^1 - (p^1)^T \Im(H) q^2 + (p^2)^T \Re(H) q^2$$

$$= \sum_{k=1}^{m} \sum_{k'=1}^{n} \Re(H)_{kk'} \left(p_k^1 q_{k'}^1 + p_k^2 q_{k'}^2 \right) + \sum_{k=1}^{m} \sum_{k'=1}^{n} \Im(H)_{kk'} \left(p_k^2 q_{k'}^1 - p_k^1 q_{k'}^2 \right).$$

It follows that $p^T \tilde{H} q$ is a Gaussian random variable with mean 0. Moreover, its variance is given by

$$\frac{1}{2} \sum_{k=1}^{m} \sum_{k'=1}^{n} \left[\left(p_k^1 q_{k'}^1 + p_k^2 q_{k'}^2 \right)^2 + \left(p_k^2 q_{k'}^1 - p_k^1 q_{k'}^2 \right)^2 \right]$$

$$= \frac{1}{2} \sum_{k=1}^{m} \sum_{k'=1}^{n} \left[(p_k^1)^2 (q_{k'}^1)^2 + (p_k^2)^2 (q_{k'}^2)^2 + (p_k^2)^2 (q_{k'}^1)^2 + (p_k^1)^2 (q_{k'}^2)^2 \right]$$

$$= \frac{1}{2} \sum_{k=1}^{m} \left[(p_k^1)^2 \left(\sum_{k'=1}^{n} (q_{k'}^1)^2 + (q_{k'}^2)^2 \right) + (p_k^2)^2 \left(\sum_{k'=1}^{n} (q_{k'}^1)^2 + (q_{k'}^2)^2 \right) \right]$$

$$= \frac{1}{2},$$

where the last equality follows from the fact that $p = (p^1, p^2) \in \mathbb{S}^{2m-1}$ and $q = (q^1, q^2) \in \mathbb{S}^{2n-1}$. This establishes the claim. In particular, for any $\theta > 0$ and $m \geq n \geq 1$, we have

$$\Pr_H \left(|p^T \tilde{H} q| > \theta \sqrt{m} \right) \leq \frac{1}{\theta \sqrt{\pi m}} \cdot \exp \left(-\theta^2 m \right).$$

It then follows from Proposition A5.2 and the union bound that

$$\Pr_H \left(\|\tilde{H}\|_\infty > \theta \sqrt{m} \right) \leq \frac{4}{\theta \sqrt{\pi m}} \cdot \left(6^4 \exp(-\theta^2) \right)^m.$$

Upon setting $\theta = \sqrt{8}$, we obtain the desired result. ∎

A5.3   The largest singular value of a real Gaussian random matrix

Let $\tilde{H} \in \mathbb{R}^{2m \times n}$ be a matrix whose entries are IID, real Gaussian random variables with mean 0 and variance $1/2$. Using the machinery developed in Section A5.1, we can establish the following result:

PROPOSITION A5.4   *Let $\tilde{H} \in \mathbb{R}^{2m \times n}$ be defined as above, with $m \geq n \geq 1$. Then, we have*

$$\Pr_{\tilde{H}} \left(\|\tilde{H}\|_\infty > \sqrt{6m} \right) = \Pr_{\tilde{H}} \left(\lambda_{max}(\tilde{H}^T \tilde{H}) > 6m \right) \leq \exp(-m/2).$$

Proof   Let N be a $(1/2)$–net of $\mathbb{S}^{n-1}$, and let M be a $(1/2)$–net of $\mathbb{S}^{2m-1}$. By Proposition A5.1, these nets can be chosen with $|M| \leq 6^{2m}$ and $|N| \leq 6^n$. Now, let $p = (p^1, p^2) \in M$ and $q \in N$, where $p^1, p^2 \in \mathbb{R}^m$. We compute

$$p^T \tilde{H} q = (p^1)^T \Re(H) q + (p^2)^T \Im(H) q$$

$$= \sum_{k=1}^{m} \sum_{k'=1}^{n} \left(\Re(H)_{kk'} p_k^1 q_{k'} + \Im(H)_{kk'} p_k^2 q_{k'} \right).$$

It follows that $p^T \tilde{H} q$ is a mean 0 Gaussian random variable whose variance is given by

$$\frac{1}{2} \sum_{k=1}^{m} \sum_{k'=1}^{n} \left[(p_k^1 q_{k'})^2 + (p_k^2 q_{k'})^2 \right] = \frac{1}{2} \sum_{k=1}^{m} \left[\left((p_k^1)^2 + (p_k^2)^2 \right) \left(\sum_{k'=1}^{n} q_{k'}^2 \right) \right] = \frac{1}{2},$$

where the last equality follows from the fact that $p = (p^1, p^2) \in \mathbb{S}^{2m-1}$ and $q \in \mathbb{S}^{n-1}$. Thus, using the argument in the proof of Proposition A5.3, we obtain

$$\Pr_{\tilde{H}} \left(\|\tilde{H}\|_\infty > \theta \sqrt{m} \right) \leq \frac{4}{\theta \sqrt{\pi m}} \cdot \left(6^3 \exp(-\theta^2) \right)^m.$$

Upon setting $\theta = \sqrt{6}$, we obtain the desired result. ∎

A5.4   The squared-norm of a complex Gaussian random vector

Recall that $y = Hx + v$, where $x \in \mathcal{S}_M^n$, $H \in \mathbb{C}^{m \times n}$ is the channel matrix whose entries are IID, standard, complex Gaussian random variables, and $v \in \mathbb{C}^m$ is a circular, symmetric, complex Gaussian random vector with covariance matrix $\sigma^2 I$ and is independent of H. It follows that each entry of y is a complex Gaussian random variable with mean 0 and variance $n + \sigma^2$, and that the entries are independent. Consequently, the random variable $\|y\|_2^2$ is simply a sum of $2m$ independent, real Gaussian random variables, each of which has mean 0 and variance $(n + \sigma^2)/2$. In particular, it should concentrate around its mean, that is, $m(n + \sigma^2)$. To make this argument precise, we need the following proposition (cf. [14, Propositions 2.1 and 2.2]):

PROPOSITION A5.5   *Let $\xi_1, \ldots, \xi_d$ be IID standard, real Gaussian random variables. Let $\alpha \in (1, \infty)$ and $\beta \in (0, 1)$ be constants, and set $U_d = \sum_{k=1}^{d} \xi_k^2$. Then, the following holds:*

$$\Pr(U_d \geq \alpha d) \leq \exp\left[\frac{d}{2} (1 - \alpha + \ln \alpha) \right]. \tag{A5.5}$$

$$\Pr(U_d \leq \beta d) \leq \exp\left[\frac{d}{2} (1 - \beta + \ln \beta) \right]. \tag{A5.6}$$

Proof   To establish (A5.5), we let $t \in [0, 1/2)$ and compute

$$\Pr(U_d \geq \alpha d) = \Pr\{\exp[t(U_d - \alpha d)] \geq 1\}$$

$$\leq \mathbb{E}\left[\exp[t(U_d - \alpha d)]\right] \quad \text{(by Markov's inequality)}$$

$$= \exp(-t\alpha d) \cdot \left(\mathbb{E}\left[\exp\left(t\xi_1^2\right)\right] \right)^d \quad \text{(by independence)}$$

$$= \exp(-t\alpha d) \cdot (1 - 2t)^{-d/2}.$$

Let $f : [0, 1/2) \to \mathbb{R}$ be given by

$$f(t) = \exp(-t\alpha d) \cdot (1 - 2t)^{-d/2}.$$

Then, we have

$$f'(t) = -\exp(-t\alpha d)\,\alpha d(1 - 2t)^{-d/2} + \exp(-t\alpha d)\,d(1 - 2t)^{-(d/2+1)},$$

and hence f is minimized at

$$t^* = \frac{1}{2}\left(1 - \frac{1}{\alpha}\right).$$

Note that $t^* \in (0, 1/2)$ whenever $\alpha \in (1, \infty)$. Thus, we conclude that

$$\Pr(U_d \geq \alpha d) \leq f(t^*) = \left[\sqrt{\alpha}\,\exp\left(\frac{1-\alpha}{2}\right)\right]^d = \exp\left[\frac{d}{2}(1 - \alpha + \ln \alpha)\right],$$

as desired. ∎

To establish (A5.6), we proceed in a similar fashion. For $t \geq 0$, we have

$$\begin{aligned}
\Pr(U_d \leq \beta d) &= \Pr\{\exp[t(\beta d - U_d)] \geq 1\} \\
&\leq \mathbb{E}[\exp[t(\beta d - U_d)]] \quad \text{(by Markov's inequality)} \\
&= \exp(t\beta d) \cdot \left(\mathbb{E}\left[\exp\left(-t\xi_1^2\right)\right]\right)^d \quad \text{(by independence)} \\
&= \exp(t\beta d) \cdot (1 + 2t)^{-d/2}.
\end{aligned}$$

Now, let $f : [0, \infty) \to \mathbb{R}$ be given by

$$f(t) = \exp(t\beta d) \cdot (1 + 2t)^{-d/2}.$$

Then, we have

$$f'(t) = \exp(t\beta d)\,\beta d(1 + 2t)^{-d/2} - \exp(t\beta d)\,d(1 + 2t)^{-(d/2+1)},$$

and hence f is minimized at

$$t^* = \frac{1}{2}\left(\frac{1}{\beta} - 1\right).$$

Moreover, we have $t^* > 0$ whenever $\beta < 1$. It follows that

$$\Pr(U_d \leq \beta d) \leq f(t^*) = \left[\sqrt{\beta}\,\exp\left(\frac{1-\beta}{2}\right)\right]^d = \exp\left[\frac{d}{2}(1 - \beta + \ln \beta)\right],$$

as desired. ∎

Now, since the random variable $\|y\|_2^2$ has the same distribution as $(n+\sigma^2)U_{2m}/2$, the following corollary of Proposition A5.5 is immediate:

COROLLARY A5.1   *Let $y \in \mathbb{C}^m$ be defined as above. Then, we have*

$$\Pr_{(H,v)}\left[\|y\|_2^2 \leq \frac{1}{2}m(n+\sigma^2)\right] \leq \exp(-m/6),$$

$$\Pr_{(H,v)}\left[\|y\|_2^2 \geq 2m(n+\sigma^2)\right] \leq \exp(-m/4). \qquad \square$$

References

[1] S. Boyd, L. El Ghaoui, E. Feron, and V. Balakrishnan, *Linear Matrix Inequalities in System and Control Theory*, ser. SIAM Studies in Applied and Numerical Mathematics. Philadelphia, PA: Society for Industrial and Applied Mathematics, 1994, vol. 15.

[2] V. Balakrishnan and F. Wang, "Semidefinite programming in systems and control theory," in *Handbook of Semidefinite Programming: Theory, Algorithms, and Applications*, ser. International Series in Operations Research and Management Science, vol. 27, H. Wolkowicz, R. Saigal, and L. Vandenberghe, eds. Boston, MA: Kluwer Academic Publishers, 2000, pp. 421–41.

[3] A. Ben-Tal and A. Nemirovski, "Structural design," in *Handbook of Semidefinite Programming: Theory, Algorithms, and Applications*, ser. International Series in Operations Research and Management Science, vol. 27, H. Wolkowicz, R. Saigal, and L. Vandenberghe, eds. Boston, MA: Kluwer Academic Publishers, 2000, pp. 443–67.

[4] P. H. Tan and L. K. Rasmussen, "The application of semidefinite programming for detection in CDMA," *IEEE Journal on Selected Areas in Communications*, vol. 19, no. 8, pp. 1442–9, 2001.

[5] W.-K. Ma, T. N. Davidson, K. M. Wong, Z.-Q. Luo, and P.-C. Ching, "Quasi-maximum-likelihood multiuser detection using semi-definite relaxation with application to synchronous CDMA," *IEEE Transactions on Signal Processing*, vol. 50, no. 4, pp. 912–22, 2002.

[6] P. Biswas, T.-C. Lian, T.-C. Wang, and Y. Ye, "Semidefinite programming based algorithms for sensor network localization," *ACM Transactions on Sensor Networks*, vol. 2, no. 2, pp. 188–220, 2006.

[7] A. M.-C. So and Y. Ye, "Theory of semidefinite programming for sensor network localization," *Mathematical Programming, Series B*, vol. 109, no. 2, pp. 367–84, 2007.

[8] ——, "A semidefinite programming approach to tensegrity theory and realizability of graphs," in *Proceedings of the 17th Annual ACM–SIAM Symposium on Discrete Algorithms (SODA 2006)*, 2006, pp. 766–75.

[9] M. X. Goemans, "Semidefinite programming in combinatorial optimization," *Mathematical Programming*, vol. 79, pp. 143–61, 1997.

[10] Yu. Nesterov, H. Wolkowicz, and Y. Ye, "Semidefinite programming relaxations of nonconvex quadratic optimization," in *Handbook of Semidefinite Programming: Theory, Algorithms, and Applications*, ser. International Series in Operations Research and Management Science, vol. 27, H. Wolkowicz, R. Saigal, and L. Vandenberghe, eds. Boston, MA: Kluwer Academic Publishers, 2000, pp. 361–419.

[11] N. Alon, K. Makarychev, Y. Makarychev, and A. Naor, "Quadratic forms on graphs," *Inventiones Mathematicae*, vol. 163, no. 3, pp. 499–522, 2006.

[12] A. M.-C. So, J. Zhang, and Y. Ye, "On approximating complex quadratic optimization problems via semidefinite programming relaxations," *Mathematical Programming, Series B*, vol. 110, no. 1, pp. 93–110, 2007.

[13] S. Arora, J. R. Lee, and A. Naor, "Euclidean distortion and the sparsest cut," *Journal of the American Mathematical Society*, vol. 21, no. 1, pp. 1–21, 2008.

[14] A. M.-C. So, Y. Ye, and J. Zhang, "A unified theorem on SDP rank reduction," *Mathematics of Operations Research*, vol. 33, no. 4, pp. 910–20, 2008.

[15] M. Laurent, "Sums of squares, moment matrices and optimization over polynomials," in *Emerging Applications of Algebraic Geometry*, ser. The IMA Volumes in Mathematics and Its Applications, vol. 149, M. Putinar and S. Sullivant, eds. New York: Springer Science & Business Media, LLC, 2009, pp. 157–270.

[16] S. Verdú, *Multiuser Detection*. Cambridge: Cambridge University Press, 1998.

[17] D. Tse and P. Viswanath, *Fundamentals of Wireless Communication*. New York: Cambridge University Press, 2005.

[18] S. Verdú, "Computational complexity of optimum multiuser detection," *Algorithmica*, vol. 4, pp. 303–12, 1989.

[19] M. Kisialiou and Z.-Q. Luo, "Performance analysis of quasi-maximum-likelihood detector based on semidefinite programming," in *Proceedings of the IEEE International Conference on Acoustics, Speech, and Signal Processing, 2005 (ICASSP 2005)*, vol. 3, 2005, pp. III-433–III-6.

[20] J. Jaldén and B. Ottersten, "The diversity order of the semidefinite relaxation detector," *IEEE Transactions on Information Theory*, vol. 54, no. 4, pp. 1406–22, 2008.

[21] A. M.-C. So, "Probabilistic analysis of the semidefinite relaxation detector in digital communications," to appear in *Proceedings of the 21st Annual ACM-SIAM Symposium on Discrete Algorithms (SODA 2010)*.

[22] A. T. James, "Distributions of matrix variates and latent roots derived from normal samples," *The Annals of Mathematical Statistics*, vol. 35, no. 2, pp. 475–501, 1964.

[23] M. Rudelson and R. Vershynin, "The smallest singular value of a random rectangular matrix," 2008, available on *arXiv*.

[24] A. M.-C. So, "On the performance of semidefinite relaxation MIMO detectors for QAM constellations," in *Proceedings of the 2009 IEEE International Conference on Acoustics, Speech, and Signal Processing (ICASSP 2009)*, 2009 pp. 2449–52.

[25] Z.-Q. Luo, X. Luo, and M. Kisialiou, "An efficient quasi-maximum likelihood decoder for PSK signals," in *Proceedings of the IEEE International Conference on Acoustics, Speech, and Signal Processing, 2003 (ICASSP 2003)*, vol. 6, 2003, pp. VI-561–IV-4.

[26] W.-K. Ma, P.-C. Ching, and Z. Ding, "Semidefinite relaxation based multiuser detection for *M*-Ary PSK multiuser systems," *IEEE Transactions on Signal Processing*, vol. 52, no. 10, pp. 2862–72, 2004.

[27] M. Grötschel, L. Lovász, and A. Schrijver, *Geometric Algorithms and Combinatorial Optimization*, second corrected ed., ser. Algorithms and Combinatorics, vol. 2. Berlin: Springer, 1993.

[28] F. Alizadeh, "Interior point methods in semidefinite programming with applications to combinatorial optimization," *SIAM Journal on Optimization*, vol. 5, no. 1, pp. 13–51, 1995.

[29] K.-C. Toh and L. N. Trefethen, "The Chebyshev polynomials of a matrix," *SIAM Journal on Matrix Analysis and Applications*, vol. 20, no. 2, pp. 400–19, 1998.

[30] M. X. Goemans and D. P. Williamson, "Approximation algorithms for MAX-3-CUT and other problems via complex semidefinite programming," *Journal of Computer and System Sciences*, vol. 68, no. 2, pp. 442–70, 2004.

[31] Y. Huang and S. Zhang, "Approximation algorithms for indefinite complex quadratic maximization problems," Department of Systems Engineering and Engineering Management, The Chinese University of Hong Kong, Shatin, N. T., Hong Kong, Tech. Rep. SEEM2005-03, 2005.

[32] C. Helmberg, "Semidefinite programming for combinatorial optimization," Konrad–Zuse–Zentrum für Informationstechnik Berlin, Takustraße 7, D–14195, Berlin, Germany, Tech. Rep. ZR–00–34, 2000.

[33] A. Wiesel, Y. C. Eldar, and S. Shamai (Shitz), "Semidefinite relaxation for detection of 16–QAM signaling in MIMO channels," *IEEE Signal Processing Letters*, vol. 12, no. 9, pp. 653–56, 2005.

[34] N. D. Sidiropoulos and Z.-Q. Luo, "A semidefinite relaxation approach to MIMO detection for high-order QAM constellations," *IEEE Signal Processing Letters*, vol. 13, no. 9, pp. 525–8, 2006.

[35] Z. Mao, X. Wang, and X. Wang, "Semidefinite programming relaxation approach for multiuser detection of QAM signals," *IEEE Transactions on Wireless Communications*, vol. 6, no. 12, pp. 4275–9, 2007.

[36] A. Mobasher, M. Taherzadeh, R. Sotirov, and A. K. Khandani, "A near-maximum-likelihood decoding algorithm for MIMO systems based on semi-definite programming," *IEEE Transactions on Information Theory*, vol. 53, no. 11, pp. 3869–86, 2007.

[37] Y. Yang, C. Zhao, P. Zhou, and W. Xu, "MIMO detection of 16-QAM signaling based on semidefinite relaxation," *IEEE Signal Processing Letters*, vol. 14, no. 11, pp. 797–800, 2007.

[38] W.-K. Ma, C.-C. Su, J. Jaldén, T.-H. Chang, and C.-Y. Chi, "The equivalence of semidefinite relaxation MIMO detectors for higher-order QAM," 2008, submitted for publication to *IEEE Journal of Selected Topics in Signal Processing*. Available on *arXiv*.

[39] D. P. Bertsekas, *Nonlinear Programming*, 2nd ed. Belmont, MA: Athena Scientific, 1999.

[40] R. A. Horn and C. R. Johnson, *Matrix Analysis*. Cambridge: Cambridge University Press, 1985.

6 Semidefinite programming, matrix decomposition, and radar code design

Yongwei Huang, Antonio De Maio, and Shuzhong Zhang

In this chapter, we study specific rank-1 decomposition techniques for Hermitian positive semidefinite matrices. Based on the semidefinite programming relaxation method and the decomposition techniques, we identify several classes of quadratically constrained quadratic programming problems that are polynomially solvable. Typically, such problems do not have too many constraints. As an example, we demonstrate how to apply the new techniques to solve an optimal code design problem arising from radar signal processing.

6.1 Introduction and notation

Semidefinite programming (SDP) is a relatively new subject of research in optimization. Its success has caused major excitement in the field. One is referred to Boyd and Vandenberghe [11] for an excellent introduction to SDP and its applications. In this chapter, we shall elaborate on a special application of SDP for solving *quadratically constrained quadratic programming* (QCQP) problems. The techniques we shall introduce are related to how a positive semidefinite matrix can be decomposed into a sum of rank-1 positive semidefinite matrices, in a specific way that helps to solve nonconvex quadratic optimization with quadratic constraints. The advantage of the method is that the convexity of the original quadratic optimization problem becomes irrelevant; only the number of constraints is important for the method to be effective. We further present a study on how this method helps to solve a radar code design problem. Through this investigation, we aim to make a case that solving nonconvex quadratic optimization by SDP is a viable approach. The featured technique is to decompose a Hermitian positive semidefinite matrix into a sum of rank-1 matrices with a desirable property. The method involves some duality theory for SDP. Our simulation study shows that the techniques are numerically efficient and stable.

The organization of the chapter is as follows. In Section 6.2, we shall present several key results about the rank-1 decomposition of a positive semidefinite matrix. In Section 6.3, we shall introduce SDP and some basic properties of SDP, which will be useful for our subsequent studies. Section 6.4 introduces nonconvex QCQP problems. Such problems have a natural connection with SDP. In fact, relaxing a QCQP problem

Convex Optimization in Signal Processing and Communication, eds. Daniel P. Palomar and Yonina C. Eldar.
Published by Cambridge University Press © Cambridge University Press, 2010.

means that one gets a convex SDP problem in the lifted (matrix) space. Section 6.5 is devoted to the solution method for some specific types of QCQP problems, where the number of constraints is relatively few. We shall show that the matrix decomposition technique can be applied to obtain a global optimal solution for the nonconvex optimization problem. In the remaining sections we shall present a case study on how the proposed framework applies to radar signal processing. In particular, in Section 6.6, we introduce a radar code design problem, and in Section 6.7 the problem is formulated as nonconvex quadratic optimization. Then, in Sections 6.8 and 6.9, we apply the solution method and test its performance. Finally, in Section 6.10, we shall conclude the study and the whole chapter.

Before proceeding, let us first introduce the notations that shall be used in this chapter. We use $\mathcal{H}^N$ to denote the space of Hermitian $N \times N$ matrices, $\mathcal{S}^N$ the space of symmetric $N \times N$ matrices, and $\mathcal{M}^{M,N}$ the space of $M \times N$ matrices with both complex-valued and real-valued entries. We adopt the notation of using boldface for vectors $\boldsymbol{a}$ (lower case), and matrices $\boldsymbol{A}$ (upper case), especially, $\boldsymbol{0}$ stands for the zero column vector, or row vector, or matrix if its dimension is clear in context. The "Frobenius" inner product is defined over $\mathcal{M}^{M,N}$ as follows:

$$\boldsymbol{A} \bullet \boldsymbol{B} = \mathrm{tr}(\boldsymbol{A}\boldsymbol{B}^H) = \sum_{k=1}^{M} \sum_{l=1}^{N} A_{kl}\overline{B_{kl}},$$

where "tr" denotes the trace of a square matrix, $\boldsymbol{B}^H$ stands for the component wise conjugate of the transpose $\boldsymbol{B}^T$ of $\boldsymbol{B}$, A_{kl}, B_{kl} are the kl-th entries of $\boldsymbol{A}$ and $\boldsymbol{B}$ respectively, and $\overline{a}$ denotes the conjugate of the complex number a.

Note that the inner product is real-valued when it is defined in $\mathcal{H}^N$ or $\mathcal{S}^N$. In this chapter, we shall always work with either $\mathcal{H}^N$ (mainly) or $\mathcal{S}^N$. For simplicity, we shall only mention $\mathcal{H}^N$ if results are similar for $\mathcal{H}^N$ and $\mathcal{S}^N$ alike. Induced by the matrix inner product, the notion of orthogonality, orthogonal complement of a linear subspace, and the (Frobenius) norm of a matrix, namely

$$\|\boldsymbol{Z}\| = \sqrt{\boldsymbol{Z} \bullet \boldsymbol{Z}}, \boldsymbol{Z} \in \mathcal{H}^N,$$

all follow.

We denote by $\boldsymbol{Z} \succeq 0$ ($\succ 0$) a positive semidefinite matrix (positive definite), in other words $z^H \boldsymbol{Z} z \geq 0$ (> 0, respectively), $\forall$ nonzero $z \in \mathbb{C}^N$, where $\mathbb{C}^N$ stands for the Euclidean space in the complex number field $\mathbb{C}$ (similarly, $\mathbb{R}^N$ for the Euclidean space in the real number field $\mathbb{R}$). We also denote the set of all positive semidefinite matrices (positive definite matrices) by $\mathcal{H}_+^N$ ($\mathcal{H}_{++}^N$, respectively), and likewise, the set of all positive semidefinite matrices (positive definite matrices) in $\mathcal{S}^N$ by $\mathcal{S}_+^N$ ($\mathcal{S}_{++}^N$, respectively).

For any complex number z, we use Re (z) and Im (z) to denote, respectively, the real and the imaginary part of z, while $|z|$ and $\arg(z)$ represent the modulus and the argument of z, respectively. For a vector z, we denote by $\|z\|$ the Euclidean norm of it. For a matrix $\boldsymbol{Z}$ (a vector z), we denote by $\overline{\boldsymbol{Z}}$ ($\overline{z}$) the component wise conjugate of $\boldsymbol{Z}$ (z, respectively).

We use $\odot$ to denote the Hadamard element-wise product [29] of two vectors (or two matrices) with the same size. In this chapter, we reserve "i" for the imaginary unit (i.e., $i = \sqrt{-1}$), while the other commonly used letter "j" shall be reserved for index in this chapter.

6.2   Matrix rank-1 decomposition

6.2.1   Rank-1 matrix decomposition schemes

Let us first introduce two useful rank-1 matrix decomposition theorems for symmetric and Hermitian positive semidefinite matrices, respectively. They first appeared in Sturm and Zhang [44], and Huang and Zhang [31], respectively. It turns out that these results are quite useful. For instance, they offer alternative proofs to the famous $\mathcal{S}$-lemma with real-valued variables and complex-valued variables [19, 42, 47]. Also, these theorems lead to some very fundamental results in the so-called joint numerical ranges [12, 27, 30]. As we will see later, they are applicable to find a rank-1 optimal solution from high rank optimal solutions for some SDP problems.

THEOREM 6.1   (**Matrix Decomposition Theorem for Symmetric Matrix.**)   *Let $Z \in \mathcal{S}^N$ be a positive semidefinite matrix of rank R, and $A \in \mathcal{S}^N$ be a given matrix. Then, there is a decomposition of Z, $Z = \sum_{k=1}^{R} z_k z_k^T$, such that, $z_k^T A z_k = A \bullet Z / R$, for all $k = 1, 2, \ldots, R$.*

Proof   Let

$$Z = \sum_{k=1}^{R} z_k z_k^T$$

be any decomposition of Z, for example, by any Cholesky factorization.

If $z_k^T A z_k = A \bullet z_k z_k^H = \frac{A \bullet Z}{R}$ for all $k = 1, \ldots, R$, then the desired decomposition is achieved. Otherwise, let us assume, without loss of generality, that

$$z_1^T A z_1 < \frac{A \bullet Z}{R}, \text{ and } z_2^T A z_2 > \frac{A \bullet Z}{R}. \tag{6.1}$$

Now, let

$$v^1 = (z_1 + \gamma z_2)/\sqrt{1 + \gamma^2} \text{ and } v_2 = (-\gamma z_1 + z_2)/\sqrt{1 + \gamma^2},$$

where γ satisfies

$$(z_1 + \gamma z_2)^T A (z_1 + \gamma z_2) = \frac{Z \bullet A}{R}(1 + \gamma^2).$$

The above equation has two distinguished roots due to (6.1). It is easy to see that

$$v_1 v_1^T + v_2 v_2^T = z_1 z_1^T + z_2 z_2^T.$$

Let $Z^{new} := Z - v_1 v_1^T$ and $R := R - 1$. Then $Z^{new} \in \mathcal{S}_+^N$ with rank $R-1$, $A \bullet v_1 v_1^T = \dfrac{Z \bullet A}{R}$ and

$$\frac{A \bullet (Z - v_1 v_1^T)}{R - 1} = \frac{A \bullet Z^{new}}{R - 1} = \frac{A \bullet Z}{R}.$$

Repeating the procedure, as described above, for at most $R - 1$ times, we are guaranteed to arrive at the desired matrix decomposition. ∎

It is simple to observe that the previous theorem remains true if all data matrices are complex-valued, namely $Z \in \mathcal{H}_+^N$ and $A \in \mathcal{H}^N$. As a matter of fact, more can be done in the case of Hermitian complex matrices. Let us recall that the rank of a complex matrix Z is the largest number of columns of Z that constitute a linearly independent set (for example, see Section 0.4 of [29]).

THEOREM 6.2   (**Matrix Decomposition Theorem for Hermitian Matrix.**)   *Suppose that $Z \in \mathcal{H}^N$ is a positive semidefinite matrix of rank R, and $A, B \in \mathcal{H}^N$ are two given matrices. Then, there is a decomposition of Z, $Z = \sum_{k=1}^R z_k z_k^H$, such that,*

$$z_k^H A z_k = \frac{A \bullet Z}{R}, \; z_k^H B z_k = \frac{B \bullet Z}{R}, \tag{6.2}$$

for all $k = 1, 2, \ldots, R$.

Proof   It follows from Theorem 6.1 that there is a decomposition of Z:

$$Z = \sum_{k=1}^R u_k u_k^H \text{ such that } u_k^H A u_k = \frac{A \bullet Z}{R}, \text{ for } k = 1, \ldots, R.$$

If $u_k^H B u_k = B \bullet Z / R$ for any $k = 1, \ldots, R$, then the intended decomposition is achieved. Otherwise, there must exist two indices, say 1 and 2, such that

$$u_1^H B u_1 > \frac{B \bullet Z}{R} \text{ and } u_2^H B u_2 < \frac{B \bullet Z}{R}.$$

Denote $u_1^H A u_2 = \gamma_1 e^{i\alpha_1}$ and $u_1^H B u_2 = \gamma_2 e^{i\alpha_2}$. Let $w = \gamma e^{i\alpha} \in \mathbb{C}$ with $\alpha = \alpha_1 + \frac{\pi}{2}$ and $\gamma > 0$ be a root of the real quadratic equation in terms of γ:

$$\left(u_1^H B u_1 - \frac{B \bullet Z}{R} \right) \gamma^2 + 2\gamma_2 \sin(\alpha_2 - \alpha_1)\gamma + u_2^H B u_2 - \frac{B \bullet Z}{R} = 0. \tag{6.3}$$

Since $u_1^H B u_1 - \frac{B \bullet Z}{R} > 0$ and $u_2^H B u_2 - \frac{B \bullet Z}{R} < 0$, the above equation has two real roots. Set

$$v_1 = (wu_1 + u_2)/\sqrt{1 + \gamma^2}, \quad v_2 = (-u_1 + \overline{w}u_2)/\sqrt{1 + \gamma^2}.$$

It is easy to verify that

$$v_1 v_1^H + v_2 v_2^H = u_1 u_1^H + u_2 u_2^H. \tag{6.4}$$

Moreover

$$
\begin{aligned}
(1 + \gamma^2) v_1^H A v_1 &= (\overline{w} u_1^H + u_2^H) A (wu_1 + u_2) \\
&= \gamma^2 u_1^H A u_1 + u_2^H A u_2 + \overline{w} u_1^H A u_2 + w u_2^H A u_1 \\
&= \gamma^2 u_1^H A u_1 + 2\gamma\gamma_1 \mathrm{Re}\,(e^{i(\alpha_1 - \alpha)}) + u_2^H A u_2 \\
&= \gamma^2 u_1^H A u_1 + u_2^H A u_2 \\
&= (\gamma^2 + 1) \frac{A \bullet Z}{R},
\end{aligned}
$$

which amounts to $v_1^H A v_1 = A \bullet Z / R$. Likewise, it holds that $v_2^H A v_2 = A \bullet Z / R$. Furthermore,

$$
\begin{aligned}
(1 + \gamma^2) v_1^H B v_1 &= (\overline{w} u_1^H + u_2^H) B (wu_1 + u_2) \\
&= \gamma^2 u_1^H B u_1 + u_2^H B u_2 + 2\mathrm{Re}\,(\overline{w} u_1^H B u_2) \\
&= \gamma^2 u_1^H B u_1 + 2\gamma\gamma_2 \sin(\alpha_2 - \alpha_1) + u_2^H B u_2 \\
&= (1 + \gamma^2) \frac{B \bullet Z}{R},
\end{aligned}
$$

where, in the last equality, we use the fact that γ solves (6.3).

Due to (6.4), by letting $z_1 = v_1$, we get

$$Z - z_1 z_1^H = Z - v_1 v_1^H = v_2 v_2^H + \sum_{k=3}^{R} u_k u_k^H \succeq 0.$$

We conclude that $z_1^H A z_1 = A \bullet Z / R$ and $z_1^H B z_1 = B \bullet Z / R$. Note that $\mathrm{rank}(Z - z_1 z_1^H) = R - 1$ and $v_2^H A v_2 = u_k^H A u_k = A \bullet Z / R$ for $k = 3, \ldots, R$. Repeating this process, if there are still rank-1 terms which do not comply with (6.2), we obtain a rank-1 matrix decomposition of Z:

$$Z = \sum_{k=1}^{R} z_k z_k^H$$

where $z_k^H A z_k = A \bullet Z / R$ and $z_k^H B z_k = B \bullet Z / R$, $k = 2, \ldots, R$. $\blacksquare$

Denote '$\trianglelefteq$' to be '$=$', '$\geq$', or '$\leq$'. An immediate corollary follows.

COROLLARY 6.1 *Let $A, B \in \mathcal{H}^N$ be two arbitrary matrices. Let $Z \in \mathcal{H}^N$ be a positive semidefinite matrix of rank R. Suppose that $A \bullet Z \trianglelefteq_1 0$ and $B \bullet Z \trianglelefteq_2 0$. Then there is a rank-1 decomposition of Z,*

$$Z = \sum_{k=1}^{R} z_k z_k^H ,$$

such that $z_k^H A z_k \trianglelefteq_1 0$ and $z_k^H B z_k \trianglelefteq_2 0$, for all $k = 1, \ldots, R$. $\square$

6.2.2 Numerical performance

Theorems 6.1 and 6.2 are not mere existence results. Clearly, their proofs are constructive, and the respective rank-1 decompositions can be found fairly quickly. In this subsection, we shall consider the computational procedures for achieving such rank-1 decompositions, and report the simulation results of our computational procedures.

Algorithm DCMP1: computing the rank-1 decomposition as assured in Theorem 6.1

Input: $A \in \mathcal{S}^N$, and $Z \in \mathcal{S}_+^N$ with $R = \text{rank}\,(Z)$.

Output: $Z = \sum_{k=1}^{R} z_k z_k^T$, a rank-1 decomposition of Z, such that $A \bullet z_k z_k^T = A \bullet Z / R, k = 1, \ldots, R$.

1. Compute $p_1, \ldots, p_R$ such that $Z = \sum_{k=1}^{R} p_k p_k^T$.
2. Let $k = 1$. Repeat the following steps until $k = R - 1$:

    (a)   i. If $\left(p_k^T A p_k - \frac{A \bullet Z}{R}\right)\left(p_j^T A p_j - \frac{A \bullet Z}{R}\right) \geq 0$ for all $j = k + 1, \ldots, R$, then $z_k := p_k$.

       ii. Otherwise, let $l \in \{k + 1, \ldots, R\}$ be such that

$$\left(p_k^T A p_k - \frac{A \bullet Z}{R}\right)\left(p_l^T A p_l - \frac{A \bullet Z}{R}\right) < 0.$$

       Determine γ such that

$$(p_k + \gamma p_l)^T A (p_k + \gamma p_l) = \frac{Z \bullet A}{R}(1 + \gamma^2).$$

       Return $z_k := (p_k + \gamma p_l)/\sqrt{1 + \gamma^2}$, and set $p_l := (-\gamma p_k + p_l)/\sqrt{1 + \gamma^2}$.

    (b) If $k = R - 1$, then $z_R = p_l$.

    (c) $k := k + 1$.

Algorithm DCMP2: computing the rank-1 decomposition as assured in Theorem 6.2

Input: $A, B \in \mathcal{H}^N$, and $Z \in \mathcal{H}^N_+$ with $R = \mathrm{rank}\,(Z)$.

Output: $Z = \sum_{k=1}^{R} z_k z_k^H$, a rank-1 decomposition of Z, such that $A \bullet z_k z_k^H = A \bullet Z/R, B \bullet z_k z_k^H = B \bullet Z/R, k = 1, \ldots, R$.

1. Call Algorithm DCMP1 to output $p_1, \ldots, p_R$ such that $Z = \sum_{k=1}^{R} p_k p_k^H$ and $A \bullet p_k p_k^H = A \bullet Z/R, k = 1, \ldots, R$.

2. Let $k = 1$. Repeat the following steps until $k = R - 1$:

    (a)   i. If $\left(p_k^T B p_k - \frac{B \bullet Z}{R}\right)\left(p_j^T B p_j - \frac{B \bullet Z}{R}\right) \geq 0$ for all $j = k+1, \ldots, R$, then $z_k := p_k$.

        ii. Otherwise, let $l \in \{k+1, \ldots, R\}$ be such that

$$\left(p_k^T B p_k - \frac{B \bullet Z}{R}\right)\left(p_l^T B p_l - \frac{B \bullet Z}{R}\right) < 0.$$

        Compute the arguments $\alpha_1 := \arg(p_k^H A p_l)$ and $\alpha_2 := \arg(p_k^H B p_l)$ and the modulus $\gamma_0 = |p_k^H B p_l|$, and determine γ such that

$$\left(p_k^H B p_k - \frac{B \bullet Z}{R}\right)\gamma^2 + 2\gamma_0 \sin(\alpha_2 - \alpha_1)\gamma + p_l^H B p_l - \frac{B \bullet Z}{R} = 0.$$

        Set $w = \gamma e^{i(\alpha_1 + \pi/2)}$. Return $z_k := (w p_k + p_l)/\sqrt{1 + \gamma^2}$, and set $p_l := (-p_k + \overline{w} p_l)/\sqrt{1 + \gamma^2}$.

    (b) If $k = R - 1$, then $z_R = p_l$.

    (c) $k := k + 1$.

It can be readily verified that the computational complexity of the above rank-1 decomposition schemes is $O(N^3)$. In fact, the required amount of operations is dominated by that of the Cholesky decomposition, which is known to be $O(N^3)$ [11, Appendix C.3.2].

We report a numerical implementation of Algorithm DCMP1 and Algorithm DCMP2. We executed Algorithm DCMP1 and Algorithm DCMP2 for 500 trial runs, respectively. At each run, data matrices A and Z for Algorithm DCMP1 (A, B, and Z for Algorithm DCMP2) were randomly generated, with the size N=10+floor(40*rand) and Z's rank R=min(2+floor((N-1)*rand),N)). The performance of each trial run is measured by the max error:

$$\max\{|z_k^T A z_k - A \bullet Z/R|, 1 \leq k \leq R\}$$

for Algorithm DCMP1, or

$$\max\{|z_k^T A z_k - A \bullet Z/R|, |z_k^T B z_k - B \bullet Z/R|, 1 \leq k \leq R\}$$

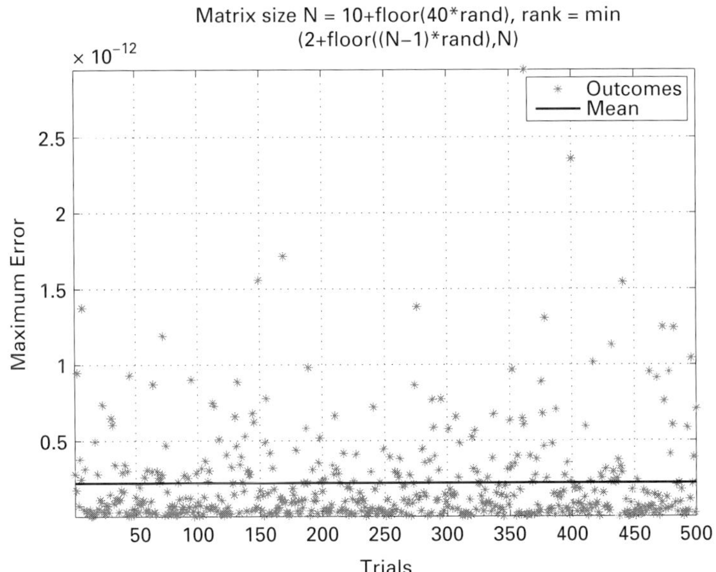

Figure 6.1 Performance of Algorithm DCMP1 for 500 trial runs; the given accuracy is $\zeta = 10^{-8}$.

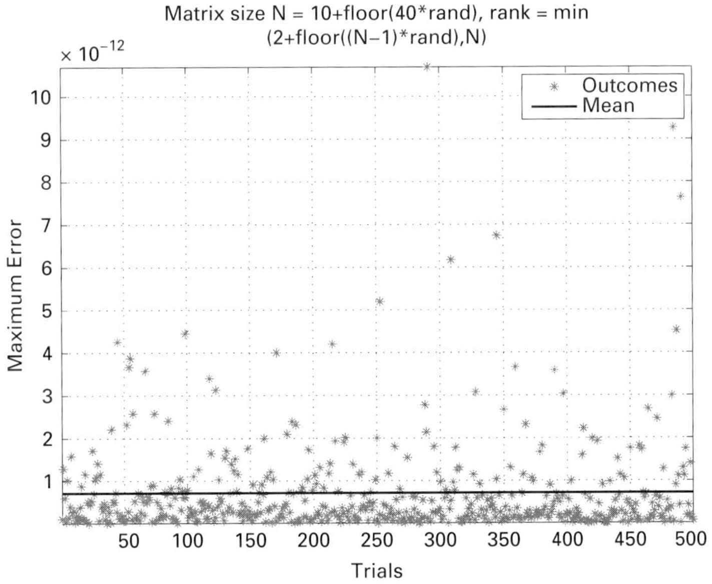

Figure 6.2 Performance of Algorithm DCMP2 for 500 trial runs; the given accuracy is $\zeta = 10^{-8}$.

for Algorithm DCMP2. Figures 6.1 and 6.2 summarize the performance of these 500 trial runs for Algorithms DCMP1 and DCMP2, respectively, where the red line in the figure is the mean of the performance measure.

6.3   Semidefinite programming

In this section, we consider the following SDP problem:

$$(\text{P}) \begin{cases} \max_X & A_0 \bullet X \\ \text{s.t.} & A_k \bullet X \trianglelefteq_k c_k, k = 1, \ldots, M, \\ & X \succeq 0, \end{cases} \tag{6.5}$$

where

$$\trianglelefteq_k \in \{\leq, =\}, \qquad k = 1, \ldots, M. \tag{6.6}$$

The dual of (P) is

$$(\text{D}) \begin{cases} \min_{y_1, \ldots, y_M} & \sum_{k=1}^{M} c_k y_k \\ \text{s.t.} & y_1 A_1 + \cdots + y_M A_M - A_0 \succeq 0, \\ & y_k \trianglelefteq_k^* 0, \end{cases} \tag{6.7}$$

where

$$\trianglelefteq_k^* \text{ is } \begin{cases} \geq, & \text{if } \trianglelefteq_k \text{ is } \leq \\ \text{unrestricted,} & \text{if } \trianglelefteq_k \text{ is } = \end{cases}, k = 1, \ldots, M, \tag{6.8}$$

with $y_k \trianglelefteq_k^* 0$ meaning that the sign of y_k is unrestricted, if $\trianglelefteq_k^*$ is "unrestricted". It is evident that the weak duality is true for (P) and (D), and it is known that the strong duality is true if some Slater conditions are satisfied. Before stating the strong-duality theorem, we denote by

$$\trianglelefteq_k' \text{ is } \begin{cases} <, & \text{if } \trianglelefteq_k \text{ is } \leq \\ =, & \text{if } \trianglelefteq_k \text{ is } = \end{cases} \tag{6.9}$$

and

$$\trianglelefteq_k^{*\prime} \text{ is } \begin{cases} >, & \text{if } \trianglelefteq_k^* \text{ is } \geq \\ \text{unrestricted,} & \text{if } \trianglelefteq_k^* \text{ is unrestricted} \end{cases} \tag{6.10}$$

for $k = 1, \ldots, M$. To sum up, we list the relation between $\trianglelefteq, \trianglelefteq', \trianglelefteq^*$ and $\trianglelefteq^{*\prime}$ in Table 6.1.

Under some suitable regularity conditions, the SDP problems (P) and (D) are solvable, that is, their respective finite optimal solutions exist. An easy verifiable regularity condition leading to the existence of optimal solutions is the so-called Slater condition, which essentially requires that both (P) and (D) are strictly feasible. By saying that (P) is strictly feasible, we mean that there is an $X_0 \succ 0$ such that $A_k \bullet X_0 \trianglelefteq_k' c_k, k = 1, \ldots, M$, and by saying that (D) is strictly feasible, we mean that there are $y_1, \ldots, y_M$ such that $y_1 A_1 + \cdots + y_M A_M - A_0 \succ 0$ and $y_k \trianglelefteq_k^{*\prime} 0$. We quote the strong-duality theorem, for example from [39], as follows:

Table 6.1. Relation between $\unlhd$, $\unlhd'$, $\unlhd^*$ and $\unlhd^{*'}$.

$\unlhd_k$	$\unlhd'_k$	$\unlhd^*_k$	$\unlhd^{*'}_k$
$\le$	$<$	$\ge$	$>$
$=$	$=$	unrestricted	unrestricted

THEOREM 6.3 **(Strong-Duality Theorem.)** *Consider the SDP problem (P) and its dual problem (D). Then,*

1. *If the primal problem (P) is bounded above and strictly feasible, then the optimal value of the dual problem (D) is attained at a feasible point, and the optimal values of (P) and (D) are equal to each other.*
2. *If the dual problem (D) is bounded below and strictly feasible, then the optimal value of the primal problem (P) is attained at a feasible point, and the optimal values of (P) and (D) are equal to each other.*
3. *If one of the problems (P) and (D) is bounded and strictly feasible, then a primal–dual feasible pair $(X, y_1, \ldots, y_M)$ is a pair of optimal solutions to the respective problems, if, and only if*

$$X \bullet (y_1 A_1 + \cdots + y_M A_M - A_0) = 0, \tag{6.11}$$

$$y_k (A_k \bullet X - c_k) = 0, \quad k = 1, \ldots, M. \tag{6.12}$$

We remark that (6.11) and (6.12) are called complementary conditions or optimality conditions for the SDP problems (P) and (D). Note that for $X_1 \succeq 0$ and $X_2 \succeq 0$, $X_1 \bullet X_2 = 0$ amounts to $X_1 X_2 = \mathbf{0}$, and thus (6.11) amounts to

$$X(y_1 A_1 + \cdots + y_M A_M - A_0) = \mathbf{0}. \tag{6.13}$$

By (6.12), it follows that for each k, either $A_k \bullet X = c_k$ or $y_k = 0$.

6.4 Quadratically constrained quadratic programming and its SDP relaxation

Let us consider the following quadratically constrained complex quadratic programming problem:

$$(\text{QCQP}) \begin{cases} \max_z & z^H A_0 z + 2\text{Re}\,(z^H b_0) \\ \text{s.t.} & z^H A_k z + 2\text{Re}\,(z^H b_k) + c_k \unlhd_k 0, \ k = 1, \ldots, M, \end{cases} \tag{6.14}$$

where the problem data is given as follows: $A_k \in \mathcal{H}^N, b_k \in \mathbb{C}^N, c_k \in \mathbb{R}, k = 0, 1, \ldots, M$. Denote by

$$
\boldsymbol{B}_0 = \begin{bmatrix} 0 & \boldsymbol{b}_0^H \\ \boldsymbol{b}_0 & \boldsymbol{A}_0 \end{bmatrix}, \; \boldsymbol{B}_k = \begin{bmatrix} c_k & \boldsymbol{b}_k^H \\ \boldsymbol{b}_k & \boldsymbol{A}_k \end{bmatrix}, k = 1, \ldots, M, \; \boldsymbol{B}_{M+1} = \begin{bmatrix} 1 & \boldsymbol{0} \\ \boldsymbol{0} & \boldsymbol{0} \end{bmatrix},
$$

and observe that

$$
z^H A_k z + 2\mathrm{Re}\,(z^H b_k) + c_k = \boldsymbol{B}_k \bullet \begin{bmatrix} 1 & z^H \\ z & zz^H \end{bmatrix}, \quad k = 1, \ldots, M.
$$

Hence, the problem (QCQP) can be rewritten in an equivalent matrix form:

$$
\text{(QCQP)} \begin{cases} \max_z & \boldsymbol{B}_0 \bullet \begin{bmatrix} 1 & z^H \\ z & zz^H \end{bmatrix} \\[2em] \text{s.t.} & \boldsymbol{B}_k \bullet \begin{bmatrix} 1 & z^H \\ z & zz^H \end{bmatrix} \trianglelefteq_k 0, \; k = 1, \ldots, M. \end{cases} \tag{6.15}
$$

Recall that a quadratic optimization problem is homogeneous if the objective and the constraint functions are all homogenous quadratic functions (i.e., there are no linear terms). A homogenized version of (QCQP) is

$$
\text{(HQ)} \begin{cases} \max_{t,z} & \boldsymbol{B}_0 \bullet \begin{bmatrix} |t|^2 & tz^H \\ \bar{t}z & zz^H \end{bmatrix} \\[2em] \text{s.t.} & \boldsymbol{B}_k \bullet \begin{bmatrix} |t|^2 & tz^H \\ \bar{t}z & zz^H \end{bmatrix} \trianglelefteq_k 0, \; k = 1, \ldots, M, \\[2em] & \boldsymbol{B}_{M+1} \bullet \begin{bmatrix} |t|^2 & tz^H \\ \bar{t}z & zz^H \end{bmatrix} = 1. \end{cases} \tag{6.16}
$$

It is easy to verify that if $[t, z^T]^T$ solves (HQ), then z/t solves (QCQP), and on the other hand, if z solves (QCQP), then $[1, z^T]^T$ also solves (HQ). Also, the optimal values of (QCQP) and (HQ) are equal to each other. Therefore, solving (QCQP) is equivalent to solving (HQ).

(QCQP) is NP-hard in general. Consider for example, $M = N, b_k = 0, c_k = -1$ for $k = 0, 1, \ldots, M$ and $A_k = e_k e_k^T$ where all entries of $e_k \in \mathbb{R}^N$ are zeros except for the kth entry, which is 1. In this case, (QCQP) becomes the problem of maximizing a homogeneous complex quadratic form over the unit hypercube, which is known to be NP-hard even in the case where A_0 is positive semidefinite [7, 20].

The SDP relaxation of (QCQP) or (HQ), by dropping the rank-1 constraint on the matrix variable, is:

$$
\text{(QSR)} \begin{cases} \max_Z & \boldsymbol{B}_0 \bullet \boldsymbol{Z} \\ \text{s.t.} & \boldsymbol{B}_k \bullet \boldsymbol{Z} \trianglelefteq_k 0, \; k = 1, \ldots, M, \\ & \boldsymbol{B}_{M+1} \bullet \boldsymbol{Z} = 1, \\ & \boldsymbol{Z} \succeq 0. \end{cases} \tag{6.17}
$$

The "SDP problem" (QSR) is standard, and the "dual problem" (DQSR) of (QSR) is given by:

$$
\text{(DQSR)} \begin{cases} \min_{y_1,\ldots,y_{M+1}} & y_{M+1} \\ \text{s.t.} & Y = \sum_{k=1}^{M} y_k B_k - B_0 + y_{M+1} B_{M+1} \succeq 0, \\ & y_k \unlhd_k^* 0, \quad k = 1,\ldots,M+1. \end{cases} \tag{6.18}
$$

Note that the dual problem (DQSR) is identical to the so-called Lagrangian dual of (QCQP), and the weak duality always holds.

We remark that in general, the SDP relaxation of a QCQP problem is not tight because of its nonconvex nature. However, for some QCQP problems with nice structures [28, 35, 40] the duality gap can be estimated. Moreover, it is interesting that in some other cases, due to some kind of hidden convexity, the SDP relaxation is tight, and the original QCQP problem can be solved in polynomial time.

6.5 Polynomially solvable QCQP problems

In this subsection, we shall identify several important classes of the QCQP problem that can be solved in polynomial time. The main tools that we need to exploit are SDP relaxation and rank-1 decomposition schemes. Indeed, a polynomially solvable QCQP problem has the same optimal value as its SDP relaxation problem. The latter is convex optimization, and can be solved in polynomial time by, for example, interior-point methods. As a consequence, an optimal solution for the underlying QCQP problem can be found by using our rank-1 decomposition schemes, hence the entire process will run in polynomial time.

Before proceeding, let us assume, throughout the subsection, that (QSR) and its dual (DQSR) satisfy the strict feasibility assumption. Namely, there is a positive definite matrix $Z_0 \succ 0$ such that

$$
B_k \bullet Z_0 \unlhd_k' 0, \ k = 1,\ldots,M, \quad B_{M+1} \bullet Z_0 = 1, \tag{6.19}
$$

and there is a vector $y \in \mathbb{R}^{M+1}$ such that $y_k \unlhd_k^* 0, k = 1,\ldots,M+1$, and

$$
Y = \sum_{k=1}^{M} y_k B_k - B_0 + y_{M+1} B_{M+1} \succ 0. \tag{6.20}
$$

It is evident that the dual strict feasibility is satisfied if one of $y_k B_k, k = 1,\ldots,M$ or $-B_0$ is positive definite.

6.5.1 QCQP problem with two constraints

The complex-valued QCQP problem with two constraints will be the first candidate of our study.

THEOREM 6.4 *Suppose (QSR) and (DQSR) are strictly feasible, and $M = 2$. Then (QCQP) can be solved in polynomial time and the SDP relaxation (QSR) and (QCQP) have the same optimal value.*

Proof By the assumption, we know that the strong duality holds for (QSR) and (DQSR). Let $\boldsymbol{Z}^* \succeq 0$ and $(y_1^*, y_2^*, y_3^*, \boldsymbol{Y}^*)$ be optimal solutions of (QSR) and (DQSR), respectively. Recall the notation "$\unlhd$" to be either "$<$" or "$=$". Then $\boldsymbol{B}_j \bullet \boldsymbol{Z}^* \unlhd_k 0$, $k = 1, 2$. By the decomposition Theorem 6.2, there are nonzero $z_j, j = 1, \ldots, R$, where R is the rank of $\boldsymbol{Z}^*$, such that

$$\boldsymbol{Z}^* = \sum_{j=1}^{R} z_j z_j^H, \ \boldsymbol{B}_k \bullet z_j z_j^H \unlhd_k 0, k = 1, 2, j = 1, \ldots, R.$$

Since $\boldsymbol{Z}_{11}^* = 1$, there is $l \in \{1, \ldots, R\}$ such that $t_l \neq 0$ where $z_l = \left[t_l, \tilde{z}_l^T \right]^T$. It is not hard to verify

$$(\tilde{z}_l / t_l)^H \boldsymbol{A}_k (\tilde{z}_l / t_l) + 2\mathrm{Re} \left((\tilde{z}_l / t_l)^H b_k \right) + c_k = \boldsymbol{B}_k \bullet \begin{bmatrix} 1 \\ \tilde{z}_l / t_l \end{bmatrix} \begin{bmatrix} 1 \\ \tilde{z}_l / t_l \end{bmatrix}^H \unlhd_k 0 \qquad (6.21)$$

for $k = 1, 2$, which implies that $\tilde{z}_l / t_l$ is a feasible point of (QCQP).

It follows from the third claim of Theorem 6.3 that the complementary conditions (6.11) and (6.12) become

$$\boldsymbol{Z}^* \bullet \boldsymbol{Y}^* = \boldsymbol{Z}^* \bullet (y_1^* \boldsymbol{B}_1 + y_2^* \boldsymbol{B}_2 - \boldsymbol{B}_0 + y_3^* \boldsymbol{B}_3) = 0,$$

and

$$y_k^*(\boldsymbol{B}_k \bullet \boldsymbol{Z}^*) = 0, \ \ k = 1, 2,$$
$$y_3^*(\boldsymbol{B}_3 \bullet \boldsymbol{Z}^* - 1) = 0.$$

Due to $\boldsymbol{Y}^* \succeq 0$, we have

$$\boldsymbol{Y}^* \bullet z_l z_l^H = 0. \qquad (6.22)$$

If $\boldsymbol{B}_k \bullet \boldsymbol{Z}^* < 0$, then by complementary conditions, it follows that $y_k^* = 0$. Otherwise, if $\boldsymbol{B}_k \bullet \boldsymbol{Z}^* = 0$, then by the rank-1 decomposition construction, we have $\boldsymbol{B}_k \bullet z_l z_l^H = 0$. Therefore, we always have

$$y_k^*(\boldsymbol{B}_k \bullet z_l z_l^H) = 0, k = 1, 2.$$

This, combined with the complementary condition (6.22) and the feasible condition (6.21), leads to the conclusion that the rank-1 matrix

$$\begin{bmatrix} 1 \\ \tilde{z}_l / t_l \end{bmatrix} \begin{bmatrix} 1 \\ \tilde{z}_l / t_l \end{bmatrix}^H$$

is an optimal solution of (QSR). Therefore $\tilde{z}_l / t_l$ is an optimal solution of (QCQP).

Note that all the computational procedures involved, including solving the SDP relaxation problem and the rank-1 decomposition algorithms, run in polynomial time. Thus, the desired result follows. ∎

Now, we consider a particular case of (QCQP), where $M = 2, A_2 = I, b_2 = 0$, and $c_2 = -1$. That is

$$
\text{(CDT1)} \begin{cases} \max_z & z^H A_0 z + 2\text{Re}\,(z^H b_0) \\ \text{s.t.} & z^H A_1 z + 2\text{Re}\,(z^H b_1) + c_1 \leq 0, \\ & z^H z \leq 1. \end{cases}
$$

This problem has its own history. In the case when all the matrices are real, the problem was proposed by Celis, Dennis, and Tapia [13] as a basic quadratic model in the so-called trust region method to solve constrained nonlinear programs. How to solve this problem efficiently has attracted much attention in optimization. That problem, now abbreviated as the "CDT problem", falls precisely into the category of nonhomogeneous QCQP with two constraints. We shall consider the CDT problem in the context that the matrices are Hermitian and complex. As we shall see, the problem becomes easy in some sense. Clearly, the feasible set of (CDT1)'s dual problem is

$$
\{(y_1, y_2, y_3, Y) : Y = y_1 \begin{bmatrix} c_1 & b_1^H \\ b_1 & A_1 \end{bmatrix} + y_2 \begin{bmatrix} -1 & 0 \\ 0 & I \end{bmatrix} - \begin{bmatrix} 0 & b_0^H \\ b_0 & A_0 \end{bmatrix}
$$
$$
+ y_3 \begin{bmatrix} 1 & 0 \\ 0 & 0 \end{bmatrix} \succeq 0,\ y_k \trianglelefteq_k^* 0, k = 1, 2, 3.\},
$$

which has an interior point. The SDP relaxation of (CDT1) is

$$
\begin{cases} \max_{v, Z} & A_0 \bullet Z + 2\text{Re}\,(v^H b_0) \\ \text{s.t.} & A_1 \bullet Z + 2\text{Re}\,(v^H b_1) + c_1 \leq 0, \\ & \begin{bmatrix} 1 & v^H \\ v & Z \end{bmatrix} \succeq 0 \text{ and } \text{tr}(Z) \leq 1. \end{cases}
$$

An immediate consequence now follows.

COROLLARY 6.2 *If (CDT1) is strictly feasible, then (CDT1) can be solved within polynomial time and its SDP relaxation admits no gap.* □

How to solve the CDT problem where all the matrices are restricted to real values remains a challenge. However, in case $A_k = A_{k1} + iA_{k2}, b_k = b_{k1} + ib_{k1}, k = 0, 1$, and $z = z_1 + iz_2$, then Corollary 6.2 states that whenever the quadratic optimization problem in the variable $[z_1^T, z_2^T]^T \in \mathbb{R}^{2N}$:

$$
\begin{cases}
\max_{z_1, z_2} & [z_1^T, z_2^T]\begin{bmatrix} A_{01} & -A_{02} \\ A_{02} & A_{01} \end{bmatrix}\begin{bmatrix} z_1 \\ z_2 \end{bmatrix} + 2(b_{01}^T z_1 + b_{02}^T z_2) \\
\text{s.t.} & [z_1^T, z_2^T]\begin{bmatrix} A_{11} & -A_{12} \\ A_{12} & A_{11} \end{bmatrix}\begin{bmatrix} z_1 \\ z_2 \end{bmatrix} + 2(b_{01}^T z_1 + b_{02}^T z_2) + c_1 \leq 0, \\
& z_1^T z_1 + z_2^T z_2 \leq 1,
\end{cases}
$$

has an interior point, then it has an exact SDP relaxation.

6.5.2   QCQP problem with three constraints

It is interesting to study another special case of QCQP where there are at most three constraints. One can show the following result:

THEOREM 6.5   *Suppose that (QCQP) has $M = 3$, and both (QSR) and (DQSR) are strictly feasible. Furthermore, suppose that the primal problem (QSR) has at least one nonbinding (inactive) constraint at optimality. Then (QCQP) can be solved in polynomial time.*

Proof   Let $\mathbf{Z}^*$ be an optimal solution of (QSR), and $(y_1^*, y_2^*, y_3^*, y_4^*, \mathbf{Y}^*)$ be an optimal solution of (DQSR) such that they satisfy the complementary conditions (6.11) and (6.12). By assumption on the nonbinding constraint at optimality, we may assume, without loss of generality, that $\mathbf{B}_3 \bullet \mathbf{Z}^* < 0$. Then $y_3^* = 0$ by the complementary condition.

Suppose $\mathbf{B}_1 \bullet \mathbf{Z}^* \lhd_1 0, \mathbf{B}_2 \bullet \mathbf{Z}^* \lhd_2 0$. Corollary 6.1 implies that there are nonzero vectors $z_j, j = 1, \ldots, R$, where R is the rank of $\mathbf{Z}^*$, such that

$$
\mathbf{Z}^* = \sum_{j=1}^{R} z_j z_j^H, \quad \mathbf{B}_k \bullet z_j z_j^H \lhd_k 0, k = 1, 2, j = 1, \ldots, R.
$$

Due to $\mathbf{B}_3 \bullet \mathbf{Z}^* < 0$, consequently, there is an $l \in \{1, \ldots, R\}$ such that $\mathbf{B}_3 \bullet z_l z_l^H < 0$. Let $z_l = \left[t_l, \tilde{z}_l^T\right]^T$, i.e., $\tilde{z}_l \in \mathbb{C}^N$. Without loss of generality, assume that $l = 1$.

We claim that $t_1 \neq 0$. In fact, suppose that $t_1 = 0$. Let $\mathbf{Z}' = \lambda^2 z_1 z_1^H + \sum_{j=2}^{R} z_j z_j^H$, for $\lambda > 0$. It is easy to verify that $\mathbf{B}_k \bullet \mathbf{Z}' \lhd_k 0, k = 1, 2$ and $\mathbf{B}_3 \bullet \mathbf{Z}' < 0$ for $\lambda \geq 1$. Then $\mathbf{Z}'$ for $\lambda \geq 1$ is feasible. Since $\mathbf{Z}^* \bullet \mathbf{Y}^* = 0$ and $\mathbf{Y}^* \succeq 0$, it follows that $\mathbf{Z}' \bullet \mathbf{Y}^* = 0$, whence $\mathbf{Z}'$ is an optimal solution of (QSR), for any $\lambda \geq 1$. However, the Frobenius norm $\|\mathbf{Z}'\|_F = \sqrt{\mathbf{Z}' \bullet \mathbf{Z}'}$ is unbounded when λ goes to infinity. That is, the optimal solution set is unbounded, which is impossible due to the assumption that (DQSR) is strictly feasible. (For a detailed account of the duality relations for conic optimization, one is referred to Chapter 2 of [21].)

Then,

$$
\mathbf{B}_3 \bullet z_1 z_1^H < 0,
$$

and

$$|t_1|^2((\tilde{z}_1/t_1)^H A_k(\tilde{z}_1/t_1) + 2\mathrm{Re}\,((\tilde{z}_1/t_1)^H b_k) + c_k) = B_k \bullet z_1 z_1^H \lhd_k 0, k = 1, 2.$$

That is, $\tilde{z}_1/t_1$ is feasible to (QCQP).

We can easily verify that the rank-1 matrix

$$\begin{bmatrix} 1 \\ \tilde{z}_1/t_1 \end{bmatrix} [1\ \ \tilde{z}_1^H/\bar{t}_1]$$

satisfies the complementarity, meaning that it is an optimal solution of (QSR), and thus $\tilde{z}_l/t_l$ is an optimal solution of (QCQP). Furthermore, all the computations involved can be run in polynomial time. The desired results follow. $\blacksquare$

6.5.3 QCQP problem with homogeneous functions

Let us now consider

$$(\mathrm{HQCQP}) \begin{cases} \max_z & z^H A_0 z \\ \text{s.t.} & z^H A_k z \lhd_k 1,\ k = 1, \ldots, M, \end{cases}$$

where we suppose that $M \geq 3$, $N \geq 2$, and $A_k \in \mathcal{H}^N$, $k = 0, \ldots, M$.

The corresponding SDP relaxation is

$$(\mathrm{HQSR}) \begin{cases} \max_Z & A_0 \bullet Z \\ \text{s.t.} & A_k \bullet Z \lhd_k 1,\ k = 1, \ldots, M, \\ & Z \succeq 0. \end{cases}$$

Its dual problem is

$$(\mathrm{DHQSR}) \begin{cases} \min_{y_1, \ldots, y_M} & \sum_{k=1}^{M} y_k \\ \text{s.t.} & Y = \sum_{k=1}^{M} y_k A_k - A_0 \succeq 0, \\ & y_k \lhd_k^* 0, k = 1, \ldots, M. \end{cases}$$

THEOREM 6.6 *Suppose that $M = 3$ and the primal and dual SDP problems (HQSR) and (DHQSR) are strictly feasible. Then the SDP relaxation of (HQCQP) has zero gap, and an optimal solution of the problem (HQCQP) can be constructed from an optimal solution of (HQSR) in polynomial time.*

Proof If there is a nonbinding constraint at Z^*, the conclusion holds due to Theorem 6.5. Suppose that the three constraints are binding at Z^*, that is, $A_1 \bullet Z^* = A_2 \bullet Z^* = A_3 \bullet Z^* = 1$. Then we have $(A_1 - A_2) \bullet Z^* = (A_2 - A_3) \bullet Z^* = 0$. By the decomposition Theorem 6.2, it follows that there is a rank-1 decomposition $Z^* = \sum_{j=1}^{R} z_j z_j^H$ such that $(A_1 - A_2) \bullet z_j z_j^H = (A_2 - A_3) \bullet z_j z_j^H = 0, j = 1, \ldots, R$, where R is the rank of Z^*. Since $A_1 \bullet Z^* = 1$, there is a $j_0 \in \{1, 2, \ldots, R\}$, say $j_0 = 1$, such that $A_1 \bullet z_1 z_1^H = s > 0$. By checking the complementary conditions (6.11) and (6.12), we conclude that

$(z_1/\sqrt{s})(z_1/\sqrt{s})^H$ is also an optimal solution of (HQSR). Then $z_1/\sqrt{s}$ is an optimal solution of (HQCQP). Then the proof is complete.   ∎

6.6   The radar code-design problem

In the next few sections we shall elaborate on one particular application of the results that we have developed so far. The application finds its root in radar code design.

6.6.1   Background

The huge advances in high-speed signal processing hardware, digital array radar technology, and the requirement of better and better radar performances has promoted, during the last two decades, the development of very sophisticated algorithms for radar waveform design [16].

Waveform optimization in the presence of colored disturbance with known covariance matrix has been addressed in [9]. Three techniques based on the maximization of the SNR are introduced and analyzed. Two of them also exploit the degrees of freedom provided by a rank-deficient disturbance covariance matrix. In [34], a signal-design algorithm relying on the maximization of the SNR under a similarity constraint with a given waveform is proposed and assessed. The solution potentially emphasizes the target contribution and de-emphasizes the disturbance. Moreover, it also preserves some characteristics of the desired waveform. In [22], a signal subspace framework, which allows the derivation of the optimal radar waveform (in the sense of maximizing the SNR at the output of the detector) for a given scenario, is presented under the Gaussian assumption for the statistics of both the target and the clutter.

A quite different signal-design approach relies on the modulation of pulse train parameters (amplitude, phase, and frequency) in order to synthesize waveforms with some specified properties. This technique is known as the "radar coding" and a substantial bulk of work is nowadays available in open literature about this topic. Here we mention Barker, Frank, and Costas codes which lead to waveforms whose ambiguity functions share good resolution properties both in range and Doppler. This list is not exhaustive, and a comprehensive treatment can be found in [5,33]. It is, however, worth pointing out that the ambiguity function is not the only relevant objective function for code design in operating situations where the disturbance is not white. This might be the case of optimum radar detection in the presence of colored disturbance, where the standard matched filter is no longer optimum and the most powerful radar receiver requires whitening operations.

In the remainder of the chapter, following [15], we shall tackle the problem of code optimization in the presence of colored Gaussian disturbance, as an application of our rank-1 matrix decomposition theorems. At the design stage, we focus on the class of coded pulse trains, and propose a code-selection algorithm which is optimum according to the following criterion: maximization of the detection performance under a control both on the region of achievable values for the Doppler estimation accuracy and on the degree of similarity with a pre-fixed radar code. Actually, this last constraint is equivalent to force a similarity between the ambiguity functions of the devised waveform and of the pulse train encoded with the pre-fixed sequence. The resulting optimization problem belongs to the family of nonconvex quadratic programs [6,11]. In order to solve it, we first resort to a relaxation of the original problem into a convex one which belongs to the SDP class. Then an optimum code is constructed through the special rank-1 decomposition theorem, Theorem 6.2, applied to an optimal solution of the relaxed problem. Thus, the entire code-search algorithm possesses a polynomial computational complexity.

At the numerical analysis stage, we assess the performance of the new encoding algorithm in terms of detection performance, region of estimation accuracies that an estimator of the Doppler frequency can theoretically achieve, and ambiguity function. The results show that it is possible to realize a trade-off among the three aforementioned performance metrics. In other words, detection capabilities can be swapped for desirable properties of the waveform ambiguity function, and/or for an enlarged region of achievable Doppler estimation accuracies.

6.6.2 System model

We consider a radar system which transmits a coherent burst of pulses

$$s(t) = a_t u(t) \exp[i(2\pi f_0 t + \phi)],$$

where a_t is the transmit signal amplitude,

$$u(t) = \sum_{j=0}^{N-1} a(j)p(t - jT_r),$$

is the signal's complex envelope (see Figure 6.3), $p(t)$ is the signature of the transmitted pulse, T_r is the "pulse repetition time" (PRT), $[a(0), a(1), \ldots, a(N-1)] \in \mathbb{C}^N$ is the radar code (assumed without loss of generality with unit norm), f_0 is the carrier frequency, and ϕ is a random phase. Moreover, the pulse waveform $p(t)$ is of duration $T_p \leq T_r$ and

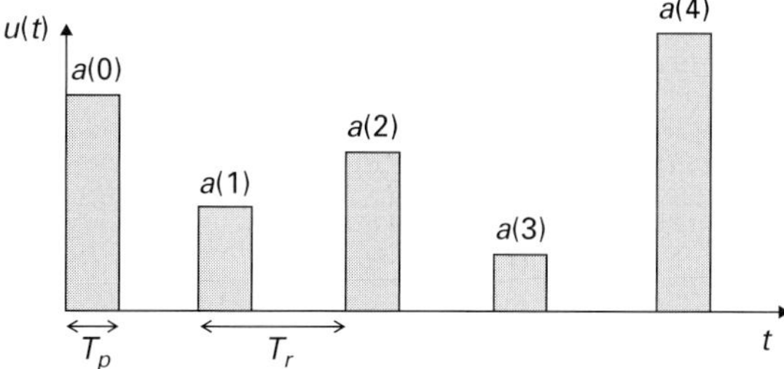

Figure 6.3 Coded pulse train $u(t)$ for $N = 5$ and $p(t)$ with rectangular shape.

has unit energy, in other words,

$$\int_0^{T_p} |p(t)|^2 dt = 1.$$

The signal backscattered by a target with a two-way time delay τ and received by the radar is

$$r(t) = \alpha_r e^{i2\pi(f_0+f_d)(t-\tau)} u(t-\tau) + n(t),$$

where α_r is the complex echo amplitude (accounting for the transmit amplitude, phase, target reflectivity, and channel's propagation effects), f_d is the target Doppler frequency, and $n(t)$ is additive disturbance due to clutter and thermal noise.

This signal is down-converted to baseband and filtered through a linear system with impulse response $h(t) = \overline{p(-t)}$. Let the filter output be

$$v(t) = \alpha_r e^{-i2\pi f_0 \tau} \sum_{j=0}^{N-1} a(j) e^{i2\pi j d T_r} \chi_p(t - jT_r - \tau, f_d) + w(t),$$

where $\chi_p(\lambda, f)$ is the pulse waveform ambiguity function [33], in other words,

$$\chi_p(\lambda, f) = \int_{-\infty}^{+\infty} p(\beta)\overline{p(\beta - \lambda)} e^{i2\pi f \beta} d\beta,$$

and $w(t)$ is the down-converted and filtered disturbance component. The signal $v(t)$ is sampled at $t_k = \tau + kT_r$, $k = 0, \ldots, N-1$, providing the observables[1]

$$v(t_k) = \alpha a(k) e^{i2\pi k f_d T_r} \chi_p(0, f_d) + w(t_k), \qquad k = 0, \ldots, N-1,$$

where $\alpha = \alpha_r e^{-i2\pi f_0 \tau}$. Assuming that the pulse waveform time-bandwidth product and the expected range of target Doppler frequencies are such that the single pulse waveform is insensitive to target Doppler shift,[2] namely $\chi_p(0, f_d) \sim \chi_p(0,0) = 1$, we can rewrite the samples $v(t_k)$ as

$$v(t_k) = \alpha a(k) e^{i2\pi k f_d T_r} + w(t_k), \qquad k = 0, \ldots, N-1.$$

Moreover, denoting by $c = [a(0), a(1), \ldots, a(N-1)]^T$ the N-dimensional column vector containing the code elements, $p = [1, e^{i2\pi f_d T_r}, \ldots, e^{i2\pi(N-1)f_d T_r}]^T$ the temporal steering vector, $v = [v(t_0), v(t_1), \ldots, v(t_{N-1})]^T$, and $w = [w(t_0), w(t_1), \ldots, w(t_{N-1})]^T$, we get the following vectorial model for the backscattered signal

$$v = \alpha c \odot p + w. \tag{6.23}$$

---

[1] We neglect range straddling losses and also assume that there are no target range ambiguities.

[2] Notice that this assumption might be restrictive for the cases of very fast moving targets such as fighters and ballistic missiles.

6.7　Performance measures for code design

In this section, we introduce some key performance measures to be optimized or controlled during the selection of the radar code. As it will be shown, they permit us to formulate the design of the code as a constrained optimization problem. The metrics considered in this paper are:

6.7.1　Detection probability

This is one of the most important performance measures which radar engineers attempt to optimize. We just remind the readers that the problem of detecting a target in the presence of observables described by the model (6.23) can be formulated in terms of the following binary hypotheses test:

$$\begin{cases} H_0 : v = w \\ H_1 : v = \alpha c \odot p + w. \end{cases} \tag{6.24}$$

Assuming that the disturbance vector w is a zero-mean, complex, circular Gaussian vector with known positive definite covariance matrix

$$\mathsf{E}[ww^H] = M$$

($\mathsf{E}[\cdot]$ denotes statistical expectation), the "generalized-likelihood ratio test" (GLRT) detector for (6.24), which coincides with the optimum test (according to the Neyman–Pearson criterion) if the phase of α is uniformly distributed in $[0, 2\pi[$ [26], is given by

$$|v^H M^{-1}(c \odot p)|^2 \underset{H_0}{\overset{H_1}{\underset{<}{\gtrless}}} G, \tag{6.25}$$

where G is the detection threshold set according to a desired value of the false alarm Probability (P_{fa}). An analytical expression of the detection Probability (P_d), for a given value of P_{fa}, is available both for the cases of "nonfluctuating" and "fluctuating target". In the former case (NFT),

$$P_d = Q\left(\sqrt{2|\alpha|^2(c \odot p)^H M^{-1}(c \odot p)}, \sqrt{-2\ln P_{fa}}\right), \tag{6.26}$$

while, for the case of the "Rayleigh fluctuating target" (RFT) with $E[|\alpha|^2] = \sigma_a^2$,

$$P_d = \exp\left(\frac{\ln P_{fa}}{1 + \sigma_a^2(c \odot p)^H M^{-1}(c \odot p)}\right), \tag{6.27}$$

where $Q(\cdot, \cdot)$ denotes the Marcum Q function of order 1. These last expressions show that, given P_{fa}, P_d depends on the radar code, the disturbance covariance matrix and the temporal steering vector only through the SNR, defined as

$$
\text{SNR} = \begin{cases} |\alpha|^2 (c \odot p)^H M^{-1} (c \odot p) & \text{NFT} \\[2ex] \sigma_a^2 (c \odot p)^H M^{-1} (c \odot p) & \text{RFT.} \end{cases}
\tag{6.28}
$$

Moreover, P_d is an increasing function of SNR and, as a consequence, the maximization of P_d can be obtained by maximizing the SNR over the radar code.

6.7.2   Doppler frequency estimation accuracy

The Doppler accuracy is bounded below by "Cramer–Rao bound" (CRB) and CRB-like techniques, which provide lower bounds for the variances of unbiased estimates. Constraining the CRB is tantamount to controlling the region of achievable Doppler estimation accuracies, referred to in the following as $\mathcal{A}$. We just highlight that a reliable measurement of the Doppler frequency is very important in radar signal processing because it is directly related to the target radial velocity useful to speed the track initiation, to improve the track accuracy [17], and to classify the dangerousness of the target. In this subsubsection, we introduce the CRB for the case of known α, the "modified CRB" (MCRB) [14], and the "Miller–Chang bound" (MCB) [36] for random α, and the "hybrid CRB" (HCRB) [45, pp. 931–932] for the case of a random zero-mean α.

PROPOSITION 6.1   *The CRB for known α (Case 1), the MCRB and the MCB for random α (Case 2 and Case 3, respectively), and the HCRB for a random zero-mean α (Case 4) are given by*

$$
\Delta_{CR}(f_d) = \frac{\Psi}{2 \dfrac{\partial h^H}{\partial f_d} M^{-1} \dfrac{\partial h}{\partial f_d}},
\tag{6.29}
$$

where $h = c \odot p$,

$$
\Psi = \begin{cases} \dfrac{1}{|\alpha|^2} & \text{Case 1} \\[3ex] \dfrac{1}{\mathrm{E}[|\alpha|^2]} & \text{Cases 2 and 4} \\[3ex] \mathrm{E}\left[\dfrac{1}{|\alpha|^2}\right] & \text{Case 3.} \end{cases}
\tag{6.30}
$$

Proof   See Appendix A6.1.

Notice that

$$
\frac{\partial h}{\partial f_d} = T_r c \odot p \odot u,
$$

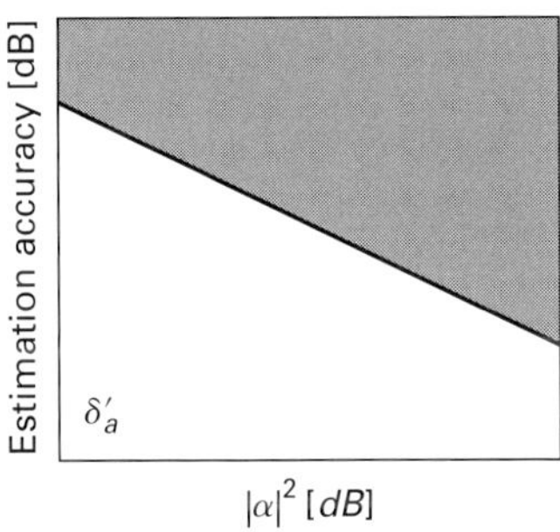
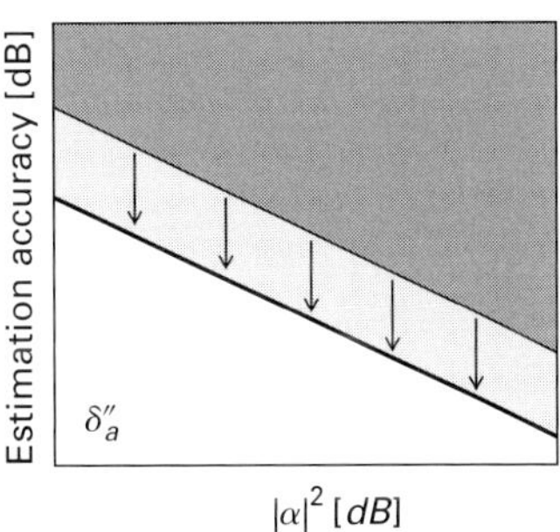

Figure 6.4 Lower bound to the size of the region $\mathcal{A}$ for two different values of δ_a ($\delta_a' < \delta_a''$).

with $u = [0, i2\pi, \ldots, i2\pi(N-1)]^T$, and (6.29) can be rewritten as

$$\Delta_{CR}(f_d) = \frac{\Psi}{2T_r^2(c \odot p \odot u)^H M^{-1}(c \odot p \odot u)}. \tag{6.31}$$

■

As already stated, forcing an upper bound to the CRB, for a specified Ψ value, results in a lower bound on the size of $\mathcal{A}$. Hence, according to this guideline, we focus on the class of radar codes complying with the condition

$$\Delta_{CR}(f_d) \leq \frac{\Psi}{2T_r^2 \delta_a}, \tag{6.32}$$

which can be equivalently written as

$$(c \odot p \odot u)^H M^{-1}(c \odot p \odot u) \geq \delta_a, \tag{6.33}$$

where the parameter δ_a rules the lower bound on the size of $\mathcal{A}$ (see Figure 6.4 for a pictorial description). Otherwise stated, suitably increasing δ_a, we ensure that new points fall in the region $\mathcal{A}$, namely new smaller values for the estimation variance can be theoretically reached by estimators of the target Doppler frequency.

6.7.3 The similarity constraint

Designing a code which optimizes the detection performance does not provide any kind of control to the shape of the resulting coded waveform. Precisely, the unconstrained optimization of P_d can lead to signals with significant modulus variations, poor range resolution, high peak sidelobe levels, and, more generally, with an undesired ambiguity function behavior. These drawbacks can be partially circumvented by imposing a further constraint to the sought radar code. Precisely, it is required that the solution be similar to a known code c_0 ($\|c_0\|^2 = 1$, that is, c_0 is assumed to be a normalized vector, without loss of generality), which shares constant modulus, reasonable range resolution, and peak sidelobe level. This is tantamount to imposing that [34]

$$\|c - c_0\|^2 \leq \epsilon, \qquad (6.34)$$

where the parameter $\epsilon \geq 0$ rules the size of the similarity region. In other words, (6.34) permits us to indirectly control the ambiguity function of the considered coded pulse train: the smaller ϵ, the higher the degree of similarity between the ambiguity functions of the designed radar code and of c_0.

6.8   Optimal code design

In this section, we propose a technique for the selection of the radar code which attempts to maximize the detection performance but, at the same time, provides a control both on the target Doppler estimation accuracy and on the similarity with a given radar code. To this end, we first observe that

$$(c \odot p)^H M^{-1}(c \odot p) = c^H \left[M^{-1} \odot \overline{(pp^H)} \right] c = c^H Rc \qquad (6.35)$$

and

$$(c \odot p \odot u)^H M^{-1}(c \odot p \odot u) = c^H \left[M^{-1} \odot \overline{(pp^H)} \odot \overline{(uu^H)} \right] c = c^H R_1 c, \qquad (6.36)$$

where $R = M^{-1} \odot \overline{(pp^H)}$ and $R_1 = M^{-1} \odot \overline{(pp^H)} \odot \overline{(uu^H)}$ are positive semidefinite [45, p. 1352, A.77]. It follows that P_d and $\Delta_{CR}(f_d)$ are competing with respect to the radar code. In fact, their values are ruled by two different quadratic forms, (6.35) and (6.36) respectively, with two different matrices R and R_1. Hence, only when they share the same eigenvector corresponding to the maximum eigenvalue are (6.35) and (6.36) maximized by the same code. In other words, maximizing P_d results in a penalization of $\Delta_{CR}(f_d)$, and vice versa.

Exploiting (6.35) and (6.36), the code-design problem can be formulated as a nonconvex optimization "quadratic problem" (QP)

$$\begin{cases} \max_c & c^H Rc \\ \text{s.t.} & c^H c = 1 \\ & c^H R_1 c \geq \delta_a \\ & \|c - c_0\|^2 \leq \epsilon, \end{cases}$$

which can be equivalently written as

$$(\text{QP}) \begin{cases} \max_c & c^H Rc \\ \text{s.t.} & c^H c = 1 \\ & c^H R_1 c \geq \delta_a \\ & \text{Re}\left(c^H c_0 \right) \geq 1 - \epsilon/2. \end{cases} \qquad (6.37)$$

The feasibility of the problem, which not only depends on the parameters δ_a and ϵ, but also on the pre-fixed code c_0, is discussed in Appendix A6.2.

In what follows, we show that an optimal solution of (6.37) can be obtained from an optimal solution of the following "enlarged quadratic problem" (EQP):

$$\text{(EQP)} \quad \begin{cases} \max_c & c^H R c \\ \text{s.t.} & c^H c = 1 \\ & c^H R_1 c \geq \delta_a \\ & \text{Re}^2\left(c^H c_0\right) + \text{Im}^2\left(c^H c_0\right) = c^H c_0 c_0{}^H c \geq \delta_\epsilon, \end{cases} \tag{6.38}$$

where $\delta_\epsilon = (1 - \epsilon/2)^2$. Since the feasibility region of (EQP) is larger than that of (QP), every optimal solution of (EQP), which is feasible for (QP), is also an optimal solution for (QP). Thus, assume that c^* is an optimal solution of (EQP) and let $\phi = \arg\left(c^{*H} c_0\right)$. It is easily seen that $c^* e^{i\phi}$ is still an optimal solution of (EQP). Now, observing that $(c^* e^{i\phi})^H c_0 = |c^{*H} c_0|$, $c^* e^{i\phi}$ is a feasible solution of (QP). In other words, $c^* e^{i \arg\left(c^{*H} c_0\right)}$ is optimal for both (QP) and (EQP).

Now, we have to find an optimal solution of (EQP) and, to this end, we exploit the equivalent matrix formulation

$$\text{(EQP)} \quad \begin{cases} \max_C & C \bullet R \\ \text{s.t.} & C \bullet I = 1 \\ & C \bullet R_1 \geq \delta_a \\ & C \bullet C_0 \geq \delta_\epsilon \\ & C = cc^H, \end{cases} \tag{6.39}$$

where I stands for the identity matrix, and $C_0 = c_0 c_0{}^H$.

Problem (6.39) can be relaxed into an SDP by neglecting the rank-1 constraint [2]. By doing so, we obtain an "enlarged quadratic problem relaxed" (EQPR)

$$\text{(EQPR)} \quad \begin{cases} \max_C & C \bullet R \\ \text{s.t.} & C \bullet I = 1 \\ & C \bullet R_1 \geq \delta_a \\ & C \bullet C_0 \geq \delta_\epsilon \\ & C \succeq 0, \end{cases} \tag{6.40}$$

where the last constraint means that C is to be Hermitian positive semidefinite. The dual problem of (6.40), is

$$\text{(EQPRD)} \quad \begin{cases} \min_{y_1, y_2, y_3} & y_1 - y_2 \delta_a - y_3 \delta_\epsilon \\ \text{s.t.} & y_1 I - y_2 R_1 - y_3 C_0 \succeq R \\ & y_2 \geq 0, y_3 \geq 0. \end{cases}$$

This problem is bounded below and is strictly feasible: it follows by the strong duality Theorem 6.3 that the optimal value is the same as the primal, and the complementary conditions, (6.11) and (6.12), are satisfied at an optimal primal–dual pair, due to the strict feasibility of the primal problem (see Appendix A6.2).

In the following, we prove that a solution of (EQP) can be obtained from an optimal solution of (EQPR) C^*, and from an optimal solution of (EQPRD) (y_1^*, y_2^*, y_3^*). Precisely, we show how to obtain a rank-1 feasible solution of (EQPR) that satisfies the complementary conditions, which specify (6.11) and (6.12) for (EQPR) and (EQPRD),

$$\left(y_1^* I - y_2^* R_1 - y_3^* C_0 - R\right) \bullet C^* = 0 \tag{6.41}$$

$$\left[C^* \bullet R_1 - \delta_a\right] y_2^* = 0 \tag{6.42}$$

$$\left[C^* \bullet C_0 - \delta_\epsilon\right] y_3^* = 0. \tag{6.43}$$

Such a rank-1 solution is also optimal for (EQP). The proof, we propose, is based on the rank-1 matrix decomposition Theorem 6.2, synthetically denoted as $\mathcal{D}(Z, A, B)$.

Moreover, in order to find a step-by-step algorithm, we distinguish four possible cases:

Case 1 $C^* \bullet R_1 - \delta_a > 0$ and $C^* \bullet C_0 - \delta_\epsilon > 0$
Case 2 $C^* \bullet R_1 - \delta_a = 0$ and $C^* \bullet C_0 - \delta_\epsilon > 0$
Case 3 $C^* \bullet R_1 - \delta_a > 0$ and $C^* \bullet C_0 - \delta_\epsilon = 0$
Case 4 $C^* \bullet R_1 - \delta_a = 0$ and $C^* \bullet C_0 - \delta_\epsilon = 0$

Case 1: Using the decomposition $\mathcal{D}(C^*, I, R_1)$ in Theorem 6.2, we can express C^* as

$$C^* = \sum_{r=1}^{R} c_r c_r^{\ H}.$$

Now, we show that there exists a $k \in \{1, \ldots, R\}$ such that $\sqrt{R} c_k$ is an optimal solution of (EQP). In fact, the decomposition $\mathcal{D}(C^*, I, R_1)$ implies that every $(\sqrt{R} c_r)(\sqrt{R} c_r)^H$, $r = 1, \ldots, R$, satisfies the first and the second constraints in (EQPR). Moreover, there must be a $k \in \{1, \ldots, R\}$ such that $(\sqrt{R} c_k)^H C_0(\sqrt{R} c_k) \geq \delta_\epsilon$. Indeed, if $(\sqrt{R} c_r)^H C_0(\sqrt{R} c_r) < \delta_\epsilon$ for every r, then

$$\sum_{r=1}^{R} (\sqrt{R} c_r)^H C_0(\sqrt{R} c_r) < R \delta_\epsilon,$$

that is,

$$C^* \bullet C_0 < \delta_\epsilon,$$

which is in contradiction to the feasibility of C^*. Thus, the rank-1 matrix $(\sqrt{R} c_k)(\sqrt{R} c_k)^H$ is feasible for (EQPR). As to fulfillment of the complementary conditions, $C^* \bullet R_1 - \delta_a > 0$ and $C^* \bullet C_0 - \delta_\epsilon > 0$ imply $y_2^* = 0$ and $y_3^* = 0$, namely (6.42) and (6.43) are verified for every $(\sqrt{R} c_r)(\sqrt{R} c_r)^H$, with $r = 1, \ldots, R$. Also, (6.41) can be recast as

$$(y_1^* \boldsymbol{I} - \boldsymbol{R}) \bullet \boldsymbol{C}^* = (y_1^* \boldsymbol{I} - \boldsymbol{R}) \bullet \left(\sum_{r=0}^{R} \boldsymbol{c}_r \boldsymbol{c}_r^H \right) = 0,$$

which, since $\boldsymbol{c}_r \boldsymbol{c}_r^H \succeq 0$, $r = 1, \dots, R$, and $y_1^* \boldsymbol{I} - \boldsymbol{R} \succeq 0$ (from the first constraint of (EQPRD)), implies

$$\left(y_1^* \boldsymbol{I} - \boldsymbol{R} \right) \bullet \left(\sqrt{R} \boldsymbol{c}_r \boldsymbol{c}_r^H \sqrt{R} \right) = 0, r = 1, \dots, R.$$

In other words, $(\sqrt{R} \boldsymbol{c}_k)(\sqrt{R} \boldsymbol{c}_k)^H$, together with (y_1^*, y_2^*, y_3^*), fulfills all the complementary conditions (6.41) – (6.43). Thus $(\sqrt{R} \boldsymbol{c}_k)(\sqrt{R} \boldsymbol{c}_k)^H$ is an optimal solution of (EQPR), and $\sqrt{R} \boldsymbol{c}_k$ is an optimal solution of (EQP).

Cases 2 and 3: The proof is completely similar to Case 1, hence we omit it.

Case 4: In this case, all the constraints of (EQPR) are active, namely $\boldsymbol{C}^* \bullet \boldsymbol{I} = 1$, $\boldsymbol{C}^* \bullet \boldsymbol{R}_1 = \delta_a$, and $\boldsymbol{C}^* \bullet \boldsymbol{C}_0 = \delta_\epsilon$. Then the proof is the same as the proof in Theorem 6.6.

In conclusion, using the decomposition of Theorem 6.2, we have shown how to construct a rank-1 optimal solution of (EQPR), which is tantamount to finding an optimal solution of (EQP). Summarizing, the optimum code can be constructed according to the following procedure:

Code Construction Procedure
Input: the datum of (QP): $\boldsymbol{R}, \boldsymbol{R}_1, \boldsymbol{c}_0, \delta_a$ and ϵ.
Output: an optimal solution of (QP).

1. Enlarge the original (QP) into the (EQP) (i.e., substitute $\mathrm{Re}(\boldsymbol{c}^H \boldsymbol{c}_0) \geq 1 - \epsilon/2$ with $|\boldsymbol{c}^H \boldsymbol{c}_0|^2 \geq \delta_\epsilon$);
2. Relax (EQP) into (EQPR) (i.e., relax $\boldsymbol{C} = \boldsymbol{c}\boldsymbol{c}^H$ into $\boldsymbol{C} \succeq 0$);
3. Solve the SDP problem (EQPR) finding an optimal solution $\boldsymbol{C}^*$;
4. Evaluate $\boldsymbol{C}^* \bullet \boldsymbol{R}_1 - \delta_a$ and $\boldsymbol{C}^* \bullet \boldsymbol{C}_0 - \delta_\epsilon$: if both are equal to 0 go to 5), else go to 7);
5. Evaluate $\mathcal{D}(\boldsymbol{C}^*, \boldsymbol{R}_1/\delta_a - \boldsymbol{I}, \boldsymbol{C}_0/\delta_\epsilon - \boldsymbol{I})$, obtaining $\boldsymbol{C}^* = \sum_{r=1}^{R} \boldsymbol{c}_r \boldsymbol{c}_r^H$;
6. Compute $\boldsymbol{c}^* = \sqrt{\gamma_1} \boldsymbol{c}_1$, with $\gamma_1 = 1/\|\boldsymbol{c}_1\|^2$, and go to 9);
7. Evaluate $\mathcal{D}(\boldsymbol{C}^*, \boldsymbol{R}_1, \boldsymbol{I})$ obtaining $\boldsymbol{C}^* = \sum_{r=1}^{R} \boldsymbol{c}_r \boldsymbol{c}_r^H$;
8. Find k such that $\boldsymbol{c}_k^H \boldsymbol{C}_0 \boldsymbol{c}_k \geq \delta_\epsilon/R$ and compute $\boldsymbol{c}^* = \sqrt{R} \boldsymbol{c}_k$;
9. Evaluate the optimal solution of the original problem (QP) as $\boldsymbol{c}^* e^{i\phi}$, with $\phi = \arg((\boldsymbol{c}^*)^H \boldsymbol{c}_0)$.

The computational complexity connected with the implementation of the algorithm is polynomial, as both the SDP problem and the decomposition of Theorem 6.2 can be performed in polynomial time. In fact, the amount of operations, involved in solving the SDP problem, is $O\left(N^{3.5} \log \frac{1}{\zeta}\right)$ [6, p. 250], where ζ is a prescribed accuracy, and the rank-1 decomposition requires $O(N^3)$ operations (as shown in Section 6.2).

6.9   Performance analysis

This section is aimed at analyzing the performance of the proposed encoding scheme. To this end, we assume that the disturbance covariance matrix is exponentially shaped with a one-lag correlation coefficient $\rho = 0.8$, in other words,

$$M(k,l) = \rho^{|k-l|},$$

and we fix the P_{fa} of the receiver (6.25) to 10^{-6}. The analysis is conducted in terms of P_d, the region of achievable Doppler estimation accuracies, and the ambiguity function of the coded pulse train, which results from exploiting the proposed algorithm (cf. Section 6.6.2), that is,

$$\chi(\lambda,f) = \int_{-\infty}^{\infty} u(\beta)\overline{u(\beta - \lambda)}e^{i2\pi f\beta}d\beta = \sum_{l=0}^{N-1}\sum_{k=0}^{N-1}(a(l))^*\overline{(a(k))^*}\chi_p[\lambda - (l-k)T_r,f],$$

where $[(a(0))^*,\ldots,(a(N-1))^*]$ is an optimum code. As to the temporal steering vector p, we set the normalized Doppler frequency[3] $f_d T_r = 0$. The convex optimization MATLAB toolbox SElf-DUal-MInimization (SeDuMi) of Jos Sturm [43] is used for solving the SDP relaxation. The decomposition $\mathcal{D}(\cdot,\cdot,\cdot)$ of the SeDuMi solution is performed using the technique described in Algorithm DCMP2. Finally, the MATLAB toolbox of [37] is used to plot the ambiguity functions of the coded pulse trains.

In the following, a generalized Barker sequence [33, pp. 109–113] is chosen as similarity code. We just highlight that generalized Barker sequences are polyphase codes whose autocorrelation function has a minimal peak-to-sidelobe ratio excluding the outermost sidelobe. Examples of such sequences were found for all $N \leq 45$ [10, 23] using numerical optimization techniques. In the simulations of this subsection, we assume $N = 7$ and set the similarity code equal to the generalized Barker sequence $c_0 = [0.3780, 0.3780, -0.1072 - 0.3624i, -0.0202 - 0.3774i, 0.2752 + 0.2591i, 0.1855 - 0.3293i, 0.0057 + 0.3779i]^T$.

In Figure 6.5a, we plot P_d of the optimum code (according to the proposed criterion) versus $|\alpha|^2$ for several values of δ_a, $\delta_\epsilon = 0.01$, and for a nonfluctuating target. In the same figure, we also represent both the P_d of the similarity code, as well as the benchmark performance, namely the maximum achievable detection rate (over the radar code), given by

$$P_d = Q\left(\sqrt{2|\alpha|^2\lambda_{\max}(R)}, \sqrt{-2\ln P_{fa}}\right), \tag{6.44}$$

where $\lambda_{\max}(\cdot)$ denotes the maximum eigenvalue of the argument.

The curves show that with increasing δ_a we get lower and lower values of P_d for a given $|\alpha|^2$ value. This was expected, since the higher the δ_a the smaller the feasibility

---

[3] We have also considered other values for the target normalized Doppler frequency. The results, not reported here, confirm the performance behavior shown in the next two subsections.

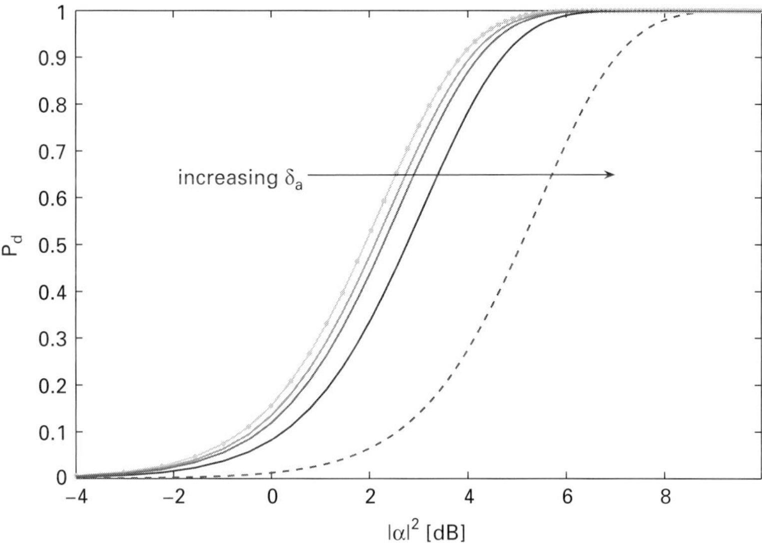

Figure 6.5a P_d versus $|\alpha|^2$ for $P_{fa} = 10^{-6}, N = 7, \delta_\epsilon = 0.01$, nonfluctuating target, and several values of $\delta_a \in \{10^{-6}, 6165.5, 6792.6, 7293.9\}$. Generalized Barker code (dashed curve). Code which maximizes the SNR for a given δ_a (solid curve). Benchmark code (dotted-marked curve). Notice that the curve for $\delta_a = 10^{-6}$ perfectly overlaps with the benchmark P_d.

region of the optimization problem to be solved for finding the code. Nevertheless, the proposed encoding algorithm usually ensures a better detection performance than the original generalized Barker code.

In Figure 6.5b, the normalized CRB ($\mathrm{CRB}_n = T_r^2 \mathrm{CRB}$) is plotted versus $|\alpha|^2$ for the same values of δ_a as in Figure 6.5a. The best value of CRB_n is plotted too, that is,

$$\mathrm{CRB}_n = \frac{1}{2|\alpha|^2 \lambda_{\max}\left(\boldsymbol{R}_1\right)} . \tag{6.45}$$

The curves highlight that by increasing δ_a, better and better CRB values can be achieved. This is in accordance with the considered criterion, because the higher the δ_a the larger the size of the region $\mathcal{A}$. Summarizing, the joint analysis of Figures 6.5a and 6.5b shows that a trade-off can be realized between the detection performance and the estimation accuracy. Moreover, there exist codes capable of outperforming the generalized Barker code, both in terms of P_d and size of $\mathcal{A}$.

The effects of the similarity constraint are analyzed in Figure 6.5c. Therein, we set $\delta_a = 10^{-6}$ and consider several values of δ_ϵ. The plots show that with increasing δ_ϵ, worse and worse P_d values are obtained; this behavior can be explained by observing that the smaller the δ_ϵ the larger the size of the similarity region. However, this detection loss is compensated for by an improvement of the coded pulse-train ambiguity function. This is shown in Figures 6.6b–6.6e, where such a function is plotted assuming rectangular

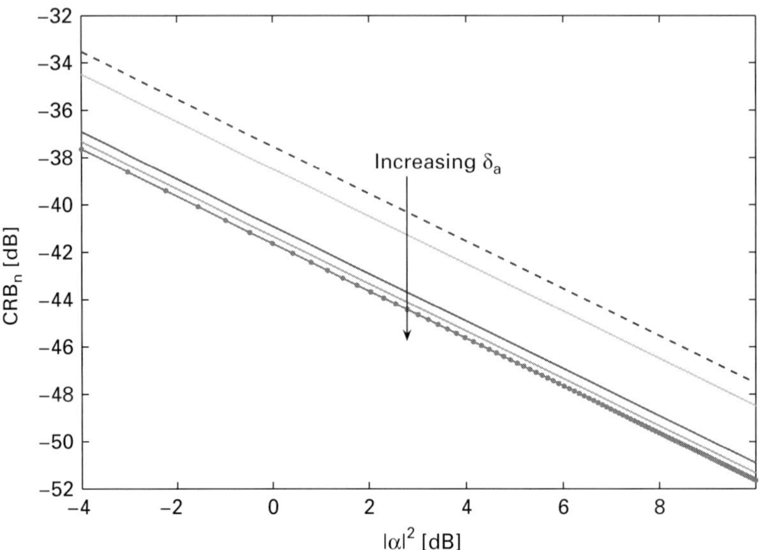

Figure 6.5b CRB_n versus $|\alpha|^2$ for $N = 7$, $\delta_\epsilon = 0.01$, and several values of $\delta_a \in \{10^{-6}, 6165.5, 6792.6, 7293.9\}$. Generalized Barker code (dashed curve). Code which maximizes the SNR for a given δ_a (solid curve). Benchmark code (dotted-marked curve). Notice that the curve for $\delta_a = 7293.9$ perfectly overlaps with the benchmark CRB_n.

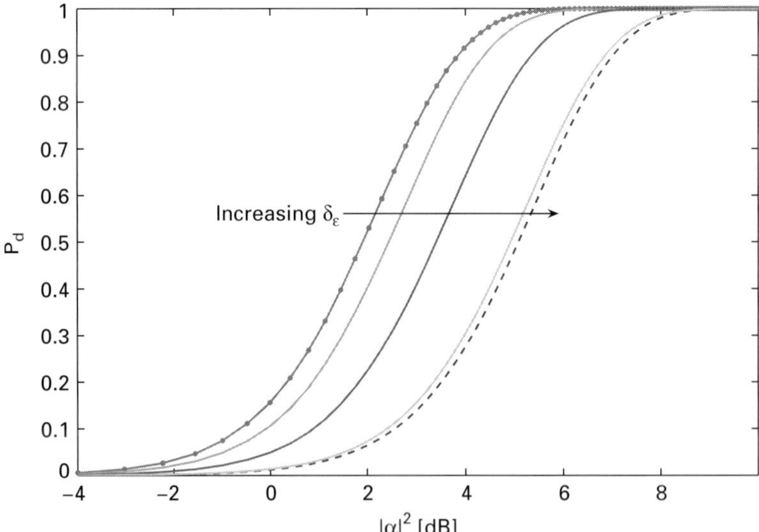

Figure 6.5c P_d versus $|\alpha|^2$ for $P_{fa} = 10^{-6}$, $N = 7$, $\delta_a = 10^{-6}$, nonfluctuating target, and several values of $\delta_\epsilon \in \{0.01, 0.6239, 0.8997, 0.9994\}$. Generalized Barker code (dashed curve). Code which maximizes the SNR for a given δ_ϵ (solid curve). Benchmark code (dotted-marked curve). Notice that the curve for $\delta_\epsilon = 0.01$ perfectly overlaps with the benchmark P_d.

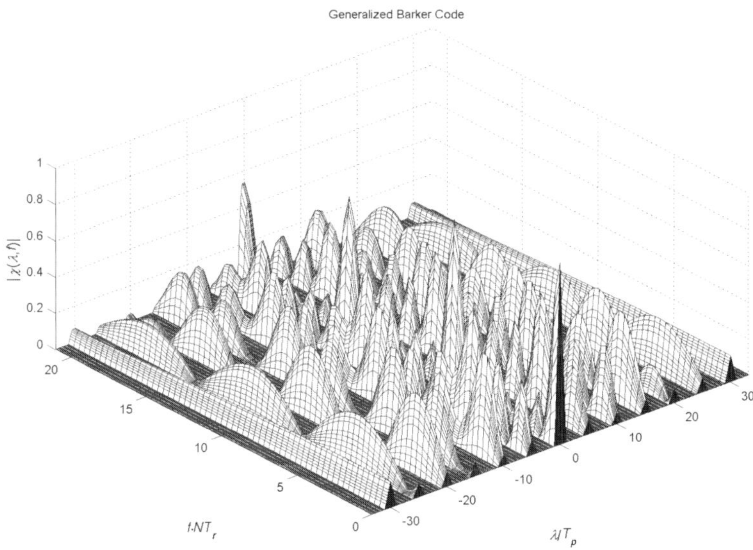

Figure 6.6a Ambiguity function modulus of the generalized Barker code $c_0 = [0.3780, 0.3780, -0.1072 - 0.3624i, -0.0202 - 0.3774i, 0.2752 + 0.2591i, 0.1855 - 0.3293i, 0.0057 + 0.3779i]^T$.

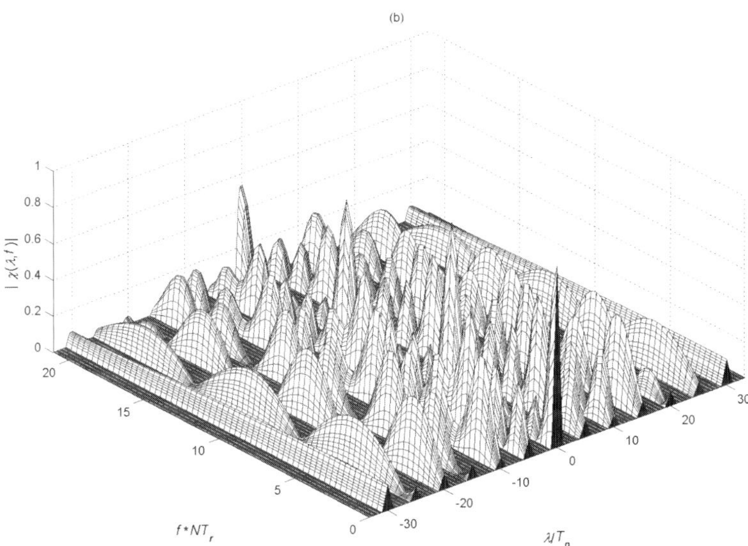

Figure 6.6b Ambiguity function modulus of code which maximizes the SNR for $N = 7$, $\delta_a = 10^{-6}$, c_0-generalized Barker code, and several values of $\delta_\epsilon = 0.9994$.

pulses, $T_r = 5T_p$ and the same values of δ_a and δ_ϵ as in Figure 6.5c. Moreover, for comparison purposes, the ambiguity function of c_0 is plotted too (Figure 6.6a). The plots highlight that the closer δ_ϵ is to 1, the higher the degree of similarity is between the ambiguity functions of the devised, and of the pre-fixed codes. This is due to the fact

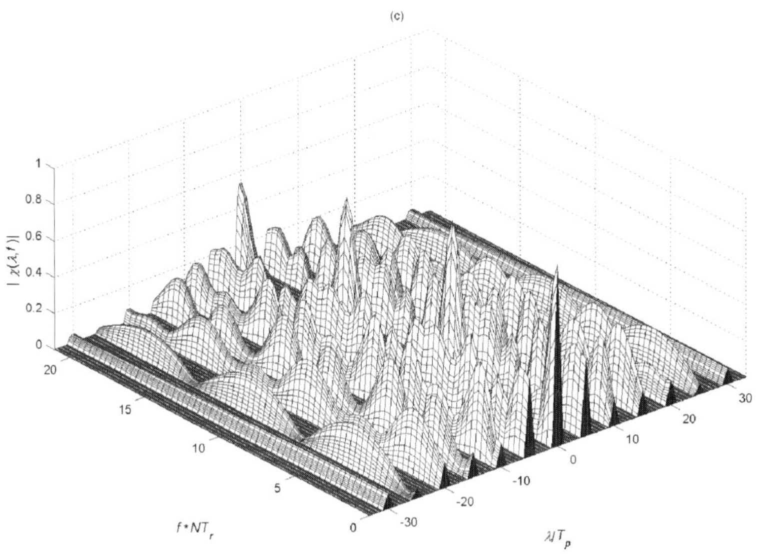

Figure 6.6c   Ambiguity function modulus of code which maximizes the SNR for $N = 7$, $\delta_a = 10^{-6}$, $\mathbf{c}_0$-generalized Barker code, and several values of $\delta_\epsilon = 0.8997$.

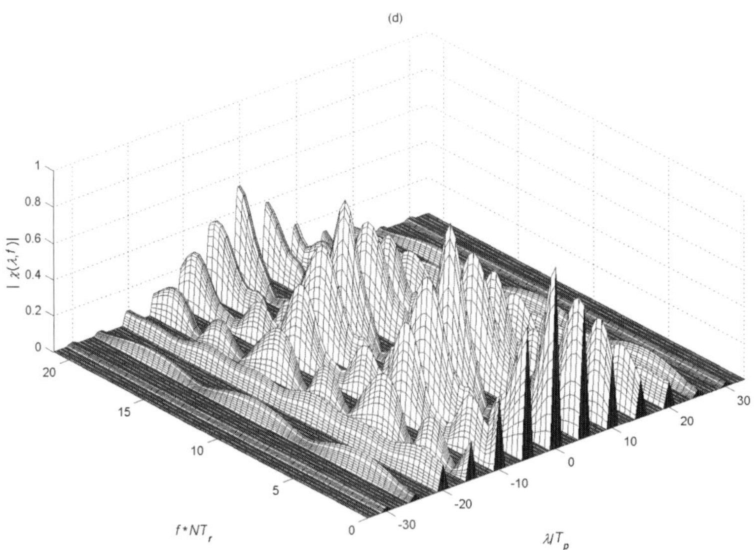

Figure 6.6d   Ambiguity function modulus of code which maximizes the SNR for $N = 7$, $\delta_a = 10^{-6}$, $\mathbf{c}_0$-generalized Barker code, and several values of $\delta_\epsilon = 0.6239$.

that increasing δ_ϵ is tantamount to reducing the size of the similarity region. In other words, we force the devised code to be more and more similar to the pre-fixed one and, as a consequence, we get more and more similar ambiguity functions.

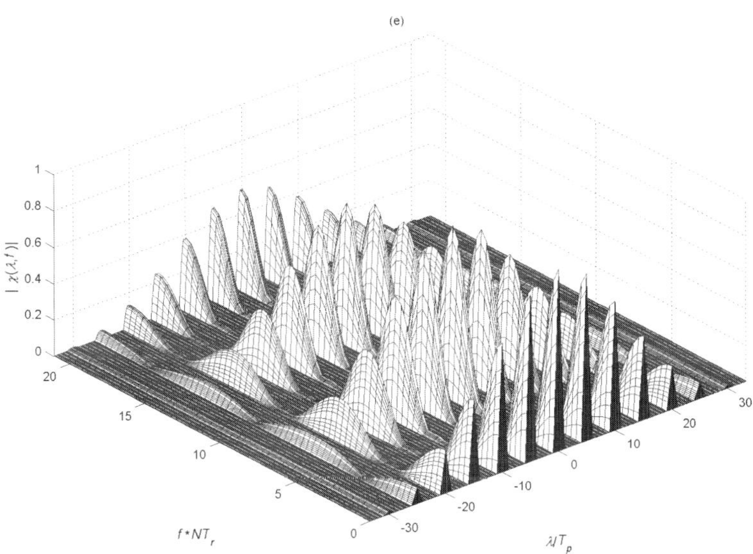

Figure 6.6e   Ambiguity function modulus of code which maximizes the SNR for $N = 7$, $\delta_a = 10^{-6}$, $\mathbf{c}_0$-generalized Barker code, and several values of $\delta_\epsilon = 0.01$.

Table 6.2. Average N_{it} and CPU time in seconds required to solve problem (6.40). Generalized Barker code as similarity sequence.

δ_a	δ_ϵ	Average N_{it}	Average CPU time
10^{-6}	0.01	21	0.30
6165.5	0.01	11	0.15
6792.6	0.01	11	0.15
7293.9	0.01	16	0.19
10^{-6}	0.6239	22	0.28
10^{-6}	0.8997	19	0.24
10^{-6}	0.9994	17	0.23

Finally, Table 6.2 provides the average number of iterations N_{it} and CPU time (in seconds) which are required to solve the SDP problem (6.40). The computer used to get these results is equipped with a 3-GHz Intel XEON processor.

6.10   Conclusions

We have presented two rank-1 decomposition theorems for Hermitian positive semidefinite matrices with some special desired properties, designed computational procedures to realize the decompositions respectively, and assessed their computational complexity. Resorting to the known SDP theory and the decomposition theorems, we have studied

QCQP problems, and identified several classes of polynomially solvable QCQP problems with a few constraints.

As an application of SDP and the matrix decomposition techniques to radar signal processing, we have considered the design of coded waveforms in the presence of colored Gaussian disturbance. Particularly, we have illustrated a polynomial-time algorithm, originally proposed in [15], which attempts to maximize the detection performance under control both on the region of achievable values for the Doppler estimation accuracy, and on the similarity with a given radar code. Finally, numerical analysis of the algorithm has been conducted.

A6   Appendix to Chapter 6

A6.1   Proof of Proposition 6.1

Cases 1, 2, and 3 follow respectively from [45, p. 928, Eq. 8.37], the definitions [14, Eq. 4], and [25, Eq. 20]. Case 4 is based on the division of the parameter vector $\boldsymbol{\theta}$ into a nonrandom and a random component, that is, $\boldsymbol{\theta} = [\theta_1, \boldsymbol{\theta}_2^T]^T$, where $\theta_1 = f_d$, $\boldsymbol{\theta}_2 = [\alpha_R, \alpha_I]^T$, and α_R and α_I are the real and the imaginary part of α. Then, according to the definition [45, pp. 931–932, Section 8.2.3.2], the "Fisher information matrix" (FIM) for the HCRB can be written as

$$\mathbf{J}_H = \mathbf{J}_D + \mathbf{J}_P,$$

where $\mathbf{J}_D$ and $\mathbf{J}_P$ are given by [45, pp. 931–932, Eq. 8.50 and 8.59]. Standard calculus implies that

$$\mathbf{J}_D = 2\mathrm{diag}\left(\mathsf{E}[|\alpha|^2]\frac{\partial \boldsymbol{h}^H}{\partial f_d}\boldsymbol{M}^{-1}\frac{\partial \boldsymbol{h}}{\partial f_d}, \boldsymbol{h}^H\boldsymbol{M}^{-1}\boldsymbol{h}, \boldsymbol{h}^H\boldsymbol{M}^{-1}\boldsymbol{h}\right),$$

and

$$\mathbf{J}_P = \begin{bmatrix} 0 & \mathbf{0} \\ \mathbf{0} & \mathbf{K} \end{bmatrix},$$

where $\mathrm{diag}(\cdot)$ denotes a diagonal matrix, and $\mathbf{K}$ is a 2×2 matrix whose elements are related to the a-priori, probability density function of α through [45, p. 930, Eq. 8.51]. The HCRB for Doppler frequency estimation can be evaluated as $[\mathbf{J}_H^{-1}](1,1)$ which provides (6.29).

A6.2 Feasibility of (QP), (EQPR), and (EQPRD)

A6.2.1 Feasibility of (QP)

The feasibility of (QP) depends on the three parameters δ_a, δ_ϵ, and c_0.

Accuracy parameter δ_a. This parameter rules the constraint $c^H R_1 c \geq \delta_a$. Since c has unitary norm, $c^H R_1 c$ ranges between the minimum ($\lambda_{\min}(R_1)$) (which is zero in this case) and the maximum ($\lambda_{\max}(R_1)$) eigenvalue of R_1. As a consequence, a necessary condition on δ_a in order to ensure feasibility is $\delta_a \in [0, \lambda_{\max}]$.

Similarity parameter δ_ϵ. This parameter rules the similarity between the sought code and the pre-fixed code, through the constraint $\|c - c_0\|^2 \leq \epsilon$. If $\epsilon < 0$ then (QP) is infeasible. Moreover, $0 \leq \epsilon \leq 2 - 2\mathrm{Re}(c^H c_0) \leq 2$, where the last inequality stems from the observation that we choose the phase of c such that $\mathrm{Re}(c^H c_0) = |c^H c_0| \geq 0$. It follows that a necessary condition on $\delta_\epsilon = (1 - \epsilon/2)^2$, in order to ensure feasibility, is $\delta_\epsilon \in [0, 1]$.

Pre-fixed code c_0. Choosing $0 \leq \delta_a \leq \lambda_{\max}(R_1)$ and $0 \leq \delta_\epsilon \leq 1$ is not sufficient in order to ensure the feasibility of (QP). A possible way to construct a strict feasible (QP) problem is to reduce the range of δ_a according to the value of c_0. In fact, denoting by $\delta_{\max} = c_0^H R_1 c_0$, the problem (QP) is strictly feasible, assuming that $0 \leq \delta_a < \delta_{\max}$ and $0 \leq \delta_\epsilon < 1$. Moreover, a strict feasible solution of (QP) is c_0 itself.

A6.2.2 Feasibility of (EQPR)

The primal problem is strictly feasible due to the strict feasibility of (QP). Precisely, assume that c_s is a strictly feasible solution of (QP), in other words, $c_s^H c_s = 1$, $c_s^H R_1 c_s > \delta_a$, and $|c_s^H c_0|^2 \geq \mathrm{Re}^2(c_s^H c_0) > \delta_\epsilon$. Then there are $u_1, u_2, \ldots, u_{N-1}$ such that $U = [c_s, u_1, u_2, \ldots, u_{N-1}]$ is a unitary matrix [29], and for a sufficiently small $\eta > 0$ the matrix C_s

$$(1 - \eta)U e_1 e_1^H U^H + \frac{\eta}{N-1} U \left(I - e_1 e_1^H \right) U^H,$$

with $e_1 = [1, 0, \ldots, 0]^T$, is a strictly feasible solution of the SDP problem (EQPR). In fact,

$$C_s \bullet I = 1$$

$$C_s \bullet R_1 = (1 - \eta)c_s^H R_1 c_s + \frac{\eta}{N-1} \sum_{n=1}^{N-1} u_n^H R_1 u_n$$

$$C_s \bullet R_1 = (1 - \eta)c_s^H C_0 c_s + \frac{\eta}{N-1} \sum_{n=1}^{N-1} u_n^H C_0 u_n,$$

which highlights that, if η is suitably chosen, then C_s is a strictly feasible solution, in other words, $C_s \bullet I = 1$, $C_s \bullet R_1 > \delta_a$ and $C_s \bullet C_0 > \delta_\epsilon$.

A6.2.3   Feasibility of (EQPRD)

The dual problem is strictly feasible because, for every finite value of y_2 and y_3, say $\hat{y}_2, \hat{y}_3$, we can choose a $\hat{y}_1$, such that $\hat{y}_1 \boldsymbol{I} - \hat{y}_2 \boldsymbol{R}_1 - \hat{y}_3 \boldsymbol{C}_0 \succ \boldsymbol{R}$. It follows that $(\hat{y}_1, \hat{y}_2, \hat{y}_3)$ is a strictly feasible solution of (EQPRD) [6].

References

[1] W. Ai, Y. Huang, and S. Zhang, "New results on Hermitian matrix rank-one decomposition," *Mathematical Programming*, 2009. DOI: 10.1007/s10107-009-0304-7.

[2] A. d'Aspremont and S. Boyd, "Relaxations and randomized methods for nonconvex QCQPs," EE392o Class Notes, Stanford University, Autumn 2003.

[3] A. Beck and Y. Eldar, "Strong duality in nonconvex optimization with two quadratic constraints," *SIAM Journal on Optimization*, vol. 17, no. 3, pp. 844–60, 2006.

[4] M.R. Bell, "Information theory and radar waveform design," *IEEE Transactions on Information Theory*, vol. 39, no. 5, pp. 1578–97, 1993.

[5] M.R. Bell, "Information theory of radar and sonar waveforms," in *Wiley Encyclopedia of Electrical and Electronic Engineering*, vol. 10, J.G. Webster, ed. New York, NY: Wiley-Interscience, 1999, pp. 180–90.

[6] A. Ben-Tal and A. Nemirovski, *Lectures on Modern Convex Optimization: Analysis, Algorithms, and Engineering Applications*, Philadelphia, PA: MPS-SIAM. 2001.

[7] A. Ben-Tal, A. Nemirovski, and C. Roos, "Extended matrix cube theorems with applications to μ-theory in control," *Mathematics of Operations Research*, vol. 28, no. 3, pp. 497–523, 2003.

[8] A. Ben-Tal and M. Teboulle, "Hidden convexity in some nonconvex quadratically constrained quadratic programming," *Mathematical Programming*, vol. 72, pp. 51–63, 1996.

[9] J.S. Bergin, P.M. Techau, J.E. Don Carlos, and J.R. Guerci, "Radar waveform optimization for colored noise mitigation," *2005 IEEE International Radar Conference*, Alexandria, VA, May 9–12, 2005, pp. 149–54.

[10] L. Bomer and M. Antweiler, "Polyphase Barker sequences," *Electronics Letters*, vol. 25, no. 23, pp. 1577–9, 1989.

[11] S. Boyd and L. Vandenberghe, *Convex Optimization*. Cambridge: Cambridge University Press, 2004.

[12] L. Brickman, "On the field of values of a matrix," in *Proceedings of the American Mathematical Society*, 1961, vol. 12, pp. 61–6.

[13] M.R. Celis, J.E. Dennis, and R.A. Tapia, "A trust region algorithm for nonlinear equality constrained optimization," in *Numerical Optimization*. R.T. Boggs, R.H. Byrd, and R.B. Schnabel, Eds. Philadelphia, PA: SIAM. 1985, pp. 71–82.

[14] A.N. D'Andrea, U. Mengali, and R. Reggiannini, "The modified Cramer–Rao bound and its application to synchronization problems," *IEEE Transactions on Communications*, vol. 42, no. 3, pp. 1391–9, 1994.

[15] A. De Maio, S. De Nicola, Y. Huang, S. Zhang, and A. Farina, "Code design to optimize radar detection performance under accuracy and similarity constraints," *IEEE Transactions on Signal Processing*, vol. 56, no. 11, pp. 5618–29, 2008.

[16] A. Farina, "Waveform Diversity: Past, Present, and Future," presented at the Third International Waveform Diversity & Design Conference, Plenary Talk, Pisa, June 2007.

[17] A. Farina and S. Pardini, "A track-while-scan algorithm using radial velocity in a clutter environment," *IEEE Transactions on Aerospace and Electronic Systems*, vol. 14, no. 5, pp. 769–79, 1978.

[18] A. Farina and F.A. Studer, "Detection with high resolution radar: great promise, big challenge," *Microwave Journal*, pp. 263–73, 1991.

[19] A.L. Fradkov and V.A. Yakubovich, "The S-procedure and duality relations in nonconvex problems of quadratic programming," *Vestnik Leningrad University*, vol. 6, pp. 101–9, 1979. (In Russian 1973.)

[20] M.Y. Fu, Z.Q. Luo, and Y. Ye, "Approximation algorithms for quadratic programming," *Journal of Combinatorial Optimization*, vol. 2, pp. 29–50, 1998.

[21] J.B.G. Frenk, K. Roos, T. Terlaky, and S. Zhang, eds., *High performance optimization*. Dordrecht: Kluwer Academic Publishers, 2000.

[22] B. Friedlander, "A Subspace Framework for Adaptive Radar Waveform Design," in *Record of the Thirty-Ninth Asilomar Conference on Signals, Systems and Computers 2005*, Pacific Grove, CA, Oct. 28–Nov. 1, 2005, pp. 1135–9.

[23] M. Friese, "Polyphase Barker sequences up to length 36," *IEEE Transactions on Information Theory*, vol. 42, no. 4, pp. 1248–50, 1996.

[24] D.A. Garren, A.C. Odom, M.K. Osborn, J.S. Goldstein, S.U. Pillai, and J.R. Guerci, "Full-polarization matched-illumination for target detection and identification," *IEEE Transactions on Aerospace and Electronic Systems*, vol. 38, no. 3, pp. 824–37, 2002.

[25] F. Gini and R. Reggiannini, "On the use of Cramer-Rao-like bounds in the presence of random nuisance parameters," *IEEE Transactions on Communications*, vol. 48, no. 12, pp. 2120–26, 2000.

[26] J.S. Goldstein, I.S. Reed, and P.A. Zulch, "Multistage partially adaptive STAP CFAR detection algorithm," *IEEE Transactions on Aerospace and Electronic Systems*, vol. 35, no. 2, pp. 645–61, 1999.

[27] F. Hausdorff, "Der wertvorrat einer bilinearform," *Mathematische Zeitschrift*, vol. 3, pp. 314–6, 1919.

[28] S.M. He, Z.-Q. Luo, J.W. Nie, and S. Zhang, "Semidefinite relaxation bounds for indefinite homogeneous quadratic optimization," *SIAM Journal on Optimization*, vol. 19, no. 2, pp. 503–23, 2008.

[29] R.A. Horn and C.R. Johnson, *Matrix Analysis*. Cambridge: Cambridge University Press, 1985.

[30] R.A. Horn and C.R. Johnson, *Topics in Matrix Analysis*. New York: Cambridge University Press, 1991.

[31] Y. Huang and S. Zhang, "Complex matrix decomposition and quadratic programming," *Mathematics of Operations Research*, vol. 32, no. 3, pp. 758–68, 2007.

[32] S. Kay, "Optimal signal design for detection of point targets in stationary gaussian clutter/reverberation," *IEEE Journal on Selected Topics in Signal Processing*, vol. 1, no. 1, pp. 31–41, 2007.

[33] N. Levanon and E. Mozeson, *Radar Signals*. Hoboken, NJ: John Wiley & Sons, 2004.

[34] J. Li, J.R. Guerci, and L. Xu, "Signal waveform's optimal-under-restriction design for active sensing," *IEEE Signal Processing Letters*, vol. 13, no. 9, pp. 565–8, 2006.

[35] Z.-Q. Luo, N.D. Sidiropoulos, P. Tseng, and S. Zhang, "Approximation bounds for quadratic optimization with homogeneous quadratic constraints," *SIAM Journal on Optimization*, vol. 18, no. 1, pp. 1–28, 2007.

[36] R.W. Miller and C.B. Chang, "A modified Cramér-Rao bound and its applications," *IEEE Transactions on Information Theory*, vol. 24, no. 3, pp. 398–400, 1978.

[37] E. Mozeson and N. Levanon, "MATLAB code for plotting ambiguity functions," *IEEE Transactions on Aerospace and Electronic Systems*, vol. 38, no. 3, pp. 1064–8, July 2002.

[38] H. Naparst, "Dense target signal processing," *IEEE Transactions on Information Theory*, vol. 37, no. 2, pp. 317–27, 1991.

[39] A. Nemirovski, "Lectures on Modern Convex Optimization," Class Notes, Georgia Institute of Technology, Fall 2005.

[40] A. Nemirovski, C. Roos, and T. Terlaky, "On maximization of quadratic form over intersection of ellipsoids with common center," *Mathematical Programming*, vol. 86, pp. 463–73, 1999.

[41] S.U. Pillai, H.S. Oh, D.C. Youla, and J.R. Guerci, "Optimum transmit-receiver design in the presence of signal-dependent interference and channel noise," *IEEE Transactions on Information Theory*, vol. 46, no. 2, pp. 577–84, 2000.

[42] I. Pólik and T. Terlaky, "S-lemma: a survey," *SIAM Review*, vol. 49, no. 3, pp. 371–418, 2007.

[43] J.F. Sturm, "Using SeDuMi 1.02, a MATLAB toolbox for optimization over symmetric cones," *Optim. Meth. Software*, vol. 11–12, pp. 625–53, 1999.

[44] J.F. Sturm and S. Zhang, "On cones of nonnegative quadratic functions," *Mathematics of Operations Research*, vol. 28, pp. 246–67, 2003.

[45] H.L. van Trees, *Optimum Array Processing. Part IV of Detection, Estimation and Modulation Theory*. New York: John Wiley & Sons, 2002.

[46] C.H. Wilcox, "The synthesis problem for radar ambiguity functions," Math Resource Center, U.S. Army, University Wisconsin, Madison, WI, MRC Tech. Summary Rep. 157, April 1960; reprinted in R.E. Blahut, W.M. Miller, and C.H. Wilcox, *Radar and Sonar*, Part 1. New York: Springer, 1991.

[47] V.A. Yakubovich, "S-procedure in nonlinear control theory," *Vestnik Leningrad University*, vol. 4, no. 1, pp. 73–93, 1977. (In Russian 1971.)

[48] Y. Ye and S. Zhang, "New results on quadratic optimization," *SIAM Journal on Optimization*, vol. 14, no. 1, pp. 245–67, 2003.

7 Convex analysis for non-negative blind source separation with application in imaging

Wing-Kin Ma, Tsung-Han Chan, Chong-Yung Chi, and Yue Wang

In recent years, there has been a growing interest in blind separation of non-negative sources, known as simply *non-negative blind source separation* (nBSS). Potential applications of nBSS include biomedical imaging, multi/hyper-spectral imaging, and analytical chemistry. In this chapter, we describe a rather new endeavor of nBSS, where convex geometry is utilized to analyze the nBSS problem. Called *convex analysis of mixtures of non-negative sources* (CAMNS), the framework described here makes use of a very special assumption called *local dominance*, which is a reasonable assumption for source signals exhibiting sparsity or high contrast. Under the locally dominant and some usual nBSS assumptions, we show that the source signals can be perfectly identified by finding the extreme points of an observation-constructed polyhedral set. Two methods for practically locating the extreme points are also derived. One is analysis-based with some appealing theoretical guarantees, while the other is heuristic in comparison, but is intuitively expected to provide better robustness against model mismatches. Both are based on linear programming and thus can be effectively implemented. Simulation results on several data sets are presented to demonstrate the efficacy of the CAMNS-based methods over several other reported nBSS methods.

7.1 Introduction

Blind source separation (BSS) is a signal-processing technique, the purpose of which is to separate source signals from observations, without information of how the source signals are mixed in the observations. BSS presents a technically very challenging topic to the signal processing community, but it has stimulated significant interest for many years due to its relevance to a wide variety of applications. BSS has been applied to wireless communications and speech processing, and recently there has been an increasing interest in imaging applications.

BSS methods are "blind" in the sense that the mixing process is not known, at least not explicitly. But what is universally true for all BSS frameworks is that we make certain presumptions on the source characteristics (and sometimes on the mixing characteristics as well), and then exploit such characteristics during the blind-separation process. For instance, independent component analysis (ICA) [1, 2], a major and very

Convex Optimization in Signal Processing and Communication, eds. Daniel P. Palomar and Yonina C. Eldar. Published by Cambridge University Press © Cambridge University Press, 2010.

representative BSS framework on which many BSS methods are based, assumes that the sources are mutually uncorrelated/independent random processes, possibly with non-Gaussian distributions. There are many other possibilities one can consider; for example, using quasi-stationarity [3,4] (speech signals are quasi-stationary), and using boundness of the source magnitudes [5–7] (suitable for digital signals). In choosing a right BSS method for a particular application, it is important to examine whether the underlying assumptions of the BSS method are a good match to the application. For instance, statistical independence is a reasonable assumption in applications such as speech signal separation, but it may be violated in certain imaging scenarios such as hyper-spectral imaging [8].

This chapter focuses on *non-negative blind source separation* (nBSS), in which the source signals are assumed to take on non-negative values. Naturally, images are non-negative signals. Potential applications of nBSS include biomedical imaging [9], hyper-spectral imaging [10], and analytical chemistry [11]. In biomedical imaging, for instance, there are realistic, meaningful problems where nBSS may serve as a powerful image analysis tool for practitioners. Such examples will be briefly described in this chapter.

In nBSS, how to cleverly utilize source non-negativity to achieve clean separation has been an intriguing subject that has received much attention recently. Presently available nBSS methods may be classified into two groups. One group is similar to ICA: assume that the sources are mutually uncorrelated or independent, but with non-negative source distributions. Methods falling in this class include non-negative ICA (nICA) [12], stochastic non-negative ICA (SNICA) [13], and Bayesian positive source separation (BPSS) [14]. In particular, in nICA, the blind-separation criterion can theoretically guarantee perfect separation of sources [15], under an additional assumption where the source distributions are nonvanishing around zero (this is called the *well-grounded condition*).

Another group of nBSS methods does not rely on statistical assumptions. Roughly speaking, these methods explicitly exploit source non-negativity or even mixing matrix non-negativity, with an attempt to achieve some kind of least-square fitting criterion. Methods falling in this group are generally known as (or may be vaguely recognized as) non-negative matrix factorizations (NMFs) [16, 17]. An advantage with NMF is that it does not operate on the premise of mutual uncorrelatedness/independence as in the first group of nBSS methods. NMF is a nonconvex constrained-optimization problem. A popular way of handling NMF is to apply gradient descent [17], but it is known to be suboptimal and slowly convergent. A projected quasi-Newton method has been incorporated in NMF to speed up its convergence [18]. Alternatively, alternating least squares (ALS) [19–22] can also be applied. Fundamentally, the original NMF [16, 17] does not always yield unique factorization, and this means that NMF may fail to provide perfect separation. Possible circumstances under which NMF draws a unique decomposition can be found in [23]. Simply speaking, unique NMF would be possible if both the source signals and mixing process exhibit some form of sparsity. Some recent works have focused on incorporating additional penalty functions or constraints, such as sparse constraints, to strengthen the NMF uniqueness [24, 25].

In this chapter we introduce an nBSS framework that is different from the two groups of nBSS approaches mentioned above. Called *convex analysis of mixtures of non-negative sources* (CAMNS) [26], this framework is deterministic, using convex geometry to analyze the relationships of the observations and sources in a vector space. Apart from source non-negativity, CAMNS adopts a special deterministic assumption called *local dominance*. We initially introduced this assumption to capture the sparse characteristics of biomedical images [27,28], but we also found that local dominance can be perfectly or approximately satisfied for high-contrast images such as human portraits. (We, however, should stress that the local dominance assumption is different from the sparsity assumption in compressive sensing.) Under the local dominance assumption and some standard nBSS assumptions, we can show using convex analysis that the true source vectors serve as the extreme points of some observation-constructed polyhedral set. This geometrical discovery is surprising, with a profound implication that perfect blind separation can be achieved by solving an extreme point-finding problem that is not seen in the other BSS approaches to our best knowledge. Then we will describe two methods for practical realizations of CAMNS. The first method is analysis-based, using LPs to locate all the extreme points systematically. Its analysis-based construction endows it with several theoretically appealing properties, as we will elaborate upon later. The second method is heuristic in comparison, but intuitively it is expected to have better robustness against a mismatch of model assumptions. In our simulation results with real images, the second method was found to exhibit further improved separation performance over the first.

In Figure 7.1 we use diagrams to give readers some impression of how CAMNS works.

The chapter is organized as follows. In Section 7.2, the problem statement is given. In Section 7.3, we review some key concepts of convex analysis, which would be useful for understanding of the mathematical derivations that follow. CAMNS and its resultant implications on nBSS criteria are developed in Section 7.4. The systematic, analysis-based LP method for implementing CAMNS is described in Section 7.5. We then introduce an alternating volume-maximization heuristic for implementing CAMNS in Section 7.6. Finally, in Section 7.7, we use simulations to evaluate the performance of the proposed CAMNS-based nBSS methods and some other existing nBSS methods.

7.2 Problem statement

Our notations are standard, following those in convex optimization for signal processing:

$\mathbb{R}, \mathbb{R}^N, \mathbb{R}^{M \times N}$	set of real numbers, N-vectors, $M \times N$ matrices;
$\mathbb{R}_+, \mathbb{R}_+^N, \mathbb{R}_+^{M \times N}$	set of non-negative real numbers, N-vectors, $M \times N$ matrices;
$\mathbf{1}$	all one vector;
$\mathbf{I}_N$	$N \times N$ identity matrix;
$\mathbf{e}_i$	unit vector of proper dimension with the ith entry equal to 1;
$\succeq$	componentwise inequality;
$\| \cdot \|$	Euclidean norm;
$\mathcal{N}(\boldsymbol{\mu}, \boldsymbol{\Sigma})$	Gaussian distribution with mean $\boldsymbol{\mu}$ and covariance $\boldsymbol{\Sigma}$;
$\mathrm{diag}(x_1, \ldots, x_N)$	diagonal matrix with diagonal entries $x_1, \ldots, x_N$;
$\det(\mathbf{X})$	determinant of a square matrix $\mathbf{X}$.

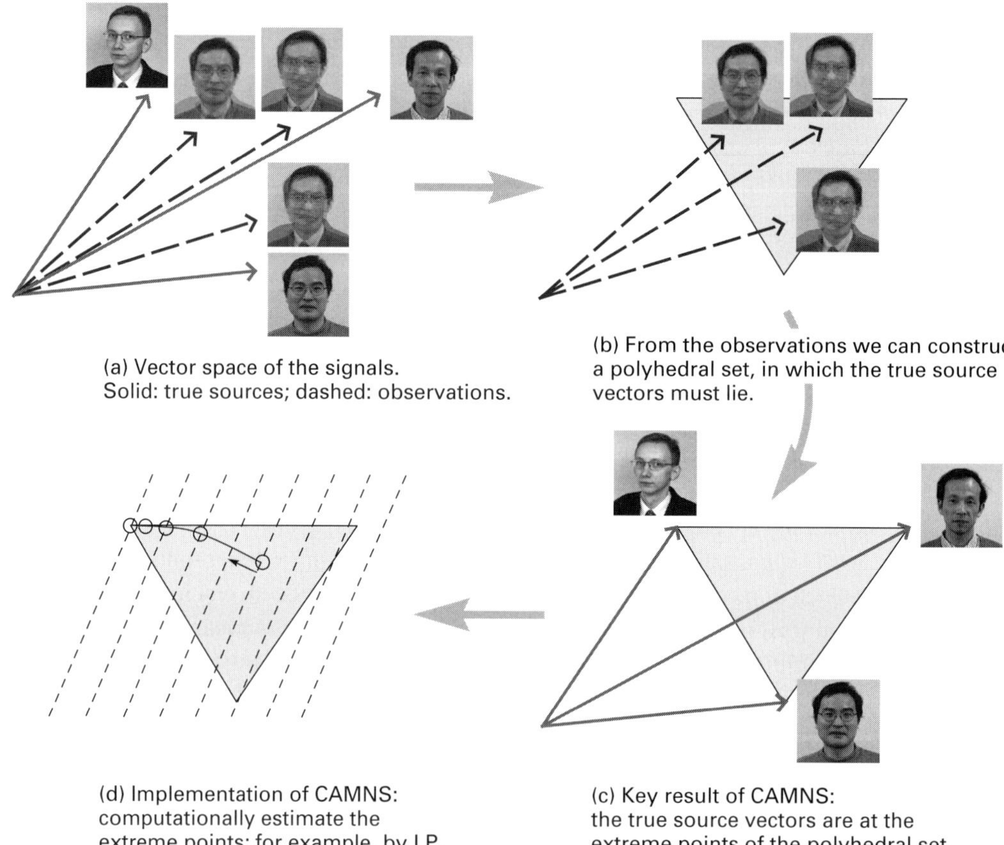

(a) Vector space of the signals. Solid: true sources; dashed: observations.

(b) From the observations we can construct a polyhedral set, in which the true source vectors must lie.

(d) Implementation of CAMNS: computationally estimate the extreme points; for example, by LP.

(c) Key result of CAMNS: the true source vectors are at the extreme points of the polyhedral set.

Figure 7.1   An intuitive illustration of how CAMNS operates.

We consider the scenario of linear instantaneous mixtures of unknown source signals, in which the signal model is

$$\mathbf{x}[n] = \mathbf{A}\mathbf{s}[n], \qquad n = 1, \ldots, L, \tag{7.1}$$

where

$\mathbf{s}[n] = [s_1[n], \ldots, s_N[n]]^T$ input or source vector sequence, with N denoting the input dimension;

$\mathbf{x}[n] = [x_1[n], \ldots, x_M[n]]^T$ output or observation vector sequence, with M denoting the output dimension;

$\mathbf{A} \in \mathbb{R}^{M \times N}$ mixing matrix describing the input–output relation;

L sequence (or data) length, with $L \gg \max\{M, N\}$ (often true in practice).

The linear instantaneous-mixture model in (7.1) can also be expressed as

$$x_i = \sum_{j=1}^{N} a_{ij} s_j, \quad i = 1, \ldots, M,$$
(7.2)

where

$$
\begin{array}{ll}
a_{ij} & (i,j)\text{th element of } \mathbf{A}; \\
s_j = [s_j[1], \ldots, s_j[L]]^T & \text{vector representing the } j\text{th source signal}; \\
x_i = [x_i[1], \ldots, x_i[L]]^T & \text{vector representing the } i\text{th observed signal}.
\end{array}
$$

In blind source separation (BSS), the problem is to retrieve the sources $s_1, \ldots, s_N$ from the observations $x_1, \ldots, x_M$, without knowledge of $\mathbf{A}$. BSS shows great potential in various applications, and here we describe two examples in biomedical imaging.

Example 7.1 Magnetic resonance imaging (MRI) *Dynamic contrast-enhanced MRI* (DCE-MRI) uses various molecular weight contrast agents to assess tumor vasculature perfusion and permeability, and has potential utility in evaluating the efficacy of angio-genesis inhibitors in cancer treatment [9]. While DCE-MRI can provide a meaningful estimation of vasculature permeability when a tumor is homogeneous, many malignant tumors show markedly heterogeneous areas of permeability, and thereby the signal at each pixel often represents a convex mixture of more than one distinct vasculature source independent of spatial resolution.

The raw DCE-MRI images of a breast tumor, for example, are given at the top of Figure 7.2, and its bottom plot illustrates the temporal mixing process of the source patterns where the tumor angiogenic activities (observations) represent the weighted summation of spatially-distributed vascular permeability associated with different perfusion rates. The BSS methods can be applied to computationally estimate the time activity curves (mixing matrix) and underlying compartment vascular permeability (sources) within the tumor site.

Example 7.2 Dynamic fluorescent imaging (DFI) DFI exploits highly specific and bio-compatible fluorescent contrast agents to interrogate small animals for drug development and disease research [29]. The technique generates a time series of images acquired after injection of an inert dye, where the dye's differential biodistribution dynamics allow precise delineation and identification of major organs. However, spatial resolution and quantitation at depth is not one of the strengths of planar optical approaches, due mainly to the malign effects of light scatter and absorption.

The DFI data acquired in a mouse study, for instance, is shown in Figure 7.3, where each DFI image (observation) is delineated as a linear mixture of the anatomical maps

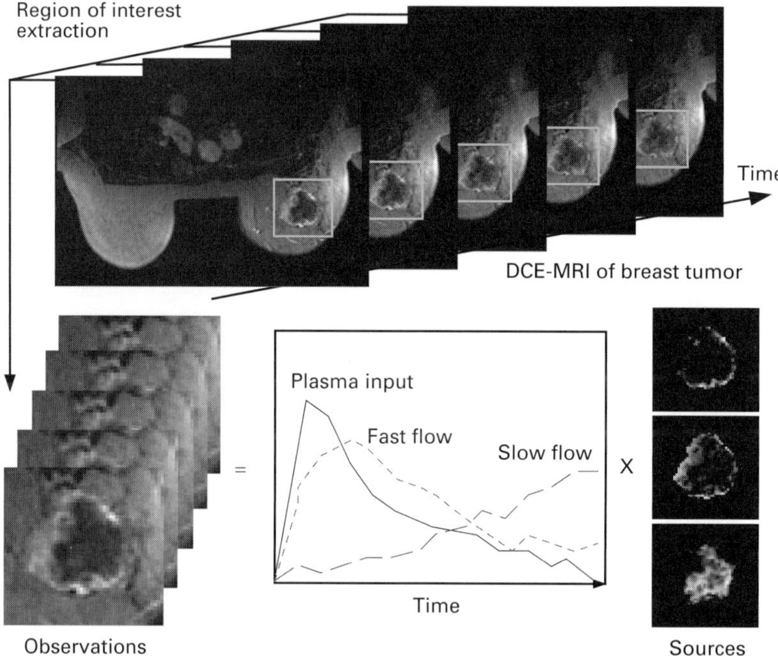

Figure 7.2 The BSS problem in DCE-MRI applications.

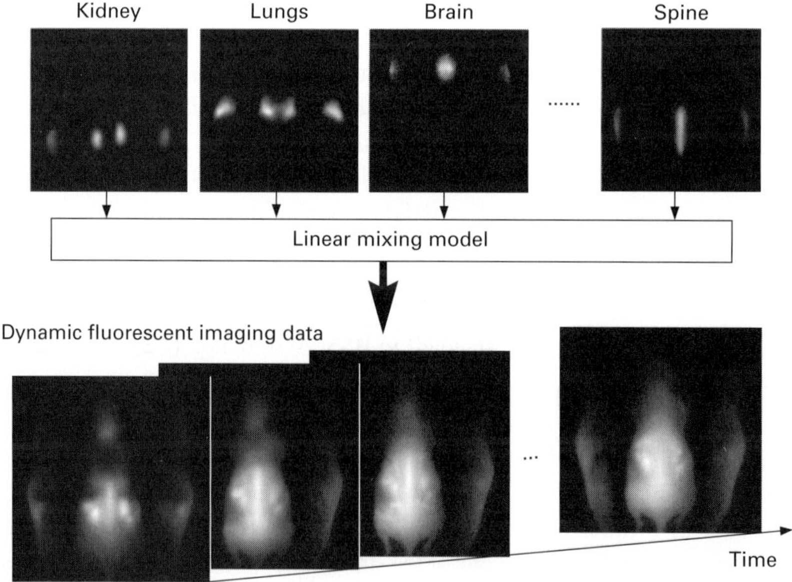

Figure 7.3 The BSS problem in DFI applications.

associated with different organs. The BSS methods can be used to numerically unmix the anatomical maps (sources) and their mixing portions (mixing matrix).

7.2.1 Assumptions

Like other non-negative BSS techniques, the CAMNS framework to be presented will make use of source signal non-negativity. Hence, we assume that

(A1) All s_j are componentwise non-negative; in other words, for each j, $s_j \in \mathbb{R}_+^L$.

What makes CAMNS special compared to the other available nBSS frameworks is the use of the *local dominance* assumption, as follows:

(A2) Each source signal vector is *locally dominant*, in the following sense: for each $i \in \{1,\ldots,N\}$, there exists an (unknown) index ℓ_i such that $s_i[\ell_i] > 0$ and $s_j[\ell_i] = 0, \forall j \neq i$.

In imaging, (A2) means that for each source, say source i, there exists at least one pixel (indexed by n) such that source i has a nonzero pixel while the other sources are zero. It may be completely satisfied, or serve as a good approximation when the source signals are sparse (or contain many zeros). In brain MRI, for instance, the non-overlapping region of the spatial distribution of fast perfusion and slow perfusion source images [9] can be higher than 95%. For high contrast images such as human portraits, we found that (A2) would also be an appropriate assumption.

We make two more assumptions:

(A3) The mixing matrix has unit row sum; that is, for all $i = 1,\ldots,M$,

$$\sum_{j=1}^{N} a_{ij} = 1. \tag{7.3}$$

(A4) $M \geq N$ and $\mathbf{A}$ is of full column rank.

Assumption (A4) is rather standard in BSS. Assumption (A3) is essential to CAMNS, but can be relaxed through a model reformulation [28]. Moreover, in MRI (e.g., in Example 7.1), (A3) is automatically satisfied due to the so-called partial-volume effect [28]. The following example shows how we can relax (A3):

Example 7.3 Suppose that the model in (7.2) does not satisfy (A3). For simplicity of exposition of the idea, assume non-negative mixing; in other words, $a_{ij} \geq 0$ for all (i,j) (extension to $a_{ij} \in \mathbb{R}$ is possible). Under such circumstances, the observations are all non-negative, and we can assume that

$$x_i^T \mathbf{1} \neq 0$$

for all $i = 1, \ldots, M$. Likewise, we can assume $s_j^T 1 \neq 0$ for all j. The idea is to enforce (A3) by normalizing the observation vectors:

$$\bar{x}_i \triangleq \frac{x_i}{x_i^T 1} = \sum_{j=1}^{N} \left(\frac{a_{ij} s_j^T 1}{x_i^T 1} \right) \left(\frac{s_j}{s_j^T 1} \right). \tag{7.4}$$

By letting $\bar{a}_{ij} = a_{ij} s_j^T 1 / x_i^T 1$ and $\bar{s}_j = s_j / s_j^T 1$, we obtain a model $\bar{x}_i = \sum_{j=1}^{N} \bar{a}_{ij} \bar{s}_j$ which is in the same form as the original signal model in (7.2). It is easy to show that the new mixing matrix, denoted by $\bar{A}$, has unit row sum (or (A3)).

It should also be noted that the model reformulation above does not damage the rank of the mixing matrix. Specifically, if the original mixing matrix A satisfies (A4), then the new mixing matrix $\bar{A}$ also satisfies (A4). To show this, we notice that the relationship of A and $\bar{A}$ can be expressed as

$$\bar{A} = D_1^{-1} A D_2, \tag{7.5}$$

where $D_1 = \mathrm{diag}(x_1^T 1, \ldots, x_M^T 1)$ and $D_2 = \mathrm{diag}(s_1^T 1, \ldots, s_N^T 1)$. Since D_1 and D_2 are of full rank, we have $\mathrm{rank}(\bar{A}) = \mathrm{rank}(A)$.

Throughout this chapter, we will assume that (A1)–(A4) are satisfied unless specified.

7.3   Review of some concepts in convex analysis

In CAMNS we analyze the geometric structures of the signal model by utilizing some fundamental convex-analysis concepts, namely *affine hull*, *convex hull*, and their associated properties. As we will see in the next section, such a convex analysis will shed light into how we can separate the sources. Here we provide a review of the essential concepts. Readers who are interested in further details of convex analysis are referred to the literature [30–32].

7.3.1   Affine hull

Given a set of vectors $\{s_1, \ldots, s_N\} \subset \mathbb{R}^L$, the *affine hull* is defined as

$$\mathrm{aff}\{s_1, \ldots, s_N\} = \left\{ x = \sum_{i=1}^{N} \theta_i s_i \,\middle|\, \theta \in \mathbb{R}^N, \sum_{i=1}^{N} \theta_i = 1 \right\}. \tag{7.6}$$

Some examples of affine hulls are illustrated in Figure 7.4. We see that for $N = 2$, an affine hull is a line passing through s_1 and s_2; and for $N = 3$, it is a plane passing through s_1, s_2, and s_3.

An affine hull can always be represented by

$$\mathrm{aff}\{s_1, \ldots, s_N\} = \left\{ x = C\alpha + d \,\middle|\, \alpha \in \mathbb{R}^P \right\} \tag{7.7}$$

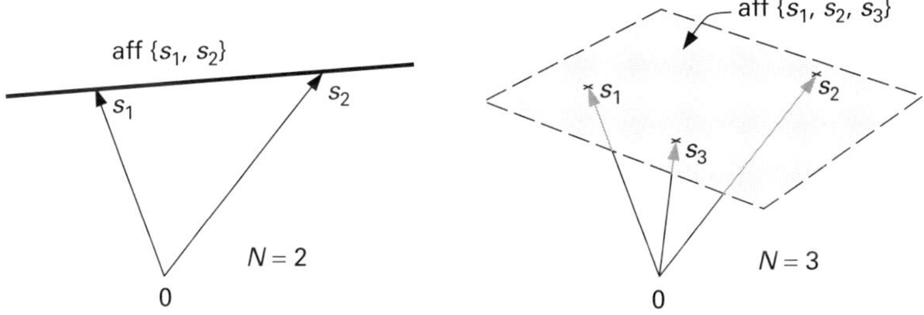

Figure 7.4   Examples of affine hulls for $N = 2$ and $N = 3$.

for some $\mathbf{d} \in \mathbb{R}^L$ (non-unique), for some full column rank $\mathbf{C} \in \mathbb{R}^{L \times P}$ (also non-unique), and for some $P \geq 1$. To understand this, consider a simple example where $\{s_1, \ldots, s_N\}$ is linearly independent. One can verify that (7.6) can be rewritten as (7.7), with

$$\mathbf{d} = s_N, \quad \mathbf{C} = [s_1 - s_N, s_2 - s_N, \ldots, s_{N-1} - s_N],$$

$$P = N - 1, \quad \boldsymbol{\alpha} = [\theta_1, \ldots, \theta_{N-1}]^T.$$

The number P in (7.7) is called the *affine dimension*, which characterizes the effective dimension of the affine hull. The affine dimension must satisfy $P \leq N - 1$. Moreover,

Property 7.1   If $\{s_1, \ldots, s_N\}$ is an affinely independent set (which means that $\{s_1 - s_N, \ldots, s_{N-1} - s_N\}$ is linearly independent), then the affine dimension is maximal; in other words, $P = N - 1$.

7.3.2   Convex hull

Given a set of vectors $\{s_1, \ldots, s_N\} \subset \mathbb{R}^L$, the *convex hull* is defined as

$$\text{conv}\{s_1, \ldots, s_N\} = \left\{ x = \sum_{i=1}^{N} \theta_i s_i \;\middle|\; \boldsymbol{\theta} \in \mathbb{R}_+^N, \sum_{i=1}^{N} \theta_i = 1 \right\}. \tag{7.8}$$

A convex hull would be a line segment for $N = 2$, and a triangle for $N = 3$. This is illustrated in Figure 7.5.

An important concept related to the convex hull is that of *extreme points*, also known as *vertices*. From a geometric perspective, extreme points are the corner points of the convex hull. A point $x \in \text{conv}\{s_1, \ldots, s_N\}$ is said to be an extreme point of $\text{conv}\{s_1, \ldots, s_N\}$ if x can never be a convex combination of $s_1, \ldots, s_N$ in a non-trivial manner; in other words,

$$x \neq \sum_{i=1}^{N} \theta_i s_i \tag{7.9}$$

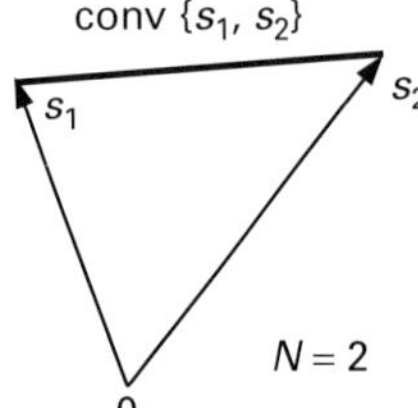

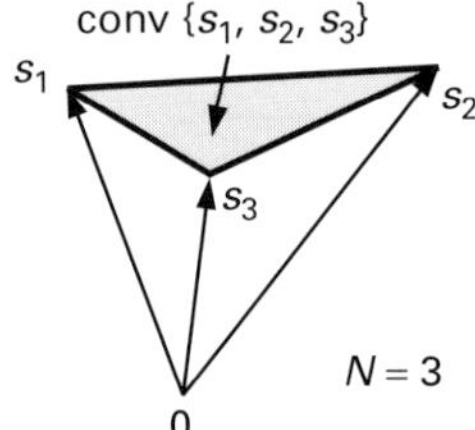

Figure 7.5   Examples of convex hulls for $N = 2$ and $N = 3$.

for all $\boldsymbol{\theta} \in \mathbb{R}^N_+$, $\sum_{i=1}^N \theta_i = 1$, and $\boldsymbol{\theta} \neq \mathbf{e}_i$ for any i. Some basic properties about extreme points are as follows:

Property 7.2   The set of extreme points of $\mathrm{conv}\{s_1,\ldots,s_N\}$ must be either the full set or a subset of $\{s_1,\ldots,s_N\}$.

Property 7.3   If $\{s_1,\ldots,s_N\}$ is affinely independent, then the set of extreme points of $\mathrm{conv}\{s_1,\ldots,s_N\}$ is exactly $\{s_1,\ldots,s_N\}$.

For example, in the illustrations in Figure 7.5 the extreme points are the corner points $\{s_1,\ldots,s_N\}$.

A special, but representative case of the convex hull is *simplex*. A convex hull is called a simplex if $L = N - 1$ and $\{s_1,\ldots,s_N\}$ is affinely independent. It follows that:

Property 7.4   The set of extreme points of a simplex $\mathrm{conv}\{s_1,\ldots,s_N\} \subset \mathbb{R}^{N-1}$ is $\{s_1,\ldots,s_N\}$.

In other words, a simplex on $\mathbb{R}^{N-1}$ is a convex hull with exactly N extreme points. A simplex for $N = 2$ is a line segment on $\mathbb{R}$, while a simplex for $N = 3$ is a triangle on $\mathbb{R}^2$.

7.4   Non-negative blind source-separation criterion via CAMNS

We now consider applying convex analysis to the nBSS problem (the model in (7.2), with assumptions (A1)–(A4)). Such a convex analysis of mixtures of non-negative sources will lead to an nBSS criterion that guarantees perfect source separation.

7.4.1   Convex analysis of the problem, and the CAMNS criterion

Recall from (7.2) that the signal model is given by

$$\boldsymbol{x}_i = \sum_{j=1}^N a_{ij}\boldsymbol{s}_j, \ i = 1,\ldots,M.$$

Since $\sum_{j=1}^N a_{ij} = 1$ ((A3)), every $\boldsymbol{x}_i$ is indeed an affine combination of $\{s_1,\ldots,s_N\}$:

$$\boldsymbol{x}_i \in \mathrm{aff}\{s_1,\ldots,s_N\} \tag{7.10}$$

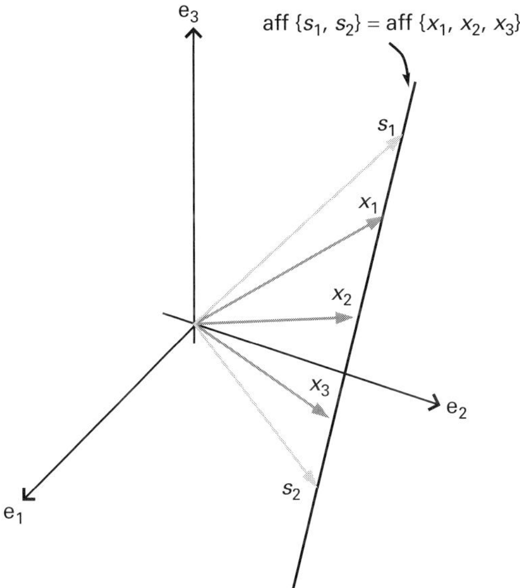

Figure 7.6 A geometric illustration of the affine hull equivalence in Lemma 7.1 for the special case of $N = 2$, $M = 3$, and $L = 3$.

for any $i = 1, \ldots, M$. Hence an interesting question is the following: can we use the observations $x_1, \ldots, x_M$ to construct the source affine hull aff $\{s_1, \ldots, s_N\}$?

To answer the question above, let us first consider the following lemma:

LEMMA 7.1 *The observation affine hull is identical to the source affine hull; that is,*

$$\text{aff}\{s_1, \ldots, s_N\} = \text{aff}\{x_1, \ldots, x_M\}. \tag{7.11}$$

An illustration is shown in Figure 7.6 to pictorially demonstrate the affine hull equivalence in Lemma 7.1. Since Lemma 7.1 represents an essential part of CAMNS, here we provide the proof to illustrate its idea.

Proof of Lemma 7.1 Any $x \in \text{aff}\{x_1, \ldots, x_M\}$ can be represented by

$$x = \sum_{i=1}^{M} \theta_i x_i, \tag{7.12}$$

where $\theta \in \mathbb{R}^M$, $\theta^T \mathbf{1} = 1$. Substituting (7.2) into (7.12), we get

$$x = \sum_{j=1}^{N} \beta_j s_j, \tag{7.13}$$

where $\beta_j = \sum_{i=1}^{M} \theta_i a_{ij}$ for $j = 1, \ldots, N$, or equivalently

$$\beta = \mathbf{A}^T \theta. \tag{7.14}$$

Since $\mathbf{A}$ has unit row sum [(A3)], we have

$$\boldsymbol{\beta}^T \mathbf{1} = \boldsymbol{\theta}^T (\mathbf{A} \mathbf{1}) = \boldsymbol{\theta}^T \mathbf{1} = 1. \tag{7.15}$$

This implies that $\boldsymbol{\beta}^T \mathbf{1} = 1$, and as a result, it follows from (7.13) that $x \in \mathrm{aff}\{s_1, \ldots, s_N\}$.

On the other hand, any $x \in \mathrm{aff}\{s_1, \ldots, s_N\}$ can be represented by (7.13) for $\boldsymbol{\beta}^T \mathbf{1} = 1$. Since $\mathbf{A}$ has full column rank [(A4)], a $\boldsymbol{\theta}$ always exist such that (7.14) holds. Substituting (7.14) into (7.13) yields (7.12). Since (7.15) implies that $\boldsymbol{\theta}^T \mathbf{1} = 1$, we conclude that $x \in \mathrm{aff}\{x_1, \ldots, x_M\}$. $\blacksquare$

Thus, by constructing the observation affine hull, the source affine hull will be obtained. Using the linear-equality representation of an affine hull, $\mathrm{aff}\{s_1, \ldots, s_N\}$ (or $\mathrm{aff}\{x_1, \ldots, x_M\}$) can be characterized as

$$\mathrm{aff}\{s_1, \ldots, s_N\} = \left\{ x = \mathbf{C}\boldsymbol{\alpha} + \mathbf{d} \mid \boldsymbol{\alpha} \in \mathbb{R}^P \right\} \tag{7.16}$$

for some $(\mathbf{C}, \mathbf{d}) \in \mathbb{R}^{L \times P} \times \mathbb{R}^L$ such that $\mathrm{rank}(\mathbf{C}) = P$, with P being the affine dimension. From **(A2)** it can be shown that (see Appendix A7.1):

LEMMA 7.2   *The set of source vectors $\{s_1, \ldots, s_N\}$ is linearly independent.*

Hence, by Property 7.1 the affine dimension of $\mathrm{aff}\{s_1, \ldots, s_N\}$ is maximal; that is, $P = N - 1$. For the special case of $M = N$ (the number of inputs being equal to the number of outputs), it is easy to obtain $(\mathbf{C}, \mathbf{d})$ from the observations $x_1, \ldots, x_M$; see the review in Section 7.3.

For $M \geq N$, a method called *affine set fitting* would be required. Since $(\mathbf{C}, \mathbf{d})$ is non-unique, without loss of generality, one can restrict $\mathbf{C}^T \mathbf{C} = \mathbf{I}$. The following affine set-fitting problem is used to find $(\mathbf{C}, \mathbf{d})$

$$(\mathbf{C}, \mathbf{d}) = \underset{\substack{\tilde{\mathbf{C}}, \tilde{\mathbf{d}} \\ \tilde{\mathbf{C}}^T \tilde{\mathbf{C}} = \mathbf{I}}}{\mathrm{argmin}} \sum_{i=1}^{M} e_{\mathcal{A}(\tilde{\mathbf{C}}, \tilde{\mathbf{d}})}(x_i), \tag{7.17}$$

where $e_{\mathcal{A}}(x)$ is the projection error of x onto $\mathcal{A}$, defined as

$$e_{\mathcal{A}}(x) = \min_{\tilde{x} \in \mathcal{A}} \|x - \tilde{x}\|^2, \tag{7.18}$$

and

$$\mathcal{A}(\tilde{\mathbf{C}}, \tilde{\mathbf{d}}) = \{\tilde{x} = \tilde{\mathbf{C}}\boldsymbol{\alpha} + \tilde{\mathbf{d}} \mid \boldsymbol{\alpha} \in \mathbb{R}^{N-1}\} \tag{7.19}$$

is an affine set parameterized by $(\tilde{\mathbf{C}}, \tilde{\mathbf{d}})$. The objective of (7.17) is to find an $(N - 1)$-dimensional affine set that has the minimum projection error with respect to the observations (which is zero for the noise-free case). Problem (7.17) is shown to have a closed-form solution:

PROPOSITION 7.1 *A solution to the affine set fitting problem in (7.17) is*

$$\mathbf{d} = \frac{1}{M} \sum_{i=1}^{M} \mathbf{x}_i \tag{7.20}$$

$$\mathbf{C} = [\mathbf{q}_1(\mathbf{U}\mathbf{U}^T), \mathbf{q}_2(\mathbf{U}\mathbf{U}^T), \dots, \mathbf{q}_{N-1}(\mathbf{U}\mathbf{U}^T)], \tag{7.21}$$

where $\mathbf{U} = [\mathbf{x}_1 - \mathbf{d}, \dots, \mathbf{x}_M - \mathbf{d}] \in \mathbb{R}^{L \times M}$, *and the notation* $\mathbf{q}_i(\mathbf{R})$ *denotes the eigenvector associated with the ith principal eigenvalue of the input matrix* $\mathbf{R}$.

The proof of the above proposition is given in Appendix A7.2. We should stress that this affine set fitting provides a best affine set in terms of minimizing the projection error. Hence, in the presence of additive noise, it has an additional advantage of noise mitigation for $M > N$.

Remember that we are dealing with non-negative sources. Hence, any source vector $\mathbf{s}_i$ must lic in

$$\begin{aligned} \mathcal{S} &\triangleq \text{aff}\{\mathbf{s}_1, \dots, \mathbf{s}_N\} \cap \mathbb{R}_+^L \\ &= \mathcal{A}(\mathbf{C}, \mathbf{d}) \cap \mathbb{R}_+^L \\ &= \{\mathbf{x} \mid \mathbf{x} = \mathbf{C}\boldsymbol{\alpha} + \mathbf{d}, \ \mathbf{x} \geq \mathbf{0}, \ \boldsymbol{\alpha} \in \mathbb{R}^{N-1}\}. \end{aligned} \tag{7.22}$$

Note that we have knowledge of $\mathcal{S}$ only through (7.22), a polyhedral set representation. The following lemma plays an important role:

LEMMA 7.3 *The polyhedral set* $\mathcal{S}$ *is identical to the source convex hull; that is,*

$$\mathcal{S} = \text{conv}\{\mathbf{s}_1, \dots, \mathbf{s}_N\}. \tag{7.23}$$

Following the illustration in Figure 7.6, in Figure 7.7 we geometrically demonstrate the equivalence of $\mathcal{S}$ and $\text{conv}\{\mathbf{s}_1, \dots, \mathbf{s}_N\}$. This surprising result is due mainly to the local dominance, and we include its proof here, considering its importance.

Proof of Lemma 7.3 Assume that $\mathbf{z} \in \text{aff}\{\mathbf{s}_1, \dots, \mathbf{s}_N\} \cap \mathbb{R}_+^L$:

$$\mathbf{z} = \sum_{i=1}^{N} \theta_i \mathbf{s}_i \succeq \mathbf{0}, \ \mathbf{1}^T \boldsymbol{\theta} = 1.$$

From (A2), it follows that $z[\ell_i] = \theta_i s_i[\ell_i] \geq 0$, $\forall i$. Since $s_i[\ell_i] > 0$, we must have $\theta_i \geq 0$, $\forall i$. Therefore, $\mathbf{z}$ lies in $\text{conv}\{\mathbf{s}_1, \dots, \mathbf{s}_N\}$. On the other hand, assume that $\mathbf{z} \in \text{conv}\{\mathbf{s}_1, \dots, \mathbf{s}_N\}$, in other words,

$$\mathbf{z} = \sum_{i=1}^{N} \theta_i \mathbf{s}_i, \ \mathbf{1}^T \boldsymbol{\theta} = 1, \ \boldsymbol{\theta} \succeq \mathbf{0},$$

implying that $\mathbf{z} \in \text{aff}\{\mathbf{s}_1, \dots, \mathbf{s}_N\}$. From (A1), we have $\mathbf{s}_i \succeq \mathbf{0}$ $\forall i$ and subsequently $\mathbf{z} \succeq \mathbf{0}$. This completes the proof for (7.23). ∎

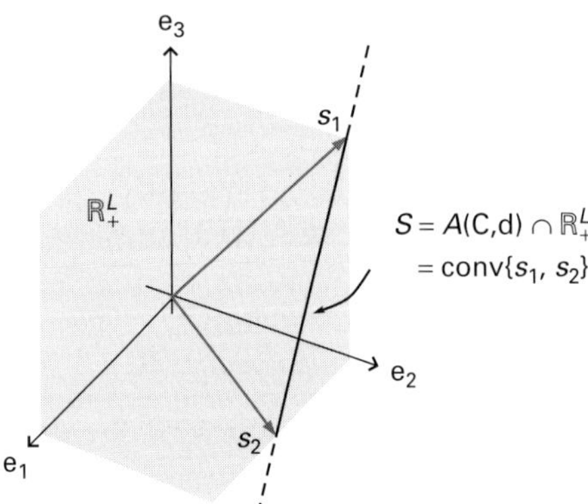

Figure 7.7   A geometric illustration of the convex-hull equivalence in Lemma 7.3, for the special case of $N = 2$, $M = 3$, and $L = 3$. Note that for each s_i there is a coordinate along which all the other sources have zero element; specifically $\mathbf{e}_3$ for s_1 and $\mathbf{e}_1$ for s_2.

Furthermore, we can deduce from Lemma 7.2 and Property 7.3 that

LEMMA 7.4   *The set of extreme points of* $\mathrm{conv}\{s_1,\ldots,s_N\}$ *is* $\{s_1,\ldots,s_N\}$.

Summarizing all the results above, we establish an nBSS criterion as follows:

CRITERION 7.1   *Use the affine set fitting solution in Proposition 7.1 to compute* $(\mathbf{C},\mathbf{d})$. *Then, find all the extreme points of the polyhedral set*

$$S = \left\{ x \in \mathbb{R}^L \mid x = \mathbf{C}\boldsymbol{\alpha} + \mathbf{d} \succeq \mathbf{0},\ \boldsymbol{\alpha} \in \mathbb{R}^{N-1} \right\} \tag{7.24}$$

and denote the obtained set of extreme points by $\{\hat{s}_1,\ldots,\hat{s}_N\}$. *Output* $\{\hat{s}_1,\ldots,\hat{s}_N\}$ *as the set of estimated source vectors.*

And it follows from the development above that:

THEOREM 7.1   *The solution to Criterion 7.1 is uniquely given by the set of true source vectors* $\{s_1,\ldots,s_N\}$, *under the premises in* (**A1**) *to* (**A4**).

The implication of Theorem 7.1 is profound: it suggests that the true source vectors can be perfectly identified by finding all the extreme points of S. This provides new opportunities in nBSS that cannot be found in the other, presently available literature to our best knowledge.

In the next section we will describe a systematic LP-based method for realizing Criterion 7.1 in practice.

7.4.2   An alternative form of the CAMNS criterion

There is an alternative form to the CAMNS criterion (Criterion 7.1, specifically). The alternative form is useful for deriving simple CAMNS algorithms in some special cases.

It will also shed light into the volume-maximization heuristics considered in the later part of this chapter.

Consider the pre-image of the observation-constructed polyhedral set $\mathcal{S}$, under the mapping $s = \mathbf{C}\alpha + \mathbf{d}$:

$$\mathcal{F} = \left\{ \alpha \in \mathbb{R}^{N-1} \mid \mathbf{C}\alpha + \mathbf{d} \succeq \mathbf{0} \right\}$$
$$= \left\{ \alpha \in \mathbb{R}^{N-1} \mid \mathbf{c}_n^T \alpha + d_n \geq 0, \ n = 1, \ldots, L \right\}, \tag{7.25}$$

where $\mathbf{c}_n^T$ is the nth row of $\mathbf{C}$. There is a direct correspondence between the extreme points of $\mathcal{S}$ and $\mathcal{F}$ [26]:

LEMMA 7.5 *The polyhedral set $\mathcal{F}$ in (7.25) is equivalent to a simplex*

$$\mathcal{F} = \mathrm{conv}\{\alpha_1, \ldots, \alpha_N\} \tag{7.26}$$

where each $\alpha_i \in \mathbb{R}^{N-1}$ satisfies

$$\mathbf{C}\alpha_i + \mathbf{d} = s_i. \tag{7.27}$$

The proof of Lemma 7.5 is given in Appendix A7.3. Since the set of extreme points of a simplex $\mathrm{conv}\{\alpha_1, \ldots, \alpha_N\}$ is $\{\alpha_1, \ldots, \alpha_N\}$ (Property 7.4), we have the following alternative nBSS criterion:

CRITERION 7.2 Alternative form of Criterion 7.1
Use the affine set fitting solution in Proposition 7.1 to compute $(\mathbf{C}, \mathbf{d})$. Then, find all the extreme points of the simplex

$$\mathcal{F} = \left\{ \alpha \in \mathbb{R}^{N-1} \mid \mathbf{C}\alpha + \mathbf{d} \succeq \mathbf{0} \right\} \tag{7.28}$$

and denote the obtained set of extreme points by $\{\hat{\alpha}_1, \ldots, \hat{\alpha}_N\}$. Output

$$\hat{s}_i = \mathbf{C}\hat{\alpha}_i + \mathbf{d}, \ i = 1, \ldots, N \tag{7.29}$$

as the set of estimated source vectors.

It follows directly from Theorem 7.1 and Lemma 7.5 that:

THEOREM 7.2 *The solution to Criterion 7.2 is uniquely given by the set of true source vectors $\{s_1, \ldots, s_N\}$, under the premises in* (A1) *to* (A4).

In [26], we have used Criterion 7.2 to develop simple nBSS algorithms for the cases of two and three sources. In the following we provide the solution for the two-source case and demonstrate its effectiveness using synthetic X-ray observations:

Example 7.4   For $N = 2$, the simplex $\mathcal{F}$ is a line segment on $\mathbb{R}$. Hence, by locating the two endpoints of the line segment, the extreme points will be found. To see how this can be done, let us examine $\mathcal{F}$ (in polyhedral form):

$$\mathcal{F} = \{\alpha \in \mathbb{R} \mid c_n\alpha + d_n \geq 0, \ n = 1,\ldots,L\}. \tag{7.30}$$

From (7.30) we see that $\alpha \in \mathcal{F}$ implies the following two conditions:

$$\alpha \geq -d_n/c_n, \ \text{for all } n \text{ such that } c_n > 0, \tag{7.31}$$

$$\alpha \leq -d_n/c_n, \ \text{for all } n \text{ such that } c_n < 0. \tag{7.32}$$

We therefore conclude from (7.31) and (7.32) that the extreme points are given by

$$\alpha_1 = \min\{-d_n/c_n \mid c_n < 0, \ n = 1, 2, \ldots, L\}, \tag{7.33}$$

$$\alpha_2 = \max\{-d_n/c_n \mid c_n > 0, \ n = 1, 2, \ldots, L\}. \tag{7.34}$$

Thus, for two sources, CAMNS blind source separation reduces to a simple closed-form solution.

We carried out a quick simulation to verify the idea. Figure 7.8(a) shows the source images, which are X-ray images. The observations are shown in Figure 7.8(b). We separate the sources using the closed-form solution in (7.33) and (7.34), and the results are shown in Figure 7.8(c). It is clear that CAMNS can successfully recover the two sources.

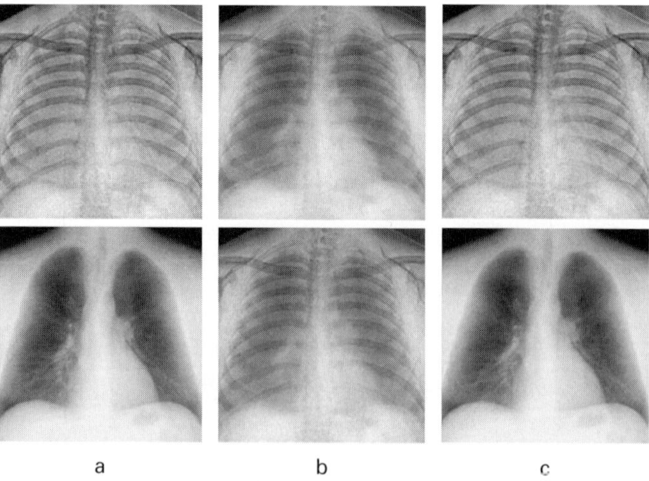

Figure 7.8   Simulation result for testing the CAMNS two-source, closed-form solution. (a) Sources, (b) observations, and (c) extracted sources by CAMNS.

7.5 Systematic linear-programming method for CAMNS

This section, as well as the next section, are dedicated to the practical implementations of CAMNS. In this section, we propose an approach that uses linear programs (LPs) to systematically fulfill the CAMNS criterion, specifically Criterion 7.1. An appealing characteristic of this CAMNS–LP method is that Criterion 7.1 does not appear to be related to convex optimization at first look, and yet it can be exactly solved by CAMNS–LP as long as the problem assumptions (**A1**)–(**A4**) are valid.

Our problem, as specified in Criterion 7.1, is to find all the extreme points of the polyhedral set $\mathcal{S}$ in (7.22). In the optimization literature this problem is known as *vertex enumeration* [33–35]. The available extreme-point finding methods are sophisticated, requiring no assumption on the extreme points. However, the complexity of those methods would increase exponentially with the number of inequalities L (note that L is also the data length in our problem, which is often large in practice). The notable difference of the development here is that we exploit the characteristic that the extreme points $s_1, \ldots, s_N$ are linearly independent (Lemma 7.2). By doing so we will establish an extreme-point finding method for CAMNS, whose complexity is polynomial in L.

Our approach is to identify one extreme point at one time. Consider the following linear minimization problem:

$$p^\star = \min_s \ \mathbf{r}^T s$$

$$\text{subject to (s.t.)} \ \ s \in \mathcal{S} \tag{7.35}$$

for some arbitrarily chosen direction $\mathbf{r} \in \mathbb{R}^L$, where $p^\star$ denotes the optimal objective value of (7.35). By the polyhedral representation of $\mathcal{S}$ in (7.22), problem (7.35) can be equivalently represented by an LP

$$p^\star = \min_\alpha \ \mathbf{r}^T (\mathbf{C}\alpha + \mathbf{d})$$

$$\text{s.t.} \ \ \mathbf{C}\alpha + \mathbf{d} \succeq \mathbf{0}, \tag{7.36}$$

which can be solved by readily available algorithms such as the polynomial-time, interior-point methods [36, 37]. Problem (7.36) is the problem we solve in practice, but (7.35) leads to important implications to extreme-point search.

A fundamental result in LP theory is that $\mathbf{r}^T s$, the objective function of (7.35), attains the minimum at a point of the boundary of $\mathcal{S}$. To provide more insights, some geometric illustrations are given in Figure 7.9. We can see that the solution of (7.35) may be uniquely given by one of the extreme points s_i (Figure 7.9(a)), or it may be any point on a face (Figure 7.9(b)). The latter case poses a difficulty in our task of identifying s_i, but it is arguably not a usual situation. For instance, in the illustration in Figure 7.9(b), $\mathbf{r}$ must be normal to $s_2 - s_3$, which may be unlikely to happen for a randomly picked $\mathbf{r}$. With this intuition in mind, we can prove the following lemma:

LEMMA 7.6 *Suppose that* $\mathbf{r}$ *is randomly generated following a distribution* $\mathcal{N}(\mathbf{0}, \mathbf{I}_L)$. *Then, with probability 1, the solution of (7.35) is uniquely given by* s_i *for some* $i \in \{1, \ldots, N\}$.

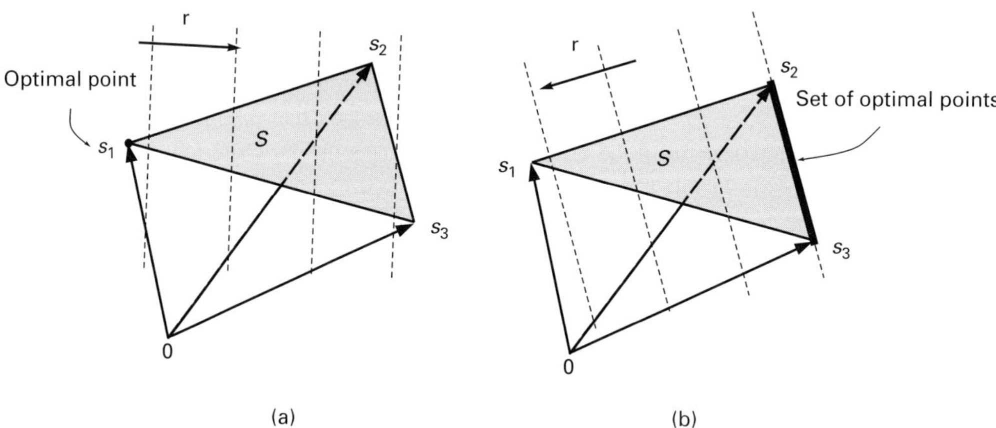

Figure 7.9   Geometric interpretation of an LP.

The proof of Lemma 7.6 is given in Appendix A7.4. The idea behind the proof is that undesired cases, such as that in Figure 7.9(b), happen with probability zero.

We may find another extreme point by solving the maximization counterpart of (7.35):

$$q^{\star} = \max_{\alpha} \ \mathbf{r}^T (\mathbf{C}\boldsymbol{\alpha} + \mathbf{d})$$

$$\text{s.t.} \ \ \mathbf{C}\boldsymbol{\alpha} + \mathbf{d} \succeq \mathbf{0}. \tag{7.37}$$

Using the same derivations as above, we can show the following: under the premise of Lemma 7.6, the solution of (7.37) is, with probability 1, uniquely given by an extreme point s_i different from that in (7.35).

Suppose that we have identified l extreme points, say, without loss of generality, $\{s_1,\ldots,s_l\}$. Our interest is in refining the above LP extreme-point finding procedure such that the search space is restricted to $\{s_{l+1},\ldots,s_N\}$. To do so, consider a thin QR decomposition [38] of $[s_1,\ldots,s_l]$

$$[s_1,\ldots,s_l] = \mathbf{Q}_1\mathbf{R}_1, \tag{7.38}$$

where $\mathbf{Q}_1 \in \mathbb{R}^{L\times l}$ is semi-unitary and $\mathbf{R}_1 \in \mathbb{R}^{l\times l}$ is upper triangular. Let

$$\mathbf{B} = \mathbf{I}_L - \mathbf{Q}_1\mathbf{Q}_1^T. \tag{7.39}$$

We assume that $\mathbf{r}$ takes the form

$$\mathbf{r} = \mathbf{B}w \tag{7.40}$$

for some $w \in \mathbb{R}^L$, and consider solving (7.36) and (7.37) with such an $\mathbf{r}$. Since $\mathbf{r}$ is orthogonal to the old extreme points $s_1,\ldots,s_l$, the intuitive expectation is that (7.36) and (7.37) should both lead to new extreme points. Interestingly, we found theoretically that such an expectation is not true, but close. It can be shown that (see Appendix A7.5):

LEMMA 7.7 *Suppose that* $\mathbf{r} = \mathbf{B}w$, *where* $\mathbf{B} \in \mathbb{R}^{L \times L}$ *is given by (7.39) and* w *is randomly generated following a distribution* $\mathcal{N}(\mathbf{0}, \mathbf{I}_L)$. *Then, with probability 1, at least one of the optimal solutions of (7.36) and (7.37) is a new extreme point; that is,* $\mathbf{s}_i$ *for some* $i \in \{l + 1, \ldots, N\}$. *The certificate of finding new extreme points is indicated by* $|p^\star| \neq 0$ *for the case of (7.36), and* $|q^\star| \neq 0$ *for (7.37).*

By repeating the above-described procedures, we can identify all the extreme points $\mathbf{s}_1, \ldots, \mathbf{s}_N$. The resultant CAMNS–LP method is summarized in Algorithm 7.1.

The CAMNS–LP method in Algorithm 7.1 is not only systematically straightforward to apply, it is also efficient due to the maturity of convex-optimization algorithms. Using a primal-dual, interior-point method, each LP problem (or the problem in (7.35) or (7.37)) can be solved with a worst-case complexity of $\mathcal{O}(L^{0.5}(L(N - 1) + (N - 1)^3)) \simeq \mathcal{O}(L^{1.5}(N - 1))$ for $L \gg N$ [37]. Since the algorithm solves $2(N - 1)$ LP problems in the worst case, we infer that its worst-case complexity is $\mathcal{O}(L^{1.5}(N - 1)^2)$.

Based on Theorem 7.1, Lemma 7.6, Lemma 7.7, and the complexity discussion above, we conclude that:

Algorithm 7.1. CAMNS–LP

Given an affine set characterization 2-tuple $(\mathbf{C}, \mathbf{d})$.

Step 1. Set $l = 0$, and $\mathbf{B} = \mathbf{I}_L$.

Step 2. Randomly generate a vector $w \sim \mathcal{N}(\mathbf{0}, \mathbf{I}_L)$, and set $\mathbf{r} := \mathbf{B}w$.

Step 3. Solve the LPs

$$p^\star = \min_{\alpha : \mathbf{C}\alpha + \mathbf{d} \geq \mathbf{0}} \mathbf{r}^T (\mathbf{C}\alpha + \mathbf{d})$$

$$q^\star = \max_{\alpha : \mathbf{C}\alpha + \mathbf{d} \geq \mathbf{0}} \mathbf{r}^T (\mathbf{C}\alpha + \mathbf{d})$$

and obtain their optimal solutions, denoted by $\alpha_1^\star$ and $\alpha_2^\star$, respectively.

Step 4. If $l = 0$

$$\hat{\mathbf{S}} = [\mathbf{C}\alpha_1^\star + \mathbf{d}, \ \mathbf{C}\alpha_2^\star + \mathbf{d}]$$

else

If $|p^\star| \neq 0$ then $\hat{\mathbf{S}} := [\hat{\mathbf{S}} \ \mathbf{C}\alpha_1^\star + \mathbf{d}]$.

If $|q^\star| \neq 0$ then $\hat{\mathbf{S}} := [\hat{\mathbf{S}} \ \mathbf{C}\alpha_2^\star + \mathbf{d}]$.

Step 5. Update l to be the number of columns of $\hat{\mathbf{S}}$.

Step 6. Apply QR decomposition

$$\hat{\mathbf{S}} = \mathbf{Q}_1 \mathbf{R}_1,$$

where $\mathbf{Q}_1 \in \mathbb{R}^{L \times l}$ and $\mathbf{R}_1 \in \mathbb{R}^{l \times l}$. Update $\mathbf{B} := \mathbf{I}_L - \mathbf{Q}_1 \mathbf{Q}_1^T$.

Step 7. Repeat **Step 2** to **Step 6** until $l = N$.

PROPOSITION 7.2   *Algorithm 7.1 finds all the true source vectors* $s_1, \ldots, s_N$ *with probability 1, under the premises of* **(A1)**–**(A4)**. *It does so with a worst-case complexity of* $\mathcal{O}(L^{1.5}(N-1)^2)$.

We have provided a practical implementation of CAMNS–LP at www.ee.cuhk.edu. hk/~wkma/CAMNS/CAMNS.htm. The source codes were written in MATLAB, and are based on the reliable convex optimization software SeDuMi [36]. Readers are encouraged to test the codes and give us some feedback.

7.6   Alternating volume-maximization heuristics for CAMNS

The CAMNS–LP method developed in the last section elegantly takes advantage of the model assumptions to sequentially track down the extreme points or the true source vectors. In particular, the local dominant assumption **(A2)** plays a key role. Our simulation experience is that CAMNS–LP can provide good separation performance on average, even when the local dominance assumption is not perfectly satisfied. In this section we consider an alternative that is also inspired by the CAMNS criterion, and that is intuitively expected to offer better robustness against model mismatch with local dominance. As we will further explain soon, the idea is to perform simplex volume maximization. Unfortunately, such an attempt will lead us to a nonconvex-optimization problem. We will propose an alternating, LP-based optimization heuristic to the simplex volume-maximization problem. Although the alternating heuristics is suboptimal, simulation results will indicate that the alternating heuristic can provide a better separation than CAMNS–LP, by a factor of about several dBs in sum-square-error performance (for data where local dominance is not perfectly satisfied).

Recall the CAMNS criterion in Criterion 7.2: find the extreme points of the polyhedral set

$$\mathcal{F} = \left\{ \boldsymbol{\alpha} \in \mathbb{R}^{N-1} \mid \mathbf{C}\boldsymbol{\alpha} + \mathbf{d} \succeq \mathbf{0} \right\}$$

which, under the model assumptions in **(A1)**–**(A4)**, is a simplex in the form of

$$\mathcal{F} = \mathrm{conv}\{\boldsymbol{\alpha}_1, \ldots, \boldsymbol{\alpha}_N\}.$$

For a simplex we can define its volume: a simplex, say denoted by $\mathrm{conv}\{\boldsymbol{\beta}_1, \ldots, \boldsymbol{\beta}_N\} \subset \mathbb{R}^{N-1}$, has its volume given by [39]

$$V(\boldsymbol{\beta}_1, \ldots, \boldsymbol{\beta}_N) = \frac{\left| \det\left(\boldsymbol{\Delta}(\boldsymbol{\beta}_1, \ldots, \boldsymbol{\beta}_N) \right) \right|}{(N-1)!}, \tag{7.41}$$

where

$$\boldsymbol{\Delta}(\boldsymbol{\beta}_1, \ldots, \boldsymbol{\beta}_N) = \begin{bmatrix} \boldsymbol{\beta}_1 & \cdots & \boldsymbol{\beta}_N \\ 1 & \cdots & 1 \end{bmatrix} \in \mathbb{R}^{N \times N}. \tag{7.42}$$

Suppose that $\{\boldsymbol{\beta}_1, \ldots, \boldsymbol{\beta}_N\} \subset \mathcal{F}$. As illustrated in the picture in Figure 7.10, the volume of $\mathrm{conv}\{\boldsymbol{\beta}_1, \ldots, \boldsymbol{\beta}_N\}$ should be no greater than that of $\mathcal{F} = \mathrm{conv}\{\boldsymbol{\alpha}_1, \ldots, \boldsymbol{\alpha}_N\}$. Hence,

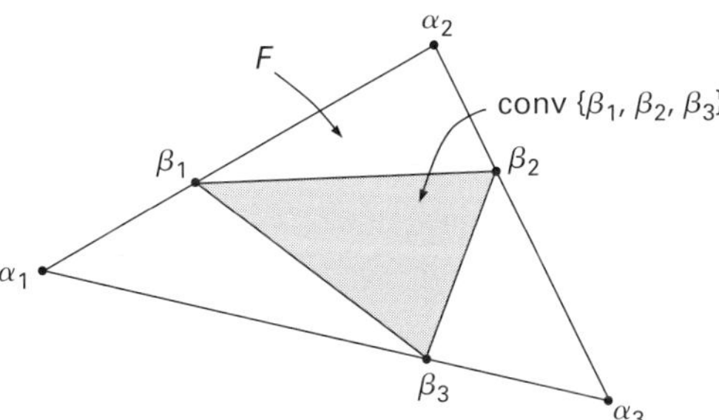

Figure 7.10 A geometric illustration for $\{\beta_1,\ldots,\beta_N\} \subset \mathcal{F}$ for $N = 3$.

by finding $\{\beta_1,\ldots,\beta_N\} \subset \mathcal{F}$ such that the respective simplex volume is maximized, we would expect that $\{\beta_1,\ldots,\beta_N\}$ is exactly $\{\alpha_1,\ldots,\alpha_N\}$; the ground truth we are seeking. This leads to the following variation of the CAMNS criterion:

CRITERION 7.3 Volume-maximization alternative to Criterion 7.2 *Use the affine set fitting solution in Proposition 7.1 to compute* $(\mathbf{C}, \mathbf{d})$. *Then, solve the volume maximization problem*

$$\{\hat{\alpha}_1,\ldots,\hat{\alpha}_N\} = \underset{\beta_1,\ldots,\beta_N}{\operatorname{argmax}} \quad V(\beta_1,\ldots,\beta_N) \tag{7.43}$$

$$\text{s.t.} \ \ \{\beta_1,\ldots,\beta_N\} \subset \mathcal{F}.$$

Output

$$\hat{s}_i = \mathbf{C}\hat{\alpha}_i + \mathbf{d}, \ i = 1,\ldots,N \tag{7.44}$$

as the set of estimated source vectors.

Like Criteria 7.1 and 7.2, Criterion 7.3 can be shown to provide the same perfect separation result:

THEOREM 7.3 *The globally optimal solution of (7.43) is uniquely given by* $\alpha_1,\ldots,\alpha_N$, *under the premises of (A1)–(A4).*

The proof of Theorem 7.3 is given in Appendix A7.6. As we mentioned in the beginning of this section, what is interesting with simplex volume maximization is when local dominance is not perfectly satisfied: the polyhedral set $\mathcal{F}$ may no longer be a simplex under such circumstances, though it would exhibit a geometric structure similar to a simplex. Simplex volume maximization would then be meaningful, because it gives a "best" simplex approximation to $\mathcal{F}$. Figure 7.11 provides an illustration of our argument above.

In the volume maximization approach, the challenge is with the simplex volume maximization problem in (7.43). To see this, we substitute (7.25) and (7.41) into (7.43)

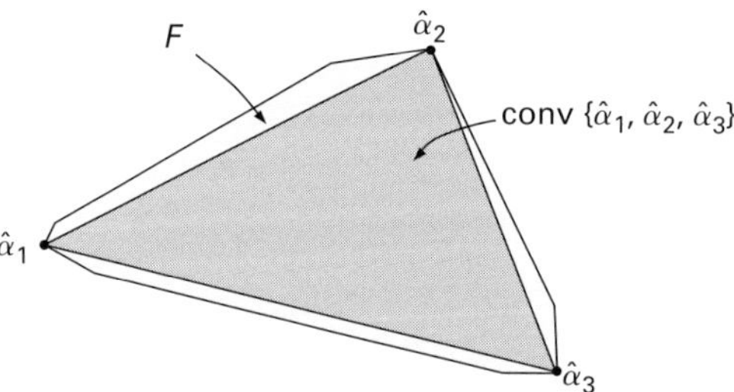

Figure 7.11   A geometric illustration for Criterion 7.3 when the local dominance assumption is not perfectly satisfied.

to obtain a more explicit formulation of the simplex volume-maximization problem:

$$\max_{\beta_1,\dots,\beta_N} \quad \left| \det\left(\Delta(\beta_1,\dots,\beta_N)\right) \right|$$
$$\text{s.t.} \quad \mathbf{C}\beta_i + \mathbf{d} \succeq \mathbf{0}, \ \forall\, i = 1,\dots,N. \tag{7.45}$$

The constraints of the problem above are affine (and convex), but the objective function is nonconvex.

Although a globally optimal solution of (7.45) may be difficult to obtain, we can approximate (7.45) in a convenient manner by using an alternating optimization heuristics proposed as follows. If we apply the cofactor expansion of $\Delta(\beta_1,\dots,\beta_N)$ along the jth column (for any j), we obtain an expression

$$\det(\Delta(\beta_1,\dots,\beta_N)) = \mathbf{b}_j^T \beta_j + (-1)^{N+j}\det(\mathcal{B}_{Nj}), \tag{7.46}$$

where $\mathbf{b}_j = [(-1)^{i+j}\det(\mathcal{B}_{ij})]_{i=1}^{N-1} \in \mathbb{R}^{N-1}$ and $\mathcal{B}_{ij} \in \mathbb{R}^{(N-1)\times(N-1)}$ is a submatrix of $\Delta(\beta_1,\dots,\beta_N)$ with the ith row and jth column being removed [39]. It is apparent from (7.46) that $\det(\Delta(\beta_1,\dots,\beta_N))$ is affine in each β_j. Now, consider partial maximization of (7.45) with respect to β_j, while fixing $\beta_1,\dots,\beta_{j-1},\beta_{j+1},\dots,\beta_N$:

$$\max_{\beta_j \in \mathbb{R}^{N-1}} \quad \left| \mathbf{b}_j^T \beta_j + (-1)^{N+j}\det(\mathcal{B}_{Nj}) \right|$$
$$\text{s.t.} \quad \mathbf{C}\beta_j + \mathbf{d} \succeq \mathbf{0}. \tag{7.47}$$

The objective function in (7.47) is still nonconvex, but (7.47) can be solved in a globally optimal manner by breaking it into two LPs:

$$p^\star = \max_{\beta_j \in \mathbb{R}^{N-1}} \quad \mathbf{b}_j^T \beta_j + (-1)^{N+j}\det(\mathcal{B}_{Nj})$$
$$\text{s.t.} \quad \mathbf{C}\beta_j + \mathbf{d} \succeq \mathbf{0}. \tag{7.48}$$

$$q^\star = \min_{\beta_j \in \mathbb{R}^{N-1}} \quad \mathbf{b}_j^T \beta_j + (-1)^{N+j}\det(\mathcal{B}_{Nj})$$
$$\text{s.t.} \quad \mathbf{C}\beta_j + \mathbf{d} \succeq \mathbf{0}. \tag{7.49}$$

The optimal solution of (7.47), denoted by $\hat{\alpha}_j$, is the optimal solution of (7.48) if $|p^\star| > |q^\star|$, and the optimal solution of (7.49) if $|q^\star| > |p^\star|$. This partial maximization is conducted alternately (i.e., $j := (j \bmod N) + 1$) until some stopping rule is satisfied.

The *CAMNS alternating volume maximization heuristics*, or simply CAMNS–AVM, is summarized in Algorithm 7.2.

Like alternating optimization in many other applications, the number of iterations required for CAMNS–AVM to terminate may be difficult to analyze. In the simulations considered in the next section, we found that for an accuracy of $\varepsilon = 10^{-13}$, CAMNS–AVM takes about 2 to 4 iterations to terminate, which is, surprisingly, quite small. Following the same complexity evaluation as in CAMNS–LP, CAMNS–AVM has a complexity of $\mathcal{O}(N^2 L^{1.5})$ per iteration. This means that CAMNS–AVM is only about 2 to 4 times more expensive than CAMNS–LP (by empirical experience).

Algorithm 7.2. CAMNS–AVM

Given a convergence tolerance $\varepsilon > 0$, an affine set characterization 2-tuple $(\mathbf{C}, \mathbf{d})$, and the observations $\boldsymbol{x}_1, \ldots, \boldsymbol{x}_M$.

Step 1. Initialize $\boldsymbol{\beta}_1, \ldots, \boldsymbol{\beta}_N \in \mathcal{F}$. (Our suggested choice: randomly choose N vectors out of the M observation-constructed vectors $\{(\mathbf{C}^T\mathbf{C})^{-1}\mathbf{C}^T(\boldsymbol{x}_i - \mathbf{d}),\ i = 1, \ldots, M\ \}$). Set

$$\boldsymbol{\Delta}(\boldsymbol{\beta}_1, \ldots, \boldsymbol{\beta}_N) = \begin{bmatrix} \boldsymbol{\beta}_1 & \cdots & \boldsymbol{\beta}_N \\ 1 & \cdots & 1 \end{bmatrix},$$

$\varrho := |\det(\boldsymbol{\Delta}(\boldsymbol{\beta}_1, \ldots, \boldsymbol{\beta}_N))|$, and $j := 1$.

Step 2. Update $\mathcal{B}_{ij}$ by a submatrix of $\boldsymbol{\Delta}(\boldsymbol{\beta}_1, \ldots, \boldsymbol{\beta}_N)$ with the ith row and jth column removed, and $\mathbf{b}_j := [(-1)^{i+j}\det(\mathcal{B}_{ij})]_{i=1}^{N-1}$.

Step 3. Solve the LPs

$$p^\star = \max_{\boldsymbol{\beta}_j : \mathbf{C}\boldsymbol{\beta}_j + \mathbf{d} \geq \mathbf{0}} \mathbf{b}_j^T \boldsymbol{\beta}_j + (-1)^{N+j}\det(\mathcal{B}_{Nj})$$

$$q^\star = \min_{\boldsymbol{\beta}_j : \mathbf{C}\boldsymbol{\beta}_j + \mathbf{d} \geq \mathbf{0}} \mathbf{b}_j^T \boldsymbol{\beta}_j + (-1)^{N+j}\det(\mathcal{B}_{Nj})$$

and obtain their optimal solutions, denoted by $\bar{\boldsymbol{\beta}}_j$ and $\underline{\boldsymbol{\beta}}_j$, respectively.

Step 4. If $|p^\star| > |q^\star|$, then update $\boldsymbol{\beta}_j := \bar{\boldsymbol{\beta}}_j$. Otherwise, update $\boldsymbol{\beta}_j := \underline{\boldsymbol{\beta}}_j$.

Step 5. If $(j \bmod N) \neq 0$, then $j := j + 1$, and go to **Step 2**,

else

If $|\max\{|p^\star|, |q^\star|\} - \varrho|/\varrho < \varepsilon$, then $\hat{\boldsymbol{\alpha}}_i = \boldsymbol{\beta}_i$ for $i = 1, \ldots, N$.
Otherwise, set $\varrho := \max\{|p^\star|, |q^\star|\}$, $j := 1$, and go to **Step 2**.

Step 6. Compute the source estimates $\hat{\boldsymbol{s}}_1, \ldots, \hat{\boldsymbol{s}}_N$ through $\hat{\boldsymbol{s}}_j = \mathbf{C}\hat{\boldsymbol{\alpha}}_j + \mathbf{d}$.

7.7   Numerical results

To demonstrate the efficacy of the CAMNS–LP and CAMNS–AVM methods, four simulation results are presented here. Section 7.7.1 is a cell image example where our task is to distinguish different types of cells. Section 7.7.2 focuses on a challenging scenario reminiscent of ghosting effects in photography. Section 7.7.3 considers a problem in which the sources are faces of five different persons. Section 7.7.4 uses Monte Carlo simulation to evaluate the performance of CAMNS-based algorithms under noisy conditions. For performance comparison, we also test three existing nBSS algorithms, namely non-negative matrix factorization (NMF) [16], non-negative independent component analysis (nICA) [12], and Ergodan's algorithm (a BSS method that exploits magnitude bounds of the sources) [6].

The performance measure used in this chapter is described as follows. Let $\mathbf{S} = [\mathbf{s}_1, \ldots, \mathbf{s}_N]$ be the true, multi-source signal matrix, and $\hat{\mathbf{S}} = [\hat{\mathbf{s}}_1, \ldots, \hat{\mathbf{s}}_N]$ be the multi-source output of a BSS algorithm. It is well known that a BSS algorithm is inherently subject to permutation and scaling ambiguities. We propose a *sum-square-error* (SSE) measure for $\mathbf{S}$ and $\hat{\mathbf{S}}$ [40, 41], given as follows:

$$
e(\mathbf{S}, \hat{\mathbf{S}}) = \min_{\boldsymbol{\pi} \in \Pi_N} \sum_{i=1}^{N} \left\| \mathbf{s}_i - \frac{\|\mathbf{s}_i\|}{\|\hat{\mathbf{s}}_{\pi_i}\|} \hat{\mathbf{s}}_{\pi_i} \right\|^2
\tag{7.50}
$$

where $\boldsymbol{\pi} = (\pi_1, \ldots, \pi_N)$, and $\Pi_N = \{ \boldsymbol{\pi} \in \mathbb{R}^N \mid \pi_i \in \{1, 2, \ldots, N\}, \ \pi_i \neq \pi_j \text{ for } i \neq j \}$ is the set of all permutations of $\{1, 2, \ldots, N\}$. The optimization of (7.50) is to adjust the permutation $\boldsymbol{\pi}$ such that the best match between true and estimated signals is yielded, while the factor $\|\mathbf{s}_i\|/\|\hat{\mathbf{s}}_{\pi_i}\|$ is to fix the scaling ambiguity. Problem (7.50) is the optimal assignment problem which can be efficiently solved by the Hungarian algorithm[1] [42].

7.7.1   Example of 3-source case: cell separation

In this example three 125×125 cell images, displayed in Figure 7.12(a), were taken as the source images. Each image is represented by a source vector $\mathbf{s}_i \in \mathbb{R}^L$, by scanning the image vertically from top left to bottom right (thereby $L = 125^2 = 15\,625$). For the three source images, we found that the local dominance assumption is not perfectly satisfied. To shed some light into this, we propose a measure called the *local dominance proximity factor* (LDPF) of the ith source, defined as follows:

$$
\kappa_i = \max_{n=1,\ldots,L} \frac{s_i[n]}{\sum_{j \neq i} s_j[n]}.
\tag{7.51}
$$

When $\kappa_i = \infty$, we have the ith source satisfying the local dominance assumption perfectly. The values of κ_is in this example are shown in Table 7.1, where we see that the LDPFs of the three sources are strong but not infinite.

---

[1] A Matlab implementation is available at http://si.utia.cas.cz/Tichavsky.html

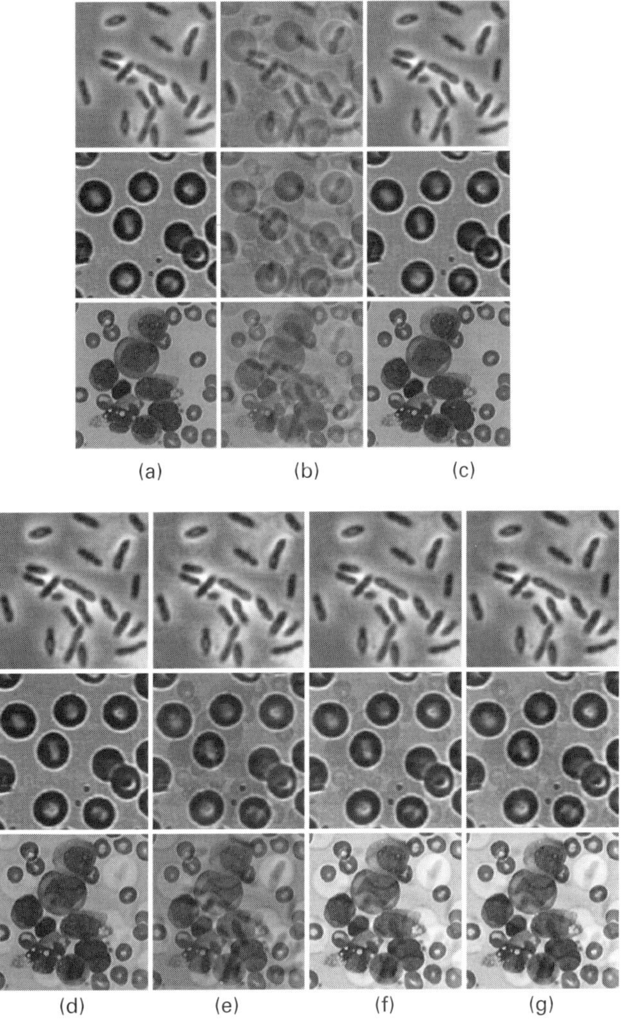

Figure 7.12 Cell separation: (a) The sources, (b) the observations, and the extracted sources obtained by (c) CAMNS–AVM method, (d) CAMNS–LP method, (e) NMF, (f) nICA, and (g) Erdogan's algorithm.

The three observation vectors are synthetically generated using a mixing matrix

$$\mathbf{A} = \begin{bmatrix} 0.20 & 0.62 & 0.18 \\ 0.35 & 0.37 & 0.28 \\ 0.40 & 0.40 & 0.20 \end{bmatrix}. \tag{7.52}$$

The mixed images are shown in Figure 7.12(b). The separated images of the various nBSS methods are illustrated in Figure 7.12(c)–(g). By visual inspection, the CAMNS-based methods provide good separation, despite the fact that the local dominance assumption is not perfectly satisfied. This result indicates that the CAMNS-based methods have some

Table 7.1. Local dominance proximity factors in the three scenarios.

	κ_i				
	source 1	source 2	source 3	source 4	source 5
Cell separation	48.667	3.821	15.200	–	–
Ghosting reduction	2.133	2.385	2.384	2.080	–
Human face separation	10.450	9.107	5.000	3.467	2.450

Table 7.2. The SSEs of the various nBSS methods in the three scenarios.

	SSE $e(\mathbf{S}, \hat{\mathbf{S}})$ (in dB)				
	CAMNS–AVM	CAMNS–LP	NMF	nICA	Erdogan's algorithm
Cell separation	3.710	12.323	23.426	19.691	19.002
Ghosting reduction	11.909	20.754	38.620	41.896	39.126
Human face separation	0.816	17.188	39.828	43.581	45.438

robustness against violation of local dominance. The SSE performance of the various methods is given in Table 7.2. We observe that the CAMNS–AVM method yields the best performance among all the methods under test, followed by CAMNS–LP. This suggests that CAMNS–AVM is more robust than CAMNS–LP, when local dominance is not exactly satisfied. This result will be further confirmed in the Monte Carlo simulation in Section 7.7.4.

7.7.2   Example of 4-source case: ghosting effect

We take a 285×285 Lena image from [2] as one source and then shift it diagonally to create three more sources; see Figure 7.13(a). Apparently, these sources are strongly correlated. Even worse, their LDPFs, shown in Table 7.1, are not too satisfactory compared to the previous example. The mixing matrix is

$$\mathbf{A} = \begin{bmatrix} 0.02 & 0.37 & 0.31 & 0.30 \\ 0.31 & 0.21 & 0.26 & 0.22 \\ 0.05 & 0.38 & 0.28 & 0.29 \\ 0.33 & 0.23 & 0.21 & 0.23 \end{bmatrix}. \tag{7.53}$$

Figure 7.13(b) displays the observations, where the mixing effect is reminiscent of the ghosting effect in analog televisions. The image separation results are illustrated in Figure 7.13(c)–(g). Clearly, only the CAMNS-based methods provide sufficiently good mitigation of the "ghosts." This result once again suggests that the CAMNS-based method is not too sensitive to the effect of local-dominance violation. The numerical results shown in Table 7.2 reflect that the SSE of CAMNS–AVM is about 9 dB smaller than that of CAMNS–LP, which can be validated by visual inspection of Figure 7.13(d)

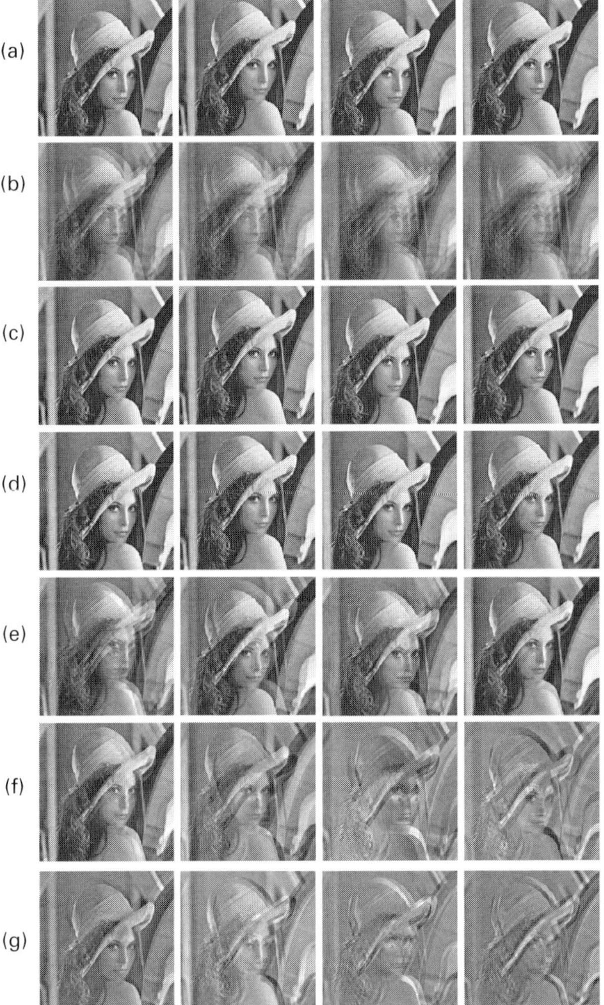

Figure 7.13 Ghosting reduction: (a) The sources, (b) the observations, and the extracted sources obtained by (c) CAMNS–AVM method, (d) CAMNS–LP method, (e) NMF, (f) nICA, and (g) Erdogan's algorithm.

where there are slight residuals on the 4th separated image. We argue that the residuals are harder to notice for the CAMNS–AVM method.

7.7.3 Example of 5-source case: human face separation

Five 240×320 photos taken from the second author and his friends are used as the source images in this example; see Figure 7.14(a). Since each human face was captured almost at the same position, the source images have some correlations. Once again, the local dominance assumption is not perfectly satisfied; as shown in Table 7.1. The five mixed

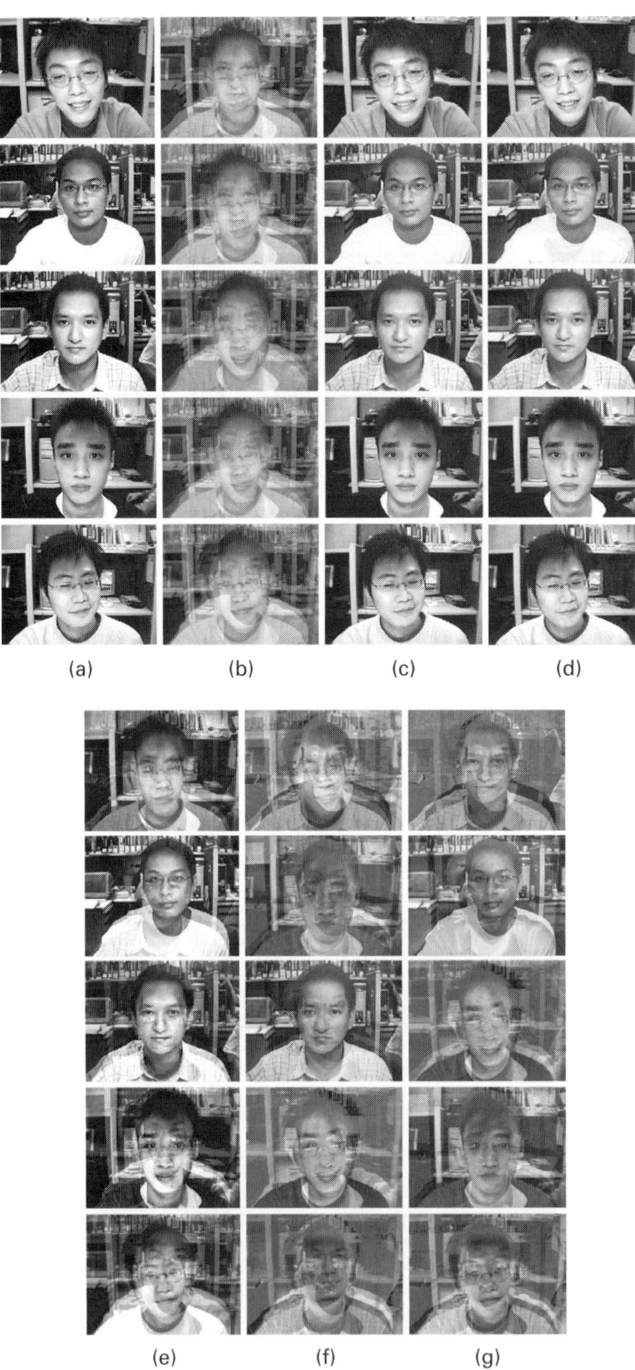

Figure 7.14   Human face separation: (a) The sources, (b) the observations, and the extracted sources obtained by (c) CAMNS–AVM method, (d) CAMNS–LP method, (e) NMF, (f) nICA, and (g) Erdogan's algorithm.

images, displayed in Figure 7.14(b), are generated through a mixing matrix given by

$$\mathbf{A} = \begin{bmatrix} 0.01 & 0.05 & 0.35 & 0.21 & 0.38 \\ 0.04 & 0.14 & 0.26 & 0.20 & 0.36 \\ 0.23 & 0.26 & 0.19 & 0.28 & 0.04 \\ 0.12 & 0.23 & 0.19 & 0.22 & 0.24 \\ 0.29 & 0.32 & 0.02 & 0.12 & 0.25 \end{bmatrix}. \tag{7.54}$$

Figures 7.14(c)–(g) show the separated images of the various nBSS methods. Apparently, one can see that the CAMNS-based methods have more accurate separation than the other methods, except for some slight residual image appearing in the 2nd CAMNS–LP separated image, by careful visual inspection. The numerical results shown in Table 7.2 indicate that the CAMNS-based methods perform better than the other methods. Moreover, comparing CAMNS–AVM and CAMNS–LP, there is a large performance gap of about 16 dB.

7.7.4 Monte Carlo simulation: noisy environment

We use Monte Carlo simulation to test the performance of the various methods when noise is present. The three cell images in Figure 7.12(a) were used to generate six noisy observations. The noise is independent and identically distributed (IID), following a Gaussian distribution with zero mean and variance σ^2. To maintain non-negativity of the observations in the simulation, we force the negative noisy observations to zero. We performed 100 independent runs. At each run the mixing matrix was IID uniformly generated on $[0,1]$ and then each row was normalized to 1 to maintain (**A3**). The average SSE for different SNRs (defined here as SNR$= \sum_{i=1}^{N} \|s_i\|^2 / LN\sigma^2$) are shown in Figure 7.15. One can see that the CAMNS-based methods perform better than the other methods.

We examine the performance of the various methods for different numbers of noisy observations with fixed SNR $= 25$ dB. The average SSEs for the various methods are shown in Figure 7.16. One can see that the performance of CAMNS-based methods becomes better when more observations are given. This phenomenon clearly validates the noise-mitigation merit of the affine set-fitting procedure (Proposition 7.1) in CAMNS.

7.8 Summary and discussion

In this chapter, we have shown how convex analysis provides a new avenue to approaching non-negative blind source separation. Using convex geometry concepts such as affine hull and convex hull, an analysis was carried out to show that under some appropriate assumptions nBSS can be boiled down to a problem of finding extreme points of a polyhedral set. We have also shown how this extreme-point finding problem can be solved by convex optimization, specifically by using LPs to systematically find all the extreme points.

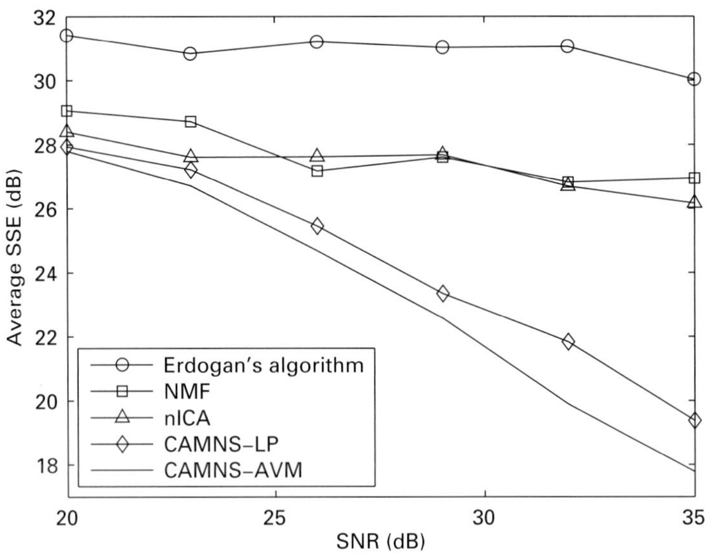

Figure 7.15   Performance evaluation of the CAMNS-based methods, NMF, nICA, and Erdogan's method for the cell images experiment for different SNRs.

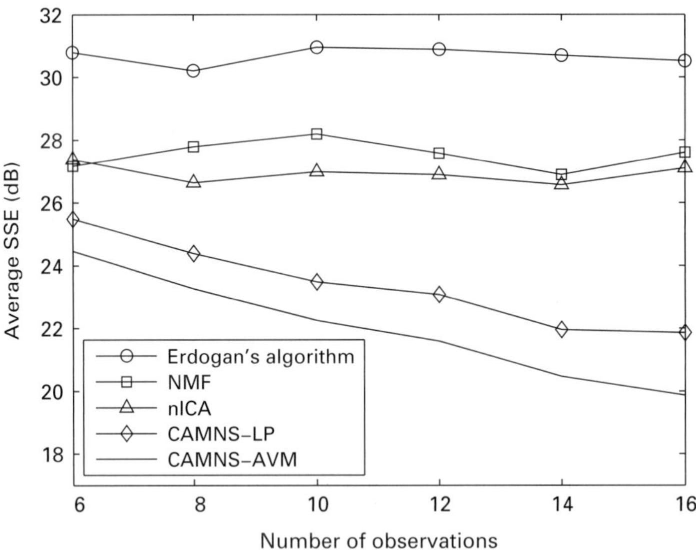

Figure 7.16   Performance evaluation of the CAMNS-based methods, NMF, nICA, and Erdogan's method for the cell images experiment for a different number of noisy observations.

The key success of this new nBSS framework is based on a deterministic signal assumption called local dominance. Local dominance is a good model assumption for sparse or high-contrast images, but it may not be perfectly satisfied sometimes. We have developed an alternative to the systematic LP method that is expected to yield better robustness against violation of local dominance. The idea is to solve a volume maximization problem. Despite the fact that the proposed algorithm uses heuristics to handle volume maximization (which is nonconvex), simulation results match with our intuitive expectation that volume maximization (done by a heuristics) can exhibit better resistance against the model mismatch.

We have carried out a number of simulations using different sets of image data, and have demonstrated that the proposed convex-analysis-based nBSS methods are promising, both by visual inspection and by the sum-square-error separation performance measure. Other methods such as nICA and NMF were also compared to demonstrate the effectiveness of the proposed methods.

A7   Appendix to Chapter 7

A7.1   Proof of Lemma 7.2

We prove the linear independence of $s_1, \ldots, s_N$ by showing that $\sum_{j=1}^{N} \theta_j s_j = \mathbf{0}$ only has the trivial solution $\theta_1 = \theta_2 = \ldots = \theta_N = 0$.

Suppose that $\sum_{j=1}^{N} \theta_j s_j = \mathbf{0}$ is true. Under (A2), for each source i we have the ℓ_ith entry (the locally dominant point) of $\sum_{j=1}^{N} \theta_j s_j$ given by

$$0 = \sum_{j=1}^{N} \theta_j s_j[\ell_i] = \theta_i s_i[\ell_i]. \tag{A7.1}$$

Since $s_i[\ell_i] > 0$, we must have $\theta_i = 0$ and this has to be satisfied for all i. As a result, Lemma 7.2 is obtained.

A7.2   Proof of Proposition 7.1

As a basic result in least-squares, each projection error in (7.17)

$$e_{\mathcal{A}(\tilde{\mathbf{C}},\tilde{\mathbf{d}})}(\mathbf{x}_i) = \min_{\boldsymbol{\alpha} \in \mathbb{R}^{N-1}} \| \tilde{\mathbf{C}}\boldsymbol{\alpha} + \tilde{\mathbf{d}} - \mathbf{x}_i \|^2 \tag{A7.2}$$

has a closed form

$$e_{\mathcal{A}(\tilde{\mathbf{C}},\tilde{\mathbf{d}})}(\mathbf{x}_i) = (\mathbf{x}_i - \tilde{\mathbf{d}})^T \mathbf{P}_{\tilde{\mathbf{C}}}^{\perp} (\mathbf{x}_i - \tilde{\mathbf{d}}), \tag{A7.3}$$

where $\mathbf{P}_{\tilde{\mathbf{C}}}^{\perp}$ is the orthogonal-complement projection of $\tilde{\mathbf{C}}$. Using (A7.3), the affine set fitting problem [in (7.17)] can be rewritten as

$$\min_{\tilde{\mathbf{C}}^T \tilde{\mathbf{C}} = \mathbf{I}} \left\{ \min_{\tilde{\mathbf{d}}} \sum_{i=1}^{M} (\mathbf{x}_i - \tilde{\mathbf{d}})^T \mathbf{P}_{\tilde{\mathbf{C}}}^{\perp} (\mathbf{x}_i - \tilde{\mathbf{d}}) \right\}. \tag{A7.4}$$

The inner-minimization problem in (A7.4) is an unconstrained, convex quadratic program, and it can be easily verified that $\mathbf{d} = \frac{1}{M}\sum_{i=1}^{M} \boldsymbol{x}_i$ is an optimal solution to the inner-minimization problem. By substituting this optimal $\mathbf{d}$ into (A7.4) and by letting $\mathbf{U} = [\boldsymbol{x}_1 - \mathbf{d}, \ldots, \boldsymbol{x}_M - \mathbf{d}]$, problem (A7.4) can be reduced to

$$\min_{\tilde{\mathbf{C}}^T \tilde{\mathbf{C}} = \mathbf{I}_{N-1}} \mathrm{Trace}\{\mathbf{U}^T \mathbf{P}_{\tilde{\mathbf{C}}}^{\perp} \mathbf{U}\}. \tag{A7.5}$$

When $\tilde{\mathbf{C}}^T \tilde{\mathbf{C}} = \mathbf{I}_{N-1}$, the projection matrix $\mathbf{P}_{\tilde{\mathbf{C}}}^{\perp}$ can be simplified to $\mathbf{I}_L - \tilde{\mathbf{C}}\tilde{\mathbf{C}}^T$. Subsequently (A7.5) can be further reduced to

$$\max_{\tilde{\mathbf{C}}^T \tilde{\mathbf{C}} = \mathbf{I}_{N-1}} \mathrm{Trace}\{\mathbf{U}^T \tilde{\mathbf{C}}\tilde{\mathbf{C}}^T \mathbf{U}\}. \tag{A7.6}$$

An optimal solution of (A7.6) is known to be the $N-1$ principal eigenvector matrix of $\mathbf{U}\mathbf{U}^T$ [38] as given by (7.21).

A7.3   Proof of Lemma 7.5

Equation (7.25) can also be expressed as

$$\mathcal{F} = \left\{ \boldsymbol{\alpha} \in \mathbb{R}^{N-1} \mid \mathbf{C}\boldsymbol{\alpha} + \mathbf{d} \in \mathrm{conv}\{s_1, \ldots, s_N\} \right\}.$$

Thus, every $\boldsymbol{\alpha} \in \mathcal{F}$ satisfies

$$\mathbf{C}\boldsymbol{\alpha} + \mathbf{d} = \sum_{i=1}^{N} \theta_i s_i \tag{A7.7}$$

for some $\boldsymbol{\theta} \succeq \mathbf{0}$, $\boldsymbol{\theta}^T \mathbf{1} = 1$. Since $\mathbf{C}$ has full column rank, (A7.7) can be re-expressed as

$$\boldsymbol{\alpha} = \sum_{i=1}^{N} \theta_i \boldsymbol{\alpha}_i, \tag{A7.8}$$

where $\boldsymbol{\alpha}_i = (\mathbf{C}^T \mathbf{C})^{-1}\mathbf{C}^T (s_i - \mathbf{d})$ (or $\mathbf{C}\boldsymbol{\alpha}_i + \mathbf{d} = s_i$). Equation (A7.8) implies that $\mathcal{F} = \mathrm{conv}\{\boldsymbol{\alpha}_1, \ldots, \boldsymbol{\alpha}_N\}$.

We now show that $\mathcal{F} = \mathrm{conv}\{\boldsymbol{\alpha}_1, \ldots, \boldsymbol{\alpha}_N\}$ is a simplex by contradiction. Suppose that $\{\boldsymbol{\alpha}_1, \ldots, \boldsymbol{\alpha}_N\}$ are not affinely independent. This means that for some $\gamma_1, \ldots, \gamma_{N-1}$, $\sum_{i=1}^{N-1} \gamma_i = 1$, $\boldsymbol{\alpha}_N = \sum_{i=1}^{N-1} \gamma_i \boldsymbol{\alpha}_i$ can be satisfied. One then has $s_N = \mathbf{C}\boldsymbol{\alpha}_N + \mathbf{d} = \sum_{i=1}^{N-1} \gamma_i s_i$, which is a contradiction to the property that $\{s_1, \ldots, s_N\}$ is linearly independent.

A7.4   Proof of Lemma 7.6

Any point in $\mathcal{S} = \mathrm{conv}\{s_1, \ldots, s_N\}$ can be equivalently represented by $s = \sum_{i=1}^{N} \theta_i s_i$, where $\boldsymbol{\theta} \succeq \mathbf{0}$ and $\boldsymbol{\theta}^T \mathbf{1} = 1$. Applying this result to (7.35), Problem (7.35) can be reformulated as

$$\min_{\boldsymbol{\theta} \in \mathbb{R}^N} \quad \sum_{i=1}^{N} \theta_i \rho_i$$
$$\text{s.t.} \quad \boldsymbol{\theta}^T \mathbf{1} = 1, \ \boldsymbol{\theta} \geq \mathbf{0}, \tag{A7.9}$$

where $\rho_i = \mathbf{r}^T s_i$. We assume without loss of generality that $\rho_1 < \rho_2 \leq \cdots \leq \rho_N$. If $\rho_1 < \rho_2 < \cdots < \rho_N$, then it is easy to verify that the optimal solution to (A7.9) is uniquely given by $\boldsymbol{\theta}^\star = \mathbf{e}_1$. In its counterpart in (7.35), this translates into $s^\star = s_1$. But when $\rho_1 = \rho_2 = \cdots = \rho_P$ and $\rho_P < \rho_{P+1} \leq \cdots \leq \rho_N$ for some P, the solution of (A7.9) is not unique. In essence, the latter case can be shown to have a solution set

$$\Theta^\star = \{\boldsymbol{\theta} \mid \boldsymbol{\theta}^T \mathbf{1} = 1, \ \boldsymbol{\theta} \geq \mathbf{0}, \ \theta_{P+1} = \cdots = \theta_N = 0\}. \tag{A7.10}$$

We now prove that the non-unique solution case happens with probability zero. Suppose that $\rho_i = \rho_j$ for some $i \neq j$, which means that

$$(s_i - s_j)^T \mathbf{r} = 0. \tag{A7.11}$$

Let $v = (s_i - s_j)^T \mathbf{r}$. Apparently, v follows a distribution $\mathcal{N}(0, \|s_i - s_j\|^2)$. Since $s_i \neq s_j$, the probability $\Pr[\rho_i = \rho_j] = \Pr[v = 0]$ is of measure zero. This in turn implies that $\rho_1 < \rho_2 < \cdots < \rho_N$ holds with probability 1.

A7.5 Proof of Lemma 7.7

The approach to proving Lemma 7.7 is similar to that in Lemma 7.6. Let

$$\rho_i = \mathbf{r}^T s_i = (\mathbf{B} w)^T s_i \tag{A7.12}$$

for which we have $\rho_i = 0$ for $i = 1, \ldots, l$. It can be shown that

$$\rho_{l+1} < \rho_{l+2} < \cdots < \rho_N \tag{A7.13}$$

holds with probability 1, as long as $\{s_1, \ldots, s_N\}$ is linearly independent. Problems (7.35) and (7.37) are respectively equivalent to

$$p^\star = \min_{\boldsymbol{\theta} \in \mathbb{R}^N} \sum_{i=l+1}^{N} \theta_i \rho_i$$
$$\text{s.t.} \ \boldsymbol{\theta} \geq \mathbf{0}, \ \boldsymbol{\theta}^T \mathbf{1} = 1, \tag{A7.14}$$

$$q^\star = \max_{\boldsymbol{\theta} \in \mathbb{R}^N} \sum_{i=l+1}^{N} \theta_i \rho_i$$
$$\text{s.t.} \ \boldsymbol{\theta} \geq \mathbf{0}, \ \boldsymbol{\theta}^T \mathbf{1} = 1. \tag{A7.15}$$

Assuming (A7.13), we have three distinct cases to consider: (C1) $\rho_{l+1} < 0$, $\rho_N < 0$, (C2) $\rho_{l+1} < 0$, $\rho_N > 0$, and (C3) $\rho_{l+1} > 0$, $\rho_N > 0$.

For (C2), we can see the following: Problem (A7.14) has a unique optimal solution $\boldsymbol{\theta}^{\star} = \mathbf{e}_{l+1}$ [and $s^{\star} = s_{l+1}$ in its counterpart in (7.35)], attaining an optimal value $p^{\star} = \rho_{l+1} < 0$. Problem (A7.15) has a unique optimal solution $\boldsymbol{\theta}^{\star} = \mathbf{e}_N$ [and $s^{\star} = s_N$ in its counterpart in (7.37)], attaining an optimal value $q^{\star} = \rho_N > 0$. In other words, both (A7.14) and (A7.15) lead to finding new extreme points. For (C1), problem (A7.15) is shown to have a solution set

$$\Theta^{\star} = \{\boldsymbol{\theta} \mid \boldsymbol{\theta}^T \mathbf{1} = 1,\ \boldsymbol{\theta} \succeq \mathbf{0},\ \theta_{l+1} = \cdots = \theta_N = 0\}, \tag{A7.16}$$

which contains convex combinations of the old extreme points, and the optimal value is $q^{\star} = 0$. Nevertheless, it is still true that (A7.14) finds a new extreme point with $p^{\star} < 0$. A similar situation happens with (C3), where (A7.14) does not find a new extreme point with $p^{\star} = 0$, but (A7.15) finds a new extreme point with $q^{\star} > 0$.

A7.6   Proof of Theorem 7.3

In Problem (7.43), the constraints $\boldsymbol{\beta}_i \in \mathcal{F} = \mathrm{conv}\{\boldsymbol{\alpha}_1, \ldots, \boldsymbol{\alpha}_N\}$ imply that

$$\boldsymbol{\beta}_i = \sum_{j=1}^{N} \theta_{ij} \boldsymbol{\alpha}_j, \tag{A7.17}$$

where $\sum_{j=1}^{N} \theta_{ij} = 1$ and $\theta_{ij} \geq 0$ for $i = 1, \ldots, N$. Hence we can write

$$\boldsymbol{\Delta}(\boldsymbol{\beta}_1, \ldots, \boldsymbol{\beta}_N) = \boldsymbol{\Delta}(\boldsymbol{\alpha}_1, \ldots, \boldsymbol{\alpha}_N) \boldsymbol{\Theta}^T, \tag{A7.18}$$

where $\boldsymbol{\Theta} = [\theta_{ij}] \in \mathbb{R}_+^{N \times N}$ and $\boldsymbol{\Theta} \mathbf{1} = \mathbf{1}$. For such a structured $\boldsymbol{\Theta}$ it was shown that (Lemma 1 in [43])

$$|\det(\boldsymbol{\Theta})| \leq 1 \tag{A7.19}$$

and that $|\det(\boldsymbol{\Theta})| = 1$ if and only if $\boldsymbol{\Theta}$ is a permutation matrix. It follows from (7.41), (A7.18), and (A7.19) that

$$V(\boldsymbol{\beta}_1, \ldots, \boldsymbol{\beta}_N) = |\det(\boldsymbol{\Delta}(\boldsymbol{\alpha}_1, \ldots, \boldsymbol{\alpha}_N) \boldsymbol{\Theta}^T)| / (N - 1)!$$

$$= |\det(\boldsymbol{\Delta}(\boldsymbol{\alpha}_1, \ldots, \boldsymbol{\alpha}_N))| |\det(\boldsymbol{\Theta})| / (N - 1)!$$

$$\leq V(\boldsymbol{\alpha}_1, \ldots, \boldsymbol{\alpha}_N), \tag{A7.20}$$

and that the equality holds if and only if $\boldsymbol{\Theta}$ is a permutation matrix, which implies $\{\boldsymbol{\beta}_1, \ldots, \boldsymbol{\beta}_N\} = \{\boldsymbol{\alpha}_1, \ldots, \boldsymbol{\alpha}_N\}$. Hence we conclude that $V(\boldsymbol{\beta}_1, \ldots, \boldsymbol{\beta}_N)$ is maximized if and only if $\{\boldsymbol{\beta}_1, \ldots, \boldsymbol{\beta}_N\} = \{\boldsymbol{\alpha}_1, \ldots, \boldsymbol{\alpha}_N\}$.

Acknowledgments

This work was supported in part by the National Science Council (R.O.C.) under Grants NSC 96-2628-E-007-003-MY3, by the U.S. National Institutes of Health under Grants EB000830 and CA109872, and by a grant from the Research Grant Council of Hong Kong (General Research Fund, Project 2150599).

References

[1] A. Hyvärinen, J. Karhunen, and E. Oja, *Independent Component Analysis*. New York: John Wiley, 2001.

[2] A. Cichocki and S. Amari, *Adaptive Blind Signal and Image Processing*. Chichester: John Wiley, Inc., 2002.

[3] L. Parra and C. Spence, "Convolutive blind separation of non-stationary sources," *IEEE Transactions on Speech & Audio Processings*, vol. 8, no. 3, pp. 320–7, 2000.

[4] D.-T. Pham and J.-F. Cardoso, "Blind separation of instantaneous mixtures of nonstationary sources," *IEEE Transactions on Signal Processing*, vol. 49, no. 9, pp. 1837–48, 2001.

[5] A. Prieto, C. G. Puntonet, and B. Prieto, "A neural learning algorithm for blind separation of sources based on geometric properties," *Signal Processing*, vol. 64, pp. 315–31, 1998.

[6] A. T. Erdogan, "A simple geometric blind source separation method for bound magnitude sources," *IEEE Transactions on Signal Processing*, vol. 54, no. 2, pp. 438–49, 2006.

[7] F. Vrins, J. A. Lee, and M. Verleysen, "A minimum-range approach to blind extraction of bounded sources," *IEEE Transactions on Neural Networks*, vol. 18, no. 3, pp. 809–22, 2006.

[8] J. M. P. Nascimento and J. M. B. Dias, "Does independent component analysis play a role in unmixing hyperspectral data?" *IEEE Transactions on Geoscience and Remote Sensing*, vol. 43, no. 1, pp. 175–87, 2005.

[9] Y. Wang, J. Xuan, R. Srikanchana, and P. L. Choyke, "Modeling and reconstruction of mixed functional and molecular patterns," *International Journal of Biomedical Imaging*, ID29707, 2006.

[10] N. Keshava and J. Mustard, "Spectral unmixing," *IEEE Signal Processing Magazine*, vol. 19, no. 1, pp. 44–57, 2002.

[11] E. R. Malinowski, *Factor Analysis in Chemistry*. New York: John Wiley, 2002.

[12] M. D. Plumbley, "Algorithms for non-negative independent component analysis," *IEEE Transactions on Neural Networks*, vol. 14, no. 3, pp. 534–43, 2003.

[13] S. A. Astakhov, H. Stogbauer, A. Kraskov, and P. Grassberger, "Monte Carlo algorithm for least dependent non-negative mixture decomposition," *Analytical Chemistry*, vol. 78, no. 5, pp. 1620–7, 2006.

[14] S. Moussaoui, D. Brie, A. Mohammad-Djafari, and C. Carteret, "Separation of non-negative mixture of non-negative sources using a Bayesian approach and MCMC sampling," *IEEE Transactions on Signal Processing*, vol. 54, no. 11, pp. 4133–45, 2006.

[15] M. D. Plumbley, "Conditions for nonnegative independent component analysis," *IEEE Signal Processing Letters*, vol. 9, no. 6, pp. 177–80, 2002.

[16] D. D. Lee and H. S. Seung, "Learning the parts of objects by non-negative matrix factorization," *Nature*, vol. 401, pp. 788–91, 1999.

[17] ——, "Algorithms for non-negative matrix factorization," in *NIPS*. MIT Press, 2001, pp. 556–562.

[18] R. Zdunek and A. Cichocki, "Nonnegative matrix factorization with constrained second-order optimization," *Signal Processing*, vol. 87, no. 8, pp. 1904–16, 2007.

[19] C. Lawson and R. J. Hanson, *Solving Least-Squares Problems*. New Jersey: Prentice-Hall, 1974.

[20] R. Tauler and B. Kowalski, "Multivariate curve resolution applied to spectral data from multiple runs of an industrial process," *Analytical Chemistry*, vol. 65, pp. 2040–7, 1993.

[21] A. Zymnis, S.-J. Kim, J. Skaf, M. Parente, and S. Boyd, "Hyperspectral image unmixing via alternating projected subgradients," Proceedings of the 41st Asilomar Conference on Signals, Systems, and Computers, Pacific Grove, CA, 2007.

[22] C.-J. Lin, "Projected gradient methods for non-negative matrix factorization," *Neural Computation*, vol. 19, no. 10, pp. 2756–79, 2007.

[23] H. Laurberg, M. G. Christensen, M. D. Plumbley, L. K. Hansen, and S. H. Jensen, "Theorems on positive data: On the uniqueness of NMF," *Computational Intelligence and Neuroscience*, ID764206, 2008.

[24] P. Hoyer, "Nonnegative sparse coding," in *IEEE Workshop on Neural Networks for Signal Processing*, Martigny, Switzerland, Sept. 4-6, 2002, pp. 557–565.

[25] H. Kim and H. Park, "Sparse non-negative matrix factorizations via alternating non-negativity-constrained least squares for microarray data analysis," *Bioinformatics*, vol. 23, no. 12, pp. 1495–502, 2007.

[26] T.-H. Chan, W.-K. Ma, C.-Y. Chi, and Y. Wang, "A convex analysis framework for blind separation of non-negative sources," *IEEE Trans. Signal Processing*, vol. 56, no. 10, pp. 5120–34, 2008.

[27] F.-Y. Wang, Y. Wang, T.-H. Chan, and C.-Y. Chi, "Blind separation of multichannel biomedical image patterns by non-negative least-correlated component analysis," in *Lecture Notes in Bioinformatics (Proc. PRIB'06), Springer-Verlag*, vol. 4146, Berlin, Dec. 9-14, 2006, pp. 151–62.

[28] F.-Y. Wang, C.-Y. Chi, T.-H. Chan, and Y. Wang, "Blind separation of positive dependent sources by non-negative least-correlated component analysis," in *IEEE International Workshop on Machine Learning for Signal Processing (MLSP'06)*, Maynooth, Ireland, Sept. 6-8, 2006, pp. 73–78.

[29] E. Hillman and A. Moore, "All-optical anatomical co-registration for molecular imaging of small animals using dynamic contrast," *Nature Photonics Letters*, vol. 1, pp. 526–530, 2007.

[30] S. Boyd and L. Vandenberghe, *Convex Optimization*. Cambridge: Cambridge University Press, 2004.

[31] D. P. Bertsekas, A. Nedić, and A. E. Ozdaglar, *Convex Analysis and Optimization*. Belmont, MA: Athena Scientific, 2003.

[32] B. Grünbaum, *Convex Polytopes*. New York: Springer, 2003.

[33] M. E. Dyer, "The complexity of vertex enumeration methods," *Mathematics of Operations Research*, vol. 8, no. 3, pp. 381–402, 1983.

[34] K. G. Murty and S.-J. Chung, "Extreme point enumeration," College of Engineering, University of Michigan, Technical Report 92-21, 1992. Available http://deepblue.lib.umich.edu/handle/2027.42/6731

[35] K. Fukuda, T. M. Liebling, and F. Margot, "Analysis of backtrack algorithms for listing all vertices and all faces of a convex polyhedron," *Computational Geometry: Theory and Applications*, vol. 8, no. 1, pp. 1–12, 1997.

[36] J. F. Sturm, "Using SeDuMi 1.02, a MATLAB toolbox for optimization over symmetric cones," *Optimization Methods and Software*, vol. 11–12, pp. 625–53, 1999.

[37] I. J. Lustig, R. E. Marsten, and D. F. Shanno, "Interior point methods for linear programming: computational state of the art," *ORSA Journal on Computing*, vol. 6, no. 1, pp. 1–14, 1994.

[38] G. H. Golub and C. F. V. Loan, *Matrix Computations*. Baltimore, MD: The Johns Hopkins University Press, 1996.

[39] G. Strang, *Linear Algebra and Its Applications*, 4th ed. CA: Thomson, 2006.

[40] P. Tichavský and Z. Koldovský, "Optimal pairing of signal components separated by blind techniques," *IEEE Signal Processing Letters*, vol. 11, no. 2, pp. 119–22, 2004.

[41] J. R. Hoffman and R. P. S. Mahler, "Multitarget miss distance via optimal assignment," *IEEE Transactions on System, Man, and Cybernetics*, vol. 34, no. 3, pp. 327–36, 2004.

[42] H. W. Kuhn, "The Hungarian method for the assignment method," *Naval Research Logistics Quarterly*, vol. 2, pp. 83–97, 1955.

[43] F.-Y. Wang, C.-Y. Chi, T.-H. Chan, and Y. Wang, "Non-negative least-correlated component analysis for separation of dependent sources by volume maximization," *IEEE Trans. Pattern Analysis and Machine Intelligence*, accepted for publication.

8 Optimization techniques in modern sampling theory

Tomer Michaeli and Yonina C. Eldar

Sampling theory has benefited from a surge of research in recent years, due in part to intense research in wavelet theory and the connections made between the two fields. In this chapter we present several extensions of the Shannon theorem, which treat a wide class of input signals, as well as nonideal-sampling and constrained-recovery procedures. This framework is based on an optimization viewpoint, which takes into account both the goodness of fit of the reconstructed signal to the given samples, as well as relevant prior knowledge on the original signal. Our exposition is based on a Hilbert-space interpretation of sampling techniques, and relies on the concepts of bases (frames) and projections. The reconstruction algorithms developed in this chapter lead to improvement over standard interpolation approaches in signal- and image-processing applications.

8.1   Introduction

Sampling theory treats the recovery of a continuous-time signal from a discrete set of measurements. This field attracted significant attention in the engineering community ever since the pioneering work of Shannon [1] (also attributed to Whitaker [2], Kotelnikov [3], and Nyquist [4]) on sampling bandlimited signals. Discrete-time signal processing (DSP) inherently relies on sampling a continuous-time signal to obtain a discrete-time representation. Therefore, with the rapid development of digital applications, the theory of sampling has gained importance.

Traditionally, sampling theories addressed the problem of perfectly reconstructing a given class of signals from their samples. During the last two decades, it has been recognized that these theories can be viewed in a broader sense of projection onto appropriate subspaces of L_2 [5–7], and also extended to arbitrary Hilbert spaces [8, 9].

The goal of this chapter is to introduce a complementary viewpoint on sampling, which is based on optimization theory. The key idea in this approach is to construct an optimization problem that takes into account both the goodness of fit of the reconstructed signal to the given samples, as well as relevant prior knowledge on the original signal, such as smoothness. This framework is rooted in the theory of spline interpolation: one of the arguments in favor of using smoothing splines for interpolation is that they minimize an energy functional. This objective accounts for both the miss-fit at the sampling

Convex Optimization in Signal Processing and Communication, eds. Daniel P. Palomar and Yonina C. Eldar.
Published by Cambridge University Press © Cambridge University Press, 2010.

locations, and for the energy of the mth derivative [10] (where m is related to the spline order). Several other interpolation techniques have been proposed in recent years, which are based on variational arguments of the same spirit [11]. These methods also have connections with Wiener's estimation theory of random processes [11–15]. This chapter provides extensions, generalizations, and rigorous proofs of a variety of optimization-based interpolation techniques. Some of these methods were recently reported (without proof) in the review paper [15].

We focus on sampling problems in an abstract Hilbert space setting. This facilitates the treatment of various sampling scenarios via an optimization framework. Since signals are generally functions over a continuous domain, the optimization problems we will encounter in this chapter are infinite-dimensional and cannot be solved numerically. Furthermore, in order to make the discussion relevant to arbitrary Hilbert spaces and not only for signals which are functions over $\mathbb{R}$, we refrain from using calculus of variations in our derivations. Therefore, most of our treatment relies on the interpretation of the resulting optimization problems in terms of projections onto appropriate spaces. Besides being infinite-dimensional, many of the optimization problems we attempt to solve are also not convex. Nevertheless, we derive closed-form solutions for many of these, by employing a geometric viewpoint. In the last section we tackle interpolation problems which do not admit a closed-form solution. In this scenario, we narrow the discussion to signals lying in $\mathbb{R}^n$ or $\mathbb{C}^n$ and employ semidefinite relaxation and saddle-point techniques to arrive at optimization problems which can be solved numerically.

The scenarios treated in this chapter differ from one another in several aspects. First, we distinguish between noiseless and noisy samples. Second, the reconstruction algorithms we consider can either be adjusted according to some objective, or constrained to be of a certain predefined structure. Third, we treat two types of prior knowledge on the original signal, which we term subspace priors and smoothness priors. Last, we treat two classes of optimization criteria for each of the scenarios: least-squares and minimax. The setups we consider are summarized in Table 8.1. Throughout the chapter we highlight the connection between the resulting reconstruction methods, demonstrate how they can be implemented efficiently, and provide concrete examples of interpolation results.

The chapter is organized as follows. In Section 8.2 we provide mathematical preliminaries needed for the derivations to follow. Section 8.3 describes in detail the sampling and reconstruction setups treated in this chapter. In particular, we elaborate on the types of prior knowledge and reconstruction approaches that are considered. Section 8.4 is devoted to the different objectives that are at the heart of the proposed recovery techniques. In Sections 8.5 and 8.6 we develop reconstruction methods for the case of subspace and smoothness priors, respectively. Each of these priors is studied in a constrained and unconstrained reconstruction setting using both the least-squares and minimax objectives. All reconstruction methods in these sections possess closed-form expressions. Section 8.7 includes comparisons between the various recovery algorithms. Finally, in Section 8.8 we treat the case in which the samples are noisy. There, we focus our attention on smoothness priors and on the minimax objective. We use semidefinite relaxation to tackle the resulting nonconvex quadratic programs. This section also includes a summary of recent results on semidefinite relaxation of nonconvex quadratic programs, which is needed for our derivations.

Table 8.1. Different scenarios treated in this chapter.

	Unconstrained Reconstruction		Constrained Reconstruction	
	Least-Squares	Minimax	Least-Squares	Minimax
Subspace Priors Noise-Free Samples	Section 8.5.1	Section 8.5.1	Section 8.5.2	Section 8.5.2
Smoothness Priors Noise-Free Samples	Section 8.6.1	Section 8.6.1	Section 8.6.2	Section 8.6.2
Smoothness Priors Noisy Samples	Section 8.8	Section 8.8	Section 8.8	Section 8.8

8.2 Notation and mathematical preliminaries

The exposition in this chapter is based on a Hilbert-space interpretation of sampling techniques, and relies on the concepts of frames and projections. In this section we introduce some notations and mathematical preliminaries which form the basis for the derivations in the sections to follow.

8.2.1 Notation

We denote vectors in an arbitrary Hilbert space $\mathcal{H}$ by lowercase letters, and the elements of a sequence $c \in \ell_2$ by $c[n]$. Traditional sampling theories deal with signals, which are defined over the real line. In this case the Hilbert space $\mathcal{H}$ of signals of interest is the space L_2 of square integrable functions, in other words every vector $x \in \mathcal{H}$ is a function $x(t)$, $t \in \mathbb{R}$. We use the notations x and $x(t)$ interchangeably according to the context. The continuous-time Fourier transform (CTFT) of a signal $x(t)$ is denoted by $X(\omega)$ and is defined by

$$X(\omega) = \int_{-\infty}^{\infty} x(t)e^{-j\omega t} dt. \tag{8.1}$$

Similarly, the discrete-time Fourier transform (DTFT) of a sequence $c[n]$ is denoted by $C(e^{j\omega})$ and is defined as

$$C(e^{j\omega}) = \sum_{n=-\infty}^{\infty} c[n]e^{-j\omega n}. \tag{8.2}$$

The inner product between vectors $x, y \in \mathcal{H}$ is denoted $\langle x, y \rangle$, and is linear in the second argument; $\|x\|^2 = \langle x, x \rangle$ is the squared norm of x. The ℓ_2 norm of a sequence is $\|c\|^2 = \sum_n |c[n]|^2$. The orthogonal complement of a subspace $\mathcal{A}$ is denoted by $\mathcal{A}^\perp$. A direct sum between two closed subspaces $\mathcal{W}$ and $\mathcal{S}$ is written as $\mathcal{W} \oplus \mathcal{S}$, and is the sum set $\{w + s; \; w \in \mathcal{W}, s \in \mathcal{S}\}$ with the property $\mathcal{W} \cap \mathcal{S} = \{0\}$. Given an operator T, T^* is its adjoint, and $\mathcal{N}(T)$ and $\mathcal{R}(T)$ are its null space and range space, respectively.

8.2.2 Projections

A projection E in a Hilbert space $\mathcal{H}$ is a linear operator from $\mathcal{H}$ onto itself that satisfies the property

$$E^2 = E. \tag{8.3}$$

A projection operator maps the entire space $\mathcal{H}$ onto the range $\mathcal{R}(E)$, and leaves vectors in this subspace unchanged. Property (8.3) implies that every vector in $\mathcal{H}$ can be uniquely decomposed into a vector in $\mathcal{R}(E)$ and a vector in $\mathcal{N}(E)$, that is, we have the direct sum decomposition $\mathcal{H} = \mathcal{R}(E) \oplus \mathcal{N}(E)$. Therefore, a projection is completely determined by its range space and null space.

An orthogonal projection P is a Hermitian projection operator. In this case the range space $\mathcal{R}(P)$ and null space $\mathcal{N}(P)$ are orthogonal, and consequently P is completely determined by its range. We use the notation $P_{\mathcal{V}}$ to denote an orthogonal projection with range $\mathcal{V} = \mathcal{R}(P_{\mathcal{V}})$. An important property of an orthogonal projection onto a closed subspace $\mathcal{V}$ is that it maps every vector in $\mathcal{H}$ to the vector in $\mathcal{V}$ which is closest to it:

$$P_{\mathcal{V}} y = \operatorname*{argmin}_{x \in \mathcal{V}} \|y - x\|. \tag{8.4}$$

This property will be useful in a variety of different sampling scenarios.

An oblique projection is a projection operator that is not necessarily Hermitian. The notation $E_{\mathcal{A}\mathcal{S}^\perp}$ denotes an oblique projection with range space $\mathcal{A}$ and null space $\mathcal{S}^\perp$. If $\mathcal{A} = \mathcal{S}$, then $E_{\mathcal{A}\mathcal{S}^\perp} = P_{\mathcal{A}}$ [16]. The oblique projection onto $\mathcal{A}$ along $\mathcal{S}^\perp$ is the unique operator satisfying

$$E_{\mathcal{A}\mathcal{S}^\perp} a = a \text{ for any } a \in \mathcal{A};$$

$$E_{\mathcal{A}\mathcal{S}^\perp} s = 0 \text{ for any } s \in \mathcal{S}^\perp. \tag{8.5}$$

Projections can be used to characterize the pseudo-inverse of a given operator. Specifically, let T be a bounded operator with closed range. The Moore–Penrose pseudo-inverse of T, denoted $T^\dagger$, is the unique operator satisfying [17]:

$$\mathcal{N}(T^\dagger) = \mathcal{R}(T)^\perp,$$

$$\mathcal{R}(T^\dagger) = \mathcal{N}(T)^\perp,$$

$$TT^\dagger x = x \quad \forall x \in \mathcal{R}(T). \tag{8.6}$$

The following is a set of properties of the pseudo-inverse operator, which will be used extensively throughout the chapter [17].

LEMMA 8.1 *Let T be a bounded operator with closed range. Then:*

1. $P_{\mathcal{R}(T)} = T^\dagger T$.
2. $P_{\mathcal{N}(T)^\perp} = TT^\dagger$.
3. T^* *has closed range and* $(T^*)^\dagger = (T^\dagger)^*$.

8.2.3   Frames

As we will see in the sequel, sampling can be viewed as the process of taking inner products of a signal x with a sequence of vectors $\{a_n\}$. To simplify the derivations associated with such sequences, we use the notion of set transforms.

DEFINITION 8.1   *A set transformation* $A : \ell_2 \rightarrow \mathcal{H}$ *corresponding to vectors* $\{a_n\}$ *is defined by* $Ab = \sum_n b[n]a_n$ *for all* $b \in \ell_2$. *From the definition of the adjoint, if* $c = A^*x$, *then* $c[n] = \langle a_n, x \rangle$.

Note that a set transform A corresponding to vectors $\{a_n\}_{n=1}^{N}$ in $\mathbb{R}^M$ is simply an $M \times N$ matrix whose columns are $\{a_n\}_{n=1}^{N}$.

To guarantee stability of the sampling theorems we develop, we concentrate on vector sets that generate frames [18, 19].

DEFINITION 8.2   *A family of vectors* $\{a_n\}$ *in a Hilbert space* $\mathcal{H}$ *is called a frame for a subspace* $\mathcal{A} \subseteq \mathcal{H}$ *if there exist constants* $\alpha > 0$ *and* $\beta < \infty$ *such that the associated set transform* A *satisfies*

$$\alpha \|x\|^2 \leq \|A^*x\|^2 \leq \beta \|x\|^2, \quad \forall x \in \mathcal{A}. \tag{8.7}$$

The norm in the middle term is the ℓ_2 *norm of sequences.*

The lower bound in (8.7) ensures that the vectors $\{a_n\}$ span $\mathcal{A}$. Therefore, the number of frame elements, which we denote by N, must be at least as large as the dimension of $\mathcal{A}$. If $N < \infty$, then the right-hand inequality of (8.7) is always satisfied with $\beta = \sum_n \|a_n\|^2$. Consequently, any finite set of vectors that spans $\mathcal{A}$ is a frame for $\mathcal{A}$. For an infinite set of frame vectors $\{a_n\}$, condition (8.7) ensures that the sum $x = \sum_n b[n]a_n$ converges for any sequence $b \in \ell_2$ and that a small change in the expansion coefficients b results in a small change in x [20]. Similarly, a slight perturbation of x will entail only a small change in the inner products with the frame elements.

8.3   Sampling and reconstruction setup

We are now ready to introduce the sampling and reconstruction setup that will be studied in this chapter. Before we elaborate on the abstract Hilbert-space exposition that will be at the heart of our derivations to follow, we first review the famous Shannon sampling theorem [1].

Shannon's theorem states that a signal $x(t)$ bandlimited to π/T can be recovered from its uniform samples at time instants nT. Reconstruction is obtained by filtering the samples with a sinc interpolation kernel:

$$\hat{x}(t) = \frac{1}{T} \sum_{n=-\infty}^{\infty} x(nT)\operatorname{sinc}(t/T - n), \tag{8.8}$$

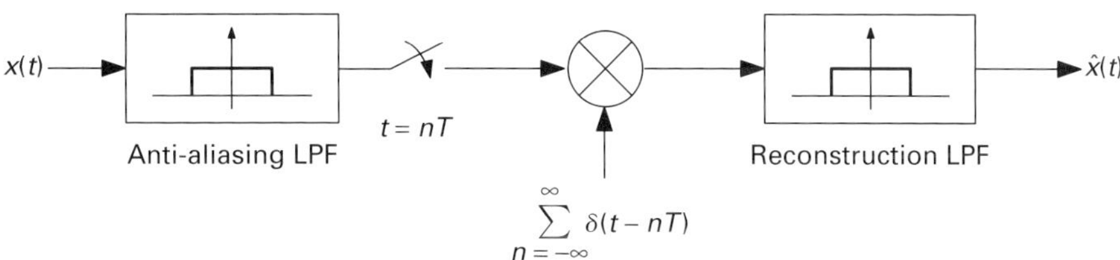

Figure 8.1 Shannon's sampling paradigm. The signal $x(t)$ passes through an ideal LPF prior to sampling. Reconstruction is obtained using the same LPF.

where $\mathrm{sinc}(t) = \sin(\pi t)/(\pi t)$. If the bandwidth of $x(t)$ exceeds π/T, then an anti-aliasing low-pass filter (LPF) with cutoff π/T can be used prior to sampling, as shown in Figure 8.1. In this case, the reconstruction formula (8.8) produces the best approximation (in an L_2 sense) to $x(t)$ within the class of π/T-bandlimited functions. To see this, note that an LPF with unit magnitude in its pass-band satisfies (8.3) and is thus a projection operator. Furthermore, the range space of such a filter comprises all signals whose CTFT vanishes outside $[-\pi/T, \pi/T]$, and its null space is the set of signals that vanish in $[-\pi/T, \pi/T]$, which is orthogonal to the range. Therefore, the anti-aliasing filter is an *orthogonal* projection. Property (8.4) then implies that its output is the best approximation to its input within all signals in its range, namely π/T-bandlimited signals.

Shannon's theorem contains four fundamental aspects that are important in any sampling theorem:

1. *Prior knowledge.* The input lies in the class of π/T-bandlimited signals;
2. *Sampling mechanism.* Pre-filtering with an LPF with cutoff π/T, followed by pointwise sampling;
3. *Reconstruction method.* Sinc interpolation kernel modulated by the sample values;
4. *Objective.* Minimization of the L_2 norm of the error $x(t) - \hat{x}(t)$.

These specific choices of prior knowledge, sampling mechanism, reconstruction method, and objective are often not met in practical scenarios. First, natural signals are rarely truly bandlimited. Second, the sampling device is usually not ideal, that is, it does not produce exact signal values at the sampling locations. A common situation is that the analog-to-digital converter (ADC) integrates the signal, usually over small neighborhoods surrounding the sampling points. Moreover, in many applications the samples are contaminated by noise due to quantization and other physical effects. Third, the use of the sinc kernel for reconstruction is often impractical because of its slow decay. Finally, when considering signal priors which are richer than the bandlimited assumption, it is usually impossible to minimize the error norm uniformly over all feasible signals. Therefore other criteria must be considered.

In this chapter we treat each of these essential components of the sampling scheme, focusing on several models which commonly arise in signal-processing, image-processing, and communication systems. For simplicity, throughout the chapter we assume a sampling period of $T = 1$. We next elaborate on the signal priors and general sampling and reconstruction processes we treat.

8.3.1   Signal priors

In essence, the Shannon sampling theorem states that if $x(t)$ is known a priori to lie in the space of bandlimited signals, then it can be perfectly recovered from ideal uniformly-spaced samples. Clearly, the question of whether $x(t)$ can be recovered from its samples depends on the prior knowledge we have on the class of input signals. In this chapter we depart from the traditional bandlimited assumption and discuss signal priors that appear more frequently in signal-processing and communication scenarios.

Subspace priors

Our first focus is on cases where the signal $x(t)$ is known to lie in a subspace $\mathcal{A}$, spanned by vectors $\{a_n\}$. Although the discussion in this chapter is valid for a wide class of such subspaces, we take special interest in subspaces of L_2 that are *shift invariant* (SI). A SI subspace $\mathcal{A}$ of L_2, is a space of signals that can be expressed as linear combinations of shifts of a generator $a(t)$ [7]:

$$x(t) = \sum_{n=-\infty}^{\infty} d[n]a(t - n), \tag{8.9}$$

where $d[n]$ is an arbitrary norm-bounded sequence. Note that $d[n]$ does not necessarily correspond to samples of the signal, that is, we can have $x(n) \neq d[n]$. More generally, $\mathcal{A}$ may be generated by several generators $a_k(t)$ so that $x(t) = \sum_{k=1}^{K} \sum_{n=-\infty}^{\infty} d_k[n]a_k(t - n)$. For simplicity we focus here on the single-generator case. Using set-transform formulation, (8.9) can be written compactly as

$$x = Ad, \tag{8.10}$$

where A is the set transform associated with the functions $\{a(t - n)\}$.

Choosing $a(t) = \mathrm{sinc}(t)$ in (8.9) results in the space of π-bandlimited signals. However, a much broader class of signal spaces can be defined, including spline functions [21]. In these cases $a(t)$ may be easier to handle numerically than the sinc function.

A popular choice of SI spaces in many image-processing applications is the class of splines. A spline $f(t)$ of degree N is a piecewise polynomial with the pieces combined at knots, such that the function is continuously differentiable $N - 1$ times. It can be shown that any spline of degree N with knots at the integers can be generated using (8.9) by a B-spline of degree N, denoted $\beta_N(t)$. The latter is the function obtained by the $(N + 1)$-fold convolution of the unit square

$$\beta_0(t) = \begin{cases} 1 & -\tfrac{1}{2} < t < \tfrac{1}{2}; \\ 0 & \text{otherwise.} \end{cases} \tag{8.11}$$

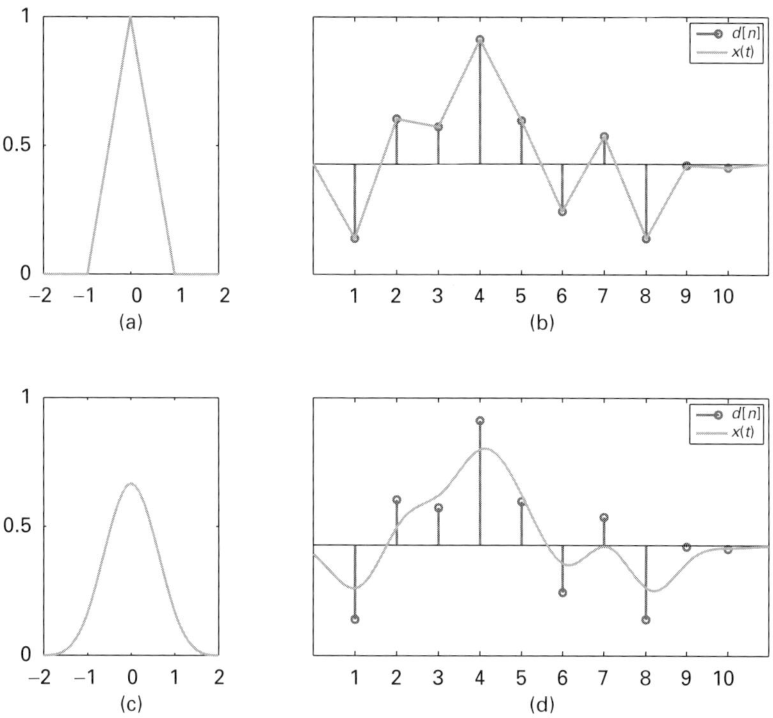

Figure 8.2 Spline functions of different orders generated using (8.9) with the same sequence $d[n]$. (a) $\beta_1(t)$. (b) A spline of degree 1. (c) $\beta_3(t)$. (d) A spline of degree 3.

As demonstrated in Figure 8.2, the sequence $d[n]$ in (8.9) is not equal to the samples $x(n)$ for splines of order greater than 1.

An important generalization of the SI subspace prior (8.9) is the class of signals that lie in a *union* of SI spaces. In this case,

$$x(t) = \sum_{k=1}^{K} \sum_{n=-\infty}^{\infty} d_k[n] a_k(t-n), \tag{8.12}$$

for a set of generators $a_k(t)$ where only $M < K$ out of the sequences $d_k[n]$ are not identically zero. However, we do not know in advance which M are chosen. This model can be used, for example, to describe multiband signals whose total number of active bands is small compared to the Nyquist rate [22,23]. The techniques developed to sample and reconstruct such classes of signals are based on ideas and results from the emerging field of compressed sensing [24, 25]. However, while the latter deals with sampling of finite vectors, the multiband problem is concerned with analog sampling. By using several tools, developed in more detail in [23,26,27], it is possible to extend the essential ideas of compressed sensing to the analog domain. These results can also be applied more generally to signals that lie in a union of subspaces [28, 29], which are not necessarily shift invariant. Unlike subspace priors, nonlinear techniques are required in order to recover signals of this type. Therefore, for simplicity we will confine the discussion in this chapter to the single-subspace case.

Smoothness priors

Subspace priors are very useful because, as we will see, they often can be used to perfectly recover $x(t)$ from its samples. However, in many practical scenarios our knowledge about the signal is much less complete and can only be formulated in very general terms. An assumption prevalent in image and signal processing is that natural signals are smooth in some sense. Here we focus on approaches that quantify the extent of smoothness using the L_2 norm $\|Lx\|$, where L is a linear operator. Specifically, we assume that

$$\|Lx\| \leq \rho \tag{8.13}$$

for some finite constant $\rho > 0$. It is common to use SI smoothness priors, resulting in a linear time-invariant (LTI) operator L. In these cases the corresponding filter $L(\omega)$ is often chosen to be a first- or second-order derivative in order to constrain the solution to be smooth and nonoscillating, that is, $L(\omega) = a_0 + a_1 j\omega + a_2 (j\omega)^2 + \cdots$ for some constants a_n. Another common choice is the filter $L(\omega) = (a_0^2 + \omega^2)^\gamma$ with some parameter γ. The latter is in use mainly in image-processing applications. The appeal of the model (8.13) stems from the fact that it leads to linear-recovery procedures. In contrast, smoothness measures such as total variation [30], result in nonlinear-interpolation techniques.

The class of smooth signals is much richer than its subspace counterpart. Consequently, it is often impossible to distinguish between one smooth signal and another based solely on their samples. In other words, in contrast to subspace priors, perfect reconstruction is typically impossible under the smoothness assumption. Instead, we develop recovery algorithms that attempt to best approximate a smooth signal from the given samples.

8.3.2   Sampling process

We now present the general sampling process we treat. As we have seen, in the Shannon sampling theorem $x(t)$ is filtered with an LPF with cutoff π prior to sampling. In practical applications the sampling is not ideal. Therefore, a more realistic setting is to let the anti-aliasing filter, which we denote by $s(-t)$, be an arbitrary sampling function, as depicted in Figure 8.3. This allows us to incorporate imperfections in the ideal sampler into the

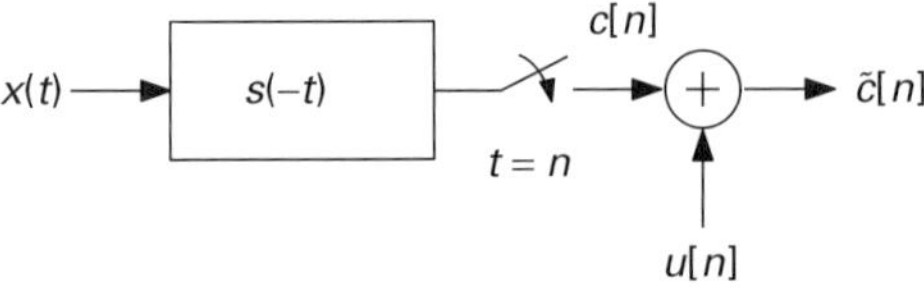

Figure 8.3   Shift-invariant sampling. Filtering the signal $x(t)$ prior to taking ideal and uniform samples, can be interpreted as L_2 inner-products between $x(t)$ and shifts of $s(t)$. In practical applications the samples are contaminated by noise $u[n]$.

function $s(t)$ [5, 8, 15, 31, 32]. As an example, typical ADCs average the signal over a small interval rather than outputting pointwise signal values. This distortion can be taken into account by modifying $s(t)$ to include the integration.

The samples $c[n]$ can be expressed as

$$c[n] = \int_{-\infty}^{\infty} x(t)s(t-n)dt = \langle s_n, x \rangle, \tag{8.14}$$

where $s_n(t) = s(t-n)$. More generally, we treat the scenario in which the samples $c[n]$ are obtained as inner products with a set of arbitrary functions $\{s_n\}$. Using set-transform notation, the samples can be written as

$$c = S^* x, \tag{8.15}$$

where S is the set transform corresponding to the sampling vectors $\{s_n\}$ and c is the sequence whose nth element is $c[n]$.

To ensure that the sampling process is stable, we concentrate on the case in which the vectors $\{s_n\}$ form a frame for their span, which we term the *sampling space* $\mathcal{S}$. It follows immediately from the upper bound in (8.7) that the sequence of samples $c[n] = \langle s_n, x \rangle$ is then in ℓ_2 for any signal x that has bounded norm.

In the case where $\mathcal{S}$ is an SI space, condition (8.7) can be stated in terms of $S(\omega)$, the CTFT of the generator $s(t)$. Specifically, it can be shown that the functions $\{s(t-n)\}$ generate a frame if, and only if

$$\alpha \le \phi_{SS}\left(e^{j\omega}\right) \le \beta, \quad \omega \in \mathcal{I}_\mathcal{S}, \tag{8.16}$$

for some constants $\alpha > 0$ and $\beta < \infty$. Here,

$$\phi_{SS}\left(e^{j\omega}\right) = \sum_{k=-\infty}^{\infty} |S(\omega - 2\pi k)|^2 \tag{8.17}$$

is the DTFT of the sampled correlation function $r_{ss}[n] = \langle s(t), s(t-n) \rangle$ and $\mathcal{I}_\mathcal{S}$ is the set of frequencies ω for which $\phi_{SS}(e^{j\omega}) \ne 0$ [33]. It is easy to see that $s(t) = \mathrm{sinc}(t)$ satisfies (8.16). Furthermore, B-splines of all orders also satisfy (8.16) [21].

In many situations the samples are perturbed by the sampling device, for example due to quantization or noise. Thus, as shown in Figure 8.3, one usually only has access to the modified samples

$$\tilde{c}[n] = c[n] + u[n], \tag{8.18}$$

where $u[n]$ is a discrete-time noise process. Clearly, the noise should be taken into consideration when designing a reconstruction algorithm.

Another setup which was treated recently is that of reconstructing a signal which has undergone nonlinear distortion prior to sampling [34]. Using optimization tools, together with frame-perturbation theory, it can be shown that under a subspace prior and several technical conditions the signal can be reconstructed perfectly despite the nonlinearity. In this chapter, we focus on linear sampling and, therefore, do not survey these results.

8.3.3   Reconstruction method

The problem at the heart of sampling theory is how to reconstruct a signal from a given set of samples. For a sampling theorem to be practical, it must take into account constraints that are imposed on the reconstruction method. One aspect of the Shannon sampling theorem, which renders it unrealizable, is the use of the sinc interpolation kernel. Due to its slow decay, the evaluation of $x(t)$ at a certain time instant t_0, requires using a large number of samples located far away from t_0. In many applications, reduction of computational load is achieved by employing much simpler methods, such as nearest-neighbor interpolation. In these cases the sampling scheme should be modified to compensate for the chosen nonideal kernel.

In this chapter we study two interpolation strategies: unconstrained and constrained reconstruction. In the former, we pose no limitation on the interpolation algorithm. The goal then is to extend Shannon's theorem to more general classes of input signals. In the latter strategy, we restrict the reconstruction to be of a predefined form in order to reduce computational load. Here too, we address a variety of input signals.

Unconstrained reconstruction

The first setup we treat is unconstrained recovery. Here, we design reconstruction methods that are best adapted to the underlying signal prior according to an appropriately defined objective, without restricting the reconstruction mechanism. In these scenarios, it is sometimes possible to obtain perfect recovery, as in the Shannon sampling theorem. When both the sampling process and signal prior are SI, the unconstrained reconstruction methods under the different scenarios treated in this chapter all have a common structure, depicted in Figure 8.4. Here $w(t)$ is the impulse response of a continuous-time filter, which serves as the interpolation kernel, while $h[n]$ represents a discrete-time filter used to process the samples prior to reconstruction. Denoting the output of the discrete-time filter by $d[n]$, the input to the analog filter $w(t)$ is a modulated impulse

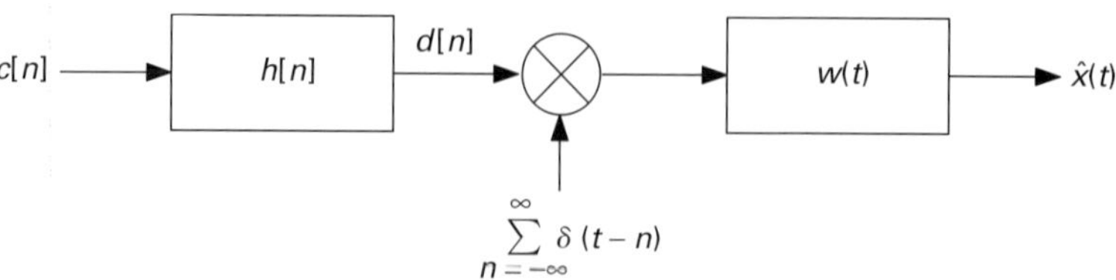

Figure 8.4   Reconstruction using a digital compensation filter $h[n]$ and interpolation kernel $w(t)$.

train $\sum_n d[n]\delta(t-n)$. The filter's output is given by

$$\hat{x}(t) = \sum_{n=-\infty}^{\infty} d[n]w(t-n). \tag{8.19}$$

If either the sampling process or the prior are not SI, then the reconstruction has the more general form $\hat{x} = \sum_n d[n]w_n$, where $d[n]$ is obtained by a linear transform of $c[n]$. Using set-transform notation, the recovered signal can be written as

$$\hat{x} = Wd = WHc, \tag{8.20}$$

where W is the set transform corresponding to the reconstruction vectors $\{w_n\}$ and H is a linear operator. Our goal is to determine both the reconstruction vectors $\{w_n\}$ and the transform H to be optimal in some sense.

We will see that in the SI case, optimal interpolation kernels are typically derived in the frequency domain but very often do not admit a closed-form in the time domain. This limits the applicability of these recovery techniques. One way to resolve this problem is to choose the signal prior so as to yield an efficient interpolation algorithm, as done, for example, in [12] in the context of exponential B-splines. Nevertheless, this approach restricts the type of priors that can be handled.

Constrained reconstruction

To overcome the difficulties in implementing the unconstrained solutions, we may resort to a system that uses a predefined interpolation kernel $w(t)$ that is easy to implement. For example, in image-processing applications kernels with small supports are often used. These include nearest neighbor (B-spline of order 0), bilinear (B-spline of order 1), bicubic (Keys kernel [35]), Lanczos, and B-spline of order 3. In this setup, the only freedom is in the design of the digital correction filter $h[n]$ in Figure 8.4, which may be used to compensate for the nonideal behavior of the pre-specified kernel [5, 8, 9, 32, 36]. The filter $h[n]$ is selected to optimize a criterion matched to the signal prior.

As a simple example demonstrating the need for constraining the reconstruction, consider the situation where we are given pointwise samples of a π-bandlimited signal and our goal is to compute $x(t_0)$ for some non-integer t_0. The sinc-interpolation formula (8.8) uses the entire sequence of samples $c[n]$, since $\mathrm{sinc}(t)$ is not compactly supported. To reduce computational load, we can replace the sinc kernel, for instance, by $\beta_3(t)$ (a B-spline of degree 3). Since the support of the latter is $[-2, 2]$, the approximation (8.19) includes only 4 summands per time instance t_0. In this example, however, using the sample values $c[n]$ as the expansion coefficients $d[n]$ in (8.19) is not desired, as demonstrated in Figure 8.2(d). To obtain a good approximation, a digital processing step is required prior to reconstruction, as depicted in Figure 8.4.

More generally, when the signal is reconstructed using an arbitrary given set of vectors $\{w_n\}$, the goal is to design the correction transform H in (8.20), which will usually not correspond to digital filtering. We restrict attention to the case where $\{w_n\}$ form a frame for their span $\mathcal{W}$, which we call the *reconstruction space*. By restricting the reconstruction to the form $\hat{x} = Wd$, we are essentially imposing that the recovered signal $\hat{x}$ lie in the predefined space $\mathcal{W}$. This space can be chosen so as to lead to highly efficient interpolation methods. For example, by appropriate choice of a generator $w(t)$, the family of splines can be described using (8.19) [37–39].

8.4   Optimization methods

The fundamental problem we wish to address in this chapter is the following. Given the (noisy) samples of a signal $\tilde{c} = S^*x + u$ and some prior knowledge of the form $x \in \mathcal{A}$, produce a reconstruction $\hat{x}$ that is close to x in some sense. The set $\mathcal{A}$ incorporates our knowledge about the typical input signals and can be a subspace, as in (8.10), or an ellipsoid, as in (8.13).

Assuming that the noise u is known to be norm bounded, the samples $\tilde{c}$ together with the set $\mathcal{A}$ can be used to determine the set of feasible signals:

$$\mathcal{G} = \{x : x \in \mathcal{A}, \|S^*x - \tilde{c}\| \leq \alpha\}. \tag{8.21}$$

Thus, the unknown signal lies in $\mathcal{G}$. To find a good approximation to x in $\mathcal{G}$, it is important to notice that the reconstruction error $\|\hat{x} - x\|$ of any recovery method generally depends on the unknown signal x. This renders comparison between different methods difficult, as one method may be better than another for certain input signals and worse for others. In other words, it is generally impossible to minimize the error uniformly over $\mathcal{G}$. The same phenomenon occurs in the case where the noise u is random and the goal is to minimize the mean-square-error (MSE) over the set $\mathcal{A}$ [40]. Two approaches to deal with this dependency are *least-squares* (LS) and *worst-case* (minimax) design.

In the LS strategy, the reconstruction error $\|\hat{x} - x\|$ is replaced by the error-in-samples objective $\|S^*\hat{x} - \tilde{c}\|$. This approach seeks a signal $\hat{x}$ that produces samples as close as possible to the measured samples $\tilde{c}$:

$$\hat{x}_{\mathrm{LS}} = \underset{x \in \mathcal{G}}{\operatorname{argmin}} \|S^*x - \tilde{c}\|^2. \tag{8.22}$$

The objective in (8.22) is convex (quadratic) in x and therefore, if $\mathcal{G}$ is a convex set, as we assume throughout the chapter, then the problem is convex. Furthermore, the LS reconstruction admits a closed-form solution for many interesting priors. Due to its simplicity, this criterion is widely used in inverse problems in general, and in sampling in particular [11, 12]. However, it is important to note that there are situations where minimization of the error-in-samples leads to a large reconstruction error. This happens, for example, when S is such that large perturbations in x lead to small perturbations in S^*x. Therefore, this method does not guarantee a small recovery error.

An alternative to the LS approach is worst-case (or minimax) design [14, 32, 40–42]. This method attempts to control the estimation error by minimizing its largest possible value. Since x is unknown, we seek the reconstruction $\hat{x}$ that minimizes the error for the worst feasible signal:

$$\hat{x}_{\text{MX}} = \arg\min_{\hat{x}} \max_{x \in \mathcal{G}} \|\hat{x} - x\|^2. \tag{8.23}$$

In contrast to (8.22), here we attempt to directly control the reconstruction error $\|x - \hat{x}\|$, which is the quantity of interest in many applications. Problem (8.23), however, is more challenging than (8.22), as we discuss next.

There are several possible approaches to solving minimax problems. In convex–concave problems we can replace the order of the minimization and the maximization [43], as incorporated in the following proposition.

PROPOSITION 8.1 *Let $\mathcal{X}$ and $\mathcal{Y}$ be convex compact sets, and let $f(x, y)$ be a continuous function which is convex in $x \in \mathcal{X}$ for every fixed $y \in \mathcal{Y}$ and concave in $y \in \mathcal{Y}$ for every fixed $x \in \mathcal{X}$. Then,*

$$\min_{x \in \mathcal{X}} \max_{y \in \mathcal{Y}} f(x, y) = \max_{y \in \mathcal{Y}} \min_{x \in \mathcal{X}} f(x, y),$$

and we can replace the order of the minimization and the maximization.

There are many variants of Proposition 8.1 under weaker conditions. In particular, it is sufficient that only one of the sets will be compact, and convexity may be replaced by quasi-convexity. In the case in which the minimization is easy to solve, the problem reduces to a convex-optimization problem of maximizing a concave function.

Unfortunately, since the objective in (8.23) is convex in both x and $\hat{x}$, we cannot employ Proposition 8.1 to solve it. An alternative strategy is to establish a lower bound on the objective and find a vector $\hat{x}$, which is not a function of x, that achieves it. Specifically, suppose that we show that $\max_{x \in \mathcal{G}} \|\hat{x} - x\|^2 \geq \max_{x \in \mathcal{G}} g(x)$ for all $\hat{x}$, where $g(x)$ is some function of x. Then every reconstruction $\hat{x}$, which is not a function of x, that achieves the bound is a solution. Although this approach is not constructive, when applicable, it leads to a closed-form solution to the infinite-dimensional minimax problem (8.23). The minimax problems of Sections 8.5 and 8.6 will be treated using this strategy.

Another approach to solve minimax problems is to replace the inner maximization by its dual function when strong duality holds. This will result in a minimization problem that can be combined with the outer minimization. In order to follow this method, we need to be able to establish strong duality. The inner maximization in (8.23) is a non-convex, constrained quadratic-optimization problem. The nonconvexity is the result of the fact that we are maximizing a convex quadratic function, rather than minimizing it. Fortunately, we will see that under the signal priors considered here, strong duality exists in some cases [44]. In other cases, the dual problem leads to an upper bound on the objective of the inner maximization, and can be used to approximate the solution. The drawback of this route is that the resulting optimization problem usually does not admit a closed-form solution and must be solved numerically. This limits this method to finite-dimensional problems. The minimax problems of Section 8.8, which involve noisy samples, will be approached using this strategy.

8.5   Subspace priors

Our first focus is on cases in which the signal $x(t)$ is known to lie in a subspace $\mathcal{A}$ spanned by frame vectors $\{a_n\}$. Given a sequence of measurements $c[n] = \langle s_n, x \rangle$, namely $c = S^*x$, where the vectors $\{s_n\}$ form a frame for the sampling space $\mathcal{S}$, our goal is to produce a reconstruction $\hat{x}$ that best approximates x in some sense.

8.5.1   Unconstrained reconstruction

We begin the discussion with the case where no constraints are imposed on the reconstruction $\hat{x}$. Interestingly, we will see that in this setting the minimax and LS solutions coincide.

Least-squares recovery

As explained in Section 8.4, in the LS strategy, the reconstruction error $\|\hat{x} - x\|$ is replaced by the error-in-samples objective $\|S^*\hat{x} - c\|$. Taking into account the prior knowledge that $x \in \mathcal{A}$, the LS recovery method can be written as

$$\hat{x}_{\mathrm{LS}} = \operatorname*{argmin}_{x \in \mathcal{A}} \|S^*x - c\|^2. \tag{8.24}$$

By assumption, there exists a signal $x \in \mathcal{A}$ for which $c = S^*x$. Therefore, the optimal value in (8.24) is 0. The signal x attaining this optimum is not unique if $\mathcal{A} \cap \mathcal{S}^\perp \neq \{0\}$. Indeed, suppose that x is a nonzero signal in $\mathcal{A} \cap \mathcal{S}^\perp$. Then $c[n] = \langle s_n, x \rangle = 0$ for all n, and clearly x cannot be reconstructed from the measurements $c[n]$. A sufficient condition ensuring the uniqueness of the solution is that $\mathcal{A} \cap \mathcal{S}^\perp = \{0\}$ and that the Hilbert space $\mathcal{H}$ of signals can be decomposed as [45]

$$\mathcal{H} = \mathcal{A} \oplus \mathcal{S}^\perp. \tag{8.25}$$

When this condition holds, we can perfectly recover x from the samples c. This condition can be easily verified in SI spaces, as we discuss below.

Since (8.24) is defined over an infinite-dimensional Hilbert space, to solve it we do not use standard techniques such as setting the derivative of the Lagrangian to 0. Instead, we rely on the properties of the relevant spaces. Specifically, to determine the set of optimal solutions to (8.24), we express x in terms of its expansion coefficients in $\mathcal{A}$. Writing $x = \sum d[n]a_n = Ad$, the optimal sequence d is the solution to

$$\hat{d}_{\mathrm{LS}} = \operatorname*{argmin}_{d} \|S^*Ad - c\|^2. \tag{8.26}$$

The set of solutions to this optimization problem is given in the following theorem.

THEOREM 8.1   *Every solution to (8.26) is of the form*

$$\hat{d}_{LS} = (S^*A)^\dagger c + v, \tag{8.27}$$

where v is some vector in $\mathcal{N}(S^*A)$. Furthermore, the minimal-norm solution is given by $\hat{d} = (S^*A)^\dagger c$.

Before proving the theorem, we first need to verify that the pseudo-inverse is well defined. If S and A have finite dimensions, say M and N respectively, then S^*A corresponds to an $M \times N$ matrix and $(S^*A)^\dagger$ is trivially a bounded operator. However, this is not necessarily true for infinite-dimensional operators. Fortunately, the fact that S and A are synthesis operators of frames, guarantees that $(S^*A)^\dagger$ is bounded, as stated in the next proposition.

PROPOSITION 8.2 *Let S and A be set transformations corresponding to frames $\{s_n\}$ and $\{a_n\}$, respectively. Then $(S^*A)^\dagger$ is a bounded operator.*

Proof The proof of the proposition relies on the fact that the pseudo-inverse of an operator is well defined if its range is closed. In other words, we need to show that every Cauchy sequence $c_n \in \mathcal{R}(S^*A)$ converges to a limit in $\mathcal{R}(S^*A)$. This can be established by using the lower frame bound of S, the fact that $\mathcal{R}(A)$ and $\mathcal{R}(S^*)$ are closed, and that S^* is continuous. $\blacksquare$

We now prove Theorem 8.1.

Proof of Theorem 8.1 To see that the set of solutions to (8.26) is given by (8.27), we substitute $\hat{d}_{\mathrm{LS}}$ of (8.27) into the objective of (8.26):

$$
\begin{aligned}
S^*A\hat{d}_{\mathrm{LS}} - c &= (S^*A)((S^*A)^\dagger c + v) - c \\
&= (S^*A)(S^*A)^\dagger c - c \\
&= P_{\mathcal{R}(S^*A)}c - c = 0.
\end{aligned}
\tag{8.28}
$$

The second equality follows from the fact that $v \in \mathcal{N}(S^*A)$, the third equality follows from Lemma 8.1, and the last equality is a result of $c \in \mathcal{R}(S^*A)$, since $c = S^*x$ for some $x \in \mathcal{A}$. Therefore, every vector described by (8.27) attains an optimal value of 0 in (8.26). It is also clear from (8.28) that any vector of the form $\hat{d}_{\mathrm{LS}} + w$, where $\hat{d}_{\mathrm{LS}}$ is given by (8.27) and $w \in \mathcal{N}(S^*A)^\perp$, is not a solution to (8.26). Thus, d solves (8.26) if, and only if it is of the form (8.27).

Among all solutions, the one with minimal norm is given by $\hat{d}_{\mathrm{LS}} = (S^*A)^\dagger c$. This follows from the fact that $(S^*A)^\dagger c \in \mathcal{N}(S^*A)^\perp$ by definition of the pseudo-inverse, and v lies in $\mathcal{N}(S^*A)$. The Pythagorean theorem therefore implies that $\|(S^*A)^\dagger c + v\|^2 = \|(S^*A)^\dagger c\|^2 + \|v\|^2 > \|(S^*A)^\dagger c\|^2$ for any nonzero v. $\blacksquare$

In the sequel, we take interest only in the minimal-norm solution. From Theorem 8.1, the LS recovery method with minimal-norm amounts to applying the transformation

$$
H = (S^*A)^\dagger
\tag{8.29}
$$

to the samples c to obtain a sequence of expansion coefficients d. This sequence is then used to synthesize $\hat{x}$ via $\hat{x} = \sum_n d[n]a_n$. Thus, $\hat{x}$ is related to x by

$$\hat{x}_{\mathrm{LS}} = Ad = AHc = A(S^*A)^{\dagger}c = A(S^*A)^{\dagger}S^*x. \tag{8.30}$$

In settings where the solution is unique, namely when $\mathcal{A} \cap \mathcal{S}^{\perp} = \{0\}$ and (8.25) holds, the LS strategy leads to perfect recovery of x. This has a simple geometric interpretation. It is easily verified that $\mathcal{N}(S^*A) = \mathcal{N}(A)$ in this case [45]. Consequently, $A(S^*A)^{\dagger}S^*$ is an oblique projection with range $\mathcal{A}$ and null space $\mathcal{S}^{\perp}$, denoted by $E_{\mathcal{AS}^{\perp}}$. To see this, note that every $x \in \mathcal{A}$ can be written as $x = Ad$ for some $d \in \ell_2$ and therefore

$$A(S^*A)^{\dagger}S^*x = A(S^*A)^{\dagger}S^*Ad = AP_{\mathcal{N}(S^*A)^{\perp}}d = AP_{\mathcal{N}(A)^{\perp}}d = Ad = x. \tag{8.31}$$

On the other hand, for any $x \in \mathcal{S}^{\perp}$ we have $A(S^*A)^{\dagger}S^*x = 0$. Thus, $\hat{x} = E_{\mathcal{AS}^{\perp}}x = x$ for any $x \in \mathcal{A}$, which implies that the LS reconstruction (8.30) coincides with the original signal x. As a special case, if $\mathcal{A} = \mathcal{S}$ then the LS solution reduces to an orthogonal projection $P_{\mathcal{S}} = S(S^*S)^{\dagger}S^*$.

The geometric explanation of $\hat{x} = E_{\mathcal{AS}^{\perp}}x$ follows from the fact that knowing the samples c is equivalent to knowing the orthogonal projection of the signal onto the sampling space, since $P_{\mathcal{S}}x = S(S^*S)^{\dagger}S^*x = S(S^*S)^{\dagger}c$. The direct-sum condition (8.25) ensures that there is a unique vector in $\mathcal{A}$ with the given projection onto $\mathcal{S}$. As depicted in Figure 8.5, in this case we can draw a vertical line from the projection until we hit the space $\mathcal{A}$, and in such a way obtain the unique vector in $\mathcal{A}$ that is consistent with the given samples. Therefore, under (8.25), x can be perfectly recovered from c by using H of (8.29). To conclude, we see that if $\mathcal{A} \cap \mathcal{S}^{\perp} = \{0\}$ then the LS approach leads to perfect recovery. If the intersection is non-trivial, on the other hand, then clearly perfect reconstruction is not possible since there are generally infinitely many signals in $\mathcal{A}$ yielding the same samples $c = S^*x$.

We now turn our attention to the case in which $\mathcal{A}$ and $\mathcal{S}$ are SI spaces with generators $a(t)$ and $s(t)$, respectively. In this setting, the operator S^*A corresponds to convolution with the sequence $r_{SA}[n] = (a(t) * s(-t))(n)$. To see this, let $c = S^*Ad$. From

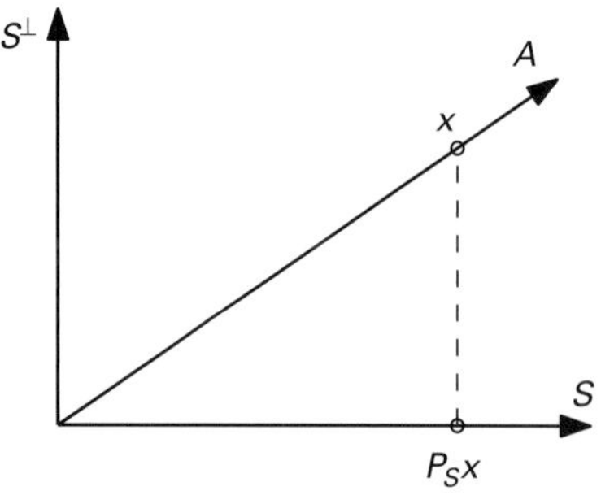

Figure 8.5   A unique element in $\mathcal{A}$ which is consistent with the samples in $\mathcal{S}$ can be recovered from the known samples.

Definition 8.1,

$$c[k] = \int_{-\infty}^{\infty} s(t-k) \sum_{n=-\infty}^{\infty} d[n]a(t-n)dt$$

$$= \sum_{n=-\infty}^{\infty} d[n](a(t) * s(-t))(k-n)$$

$$= (d[n] * r_{SA}[n])[k]. \tag{8.32}$$

Therefore, the correction $H = (S^*A)^{\dagger}$ is a digital filter with frequency response

$$H(e^{j\omega}) = \begin{cases} \frac{1}{\phi_{SA}(e^{j\omega})}, & \phi_{SA}\left(e^{j\omega}\right) \neq 0; \\ 0, & \phi_{SA}\left(e^{j\omega}\right) = 0, \end{cases} \tag{8.33}$$

where $\phi_{SA}(e^{j\omega})$ is the DTFT of $r_{SA}[n]$, and is given by

$$\phi_{SA}(e^{j\omega}) = \sum_{k=-\infty}^{\infty} S^*(\omega + 2\pi k)A(\omega + 2\pi k). \tag{8.34}$$

The overall scheme fits that depicted in Figure 8.4, with $w(t) = a(t)$ and $h[n]$ given by (8.33).

The direct-sum condition (8.25) ensuring perfect recovery, can be verified easily in SI spaces. Specifically, (8.25) is satisfied, if and only if [46] the supports $\mathcal{I}_A$ and $\mathcal{I}_S$ of $\phi_{SS}(e^{j\omega})$ and $\phi_{AA}(e^{j\omega})$ respectively coincide, and there exists a constant $\alpha > 0$ such that $|\phi_{SA}(e^{j\omega})| > \alpha$, for all $\omega \in \mathcal{I}_A$.

We conclude the discussion with a non-intuitive example, in which a signal that is not bandlimited is filtered with an LPF prior to sampling, and still can be perfectly reconstructed from the resulting samples.

Example 8.1 Consider a signal $x(t)$ formed by exciting an RC circuit with a modulated impulse train $\sum_n d[n]\delta(t-n)$, as shown in Figure 8.6(a). The impulse response of the RC circuit is known to be $a(t) = \tau^{-1} \exp\{-t/\tau\}u(t)$, where $u(t)$ is the unit step function and $\tau = RC$ is the time constant. Therefore

$$x(t) = \frac{1}{\tau} \sum_{n=-\infty}^{\infty} d[n] \exp\{-(t-n)/\tau\}u(t-n). \tag{8.35}$$

Clearly, $x(t)$ is not bandlimited. Now, suppose that $x(t)$ is filtered by an ideal LPF $s(t) = \mathrm{sinc}(t)$ and then sampled at times $t = n$ to obtain the sequence $c[n]$. The signal $x(t)$ and its samples are depicted in Figure 8.6(b). Intuitively, there seems to be information loss in the sampling process since the entire frequency content of $x(t)$ outside $[-\pi, \pi]$ is zeroed out, as shown in Figure 8.6(c). However, it is easily verified that if $\tau < \pi^{-1}$, then $|\phi_{SA}(e^{j\omega})| > (1 - \pi^2\tau^2)^{-1/2} > 0$ for all $\omega \in [-\pi, \pi]$ so that condition (8.25) is

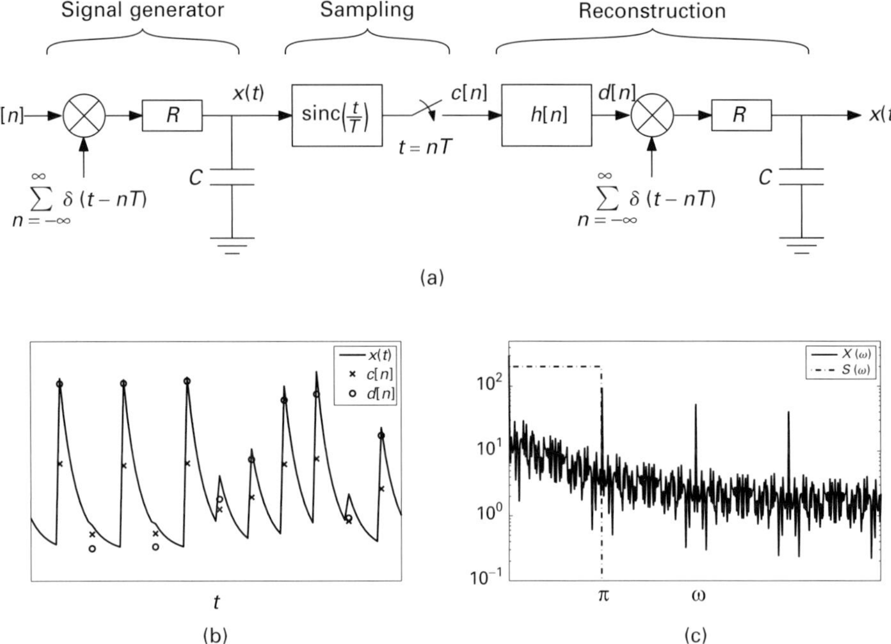

Figure 8.6   A non-bandlimited signal $x(t)$, formed by exciting an RC-circuit with a modulated impulse train, is sampled after passing through an ideal LPF and then perfectly reconstructed using the LS approach. (a) Sampling and reconstruction setup. (b) The signal $x(t)$, its samples $c[n]$ and expansion coefficients $d[n]$. (c) The signal $X(\omega)$ and the sampling filter $S(\omega)$.

satisfied. Therefore, perfect recovery is possible in this setup using the LS approach. The digital correction filter (8.33) in this case can be shown to be

$$h[n] = \begin{cases} 1 & n = 0; \\ \frac{\tau}{n}(-1)^n & n \neq 0. \end{cases} \tag{8.36}$$

Thus, to reconstruct $x(t)$ we need to excite an identical RC circuit with an impulse train modulated by the sequence $d[n] = h[n] * c[n]$. The entire sampling-reconstruction setup is depicted in Figure 8.6(a).

Minimax recovery

We now treat the recovery of x via a minimax framework:

$$\hat{x}_{\mathrm{MX}} = \arg \min_{\hat{x}} \max_{x \in \mathcal{G}} \|\hat{x} - x\|^2, \tag{8.37}$$

where $\mathcal{G}$ is the set of signals $x \in \mathcal{A}$ satisfying $S^* x = c$.

To approach this problem, notice that $\hat{x}_{\mathrm{MX}}$ must lie in $\mathcal{A}$, as any $\hat{x} \notin \mathcal{A}$ can be improved upon by projecting it onto $\mathcal{A}$: $\|\hat{x} - x\|^2 \geq \|P_{\mathcal{A}}\hat{x} - x\|^2$ for any $\hat{x}$ and $x \in \mathcal{A}$. Therefore, we can express both $\hat{x}$ and x in terms of their expansion coefficients in $\mathcal{A}$, by writing $\hat{x} = A\hat{d}$ and $x = Ad$. To guarantee that the error in (8.37) cannot grow without bound, the sequence d should be constrained to lie in some bounded set. We therefore impose the additional requirement that $\|d\| \leq \rho$, for some $\rho > 0$. Problem (8.37) can then be reformulated as

$$\min_{\hat{d}} \max_{d \in \mathcal{D}} \|A\hat{d} - Ad\|^2, \tag{8.38}$$

where $\mathcal{D} = \{d : S^*Ad = c, \|d\| \leq \rho\}$. As we now show, the choice of ρ does not affect the solution, as long as $\mathcal{D}$ is a nonempty set.

THEOREM 8.2   *A solution to problem (8.38) is* $\hat{d} = (S^*A)^{\dagger}c$.

Proof   As we have seen in the proof of Theorem 8.1, assuming that $c \in \mathcal{R}(S^*A)$, a sequence d satisfies $S^*Ad = c$ if, and only if it is of the form $d = (S^*A)^{\dagger}c + v$, where v is some vector in $\mathcal{N}(S^*A)$. Furthermore, $(S^*A)^{\dagger}c \in \mathcal{N}(S^*A)^{\perp}$ so that $\|v\|^2 = \|d\|^2 - \|(S^*A)^{\dagger}c\|^2$. Therefore, the inner maximization in (8.38) becomes

$$\|A(\hat{d} - (S^*A)^{\dagger}c)\|^2 + \max_{v \in \mathcal{V}}\{\|Av\|^2 - 2v^*A^*A(\hat{d} - (S^*A)^{\dagger}c)\}, \tag{8.39}$$

where

$$\mathcal{V} = \left\{v : v \in \mathcal{N}(S^*A), \|v\|^2 \leq \rho^2 - \|(S^*A)^{\dagger}c\|^2\right\}. \tag{8.40}$$

Since $\mathcal{V}$ is a symmetric set, the vector v attaining the maximum in (8.39) must satisfy $v^*A^*A(\hat{d} - (S^*A)^{\dagger}c) \leq 0$, as we can change the sign of v without effecting the constraint. Therefore,

$$\max_{v \in \mathcal{V}}\{\|Av\|^2 - 2v^*A^*A(\hat{d} - (S^*A)^{\dagger}c)\} \geq \max_{v \in \mathcal{V}} \|Av\|^2. \tag{8.41}$$

Combining (8.41) and (8.39) we have that

$$\min_{\hat{d}} \max_{d \in \mathcal{D}} \|A\hat{d} - Ad\|^2 \geq \min_{\hat{d}}\{\|A(\hat{d} - (S^*A)^{\dagger}c)\|^2 + \max_{v \in \mathcal{V}} \|Av\|^2\}$$

$$= \max_{v \in \mathcal{V}} \|Av\|^2, \tag{8.42}$$

where the equality is a result of solving the minimization, which is obtained, for example, at $\hat{d} = (S^*A)^{\dagger}c$.

We now show that the inequality in (8.42) can be achieved with $\hat{d} = (S^*A)^{\dagger}c$. Indeed, substituting this choice of $\hat{d}$ in (8.39), we have that

$$\max_{d \in \mathcal{D}} \|A\hat{d} - Ad\|^2 = \max_{v \in \mathcal{V}}\{\|Av\|^2 - 2v^*A(\hat{d} - (S^*A)^{\dagger}c)\} = \max_{v \in \mathcal{V}} \|Av\|^2, \tag{8.43}$$

concluding the proof.   ∎

We conclude that a solution to the minimax problem (8.37) is given by

$$\hat{x}_{\mathrm{MX}} = A(S^*A)^\dagger c, \tag{8.44}$$

coinciding with the LS solution (8.30). We also see that, as in the LS strategy, the expansion coefficients of the recovery $\hat{x}$ in $\mathcal{A}$ are obtained by applying $H = (S^*A)^\dagger$ on the samples c.

Although the minimax and LS approaches coincide in the unconstrained subspace setting discussed thus far, we will see that these strategies lead to quite different reconstruction methods when the reconstruction process is constrained. In Section 8.6 we will also show that the results differ under a smoothness prior.

8.5.2   Constrained reconstruction

Up until now we specified the sampling process, but did not restrict the reconstruction or interpolation kernel $w(t)$ in Figure 8.4. We now address the problem of approximating x using a predefined set of reconstruction functions $\{w_n\}$, which form a frame for the reconstruction space $\mathcal{W}$. Given sampling functions $\{s_n\}$ and a fixed set of reconstruction functions $\{w_n\}$, an important question is how to design the correction transform H so that the output $\hat{x}$ is a good approximation of the input signal x in some sense. To handle this problem, we extend the two approaches discussed in Section 8.5.1 to the constrained setup. However, in contrast to the previous section, where perfect recovery was guaranteed under a direct-sum assumption, here $\hat{x}$ must lie in the space $\mathcal{W}$. Therefore, if x does not lie in $\mathcal{W}$ to begin with, then $\hat{x}$ cannot be equal to x.

Least-squares recovery

To obtain a reconstruction in $\mathcal{W}$ within the LS methodology we reformulate (8.24) as

$$\hat{x}_{\mathrm{CLS}} = \underset{x \in \mathcal{W}}{\mathrm{argmin}}\, \|S^*x - c\|^2. \tag{8.45}$$

Namely, the reconstruction $\hat{x}_{\mathrm{CLS}} \in \mathcal{W}$ should yield samples as close as possible to the measured sequence c. Note that (8.45) actually ignores our prior knowledge that $x \in \mathcal{A}$, a problem which is inevitable when working with the error-in-samples criterion.

Problem (8.45) is similar to (8.24) with $\mathcal{A}$ replaced by $\mathcal{W}$. However, here c does not necessarily lie in $\mathcal{R}(S^*W)$. Therefore, there does not necessarily exist an $x \in \mathcal{W}$ giving rise to the measured samples c, and consequently, the minimal distance is generally not 0.

THEOREM 8.3   *A solution to (8.45) is* $\hat{x}_{\mathrm{CLS}} = W(S^*W)^\dagger c$.

Proof Let $\hat{d}$ denote the expansion coefficients of the reconstruction, so that $\hat{x} = W\hat{d}$, and let $\hat{c} = S^*W\hat{d}$ be the samples it produces. Then, (8.45) can be written as

$$\underset{\hat{c} \in \mathcal{R}(S^*W)}{\min}\, \|\hat{c} - c\|^2. \tag{8.46}$$

This formulation shows that the optimal $\hat{c}$ is the projection of c onto $\mathcal{R}(S^*W)$:

$$\hat{c} = S^*W\hat{d} = P_{\mathcal{R}(S^*W)}c = (S^*W)(S^*W)^{\dagger}c, \tag{8.47}$$

from which the result follows. ∎

The solution of Theorem 8.3 has the same structure as the unconstrained LS reconstruction (8.30) with A replaced by W. Furthermore, as in Section 8.5.1, this solution is not unique if $\mathcal{W} \cap \mathcal{S}^{\perp} \neq \{0\}$.

It is interesting to study the relation between the unconstrained and constrained solutions. As we have seen, $\hat{x}_{\mathrm{LS}}$ of (8.30) is consistent, namely $S^*\hat{x}_{\mathrm{LS}} = c$. Therefore, $\hat{x}_{\mathrm{CLS}}$ can be expressed in terms of $\hat{x}_{\mathrm{LS}}$:

$$\hat{x}_{\mathrm{CLS}} = W(S^*W)^{\dagger}c = W(S^*W)^{\dagger}S^*\hat{x}_{\mathrm{LS}}. \tag{8.48}$$

The geometric meaning of this relation is best understood when $\mathcal{H} = \mathcal{W} \oplus \mathcal{S}^{\perp}$. Then, $\hat{x}_{\mathrm{CLS}}$ is the oblique projection of $\hat{x}_{\mathrm{LS}}$ onto $\mathcal{W}$ along $\mathcal{S}^{\perp}$:

$$\hat{x}_{\mathrm{CLS}} = E_{\mathcal{W}\mathcal{S}^{\perp}}\hat{x}_{\mathrm{LS}}. \tag{8.49}$$

Figure 8.7 depicts $\hat{x}_{\mathrm{LS}}$ and $\hat{x}_{\mathrm{CLS}}$ in a situation where $\mathcal{A}$ and $\mathcal{S}^{\perp}$ satisfy the direct-sum condition (8.25) so that $\hat{x}_{\mathrm{LS}} = \hat{x}_{\mathrm{MX}} = x$, and also $\mathcal{H} = \mathcal{W} \oplus \mathcal{S}^{\perp}$, implying, that (8.49) holds. This example highlights the disadvantage of the LS formulation. In this setting we are constrained to yield $\hat{x} \in \mathcal{W}$. But since x can be determined from the samples c in this case, so can its best approximation in $\mathcal{W}$, which is given by $P_{\mathcal{W}}x = P_{\mathcal{W}}\hat{x}_{\mathrm{LS}}$. This alternative is also shown in Figure 8.7, and is clearly advantageous to $\hat{x}_{\mathrm{CLS}}$ in terms of squared error *for every* x. We will see in Section 8.5.2 that orthogonally projecting $\hat{x}_{\mathrm{LS}}$ onto the reconstruction space $\mathcal{W}$ can be motivated also when condition (8.25) does not hold.

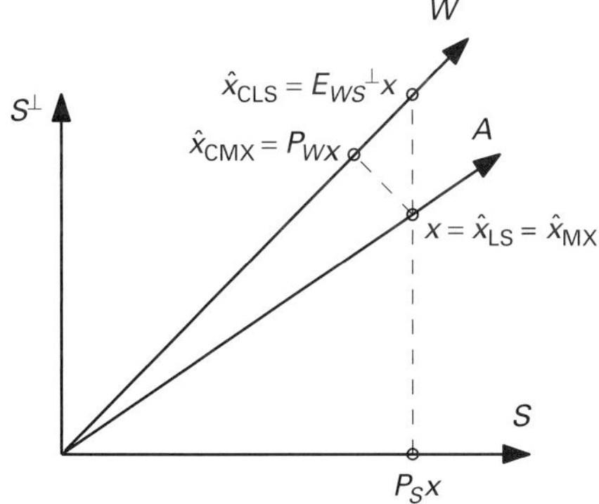

Figure 8.7 When condition (8.25) holds, $x = x_{\mathrm{LS}} = x_{\mathrm{MX}}$ and thus the signal $x \in \mathcal{A}$ can be recovered from the samples $c[n]$, allowing us to compute its projection onto $\mathcal{W}$. The constrained-minimax approach indeed yields $\hat{x}_{\mathrm{CMX}} = P_{\mathcal{W}}x$, whereas the constrained least-squares criterion leads to $\hat{x}_{\mathrm{CLS}} = E_{\mathcal{W}\mathcal{S}^{\perp}}x$.

The constrained LS method can be easily implemented in situations where $\mathcal{W}$ and $\mathcal{S}$ are SI spaces with generators $w(t)$ and $s(t)$, respectively. As we have seen, the operator S^*W corresponds to convolution with the sequence $r_{SW}[n] = (w(t) * s(-t))(n)$, and thus the correction transform $H = (S^*W)^\dagger$ is a digital filter whose frequency response is

$$H(e^{j\omega}) = \begin{cases} \dfrac{1}{\phi_{SW}\left(e^{j\omega}\right)}, & \phi_{SW}\left(e^{j\omega}\right) \neq 0; \\ 0, & \phi_{SW}\left(e^{j\omega}\right) = 0, \end{cases} \tag{8.50}$$

where $\phi_{SW}(e^{j\omega})$ is the DTFT of $r_{SW}[n]$, which is given by (8.34) with $A(\omega)$ replaced by $W(\omega)$. To conclude, reconstruction is performed by the scheme depicted in Figure 8.4, where the reconstruction kernel is $w(t)$ and the digital correction filter is given by (8.50).

Minimax recovery

We now treat the constrained recovery setting via a worst-case design strategy. The constraint $\hat{x} \in \mathcal{W}$ leads to an inherent limitation on the minimal achievable reconstruction error. Indeed, since $x \in \mathcal{A}$, the reconstruction error cannot be 0 unless $\mathcal{W} \subseteq \mathcal{A}$ [32]. From (8.4), we know that the best approximation in $\mathcal{W}$ of any signal x is given by $\hat{x} = P_\mathcal{W} x$, which, in general, cannot be computed from the sequence of samples $c[n]$. Therefore, we consider here the minimization of the regret, which is defined by $\|\hat{x} - P_\mathcal{W} x\|^2$. Since the regret is a function of the unknown signal x, we seek the reconstruction $\hat{x} \in \mathcal{W}$ minimizing the worst-case regret [14, 32, 47, 48]. Our problem is thus

$$\hat{x}_{\mathrm{CMX}} = \min_{\hat{x}\in\mathcal{W}} \max_{x\in\mathcal{G}} \|\hat{x} - P_\mathcal{W} x\|^2, \tag{8.51}$$

where $\mathcal{G}$ is the set of signals $x \in \mathcal{A}$ satisfying $S^*x = c$.

To solve (8.51) we express $\hat{x}$ and x in terms of their expansion coefficients in $\mathcal{W}$ and $\mathcal{A}$ respectively, by writing $\hat{x} = W\hat{d}$ and $x = Ad$. As in the unconstrained setting, we require that $\|d\| \leq \rho$ for some $\rho > 0$, in order for the inner maximization to be bounded. Therefore, problem (8.51) can be written as

$$\min_{\hat{d}} \max_{d\in\mathcal{D}} \|W\hat{d} - P_\mathcal{W} Ad\|^2, \tag{8.52}$$

where $\mathcal{D} = \{d : S^*Ad = c, \|d\| \leq \rho\}$.

THEOREM 8.4   *A solution to problem (8.51) is* $\hat{d} = (W^*W)^\dagger W^*A(S^*A)^\dagger c.$

Proof   The set $\mathcal{D}$ consists of all sequences of the form $d = (S^*A)^\dagger c + v$, where v is some vector in $\mathcal{N}(S^*A)$, with $\|v\|^2 \leq \|d\|^2 - \|(S^*A)^\dagger c\|^2$. Therefore, the inner maximization in (8.52) becomes

$$\|W\hat{d} - P_\mathcal{W} A(S^*A)^\dagger c\|^2 + \max_{v\in\mathcal{V}}\{\|P_\mathcal{W} Av\|^2 - 2(P_\mathcal{W} Av)^*(W\hat{d} - P_\mathcal{W} A(S^*A)^\dagger c)\}, \tag{8.53}$$

where $\mathcal{V}$ is given by (8.40). Since $\mathcal{V}$ is a symmetric set, the vector v attaining the maximum in (8.53) must satisfy $(P_{\mathcal{W}}Av)^*(W\hat{d} - P_{\mathcal{W}}(S^*A)^{\dagger}c) \leq 0$, as we can change the sign of v without effecting the constraint. Consequently,

$$\max_{v\in\mathcal{V}}\{\|P_{\mathcal{W}}Av\|^2 - 2(P_{\mathcal{W}}Av)^*(W\hat{d} - P_{\mathcal{W}}A(S^*A)^{\dagger}c)\} \geq \max_{v\in\mathcal{V}}\|P_{\mathcal{W}}Av\|^2. \tag{8.54}$$

Combining (8.54) and (8.53) we have that

$$\min_{\hat{d}}\max_{d\in\mathcal{D}}\|W\hat{d} - P_{\mathcal{W}}Ad\|^2 \geq \min_{\hat{d}}\{\|W\hat{d} - P_{\mathcal{W}}A(S^*A)^{\dagger}c\|^2 + \max_{v\in\mathcal{V}}\|P_{\mathcal{W}}Av\|^2\}$$

$$= \max_{v\in\mathcal{V}}\|P_{\mathcal{W}}Av\|^2, \tag{8.55}$$

where the equality is a result of solving the minimization, which is obtained, for example, at

$$\hat{d} = (W^*W)^{\dagger}W^*A(S^*A)^{\dagger}c. \tag{8.56}$$

We now show that the inequality can be achieved with $\hat{d}$ given by (8.56). Substituting this into (8.53), we have that

$$\max_{d\in\mathcal{D}}\|W\hat{d} - P_{\mathcal{W}}Ad\|^2 = \max_{v\in\mathcal{V}}\{\|P_{\mathcal{W}}Av\|^2 - 2(P_{\mathcal{W}}Av)^*(W\hat{d} - P_{\mathcal{W}}A(S^*A)^{\dagger}c)\}$$

$$= \max_{v\in\mathcal{V}}\|P_{\mathcal{W}}Av\|^2, \tag{8.57}$$

from which the proof follows. ∎

We conclude that the solution to the minimax problem (8.51) is given by

$$\hat{x}_{\mathrm{CMX}} = W\hat{d} = W(W^*W)^{\dagger}W^*A(S^*A)^{\dagger}c = P_{\mathcal{W}}A(S^*A)^{\dagger}c. \tag{8.58}$$

In contrast to the constrained LS reconstruction of Theorem 8.3, the minimax regret solution of Theorem 8.4 explicitly depends on A. Hence, the prior knowledge that $x \in \mathcal{A}$ plays a role, as one would expect. It is also readily observed that the relation between the unconstrained and constrained minimax solutions is different than in the LS approach. Identifying in (8.58) the expression $A(S^*A)^{\dagger}c = \hat{x}_{\mathrm{MX}} = \hat{x}_{\mathrm{LS}}$, the constrained minimax recovery can be written as

$$\hat{x}_{\mathrm{CMX}} = P_{\mathcal{W}}\hat{x}_{\mathrm{MX}}, \tag{8.59}$$

so that the constrained solution is the orthogonal projection onto $\mathcal{W}$ of the unconstrained reconstruction. In Section 8.5.2 we discussed the superiority of this approach in situations where the spaces $\mathcal{S}$ and $\mathcal{A}$ satisfy the direct-sum condition (8.25), as shown in Figure 8.7. We now see that this strategy stems from the minimization of the worst-case regret, for *any two spaces* $\mathcal{S}$ and $\mathcal{A}$.

Let us now examine the case where $\mathcal{S}$, $\mathcal{A}$, and $\mathcal{W}$ are SI spaces with generators $s(t)$, $a(t)$, and $w(t)$, respectively. As shown in Section 8.5.1, each of the operators $(W^*W)^\dagger$, W^*A, and $(S^*A)^\dagger$ corresponds to a digital filter. Therefore, the overall reconstruction scheme is that depicted in Figure 8.4 with a digital correction filter $H(e^{j\omega})$ given by

$$H\left(e^{j\omega}\right) = \begin{cases} \dfrac{\phi_{WA}(e^{j\omega})}{\phi_{SA}(e^{j\omega})\phi_{WW}(e^{j\omega})}, & \phi_{SA}(e^{j\omega})\phi_{WW}(e^{j\omega}) \neq 0; \\ 0, & \phi_{SA}(e^{j\omega})\phi_{WW}(e^{j\omega}) = 0, \end{cases} \tag{8.60}$$

where $\phi_{WA}(e^{j\omega})$, $\phi_{SA}(e^{j\omega})$, and $\phi_{WW}(e^{j\omega})$ follow from (8.34) with the corresponding substitution of the filters $W(\omega)$, $A(\omega)$, and $S(\omega)$.

To demonstrate the minimax-regret recovery procedure, we now revisit Example 8.1 imposing a constraint on the recovery mechanism.

Example 8.2   Suppose that the signal $x(t)$ of (8.35) is sampled at the integers after passing through the anti-aliasing filter $s(t) = \mathrm{sinc}(t)$, as in Example 8.1. We would now like to recover $x(t)$ from the samples $c[n]$ using a standard, zero-order-hold, digital-to-analog convertor. The corresponding reconstruction filter is therefore $w(t) = u(t) - u(t-1)$, where $u(t)$ is the unit step function. To compute the digital compensation filter (8.60), we note that $\phi_{WW}(e^{j\omega}) = 1$ in our case. Furthermore, the filter $1/\phi_{SA}(e^{j\omega})$ is given by (8.36), as we have already seen in Example 8.1. It can be easily shown that the remaining term, $\phi_{WA}(e^{j\omega})$, corresponds to the filter

$$h_{WA}[n] = \begin{cases} e^{\frac{n}{\tau}}\left(1 - e^{-\frac{1}{\tau}}\right), & n \leq 0; \\ 0, & n > 0. \end{cases} \tag{8.61}$$

Therefore, the sequence $d[n]$, feeding the DAC, is obtained by convolving the samples $c[n]$ with $h[n]$ of (8.36) and then by $h_{WA}[n]$ of (8.61).

To summarize, we have seen that treating the constrained-reconstruction scenario within the minimax-regret framework leads to a simple and plausible recovery method. In contrast, the constrained LS approach does not take the prior into account, and is thus often inferior in terms of squared error in this setting.

8.6   Smoothness priors

We now treat the problem of approximating $x(t)$ from its samples, based on the knowledge that it is smooth. Specifically, here x is assumed to obey (8.13) with some $\rho > 0$.

8.6.1 Unconstrained reconstruction

Least-squares approximation

We begin by approximating a smooth signal x via the minimization of the error-in-samples criterion. To take the smoothness prior into account, we define the set $\mathcal{G}$ of feasible signals as $\mathcal{G} = \{x : \|Lx\| \le \rho\}$. The LS problem is then

$$\hat{x}_{\mathrm{LS}} = \underset{x \in \mathcal{G}}{\mathrm{argmin}} \ \|S^*x - c\|^2. \tag{8.62}$$

Since, by assumption, there exists an x in $\mathcal{G}$ giving rise to the measured samples c, the optimal value in (8.62) is 0. Furthermore, there may be infinitely many solutions in $\mathcal{G}$ yielding 0 error-in-samples, as demonstrated in Figure 8.8(b). In this figure, the solid vertical segment is the set of signals satisfying $S^*x = c$ and $\|Lx\| \le \rho$. To resolve this ambiguity, we seek the smoothest reconstruction among all possible solutions:

$$\hat{x}_{\mathrm{LS}} = \arg \underset{x \in \mathcal{G}}{\min} \ \|Lx\|, \tag{8.63}$$

where now $\mathcal{G} = \{x : S^*x = c\}$.

Problem (8.63) is a linearly-constrained quadratic program with a convex objective. In finite dimensions there always exists a solution to this kind of problem. However, in infinite dimensions this is no longer true [49, Chapter 11]. To guarantee the existence of a solution, we focus on situations in which the operator L^*L is bounded from above and below, so that there exist constants $0 < \alpha_L \le \beta_L < \infty$ such that

$$\alpha_L \|x\|^2 \le \|L^*Lx\|^2 \le \beta_L \|x\|^2 \tag{8.64}$$

for any $x \in H$. Since L^*L is Hermitian, this condition also implies that $(L^*L)^{-1}$ is bounded and that $\beta_L^{-1} \|x\|^2 \le \|(L^*L)^{-1}x\|^2 \le \alpha_L^{-1} \|x\|^2$ for any $x \in \mathcal{H}$.

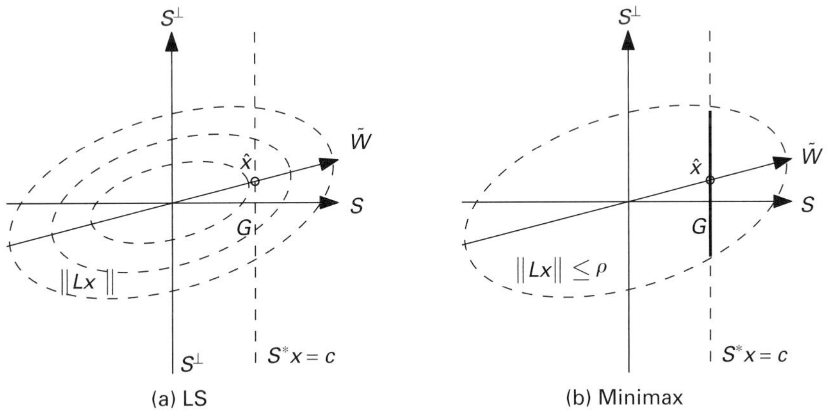

Figure 8.8 Geometric interpretation of the LS (a) and minimax (b) recoveries.

THEOREM 8.5   *Assume that the operator L satisfies condition (8.64). Then the solution (8.63) is given by*

$$\hat{x}_{LS} = \tilde{W}(S^*\tilde{W})^\dagger c, \tag{8.65}$$

where

$$\tilde{W} = (L^*L)^{-1}S. \tag{8.66}$$

Proof   Since $(L^*L)^{-1}$ is upper- and lower-bounded, and S satisfies the frame condition (8.7), $\tilde{W}$ is a synthesis operator of a frame. From Proposition 8.2, it then follows that $(S^*\tilde{W})^\dagger$ is bounded.

To solve (8.63), define the operator

$$E = \tilde{W}(S^*\tilde{W})^\dagger S^*, \tag{8.67}$$

so that $\hat{x}_{LS}$ of (8.65) is given by $\hat{x}_{LS} = Ex$. Now, any x can be decomposed as

$$x = Ex + (I - E)x = Ex + v, \tag{8.68}$$

where $v = (I - E)x$. In addition,

$$S^*E = S^*(L^*L)^{-1}S(S^*(L^*L)^{-1}S)^\dagger S^* = P_{\mathcal{N}(S^*(L^*L)^{-1}S)^\perp}S^* = S^*, \tag{8.69}$$

where we used the fact that $\mathcal{N}(S^*(L^*L)^{-1}S) = \mathcal{N}(S) = \mathcal{R}(S^*)^\perp$. Therefore, $S^*x = c$ and $S^*v = 0$. Next, using the identity $U^\dagger UU^\dagger = U^\dagger$, it can be verified that

$$E^*L^*L(I - E) = 0. \tag{8.70}$$

Consequently,

$$\|Lx\|^2 = \|LEx\|^2 + \|L(I - E)x\|^2 = \|LEx\|^2 + \|Lv\|^2, \tag{8.71}$$

and thus $\|Lx\|^2$ is minimized by choosing $Lv = 0$. Finally, if $S^*x = c$ then $Ex = \hat{x}_{LS}$ of (8.65). ∎

In Section 8.5, we have seen that (8.65) corresponds to the LS and minimax reconstructions when we have prior knowledge that x lies in the range of $\tilde{W}$, which we denote by $\tilde{\mathcal{W}}$. Thus, this approach can be viewed as first determining the optimal reconstruction space given by (8.66), and then computing the LS (or minimax) reconstruction under the subspace prior $x \in \tilde{\mathcal{W}}$.

Figure 8.8(a) shows a geometric interpretation of the LS solution. The set of feasible signals is the subspace $S^*x = c$, which is orthogonal to $\mathcal{S}$ (vertical dashed line). The dashed ellipsoids are the level sets of the objective function $\|Lx\|$. The LS solution is the intersection between the vertical line and the ellipsoid for which it constitutes a tangent.

The reconstruction space $\tilde{\mathcal{W}}$, is the line connecting all possible reconstructions (for all possible sample sequences c).

As a special case, we may choose to produce the minimal-norm consistent reconstruction $\hat{x}$ by letting L be the identity operator I. This leads to $\tilde{\mathcal{W}} = \mathcal{S}$ and consequently, $\hat{x}$ is the orthogonal projection onto the sampling space, $\hat{x} = S(S^*S)^{\dagger}S^*x = P_{\mathcal{S}}x$. This can also be seen by noting that any reconstruction $\hat{x}$ which yields the samples c has the form $\hat{x} = P_{\mathcal{S}}x + v$, where v is an arbitrary vector in $\mathcal{S}^{\perp}$. The minimal-norm approximation corresponds to the choice $v = 0$.

If L is an LTI operator corresponding to convolution with a kernel whose CTFT is $L(\omega)$, then $(L^*L)^{-1}$ corresponds to filtering with $1/|L(\omega)|^2$. In this case, if the sampling space $\mathcal{S}$ is SI, then $\tilde{\mathcal{W}}$ is an SI space with generator $\tilde{w}(t)$ whose CTFT is $\tilde{W}(\omega) = S(\omega)/|L(\omega)|^2$. This is because the nth reconstruction function $w_n(t)$ is a filtered version of the corresponding sampling function $s_n(t) = s(t-n)$, namely $\tilde{W}_n(\omega) = S_n(\omega)/|L(\omega)|^2$. As shown in the previous section, the correction transform H yielding the expansion coefficients $d[n]$ is also LTI in this case, in other words it corresponds to digital filtering. Therefore, the overall reconstruction scheme is that shown in Figure 8.4, where now the reconstruction kernel is

$$\tilde{W}(\omega) = \frac{S(\omega)}{|L(\omega)|^2}, \tag{8.72}$$

and the digital correction filter is

$$H(e^{j\omega}) = \begin{cases} \dfrac{1}{\phi_{S\tilde{w}}(e^{j\omega})}, & \phi_{S\tilde{w}}(e^{j\omega}) \neq 0; \\ 0, & \phi_{S\tilde{w}}(e^{j\omega}) = 0. \end{cases} \tag{8.73}$$

Here, the filter $\phi_{S\tilde{w}}(e^{j\omega})$ follows from (8.34) with $A(\omega)$ replaced by $\tilde{W}(\omega)$.

Minimax recovery

We now treat the problem of reconstructing a smooth signal from its samples via a worst-case design approach. The prior information we have can be used to construct a set $\mathcal{G}$ of all possible input signals:

$$\mathcal{G} = \left\{ x : S^*x = c, \|Lx\| \leq \rho \right\}. \tag{8.74}$$

As in Section 8.6.1, $\rho > 0$ is assumed to be large enough so that $\mathcal{G}$ is nonempty. The set consists of signals that are consistent with the samples and are relatively smooth. We now seek the reconstruction that minimizes the worst-case error over $\mathcal{G}$:

$$\hat{x}_{\mathrm{MX}} = \min_{\hat{x}} \max_{x \in \mathcal{G}} \|\hat{x} - x\|^2. \tag{8.75}$$

THEOREM 8.6 *The solution to problem (8.75) coincides with the LS approach, namely $\hat{x}_{MX}$ equals $\hat{x}_{LS}$ of (8.65).*

Proof   For any signal x satisfying the consistency constraint $S^*x = c$, the norm $\|Lx\|^2$ is given by (8.71), with E of (8.67). Therefore, we can write the inner maximization in (8.75) as

$$\|\hat{x} - \tilde{W}(S^*\tilde{W})^\dagger c\|^2 + \max_{v \in \mathcal{V}} \left\{ \|v\|^2 - 2(\hat{x} - \tilde{W}(S^*\tilde{W})^\dagger c)^*v \right\}, \qquad (8.76)$$

where

$$\mathcal{V} = \left\{ v : \|Lv\|^2 \le \rho^2 - \|L\tilde{W}(S^*\tilde{W})^\dagger c\|^2 \right\}. \qquad (8.77)$$

Clearly, at the maximum value of v we have that $(\hat{x} - \tilde{W}(S^*\tilde{W})^\dagger c)^*v \le 0$ since we can change the sign of v without effecting the constraint. Therefore,

$$\max_{v \in \mathcal{V}} \left\{ \|v\|^2 - 2(\hat{x} - \tilde{W}(S^*\tilde{W})^\dagger c)^*v \right\} \ge \max_{v \in \mathcal{V}} \|v\|^2. \qquad (8.78)$$

Combining (8.78) and (8.76),

$$\min_{\hat{x}} \max_{x \in \mathcal{G}} \|\hat{x} - x\|^2 \ge \min_{\hat{x}} \left\{ \|\hat{x} - \tilde{W}(S^*\tilde{W})^\dagger c\|^2 + \max_{v \in \mathcal{V}} \|v\|^2 \right\} = \max_{v \in \mathcal{V}} \|v\|^2, \qquad (8.79)$$

where the equality is a result of solving the inner minimization, obtained at $\hat{x} = \tilde{W}(S^*\tilde{W})^\dagger c$. We now show that the inequality can be achieved with $\hat{x} = \tilde{W}(S^*\tilde{W})^\dagger c$. Indeed, with this choice of $\hat{x}$, (8.76) implies that

$$\max_{x \in \mathcal{G}} \|\hat{x} - x\|^2 = \max_{v \in \mathcal{V}} \left\{ \|v\|^2 - 2(\hat{x} - \tilde{W}(S^*\tilde{W})^\dagger c)^*v \right\} = \max_{v \in \mathcal{V}} \|v\|^2, \qquad (8.80)$$

from which the theorem follows.   ∎

Figure 8.8(b) shows a geometric interpretation of the minimax solution. The set $\mathcal{G}$ (solid segment) of feasible signals is an intersection of the ellipsoid defined by $\|Lx\| \le \rho$ and the subspace $S^*x = c$, which is orthogonal to $\mathcal{S}$. Clearly, for any reconstruction $\hat{x} \in \mathcal{G}$, the worst-case signal x lies on the boundary of $\mathcal{G}$. Therefore, to minimize the worst-case error, $\hat{x}$ must be the midpoint of the solid segment, as shown in the figure. The optimal reconstruction space $\tilde{\mathcal{W}}$ connects the recoveries corresponding to all possible sequences of samples c. This is equivalent to horizontally swapping the vertical dashed line in the figure and connecting the midpoints of the corresponding feasible sets $\mathcal{G}$.

Although the two approaches we discussed are equivalent in the unrestricted setting, the minimax strategy allows more flexibility in incorporating constraints on the reconstruction, as we show in the next subsection. Furthermore, it tends to outperform the consistency approach when further restrictions are imposed, as we will demonstrate via several examples.

Example 8.3   Figure 8.9 compares the minimax (and LS) approach with bicubic interpolation in the context of image enlargement. The bicubic kernel of [35] is one of the

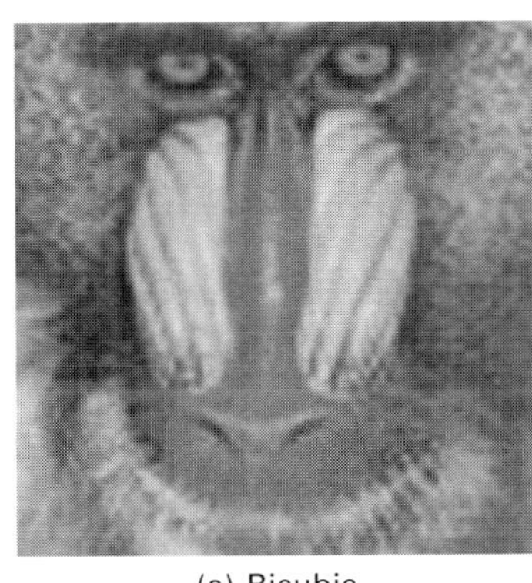
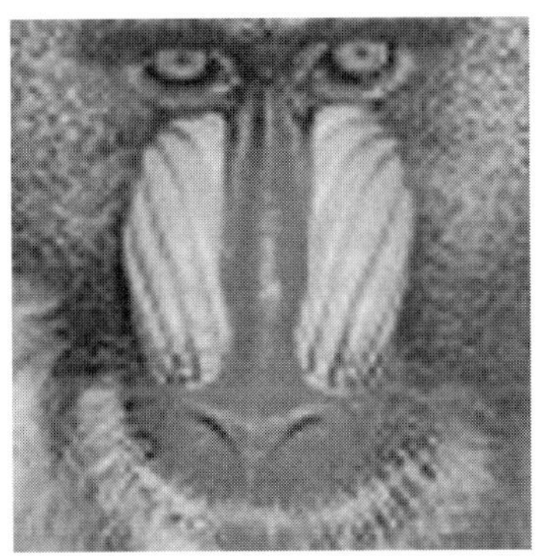

(a) Bicubic   (b) Minimax

Figure 8.9 Mandrill image rescaling: down-sampling by a factor of 3 using a rectangular sampling filter followed by upsampling back to the original dimensions using two interpolation methods. (a) The bicubic interpolation kernel leads to a blurry reconstruction with PSNR of 24.18 dB. (b) The minimax method leads to a sharper reconstruction with PSNR of 24.39 dB.

most popular image re-sampling methods. In our experiment, a high-resolution image was down-sampled by a factor of 3. A continuous model was then fitted to the samples using both the minimax solution and the bicubic kernel of [35]. These models were re-sampled on a grid with $1/3$ spacings to produce an image of the original size. The regularization operator was taken to be $L(\boldsymbol{\omega}) = \left((0.1\pi)^2 + \|\boldsymbol{\omega}\|^2\right)^{1.3}$, where $\boldsymbol{\omega}$ denotes the 2D frequency vector. In this example, minimax recovery is superior to the commonly used bicubic method in terms of peak signal-to-noise ratio (PSNR), defined as $\text{PSNR} = 10\log_{10}(255^2/\text{MSE})$ with MSE denoting the empirical squared-error average over all pixel values. In terms of visual quality, the minimax reconstruction is sharper and contains enhanced textures.

8.6.2 Constrained reconstruction

We next treat the problem of approximating x from its samples using a pre-specified set of reconstruction functions $\{w_n\}$. We will see that in this setup the LS and minimax recovery methods no longer coincide.

Least-squares approximation
In order to produce a solution $\hat{x} \in \mathcal{W}$, we modify the feasible set $\mathcal{G}$ of (8.62) to include only signals in $\mathcal{W}$:

$$\hat{x}_{\text{CLS}} = \arg\min_{x \in \tilde{\mathcal{G}}} \|S^*x - c\|^2, \tag{8.81}$$

where $\tilde{\mathcal{G}} = \{x : \|Lx\| \leq \rho, x \in \mathcal{W}\}$.

We have seen in Section 8.5 that without the constraint $\|Lx\| \leq \rho$, the set of solutions to (8.81) is given by

$$\mathcal{G} = \{x : x \in \mathcal{W}, S^*x = P_{\mathcal{R}(S^*W)}c\}. \tag{8.82}$$

We assume here that ρ is sufficiently large so that $\mathcal{G}$ contains at least one x. To choose one solution to (8.81) we minimize the smoothness measure $\|Lx\|$ over the set $\mathcal{G}$:

$$\hat{x}_{\mathrm{CLS}} = \underset{x \in \mathcal{G}}{\mathrm{argmin}} \ \|Lx\|. \tag{8.83}$$

THEOREM 8.7 *The solution to problem (8.83) is given by*

$$\hat{x}_{\mathrm{CLS}} = \hat{W}(S^*\hat{W})^{\dagger}c, \tag{8.84}$$

where now

$$\hat{W} = W(W^*L^*LW)^{\dagger}W^*S. \tag{8.85}$$

Proof The proof of the theorem follows similar steps as in Section 8.6.1 and utilizes the fact that every signal in $\mathcal{G}$ is of the form $x = W((S^*W)^{\dagger}c+v)$, where $v \in \mathcal{N}(S^*W)$. ∎

Note that this solution is feasible, namely $\hat{x}_{\mathrm{CLS}} \in \mathcal{G}$:

$$S^*x_{\mathrm{CLS}} = S^*W(W^*L^*LW)^{\dagger}W^*S(S^*W(W^*L^*LW)^{\dagger}W^*S)^{\dagger}c$$

$$= P_{\mathcal{R}(S^*W(W^*L^*LW)^{\dagger}W^*S)}c$$

$$= P_{\mathcal{R}(S^*W)}c. \tag{8.86}$$

The last equality follows from the fact that $\mathcal{R}((W^*L^*LW)^{\dagger}) = \mathcal{N}(W^*L^*LW)^{\perp} = \mathcal{N}(LW)^{\perp} = \mathcal{N}(W)^{\perp}$ and similarly $\mathcal{N}((W^*L^*LW)^{\dagger})^{\perp} = \mathcal{R}(W^*)$.

In contrast to subspace priors, here the constrained LS recovery does not generally relate to the unconstrained solution via an oblique projection. An exception is the case where $\mathcal{W} \oplus \mathcal{S}^{\perp} = \mathcal{H}$. As we have seen, in this case there exists a unique $x \in \mathcal{W}$ satisfying $S^*x = c$, which is equal to the oblique projection $E_{\mathcal{W}\mathcal{S}^{\perp}}x$. Since there is only one signal in the constraint set of problem (8.83), the smoothness measure in the objective does not play a role and the solution becomes $\hat{x}_{\mathrm{CLS}} = W(S^*W)^{\dagger}c$. In this setting, we can also use the fact that the unconstrained solution (8.65) satisfies $S^*x_{\mathrm{LS}} = c$, to write $\hat{x}_{\mathrm{CLS}} = W(S^*W)^{\dagger}S^*x_{\mathrm{LS}}$, recovering the relation we had in Section 8.5.2.

Another interesting scenario where L does not affect the solution is the case where $\mathcal{W}$ and $\mathcal{S}$ are SI spaces with generator $w(t)$ and $s(t)$, respectively, and L is an LTI operator with frequency response $L(\omega)$. The operator $(W^*L^*LW)^{\dagger}$ then corresponds to the digital filter

$$\begin{cases} \dfrac{1}{\phi_{(LW)(LW)}(e^{j\omega})}, & \phi_{(LW)(LW)}(e^{j\omega}) \neq 0; \\ 0, & \phi_{(LW)(LW)}(e^{j\omega}) = 0 \end{cases} = \begin{cases} \dfrac{1}{\phi_{(LW)(LW)}(e^{j\omega})}, & \phi_{WW}(e^{j\omega}) \neq 0; \\ 0, & \phi_{WW}(e^{j\omega}) = 0, \end{cases} \tag{8.87}$$

where $\phi_{(LW)(LW)}(e^{j\omega})$ is given by (8.34) with $S(\omega)$ and $A(\omega)$ both replaced by $L(\omega)W(\omega)$, and $\phi_{WW}(e^{j\omega})$ is given by (8.34), with $S(\omega)$ and $A(\omega)$ both replaced by $W(\omega)$. The equality follows from the fact that L is assumed to satisfy (8.64), and thus $L(\omega)$ does not vanish anywhere. Therefore, it can be verified that $\hat{x}_{CLS}(t)$ of (8.84) can be produced by filtering the sequence of samples $c[n]$ with

$$H(e^{j\omega}) = \begin{cases} \dfrac{1}{\phi_{SW}(e^{j\omega})}, & \phi_{SW}(e^{j\omega}) \neq 0, \phi_{WW}(e^{j\omega}) \neq 0; \\ 0, & \text{else,} \end{cases} \tag{8.88}$$

prior to reconstruction with $W(\omega)$. Here $\phi_{SW}(e^{j\omega})$ is given by (8.34) with $A(\omega)$ replaced by $W(\omega)$. It can be seen that (8.88) does not depend on $L(\omega)$; namely the smoothness prior does not affect the solution in the SI setting.

The situation where $\mathcal{W} \oplus \mathcal{S}^{\perp} = \mathcal{H}$, happens if and only if the supports of $S(\omega)$ and $W(\omega)$ are the same [32]. In this case it can be seen that (8.88) becomes

$$H(e^{j\omega}) = \begin{cases} \dfrac{1}{\phi_{SW}(e^{j\omega})}, & \phi_{SW}\left(e^{j\omega}\right) \neq 0; \\ 0, & \phi_{SW}\left(e^{j\omega}\right) = 0. \end{cases} \tag{8.89}$$

The resulting scheme is identical to the constrained LS reconstruction discussed in Section 8.6.2 in the context of subspace priors.

Minimax-regret recovery

We next consider the extension of the minimax approach of Section 8.6.1 to the setup where $\hat{x}$ is constrained to lie in $\mathcal{W}$. Similar to the case of subspace priors, treated in Section 8.5.2, we consider here the minimization of the worst-case regret:

$$\hat{x}_{CMX} = \arg \min_{\hat{x} \in \mathcal{W}} \max_{x \in \mathcal{G}} \left\| \hat{x} - P_{\mathcal{W}} x \right\|^2, \tag{8.90}$$

where $\mathcal{G}$ is given by (8.74).

THEOREM 8.8 *The solution to (8.90) is given by*

$$\hat{x}_{CMX} = P_{\mathcal{W}} \tilde{W} (S^* \tilde{W})^{\dagger} c = P_{\mathcal{W}} \hat{x}_{MX}, \tag{8.91}$$

where $\tilde{W}$ is given by (8.66) and $\hat{x}_{MX} = \hat{x}_{LS}$ is the unconstrained solution given by (8.65).

Proof The proof follows the exact same steps as in Section 8.6.1. ∎

This result is intuitive: when the output is constrained to the subspace $\mathcal{W}$, the minimax recovery is the orthogonal projection onto $\mathcal{W}$ of the minimax solution without the restriction. Recall that relation (8.91) is also true for subspace priors, as we have seen in Section 8.5.

Figure 8.10 shows a geometric interpretation of the minimax-regret solution. As in the unconstrained scenario of Figure 8.9, the feasible set of signals $\mathcal{G}$ is the vertical solid segment. Here, however, the reconstruction $\hat{x}$ is constrained to lie in the predefined space

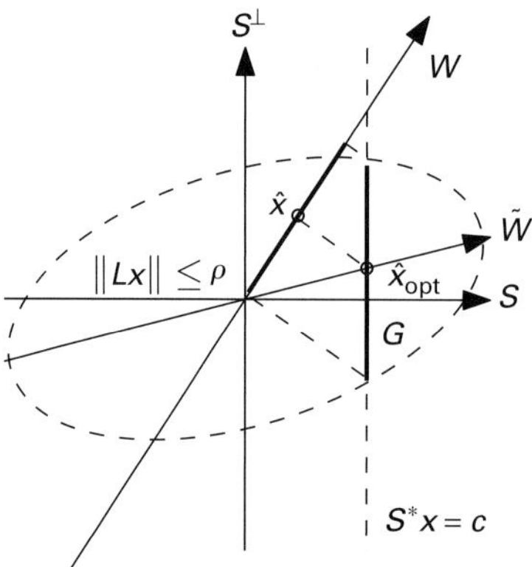

Figure 8.10   Geometric interpretation of minimax-regret recovery in a predefined reconstruction space $\mathcal{W}$.

$\mathcal{W}$. The regret criterion (8.90) measures the deviation of $\hat{x}$ from $P_{\mathcal{W}}x$. The tilted solid segment is the projection of the feasible set $\mathcal{G}$ onto $\mathcal{W}$. For every reconstruction $\hat{x}$ in $\mathcal{W}$, the signal x leading to the worst regret corresponds to one of the endpoints of this set. Therefore, the minimal regret is attained if we choose $\hat{x}$ to be the midpoint of this segment. This solution is also the projection of the midpoint of $\mathcal{G}$ onto $\mathcal{W}$, that is, the projection of the unconstrained minimax reconstruction (8.65) onto $\mathcal{W}$.

When $\mathcal{S}$ and $\mathcal{W}$ are SI spaces and L is an LTI operator, the correction transform H corresponds to a digital filter $H(e^{j\omega})$. This filter can be determined by writing $H = (W^*W)^\dagger W^*\tilde{W}(S^*\tilde{W})^\dagger$, where $\tilde{W} = (L^*L)^{-1}S$ is the set transform corresponding to the unrestricted minimax solution. The operators W^*W, $W^*\tilde{W}$, and $S^*\tilde{W}$ correspond to the digital filters $\phi_{WW}(e^{j\omega})$, $\phi_{W\tilde{W}}(e^{j\omega})$, and $\phi_{S\tilde{W}}(e^{j\omega})$, respectively. The digital correction filter of Figure 8.4 then becomes

$$H(e^{j\omega}) = \begin{cases} \dfrac{\phi_{W\tilde{W}}(e^{j\omega})}{\phi_{S\tilde{W}}(e^{j\omega})\phi_{WW}(e^{j\omega})}, & \phi_{S\tilde{W}}(e^{j\omega})\phi_{WW}(e^{j\omega}) \neq 0; \\ 0, & \text{else.} \end{cases} \tag{8.92}$$

In contrast to the constrained LS method, this filter depends on $L(\omega)$ so that the prior does affect the solution. The next example demonstrates the effectiveness of this filter in an image-processing task.

Example 8.4   In Figure 8.11 we demonstrate the difference between the LS and minimax-regret methods in an image-enlargement task. The setup is the same as that of Figure 8.9, only now the reconstruction filter is constrained to be a triangular kernel corresponding to linear interpolation. With this interpolation kernel, the direct-sum condition $L_2 = \mathcal{W} \oplus \mathcal{S}^\perp$ is satisfied. It can be seen that the error of the minimax-regret recovery is only 0.7 dB less than the unconstrained minimax shown in Figure 8.9. The constrained LS approach, on the other hand, is much worse both in terms of PSNR, and in terms of

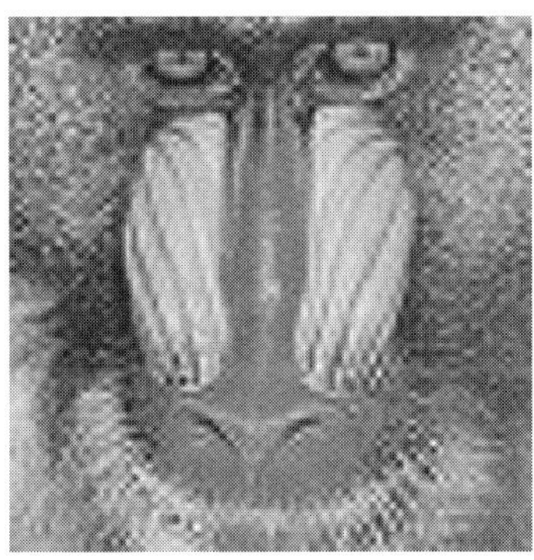
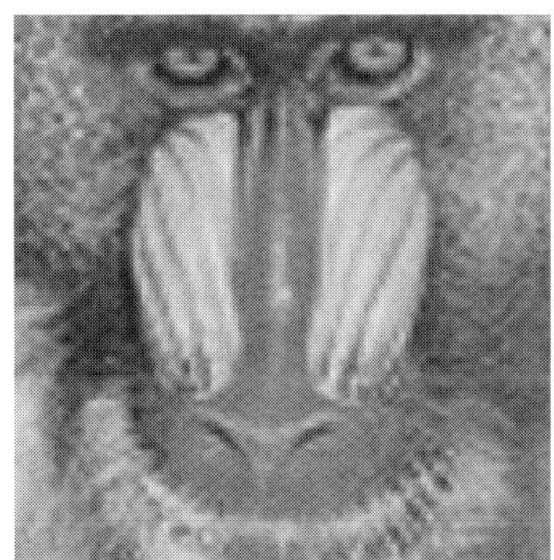

(a) Consistent   (b) Minimax regret

Figure 8.11   Mandrill image rescaling: down-sampling by a factor of 3 using a rectangular sampling filter followed by upsampling back to the original dimensions using the LS and minimax-regret methods. (a) The LS approach over-enhances the high frequencies and results in a PSNR of 22.51 dB. (b) The minimax-regret method leads to a smoother reconstruction with PSNR of 23.69 dB.

---

visual quality. Its tendency to over-enhance high frequencies stems from the fact that it ignores the smoothness prior.

Many of the interesting properties of the minimax-regret recovery (8.92) can be best understood by examining the case where our only prior on the signal is that it is norm-bounded, that is, when L is the identity operator I. This scenario was thoroughly investigated in [32]. Setting $L(\omega) = 1$ in (8.92), the correction filter becomes

$$H(e^{j\omega}) = \begin{cases} \dfrac{\phi_{WS}(e^{j\omega})}{\phi_{SS}(e^{j\omega})\phi_{WW}(e^{j\omega})}, & \phi_{SS}(e^{j\omega})\phi_{WW}(e^{j\omega}) \neq 0; \\ 0, & \text{else,} \end{cases} \tag{8.93}$$

since from (8.72), $\tilde{w}(t) = s(t)$. Applying the Cauchy–Schwartz inequality to the numerator of (8.93) and to the denominator of (8.88), it is easy to see that the magnitude of the minimax-regret filter (8.93) is smaller than that of the constrained LS filter (8.88) at all frequencies. This property renders the minimax-regret approach more resistant to noise in the samples $c[n]$, since perturbations in $\hat{x}(t)$ caused by errors in $c[n]$ are always smaller in the minimax-regret method than in the consistent approach.

In Figure 8.12 we illustrate the minimax-regret reconstruction geometrically for the case $L = I$. We have seen already that knowing the samples $c[n]$ is equivalent to knowing $P_{S}x$. In addition, our recovery is constrained to lie in the space $\mathcal{W}$. As illustrated in the figure, the minimax-regret solution is a robust recovery scheme by which the signal is first orthogonally projected onto the sampling space, and then onto the reconstruction space.

When x is known to lie in $\mathcal{S}$, it follows from the previous section that the minimal error can be obtained by using (8.60) with $A = S$. The resulting filter coincides with the minimax-regret filter of (8.89), implying that the regret approach minimizes the squared-error over all $x \in \mathcal{S}$.

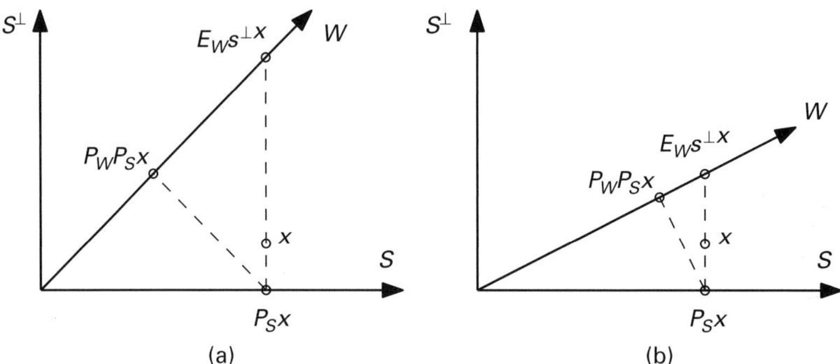

Figure 8.12 Comparison of minimax-regret reconstruction and constrained LS reconstruction for two different choices of W satisfying $\mathcal{H} = W \oplus \mathcal{S}^{\perp}$. (a) The minimax strategy ($P_W P_S x$) is preferable to LS ($E_W \mathcal{S}^{\perp} x$) when W is "far" from $\mathcal{S}$. (b) Both methods lead to errors on the same order of magnitude when W is "close" to $\mathcal{S}$.

In [32] tight bounds on the error resulting from the constrained LS and minimax-regret methods are developed for the case where $\mathcal{H} = W \oplus \mathcal{S}^{\perp}$. We omit the technical details here and only summarize the main conclusions. We first recall that if we know a priori that x lies in a subspace $\mathcal{A}$ such that $\mathcal{H} = \mathcal{A} \oplus \mathcal{S}^{\perp}$, then the filter (8.60) will yield the minimal-error approximation of x and therefore is optimal in the squared-norm sense. When $\mathcal{A} = \mathcal{S}$ this strategy reduces to the minimax-regret method, while if $\mathcal{A} = W$, then we obtain the constrained LS reconstruction.

When no prior subspace knowledge is given, the regret approach is preferable if the spaces $\mathcal{S}$ and W are sufficiently far apart, or if x has enough energy in $\mathcal{S}$. These results are intuitive as illustrated geometrically in Figure 8.12. In Figure 8.12(a) we depict the constrained LS and regret reconstruction when W is far from $\mathcal{S}$. As can be seen in the figure, in this case the error resulting from the LS solution is large with respect to the regret-approximation error. In Figure 8.12(b), W and $\mathcal{S}$ are close, and the errors have roughly the same magnitude.

8.7   Comparison of the various scenarios

Table 8.2 summarizes the reconstruction techniques developed in Sections 8.5 and 8.6. We use the superscripts "sub" and "smo" to signify whether a solution corresponds to a subspace or a smoothness prior. The transformations $\tilde{W}$ and $\hat{W}$ are given by

$$\tilde{W} = (L^*L)^{-1} S,$$

and

$$\hat{W} = W (W^* L^* L W)^{\dagger} W^* S,$$

respectively.

Table 8.2. Reconstruction from noiseless samples.

Prior	Unconstrained ($\hat{x} \in \mathcal{H}$)		Constrained ($\hat{x} \in \mathcal{W}$)	
	Least-Squares	Minimax	Least-Squares	Minimax
$x \in \mathcal{A}$	$\hat{x}_{\mathrm{LS}}^{\mathrm{sub}} = A(S^*A)^\dagger c$	$\hat{x}_{\mathrm{MX}}^{\mathrm{sub}} = \hat{x}_{\mathrm{LS}}^{\mathrm{sub}}$	$\hat{x}_{\mathrm{CLS}}^{\mathrm{sub}} = W(S^*W)^\dagger S^*\hat{x}_{\mathrm{LS}}^{\mathrm{sub}}$	$\hat{x}_{\mathrm{CMX}}^{\mathrm{sub}} = P_{\mathcal{W}}\hat{x}_{\mathrm{MX}}^{\mathrm{sub}}$
$\|Lx\| \leq \rho$	$\hat{x}_{\mathrm{LS}}^{\mathrm{smo}} = \tilde{W}(S^*\tilde{W})^\dagger c$	$\hat{x}_{\mathrm{MX}}^{\mathrm{smo}} = \hat{x}_{\mathrm{LS}}^{\mathrm{smo}}$	$\hat{x}_{\mathrm{CLS}}^{\mathrm{smo}} = \hat{W}(S^*\hat{W})^\dagger S^*\hat{x}_{\mathrm{LS}}^{\mathrm{smo}}$	$\hat{x}_{\mathrm{CMX}}^{\mathrm{smo}} = P_{\mathcal{W}}\hat{x}_{\mathrm{MX}}^{\mathrm{smo}}$

This table highlights the key observations discussed in the previous sections. We begin by examining the case in which no constraint is imposed on $\hat{x}$, shown in columns 1 and 2. First, we see that in this situation both the LS and minimax reconstructions coincide. This property holds true both for subspace and smoothness priors. In Section 8.8 we show that this is not the case when the samples are noisy. Second, smoothness-prior recovery (row 2) has the same structure as subspace-prior recovery (row 1) with $\tilde{W}$ replacing A. Therefore, we can interpret $\tilde{\mathcal{W}} = \mathcal{R}(\tilde{W})$ as the optimal reconstruction space associated with the smoothness prior. Finally, we note that for certain subspace priors, perfect recovery can be achieved, leading to $\hat{x}_{\mathrm{LS}}^{\mathrm{sub}} = x$. Specifically, this happens if the sampling space $\mathcal{S}$ and the prior space $\mathcal{A}$ satisfy the direct-sum condition $\mathcal{H} = \mathcal{A} \oplus \mathcal{S}^\perp$. In the smoothness-prior case, the direct-sum condition $\mathcal{H} = \tilde{\mathcal{W}} \oplus \mathcal{S}^\perp$ does not imply perfect recovery because the original x does not necessarily lie in $\tilde{\mathcal{W}}$. When the direct sum holds, however, recovery can be interpreted as an oblique projection of the (unknown) signal x onto the optimal reconstruction space $\tilde{\mathcal{W}}$, namely $\hat{x}_{\mathrm{LS}}^{\mathrm{smo}} = E_{\tilde{\mathcal{W}}\tilde{\mathcal{S}}^\perp}x$.

We now examine the case in which the recovery is constrained to lie in $\mathcal{W}$ (columns 3 and 4). These solutions are expressed in Table 8.2 in terms of the unconstrained reconstructions (columns 1 and 2). Here the minimax-regret solutions (column 4) are related to the unconstrained recoveries (column 2) via an orthogonal projection onto the reconstruction space $\mathcal{W}$. This implies that $\|\hat{x}_{\mathrm{CMX}}\| \leq \|\hat{x}_{\mathrm{MX}}\|$. The constrained LS solutions, on the other hand, possess a different structure. When the sampling and reconstruction spaces satisfy the direct-sum $\mathcal{H} = \mathcal{W} \oplus \mathcal{S}^\perp$, both constrained LS solutions of column 3 become $\hat{x}_{\mathrm{CLS}} = E_{\mathcal{W}\mathcal{S}^\perp}\hat{x}_{\mathrm{LS}}$. This has several implications. First, in contrast to an orthogonal projection, an oblique projection may lead to solutions with arbitrary large norm, given that $\mathcal{W}$ is sufficiently far apart from $\mathcal{S}$. Therefore, the error in the constrained LS framework is not guaranteed to be bounded, unless a bound on the "distance" between $\mathcal{S}$ and $\mathcal{W}$ is known a-priori. Second, this implies that the recovery does not depend on the prior, in other words, $\hat{x}_{\mathrm{CLS}}^{\mathrm{sub}}$ is not a function of A and $\hat{x}_{\mathrm{CLS}}^{\mathrm{smo}}$ does not depend on L. These properties are clearly undesirable and can lead to unsatisfactory results in practical applications, as demonstrated in Example 8.4.

Table 8.3 summarizes the recovery formulae obtained under the direct-sum assumptions discussed above. The expressions in column 1, rows 1 and 2, are true when $\mathcal{H} = \mathcal{A} \oplus \mathcal{S}^\perp$ and $\mathcal{H} = \tilde{\mathcal{W}} \oplus \mathcal{S}^\perp$, respectively. The recoveries of column 3 are obtained under the assumption that $\mathcal{H} = \mathcal{W} \oplus \mathcal{S}^\perp$.

Finally, we note that all the recovery techniques discussed thus far can be easily implemented in SI spaces. Specifically, suppose that $\mathcal{S}$, $\mathcal{A}$, and $\mathcal{W}$ are SI spaces with

Table 8.3. Reconstruction from noiseless samples under direct-sum assumptions.

Prior	Unconstrained ($\hat{x} \in \mathcal{H}$)		Constrained ($\hat{x} \in \mathcal{W}$)	
	Least-Squares	Minimax	Least-Squares	Minimax
$x \in \mathcal{A}$	$\hat{x}_{\mathrm{LS}}^{\mathrm{sub}} = x$	$\hat{x}_{\mathrm{MX}}^{\mathrm{sub}} = \hat{x}_{\mathrm{LS}}^{\mathrm{sub}}$	$\hat{x}_{\mathrm{CLS}}^{\mathrm{sub}} = E_{\mathcal{WS}^{\perp}}\hat{x}_{\mathrm{LS}}^{\mathrm{sub}}$	$\hat{x}_{\mathrm{CMX}}^{\mathrm{sub}} = P_{\mathcal{W}}\hat{x}_{\mathrm{MX}}^{\mathrm{sub}}$
$\|Lx\| \leq \rho$	$\hat{x}_{\mathrm{LS}}^{\mathrm{smo}} = E_{\tilde{\mathcal{W}}\mathcal{S}^{\perp}}x$	$\hat{x}_{\mathrm{MX}}^{\mathrm{smo}} = \hat{x}_{\mathrm{LS}}^{\mathrm{smo}}$	$\hat{x}_{\mathrm{CLS}}^{\mathrm{smo}} = E_{\mathcal{WS}^{\perp}}\hat{x}_{\mathrm{LS}}^{\mathrm{smo}}$	$\hat{x}_{\mathrm{CMX}}^{\mathrm{smo}} = P_{\mathcal{W}}\hat{x}_{\mathrm{MX}}^{\mathrm{smo}}$

Table 8.4. Reconstruction from noiseless samples in SI spaces.

Prior	Unconstrained ($\hat{x} \in \mathcal{H}$)		Constrained ($\hat{x} \in \mathcal{W}$)	
	Least-Squares	Minimax	Least-Squares	Minimax
$x \in \mathcal{A}$	(8.33)	(8.33)	(8.50)	(8.60)
$\|Lx\| \leq \rho$	(8.73), (8.72)	(8.73), (8.72)	(8.88)	(8.92)

generators $s(t)$, $a(t)$, and $w(t)$, respectively. Moreover, assume that L is an LTI operator corresponding to the filter $L(\omega)$. Then all the reconstruction methods of Table 8.2 can be implemented by digitally filtering the samples $c[n]$ prior to reconstruction, as depicted in Figure 8.4. The resulting interpolation methods are summarized in Table 8.4. The numbers in the table indicate the equation numbers containing the reconstruction formulae of the digital correction filter and reconstruction kernel. The optimal kernel corresponding to unconstrained recovery with a subspace prior (row 1, columns 1 and 2) is $a(t)$, while the interpolation kernel in the constrained case (columns 3 and 4) is $w(t)$.

The direct-sum conditions, under which Table 8.3 was constructed, can be easily verified in SI spaces, as explained in Section 8.5.1. Specifically, for two SI spaces $\mathcal{A}$ and $\mathcal{S}$, the condition $\mathcal{H} = \mathcal{A} \oplus \mathcal{S}^{\perp}$ is satisfied if, and only if [46] the supports $\mathcal{I}_{\mathcal{A}}$ and $\mathcal{I}_{\mathcal{S}}$ of $\phi_{SS}(e^{j\omega})$ and $\phi_{AA}(e^{j\omega})$ respectively coincide, and there exists a constant $\alpha > 0$ such that $|\phi_{SA}(e^{j\omega})| > \alpha$, for all ω in $\mathcal{I}_{\mathcal{A}}$. The filters $\phi_{SS}(e^{j\omega})$, $\phi_{AA}(e^{j\omega})$, and $\phi_{SA}(e^{j\omega})$ are defined in (8.34).

8.8 Sampling with noise

We now extend the approaches of the previous sections to the case in which the samples are perturbed by noise. Specifically, we assume that $c = S^{*}x + u$, where $u[n]$ is an unknown noise sequence.

To approach the noisy setup within the LS framework, we need to minimize the error-in-samples $\|S^{*}x - c\|^{2}$ over the set of feasible signals. Thus, the optimization problems (8.24), (8.45), (8.62), and (8.81), which correspond to the unconstrained and constrained subspace and smoothness scenarios, remain valid here too. However, note that to solve these problems we assumed that signals x for which $S^{*}x = c$ (or $S^{*}x = P_{\mathcal{R}(W^{*}S)}c$ in the constrained setting) are included in the feasible set. When the samples are noisy, this is not necessarily true so that, for example, the optimal value of the unconstrained problem $\min_{x \in \mathcal{A}} \|S^*x - c\|^2$ is no longer 0. Nevertheless, it can be easily shown that the solutions we obtained under the subspace-prior assumption (Problems (8.24) and (8.45)) and in the constrained smoothness setting (Problem (8.81)) remain the same. Furthermore, it can be shown that in the unconstrained smoothness scenario (Problem (8.62)), this fact does not change the optimal reconstruction space $\tilde{\mathcal{W}}$ of (8.66), but only the expansion coefficients of $\hat{x}$ in $\tilde{\mathcal{W}}$. Interestingly, this property holds even when the ℓ_2-norm in the error-in-samples term $\|S^*x - c\|$ is replaced by an ℓ_p-norm with arbitrary $p \in [1, \infty]$ [11].

Therefore, we focus our attention on the extension of the *minimax*-recovery techniques to the noisy case. To keep the exposition simple, we will thoroughly examine only the smoothness-prior scenarios of Section 8.6. The subspace-prior problems of Section 8.5 can be treated in the exact same manner. We thus assume in the sequel that $\|Lx\| \leq \rho$ for some $\rho \geq 0$. In this setting the solution no longer lies in the reconstruction space $\tilde{\mathcal{W}}$ of (8.66). Moreover, the resulting problems generally do not admit a closed-form solution and must be solved using numerical-optimization methods. Consequently, we will narrow the discussion from signals in an arbitrary Hilbert space $\mathcal{H}$ to signals lying in $\mathbb{R}^n$ or $\mathbb{C}^n$.

Assume that the samples c are noisy so that our only information is that $\|S^*x - c\| \leq \alpha$ for some value of α. The extension of the minimax problem (8.75) in the unconstrained scenario to the noisy case is

$$\hat{x}_{\mathrm{MX}} = \arg \min_{\hat{x}} \max_{x \in \mathcal{G}} \|\hat{x} - x\|^2, \tag{8.94}$$

where

$$\mathcal{G} = \{x : \|S^*x - c\| \leq \alpha, \|Lx\| \leq \rho\}. \tag{8.95}$$

Similarly, the counterpart of (8.90), where the solution is constrained to lie in $\mathcal{W}$, is $\hat{x}_{\mathrm{CMX}} = W\hat{d}_{\mathrm{CMX}}$, with

$$\hat{d}_{\mathrm{CMX}} = \arg \min_{d} \max_{x \in \mathcal{G}} \|Wd - P_{\mathcal{W}}x\|^2. \tag{8.96}$$

To solve (8.94) and (8.96) we replace the inner maximization by its dual function. This will result in a minimization problem that can be combined with the outer minimization. In order to follow this method, we need to be able to establish strong duality of the inner maximization with its dual function. The maximization in (8.94) and (8.96) is a special case of a nonconvex quadratic-optimization problem with two quadratic constraints. The nonconvexity is the result of the fact that we are maximizing a convex quadratic function, rather than minimizing it. Nonconvex quadratic-optimization problems have been studied extensively in the optimization literature. Below, we first briefly survey some of the main results on quadratic optimization, relevant to our problem. We then show how they can be used to develop a robust recovery method.

8.8.1 Quadratic-optimization problems

This simplest class of quadratic-optimization problems is the minimization of a single (possibly nonconvex) quadratic function subject to one quadratic constraint. A special well-studied case is that of the trust-region algorithm for unconstrained optimization, which has the form [50–55]:

$$\min_{x\in\mathbb{R}^n}\{x^T Bx + 2g^T x : \|x\|^2 \leq \delta\}, \tag{8.97}$$

where B is not necessarily non-negative definite so that the problem is not generally convex. The dual of (8.97) is given by the semidefinite program (SDP) [56]

$$\max_{\alpha,\lambda}\left\{\lambda \; : \; \begin{pmatrix} B + \alpha I & g \\ g^T & -\alpha\delta - \lambda \end{pmatrix} \succeq 0, \alpha \geq 0\right\}. \tag{8.98}$$

Problem (8.97) enjoys many useful and attractive properties. It is known that it admits no duality gap and that its semidefinite relaxation (SDR) is tight. Moreover, there exist a set of necessary and sufficient conditions that guarantee optimality of a solution to (8.97), which can be extracted from the dual solution $\bar{\alpha}$. These results all extend to the case in which the problem is to optimize an arbitrary quadratic function subject to a single quadratic constraint, where both quadratic forms are not necessarily convex.

Unfortunately, in general these results cannot be generalized to the case in which the constraint set consists of two quadratic restrictions. More specifically, consider the following quadratic problems:

$$(Q2P_{\mathbb{C}}) \quad \min_{z\in\mathbb{C}^n}\{f_3(z) : f_1(z) \geq 0, f_2(z) \geq 0\}, \tag{8.99}$$

$$(Q2P_{\mathbb{R}}) \quad \min_{x\in\mathbb{R}^n}\{f_3(x) : f_1(x) \geq 0, f_2(x) \geq 0\}. \tag{8.100}$$

In the real case each function $f_j : \mathbb{R}^n \to \mathbb{R}$ is defined by $f_j(x) = x^T A_j x + 2b_j^T x + c_j$ with $A_j = A_j^T \in \mathbb{R}^{n\times n}, b_j \in \mathbb{R}^n$, and $c_j \in \mathbb{R}$. In the complex setting, $f_j : \mathbb{C}^n \to \mathbb{R}$ is given by $f_j(z) = z^* A_j z + 2\Re(b_j^* z) + c_j$, where A_j are Hermitian matrices, that is, $A_j = A_j^*, b_j \in \mathbb{C}^n$, and $c_j \in \mathbb{R}$. We distinguish between the real and complex cases since we will see that different strong-duality results apply in both settings. In particular, there are stronger results for complex quadratic problems than for their real counterparts.

The problem $(Q2P_{\mathbb{R}})$ appears as a subproblem in some trust-region algorithms for constrained optimization [57–61] where the original problem is to minimize a general nonlinear function subject to equality constraints. Unfortunately, in general the strong duality results in the case of a single constraint cannot be extended to the case of two quadratic restrictions $(Q2P_{\mathbb{R}})$. Indeed, it is known that the SDR of $(Q2P_{\mathbb{R}})$ is not necessarily tight [60, 62]. An exception is the case in which the functions f_1, f_2, and f_3, are all homogenous quadratic functions and there exists a positive definite, linear combination of the matrices A_j [62]. Another setting in which strong duality is guaranteed is derived in [44] and will be discussed below.

Quadratic optimization in the complex domain is simpler. In [44] it is shown that under some mild conditions strong duality holds for the complex-valued problem $(Q2P_\mathbb{C})$ and that its semidefinite relaxation is tight. This result is based on the extended version of the S-lemma derived by Fradkov and Yakubovich [63]. The standard Lagrangian dual of $(Q2P_\mathbb{C})$ is given by

$$(D_\mathbb{C}) \quad \max_{\alpha \geq 0, \beta \geq 0, \lambda} \left\{ \lambda \ \middle| \ \begin{pmatrix} A_3 & b_3 \\ b_3^* & c_3 - \lambda \end{pmatrix} \succeq \alpha \begin{pmatrix} A_1 & b_1 \\ b_1^* & c_1 \end{pmatrix} + \beta \begin{pmatrix} A_2 & b_2 \\ b_2^* & c_2 \end{pmatrix} \right\}. \quad (8.101)$$

Problem $(D_\mathbb{C})$ is sometimes called Shor's relaxation [64]. Theorem 8.9 below states that if problem $(Q2P_\mathbb{C})$ is strictly feasible then $\mathrm{val}(Q2P_\mathbb{C}) = \mathrm{val}(D_\mathbb{C})$ *even* in the case where the value is equal to $-\infty$.

THEOREM 8.9 *[44] Suppose that problem $(Q2P_\mathbb{C})$ is strictly feasible, in other words, there exists $\tilde{z} \in \mathbb{C}^n$ such that $f_1(\tilde{z}) > 0, f_2(\tilde{z}) > 0$. Then,*

1. *If $\mathrm{val}(Q2P_\mathbb{C})$ is finite then the maximum of problem $(D_\mathbb{C})$ is attained and $\mathrm{val}(Q2P_\mathbb{C}) = \mathrm{val}(D_\mathbb{C})$.*
2. *$\mathrm{val}(Q2P_\mathbb{C}) = -\infty$ if, and only if $(D_\mathbb{C})$ is not feasible.*

Necessary and sufficient optimality conditions similar to those known for (8.97) were also derived in [44]. These conditions can be used to calculate the optimal solution of $(Q2P_\mathbb{C})$ from the dual solution.

It is interesting to note that the dual problem to $(D_\mathbb{C})$ is the so-called SDR of $(Q2P_\mathbb{C})$:

$$(SDR_\mathbb{C}) \quad \min_{Z} \{ \mathrm{Tr}(ZM_3) : \mathrm{Tr}(ZM_1) \geq 0, \mathrm{Tr}(ZM_2) \geq 0, Z_{n+1,n+1} = 1, Z \succeq 0 \},$$

$$(8.102)$$

where

$$M_j = \begin{pmatrix} A_j & b_j \\ b_j^* & c_j \end{pmatrix}. \quad (8.103)$$

If both problems $(Q2P_\mathbb{C})$ and $(D_\mathbb{C})$ are strictly feasible, then problems $(Q2P_\mathbb{C})$,$(D_\mathbb{C})$, and $(SDR_\mathbb{C})$ (problems (8.99),(8.101), and (8.102) respectively) attain their solutions and

$$\mathrm{val}(Q2P_\mathbb{C}) = \mathrm{val}(D_\mathbb{C}) = \mathrm{val}(SDR_\mathbb{C}). \quad (8.104)$$

The real-valued problem $(Q2P_\mathbb{R})$ is more difficult to handle. In contrast to the complex case, strong duality results are, generally speaking, not true for $(Q2P_\mathbb{R})$. It is not known whether $(Q2P_\mathbb{R})$ is a tractable problem or not and in that respect, if there is an efficient algorithm for finding its solution. If the constraints of $(Q2P_\mathbb{R})$ are convex then the complex-valued problem $(Q2P_\mathbb{C})$, considered as a relaxation of $(Q2P_\mathbb{R})$, can produce an approximate solution. Although strong duality results do not hold generally in the real case, a sufficient condition can be developed to ensure zero duality gap (and tightness of the semidefinite relaxation) for $(Q2P_\mathbb{R})$ [44]. This result is based on the connection between the image of the real and complex spaces under a quadratic mapping, and is given in terms of the dual-optimal values.

The dual problem to $(QP_\mathbb{R})$ is

$$(D_\mathbb{R}) \quad \max_{\alpha \geq 0, \beta \geq 0, \lambda} \left\{ \lambda \ \middle| \ \begin{pmatrix} A_3 & b_3 \\ b_3^T & c_3 - \lambda \end{pmatrix} \succeq \alpha \begin{pmatrix} A_1 & b_1 \\ b_1^T & c_1 \end{pmatrix} + \beta \begin{pmatrix} A_2 & b_2 \\ b_2^T & c_2 \end{pmatrix} \right\}. \quad (8.105)$$

Note that this is exactly the same as problem $(D_\mathbb{C})$ in (8.101), where here we used the fact that the data is real and therefore $b_j^* = b_j^T$. The SDR in this case is given by

$$(SDR_\mathbb{R}) \quad \min_X \{\mathrm{Tr}(XM_3) : \mathrm{Tr}(XM_1) \geq 0, \mathrm{Tr}(XM_2) \geq 0, X_{n+1,n+1} = 1, X \succeq 0\}. \quad (8.106)$$

Suppose that both problems $(Q2P_\mathbb{R})$ and $(D_\mathbb{R})$ are strictly feasible and there exists real values $\hat{\alpha}, \hat{\beta}$ such that

$$\hat{\alpha}A_1 + \hat{\beta}A_2 \succ 0. \quad (8.107)$$

Let $(\bar{\lambda}, \bar{\alpha}, \bar{\beta})$ be an optimal solution of the dual problem $(D_\mathbb{R})$. If

$$\dim\left(\mathcal{N}(A_3 - \bar{\alpha}A_1 - \bar{\beta}A_2)\right) \neq 1, \quad (8.108)$$

then $\mathrm{val}(Q2P_\mathbb{R}) = \mathrm{val}(D_\mathbb{R}) = \mathrm{val}(SDR_\mathbb{R})$ and there exists a real-valued solution to the complex-valued problem $(Q2P_\mathbb{C})$.

8.8.2   Minimax recovery using SDP relaxation

We now show how the strong duality results developed in the previous section can be used to solve the recovery problems (8.94) and (8.96). Our general approach is to replace the inner maximization by its dual [65]. Over the complex domain, strong duality holds, and the resulting problems are exact representations of (8.94) and (8.96). Over the reals, this leads to an approximation, however, in practice it is pretty tight and yields good recovery results. Alternatively, we can use an SDR approach to replace the inner maximization by its relaxation. The resulting problem is a convex–concave saddle-point program which can be further simplified by relying on Proposition 8.1 [66]. Both derivations are equivalent, since the dual problem of the inner maximization is also the dual of the (convex) SDR [56]. Here we follow the relaxation approach since its derivation is simpler.

Instead of focusing on our particular problem, we treat a general minimax formulation with a quadratic objective, and two quadratic constraints:

$$\min_{\hat{x}} \max_x \{\|A\hat{x} - Qx\|^2 : f_i(x) \leq 0, 1 \leq i \leq 2\}, \quad (8.109)$$

where

$$f_i(x) \overset{\triangle}{=} x^* A_i x + 2\Re\{b_i^* x\} + c_i. \quad (8.110)$$

Clearly (8.94) and (8.96) are special cases of (8.109) with

$$A_1 = SS^*, b_1 = -Sc, c_1 = \|c\|^2 - \alpha^2, A_2 = L^*L, b_2 = 0, c_2 = -\rho^2. \qquad (8.111)$$

The difference between the two problems is in the matrices A and Q. In (8.94) we have $A = Q = I$, whereas in (8.96) $A = W$ and $Q = P_{\mathcal{W}} = W(W^*W)^\dagger W^*$.

In order to develop a solution to (8.109) we first consider the inner maximization:

$$\max_{x}\{\|A\hat{x} - Qx\|^2 : f_i(x) \leq 0, 1 \leq i \leq 2\}, \qquad (8.112)$$

which is a special case of quadratic optimization with 2 quadratic constraints. Denoting $\Delta = xx^*$, (8.112) can be written equivalently as

$$\max_{(\Delta,x)\in\mathcal{G}}\{\|A\hat{x}\|^2 - 2\Re\{\hat{x}^*A^*Qx\} + \mathrm{Tr}(Q^*Q\Delta)\}, \qquad (8.113)$$

where

$$\mathcal{G} = \{(\Delta,x) : f_i(\Delta,x) \leq 0, \ 1 \leq i \leq 2, \Delta = xx^*\}, \qquad (8.114)$$

and we defined

$$f_i(\Delta,x) = \mathrm{Tr}(A_i\Delta) + 2\Re\{b_i^*x\} + c_i, \quad 1 \leq i \leq 2. \qquad (8.115)$$

The objective in (8.113) is concave (linear) in (Δ,x), but the set $\mathcal{G}$ is not convex. To obtain a relaxation of (8.113) we may replace $\mathcal{G}$ by the convex set

$$\mathcal{T} = \{(\Delta,x) : f_i(\Delta,x) \leq 0, \ 1 \leq i \leq 2, \Delta \succeq xx^*\}. \qquad (8.116)$$

Indeed, using Schur's lemma [67, p. 28] $\Delta \succeq xx^*$ can be written as a linear-matrix inequality. Our relaxation of (8.109) is the solution to the resulting minimax problem:

$$\min_{\hat{x}} \max_{(\Delta,x)\in\mathcal{T}}\{\|A\hat{x}\|^2 - 2\Re\{\hat{x}^*A^*Qx\} + \mathrm{Tr}(Q^*Q\Delta)\}. \qquad (8.117)$$

The objective in (8.117) is concave (linear) in Δ and x and convex in $\hat{x}$. Furthermore, the set $\mathcal{T}$ is bounded. Therefore, from Proposition 8.1 we can replace the order of the minimization and maximization, resulting in the equivalent problem

$$\max_{(\Delta,x)\in\mathcal{T}} \min_{\hat{x}}\{\|A\hat{x}\|^2 - 2\Re\{\hat{x}^*A^*Qx\} + \mathrm{Tr}(Q^*Q\Delta)\}. \qquad (8.118)$$

The inner minimization is a simple quadratic problem. Expressing the objective as $\|A\hat{x} - Qx\|^2$, it can be seen that its solution satisfies $A\hat{x} = P_{\mathcal{A}}Qx$, where $\mathcal{A} = \mathcal{R}(A)$. Substituting this result into (8.118), our problem reduces to

$$\max_{(\Delta,x)\in\mathcal{T}}\{-\|P_{\mathcal{A}}Qx\|^2 + \mathrm{Tr}(Q^*Q\Delta)\}. \qquad (8.119)$$

Problem (8.119) is a convex-optimization problem with a concave objective and linear matrix-inequality constraints and can therefore be solved easily using standard software packages. The approximate minimax solution to (8.109) is the x-part of the solution to (8.119). When (8.109) is defined over the complex domain, then this solution is exact. In the real case, it will be exact when condition (8.108) is satisfied.

Instead of solving (8.119) we may consider its dual function. Since (8.119) is convex, strong duality holds. For simplicity, we will assume that L^*L is invertible so that $A_2 \succ 0$ in our case. We will also use the fact that in our setting, $Q = P_{\mathcal{A}}$, where $\mathcal{A} = \mathcal{R}(A)$. This leads to a simple explicit expression for the solution $\hat{x}$:

THEOREM 8.10 *Assume that at least one of the matrices $\{A_i\}$ is strictly positive definite and that $Q = P_{\mathcal{A}}$. Then the solution to (8.119) is given by*

$$\hat{x} = -\left(\sum_{i=1}^{2} \alpha_i A_i\right)^{-1}\left(\sum_{i=1}^{2} \alpha_i b_i\right), \tag{8.120}$$

where (α_1, α_2) is an optimal solution of the following convex-optimization problem in 2 variables:

$$\min_{\alpha_i} \left\{ \sum_{i=1}^{2} \alpha_i b_i^* \left(\sum_{i=1}^{2} \alpha_i A_i\right)^{-1} \sum_{i=1}^{2} \alpha_i b_i - \sum_{i=1}^{2} c_i \alpha_i \right\}$$

$$\text{s.t.} \quad \sum_{i=1}^{2} \alpha_i A_i \succeq Q^*Q,$$

$$\alpha_i \geq 0, \quad 1 \leq i \leq 2. \tag{8.121}$$

Proof To prove the theorem we show that (8.121) is the dual of (8.119). Since (8.119) is convex and strictly feasible, its optimal value is equal to that of its dual problem. To compute the dual, we first form the Lagrangian:

$$\mathcal{L} = -\|P_{\mathcal{A}}Qx\|^2 + \mathrm{Tr}(Q^*Q\Delta) + \mathrm{Tr}(\Pi(\Delta - xx^T))$$

$$- \sum_{i=1}^{2} \alpha_i (\mathrm{Tr}(A_i\Delta) + 2\Re\{b_i^*x\} + c_i), \tag{8.122}$$

where $\alpha_i \geq 0$ and $\Pi \succeq 0$ are the dual variables. The maximization of $\mathcal{L}$ with respect to x yields

$$x = -(Q^*P_{\mathcal{A}}Q + \Pi)^{-1} \sum_{i=1}^{2} \alpha_i b_i. \tag{8.123}$$

The derivative with respect to Δ yields,

$$Q^*Q + \Pi = \sum_{i=1}^{2} \alpha_i A_i. \tag{8.124}$$

Using the fact that $Q^*P_{\mathcal{A}}Q = Q^*Q$ and that $Q^*(P_{\mathcal{A}} - I)Q = 0$, the combination of (8.123) and (8.124) yields (8.120).

Next we note that since $\Pi \succeq 0$, we must have from (8.124) that $\sum_{i=0}^{k} \alpha_i A_i \succeq Q^*Q$. Finally, substituting (8.123) and (8.124) into (8.122), we obtain the dual problem (8.121). $\blacksquare$

Returning to our reconstruction problems, the substitution of (8.111) in Theorem 8.10 implies that $\hat{x}_{\text{MX}}$ of (8.94) is given by

$$\hat{x}_{\text{MX}} = (SS^* + \lambda L^*L)^{-1}Sc, \tag{8.125}$$

where $\lambda = \alpha_2/\alpha_1$. If $\alpha_1 = 0$ then $\hat{x}_{\text{MX}} = 0$. Similarly, $\hat{x}_{\text{CMX}}$ corresponding to (8.96) is given by

$$\hat{x}_{\text{CMX}} = P_{\mathcal{W}}\hat{x}_{\text{MX}}, \tag{8.126}$$

where we used the fact that $A = W$ and $Q = P_{\mathcal{W}}$ in this case. This shows that, as in the noiseless scenarios of Sections 8.5 and 8.6, here too the constrained minimax-regret solution relates to the unconstrained minimax recovery via an orthogonal projection.

Problem (8.121) can be cast as an SDP:

$$\min_{\alpha_i} \quad \left\{ t - \sum_{i=1}^{2} c_i\alpha_i \right\}$$

$$\text{s.t.} \quad \begin{pmatrix} Q^*(P_{\mathcal{A}} - I)Q + \sum_{i=1}^{2}\alpha_i A_i & \sum_{i=1}^{2}\alpha_i b_i \\ \sum_{i=1}^{2}\alpha_i b_i^* & t \end{pmatrix} \succeq 0, \tag{8.127}$$

$$\sum_{i=1}^{2}\alpha_i A_i \succeq Q^*Q,$$

$$\alpha_i \geq 0, \quad 1 \leq i \leq 2.$$

This SDP can be solved by one of the many available SDP solvers such as the Self-Dual-Minimization (SeDuMi) package [68] or CVX [69].

It can be shown that both the true minimax solution and its approximation are feasible, namely they satisfy the quadratic constraints [66]. This approach can also be extended to the case when there are more than 2 constraints.

Example 8.5 Figure 8.13 shows a concrete example of the approximation discussed above for the unconstrained minimax reconstruction of (8.94). This reconstruction is compared with the LS solution $\hat{x}_{\text{LS}} = \min_{\|Lx\|\leq\rho} \|S^*x - c\|^2$. In this experiment, the sampling filter is a rectangular window whose support is equal to the sampling interval,

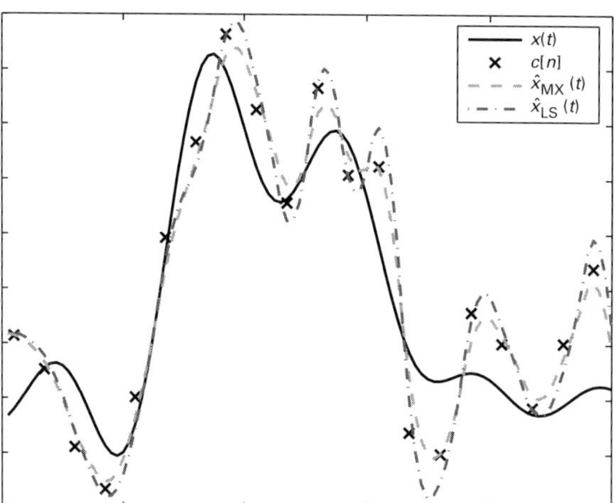

Figure 8.13   Comparison of minimax and LS reconstruction. The error norm of $\hat{x}_{LS}$ is 30% higher than that of $\hat{x}_{MX}$.

the noise is white and Gaussian with variance σ^2, and the signal $x = (x(1), \ldots, x(100))^T$ is given by $x(n) = \mathrm{sinc}(0.1(n - 33.3)) + \exp\{-0.005(n - 50.5)^2\}$. The regularization operator L was taken to be a (discrete approximation of) first-order derivative. The parameter α was chosen as $3\sqrt{K}\sigma$, where K is the number of samples. Both the minimax and LS reconstructions were produced with the same ρ. It can be seen that $\hat{x}_{LS}$ tends to oscillate more than $\hat{x}_{MX}$. Consequently, its reconstruction error is larger than that of $\hat{x}_{MX}$ by 30%.

To conclude, we have seen that when the samples are perturbed by noise, we can obtain an approximation of the minimax recovery by numerically solving an SDP. In some cases, this strategy leads to the exact minimax solution. The minimax method often yields improved reconstructions over LS. This is especially true in the constrained setting, where the minimax recovery is the orthogonal projection of the unconstrained solution onto the reconstruction space, whereas the LS recovery may deviate substantially from the unconstrained method.

8.9   Conclusions

In this chapter we revisited the fundamental problem of reconstructing signals from their samples. We considered several models for each of the essential ingredients of the sampling problem: the sampling mechanism (general pre-filter, noise), the reconstruction kernel (pre-specified, unrestricted), and the signal prior (subspace, smoothness). Our approach was to define an optimization problem that takes into account both the fit of the reconstructed signal to the given samples and the prior knowledge we have about the signal. Each of the settings studied in this chapter was treated using two optimization strategies: LS and minimax. We showed that when the samples are noise-free, both strategies coincide if the reconstruction mechanism is unrestricted. In this case, perfect recovery is often possible under a subspace prior. In contrast, when the recovery is constrained, the minimax strategy leads to solutions that are closer to the original signal. The last part of this chapter was devoted to the challenging task of treating smoothness priors via the minimax strategy in the case in which the samples are noisy. Since closed-form solutions are unavailable in this setting, we restricted our attention to signals lying in $\mathbb{R}^n$ and $\mathbb{C}^n$ and showed how the resulting problems can be solved numerically using standard optimization packages. This was made possible by relying on recent results in optimization theory, regarding the tightness of SDP relaxations in quadratic problems.

Acknowledgments

We thank Dr. Ewa Matusiak for fruitful discussions.

This work was supported in part by the Israel Science Foundation under Grant no. 1081/07 and by the European Commission in the framework of the FP7 Network of Excellence in Wireless COMmunications NEWCOM++ (contract no. 216715).

References

[1] C. E. Shannon, "Communications in the presence of noise," *Proceedings of the IRE*, vol. 37, pp. 10–21, 1949.

[2] E. T. Whitaker, "On the functions which are represented by the expansion of interpolating theory," in *Proceedings of the Royal Society of Edinburgh*, vol. 35, 1915, pp. 181–94.

[3] V. A. Kotelnikov, "On the transmission capacity of 'ether'," in *Proceedings of the First All-union Conference on Questions of Communications*, 1933.

[4] H. Nyquist, "Certain topics in telegraph transmission theory," *AIEE Transactions*, vol. 47, pp. 617–44, 1928.

[5] M. Unser and A. Aldroubi, "A general sampling theory for nonideal acquisition devices," *IEEE Transactions on Signal Processing*, vol. 42, no. 11, pp. 2915–25, 1994.

[6] A. Aldroubi and M. Unser, "Sampling procedures in function spaces and asymptotic equivalence with Shannon's sampling theory," *Numerical Functional Analysis & Optimization*, vol. 15, pp. 1–21, 1994.

[7] A. Aldroubi and K. Gröchenig, "Non-uniform sampling and reconstruction in shift-invariant spaces," *Siam Review*, vol. 43, pp. 585–620, 2001.

[8] Y. C. Eldar, "Sampling and reconstruction in arbitrary spaces and oblique dual frame vectors," *Journal of Fourier Analysis & Applications*, vol. 1, no. 9, pp. 77–96, 2003.

[9] ——, "Sampling without input constraints: consistent reconstruction in arbitrary spaces," in *Sampling, Wavelets and Tomography*, A. I. Zayed and J. J. Benedetto, eds. Boston, MA: Birkhäuser, 2004, pp. 33–60.

[10] I. J. Schoenberg, "Spline functions and the problem of graduation," *Proceedings of the National Academy of Sciences*, vol. 52, no. 4, pp. 947–50, 1964.

[11] S. Ramani, D. Van De Ville, T. Blu, and M. Unser, "Nonideal sampling and regularization theory," *IEEE Transactions on Signal Processing*, vol. 56, no. 3, pp. 1055–70, 2008.

[12] M. Unser and T. Blu, "Generalized smoothing splines and the optimal discretization of the Wiener filter," *IEEE Transactions on Signal Processing*, vol. 53, no. 6, pp. 2146–59, 2005.

[13] T. Michaeli and Y. C. Eldar, "High rate interpolation of random signals from nonideal samples," *IEEE Transactions on Signal Processing*, vol. 57, no. 3, pp. 977–92, 2009.

[14] Y. C. Eldar and M. Unser, "Nonideal sampling and interpolation from noisy observations in shift-invariant spaces," *IEEE Transactions on Signal Processing*, vol. 54, no. 7, pp. 2636–51, 2006.

[15] Y. C. Eldar and T. Michaeli, "Beyond bandlimited sampling," *IEEE Signal Processing Magazine*, vol. 26, no. 3, pp. 48–68, 2009.

[16] S. Kayalar and H. L. Weinert, "Oblique projections: formulas, algorithms, and error bounds," *Mathematics of Control, Signals & Systems*, vol. 2, no. 1, pp. 33–45, 1989.

[17] O. Christensen, *Frames and Bases. An Introductory Course*. Boston, MA: Birkhäuser, 2008.

[18] I. Daubechies, "The wavelet transform, time-frequency localization and signal analysis," *IEEE Transactions on Information*, vol. 36, pp. 961–1005, 1990.

[19] ——, *Ten Lectures on Wavelets*. Philadelphia, PA: SIAM, 1992.

[20] R. M. Young, *An Introduction to Nonharmonic Fourier Series*. New York: Academic Press, 1980.

[21] M. Unser, "Sampling—50 years after Shannon," *IEEE Proceedings*, vol. 88, pp. 569–87, 2000.

[22] M. Mishali and Y. C. Eldar, "Blind multi-band signal reconstruction: compressed sensing for analog signals," *IEEE Transactions on Signal Processing*, vol. 57, no. 3, pp. 993–1009, 2009.

[23] Y. C. Eldar, "Compressed sensing of analog signals in shift invariant spaces," *IEEE Transactions on Signal Processing*, vol. 57, no. 8, pp. 2986–97, 2009.

[24] D. L. Donoho, "Compressed sensing," *IEEE Transactions on Information Theory*, vol. 52, no. 4, pp. 1289–306, 2006.

[25] E. Candes, J. Romberg, and T. Tao, "Robust uncertainty principles: exact signal reconstruction from highly incomplete frequency information," *IEEE Transactions on Information Theory*, vol. 52, no. 2, pp. 489–509, 2006.

[26] M. Mishali and Y. C. Eldar, "Reduce and boost: recovering arbitrary sets of jointly sparse vectors," *IEEE Transactions on Signal Processing*, vol. 56, no. 10, pp. 4692–702, 2008.

[27] Y. C. Eldar, "Uncertainty relations for shift-invariant analog signals," to appear in *IEEE Transactions on Information Theory*.

[28] Y. M. Lu and M. N. Do, "A theory for sampling signals from a union of subspaces," *IEEE Transactions on Signal Processing*, vol. 56, no. 6, pp. 2334–45, 2008.

[29] Y. C. Eldar and M. Mishali, "Robust recovery of signals from a structured union of subspaces," to appear in *IEEE Transactions on Information Theory*.

[30] L. Rudin, S. Osher, and E. Fatemi, "Nonlinear total variation based noise removal algorithms," *Physica D*, vol. 60, no. 1–4, pp. 259–68, 1992.

[31] P. P. Vaidyanathan, "Generalizations of the sampling theorem: seven decades after Nyquist," *IEEE Transactions on Circuit & Systems I*, vol. 48, no. 9, pp. 1094–109, 2001.

[32] Y. C. Eldar and T. G. Dvorkind, "A minimum squared-error framework for generalized sampling," *IEEE Transactions on Signal Processing*, vol. 54, no. 6, pp. 2155–67, 2006.

[33] A. Aldroubi, "Oblique projections in atomic spaces," *Proceedings of the American Mathematical Society*, vol. 124, no. 7, pp. 2051–60, 1996.

[34] T. G. Dvorkind, Y. C. Eldar, and E. Matusiak, "Nonlinear and non-ideal sampling: theory and methods," *IEEE Transactions on Signal Processing*, vol. 56, no. 12, pp. 5874–90, 2008.

[35] R. G. Keys, "Cubic convolution interpolation for digital image processing," *IEEE Transactions on Acoustics Speech & Signal Processing*, vol. 29, no. 6, pp. 1153–60, 1981.

[36] Y. C. Eldar and T. Werther, "General framework for consistent sampling in Hilbert spaces," *International Journal of Wavelets, Multiresolution, and Information Processing*, vol. 3, no. 3, pp. 347–59, 2005.

[37] I. J. Schoenberg, *Cardinal Spline Interpolation*. Philadelphia, PA: SIAM, 1973.

[38] M. Unser, A. Aldroubi, and M. Eden, "B-Spline signal processing: Part I - Theory," *IEEE Transactions on Signal Processing*, vol. 41, no. 2, pp. 821–33, 1993.

[39] ——, "B-Spline signal processing: Part II - Efficient design and applications," *IEEE Transactions on Signal Processing*, vol. 41, no. 2, pp. 834–848, 1993.

[40] Y. C. Eldar, "Rethinking biased estimation: improving maximum likelihood and the Cramer–Rao bound," *Foundations and Trends in Signal Processing*, vol. 1, no. 4, pp. 305–449, 2007.

[41] Y. C. Eldar, A. Ben-Tal, and A. Nemirovski, "Robust mean-squared error estimation in the presence of model uncertainties," *IEEE Transactions on Signal Processing*, vol. 53, no. 1, pp. 168–81, 2005.

[42] T. G. Dvorkind, H. Kirshner, Y. C. Eldar, and M. Porat, "Minimax approximation of representation coefficients from generalized samples," *IEEE Transactions on Signal Processing*, vol. 55, pp. 4430–43, 2007.

[43] M. Sion, "On general minimax theorems," *Pacific Journal of Mathematics*, vol. 8, pp. 171–6, 1958.

[44] A. Beck and Y. C. Eldar, "Strong duality in nonconvex quadratic optimization with two quadratic constraints," *Siam Journal on Optimization*, vol. 17, no. 3, pp. 844–60, 2006.

[45] Y. C. Eldar and O. Christensen, "Characterization of oblique dual frame pairs," *Journal of Applied Signal Processing*, pp. 1–11, 2006, article ID 92674.

[46] O. Christensen and Y. C. Eldar, "Oblique dual frames and shift-invariant spaces," *Applied & Computational Harmonic Analysis*, vol. 17, no. 1, pp. 48–68, 2004.

[47] Y. C. Eldar, A. Ben-Tal, and A. Nemirovski, "Linear minimax regret estimation of deterministic parameters with bounded data uncertainties," *IEEE Transactions on Signal Processing*, vol. 52, pp. 2177–88, 2004.

[48] Y. C. Eldar and N. Merhav, "A competitive minimax approach to robust estimation of random parameters," *IEEE Transactions on Signal Processing*, vol. 52, pp. 1931–46, 2004.

[49] E. Lieb and M. Loss, *Analysis*. Providence, RI: American Mathematical Society, 2001.

[50] A. Ben-Tal and M. Teboulle, "Hidden convexity in some nonconvex quadratically constrained quadratic programming," *Mathematical Programming*, vol. 72, no. 1, pp. 51–63, 1996.

[51] H. G. Feichtinger and T. Werther, "Robustness of minimal norm interpolation in Sobolev algebras," in *Sampling, Wavelets and Tomography*, A. I. Zayed and J. J. Benedetto, eds. Boston, MA: Birkhäuser, 2004.

[52] J. M. Martínez, "Local minimizers of quadratic functions on Euclidean balls and spheres," *SIAM Journal on Optimization*, vol. 4, no. 1, pp. 159–76, 1994.

[53] J. J. Moré and D. C. Sorensen, "Computing a trust region step," *SIAM Journal on Scientific & Statistical Computing*, vol. 4, no. 3, pp. 553–72, 1983.

[54] D. C. Sorensen, "Newton's method with a model trust region modification," *SIAM Journal on Numerical Analysis*, vol. 19, no. 2, pp. 409–26, 1982.

[55] R. J. Stern and H. Wolkowicz, "Indefinite trust region subproblems and nonsymmetric eigenvalue perturbations," *SIAM Journal on Optimization*, vol. 5, no. 2, pp. 286–313, 1995.

[56] L. Vandenberghe and S. Boyd, "Semidefinite programming," *SIAM Review*, vol. 38, no. 1, pp. 40–95, 1996.

[57] M. R. Celis, J. E. Dennis, and R. A. Tapia, "A trust region strategy for nonlinear equality constrained optimization," in *Numerical Optimization, 1984 (Boulder, Colo., 1984)*. Philadelphia, PA: SIAM, 1985, pp. 71–82.

[58] A. R. Conn, N. I. M. Gold, and P. L. Toint, *Trust-Region Methods*, MPS/SIAM Series on Optimization. Philadelphia, PA: Society for Industrial and Applied Mathematics (SIAM), 2000.

[59] M. J. D. Powell and Y. Yuan, "A trust region algorithm for equality constrained optimization," *Mathematical Programming*, vol. 49, no. 2, (Ser. A), pp. 189–211, 1990.

[60] Y. Yuan, "On a subproblem of trust region algorithms for constrained optimization," *Mathematical Programming*, vol. 47, pp. 53–63, 1990.

[61] Y. M. Zhu, "Generalized sampling theorem," *IEEE Transactions on Circuits & Systems II*, vol. 39, pp. 587–8, 1992.

[62] Y. Ye and S. Zhang, "New results on quadratic minimization," *SIAM Journal on Optimization*, vol. 14, no. 1, pp. 245–67, 2003.

[63] A. L. Fradkov and V. A. Yakubovich, "The S-procedure and the duality relation in convex quadratic programming problems," *Vestnik Leningrad Univ.*, vol. 1, pp. 81–87, 1973.

[64] N. Z. Shor, "Quadratic optimization problems," *Izvestiya Akademii Tekhnicheskaya Nauk SSSR Tekhnicheskaya Kibernetika*, no. 1, pp. 128–39, 222, 1987.

[65] A. Beck and Y. C. Eldar, "Regularization in regression with bounded noise: a Chebyshev center approach," *SIAM Journal on Matrix Analysis Applications*, vol. 29, no. 2, pp. 606–25, 2007.

[66] Y. C. Eldar, A. Beck, and M. Teboulle, "A minimax Chebyshev estimator for bounded error estimation," *IEEE Transactions on Signal Processing*, vol. 56, no. 4, pp. 1388–97, 2008.

[67] S. Boyd, L. E. Ghaoui, E. Feron, and V. Balakrishnan, *Linear Matrix Inequalities in System and Control Theory*. Philadelphia, PA: SIAM, 1994.

[68] J. F. Sturm, "Using SeDuMi 1.02, a MATLAB toolbox for optimization over symmetric cones," *Optimization Methods and Software*, vol. 11–12, pp. 625–53, 1999.

[69] M. Grant and S. Boyd, (2008, March). "CVX: Matlab software for disciplined convex programming," Available: http://stanford.edu/~boyd/cvx

9 Robust broadband adaptive beamforming using convex optimization

Michael Rübsamen, Amr El-Keyi, Alex B. Gershman, and Thia Kirubarajan

Several worst-case performance optimization-based broadband adaptive beamforming techniques with an improved robustness against array manifold errors are developed. The proposed beamformers differ from the existing broadband robust techniques in that their robustness is directly matched to the amount of uncertainty in the array manifold, and the suboptimal subband decomposition step is avoided. Convex formulations of the proposed beamformer designs are derived based on *second-order cone programming* (SOCP) and *semidefinite programming* (SDP). Simulation results validate an improved robustness of the proposed robust beamformers relative to several state-of-the-art robust broadband techniques.

9.1 Introduction

Adaptive array processing has received considerable attention during the last four decades, particularly in the fields of sonar, radar, speech acquisition and, more recently, wireless communications [1,2]. The main objective of adaptive beamforming algorithms is to suppress the interference and noise while preserving the desired signal components. One of the early adaptive beamforming algorithms for broadband signals is the *linearly constrained minimum variance* (LCMV) algorithm developed by Frost in [3] and extensively studied in the follow-up literature [1, 4, 5]. Frost's broadband array processor includes a presteering delay front-end whose function is to steer the array towards the desired signal so that each of its frequency components appears in-phase across the array after the presteering delays. Each presteering delay is then followed by a *finite impulse response* (FIR) adaptive filter and the outputs of all these filters are summed together to obtain the array output. The LCMV algorithm optimizes the filter weight coefficients to minimize the array output power subject to distortionless response constraints. These constraints are used to preserve the desired signal components that have to appear in-phase after the presteering stage. However, in practical cases certain errors in the array manifold are inevitable. Such errors may be caused, for example, by signal wavefront distortions, look direction errors, array imperfections, quantization errors in presteering delays, etc. As a result, the desired signal components cannot be perfectly phase-aligned by the presteering delays and, therefore, they may be erroneously suppressed by the

Convex Optimization in Signal Processing and Communication, eds. Daniel P. Palomar and Yonina C. Eldar.
Published by Cambridge University Press © Cambridge University Press, 2010.

subsequent adaptive beamforming stage. This effect is commonly referred to as *signal self-nulling*.

Several approaches have been developed to improve the robustness of the broadband LCMV technique against look direction errors [6–10]. These approaches use the point or derivative constraints to stabilize the broadband array response. However, they do not provide robustness against arbitrary manifold errors. Moreover, it is rather difficult to match the choice of the robustness parameters of these algorithms to the amount of uncertainty in the array manifold.

Another popular approach to incorporate robustness in the broadband LCMV beam-forming scheme is the diagonal loading technique [11–13]. Diagonal loading is well known to be equivalent to penalizing large values of the beamformer weight vector [11, 14]. However, in the conventional *fixed diagonal loading* techniques, the choice of the diagonal loading factor (that determines the resulting beamformer robustness) is set up in an ad hoc way, without matching it to the amount of uncertainty in the array manifold [12–14].

Recently, several theoretically rigorous algorithms have been proposed to add robustness to the narrowband *minimum variance distortionless response* (MVDR) beam-formers so that the amount of robustness is optimally (in the worst-case sense) matched to the amount of the uncertainty in the array manifold [15–20]. It has been shown in [15] and [18] that these algorithms can be viewed as *adaptive diagonal loading* techniques.

Two subband decomposition-based extensions of the robust MVDR beamformer of [18] to the broadband case have been developed in [21]: the *constant power width* (CPW) and the *constant beam width* (CBW) MVDR beamformers. However, the CBW beamformer has a rather limited application because it requires quite a specific array structure [21]. Moreover, both the CBW and CPW beamformers optimize each subband weight vector independently and, hence, completely ignore inter-subband relationships of the beamformer phase response. This, however, may cause a substantial performance degradation of the latter beamforming schemes.

In this chapter (see also [22–24]), several alternative approaches to robust broadband beamforming are developed using worst-case performance optimization. The proposed beamformers extend the narrowband robust approaches of [15] and [17]. In contrast to the approaches of [21], these techniques avoid the suboptimal subband decomposition step. Our beamformers are designed to protect not only the desired signal appearing in-phase after the presteering delays, but also its mismatched components whose norm of the error vector does not exceed a certain preselected value. The resulting optimization problems are converted to convex *second-order cone programming* (SOCP) and *semidefinite programming* (SDP) forms and, therefore, they can be efficiently solved in polynomial time using interior-point methods. Simulation results compare the proposed robust broadband beamformer designs with several state-of-the-art wideband techniques and demonstrate their enhanced robustness relative to these existing techniques.

The remainder of this chapter is organized as follows. Some necessary background on robust broadband beamforming is presented in Section 9.2. Section 9.3 contains the formulation of our robust broadband beamformers. Simulation results are presented in Section 9.4. Conclusions are drawn in Section 9.5.

9.2 Background

Consider an M-sensor, L-tap broadband adaptive beamformer shown in Figure 9.1. The sensor locations are chosen to meet the Nyquist spatial sampling criterion, that is, the sensors are placed densely enough to avoid spatial aliasing effects at all frequencies. For a *uniform linear array* (ULA), this corresponds to the condition that the sensors are placed at a distance less than or equal to $c/2f_u$, where c is the propagation speed and f_u is the maximal frequency in the frequency band. The output of the mth sensor is passed through a presteering delay T_m and, subsequently, an FIR filter with $(L-1)$ tap delays T_s, where the sampling frequency $f_s = 1/T_s$ is selected greater than or equal to $2f_u$ to avoid aliasing in the frequency domain.

The $M \times L$ matrix W is composed of the FIR filter weight coefficients where $W_{m,l}$ is the weight coefficient in the mth sensor after the $(l-1)$th tap delay. Vectorizing W yields the $ML \times 1$ weight vector

$$w \triangleq \mathrm{vec}(W) = \left[W_{:,1}^T, W_{:,2}^T, \ldots, W_{:,L}^T \right]^T,$$

where $W_{:,l}$, $(\cdot)^T$, and $\mathrm{vec}(\cdot)$ denote the lth column of W, the transpose, and the vectorization operator, respectively.

The $ML \times 1$ beamformer snapshot vector at the kth time instant is given by

$$x(k) \triangleq \left[x_1^T(k), x_2^T(k), \ldots, x_L^T(k) \right]^T,$$

where the $M \times 1$ vector $x_l(k)$ contains the array data after the $(l-1)$th tap delay. As we consider broadband signals without down-conversion, both the array snapshot vector

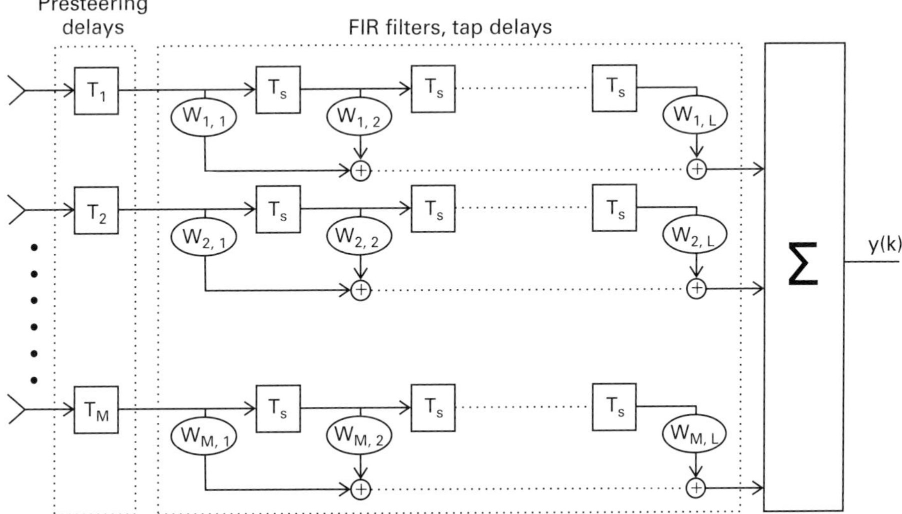

Figure 9.1 Broadband adaptive beamformer.

and the weight vector are real-valued. We can write

$$x(k) = x_s(k) + x_i(k) + x_n(k),$$

where $x_s(k)$, $x_i(k)$, and $x_n(k)$ are the signal, interference, and noise components, respectively. Then the beamformer output can be expressed as

$$y(k) = w^T x(k),$$

and the output power is given by

$$P_{\text{out}} = w^T R_x w,$$

where $R_x = \mathrm{E}\{x(k)x^T(k)\}$ is the $ML \times ML$ array covariance matrix, and $\mathrm{E}\{\cdot\}$ stands for the statistical expectation operator.

The array response for the desired signal can be expressed as

$$H_s(f) = w^T \left[d(f) \otimes (T(f)a_s(f)) \right], \tag{9.1}$$

where f is the frequency, $a_s(f)$ is the $M \times 1$ desired signal steering vector,

$$d(f) \triangleq [1, e^{-j2\pi f T_s}, \ldots, e^{-j2\pi f (L-1)T_s}]^T,$$

$$T(f) \triangleq \mathrm{diag}\{e^{-j2\pi f T_1}, \ldots, e^{-j2\pi f T_M}\},$$

and $\otimes$ denotes the Kronecker product.

In the sequel, we assume for notational simplicity that all the array sensors are omnidirectional with unit gain and that the signal sources are located in the far-field. Then the array-steering vector for the desired signal can be written as

$$a_s(f) = [e^{-j2\pi f \tau_1}, \ldots, e^{-j2\pi f \tau_M}]^T,$$

where τ_m is the propagation delay of the desired signal to the mth sensor. If the azimuth and elevation *directions-of-arrival* (DOAs) of the desired signal are ϕ_s and θ_s, respectively, then

$$\tau_m = -\frac{x_m}{c} \sin(\phi_s) \cos(\theta_s) - \frac{y_m}{c} \cos(\phi_s) \cos(\theta_s) - \frac{z_m}{c} \sin(\theta_s), \quad m = 1, \ldots, M,$$

where (x_m, y_m, z_m) are the coordinates of the mth sensor. The function of the presteering delays is to align the desired signal components in the sensors. Hence, the presteering delays should be ideally chosen as

$$T_m = T_0 - \tau_m, \quad m = 1, \ldots, M,$$

where T_0 ensures that the presteering delays are non-negative. Then, the beamformer-frequency response for the desired signal becomes

$$H_s(f) = w^T (d(f) \otimes 1_M) e^{-j2\pi f T_0} = w^T C_0 d(f) e^{-j2\pi f T_0}, \tag{9.2}$$

where $\mathbf{1}_M$ is the $M \times 1$ vector of ones, $\mathbf{C}_0 \triangleq \mathbf{I}_L \otimes \mathbf{1}_M$, and $\mathbf{I}_L$ is the $L \times L$ identity matrix.

In practice, the steering vector is known imperfectly and presteering delays contain quantization errors. Hence, we can write

$$\mathbf{T}(f)\mathbf{a}_s(f) = e^{-j2\pi f T_0}\mathbf{1}_M + \boldsymbol{\Delta}(f), \tag{9.3}$$

where the error vector $\boldsymbol{\Delta}(f)$ combines the effects of the steering vector and delay quantization errors. Then, inserting (9.3) into (9.1), the array response to the desired signal can be expressed as

$$\begin{aligned}
H_s(f) &= e^{-j2\pi f T_0}\mathbf{w}^T\left(\mathbf{d}(f) \otimes \mathbf{1}_M\right) + \mathbf{w}^T\left(\mathbf{d}(f) \otimes \boldsymbol{\Delta}(f)\right) \\
&= e^{-j2\pi f T_0}\mathbf{w}^T\mathbf{C}_0\mathbf{d}(f) + \mathbf{w}^T\mathbf{Q}(f)\boldsymbol{\Delta}(f),
\end{aligned}$$

where $\mathbf{Q}(f) \triangleq \mathbf{d}(f) \otimes \mathbf{I}_M$.

9.2.1 Linearly constrained minimum variance beamformer

The LCMV beamformer minimizes the output power subject to linear constraints on the weight vector $\mathbf{w}$. In general, the LCMV beamforming problem is formulated as

$$\min_{\mathbf{w}} \mathbf{w}^T\mathbf{R}_x\mathbf{w} \quad \text{s.t.} \quad \mathbf{D}^T\mathbf{w} = \mathbf{h}, \tag{9.4}$$

where $\mathbf{D}$ is the $ML \times J$ constraint matrix, $\mathbf{h}$ is the $J \times 1$ constraint vector, and J is the number of linear constraints used. If $\mathbf{R}_x$ is nonsingular and $\mathbf{D}$ has full-column rank, the solution to (9.4) can be written as

$$\mathbf{w}_{\text{LCMV}} = \mathbf{R}_x^{-1}\mathbf{D}\left(\mathbf{D}^T\mathbf{R}_x^{-1}\mathbf{D}\right)^{-1}\mathbf{h}.$$

The broadband MVDR beamformer [1] is an example of an LCMV beamformer which ensures that the desired signal passes through the broadband scheme without distortion when the frequency response is given by (9.2), that is, when there are no model errors. This is achieved by choosing $\mathbf{D} = \mathbf{C}_0$ and $\mathbf{h} = \mathbf{e}_{L_0}^{(L)}$, where $\mathbf{e}_{L_0}^{(L)}$ is the $L \times 1$ vector containing one in its L_0th position and zeros elsewhere, and L_0 determines the time delay of the desired signal after passing through the FIR filters. If there are no model errors, then the resulting frequency response for the desired signal becomes

$$H_s(f) = e^{-j2\pi f (T_0 + (L_0-1)T_s)} \quad \forall f \in \left[-\frac{f_s}{2}, \frac{f_s}{2}\right]. \tag{9.5}$$

In the sequel, we assume that L is odd and that $L_0 = (L+1)/2$.

The broadband MVDR beamformer is known to suffer from its sensitivity to errors in the array manifold. Several techniques have been proposed to improve the robustness by introducing additional linear constraints on the weight vector [6, 8–10]. For example, the presteering derivative constraints proposed in [10] ensure that the partial derivatives of the power response $\rho(f) = |H_s(f)|^2$ with respect to the presteering delays up to the N_dth order are equal to zero for all frequencies, that is

$$\frac{\partial^n \rho(f)}{\partial T_1^{n_1} \partial T_2^{n_2} \ldots \partial T_M^{n_M}} = 0 \quad \forall f \in \left[-\frac{f_s}{2}, \frac{f_s}{2}\right]; \ n = 1, \ldots, N_d;$$

$$n_m = 0, \ldots, n; \ \sum_{m=1}^{M} n_m = n.$$

These constraints improve the beamformer robustness against errors in the presteering delays T_m, $m = 1, \ldots, M$ (which are equivalent to phase errors in $a_s(f)$). However, the presteering derivative constraints do not provide robustness against magnitude errors in $a_s(f)$, and it is not possible to match the amount of robustness to the presumed amount of modeling errors.

It has been shown in [10] that if $C_0^T w = h$, then the first-order presteering derivative constraints are equivalent to

$$\left(\left(h^T(T_l^{(L)} - T_{-l}^{(L)})\right) \otimes e_m^{(M)^T}\right) w = 0 \quad \forall \, m = 1, \ldots, M; \ l = 1, \ldots, L_0 - 1, \quad (9.6)$$

where the $L \times L$ matrix $T_l^{(L)}$ is defined as

$$\left[T_l^{(L)}\right]_{s,t} \triangleq \begin{cases} 1 & \text{if } s = l + t, \\ 0 & \text{otherwise.} \end{cases}$$

If $h = e_{L_0}^{(L)}$, then it can be readily shown that the first-order derivative constraints in (9.6) can be simplified as

$$W_{m,L_0+l} = W_{m,L_0-l}, \quad \forall \, m = 1, \ldots, M; \ l = 1, \ldots, L_0 - 1. \quad (9.7)$$

Therefore, the first-order presteering derivative constraints are equivalent to linear phase constraints on each of the M FIR filters of the array processor.

9.2.2   Diagonal loading

Another popular technique to improve the robustness of the beamformer is to regularize the array covariance matrix by adding a diagonal loading term [11–13]. The LCMV beamformer problem with diagonal loading is given by

$$\min_{w} w^T (R_x + \eta I_{ML}) w \quad \text{s.t.} \quad D^T w = h, \quad (9.8)$$

and if D has full-column rank, then the solution to this problem is given by

$$w_{\text{DL}} = (R_x + \eta I_{ML})^{-1} D \left(D^T (R_x + \eta I_{ML})^{-1} D\right)^{-1} h. \quad (9.9)$$

However, the amount of diagonal loading η cannot be directly matched to the amount of uncertainty in the array manifold. Therefore, it is often set in an ad hoc way [11–13]. This parameter is typically chosen as $10\,\sigma_n^2$, where σ_n^2 is the sensor noise power.

9.3 Robust broadband beamformers

In this section, we derive four different robust adaptive broadband beamformers whose robustness is explicitly matched to the presumed amount of uncertainty in the array manifold. The proposed beamformers provide robustness against general array calibration errors, signal pointing errors, and presteering delay quantization effects. They offer different trade-offs in terms of interference suppression capability, robustness against signal self-nulling, and computational complexity. All these beamformers are formulated as convex optimization problems and can be solved in polynomial time using interior-point methods [25, 26]. All these beamformers minimize the output power subject to robust constraints that prevent the suppression of the desired signal.

9.3.1 Robust beamformer with separate magnitude and phase constraints

Let the 2-norm of the error vector $\mathbf{\Delta}(f)$ be bounded as

$$\|\mathbf{\Delta}(f)\|_2 \le \epsilon_2(f). \tag{9.10}$$

The upper bound $\epsilon_2(f)$ may be determined numerically based on assumptions on the array calibration errors, the signal pointing errors, and the presteering delay quantization errors. In order to prevent cancelation of the desired signal components, we impose the constraint

$$|H_s(f)| \ge 1 \quad \forall\, \|\mathbf{\Delta}(f)\|_2 \le \epsilon_2(f),\ f \in [f_l, f_u], \tag{9.11}$$

where f_l and f_u are lower and upper bounds on the frequency range of the desired signal, respectively. The constraints in (9.11) have to be satisfied for the worst-case mismatch vectors, in other words, (9.11) is equivalent to

$$\min_{\|\mathbf{\Delta}(f)\|_2 \le \epsilon_2(f)} |H_s(f)| \ge 1 \quad \forall f \in [f_l, f_u]. \tag{9.12}$$

Let us assume that $\epsilon_2(f)$ is sufficiently small such that $|\boldsymbol{w}^T \boldsymbol{C}_0 \boldsymbol{d}(f)| > |\boldsymbol{w}^T \boldsymbol{Q}(f)\mathbf{\Delta}(f)|$ for all anticipated mismatch vectors. This assumption guarantees that the magnitude of the beamformer frequency response is strictly larger than zero. Then the minimum in (9.12) is given by

$$\min_{\|\mathbf{\Delta}(f)\|_2 \le \epsilon_2(f)} |H_s(f)| = |\boldsymbol{w}^T \boldsymbol{C}_0 \boldsymbol{d}(f)| - \|\boldsymbol{Q}^T(f)\boldsymbol{w}\|_2 \epsilon_2(f), \tag{9.13}$$

which is achieved by using the worst-case mismatch vector

$$\mathbf{\Delta}(f) = -\frac{\mathbf{Q}^H(f)\mathbf{w}}{\|\mathbf{Q}^H(f)\mathbf{w}\|_2}\,\epsilon_2(f)e^{j\beta(f)},$$

where $\beta(f) = -2\pi f T_0 + \arg(\mathbf{w}^T\mathbf{C}_0\mathbf{d}(f))$, and $(\cdot)^H$ denotes the Hermitian transpose.

The gain constraint in (9.12) limits only the magnitude of the beamformer frequency response. The phase of the frequency response stays unconstrained, but there are several possibilities to control the phase of the frequency response. For example, the same linear phase as in (9.5) is achieved by using the constraints

$$\sum_{m=1}^{M} W_{m,L_0+l} = \sum_{m=1}^{M} W_{m,L_0-l} \quad \forall\, l = 1,\dots,L_0 - 1. \tag{9.14}$$

Note that these constraints result in the linear phase property when $\mathbf{\Delta}(f) = \mathbf{0}_M$, but the phase of the frequency response is in general nonlinear for arbitrary mismatch vectors $\mathbf{\Delta}(f) \neq \mathbf{0}_M$. Interestingly, if the first-order presteering derivative constraints in (9.7) are fulfilled, then the linear phase constraints in (9.14) are also satisfied. Although the derivative constraints are more restrictive than the linear phase constraints in (9.14), they improve the robustness of the beamformer against phase errors in the entries of $\mathbf{a}_s(f)$ (which is not the case for the linear phase constraints). Therefore, we add these derivative constraints to our first robust broadband beamforming problem [22–24]

$$\begin{aligned}
\min \quad & \mathbf{w}^T \mathbf{R}_x \mathbf{w} \\
\text{s.t.} \quad & |\mathbf{w}^T \mathbf{C}_0 \mathbf{d}(f)| - \|\mathbf{Q}^T(f)\mathbf{w}\|_2 \epsilon_2(f) \geq 1 \quad \forall f \in [f_l, f_u]; \\
& W_{m,L_0+l} = W_{m,L_0-l} \quad \forall\, m = 1,\dots,M;\ l = 1,\dots,L_0 - 1.
\end{aligned} \tag{9.15}$$

Note that the gain constraints in (9.15) are nonconvex due to the absolute value operator. However, (9.15) can be transformed to an equivalent convex optimization problem as follows. The presteering derivative constraints are fulfilled if

$$\mathbf{w} = \begin{bmatrix} \mathbf{W}_{:,1} \\ \mathbf{W}_{:,2} \\ \vdots \\ \mathbf{W}_{:,L_0} \\ \vdots \\ \mathbf{W}_{:,2} \\ \mathbf{W}_{:,1} \end{bmatrix} = \begin{bmatrix} \mathbf{I}_M & \mathbf{0}_{M\times M} & \cdots & \mathbf{0}_{M\times M} \\ \mathbf{0}_{M\times M} & \mathbf{I}_M & & \\ & & \ddots & \\ \vdots & & & \mathbf{I}_M \\ & & & \reflectbox{$\ddots$} \\ \mathbf{0}_{M\times M} & \mathbf{I}_M & & \\ \mathbf{I}_M & \mathbf{0}_{M\times M} & \cdots & \mathbf{0}_{M\times M} \end{bmatrix} \begin{bmatrix} \mathbf{W}_{:,1} \\ \mathbf{W}_{:,2} \\ \vdots \\ \mathbf{W}_{:,L_0} \end{bmatrix} = \mathbf{B}\tilde{\mathbf{w}}, \tag{9.16}$$

where

$$\tilde{\mathbf{w}} = \left[\mathbf{W}_{:,1}^T, \mathbf{W}_{:,2}^T, \cdots, \mathbf{W}_{:,L_0}^T \right]^T, \tag{9.17}$$

and $\mathbf{0}_{M \times M}$ stands for the $M \times M$ matrix of zeros. Using (9.16), we obtain

$$w^T C_0 d(f) = \tilde{w}^T \left(g(f) \otimes \mathbf{1}_M \right) e^{-j2\pi f(L_0 - 1)T_s},$$

where the $L_0 \times 1$ vector $g(f)$ is defined as

$$g(f) \triangleq [2\cos(2\pi f(L_0 - 1)T_s), 2\cos(2\pi f(L_0 - 2)T_s), \ldots, 2\cos(2\pi f T_s), 1]^T. \quad (9.18)$$

Hence,

$$|w^T C_0 d(f)| = |\tilde{w}^T (g(f) \otimes \mathbf{1}_M)| \geq 1 + \|Q^T(f)w\|_2 \epsilon_2(f) \quad \forall f \in [f_l, f_u],$$

where $\tilde{w}^T (g(f) \otimes \mathbf{1}_M)$ is a real-valued continuous function of f which is either larger than 1 or smaller than -1 for all frequencies. As the optimization problem (9.15) is invariant to multiplication of the weight vector by -1 and the function $\tilde{w}^T (g(f) \otimes \mathbf{1}_M)$ is continuous, we can assume, without loss of generality, that $\tilde{w}^T (g(f) \otimes \mathbf{1}_M)$ is positive for all values of f. Consequently, the robust beamforming problem (9.15) can be expressed in the following convex form:

$$
\begin{aligned}
\min \quad & \tilde{w}^T B^T R_x B \tilde{w} \\
\text{s.t.} \quad & \tilde{w}^T (g(f) \otimes \mathbf{1}_M) - \|Q^T(f)B\tilde{w}\|_2 \epsilon_2(f) \geq 1 \quad \forall f \in [f_l, f_u].
\end{aligned}
\quad (9.19)
$$

The optimization problem (9.19) contains an infinite number of second-order cone constraints [25]. In practice, we have to discretize the frequency range $[f_l, f_u]$ so that the gain constraints are maintained for a finite set of frequencies $f_1, \ldots, f_{N_f} \in [f_l, f_u]$. Then, we have

$$
\begin{aligned}
\min \quad & \tilde{w}^T B^T R_x B \tilde{w} \\
\text{s.t.} \quad & \tilde{w}^T (g(f_k) \otimes \mathbf{1}_M) - \|Q^T(f_k)B\tilde{w}\|_2 \epsilon_2(f_k) \geq 1 \quad \forall k = 1, \ldots, N_f.
\end{aligned}
\quad (9.20)
$$

The problem (9.20) is an SOCP problem that can be efficiently solved using modern interior-point methods [25]. According to the guidelines in [27], the computational complexity of solving (9.20) is $O((L^3 M^3 + L^2 M^3 N_f)\sqrt{N_f})$. Using the uniform frequency sampling, the standard choice for the number of frequency points is $N_f \approx 10L$ [33], but in our simulations we have observed that $N_f = 2L$ is already sufficient, in other words, any larger number of frequency samples gives only negligible performance improvements. Hence, the computational complexity of the robust beamformer (9.20) is only moderately higher than the computational complexity $O(L^3 M^3)$ of the conventional (non-robust) Frost beamformer.

9.3.2 Robust beamformer with combined magnitude and phase constraints

In (9.15), separate constraints have been used for the magnitude and phase of the beamformer frequency response to the desired signal. The key idea of our second robust beamformer is to combine these types of constraints together. The resulting beamforming problem can be written as [24]

$$
\begin{aligned}
\min \quad & w^T R_x w \\
\text{s.t.} \quad & \Re\left(H_s(f_k)e^{j2\pi f_k(T_0+(L_0-1)T_s)}\right) \geq 1 \quad \forall\ \|\boldsymbol{\Delta}(f_k)\|_2 \leq \epsilon_2(f_k),\ k = 1,\ldots,N_f,
\end{aligned}
\tag{9.21}
$$

where $\Re(\cdot)$ denotes the real-part operator. Note that the constraints in (9.21), together with the output power minimization, jointly limit the magnitude and the phase of the beamformer frequency response. Therefore, additional linear phase constraints are not required.

The constraints in (9.21) are less restrictive than the constraints in (9.15), since the former constraints do not guarantee that the phase of the frequency response is linear for $\boldsymbol{\Delta}(f) = \mathbf{0}_M$, and the elimination of $M(L_0-1)$ equality constraints results in more degrees of freedom for the weight vector w. Furthermore, the constraints in (9.15) ensure the linear phase property for the total frequency range $f \in [-f_s/2, f_s/2]$, whereas in (9.21) we only have constraints for the frequency domain of the desired signal. Therefore, the constraints in (9.21) can be expected to lead to an improved interferer suppression capability as compared to the robust beamformer (9.15). At the same time, the beamformer (9.21) can be more sensitive to signal self-nulling.

Similar steps as in Section 9.3.1 enable us to transform the problem (9.21) to an equivalent convex optimization problem with a finite number of constraints. In particular, by determining the worst-case mismatch vector we obtain

$$
\begin{aligned}
\min \Re\left(H_s(f_k)e^{j2\pi f_k(T_0+(L_0-1)T_s)}\right) \quad & \text{s.t.} \quad \|\boldsymbol{\Delta}(f_k)\|_2 \leq \epsilon_2(f_k) \\
= w^T \Re\left(C_0 d(f_k)e^{j2\pi f_k(L_0-1)T_s}\right) & - \|Q^T(f_k)w\|_2 \epsilon_2(f_k),
\end{aligned}
$$

and (9.21) can be written as

$$
\begin{aligned}
\min \quad & w^T R_x w \\
\text{s.t.} \quad & w^T b(f_k) - \|Q^T(f_k)w\|_2 \epsilon_2(f_k) \geq 1 \qquad \forall\ k = 1,\ldots,N_f,
\end{aligned}
\tag{9.22}
$$

where the $ML \times 1$ vector $b(f_k)$ is defined as $b(f_k) \triangleq \Re\left(C_0 d(f_k)e^{j2\pi f_k(L_0-1)T_s}\right)$.

The structure of the robust beamformers (9.20) and (9.22) is mathematically similar. In particular, both these beamformers can be computed by solving SOCP problems with the computational complexity $O((L^3 M^3 + L^2 M^3 N_f)\sqrt{N_f})$.

9.3.3   Robust beamformer with Chebychev constraints

Our first two robust beamformers only set a lower bound for the magnitude of the frequency response. However, for some frequencies the magnitude of the frequency response may be substantially larger than its lower bound. This may be especially true when the beamformer *input signal-to-interference-plus-noise ratio* (ISINR) is low. To prevent this phenomenon, our third beamformer uses the so-called Chebychev constraints that limit the deviation of the frequency response from its nominal value. Hence, such Chebychev constraints enforce small signal distortions even if the power of the signal is low. This beamformer can be formulated as

$$\min \quad w^T R_x w$$
$$\text{s.t.} \quad \left| H_s(f_k) - e^{-j2\pi f_k(T_0 + (L_0 - 1)T_s)} \right| \le \gamma(f_k) \tag{9.23}$$
$$\forall \, \|\Delta(f_k)\|_2 \le \epsilon_2(f_k); \quad k = 1, \ldots, N_f.$$

The parameters $\gamma(f_k)$, $k = 1, \ldots, N_f$ control the maximum deviation of the beamformer frequency response from the optimum value $\exp(-j2\pi f_k(T_0 + (L_0 - 1)T_s))$, $k = 1, \ldots, N_f$. The values of $\gamma(f_k)$ should be small enough to warrant that the distortion of the desired signal is small, and they should be large enough to enable efficient suppression of interferers and noise. If the values of $\gamma(f_k)$ are too small, then (9.23) may be infeasible. Therefore, the choice of $\gamma(f_k)$ is dictated by a trade-off between interference suppression and small signal distortions.

It can be readily shown that

$$\max \, \left| H_s(f_k) - e^{-j2\pi f_k(T_0 + (L_0 - 1)T_s)} \right| \quad \text{s.t.} \quad \|\Delta(f_k)\|_2 \le \epsilon_2(f_k)$$
$$= \left| w^T C_0 d(f_k) - e^{-j2\pi f_k(L_0 - 1)T_s} \right| + \|Q^T(f_k)w\|_2 \epsilon_2(f_k). \tag{9.24}$$

Using (9.24), we can rewrite our third beamformer as the following finite convex-optimization problem

$$\min \quad w^T R_x w$$
$$\text{s.t.} \quad \left| w^T C_0 d(f_k) - e^{-j2\pi f_k(L_0 - 1)T_s} \right| \le \gamma(f_k) - \epsilon_2(f_k)\|Q^T(f_k)w\|_2 \tag{9.25}$$
$$\forall \, k = 1, \ldots, N_f.$$

Similar to (9.20) and (9.22), (9.25) is an SOCP problem. From [27], we obtain that its computational complexity is upper bounded by $O((LM + N_f)^2 M (L + N_f)\sqrt{N_f})$. Hence, the beamformer (9.25) has somewhat higher computational complexity than the beamformers (9.20) and (9.22).

To choose $\gamma(f_k)$ such that (9.25) is feasible, we compute the minimum of (9.24) with respect to w. Note that (9.24) is a positive convex function of w. By setting its derivative to zero, we obtain that the necessary condition for the minimum of this function is

$$w^T C_0 d(f_k) = e^{-j2\pi f_k(L_0 - 1)T_s}. \tag{9.26}$$

As

$$w^T C_0 d(f_k) = \sum_{m=1}^{M} W_{m,:} d(f_k),$$

and

$$Q^T(f_k)w = \left[W_{1,:} d(f_k), \ldots, W_{M,:} d(f_k) \right]^T,$$

it can be shown that under (9.26),

$$\|\boldsymbol{Q}^T(f_k)\boldsymbol{w}\|_2 \geq \frac{1}{\sqrt{M}},$$

where we used the fact that $\|\boldsymbol{x}\|_2 \geq \|\boldsymbol{x}\|_1/\sqrt{M}$ for all $M \times 1$ vectors $\boldsymbol{x}$ and $\|\cdot\|_1$ denotes the vector 1-norm. Hence,

$$\left| \boldsymbol{w}^T \boldsymbol{C}_0 \boldsymbol{d}(f_k) - e^{-j2\pi f_k (L_0-1)T_s} \right| + \|\boldsymbol{Q}^T(f_k)\boldsymbol{w}\|_2 \epsilon_2(f_k) \geq \epsilon_2(f_k)\frac{1}{\sqrt{M}}.$$

It can be easily verified that the weight vector $\boldsymbol{w} = (\boldsymbol{e}_{L_0}^{(L)} \otimes \boldsymbol{1}_M)/M$ satisfies all constraints in (9.25) for $\gamma(f_k) = \epsilon_2(f_k)/\sqrt{M}$; $k = 1,\ldots,N_f$. Therefore, $\gamma(f_k) \geq \epsilon_2(f_k)/\sqrt{M}$ is a tight lower bound for the feasibility of (9.25).

9.3.4 Robust beamformer without frequency discretization

The previous three beamformers developed in Sections 9.3.1–9.3.3 are based on the discretization of the frequency range of the desired signal. Our fourth beamformer employs the technique presented in [28–31] to avoid such a discretization and related approximation errors.

Let us assume that the mismatch vectors are bounded by

$$\|\boldsymbol{\Delta}(f)\|_1 \leq \epsilon_1(f) \quad \forall f \in [f_l, f_u],$$

where $\|\cdot\|_1$ denotes the vector 1-norm. Note that in contrast to the robust beamformers (9.20), (9.22), and (9.25), the 1-norm characterization of uncertainty is used here for the sake of tractability. To protect the desired signal from being canceled, we require that

$$|H_s(f)| \geq 1 \quad \forall \, \|\boldsymbol{\Delta}(f)\|_1 \leq \epsilon_1(f), \, f \in [f_l, f_u]. \tag{9.27}$$

Similar steps as in Section 9.3.1 show that the magnitude of the frequency response for the worst-case mismatch vector is given by

$$\min_{\|\boldsymbol{\Delta}(f)\|_1 \leq \epsilon_1(f)} |H_s(f)| = |\boldsymbol{w}^T \boldsymbol{C}_0 \boldsymbol{d}(f)| - \|\boldsymbol{Q}^T(f)\boldsymbol{w}\|_\infty \epsilon_1(f),$$

where $\|\cdot\|_\infty$ denotes the vector ∞-norm. If we combine the gain constraints in (9.27) with the presteering derivative constraints in (9.7), we obtain

$$\begin{aligned}
\min \quad & \boldsymbol{w}^T \boldsymbol{R}_x \boldsymbol{w} \\
\text{s.t.} \quad & |\boldsymbol{w}^T \boldsymbol{C}_0 \boldsymbol{d}(f)| - \|\boldsymbol{Q}^T(f)\boldsymbol{w}\|_\infty \epsilon_1(f) \geq 1 \quad \forall f \in [f_l, f_u]; \\
& W_{m,L_0+l} = W_{m,L_0-l} \quad \forall \, m = 1,\ldots,M; \, l = 1,\ldots,L_0-1.
\end{aligned} \tag{9.28}$$

Using (9.16) to take into account the equality constraints of (9.28) and exploiting the fact that (9.28) is invariant to changing the sign of the weight vector, we can rewrite the latter problem as

$$
\begin{aligned}
\min \quad & \tilde{w}^T B^T R_x B \tilde{w} \\
\text{s.t.} \quad & \tilde{w}^T (g(f) \otimes \mathbf{1}_M) - \|Q^T(f) B \tilde{w}\|_\infty \epsilon_1(f) \geq 1 \quad \forall f \in [f_l, f_u],
\end{aligned}
\tag{9.29}
$$

where the notations of (9.16)–(9.18) are used. Note that (9.29) is a convex SOCP problem that can be approximately solved by discretizing the frequency range $[f_l, f_u]$. However, such discretization (and the related approximation errors) can be avoided by tightening the constraints in (9.29) as

$$
\begin{aligned}
\min \quad & \tilde{w}^T B^T R_x B \tilde{w} \\
\text{s.t.} \quad & \tilde{w}^T (g(f) \otimes \mathbf{1}_M) - \left[\max_{f \in [f_l, f_u]} \|Q^T(f) B \tilde{w}\|_\infty \right] \epsilon_1(f) \geq 1 \quad \forall f \in [f_l, f_u].
\end{aligned}
\tag{9.30}
$$

Note that (9.30) is not equivalent to (9.29) and, hence, can only provide an approximate solution to the original optimization problem. The solution provided by (9.30) is close to the solution of (9.29) if $\|Q^T(f) B \tilde{w}_{\text{opt}}\|_\infty$ is approximately constant over the frequency range of the desired signal. The tightened problem (9.30) can be written as

$$
\begin{aligned}
\min \quad & \tilde{w}^T B^T R_x B \tilde{w} \\
\text{s.t.} \quad & \tilde{w}^T (g(f) \otimes \mathbf{1}_M) - \nu \epsilon_1(f) \geq 1 \quad \forall f \in [f_l, f_u]; \\
& W_{m,1:L_0} g(f) \leq \nu \quad \forall f \in [f_l, f_u], \ m = 1, \dots, M; \\
& W_{m,1:L_0} g(f) \geq -\nu \quad \forall f \in [f_l, f_u], \ m = 1, \dots, M,
\end{aligned}
\tag{9.31}
$$

where we used the fact that

$$
Q^T(f) B \tilde{w} = W_{:,1:L_0} g(f) e^{-j 2\pi f (L_0 - 1) T_s}.
$$

Here, the row-vector $W_{m,1:L_0}$ contains the first L_0 entries of the mth row of W, and the $M \times L_0$ matrix $W_{:,1:L_0}$ contains the first L_0 columns of W.

Let us now approximate $\epsilon_1(f)$ over the frequency interval of the desired signal by a linear superposition of trigonometric functions, that is,

$$
\epsilon_1(f) \approx b_0 + \sum_{k=1}^{L_0 - 1} (a_k \sin(2\pi f k T_s) + b_k \cos(2\pi f k T_s)) \quad \forall f \in [f_l, f_u].
\tag{9.32}
$$

For simplicity, in (9.32) only the terms up to the order $k = L_0 - 1$ are used. However, the subsequently developed algorithm can be directly modified to take into account higher orders of the trigonometric functions. The coefficients $\{a_k\}_{k=1}^{L_0-1}$ and $\{b_k\}_{k=0}^{L_0-1}$ can be determined by a least-squares fit of $\epsilon_1(f)$ for $f \in [f_l, f_u]$.

The approximation in (9.32) allows us to rewrite each of the $2M + 1$ constraints in (9.31) in the form

$$\Re\left(\boldsymbol{v}^H(f)\boldsymbol{p}(\tilde{\boldsymbol{w}}, v)\right) \geq 0 \quad \forall f \in [f_l, f_u], \tag{9.33}$$

where

$$\boldsymbol{v}(f) = \left[1, e^{j2\pi f T_s}, \ldots, e^{j2\pi f T_s(L_0-1)}\right]^T,$$

and the first element of $\boldsymbol{p}(\tilde{\boldsymbol{w}}, v)$ is real, while all other elements may be complex. Note that the left-hand side in (9.33) can be written as a superposition of trigonometric functions

$$\Re\left(\boldsymbol{v}^H(f)\boldsymbol{p}(\tilde{\boldsymbol{w}}, v)\right)$$

$$= p_1(\tilde{\boldsymbol{w}}, v) + \sum_{k=1}^{L_0-1} \left(\Re(p_{k+1}(\tilde{\boldsymbol{w}}, v)) \cos(2\pi f k T_s) + \Im(p_{k+1}(\tilde{\boldsymbol{w}}, v)) \sin(2\pi f k T_s)\right),$$

where $\Im(\cdot)$ denotes the imaginary-part operator. More precisely, it can be easily verified that

$$\pm \boldsymbol{W}_{m,1:L_0}\boldsymbol{g}(f) \leq v$$
$$\Longleftrightarrow \quad \Re\left(\boldsymbol{v}^H(f)\left(\mp \tilde{\boldsymbol{J}}\boldsymbol{W}_{m,1:L_0}{}^T + v\boldsymbol{e}_1^{(L_0)}\right)\right) \geq 0, \tag{9.34}$$

and

$$\tilde{\boldsymbol{w}}^T\left(\boldsymbol{g}(f) \otimes \boldsymbol{1}_M\right) - v\epsilon_1(f) \geq 1$$
$$\Longleftrightarrow \quad \Re\left(\boldsymbol{v}^H(f)\left(\tilde{\boldsymbol{J}}\left(\boldsymbol{I}_{L_0} \otimes \boldsymbol{1}_M^T\right)\tilde{\boldsymbol{w}} - v\boldsymbol{q} - \boldsymbol{e}_1^{(L_0)}\right)\right) \geq 0, \tag{9.35}$$

where the $L_0 \times L_0$ matrix $\tilde{\boldsymbol{J}}$ is defined as

$$\tilde{\boldsymbol{J}} \triangleq \begin{bmatrix} 0 & \cdots & 0 & 1 \\ \vdots & & 2 & 0 \\ 0 & \cdot^{\cdot^{\cdot}} & & \vdots \\ 2 & 0 & \cdots & 0, \end{bmatrix},$$

and the $L_0 \times 1$ vector $\boldsymbol{q}$ is given by

$$\boldsymbol{q} \triangleq \left[b_0, b_1 + ja_1, \ldots, b_{L_0-1} + ja_{L_0-1}\right]^T.$$

Let us use the notation $(\cdot)^*$ for the complex conjugate, $\mathcal{H}_+^{(n)}$ for the convex cone of positive semidefinite Hermitian $n \times n$ matrices, and $\langle \boldsymbol{M}_1, \boldsymbol{M}_2 \rangle = \text{trace}(\boldsymbol{M}_1^H \boldsymbol{M}_2)$ for the inner product of two matrices M_1 and M_2 with equal dimensions. It has been shown in [28] that the set of vectors p that satisfy (9.33) can be written as

$$\left\{ p \in \mathbb{R} \times \mathbb{C}^{L_0-1} \,\middle|\, \Re\left(v^H(f)p\right) \geq 0 \quad \forall f \in [f_l, f_u] \right\}$$
$$= \left\{ p \in \mathbb{R} \times \mathbb{C}^{L_0-1} \,\middle|\, p = u(X, Z, \xi, f_l, f_u), \; X \in \mathcal{H}_+^{(L_0)}, \; Z \in \mathcal{H}_+^{(L_0-1)}, \; \xi \in \mathbb{R} \right\},$$
$$\tag{9.36}$$

where

$$u(X, Z, \xi, f_l, f_u) = \begin{bmatrix} \langle T_0^{(L_0)}, X \rangle + \langle \mu_1^* T_0^{(L_0-1)} + \mu_2 T_1^{(L_0-1)}, Z \rangle + j\xi \\[2ex] 2\langle T_1^{(L_0)}, X \rangle + \langle 2\mu_1^* T_1^{(L_0-1)} + \mu_2^* T_0^{(L_0-1)} + \mu_2 T_2^{(L_0-1)}, Z \rangle \\[1ex] \vdots \\[1ex] 2\langle T_{L_0-3}^{(L_0)}, X \rangle + \langle 2\mu_1^* T_{L_0-3}^{(L_0-1)} + \mu_2^* T_{L_0-4}^{(L_0-1)} + \mu_2 T_{L_0-2}^{(L_0-1)}, Z \rangle \\[1ex] 2\langle T_{L_0-2}^{(L_0)}, X \rangle + \langle 2\mu_1^* T_{L_0-2}^{(L_0-1)} + \mu_2^* T_{L_0-3}^{(L_0-1)}, Z \rangle \\[1ex] 2\langle T_{L_0-1}^{(L_0)}, X \rangle + \mu_2 \langle T_{L_0-2}^{(L_0-1)}, Z \rangle, \end{bmatrix}$$

and

$$\begin{bmatrix} \mu_1 \\ \mu_2 \end{bmatrix} = \begin{cases} \begin{bmatrix} \cos(2\pi f_l T_s) + \cos(2\pi f_u T_s) - \cos(2\pi(f_u - f_l)T_s) - 1 \\ (1 - \exp(j2\pi f_l T_s))\,(\exp(j2\pi f_u T_s) - 1) \end{bmatrix}, & \text{if } f_l > 0 \\[3ex] \begin{bmatrix} -\sin(2\pi f_u T_s) \\ j(1 - \exp(j2\pi f_u T_s)) \end{bmatrix}, & \text{if } f_l = 0. \end{cases}$$

Using (9.34), (9.35), and (9.36) we can reformulate (9.31) as

$$\min \quad \tilde{w}^T B^T R_x B \tilde{w}$$
$$\text{s.t.} \quad \tilde{J}\left(I_{L_0} \otimes 1_M^T\right)\tilde{w} - vq - e_1^{(L_0)} = u(X_1, Z_1, \xi_1, f_l, f_u);$$
$$-\tilde{J}\left(W_{m,1:L_0}\right)^T + ve_1^{(L_0)} = u(X_{m+1}, Z_{m+1}, \xi_{m+1}, f_l, f_u),$$
$$m = 1, \ldots, M;$$
$$\tilde{J}\left(W_{m,1:L_0}\right)^T + ve_1^{(L_0)} = u(X_{m+M+1}, Z_{m+M+1}, \xi_{m+M+1}, f_l, f_u),$$
$$m = 1, \ldots, M;$$
$$X_k \in \mathcal{H}_+^{(L_0)}, \; Z_k \in \mathcal{H}_+^{(L_0-1)}, \; \xi_k \in \mathbb{R}, \quad k = 1, \ldots, 2M + 1. \tag{9.37}$$

Note that (9.37) is a convex SDP optimization problem with a finite number of constraints which can be efficiently solved using modern interior-point methods [32]. The optimization variables are $\tilde{w}$, v, $\{X_k\}_{k=1}^{2M+1}$, $\{Z_k\}_{k=1}^{2M+1}$, and $\{\xi_k\}_{k=1}^{2M+1}$. An upper bound for the asymptotic growth of the computational complexity is $O(M^{4.5} L^{6.5})$ [27]. Thus, the computational complexity of the robust beamformer (9.37) is substantially higher than that of the beamformers developed in Sections 9.3.1–9.3.3.

9.3.5   Summary of the proposed techniques

The beamformer with separate magnitude and phase constraints uses more restrictive constraints as compared to the beamformer with joint magnitude and phase constraints. Hence, the latter beamformer can be expected to have an improved interferer suppression capability, whereas the former beamformer is likely more robust against signal self-nulling. The beamformer with Chebychev constraints allows us to control the trade-off between interference suppression and signal distortion, but the optimum trade-off parameters $\gamma(f_k)$ depend on the scenario and, therefore, it is a non-trivial task to choose these parameters in an optimum way. The beamformer presented in Section 9.3.4 avoids the approximation errors due to the discretization of the frequency range. At the same time, this beamformer requires more restrictive constraints as compared to all other beamformers and its computational complexity substantially exceeds those of the other beamformers.

9.4   Simulations

To evaluate the performance of the proposed robust broadband beamformers, the following performance measures are used:

- The *output signal-to-interference-plus-noise-ratio* (OSINR) is defined as

$$\mathrm{OSINR} = \frac{\boldsymbol{w}^T \mathrm{E}(\boldsymbol{x}_s(k)\boldsymbol{x}_s^T(k))\boldsymbol{w}}{\boldsymbol{w}^T \mathrm{E}\left((\boldsymbol{x}_i(k) + \boldsymbol{x}_n(k))(\boldsymbol{x}_i(k) + \boldsymbol{x}_n(k))^T\right)\boldsymbol{w}} = \frac{\boldsymbol{w}^T \boldsymbol{R}_s \boldsymbol{w}}{\boldsymbol{w}^T \boldsymbol{R}_{i+n} \boldsymbol{w}},$$

where $\boldsymbol{R}_s = \mathrm{E}\left(\boldsymbol{x}_s(k)\boldsymbol{x}_s^T(k)\right)$ is the covariance matrix of the signal component, and $\boldsymbol{R}_{i+n} = \mathrm{E}\left(\boldsymbol{x}_i(k)\boldsymbol{x}_i^T(k) + \boldsymbol{x}_n(k)\boldsymbol{x}_n^T(k)\right)$ is the covariance matrix of the interference-plus-noise component. The *optimum OSINR* (OOSINR) is given by [1]

$$\mathrm{OOSINR} = \mathcal{P}(\boldsymbol{R}_{i+n}^{-1}\boldsymbol{R}_s),$$

where $\mathcal{P}(\cdot)$ denotes the principal eigenvalue operator.

- As the OSINR criterion does not characterize the signal distortion after broadband beamforming, the *normalized signal distortion* (NSD) is defined as

$$\mathrm{NSD} = \min_{\zeta} \mathrm{E}\left(\left(\zeta \boldsymbol{w}^T \boldsymbol{x}_s(k) - s(k - L_0 + 1)\right)^2\right)/P_s, \tag{9.38}$$

where $s(k)$ is the waveform of the desired signal at the time instant k, $P_s = \mathrm{E}(s^2(k))$ denotes the power of the desired signal, and it is assumed that $T_0 = 0$. The minimization with respect to ζ in (9.38) is used to compensate for scaling errors in $\boldsymbol{w}$.

- Both the OSINR and NSD are only particular criteria, and the overall quality of broadband beamforming is determined by their trade-off in each particular case. Therefore, to characterize the overall beamforming quality through the combined effect of interference-plus-noise suppression and signal distortion, we will use the *normalized mean-square-error* (NMSE) criterion:

$$\text{NMSE} = \min_{\zeta} \text{E}\left(\left(\zeta w^T x(k) - s(k - L_0 + 1)\right)^2\right)/P_s.$$

The NMSE is lower-bounded by the *normalized minimum mean-square-error* (NMMSE)

$$\text{NMMSE} = 1 - \text{E}\left(x^T(k)s(k - L_0 + 1)\right) R_x^{-1} \text{E}\left(x(k)s(k - L_0 + 1)\right)/P_s.$$

Unless specified otherwise, we use the following settings in our simulations. The nominal (presumed) array is a ULA of $M = 10$ sensors. This array is aligned along the z-axis of the coordinate system and its interelement displacement is c/f_s where the sampling frequency $f_s = 1$. The actual array is a perturbed version of the nominal ULA. The perturbations of the sensor locations along the z-axis are generated randomly in each Monte-Carlo run as Gaussian zero-mean variables with the standard deviation of $0.05c/f_s$.

The number of FIR filter sections is $L = 15$. It is assumed that $T_0 = 0$ and that the signal impinges on the array from the DOA $\theta_s = -20°$ relative to the array broadside. The input *signal-to-noise-ratio* (SNR) is 10 dB. Besides the signal, there is one interferer with input *interference-to-noise-ratio* (INR) equal to 30 dB. The interferer DOA is $10°$ relative to the array broadside. The waveforms of the signal and the interferer are generated by filtering independent, white Gaussian waveforms using a band-pass filter with the cut-off frequencies $f_l = 0.05$ and $f_u = 0.45$. The additive noise in each sensor is temporally and spatially IID Gaussian. The signal, the interferer, and the noise are assumed to be statistically independent. We use $K = 1000$ snapshots to estimate the array covariance matrix R_x and 100 independent Monte-Carlo runs are performed for each simulation point. We use a uniform grid of $N_f = 2L = 30$ frequency samples. Signal look direction errors are generated randomly and independently in each Monte-Carlo run as zero-mean Gaussian variables with the standard deviation of $1°$.

Throughout our simulations, we compare our robust beamformers (9.20), (9.22), (9.25), and (9.37) (which are hereafter referred to as beamformers 1, 2, 3, and 4, respectively) with the diagonal loading-based MVDR beamformer and the diagonal loading-based LCMV beamformer with first-order presteering derivative constraints.

The uncertainty regions of beamformers 1–4 are set to match the look direction error $\pm 2°$ and sensor location errors $\pm 0.1c/f_s$. In both the diagonal loading and first-order presteering derivative constraints beamformers, the standard choice $\eta = 10\sigma_n^2$ of the diagonal loading factor is used. In beamformer 3, the parameters $\gamma(f_k) = 2\epsilon_2(f_k)/\sqrt{M}$ $(k = 1, \ldots, N_f)$ are taken. To solve the problems (9.20), (9.22), (9.25), and (9.37) we use CVX, a MATLAB package for solving convex optimization problems [34, 35].

Example 9.1   In our first example we demonstrate the beamformer performance as a function of the input SNR. Figure 9.2 depicts results for the beamformer NSDs, OSINRs, and the NMSEs. For the diagonal loading and the presteering derivative constraints

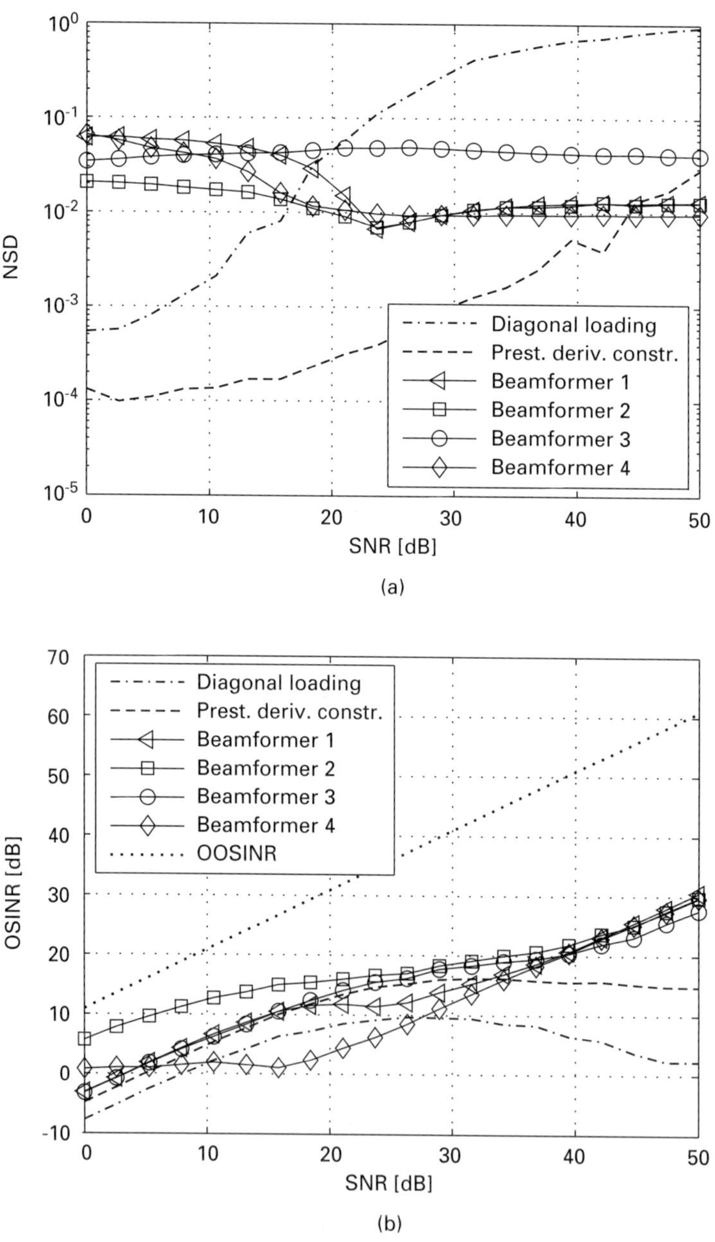

Figure 9.2   Beamformer performance versus input SNR.

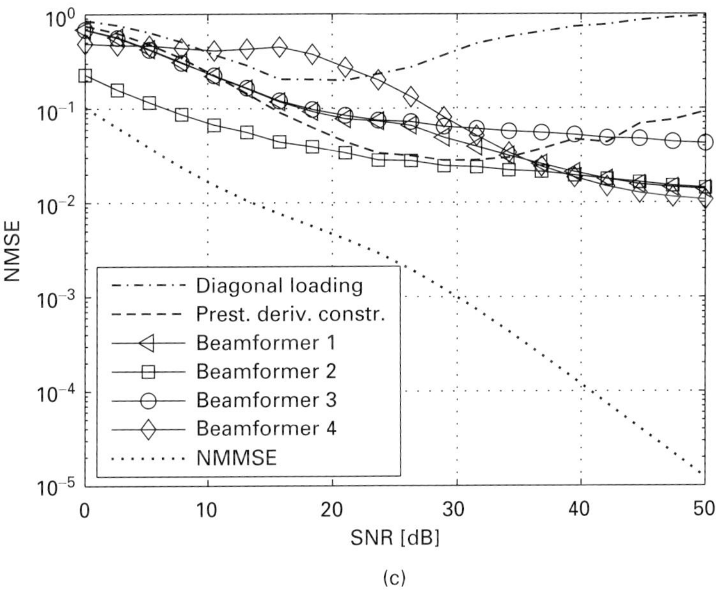

(c)

Figure 9.2   Continued.

beamformers, we can observe that the distortion of the desired signal increases with increasing input SNR. Furthermore, the OSINR of the diagonal loading beamformer degrades and the OSINR of the presteering derivative constraints beamformer saturates in the high SNR regime. Hence, for both beamformers the overall performance, expressed by the NMSE, degrades for large values of the input SNR. These two beamformers do not gain from increasing SNR, because of the negative effect of signal self-nulling. The beamformers 1–4 are more robust against signal self-nulling and, as a result, their NSD curves are approximately constant over the whole range of the input SNR values, whereas their OSINR improves with increasing input SNR. Consequently, the NMSE performance of the beamformers 1–4 improves with the input SNR. As follows from Figure 9.2c, at low and moderate SNRs, beamformer 2 achieves the best performance among the methods tested. This can be attributed to its improved interferer suppression capability as compared to beamformers 1 and 4. Since beamformer 2 is more sensitive to signal self-nulling as compared to the beamformers 1 and 4, its superiority vanishes for high values of the input SNR.

Example 9.2   In our second example, we study the impact of the angular distance between the signal and interferer. Therefore, we keep the DOA of the desired signal $\theta_s = -20°$ fixed, and the DOA of the interferer changes from $\theta_i = -17°$ to $\theta_i = 70°$. Figure 9.3 shows that the smallest signal distortions are achieved by the diagonal loading and presteering derivative constraints-based beamformers. The NSD performances of these beamformers do not degrade as in Figure 9.2 since the input SINR in each sensor is below -20 dB, so that signal self-nulling does not play a major role. Figure 9.3b shows that for a wide range of angular spacings between the signal and interferer, beamformer 2 achieves the best OSINR performance among the methods tested. The reason is, again, its improved interferer suppression capability as compared to the beamformers that use first-order presteering derivative constraints. Note that the diagonal loading beamformer

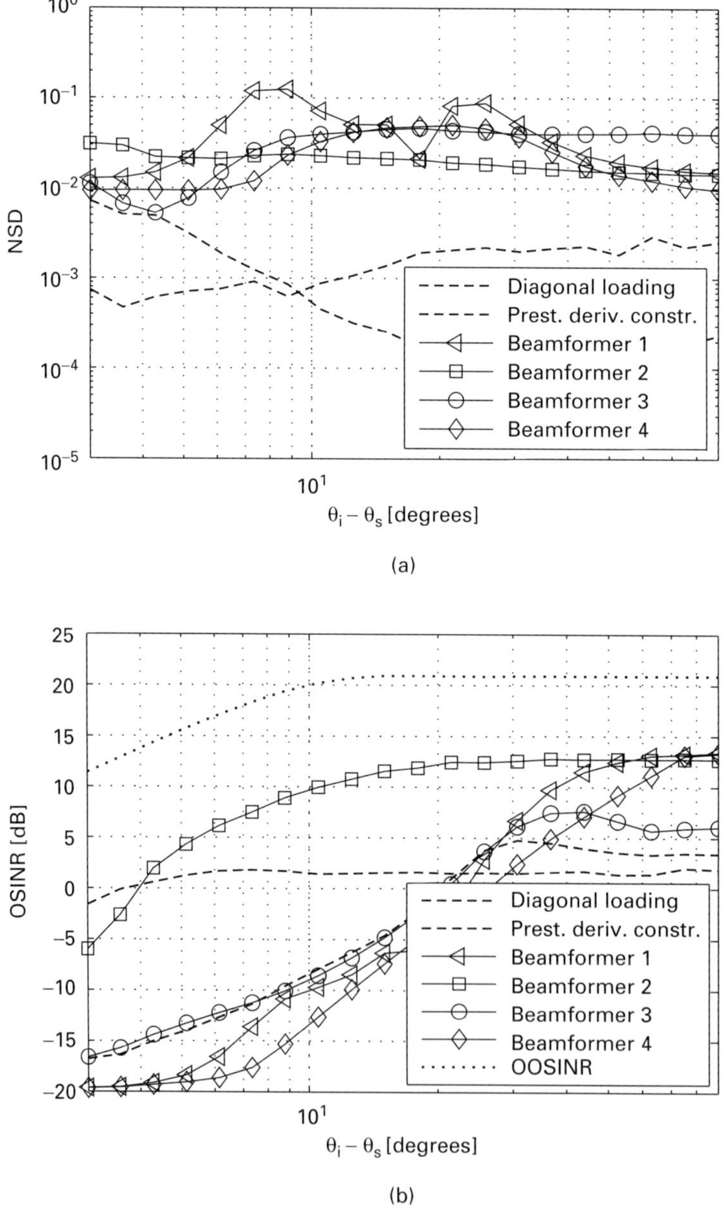

(a)

(b)

Figure 9.3   Beamformer performance versus the angular distance between the signal and the interferer.

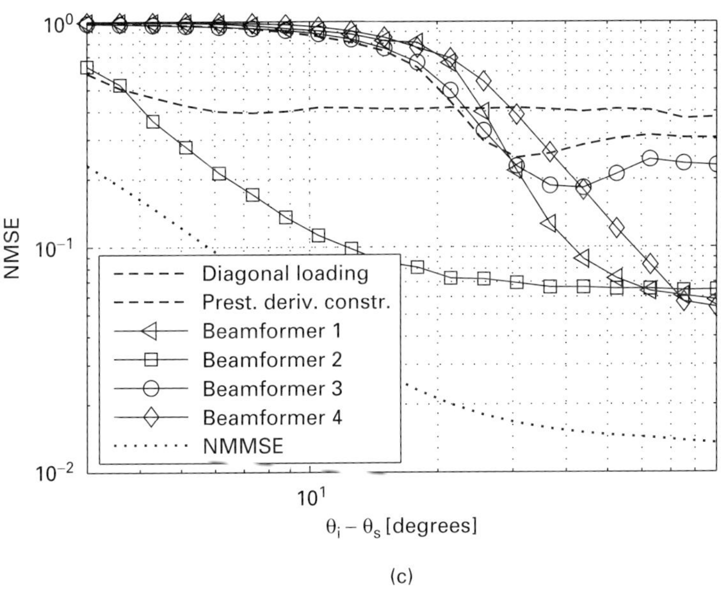

Figure 9.3   Continued.

achieves better OSINR results than beamformer 2 if the angular distance between the two sources is below $4°$, but for larger spacings between the two sources, the diagonal loading beamformer suffers from modeling mismatch errors. Since the signal distortions for all the beamformers tested are rather small, the overall beamformer NMSE performance is dominated by OSINR. Consequently, beamformer 2 mostly outperforms the other methods.

Example 9.3   In this example, we analyze the impact of the number of frequency samples N_f on the NMSE performance of the beamformers 1–3. Figure 9.4 shows that their NMSE saturates when increasing N_f. This saturation is achieved already if $N_f \gtrsim 20$.

Example 9.4   We have shown above that in some scenarios beamformer 2 outperforms the beamformers 1 and 4, and we have explained this observation with the improved interferer suppression capability of beamformer 2 that is achieved by avoiding the first-order presteering derivative constraints. These constraints are expressed in (9.14) by a set of $(L_0 - 1)M$ linear equalities that require approximately one-half of the LM degrees of freedom of the weight vector w. An important question that arises is whether beamformers 1 and 4 achieve similar performance as beamformer 2 if we double their number of FIR filter taps. To answer this question, the beamformer NMSEs are displayed in Figure 9.5 as a function of the FIR filter length L. In this example, we set $N_f = 2L$.

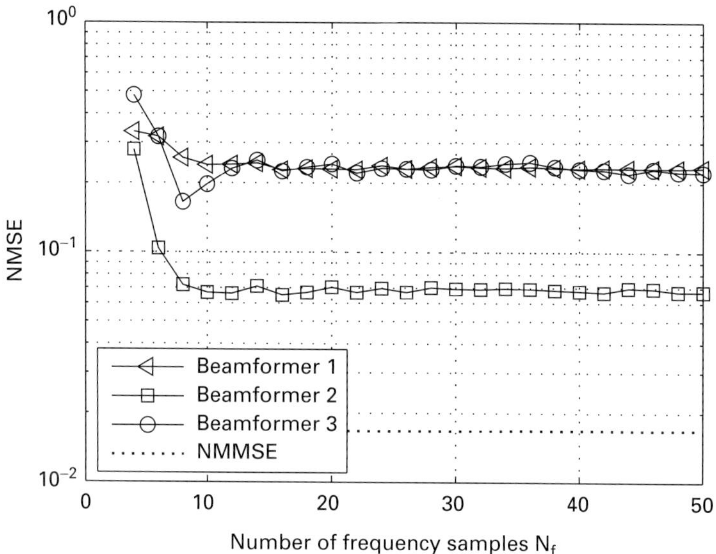

Figure 9.4   NMSE as a function of the number of frequency samples N_f.

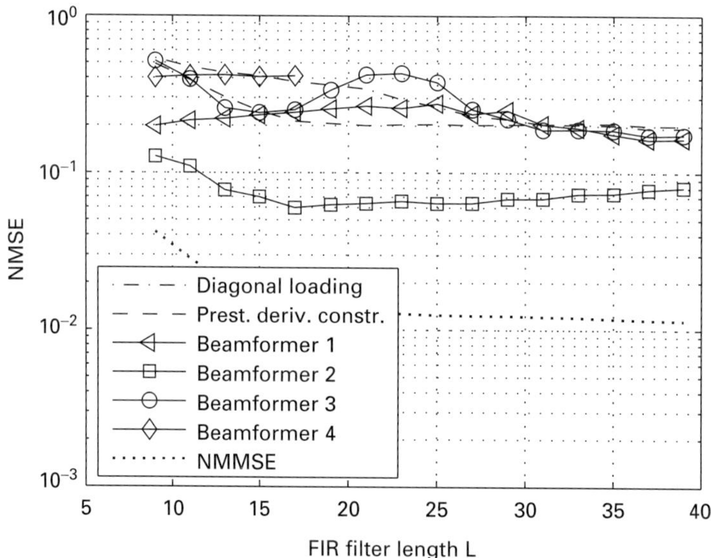

Figure 9.5   NMSE as a function of the FIR filter length L.

Figure 9.5 demonstrates that beamformers 1 and 4 with the filter length of $2L$ do not achieve the performance of beamformer 2 with the filter length of L. This may be explained by the fact that beamformers 1 and 4 satisfy the linear phase constraints for each sensor, independently from the value of L, which is not the case for beamformer 2.

Figure 9.5 also shows that, as expected, the performances of the beamformers saturate for large values of L. The curve for beamformer 4 ends at $L = 17$, because for higher values of L the CVX software cannot find a feasible point in some Monte-Carlo runs (even though the problems are all feasible). It should be noted here that the number of optimization variables increases for beamformer 4 quadratically with the FIR filter length L, whereas for all other methods the number of variables grows only linearly.

9.5 Conclusions

In this chapter, several robust, worst-case broadband adaptive beamforming techniques have been developed that offer different trade-offs in terms of interference suppression capability, robustness against signal self-nulling, and computational complexity. The proposed beamformers differ from the existing broadband robust methods in that their robustness is matched to the presumed amount of uncertainty in the array manifold. Convex SOCP and SDP formulations of the proposed beamformer designs have been obtained. Simulation results validate an improved robustness of the developed robust beamformers with respect to the earlier state-of-the-art broadband techniques.

Acknowledgments

This work was supported by the German Research Foundation (DFG) under Grant GE 1881/1-1 and the Advanced Investigator Grants Program of the European Research Council (ERC) under Grant 227477-ROSE.

References

[1] H. L. Van Trees, *Optimum Array Processing*. New York: Wiley, 2002.

[2] L. C. Godara, "Application of antenna arrays to mobile communications, Part II: Beamforming and direction-of-arrival considerations," *Proceedings of the IEEE*, vol. 85, pp. 1195–45, 1997.

[3] O. L. Frost III, "An algorithm for linearly constrained adaptive antenna array processing," *Proceedings of the IEEE*, vol. 60, pp. 926–35, 1972.

[4] L. J. Griffths and C. W. Jim, "An alternative approach to linearly constrained adaptive beamforming," *IEEE Transactions on Antennas & Propagation*, vol. 30, pp. 27–34, 1982.

[5] Z. Tian, K. L. Bell, and H. L. Van Trees, "A recursive least squares implementation for LCMP beamforming under quadratic constraint," *IEEE Transactions on Signal Processing*, vol. 49, pp. 1138–45, 2001.

[6] M. H. Er and A. Cantoni, "Derivative constraints for broad-band element space antenna array processors," *IEEE Transactions on Acoustics Speech, & Signal Processing*, vol. 31, pp. 1378–93, 1983.

[7] I. Thng, A. Cantoni, and Y. Leung, "Derivative constrained optimum broad-band antenna arrays," *IEEE Transactions on Signal Processing*, vol. 41, pp. 2376–88, 1993.

[8] I. Thng, A. Cantoni, and Y. Leung, "Constraints for maximally flat optimum broadband antenna arrays," *IEEE Transactions on Signal & Propagation*, vol. 43, pp. 1334–47, 1995.

[9] C. Y. Tseng, "Minimum variance beamforming with phase-independent derivative constraints," *IEEE Transactions on Antennas & Propagations*, vol. 40, pp. 285–94, Mar. 1992.

[10] S. Zhang and I. Thng, "Robust presteering derivative constraints for broadband antenna arrays," *IEEE Transactions on Signal Processing*, vol. 50, pp. 1–10, 2002.

[11] Y. I. Abramovich, "Controlled method for adaptive optimization of filters using the criterion of maximum SNR," *Radio Engineering & Electronic Physics*, vol. 26, pp. 87–95, March 1981.

[12] H. Cox, R. M. Zeskind, and M. H. Owen, "Robust adaptive beamforming," *IEEE Transactions on Acoustics Speech & Signal Processing*, vol. 35, pp. 1365–76, 1987.

[13] B. D. Carlson, "Covariance matrix estimation errors and diagonal loading in adaptive arrays," *IEEE Transactions on Aerospace & Electronics Systems*, vol. 24, pp. 397–401, July 1988.

[14] A. B. Gershman, "Robust adaptive beamforming in sensor arrays," *International Journal on Electronics and Communications*, vol. 53, pp. 305–14, 1999.

[15] S. A. Vorobyov, A. B. Gershman, and Z.-Q. Luo, "Robust adaptive beamforming using worst-case performance optimization: a solution to the signal mismatch problem," *IEEE Transactions on Signal Processing*, vol. 2, pp. 313–324, 2003.

[16] S. Shahbazpanahi, A. B. Gershman, Z.-Q. Luo, and K. M. Wong, "Robust adaptive beamforming for general-rank signal models," *IEEE Transactions on Signal Processing*, vol. 51, pp. 2257–69, 2003.

[17] R. Lorenz and S. P. Boyd, "Robust minimum variance beamforming," *IEEE Transactions on Signal Processing*, vol. 53, pp. 1684–96, 2005.

[18] J. Li, P. Stoica, and Z. Wang, "On robust Capon beamforming and diagonal loading," *IEEE Transactions on Signal Processing*, vol. 51, pp. 1702–15, 2003.

[19] S. A. Vorobyov, A. B. Gershman, Z.-Q. Luo, and N. Ma, "Adaptive beamforming with joint robustness against mismatched signal steering vector and interference nonstationarity," *IEEE Signal Processing Letters*, vol. 11, pp. 108–11, 2004.

[20] Y. C. Eldar, A. Nehorai, and P. S. La Rosa, "A competitive mean-squared error approach to beamforming," *IEEE Transactions on Signal Processing*, vol. 55, pp. 5143–54, 2007.

[21] Z. Wang, J. Li, P. Stoica, T. Nishida, and M. Sheplak, "Constant-beamwidth and constant-powerwidth wideband robust Capon beamformers for acoustic imaging," *Journal of the Acoustical Society of*, vol. 116, pp. 1621–31, 2004.

[22] A. El-Keyi, T. Kirubarajan, and A. B. Gershman, "Wideband robust beamforming based on worst-case performance optimization," in *Proceedings of the IEEE Workshop on Statistical Signal Processing*, Bordeaux, France, July 2005, pp. 265–70.

[23] A. El-Keyi, A. B. Gershman, and T. Kirubarajan, "Adaptive wideband beamforming with robustness against steering errors," in *Proceedings of the IEEE Sensor Array and Multichannel (SAM) Signal Processing Workshop*, Waltham, MA, 2006, pp. 11–15.

[24] M. Rübsamen and A. B. Gershman, "Robust presteered broadband beamforming based on worst-case performance optimization," in *Proceedings of the IEEE Sensor Array and Multichannel (SAM) Signal Processing Workshop*, Darmstadt, Germany, July 2008, pp. 340–44.

[25] S. Boyd and L. Vandenberghe, *Convex Optimization*. Cambridge: Cambridge University Press, 2004.

[26] Y. Nesterov and A. Nemirovsky, Algorithms "Interior-Point Polynomial in Convex Programming". *Studies in Applied Mathematics* (SIAM), Vol. 13, 1994.

[27] M. Lobo, L. Vandenberghe, S. Boyd, and H. Lebret, "Applications of second-order cone programming," *Linear Algebra & Its Applications*, vol. 284, pp. 193–228, 1998.

[28] T. N. Davidson, Z. Q. Luo, and J. F. Sturm, "Linear matrix inequality formulation of spectral mask constraints with applications to FIR filter design," *IEEE Transactions on Signal Processing*, vol. 50, pp. 2702–15, 2002.

[29] Y. Genin, Y. Hachez, Yu. Nesterov, and P. Van Dooren, "Optimization problems over positive pseudopolynomial matrices," *SIAM Journal on Matrix Analysis & Applications*, vol. 25, pp. 57–79, 2003.

[30] T. Roh and L. Vandenberghe, "Discrete transforms, semidefinite programming, and sum-of-squares representations of nonnegative polynomials," *SIAM Journal on Optimization*, vol. 16, pp. 939–64, 2006.

[31] B. Dumitrescu, *Positive Trigonometric Polynomials and Signal Processing Applications*. Dordrecht: Springer, 2007.

[32] J. F. Sturm, "Using SeDuMi 1.02, a MATLAB toolbox for optimization over symmetric cones," *Optim. Meth. Softw.*, vol. 11-12, pp. 625–653, 1999.

[33] S.-P. Wu, S. Boyd, and L. Vandenberghe, "FIR filter design via spectral factorization and convex optimization," in *Applied and Computational Control, Signals, and Circuits*, B. N. Datta, ed., Boston, MA: Birkhäuser, 1999.

[34] M. Grant and S. Boyd, "CVX: Matlab software for disciplined convex programming," http://stanford.edu/ boyd/cvx, 2008.

[35] M. Grant and S. Boyd, "Graph implementations for nonsmooth convex programs," in *Recent Advances in Learning and Control*, V. Blondel, S. Boyd, and H. Kimura, eds., London: Springer, 2008.

10 Cooperative distributed multi-agent optimization

Angelia Nedić and Asuman Ozdaglar

This chapter presents distributed algorithms for cooperative optimization among multiple agents connected through a network. The goal is to optimize a global-objective function which is a combination of local-objective functions known by the agents only. We focus on two related approaches for the design of distributed algorithms for this problem. The first approach relies on using Lagrangian-decomposition and dual-subgradient methods. We show that this methodology leads to distributed algorithms for optimization problems with special structure. The second approach involves combining consensus algorithms with subgradient methods. In both approaches, our focus is on providing convergence-rate analysis for the generated solutions that highlight the dependence on problem parameters.

10.1 Introduction and motivation

There has been much recent interest in distributed control and coordination of networks consisting of multiple agents, where the goal is to collectively optimize a global objective. This is motivated mainly by the emergence of large-scale networks and new networking applications such as mobile ad hoc networks and wireless-sensor networks, characterized by the lack of centralized access to information and time-varying connectivity. Control and optimization algorithms deployed in such networks should be completely distributed, relying only on local observations and information, robust against unexpected changes in topology, such as link or node failures, and scalable in the size of the network.

This chapter studies the problem of distributed optimization and control of multiagent networked systems. More formally, we consider a multiagent network model, where m agents exchange information over a connected network. Each agent i has a "local convex-objective function" $f_i(x)$, with $f_i : \mathbb{R}^n \to \mathbb{R}$, and a nonempty "local convex-constraint set" X_i, with $X_i \subset \mathbb{R}^n$, known by this agent only. The vector $x \in \mathbb{R}^n$ represents a global decision vector that the agents are collectively trying to decide on.

The goal of the agents is to "cooperatively optimize" a global-objective function, denoted by $f(x)$, which is a combination of the local-objective functions, that is,

$$f(x) = T\big(f_1(x), \ldots, f_m(x)\big),$$

Convex Optimization in Signal Processing and Communication, eds. Daniel P. Palomar and Yonina C. Eldar. Published by Cambridge University Press © Cambridge University Press, 2010.

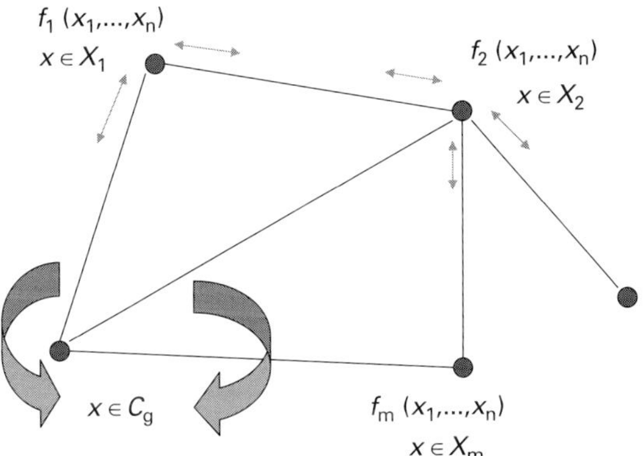

Figure 10.1   Multiagent cooperative-optimization problem.

where $T : \mathbb{R}^m \to \mathbb{R}$ is an increasing convex function.[1] The decision vector x is constrained to lie in a set, $x \in C$, which is a combination of local constraints and additional global constraints that may be imposed by the network structure, that is,

$$C = \left(\cap_{i=1}^m X_i \right) \cap C_g,$$

where C_g represents the global constraints. This model leads to the following optimization problem:

$$\text{minimize } f(x)$$
$$\text{subject to } x \in C, \tag{10.1}$$

where the function $f : \mathbb{R}^n \to \mathbb{R}$ is a "convex-objective function" and the set C is a "convex-constraint set" (see Figure 10.1). The decision vector x in problem (10.1) can be viewed as either a resource vector whose components correspond to resources allocated to each agent, or a global parameter vector to be estimated by the agents using local information.

Our goal in this chapter is to develop optimization methods that the agents can use to solve problem (10.1) within the informational structure available to them. Our development relies on using "first-order methods", in other words, gradient-based methods (or subgradient methods for the case when the local objective functions f_i are nonsmooth). Due to simplicity of computations per iteration, first-order methods have gained popularity in the last few years as low-overhead alternatives to interior-point methods, that may lend themselves to distributed implementations. Despite the fact that first-order methods have a slower convergence rate (compared to interior-point methods) in finding high-accuracy solutions, they are particularly effective in large-scale multiagent-optimization

---

[1] By an increasing function, we mean for any $w, y \in \mathbb{R}^m$ with $w \geq y$ with respect to the usual vector order (i.e., the inequality holds componentwise), we have $T(w) \geq T(y)$.

problems where the goal is to generate near-optimal *"approximate solutions"* in a relatively small number of iterations.

This chapter will present both classical results and recent advances on the design and analysis of distributed-optimization algorithms. The theoretical development will be complemented with recent application areas for these methods. Our development will focus on two major methodologies.

The first approach relies on using Lagrangian dual-decomposition and dual methods for solving problem (10.1). We will show that this approach leads to distributed-optimization algorithms when problem (10.1) is "separable" (i.e., problems where local-objective functions and constraints decompose over the components of the decision vector). This methodology has been used extensively in the networking literature to design cross-layer resource-allocation mechanisms [14, 23, 27, 48, 50]. Our focus in this chapter will be on generating approximate (primal) solutions from the dual algorithm and providing convergence-rate estimates. Despite the fact that duality yields distributed methods primarily for separable problems, our methods and rate analysis are applicable for general convex problems and will be covered here in their general version.

When problem (10.1) is not separable, the dual-decomposition approach will not lead to distributed methods. For such problems, we present optimization methods that use "consensus algorithms" as a building block. Consensus algorithms involve each agent maintaining estimates of the decision vector x and updating it based on local information that becomes available through the communication network. These algorithms have attracted much attention in the cooperative-control literature for distributed coordination of a system of dynamic agents [6–13, 20–22, 29, 44–46, 54, 55]. These works mainly focus on the "canonical-consensus problem", where the goal is to design distributed algorithms that can be used by a group of agents to agree on a common value. Here, we show that consensus algorithms can be combined with first-order methods to design distributed methods that can optimize general convex local-objective functions over a time-varying network topology.

The chapter is organized into four sections. In Section 10.2, we present distributed algorithms designed using Lagrangian duality and subgradient methods. We show that for (separable) network resource-allocation problems, this methodology yields distributed-optimization methods. We present recent results on generating approximate primal solutions from dual-subgradient methods and provide convergence-rate analysis. In Section 10.3, we develop distributed methods for optimizing the sum of general (non-separable) convex-objective functions corresponding to multiple agents connected over a time-varying topology. These methods will involve a combination of first-order methods and consensus algorithms. Section 10.4 focuses on extensions of the distributed methods to handle local constraints and imperfections associated with implementing optimization algorithms over networked systems, such as delays, asynchronism, and quantization effects, and studies the implications of these considerations on the network-algorithm performance. Section 10.5 suggests a number of areas for future research.

10.2 Distributed-optimization methods using dual decomposition

This section focuses on subgradient methods for solving the dual problem of a convex-constrained optimization problem obtained by Lagrangian relaxation of some of the constraints. For separable problems, this method leads to decomposition of the computations at each iteration into subproblems that each agent can solve using his local information and the prices (or dual variables).

In the first part of the section, we formally define the dual problem of a (primal) convex-constrained optimization problem. We establish relations between the primal and the dual optimal values, and investigate properties of the dual optimal-solution set. In Section 10.2.3, we introduce the utility-based, network resource-allocation problem and show that Lagrangian-decomposition and dual-subgradient methods yield distributed-optimization methods for solving this problem. Since the main interest in most practical applications is to obtain near-optimal solutions to problem (10.1), the remainder of the section focuses on obtaining approximate primal solutions using information directly available from dual-subgradient methods and presents the corresponding rate analysis.

We start by defining the basic notation and terminology used throughout the chapter.

10.2.1 Basic notation and terminology

We consider the n-dimensional vector space $\mathbb{R}^n$ and the m-dimensional vector space $\mathbb{R}^m$. We view a vector as a column vector, and we denote by $x'y$ the inner product of two vectors x and y. We use $\|y\|$ to denote the standard Euclidean norm, $\|y\| = \sqrt{y'y}$. We write $dist(\bar{y}, Y)$ to denote the standard Euclidean distance of a vector $\bar{y}$ from a set Y, in other words,

$$dist(\bar{y}, Y) = \inf_{y \in Y} \|\bar{y} - y\|.$$

For a vector $u \in \mathbb{R}^m$, we write u^+ to denote the projection of u on the non-negative orthant in $\mathbb{R}^m$, that is, u^+ is the component wise maximum of the vector u and the zero vector:

$$u^+ = (\max\{0, u_1\}, \cdots, \max\{0, u_m\})' \quad \text{for } u = (u_1, \cdots, u_m)'.$$

For a convex function $F : \mathbb{R}^n \to [-\infty, \infty]$, we denote the domain of F by $\text{dom}(F)$, where

$$\text{dom}(F) = \{x \in \mathbb{R}^n \mid F(x) < \infty\}.$$

We use the notion of a subgradient of a convex function $F(x)$ at a given vector $\bar{x} \in \text{dom}(F)$. A subgradient $s_F(\bar{x})$ of a convex function $F(x)$ at any $\bar{x} \in \text{dom}(F)$ provides a linear underestimate of the function F. In particular, $s_F(\bar{x}) \in \mathbb{R}^n$ is a "subgradient of a convex function" $F : \mathbb{R}^n \to \mathbb{R}$ "at a given vector" $\bar{x} \in \mathrm{dom}(F)$ when the following relation holds:

$$F(\bar{x}) + s_F(\bar{x})'(x - \bar{x}) \leq F(x) \qquad \text{for all } x \in \mathrm{dom}(F). \tag{10.2}$$

The set of all subgradients of F at $\bar{x}$ is denoted by $\partial F(\bar{x})$.

Similarly, for a concave function $q : \mathbb{R}^m \to [-\infty, \infty]$, we denote the domain of q by $\mathrm{dom}(q)$, where

$$\mathrm{dom}(q) = \{\mu \in \mathbb{R}^m \mid q(\mu) > -\infty\}.$$

A subgradient of a concave function is defined through a subgradient of a convex function $-q(\mu)$. In particular, $s_q(\bar{\mu}) \in \mathbb{R}^m$ is a "subgradient of a concave function" $q(\mu)$ "at a given vector" $\bar{\mu} \in \mathrm{dom}(q)$ when the following relation holds:

$$q(\bar{\mu}) + s_q(\bar{\mu})'(\mu - \bar{\mu}) \geq q(\mu) \qquad \text{for all } \mu \in \mathrm{dom}(q). \tag{10.3}$$

The set of all subgradients of q at $\bar{\mu}$ is denoted by $\partial q(\bar{\mu})$.

10.2.2   Primal and dual problem

We consider the following constrained-optimization problem:

$$\begin{aligned}
\text{minimize} \quad & f(x) \\
\text{subject to} \quad & g(x) \leq 0 \\
& x \in X,
\end{aligned} \tag{10.4}$$

where $f : \mathbb{R}^n \to \mathbb{R}$ is a convex function, $g = (g_1, \ldots, g_m)'$ and each $g_j : \mathbb{R}^n \to \mathbb{R}$ is a convex function, and $X \subset \mathbb{R}^n$ is a nonempty, closed convex set. We refer to problem (10.4) as the "primal problem". We denote the primal-optimal value by f^* and the primal-optimal set by X^*. Throughout this section, we assume that the value f^* is finite.

We next define the "dual problem" for problem (10.4). The dual problem is obtained by first relaxing the inequality constraints $g(x) \leq 0$ in problem (10.4), which yields the "dual function" $q : \mathbb{R}^m \to \mathbb{R}$ given by

$$q(\mu) = \inf_{x \in X} \{f(x) + \mu'g(x)\}. \tag{10.5}$$

The dual problem is then given by

$$\begin{aligned}
\text{maximize} \quad & q(\mu) \\
\text{subject to} \quad & \mu \geq 0 \\
& \mu \in \mathbb{R}^m.
\end{aligned} \tag{10.6}$$

We denote the dual-optimal value by q^* and the dual-optimal set by M^*.

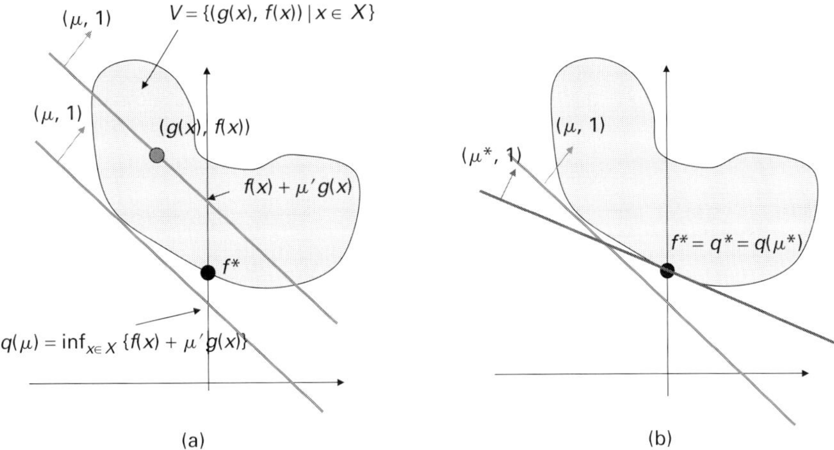

Figure 10.2   Illustration of the (a) primal, and the (b) dual problem.

The primal and the dual problem can be visualized geometrically by considering the set V of constraint-cost function values as x ranges over the set X, that is,

$$V = \{(g(x), f(x)) \mid x \in X\},$$

(see Figure 10.2). In this figure, the primal-optimal value f^* corresponds to the minimum vertical-axis value of all points on the left-half plane, that is, all points of the form $\{g(x) \leq 0 \mid x \in X\}$. Similarly, for a given dual-feasible solution $\mu \geq 0$, the dual-function value $q(\mu)$ corresponds to the vertical-intercept value of all hyperplanes with normal $(\mu, 1)$ and support the set V from below.[2] The dual-optimal value q^* then corresponds to the maximum-intercept value of such hyperplanes over all $\mu \geq 0$ [see Figure 10.2(b)]. This figure provides much insight about the relation between the primal and the dual problems and the structure of dual-optimal solutions, and has been used recently to develop a duality theory based on geometric principles (see [5] for convex-constrained optimization problems, and [34, 38, 39] for nonconvex-constrained optimization problems).

Duality gap and dual solutions

It is clear from the geometric picture that the primal- and dual-optimal values satisfy $q^* \leq f^*$, which is the well-known "weak-duality" relation (see Bertsekas *et al.* [2]).

---

[2] A *hyperplane* $H \subset \mathbb{R}^n$ is an $(n-1)$-dimensional affine set, which is defined through its nonzero normal vector $a \in \mathbb{R}^n$ and a scalar b as

$$H = \{x \in \mathbb{R}^n \mid a'x = b\}.$$

Any vector $\bar{x} \in H$ can be used to determine the constant b as $a'\bar{x} = b$, thus yielding an equivalent representation of the hyperplane H as

$$H = \{x \in \mathbb{R}^n \mid a'x = a'\bar{x}\}.$$

Here, we consider hyperplanes in $\mathbb{R}^{r+1}$ with normal vectors given by $(\mu, 1) \in \mathbb{R}^{m+1}$.

When $f^* = q^*$, we say that there is *no duality gap* or *strong duality holds*. The next condition guarantees that there is no duality gap.

Assumption 1   (Slater condition) *There exists a vector $\bar{x} \in X$ such that*

$$g_j(\bar{x}) < 0 \quad \text{for all} \quad j = 1, \ldots, m.$$

We refer to a vector $\bar{x}$ satisfying the Slater condition as a Slater vector.

Under the convexity assumptions on the primal problem (10.4) and the assumption that f^* is finite, it is well-known that the Slater condition is sufficient for no duality gap, as well as for the existence of a dual-optimal solution (see for example Bertsekas [4] or Bertsekas *et al.* [2]). Furthermore, under the Slater condition, the dual-optimal set is bounded (see Uzawa [53] and Hiriart-Urruty and Lemaréchal [19]). Figure 10.3 provides some intuition for the role of convexity and the Slater condition in establishing no duality gap and the boundedness of the dual-optimal solution set.

The following lemma extends the result on the optimal dual-set boundedness under the Slater condition. In particular, it shows that the Slater condition also guarantees the boundedness of the (level) sets $\{\mu \geq 0 \mid q(\mu) \geq q(\bar{\mu})\}$.

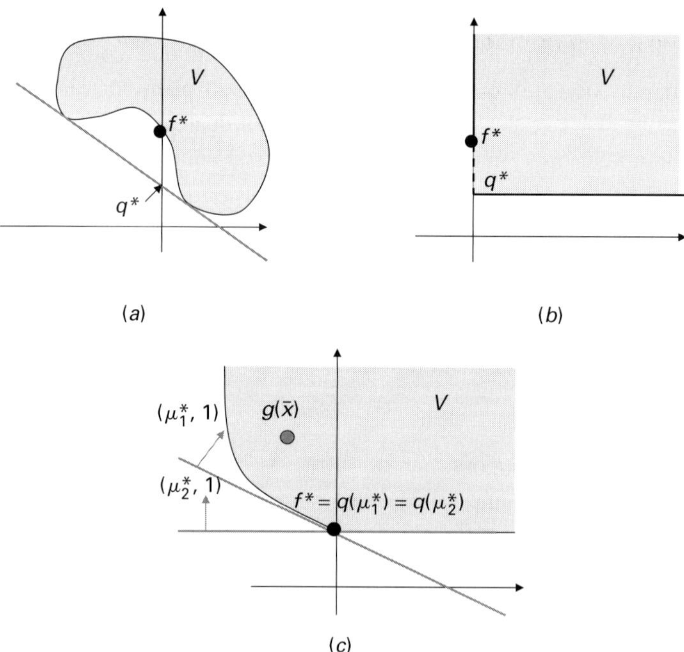

Figure 10.3   Parts (a) and (b) provide two examples where there is a duality gap [due to lack of convexity in (a) and lack of "continuity around origin" in (b)]. Part (c) illustrates the role of the Slater condition in establishing no duality gap and boundedness of the dual-optimal solutions. Note that dual-optimal solutions correspond to the normal vectors of the (nonvertical) hyperplanes supporting set V from below at the point $(0, q^*)$.

LEMMA 10.1 *Let the Slater condition hold [cf. Assumption 1]. Then, the set* $Q_{\bar{\mu}} = \{\mu \geq 0 \mid q(\mu) \geq q(\bar{\mu})\}$ *is bounded and, in particular, we have*

$$\max_{\mu \in Q_{\bar{\mu}}} \|\mu\| \leq \frac{1}{\gamma}\left(f(\bar{x}) - q(\bar{\mu})\right),$$

where $\gamma = \min_{1 \leq j \leq m}\{-g_j(\bar{x})\}$ *and* $\bar{x}$ *is a Slater vector.*

Proof　We have for any $\mu \in Q_{\bar{\mu}}$,

$$q(\bar{\mu}) \leq q(\mu) = \inf_{x \in X}\left\{f(x) + \mu' g(x)\right\} \leq f(\bar{x}) + \mu' g(\bar{x}) = f(\bar{x}) + \sum_{j=1}^{m} \mu_j g_j(\bar{x}),$$

implying that

$$-\sum_{j=1}^{m} \mu_j g_j(\bar{x}) \leq f(\bar{x}) - q(\bar{\mu}).$$

Because $g_j(\bar{x}) < 0$ and $\mu_j \geq 0$ for all j, it follows that

$$\min_{1 \leq j \leq m}\left\{-g_j(\bar{x})\right\} \sum_{j=1}^{m} \mu_j \leq -\sum_{j=1}^{m} \mu_j g_j(\bar{x}) \leq f(\bar{x}) - q(\bar{\mu}).$$

Therefore,

$$\sum_{j=1}^{m} \mu_j \leq \frac{f(\bar{x}) - q(\bar{\mu})}{\min_{1 \leq j \leq m}\left\{-g_j(\bar{x})\right\}}.$$

Since $\mu \geq 0$, we have $\|\mu\| \leq \sum_{j=1}^{m} \mu_j$ and the estimate follows.　　■

It can be seen from the preceding lemma that under the Slater condition, the dual-optimal set M^* is nonempty. In particular, by noting that $M^* = \{\mu \geq 0 \mid q(\mu) \geq q^*\}$ and by using Lemma 10.1, we see that

$$\max_{\mu^* \in M^*} \|\mu^*\| \leq \frac{1}{\gamma}\left(f(\bar{x}) - q^*\right), \tag{10.7}$$

with $\gamma = \min_{1 \leq j \leq m}\{-g_j(\bar{x})\}$.

Dual-subgradient method

Since the dual function $q(\mu)$ given by (10.5) is the infimum of a collection of affine functions, it is a concave function [2]. Hence, we can use a subgradient method to solve the dual problem (10.6). In view of its implementation simplicity, we consider the classical subgradient algorithm with a constant stepsize:

$$\mu_{k+1} = [\mu_k + \alpha g_k]^+ \qquad \text{for } k = 0, 1, \ldots, \tag{10.8}$$

where the vector $\mu_0 \geq 0$ is an initial iterate, the scalar $\alpha > 0$ is a stepsize, and the vector g_k is a subgradient of q at μ_k. Due to the form of the dual function q, the subgradients of q at a vector μ are related to the primal vectors x_μ attaining the minimum in (10.5). Specifically, the set $\partial q(\mu)$ of subgradients of q at a given $\mu \geq 0$ is given by

$$\partial q(\mu) = \mathrm{conv}\left(\{g(x_\mu) \mid x_\mu \in X_\mu\}\right), \quad X_\mu = \{x_\mu \in X \mid q(\mu) = f(x_\mu) + \mu' g(x_\mu)\}, \tag{10.9}$$

where $\mathrm{conv}(Y)$ denotes the convex hull of a set Y [2].

10.2.3   Distributed methods for utility-based network-resource allocation

In this section, we consider a utility-based network resource-allocation problem and briefly discuss how dual-decomposition and subgradient methods lead to decentralized-optimization methods that can be used over a network. This approach was proposed in the seminal work of Kelly *et al.* [23] and further developed by Low and Lapsley [27], Shakkottai and Srikant [48], and Srikant [50].

Consider a network that consists of a set $S = \{1, \ldots, S\}$ of sources and a set $\mathcal{L} = \{1, \ldots, L\}$ of undirected links, where a link l has capacity c_l. Let $\mathcal{L}(i) \subset \mathcal{L}$ denote the set of links used by source i. The application requirements of source i is represented by a concave, increasing utility function $u_i : [0, \infty) \to [0, \infty)$, in other words, each source i gains a utility $u_i(x_i)$ when it sends data at a rate x_i. We further assume that rate x_i is constrained to lie in the interval $I_i = [0, M_i]$ for all $i \in S$, where the scalar M_i denotes the maximum allowed rate for source i. Let $S(l) = \{i \in S \mid l \in \mathcal{L}(i)\}$ denote the set of sources that use link l. The goal of the "network utility-maximization problem" is to allocate the source rates as the optimal solution of the problem

$$\text{maximize} \quad \sum_{i \in S} u_i(x_i) \tag{10.10}$$

$$\text{subject to} \quad \sum_{i \in S(l)} x_i \leq c_l \quad \text{for all } l \in \mathcal{L}$$

$$x_i \in I_i \quad \text{for all } i \in S.$$

This problem is a special case of the multiagent-optimization problem (10.1), where the local objective function of each agent (or source) $f_i(x)$ is given by $f_i(x) = -u_i(x_i)$, that is, the local-objective function of each agent depends only on one component of the decision vector x and the overall-objective function $f(x)$ is the sum of the local-objective functions, $f(x) = -\sum_i u_i(x_i)$, that is, the global-objective function $f(x)$ is separable in the components of the decision vector. Moreover, the global-constraint set C_g is given by the link capacity constraints, and the local-constraint set of each agent X_i is given by the interval I_i. Note that only agent i knows his utility function $u_i(x_i)$ and his maximum allowed rate M_i, which specifies the local constraint $x_i \in I_i$.

Solving problem (10.10) directly by applying existing subgradient methods requires coordination among sources, and therefore may be impractical for real networks. This is in view of the fact that in large-scale networks, such as the Internet, there is no central entity that has access to both the source-utility functions and constraints, and the capacity of all the links in the network. Despite this information structure, in view of the separable structure of the objective and constraint functions, the dual problem can be evaluated exactly using decentralized information. In particular, the dual problem of (10.10) is given by (10.6), where the dual function takes the form

$$q(\mu) = \max_{x_i \in I_i,\, i \in \mathcal{S}} \sum_{i \in \mathcal{S}} u_i(x_i) - \sum_{l \in \mathcal{L}} \mu_l \left(\sum_{i \in \mathcal{S}(l)} x_i - c_l \right)$$

$$= \max_{x_i \in I_i,\, i \in \mathcal{S}} \sum_{i \in \mathcal{S}} \left(u_i(x_i) - x_i \sum_{l \in \mathcal{L}(i)} \mu_l \right) + \sum_{l \in \mathcal{L}} \mu_l c_l.$$

Since the optimization problem on the right-hand side of the preceding relation is separable in the variables x_i, the problem decomposes into subproblems for each source i. Letting $\mu_i = \sum_{l \in \mathcal{L}(i)} \mu_l$ for each i (i.e., μ_i is the sum of the multipliers corresponding to the links used by source i), we can write the dual function as

$$q(\mu) = \sum_{i \in \mathcal{S}} \max_{x_i \in I_i} \{ u_i(x_i) - x_i \mu_i \} + \sum_{l \in \mathcal{L}} \mu_l c_l.$$

Hence, to evaluate the dual function, each source i needs to solve the one-dimensional optimization problem $\max_{x_i \in I_i} \{ u_i(x_i) - x_i \mu_i \}$. This involves only its own utility function u_i and the value μ_i, which is available to source i in practical networks (through a direct feedback mechanism from its destination).

Using a subgradient method to solve the dual problem (10.6) yields the following distributed-optimization method, where at each iteration $k \geq 0$, links and sources update their prices (or dual-solution values) and rates respectively in a decentralized manner:
Link-Price Update. Each link l updates its price μ_l according to

$$\mu_l(k+1) = [\mu_l(k) + \alpha g_l(k)]^+,$$

where $g_l(k) = \sum_{i \in \mathcal{S}(l)} x_i(k) - c_l$, that is, $g_l(k)$ is the value of the link l capacity constraint (10.10) at the primal vector $x(k)$ [see the relation for the subgradient of the dual function $q(\mu)$ in (10.9)].
Source-Rate Update. Each source i updates its rate x_i according to

$$x_i(k+1) = \operatorname{argmax}_{x_i \in I_i} \{ u_i(x_i) - x_i \mu_i \}.$$

The preceding methodology has motivated much interest in using dual-decomposition and subgradient methods to solve network resource-allocation problems in an iterative decentralized manner (see Chiang *et al.* [14]). Other problems where the dual problem has a structure that allows exact evaluation of the dual function using local information include the problem of processor-speed control considered by Mutapcic *et al.* [30], and the traffic-equilibrium and road-pricing problems considered by Larsson *et al.* [24–26].

10.2.4   Approximate primal solutions and rate analysis

We first establish some basic relations that hold for a sequence $\{\mu_k\}$ obtained by the subgradient algorithm of (10.8).

LEMMA 10.2   *Let the sequence $\{\mu_k\}$ be generated by the subgradient algorithm (10.8). For any $\mu \geq 0$, we have*

$$\|\mu_{k+1} - \mu\|^2 \leq \|\mu_k - \mu\|^2 - 2\alpha\,(q(\mu) - q(\mu_k)) + \alpha^2 \|g_k\|^2 \qquad \textit{for all } k \geq 0.$$

Proof   By using the nonexpansive property of the projection operation, from relation (10.8) we obtain for any $\mu \geq 0$ and all k,

$$\|\mu_{k+1} - \mu\|^2 = \left\| [\mu_k + \alpha g_k]^+ - \mu \right\|^2 \leq \|\mu_k + \alpha g_k - \mu\|^2.$$

Therefore,

$$\|\mu_{k+1} - \mu\|^2 \leq \|\mu_k - \mu\|^2 + 2\alpha g_k'(\mu_k - \mu) + \alpha^2 \|g_k\|^2 \qquad \text{for all } k.$$

Since g_k is a subgradient of q at μ_k [cf. (10.3)], we have

$$g_k'(\mu - \mu_k) \geq q(\mu) - q(\mu_k),$$

implying that

$$g_k'(\mu_k - \mu) \leq -\,(q(\mu) - q(\mu_k))\,.$$

Hence, for any $\mu \geq 0$,

$$\|\mu_{k+1} - \mu\|^2 \leq \|\mu_k - \mu\|^2 - 2\alpha\,(q(\mu) - q(\mu_k)) + \alpha^2 \|g_k\|^2 \qquad \text{for all } k.$$

$\blacksquare$

Boundedness of dual iterates

Here, we show that the dual sequence $\{\mu_k\}$ generated by the subgradient algorithm is bounded under the Slater condition and a boundedness assumption on the subgradient sequence $\{g_k\}$. We formally state the latter requirement in the following.

Assumption 2   (Bounded subgradients) *The subgradient sequence $\{g_k\}$ is bounded, that is, there exists a scalar $L > 0$ such that*

$$\|g_k\| \leq L \qquad \textit{for all } k \geq 0.$$

This assumption is satisfied, for example, when the primal-constraint set X is compact. Due to the convexity of the constraint functions g_j over $\mathbb{R}^n$, each g_j is continuous over

$\mathbb{R}^n$. Thus, $\max_{x \in X} \|g(x)\|$ is finite and provides an upper bound on the norms of the subgradients g_k, in other words,

$$\|g_k\| \le L \qquad \text{for all } k \ge 0, \qquad \text{with } L = \max_{x \in X} \|g(x)\|.$$

In the following lemma, we establish the boundedness of the dual sequence generated by the subgradient method.

LEMMA 10.3   *Let the dual sequence $\{\mu_k\}$ be generated by the subgradient algorithm of (10.8). Also, let the Slater condition and the bounded-subgradients assumption hold [cf. Assumptions 1 and 2]. Then, the sequence $\{\mu_k\}$ is bounded and, in particular, we have*

$$\|\mu_k\| \le \frac{2}{\gamma} \left(f(\bar{x}) - q^* \right) + \max \left\{ \|\mu_0\|, \frac{1}{\gamma} \left(f(\bar{x}) - q^* \right) + \frac{\alpha L^2}{2\gamma} + \alpha L \right\},$$

where $\gamma = \min_{1 \le j \le m} \{-g_j(\bar{x})\}$, $\bar{x}$ is a Slater vector, L is the subgradient norm bound, and $\alpha > 0$ is the stepsize.

Proof   Under the Slater condition the optimal-dual set M^* is nonempty. Consider the set Q_α defined by

$$Q_\alpha = \left\{ \mu \ge 0 \mid q(\mu) \ge q^* - \frac{\alpha L^2}{2} \right\},$$

which is nonempty in view of $M^* \subset Q_\alpha$. We fix an arbitrary $\mu^* \in M^*$ and we first prove that for all $k \ge 0$,

$$\|\mu_k - \mu^*\| \le \max \left\{ \|\mu_0 - \mu^*\|, \frac{1}{\gamma} \left(f(\bar{x}) - q^* \right) + \frac{\alpha L^2}{2\gamma} + \|\mu^*\| + \alpha L \right\}, \qquad (10.11)$$

where $\gamma = \min_{1 \le j \le m} \{-g_j(\bar{x})\}$ and L is the bound on the subgradient norms $\|g_k\|$. Then, we use Lemma 10.1 to prove the desired estimate.

We show that relation (10.11) holds by induction on k. Note that the relation holds for $k = 0$. Assume now that it holds for some $k > 0$, that is,

$$\|\mu_k - \mu^*\| \le \max \left\{ \|\mu_0 - \mu^*\|, \frac{1}{\gamma} \left(f(\bar{x}) - q^* \right) + \frac{\alpha L^2}{2\gamma} + \|\mu^*\| + \alpha L \right\} \text{ for some } k > 0.$$

$$(10.12)$$

We now consider two cases: $q(\mu_k) \ge q^* - \alpha L^2/2$ and $q(\mu_k) < q^* - \alpha L^2/2$.

Case 1: $q(\mu_k) \ge q^* - \alpha L^2/2$.   By using the definition of the iterate μ_{k+1} in (10.8) and the subgradient boundedness, we obtain

$$\|\mu_{k+1} - \mu^*\| \le \|\mu_k + \alpha g_k - \mu^*\| \le \|\mu_k\| + \|\mu^*\| + \alpha L.$$

Since $q(\mu_k) \geq q^* - \alpha L^2/2$, it follows that $\mu_k \in Q_\alpha$. According to Lemma 10.1, the set Q_α is bounded and, in particular, $\|\mu\| \leq \frac{1}{\gamma}\left(f(\bar{x}) - q^* + \alpha L^2/2\right)$ for all $\mu \in Q_\alpha$. Therefore

$$\|\mu_k\| \leq \frac{1}{\gamma}\left(f(\bar{x}) - q^*\right) + \frac{\alpha L^2}{2\gamma}.$$

By combining the preceding two relations, we obtain

$$\|\mu_{k+1} - \mu^*\| \leq \frac{1}{\gamma}\left(f(\bar{x}) - q^*\right) + \frac{\alpha L^2}{2\gamma} + \|\mu^*\| + \alpha L,$$

thus showing that the estimate in (10.11) holds for $k + 1$.

Case 2: $q(\mu_k) < q^* - \alpha L^2/2$. By using Lemma 10.2 with $\mu = \mu^*$, we obtain

$$\|\mu_{k+1} - \mu^*\|^2 \leq \|\mu_k - \mu^*\|^2 - 2\alpha\left(q^* - q(\mu_k)\right) + \alpha^2 \|g_k\|^2.$$

By using the subgradient boundedness, we further obtain

$$\|\mu_{k+1} - \mu^*\|^2 \leq \|\mu_k - \mu^*\|^2 - 2\alpha\left(q^* - q(\mu_k) - \frac{\alpha L^2}{2}\right).$$

Since $q(\mu_k) < q^* - \alpha L^2/2$, it follows that $q^* - q(\mu_k) - \alpha L^2/2 > 0$, which when combined with the preceding relation yields

$$\|\mu_{k+1} - \mu^*\| < \|\mu_k - \mu^*\|.$$

By the induction hypothesis [cf. (10.12)], it follows that the estimate in (10.11) holds for $k + 1$ as well. Hence, the estimate in (10.11) holds for all $k \geq 0$.

From (10.11) we obtain for all $k \geq 0$,

$$\|\mu_k\| \leq \|\mu_k - \mu^*\| + \|\mu^*\|$$

$$\leq \max\left\{\|\mu_0 - \mu^*\|, \frac{1}{\gamma}\left(f(\bar{x}) - q^*\right) + \frac{\alpha L^2}{2\gamma} + \|\mu^*\| + \alpha L\right\} + \|\mu^*\|.$$

By using $\|\mu_0 - \mu^*\| \leq \|\mu_0\| + \|\mu^*\|$, we further have for all $k \geq 0$,

$$\|\mu_k\| \leq \max\left\{\|\mu_0\| + \|\mu^*\|, \frac{1}{\gamma}\left(f(\bar{x}) - q^*\right) + \frac{\alpha L^2}{2\gamma} + \|\mu^*\| + \alpha L\right\} + \|\mu^*\|$$

$$= 2\|\mu^*\| + \max\left\{\|\mu_0\|, \frac{1}{\gamma}\left(f(\bar{x}) - q^*\right) + \frac{\alpha L^2}{2\gamma} + \alpha L\right\}.$$

Since $M^* = \{\mu \geq 0 \mid q(\mu) \geq q^*\}$, according to Lemma 10.1, we have the following bound on the dual-optimal solutions

$$\max_{\mu^* \in M^*} \|\mu^*\| \leq \frac{1}{\gamma}\left(f(\bar{x}) - q^*\right),$$

implying that for all $k \geq 0$,

$$\|\mu_k\| \leq \frac{2}{\gamma}\left(f(\bar{x}) - q^*\right) + \max\left\{\|\mu_0\|, \frac{1}{\gamma}\left(f(\bar{x}) - q^*\right) + \frac{\alpha L^2}{2\gamma} + \alpha L\right\}.$$

■

Convergence-rate estimates

In this section, we generate approximate primal solutions by considering the running averages of the primal sequence $\{x_k\}$ obtained in the implementation of the subgradient method. We show that under the Slater condition, we can provide bounds for the number of subgradient iterations needed to generate a primal solution within a given level of constraint violation. We also derive upper and lower bounds on the gap from the optimal primal value.

To define the approximate primal solutions, we consider the dual sequence $\{\mu_k\}$ generated by the subgradient algorithm in (10.8), and the corresponding sequence of primal vectors $\{x_k\} \subset X$ that provide the subgradients g_k in the algorithm, that is,

$$g_k = g(x_k), \qquad x_k \in \underset{x \in X}{\operatorname{argmax}}\{f(x) + \mu_k' g(x)\} \qquad \text{for all } k \geq 0, \tag{10.13}$$

[see the subdifferential relation in (10.9)]. We define $\hat{x}_k$ as the average of the vectors $x_0, \ldots, x_{k-1}$, that is,

$$\hat{x}_k = \frac{1}{k}\sum_{i=0}^{k-1} x_i \qquad \text{for all } k \geq 1. \tag{10.14}$$

The average vectors $\hat{x}_k$ lie in the set X because X is convex and $x_i \in X$ for all i. However, these vectors need not satisfy the primal-inequality constraints $g_j(x) \leq 0, j = 1, \ldots, m$, and therefore, they can be primal infeasible.

The next proposition provides a bound on the amount of feasibility violation of the running averages $\hat{x}_k$. It also provides upper and lower bounds on the primal cost of these vectors. These bounds are given per iteration, as seen in the following.

THEOREM 10.1 Let the sequence $\{\mu_k\}$ be generated by the subgradient algorithm (10.8). Let the Slater condition and bounded-subgradients assumption hold [cf. Assumptions 1 and 2]. Also, let

$$B^* = \frac{2}{\gamma}\left(f(\bar{x}) - q^*\right) + \max\left\{\|\mu_0\|, \frac{1}{\gamma}\left(f(\bar{x}) - q^*\right) + \frac{\alpha L^2}{2\gamma} + \alpha L\right\}. \tag{10.15}$$

Let the vectors $\hat{x}_k$ for $k \geq 1$ be the averages given by (10.14). Then, the following holds for all $k \geq 1$:

(a) An upper bound on the amount of constraint violation of the vector $\hat{x}_k$ is given by

$$\|g(\hat{x}_k)^+\| \leq \frac{B^*}{k\alpha}.$$

(b) An upper bound on the primal cost of the vector $\hat{x}_k$ is given by

$$f(\hat{x}_k) \leq f^* + \frac{\|\mu_0\|^2}{2k\alpha} + \frac{\alpha L^2}{2}.$$

(c) A lower bound on the primal cost of the vector $\hat{x}_k$ is given by

$$f(\hat{x}_k) \geq f^* - \frac{1}{\gamma}\left[f(\bar{x}) - q^*\right]\|g(\hat{x}_k)^+\|.$$

Proof

(a) By using the definition of the iterate μ_{k+1} in (10.8), we obtain

$$\mu_k + \alpha g_k \leq [\mu_k + \alpha g_k]^+ = \mu_{k+1} \qquad \text{for all } k \geq 0.$$

Since $g_k = g(x_k)$ with $x_k \in X$, it follows that

$$\alpha g(x_k) \leq \mu_{k+1} - \mu_k \qquad \text{for all } k \geq 0.$$

Therefore,

$$\sum_{i=0}^{k-1} \alpha g(x_i) \leq \mu_k - \mu_0 \leq \mu_k \qquad \text{for all } k \geq 1,$$

where the last inequality in the preceding relation follows from $\mu_0 \geq 0$. Since $x_k \in X$ for all k, by the convexity of X, we have $\hat{x}_k \in X$ for all k. Hence, by the convexity of each of the functions g_j, it follows that

$$g(\hat{x}_k) \leq \frac{1}{k}\sum_{i=0}^{k-1} g(x_i) = \frac{1}{k\alpha}\sum_{i=0}^{k-1} \alpha g(x_i) \leq \frac{\mu_k}{k\alpha} \qquad \text{for all } k \geq 1.$$

Because $\mu_k \geq 0$ for all k, we have $g(\hat{x}_k)^+ \leq \mu_k/(k\alpha)$ for all $k \geq 1$ and, therefore,

$$\|g(\hat{x}_k)^+\| \leq \frac{\|\mu_k\|}{k\alpha} \qquad \text{for all } k \geq 1. \tag{10.16}$$

By Lemma 10.3 we have

$$\|\mu_k\| \leq \frac{2}{\gamma}\left(f(\bar{x}) - q^*\right) + \max\left\{\|\mu_0\|, \frac{1}{\gamma}\left(f(\bar{x}) - q^*\right) + \frac{\alpha L^2}{2\gamma} + \alpha L\right\} \qquad \text{for all } k \geq 0.$$

By the definition of B^* in (10.15), the preceding relation is equivalent to

$$\|\mu_k\| \leq B^* \qquad \text{for all } k \geq 0.$$

Combining this relation with (10.16), we obtain

$$\|g(\hat{x}_k)^+\| \le \frac{\|\mu_k\|}{k\alpha} \le \frac{B^*}{k\alpha} \qquad \text{for all } k \ge 1.$$

(b) By the convexity of the primal cost $f(x)$ and the definition of x_k as a minimizer of the Lagrangian function $f(x) + \mu_k' g(x)$ over $x \in X$ [cf. (10.13)], we have

$$f(\hat{x}_k) \le \frac{1}{k}\sum_{i=0}^{k-1} f(x_i) = \frac{1}{k}\sum_{i=0}^{k-1}\{f(x_i) + \mu_i' g(x_i)\} - \frac{1}{k}\sum_{i=0}^{k-1}\mu_i' g(x_i).$$

Since $q(\mu_i) = f(x_i) + \mu_i' g(x_i)$ and $q(\mu_i) \le q^*$ for all i, it follows that for all $k \ge 1$,

$$f(\hat{x}_k) \le \frac{1}{k}\sum_{i=0}^{k-1} q(\mu_i) - \frac{1}{k}\sum_{i=0}^{k-1}\mu_i' g(x_i) \le q^* - \frac{1}{k}\sum_{i=0}^{k-1}\mu_i' g(x_i). \qquad (10.17)$$

From the definition of the algorithm in (10.8), by using the nonexpansive property of the projection, and the facts $0 \in \{\mu \in \mathbb{R}^m \mid \mu \ge 0\}$ and $g_i = g(x_i)$, we obtain

$$\|\mu_{i+1}\|^2 \le \|\mu_i\|^2 + 2\alpha\mu_i' g(x_i) + \alpha^2\|g(x_i)\|^2 \qquad \text{for all } i \ge 0,$$

implying that

$$-\mu_i' g(x_i) \le \frac{\|\mu_i\|^2 - \|\mu_{i+1}\|^2 + \alpha^2\|g(x_i)\|^2}{2\alpha} \qquad \text{for all } i \ge 0.$$

By summing over $i = 0,\ldots,k-1$ for $k \ge 1$, we have

$$-\frac{1}{k}\sum_{i=0}^{k-1}\mu_i' g(x_i) \le \frac{\|\mu_0\|^2 - \|\mu_k\|^2}{2k\alpha} + \frac{\alpha}{2k}\sum_{i=0}^{k-1}\|g(x_i)\|^2 \qquad \text{for all } k \ge 1.$$

Combining the preceding relation and (10.17), we further have

$$f(\hat{x}_k) \le q^* + \frac{\|\mu_0\|^2 - \|\mu_k\|^2}{2k\alpha} + \frac{\alpha}{2k}\sum_{i=0}^{k-1}\|g(x_i)\|^2 \qquad \text{for all } k \ge 1.$$

Under the Slater condition, there is zero duality gap, in other words, $q^* = f^*$. Furthermore, the subgradients are bounded by a scalar L [cf. Assumption 2], so that

$$f(\hat{x}_k) \le f^* + \frac{\|\mu_0\|^2}{2k\alpha} + \frac{\alpha L^2}{2} \qquad \text{for all } k \ge 1,$$

yielding the desired estimate.

(c) Given a dual-optimal solution μ^*, we have

$$f(\hat{x}_k) = f(\hat{x}_k) + (\mu^*)' g(\hat{x}_k) - (\mu^*)' g(\hat{x}_k) \ge q(\mu^*) - (\mu^*)' g(\hat{x}_k).$$

Because $\mu^* \geq 0$ and $g(\hat{x}_k)^+ \geq g(\hat{x}_k)$, we further have

$$-(\mu^*)'g(\hat{x}_k) \geq -(\mu^*)'g(\hat{x}_k)^+ \geq -\|\mu^*\|\|g(\hat{x}_k)^+\|.$$

From the preceding two relations and the fact $q(\mu^*) = q^* = f^*$ it follows that

$$f(\hat{x}_k) \geq f^* - \|\mu^*\|\|g(\hat{x}_k)^+\|.$$

By using Lemma 10.1 with $\bar{\mu} = \mu^*$, we see that the dual set is bounded and, in particular, $\|\mu^*\| \leq \frac{1}{\gamma}(f(\bar{x}) - q^*)$ for all dual-optimal vectors μ^*. Hence,

$$f(\hat{x}_k) \geq f^* - \frac{1}{\gamma}\left[f(\bar{x}) - q^*\right]\|g(\hat{x}_k)^+\| \qquad \text{for all } k \geq 1.$$

$\blacksquare$

10.2.5   Numerical example

In this section, we study a numerical example to illustrate the performance of the dual-subgradient method with primal averaging for the utility-based network resource-allocation problem described in Section 10.2.3. Consider the network illustrated in Figure 10.4 with 2 serial links and 3 users each sending data at a rate x_i for $i = 1, 2, 3$. Link 1 has a capacity $c_1 = 1$ and link 2 has a capacity $c_2 = 2$. Assume that each user has an identical concave-utility function $u_i(x_i) = \sqrt{x_i}$, which represents the utility gained from sending rate x_i. We consider allocating rates among the users as the optimal solution of the problem

$$\text{maximize} \quad \sum_{i=1}^{3} \sqrt{x_i}$$

$$\text{subject to} \quad x_1 + x_2 \leq 1, \quad x_1 + x_3 \leq 2,$$

$$x_i \geq 0, \quad i = 1, 2, 3.$$

The optimal solution of this problem is $x^* = [0.2686, 0.7314, 1.7314]$ and the optimal value is $f^* \approx 2.7$. We consider solving this problem using the dual-subgradient method of (10.8) (with a constant stepsize $\alpha = 1$) combined with primal averaging. In particular,

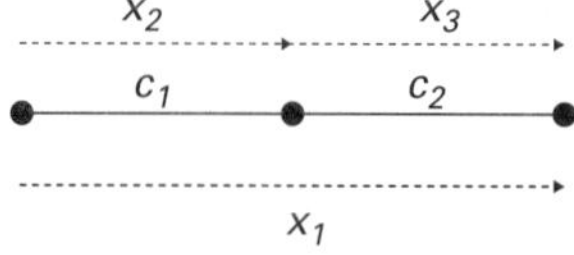

Figure 10.4   A simple network with two links of capacities $c_1 = 1$ and $c_2 = 2$, and three users, each sending data at a rate x_i.

when evaluating the subgradients of the dual function in (10.13), we obtain the primal sequence $\{x_k\}$. We generate the sequence $\{\hat{x}_k\}$ as the running average of the primal sequence [cf. (10.14)].

Figure 10.5 illustrates the behavior of the sequences $\{x_{ik}\}$ and $\{\hat{x}_{ik}\}$ for each user $i = 1, 2, 3$. As seen in this figure, for each user i, the sequences $\{x_{ik}\}$ exhibit oscillations, whereas the average sequences $\{\hat{x}_{ik}\}$ converge smoothly to near-optimal solutions within 60 subgradient iterations.

Figure 10.6 illustrates the results for the constraint-violation and the primal-objective value for the sequences $\{x_k\}$ and $\{\hat{x}_k\}$. The plot to the left in Figure 10.6 shows the

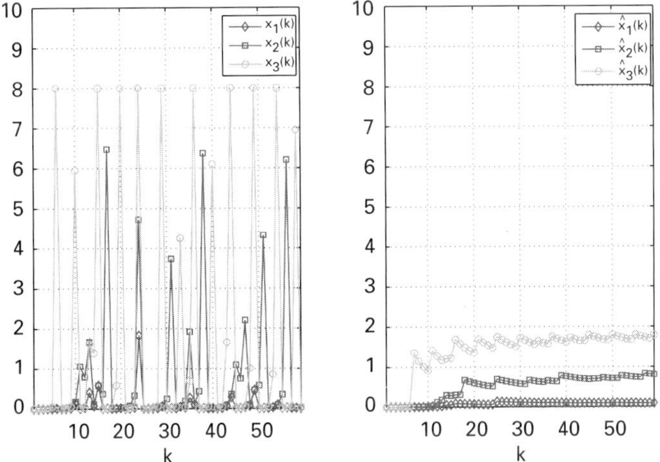

Figure 10.5 The convergence behavior of the primal sequence $\{x_k\}$ (on the left) and $\{\hat{x}_k\}$ (on the right).

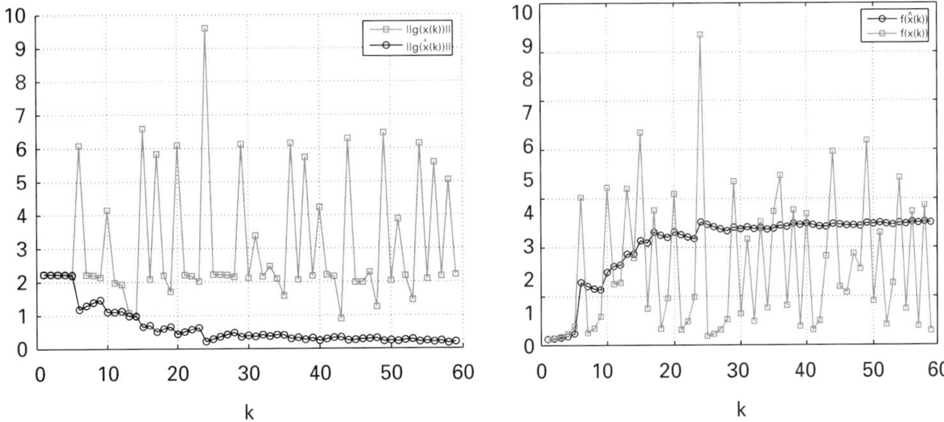

Figure 10.6 The figure on the left shows the convergence behavior of the constraint violation for the two primal sequences, $\{x_k\}$ and $\{\hat{x}_k\}$. Similarly, the figure on the right shows the convergence of the corresponding primal objective-function values.

convergence behavior of the constraint violation $\|g(x)^+\|$ for the two sequences, in other words, $\|g(x_k)^+\|$ and $\|g(\hat{x}_k)^+\|$. Note that the constraint violation for the sequence $\{x_k\}$ oscillates within a large range, while the constraint violation for the average sequence $\{\hat{x}_k\}$ rapidly converges to 0. The plot to the right in Figure 10.6 shows a similar convergence behavior for the primal objective-function values $f(x)$ along the sequences $\{x_k\}$ and $\{\hat{x}_k\}$.

10.3   Distributed-optimization methods using consensus algorithms

In this section, we develop distributed methods for minimizing the sum of non-separable convex functions corresponding to multiple agents connected over a network with time-varying topology. These methods combine first-order methods and a consensus algorithm. The consensus part serves as a basic mechanism for distributing the computations among the agents and allowing us to solve the problem in a decentralized fashion.

In contrast with the setting considered in Section 10.2.3 where each agent has an objective function that depends only on the resource-allocated to that agent, the model discussed in this section allows for the individual cost functions to depend on the entire resource-allocation vector. In particular, the focus here is on a distributed-optimization problem in a network consisting of m agents that communicate locally. The global objective is to cooperatively minimize the cost function $\sum_{i=1}^{m} f_i(x)$, where the function $f_i : \mathbb{R}^n \to \mathbb{R}$ represents the cost function of agent i, known by this agent only, and $x \in \mathbb{R}^n$ is a decision vector. The decision vector can be viewed as either a resource vector where sub-components correspond to resources allocated to each agent, or a global-decision vector which the agents are trying to compute using local information.

The approach presented here builds on the seminal work of Tsitsiklis [52] (see also Tsitsiklis *et al.* [51], Bertsekas and Tsitsiklis [3]), who developed a framework for the analysis of distributed-computation models.[3] As mentioned earlier, the approach here is to use the consensus as a mechanism for distributing the computations among the agents. The problem of reaching a consensus on a particular scalar value, or computing exact averages of the initial values of the agents, has attracted much recent attention as natural models of cooperative behavior in networked-systems (see Vicsek *et al.* [54], Jadbabaie *et al.* [20], Boyd *et al.* [8], Olfati-Saber and Murray [44], Cao *et al.* [11], and Olshevsky and Tsitsiklis [45]). Exploiting the consensus idea, recent work [42] (see also the short paper [35]) has proposed a distributed model for optimization over a network.

10.3.1   Problem and algorithm

In this section, we formulate the problem of interest and present a distributed algorithm for solving the problem.

---

[3] This framework focuses on the minimization of a (smooth) function $f(x)$ by distributing the processing of the components of vector $x \in \mathbb{R}^n$ among n agents.

Problem

We consider the problem of optimizing the sum of convex objective functions corresponding to m agents connected over a time-varying topology. The goal of the agents is to cooperatively solve the unconstrained-optimization problem

$$\text{minimize} \quad \sum_{m}^{m} f_i(x)$$

$$\text{subject to} \quad x \in \mathbb{R}^n, \tag{10.18}$$

where each $f_i : \mathbb{R}^n \to \mathbb{R}$ is a convex function, representing the local-objective function of agent i, which is known only to this agent. This problem is an unconstrained version of the multiagent-optimization problem (10.1), where the global-objective function $f(x)$ is given by the sum of the individual local-objective functions $f_i(x)$, in other words,

$$f(x) = \sum_{j=1}^{m} f_i(x),$$

(see Figure 10.7). We denote the optimal value of problem (10.18) by f^* and the set of optimal solutions by X^*.

To keep our discussion general, we do not assume differentiability of any of the functions f_i. Since each f_i is convex over the entire $\mathbb{R}^n$, the function is differentiable almost everywhere (see [2] or [47]). At the points where the function fails to be differentiable, a subgradient exists [as defined in (10.2)] and it can be used in "the role of a gradient."

Algorithm

We next introduce a distributed-subgradient algorithm for solving problem (10.18). The main idea of the algorithm is the use of consensus as a mechanism for distributing the computations among the agents. In particular, each agent starts with an initial estimate $x_i(0) \in \mathbb{R}^n$ and updates its estimate at discrete times $t_k, k = 1, 2, \ldots$. We denote by $x_i(k)$ the vector estimate maintained by agent i at time t_k. When updating, an agent

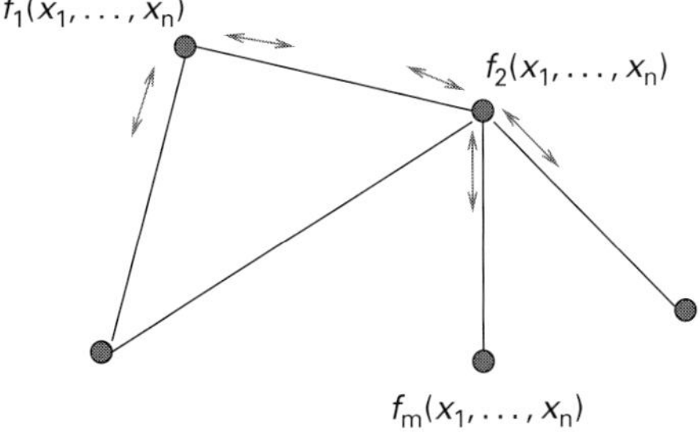

Figure 10.7   Illustration of the network with each agent having its local objective and communicating locally with its neighbors.

i combines its current estimate x_i with the estimates x_j received from its neighboring agents j. Specifically, agent i updates its estimates by setting

$$x_i(k+1) = \sum_{j=1}^{m} a_{ij}(k)x_j(k) - \alpha d_i(k), \qquad (10.19)$$

where the scalar $\alpha > 0$ is a stepsize and the vector $d_i(k)$ is a subgradient of the agent i cost function $f_i(x)$ at $x = x_i(k)$. The scalars $a_{i1}(k), \ldots, a_{im}(k)$ are non-negative weights that agent i gives to the estimates $x_1(k), \ldots, x_m(k)$. These weights capture two aspects:

1. The active links (j, i) at time k. In particular, the neighbors j that communicate with agent i at time k, will be captured by assigning $a_{ij}(k) > 0$ (including i itself). The neighbors j that do not communicate with i at time k, as well as those that are not neighbors of i, are captured by assigning $a_{ij}(k) = 0$.
2. The weight that agent i gives to the estimates received from its neighbors.

When all objective functions are zero, that is, $f_i(x) = 0$ for all x and i, the method in (10.19) reduces to

$$x_i(k+1) = \sum_{j=1}^{m} a_{ij}(k)x_j(k),$$

which is the consensus algorithm. In view of this, algorithm (10.19) can be seen as a combination of the "consensus step" $\sum_{j=1}^{m} a_{ij}(k)x_j(k)$ and the subgradient step $-\alpha d_i(k)$. The subgradient step is taken by the agent to minimize its own objective $f_i(x)$, while the consensus step serves to align its decision x_i with the decisions of its neighbors. When the network is sufficiently often connected in time to ensure the proper mixing of the agents' estimates, one would expect that all agents have the same estimate after some time, at which point the algorithm would start behaving as a "centralized" method. This intuition is behind the construction of the algorithm and also behind the analysis of its performance.

Representation using transition matrices

In the subsequent development, we find it useful to introduce $A(k)$ to denote the "weight matrix" $[a_{ij}(k)]_{i,j=1,\ldots,m}$. Using these matrices, we can capture the evolution of the estimates $x_i(k)$ generated by (10.19) over a window of time. In particular, we define a "transition matrix" $\Phi(k, s)$ for any s and k with $k \geq s$, as follows:

$$\Phi(k, s) = A(k)A(k-1) \cdots A(s+1)A(s).$$

Through the use of transition matrices, we can relate the estimate $x_i(k+1)$ to the estimates $x_1(s), \ldots, x_m(s)$ for any $s \leq k$. Specifically, for the iterates generated by (10.19), we have for any i, and any s and k with $k \geq s$,

$$x_i(k+1) = \sum_{j=1}^{m} [\Phi(k,s)]_{ij} x_j(s) - \alpha \sum_{r=s}^{k-1} \sum_{j=1}^{m} [\Phi(k,r+1)]_{ij} d_j(r) - \alpha\, d_i(k). \quad (10.20)$$

As seen from the preceding relation, to study the asymptotic behavior of the estimates $x_i(k)$, we need to understand the behavior of the transition matrices $\Phi(k,s)$. We do this under some assumptions on the agent interactions that translate into some properties of transition matrices, as seen in the next section.

10.3.2 Information-exchange model

The agent interactions and information aggregation at time k are modeled through the use of the matrix $A(k)$ of agent weights $a_{ij}(k)$. At each time k, this weight matrix captures the information flow (or the communication pattern) among the agents, as well as how the information is aggregated by each agent, in other words, how much actual weight each agent i assigns to its own estimate $x_i(k)$ and the estimates $x_j(k)$ received from its neighbors.

For the proper mixing of the agent information, we need some assumptions on the weights $a_{ij}(k)$ and the agent connectivity in time. When discussing these assumptions, we use the notion of a stochastic vector and a stochastic matrix, defined as follows. A vector a is said to be a "stochastic vector" when its components a_i are non-negative and $\sum_i a_i = 1$. A square matrix A is said to be "stochastic" when each row of A is a stochastic vector, and it is said to be "doubly stochastic" when both A and its transpose A' are stochastic matrices.

The following assumption puts conditions on the weights $a_{ij}(k)$ in (10.19).

Assumption 3 For all $k \geq 0$, the weight matrix $A(k)$ is doubly stochastic with positive diagonal. Additionally, there is a scalar $\eta > 0$ such that if $a_{ij}(k) > 0$, then $a_{ij}(k) \geq \eta$.

The doubly stochasticity assumption on the weight matrix will guarantee that the function f_i of every agent i receives the same weight in the long run. This ensures that the agents optimize the sum of the functions f_i as opposed to some weighted sum of these functions. The second part of the assumption states that each agent gives significant weight to its own value and to the values of its neighbors. This is needed to ensure that new information is aggregated into the agent system persistently in time.

We note that the lower bound η on weights in Assumption 3 need not be available to any of the agents. The existence of such a bound is merely used in the analysis of the system behavior and the algorithm's performance. Note also that such a bound η exists when each agent has a lower bound η_i on its own weights $a_{ij}(k), j = 1, \ldots, m$, in which case we can define $\eta = \min_{1 \leq i \leq m} \eta_i$.

The following are some examples of how to ensure, in a distributed manner, that the weight matrix $A(k)$ satisfies Assumption 3 when the agent communications are bidirectional.

Example 10.1   Metropolis-based weights [55] are given by for all i and j with $j \neq i$,

$$a_{ij}(k) = \begin{cases} \dfrac{1}{1 + \max\{n_i(k), n_j(k)\}} & \text{if } j \text{ communicates with } i \text{ at time } k, \\ 0 & \text{otherwise}, \end{cases}$$

with $n_i(k)$ being the number of neighbors communicating with agent i at time k. Using these, the weights $a_{ii}(k)$ for all $i = 1, \ldots, m$ are as follows

$$a_{ii}(k) = 1 - \sum_{j \neq i} a_{ij}(k).$$

The next example can be viewed as a generalization of the Metropolis weights.

Example 10.2   Each agent i has planned weights $\tilde{a}_{ij}(k), j = 1 \ldots, m$ that the agent communicates to its neighbors together with the estimate $x_i(k)$, where the matrix $\tilde{A}(k)$ of planned weights is a (row) stochastic matrix satisfying Assumption 3, except for doubly stochasticity. In particular, at time k, if agent j communicates with agent i, then agent i receives $x_j(k)$ and the planned weight $\tilde{a}_{ji}(k)$ from agent j. At the same time, agent j receives $x_i(k)$ and the planned weight $\tilde{a}_{ij}(k)$ from agent i. Then, the actual weights that an agent i uses are given by

$$a_{ij}(k) = \min\{\tilde{a}_{ij}(k),\ \tilde{a}_{ji}(k)\},$$

if i and j talk at time k, and $a_{ij}(k) = 0$ otherwise; while

$$a_{ii}(k) = 1 - \sum_{\{j \mid j \leftrightarrow i \text{ at time } k\}} a_{ij}(k),$$

where the summation is over all j communicating with i at time k. It can be seen that the weights $a_{ij}(k)$ satisfy Assumption 3.

We need the agent network to be connected to ensure that the information state of each and every agent influences the information state of other agents. However, the network need not be connected at every time instance, but rather frequently enough to persistently influence each other. To formalize this assumption, we introduce the index set $\mathcal{N} = \{1, \ldots, m\}$ and we view the agent network as a directed graph with node set $\mathcal{N}$ and time-varying link set. We define $\mathcal{E}(A(k))$ to be the set of directed links at time k induced by the weight matrix $A(k)$. In particular, the link set $\mathcal{E}(A(k))$ is given by

$$\mathcal{E}(A(k)) = \{(j, i) \mid a_{ij}(k) > 0,\ i, j = 1 \ldots, m\} \qquad \text{for all } k.$$

Note that each set $\mathcal{E}(A(k))$ includes self-edges (i, i) for all i. Now, the agents connectivity can be represented by a directed graph $G(k) = (\mathcal{N}, \mathcal{E}(A(k)))$.

The next assumption states that the agent network is frequently connected.

Assumption 4 There exists an integer $B \geq 1$ such that the directed graph

$$\Big(\mathcal{N}, \mathcal{E}(A(kB)) \cup \cdots \cup \mathcal{E}(A((k+1)B - 1))\Big)$$

is strongly connected for all $k \geq 0$.

Note that the bound B need not be known by any of the agents. This is another parameter that is used in the analysis of the network properties and the algorithm.

10.3.3 Convergence of transition matrices

Here, we study the behavior of the transition matrices $\Phi(k, s) = A(k) \cdots A(s)$ that govern the evolution of the estimates over a window of time, as seen from (10.20). Under Assumptions 3 and 4, we provide some results that we use later in the convergence analysis of method (10.19). These results are of interest on their own for consensus problems and distributed averaging.[4]

To understand the convergence of the transition matrices $\Phi(k, s)$, we start by considering a related "consensus-type" update rule of the form

$$z(k + 1) = A(k)z(k), \tag{10.21}$$

where $z(0) \in \mathbb{R}^m$ is an initial vector. This update rule captures the averaging part of (10.19), as it operates on a particular component of the agent estimates, with the vector $z(k) \in \mathbb{R}^m$ representing the estimates of the different agents for that component.

We define

$$V(k) = \sum_{j=1}^{m} (z_j(k) - \bar{z}(k))^2 \qquad \text{for all } k \geq 0,$$

where $\bar{z}(k)$ is the average of the entries of the vector $z(k)$. Under the doubly stochasticity of $A(k)$, the initial average $\bar{z}(0)$ is preserved by the update rule (10.21), in other words, $\bar{z}(k) = \bar{z}(0)$ for all k. Hence, the function $V(k)$ measures the "disagreement" in agent values.

In the next lemma, we give a bound on the decrease of the agent disagreement $V(kB)$, which is linear in η and quadratic in m^{-1}. This bound plays a crucial role in establishing the convergence and rate of convergence of transition matrices, which subsequently are used to analyze the method.

---

[4] More detailed development of these results for distributed averaging can be found in [32].

LEMMA 10.4 *Let Assumptions 3 and 4 hold. Then, $V(k)$ is nonincreasing in k. Furthermore,*

$$V((k+1)B) \leq \left(1 - \frac{\eta}{2m^2}\right) V(kB) \quad \text{for all } k \geq 0.$$

Proof This is an immediate consequence of Lemma 5 in [32], stating that[5] under Assumptions 3 and 4, for all k with $V(kB) > 0$,

$$\frac{V(kB) - V((k+1)B)}{V(kB)} \geq \frac{\eta}{2m^2}.$$

∎

Using Lemma 10.4, we establish the convergence of the transition matrices $\Phi(k,s)$ of (10.20) to the matrix with all entries equal to $\frac{1}{m}$, and we provide a bound on the convergence rate. In particular, we show that the difference between the entries of $\Phi(k,s)$ and $\frac{1}{m}$ converges to zero with a geometric rate.

THEOREM 10.2 *Let Assumptions 3 and 4 hold. Then, for all i,j and all k,s with $k \geq s$, we have*

$$\left| [\Phi(k,s)]_{ij} - \frac{1}{m} \right| \leq \left(1 - \frac{\eta}{4m^2}\right)^{\lceil \frac{k-s+1}{B} \rceil - 2}.$$

Proof By Lemma 10.4, we have for all $k \geq s$,

$$V(kB) \leq \left(1 - \frac{\eta}{2m^2}\right)^{k-s} V(sB).$$

Let k and s be arbitrary with $k \geq s$, and let

$$\tau B \leq s < (\tau + 1)B, \quad tB \leq k < (t+1)B,$$

with $\tau \leq t$. Hence, by the nonincreasing property of $V(k)$, we have

$$V(k) \leq V(tB)$$

$$\leq \left(1 - \frac{\eta}{2m^2}\right)^{t-\tau-1} V((\tau+1)B)$$

$$\leq \left(1 - \frac{\eta}{2m^2}\right)^{t-\tau-1} V(s).$$

Note that $k - s < (t - \tau)B + B$ implying that $\frac{k-s+1}{B} \leq t - \tau + 1$, where we used the fact that both sides of the inequality are integers. Therefore $\lceil \frac{k-s+1}{B} \rceil - 2 \leq t - \tau - 1$, and we have for all k and s with $k \geq s$,

$$V(k) \leq \left(1 - \frac{\eta}{2m^2}\right)^{\lceil \frac{k-s+1}{B} \rceil - 2} V(s). \tag{10.22}$$

---

[5] The assumptions in [32] are actually weaker.

By (10.21), we have $z(k+1) = A(k)z(k)$, and therefore $z(k+1) = \Phi(k,s)z(s)$ for all $k \geq s$. Let $e_i \in \mathbb{R}^m$ denote the vector with entries all equal to 0, except for the ith entry which is equal to 1. Letting $z(s) = e_i$ we obtain $z(k+1) = [\Phi(k,s)]'_i$, where $[\Phi(k,s)]'_i$ denotes the transpose of the ith row of the matrix. Using the inequalities (10.22) and $V(e_i) \leq 1$, we obtain

$$V([\Phi(k,s)]'_i) \leq \left(1 - \frac{\eta}{2m^2}\right)^{\lceil \frac{k-s+1}{B} \rceil - 2}.$$

The matrix $\Phi(k,s)$ is doubly stochastic because it is the product of doubly stochastic matrices. Thus, the average entry of $[\Phi(k,s)]_i$ is $1/m$, implying that for all i and j,

$$\left([\Phi(k,s)]_{ij} - \frac{1}{m}\right)^2 \leq V([\Phi(k,s)]'_i)$$

$$\leq \left(1 - \frac{\eta}{2m^2}\right)^{\lceil \frac{k-s+1}{B} \rceil - 2}.$$

From the preceding relation and $\sqrt{1 - \eta/(2m^2)} \leq 1 - \eta/(4m^2)$, we obtain

$$\left| [\Phi(k,s)]_{ij} - \frac{1}{m} \right| \leq \left(1 - \frac{\eta}{4m^2}\right)^{\lceil \frac{k-s+1}{B} \rceil - 2}.$$

$\blacksquare$

10.3.4 Convergence analysis of the subgradient method

Here, we study the convergence properties of the subgradient method (10.19) and, in particular, we obtain a bound on the performance of the algorithm. In what follows, we assume the uniform boundedness of the set of subgradients of the cost functions f_i at all points: for some scalar $L > 0$, we have for all $x \in \mathbb{R}^n$ and all i,

$$\|g\| \leq L \qquad \text{for all } g \in \partial f_i(x), \tag{10.23}$$

where $\partial f_i(x)$ is the set of all subgradients of f_i at x.

The analysis combines the proof techniques used for consensus algorithms and approximate subgradient methods. The consensus analysis rests on the convergence-rate result of Theorem 10.2 for transition matrices, which provides a tool for measuring the "agent disagreements" $\|x_i(k) - x_j(k)\|$ in time. Equivalently, we can measure $\|x_i(k) - x_j(k)\|$ in terms of the disagreements $\|x_i(k) - y(k)\|$ with respect to an auxiliary sequence $\{y(k)\}$, defined appropriately. The sequence $\{y_k\}$ will also serve as a basis for understanding the effects of subgradient steps in the algorithm. In fact, we will establish the suboptimality property of the sequence $\{y_k\}$, and then using the estimates for the disagreements $\|x_i(k) - y(k)\|$, we will provide a performance bound for the algorithm.

Disagreement estimate

To estimate the agent "disagreements", we use an auxiliary sequence $\{y(k)\}$ of reference points, defined as follows:[6]

$$y(k+1) = y(k) - \frac{\alpha}{m} \sum_{i=1}^{m} d_i(k), \qquad (10.24)$$

where $d_i(k)$ is the same subgradient of $f_i(x)$ at $x = x_i(k)$ that is used in the method (10.19), and

$$y(0) = \frac{1}{m} \sum_{i=1}^{m} x_i(0).$$

In the following lemma, we estimate the norms of the differences $x_i(k) - y(k)$ at each time k. The result relies on Theorem 10.2.

LEMMA 10.5   *Let Assumptions 3 and 4 hold. Assume also that the subgradients of each f_i are uniformly bounded by some scalar L [cf. (10.23)]. Then for all i and $k \geq 1$,*

$$\|x_i(k) - y(k)\| \leq \beta^{\lceil \frac{k}{B} \rceil - 2} \sum_{j=1}^{m} \|x_j(0)\| + \alpha L \left(2 + \frac{mB}{\beta(1-\beta)} \right),$$

where $\beta = 1 - \frac{\eta}{4m^2}$.

Proof   From the definition of the sequence $\{y(k)\}$ in (10.24) it follows for all k,

$$y(k) = \frac{1}{m} \sum_{i=1}^{m} x_i(0) - \frac{\alpha}{m} \sum_{r=0}^{k-1} \sum_{i=1}^{m} d_i(r). \qquad (10.25)$$

As given in equation (10.20), for the agent estimates $x_i(k)$ we have for all k,

$$x_i(k+1) = \sum_{j=1}^{m} [\Phi(k,s)]_{ij} x_j(s) - \alpha \sum_{r=s}^{k-1} \sum_{j=1}^{m} [\Phi(k,r+1)]_{ij} d_j(r) - \alpha \, d_i(k).$$

From this relation (with $s = 0$), we see that for all $k \geq 1$,

$$x_i(k) = \sum_{j=1}^{m} [\Phi(k-1,0)]_{ij} x_j(0) - \alpha \sum_{r=0}^{k-2} \sum_{j=1}^{m} [\Phi(k-1,r+1)]_{ij} d_j(r) - \alpha \, d_i(k-1).$$

$$(10.26)$$

---

[6] The iterates $y(k)$ can be associated with a stopped process related to algorithm (10.19), as discussed in [42].

By using the relations (10.25) and (10.26), we obtain for all $k \geq 1$,

$$x_i(k) - y(k) = \sum_{j=1}^{m} \left([\Phi(k-1,0)]_{ij} - \frac{1}{m} \right) x_j(0)$$
$$- \alpha \sum_{r=0}^{k-2} \sum_{j=1}^{m} \left([\Phi(k-1,r+1)]_{ij} - \frac{1}{m} \right) d_j(r)$$
$$- \alpha\, d_i(k-1) + \frac{\alpha}{m} \sum_{i=1}^{m} d_i(k-1).$$

Using the subgradient boundedness, we obtain for all $k \geq 1$,

$$\|x_i(k) - y(k)\| \leq \sum_{j=1}^{m} \left| [\Phi(k-1,0)]_{ij} - \frac{1}{m} \right| \|x_j(0)\|$$
$$+ \alpha L \sum_{s=1}^{k-1} \sum_{j=1}^{m} \left| [\Phi(k-1,s)]_{ij} - \frac{1}{m} \right| + 2\alpha L.$$

By using Theorem 10.2, we can bound the terms $\left| [\Phi(k-1,s)]_{ij} - \frac{1}{m} \right|$ and obtain for all i and any $k \geq 1$,

$$\|x_i(k) - y(k)\| \leq \sum_{j=1}^{m} \beta^{\lceil \frac{k}{B} \rceil - 2} \|x_j(0)\| + \alpha L \sum_{s=1}^{k-1} \sum_{j=1}^{m} \beta^{\lceil \frac{k-s}{B} \rceil - 2} + 2\alpha L$$
$$= \beta^{\lceil \frac{k}{B} \rceil - 2} \sum_{j=1}^{m} \|x_j(0)\| + \alpha L m \sum_{s=1}^{k-1} \beta^{\lceil \frac{k-s}{B} \rceil - 2} + 2\alpha L.$$

By using $\sum_{s=1}^{k-1} \beta^{\lceil \frac{k-s}{B} \rceil - 2} \leq \sum_{r=1}^{\infty} \beta^{\lceil \frac{r}{B} \rceil - 2} = \frac{1}{\beta} \sum_{r=1}^{\infty} \beta^{\lceil \frac{r}{B} \rceil - 1}$, and

$$\sum_{r=1}^{\infty} \beta^{\lceil \frac{r}{B} \rceil - 1} = \sum_{r=1}^{\infty} \beta^{\lceil \frac{r}{B} \rceil - 1} \leq B \sum_{t=0}^{\infty} \beta^{t} = \frac{B}{1-\beta},$$

we obtain

$$\sum_{s=1}^{k-1} \beta^{\lceil \frac{k-s}{B} \rceil - 2} \leq \frac{B}{\beta(1-\beta)}.$$

Therefore, it follows that for all $k \geq 1$,

$$\|x_i(k) - y(k)\| \leq \beta^{\lceil \frac{k}{B} \rceil - 2} \sum_{j=1}^{m} \|x_j(0)\| + \alpha L \left(2 + \frac{mB}{\beta(1-\beta)} \right).$$

Estimate for the auxiliary sequence

We next establish a result that estimates the objective function $f = \sum_{i=1}^{m} f_i$ at the running averages of the vectors $y(k)$ of (10.24). Specifically, we define

$$\hat{y}(k) = \frac{1}{k} \sum_{h=1}^{k} y(h) \qquad \text{for all } k \geq 1,$$

and we estimate the function values $f(\hat{y}(k))$. We have the following result.

LEMMA 10.6 *Let Assumptions 3 and 4 hold, and assume that the subgradients are uniformly bounded as in (10.23). Also, assume that the set X^* of optimal solutions of problem (10.18) is nonempty. Then, the average vectors $\hat{y}(k)$ satisfy for all $k \geq 1$,*

$$f(\hat{y}(k)) \leq f^* + \frac{\alpha L^2 C}{2} + \frac{2mLB}{k\beta(1-\beta)} \sum_{j=1}^{m} \|x_j(0)\| + \frac{m}{2\alpha k} \left(\operatorname{dist}(y(0), X^*) + \alpha L \right)^2,$$

where $y(0) = \frac{1}{m} \sum_{j=1}^{m} x_j(0)$, $\beta = 1 - \frac{\eta}{4m^2}$ and

$$C = 1 + 4m \left(2 + \frac{mB}{\beta(1-\beta)} \right).$$

Proof From the definition of the sequence $y(k)$ it follows for any $x^* \in X^*$ and all k,

$$\|y(k+1) - x^*\|^2 = \|y(k) - x^*\|^2 + \frac{\alpha^2}{m^2} \left\| \sum_{i=1}^{m} d_i(k) \right\|^2 - 2\frac{\alpha}{m} \sum_{i=1}^{m} d_i(k)'(y(k) - x^*).$$

$$(10.27)$$

We next estimate the terms $d_i(k)'(y(k) - x^*)$ where $d_i(k)$ is a subgradient of f_i at $x_i(k)$. For any i and k, we have

$$d_i(k)'(y(k) - x^*) = d_i(k)'(y(k) - x_i(k)) + d_i(k)'(x_i(k) - x^*).$$

By the subgradient property in (10.2), we have $d_i(k)(x_i(k) - x^*) \geq f_i(x_i(k)) - f_i(x^*)$ implying

$$d_i(k)'(y(k) - x^*) \geq d_i(k)'(y(k) - x_i(k)) + f_i(x_i(k)) - f_i(x^*)$$

$$\geq -L\|y(k) - x_i(k)\| + [f_i(x_i(k)) - f_i(y(k))] + [f_i(y(k)) - f_i(x^*)],$$

where the last inequality follows from the subgradient boundedness. We next consider $f_i(x_i(k)) - f_i(y(k))$, for which by subgradient property (10.2) we have

$$f_i(x_i(k)) - f_i(y(k)) \geq \tilde{d}_i(k)'(x_i(k) - y(k)) \geq -L\|x_i(k) - y(k)\|,$$

where $\tilde{d}_i(k)$ is a subgradient of f_i at $y(k)$, and the last inequality follows from the subgradient boundedness. Thus, by combining the preceding two relations, we have for all i and k,

$$d_i(k)'(y(k) - x^*) \geq -2L\|y(k) - x_i(k)\| + f_i(y(k)) - f_i(x^*).$$

By substituting the preceding estimate for $d_i(k)'(y(k) - x^*)$ in relation (10.27), we obtain

$$\|y(k+1) - x^*\|^2 \leq \|y(k) - x^*\|^2 + \frac{\alpha^2}{m^2} \left\| \sum_{i=1}^{m} d_i(k) \right\|^2 + \frac{4L\alpha}{m} \sum_{i=1}^{m} \|y(k) - x_i(k)\|$$

$$- \frac{2\alpha}{m} \sum_{i=1}^{m} f_i(y(k)) - f_i(x^*).$$

By using the subgradient boundedness and noting that $f = \sum_{i=1}^{m} f_i$ and $f(x^*) = f^*$, we can write

$$\|y(k+1) - x^*\|^2 \leq \|y(k) - x^*\|^2 + \frac{\alpha^2 L^2}{m} + \frac{4\alpha L}{m} \sum_{i=1}^{m} \|y(k) - x_i(k)\|$$

$$- \frac{2\alpha}{m} \left(f(y(k)) - f^* \right).$$

Taking the minimum over $x^* \in X^*$ in both sides of the preceding relation, we obtain

$$\text{dist}^2(y(k+1), X^*) \leq \text{dist}^2(y(k), X^*) + \frac{\alpha^2 L^2}{m} + \frac{4\alpha L}{m} \sum_{i=1}^{m} \|y(k) - x_i(k)\|$$

$$- \frac{2\alpha}{m} \left(f(y(k)) - f^* \right).$$

Using Lemma 10.5 to bound each of the terms $\|y(k) - x_i(k)\|$, we further obtain

$$\text{dist}^2(y(k+1), X^*) \leq \text{dist}^2(y(k), X^*) + \frac{\alpha^2 L^2}{m} + 4\alpha L \beta^{\left\lceil \frac{k}{B} \right\rceil - 2} \sum_{j=1}^{m} \|x_j(0)\|$$

$$+ 4\alpha^2 L^2 \left(2 + \frac{mB}{\beta(1-\beta)} \right) - \frac{2\alpha}{m} \left[f(y(k)) - f^* \right].$$

Therefore, by regrouping the terms and introducing

$$C = 1 + 4m \left(2 + \frac{mB}{\beta(1-\beta)} \right),$$

we have for all $k \geq 1$,

$$f(y(k)) \leq f^* + \frac{\alpha L^2 C}{2} + 2mL\beta^{\left\lceil \frac{k}{B} \right\rceil - 2} \sum_{j=1}^{m} \|x_j(0)\|$$

$$+ \frac{m}{2\alpha} \left(\text{dist}^2(y(k), X^*) - \text{dist}^2(y(k+1), X^*) \right).$$

By adding these inequalities for different values of k, we obtain

$$\frac{1}{k} \sum_{h=1}^{k} f(y(h)) \leq f^* + \frac{\alpha L^2 C}{2} + \frac{2mLB}{k\beta(1-\beta)} \sum_{j=1}^{m} \|x_j(0)\|$$

$$+ \frac{m}{2\alpha k} \left(\text{dist}^2(y(1), X^*) - \text{dist}^2(y(k), X^*) \right), \tag{10.28}$$

where we use the following inequality for $t \geq 1$,

$$\sum_{k=1}^{t} \beta^{\left\lceil \frac{k}{B} \right\rceil - 2} \leq \frac{1}{\beta} \sum_{k=1}^{\infty} \beta^{\left\lceil \frac{k}{B} \right\rceil - 1} \leq \frac{B}{\beta} \sum_{s=1}^{\infty} \beta^s = \frac{B}{\beta(1-\beta)}.$$

By discarding the nonpositive term on the right-hand side in relation (10.28) and by using the convexity of f,

$$\frac{1}{k} \sum_{h=1}^{k} f(y(h)) \leq f^* + \frac{\alpha L^2 C}{2} + \frac{2mLB}{k\beta(1-\beta)} \sum_{j=1}^{m} \|x_j(0)\|$$

$$+ \frac{m}{2\alpha k} \, \text{dist}^2(y(1), X^*).$$

Finally, by using the definition of $y(k)$ in (10.24) and the subgradient boundedness, we see that

$$\text{dist}^2(y(1), X^*) \leq \left(\text{dist}(y(0), X^*) + \alpha L \right)^2,$$

which when combined with the preceding relation yields the desired inequality.   ∎

Performance bound for the algorithm

We establish a bound on the performance of the algorithm at the time-average of the vectors $x_i(k)$ generated by method (10.19). In particular, we define the vectors $\hat{x}_i(k)$ as follows:

$$\hat{x}_i(k) = \frac{1}{k} \sum_{h=1}^{k} x_i(h).$$

The use of these vectors allows us to bound the objective-function improvement at every iteration, by combining the estimates for $\|x_i(k) - y(k)\|$ of Lemma 10.5 and the estimates for $f(\hat{y}(k))$ of Lemma 10.6. We have the following.

THEOREM 10.3   *Let Assumptions 3 and 4 hold, and assume that the set X^* of optimal solutions of problem (10.18) is nonempty. Let the subgradients be bounded as in (10.23). Then, the averages $\hat{x}_i(k)$ of the iterates obtained by the method (10.19) satisfy for all i and $k \geq 1$,*

$$f(\hat{x}_i(k)) \leq f^* + \frac{\alpha L^2 C_1}{2} + \frac{4mLB}{k\beta(1-\beta)} \sum_{j=1}^{m} \|x_j(0)\| + \frac{m}{2\alpha k} \left(\operatorname{dist}(y(0), X^*) + \alpha L\right)^2,$$

where $y(0) = \frac{1}{m} \sum_{j=1}^{m} x_j(0)$, $\beta = 1 - \frac{\eta}{4m^2}$ *and*

$$C_1 = 1 + 8m \left(2 + \frac{mB}{\beta(1-\beta)}\right).$$

Proof   By the convexity of the functions f_j, we have, for any i and $k \geq 1$,

$$f(\hat{x}_i(k)) \leq f(\hat{y}(k)) + \sum_{j=1}^{m} g_{ij}(k)'(\hat{x}_i(k) - \hat{y}(k)),$$

where $g_{ij}(k)$ is a subgradient of f_j at $\hat{x}_i(k)$. Then, by using the subgradient boundedness, we obtain for all i and $k \geq 1$,

$$f(\hat{x}_i(k)) \leq f(\hat{y}(k)) + \frac{2L}{k} \sum_{i=1}^{m} \left(\sum_{t=1}^{k} \|x_i(t) - y(t)\|\right). \tag{10.29}$$

By using the bound for $\|x_i(k) - y(k)\|$ of Lemma 10.5, we have for all i and $k \geq 1$,

$$\sum_{t=1}^{k} \|x_i(t) - \hat{y}(t)\| \leq \left(\sum_{t=1}^{k} \beta^{\lceil \frac{t}{B} \rceil - 2}\right) \sum_{j=1}^{m} \|x_j(0)\| + \alpha k L \left(2 + \frac{mB}{\beta(1-\beta)}\right).$$

Noting that

$$\sum_{t=1}^{k} \beta^{\lceil \frac{t}{B} \rceil - 2} \leq \frac{1}{\beta} \sum_{t=1}^{\infty} \beta^{\lceil \frac{t}{B} \rceil - 1} \leq \frac{B}{\beta} \sum_{s=1}^{\infty} \beta^s = \frac{B}{\beta(1-\beta)},$$

we obtain for all i and $k \geq 1$,

$$\sum_{t=1}^{k} \|x_i(t) - \hat{y}(t)\| \leq \frac{B}{\beta(1-\beta)} \sum_{j=1}^{m} \|x_j(0)\| + \alpha k L \left(2 + \frac{mB}{\beta(1-\beta)}\right).$$

Hence, by summing these inequalities over all i and by substituting the resulting estimate in relation (10.29), we obtain

$$f(\hat{x}_i(k)) \leq f(\hat{y}(k)) + \frac{2mLB}{k\beta(1-\beta)} \sum_{j=1}^{m} \|x_j(0)\| + 2m\alpha L^2 \left(2 + \frac{mB}{\beta(1-\beta)}\right).$$

The result follows by using the estimate for $f(\hat{y}(k))$ of Lemma 10.6.   ∎

The result of Theorem 10.3 provides an estimate on the values $f(\hat{x}_i(k))$ per iteration k. As the number of iterations increases to infinity, the last two terms of the estimate diminish, resulting with

$$\limsup_{k\to\infty} f(\hat{x}_i(k)) \leq f^* + \frac{\alpha L^2 C_1}{2} \qquad \text{for all } i.$$

As seen from Theorem 10.3, the constant C_1 increases only polynomially with m. When α is fixed and the parameter η is independent of m, the largest error is of the order of m^4, indicating that for high accuracy, the stepsize needs to be very small. However, our bound is for general convex functions and network topologies, and further improvements of the bound are possible for special classes of convex functions and special topologies.

10.4   Extensions

Here, we consider extensions of the distributed model of Section 10.3 to account for various network effects. We focus on two such extensions. The first is an extension of the optimization model (10.19) to a scenario where the agents communicate over a network with finite bandwidth-communication links, and the second is an extension of the consensus problem to the scenario where each agent is facing some constraints on its decisions. We discuss these extensions in the following sections.

10.4.1   Quantization effects on optimization

Here, we discuss the agent system where the agents communicate over a network consisting of finite-bandwidth links. Thus, the agents cannot exchange continuous-valued information (real numbers), but instead can only send quantized information. This problem has recently gained interest in the networking literature [12, 13, 22, 32]. In what follows, we present recent results dealing with the effects of quantization on the multiagent-distributed optimization over a network. More specifically, we discuss a "quantized" extension of the subgradient method (10.19) and provide a performance bound.[7]

We consider the case where the agents exchange quantized data, but they can store continuous data. In particular, we assume that each agent receives and sends only quantized estimates, that is, vectors whose entries are integer multiples of $1/Q$, where Q is some positive integer. At time k, an agent receives quantized estimates $x_j^Q(k)$ from some of its neighbors and updates according to the following rule:

$$x_i^Q(k+1) = \left\lfloor \sum_{j=1}^{m} a_{ij}(k)x_j^Q(k) - \alpha\tilde{d}_i(k) \right\rfloor, \tag{10.30}$$

---

[7] The result presented here can be found with more details in [31].

where $\tilde{d}_i(k)$ is a subgradient of f_i at $x_i^Q(k)$, and $\lfloor y \rfloor$ denotes the operation of (component-wise) rounding the entries of a vector y to the nearest multiple of $1/Q$. We also assume that the agents initial estimates $x_j^Q(0)$ are quantized.

We can view the agent estimates in (10.30) as consisting of a consensus part $\sum_{j=1}^m a_{ij}(k)x_j^Q(k)$, and the term due to the subgradient step and an error (due to extracting the consensus part). Specifically, we rewrite (10.30) as follows:

$$x_i^Q(k+1) = \sum_{j=1}^m a_{ij}(k)x_j^Q(k) - \alpha\tilde{d}_i(k) - \epsilon_i(k+1), \qquad (10.31)$$

where the error vector $\epsilon_i(k+1)$ is given by

$$\epsilon_i(k+1) = \sum_{j=1}^m a_{ij}(k)x_j^Q(k) - \alpha\tilde{d}_i(k) - x_i^Q(k+1).$$

Thus, the method can be viewed as a subgradient method using consensus and with external (possibly persistent) noise, represented by $\epsilon_i(k+1)$. Due to the rounding down to the nearest multiple of $1/Q$, the error vector $\epsilon_i(k+1)$ satisfies

$$0 \le \epsilon_i(k+1) \le \frac{1}{Q}\mathbf{1} \qquad \text{for all } i \text{ and } k,$$

where the inequalities above hold componentwise and $\mathbf{1}$ denotes the vector in $\mathbb{R}^n$ with all entries equal to 1. Therefore, the error norms $\|\epsilon_i(k)\|$ are uniformly bounded in time and across agents. In fact, it turns out that these errors converge to 0 as k increases. These observations are guiding the analysis of the algorithm.

Performance bound for the quantized method

We next give a performance bound for the method (10.30) assuming that the agents can store perfect information (infinitely many bits). We consider the time-average of the iterates $x_i^Q(k)$, defined by

$$\hat{x}_i^Q(k) = \frac{1}{k}\sum_{h=1}^k x_i^Q(h) \qquad \text{for } k \ge 1.$$

We have the following result (see [31] for the proof).

THEOREM 10.4 *Let Assumptions 3 and 4 hold, and assume that the optimal set X^* of problem (10.18) is nonempty. Let subgradients be bounded as in (10.23). Then, for the averages $\hat{x}_i^Q(k)$ of the iterates obtained by the method (10.30), satisfy for all i and* *all $k \geq 1$,*

$$f(\hat{x}_i^Q(k)) \leq f^* + \frac{\alpha L^2 \tilde{C}_1}{2} + \frac{4mLB}{k\beta(1-\beta)} \sum_{j=1}^{m} \|x_j^Q(0)\|$$

$$+ \frac{m}{2\alpha k} \left(\text{dist}(\tilde{y}(0), X^*) + \alpha L + \frac{\sqrt{n}}{Q} \right)^2,$$

where $\tilde{y}(0) = \frac{1}{m} \sum_{j=1}^{m} x_j^Q(0)$, $\beta = 1 - \frac{\eta}{4m^2}$ and

$$\tilde{C}_1 = 1 + 8m \left(1 + \frac{\sqrt{n}}{\alpha LQ} \right) \left(2 + \frac{mB}{\beta(1-\beta)} \right).$$

Theorem 10.4 provides an estimate on the values $f(\hat{x}_i^Q(k))$ per iteration k. As the number of iterations increases to infinity the last two terms of the estimate vanish, yielding

$$\limsup_{k \to \infty} f(\hat{x}_i^Q(k)) \leq f^* + \frac{\alpha L^2 \tilde{C}_1}{2} \qquad \text{for all } i,$$

with

$$\tilde{C}_1 = 1 + 8m \left(1 + \frac{\sqrt{n}}{\alpha LQ} \right) \left(2 + \frac{mB}{\beta(1-\beta)} \right).$$

The constant $\tilde{C}_1$ increases only polynomially with m. In fact, the growth with m is the same as that of the bound given in Theorem 10.3, since the result in Theorem 10.3 follows from Theorem 10.4. In particular, by letting the quantization level Q be increasingly finer (i.e., $Q \to \infty$), we see that the constant $\tilde{C}_1$ satisfies

$$\lim_{Q \to \infty} \tilde{C}_1 = 1 + 8m \left(2 + \frac{mB}{\beta(1-\beta)} \right),$$

which is the same as the constant C_1 in Theorem 10.3. Hence, in the limit as $Q \to \infty$, the estimate in Theorem 10.4 yields the estimate in Theorem 10.3.

10.4.2   Consensus with local constraints

Here, we focus only on the problem of reaching a consensus when the estimates of different agents are constrained to lie in different constraint sets and each agent only knows its own constraint set. Such constraints are significant in a number of applications including signal processing within a network of sensors, network motion planning and alignment, rendezvous problems, and distributed, constrained multiagent-optimization problems,[8] where each agent's position is limited to a certain region or range.

---

[8] See also [28] for constrained consensus arising in connection with potential games.

As in the preceding, we denote by $x_i(k)$ the estimate generated and stored by agent i at time slot k. The agent estimate $x_i(k) \in \mathbb{R}^n$ is constrained to lie in a nonempty closed convex set $X_i \subseteq \mathbb{R}^n$ known only to agent i. The agent's objective is to cooperatively reach a consensus on a common vector through a sequence of local-estimate updates (subject to the local-constraint set) and local information exchanges (with neighboring agents only).

To generate the estimate at time $k + 1$, agent i forms a convex combination of its estimate $x_i(k)$ with the estimates received from other agents at time k, and takes the projection of this vector on its constraint set X_i. More specifically, agent i at time $k + 1$ generates its new estimate according to the following relation:

$$x_i(k + 1) = P_{X_i} \left[\sum_{j=1}^{m} a_{ij}(k) x_j(k) \right]. \tag{10.32}$$

Through the rest of the discussion, the constraint sets $X_1, \ldots, X_m$ are assumed to be "closed convex" subsets of $\mathbb{R}^n$.

The relation in (10.32) defines the "projected-consensus algorithm". The method can be viewed as a distributed algorithm for finding a point in common to the closed convex sets $X_1, \ldots, X_m$. This problem can be formulated as an unconstrained convex minimization, as follows

$$\text{minimize} \; \frac{1}{2} \sum_{i=1}^{m} \left\| x - P_{X_i}[x] \right\|^2$$

$$\text{subject to } x \in \mathbb{R}^n. \tag{10.33}$$

In view of this optimization problem, the method in (10.32) can be interpreted as a distributed algorithm where an agent i is assigned an objective function $f_i(x) = \frac{1}{2} \left\| x - P_{X_i}[x] \right\|^2$. Each agent updates its estimate by taking a step (with step-length equal to 1) along the negative gradient of its own objective function $f_i = \frac{1}{2} \left\| x - P_{X_i} \right\|^2$ at $x = \sum_{j=1}^{m} a_{ij}(k) x_j(k)$. This interpretation of the update rule motivates our line of analysis of the projected-consensus method. In particular, we use $\sum_{i=1}^{m} \left\| x_i(k) - x \right\|^2$ with $x \in \cap_{i=1}^{m} X_i$ as a function measuring the progress of the algorithm.

Let us note that the method of (10.32), with the right choice of the weights $a_{ij}(k)$, corresponds to the classical "alternating" or "cyclic-projection method." These methods generate a sequence of vectors by projecting iteratively on the sets (either cyclically or with some given order); see Figure 10.8(a). The convergence behavior of these methods has been established by Von Neumann [43], Aronszajn [1], Gubin *et al.* [17], Deutsch [16], and Deutsch and Hundal [15]. The projected-consensus algorithm can be viewed as a version of the alternating-projection algorithm, where the iterates are combined with the weights varying over time and across agents, and then projected on the individual constraint sets.

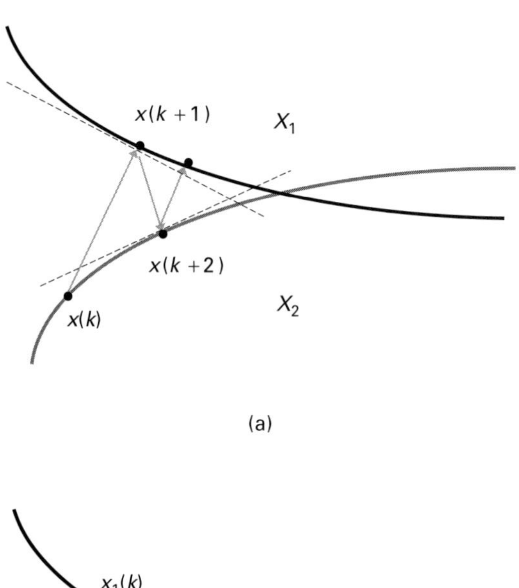

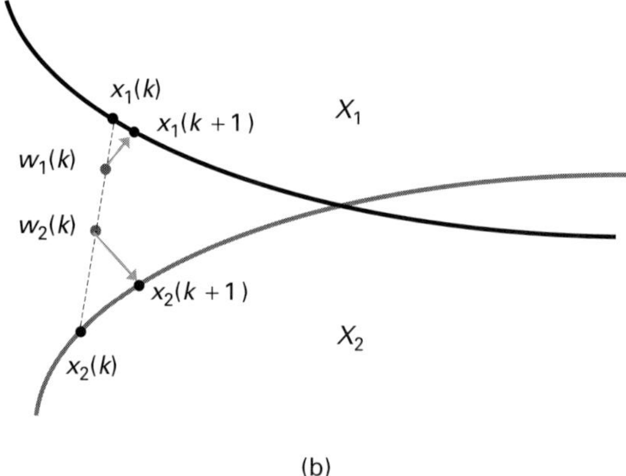

Figure 10.8 The connection between the alternating/cyclic projection method and the projected-consensus algorithm for two closed convex sets X_1 and X_2. In plot (a), the alternating-projection algorithm generates a sequence $\{x(k)\}$ by iteratively projecting onto sets X_1 and X_2, that is, $x(k+1) = P_{X_1}[x(k)], x(k+2) = P_{X_2}[x(k+1)]$. In plot (b), the projected-consensus algorithm generates sequences $\{x_i(k)\}$ for agents $i = 1, 2$ by first combining the iterates with different weights and then projecting on respective sets X_i, that is, $w_i(k) = \sum_{j=1}^{m} a_{ij}(k)x_j(k)$ and $x_i(k+1) = P_{X_i}[w_i(k)]$ for $i = 1, 2$.

To study the convergence behavior of the agent estimates $\{x_i(k)\}$ defined in (10.32), we find it useful to decompose the representation of the estimates into a linear part (corresponding to nonprojected consensus) and a nonlinear part (corresponding to the difference between the projected and nonprojected consensus). Specifically, we rewrite the update rule in (10.32) as

$$x_i(k+1) = \sum_{j=1}^{m} a_{ij}(k)x_j(k) + e^i(k), \qquad (10.34)$$

where $e^i(k)$ represents the error due to the projection operation, given by

$$e^i(k) = P_{X_i}\left[\sum_{j=1}^{m} a_{ij}(k)x_j(k)\right] - \sum_{j=1}^{m} a_{ij}(k)x_j(k). \tag{10.35}$$

As indicated by the preceding two relations, the evolution dynamics of the estimates $x_i(k)$ for each agent is decomposed into a sum of a linear (time-varying) term $\sum_{j=1}^{m} a_{ij}(k)x_j(k)$ and a nonlinear term $e^i(k)$. The linear term captures the effects of mixing the agent estimates, while the nonlinear term captures the nonlinear effects of the projection operation. This decomposition can be exploited to analyze the behavior and estimate the performance of the algorithm. It can be seen [33] that, under the doubly stochasticity assumption on the weights, the nonlinear terms $e^i(k)$ are diminishing in time for each i, and therefore, the evolution of agent estimates is "almost linear." Thus, the nonlinear term can be viewed as a non-persistent disturbance in the linear evolution of the estimates.

Convergence and rate of convergence results

We show that the projected-consensus algorithm converges to some vector that is common to all constraint sets X_i, under Assumptions 3 and 4. Under an additional assumption that the sets X_i have an interior point in common, we provide a convergence-rate estimate.

The following result shows that the agents reach a consensus asymptotically, in other words, the agent estimates $x_i(k)$ converge to the same point as k goes to infinity [33].

THEOREM 10.5   *Let the constraint sets $X_1, \ldots, X_m$ be closed convex subsets of $\mathbb{R}^n$, and let the set $X = \cap_{i=1}^{m} X_i$ be nonempty. Also, let Assumptions 3 and 4 hold. Let the sequences $\{x_i(k)\}$, $i = 1\ldots, m$, be generated by the projected-consensus algorithm (10.32). We then have for some $\tilde{x} \in X$,*

$$\lim_{k \to \infty} \|x_i(k) - \tilde{x}\| = 0 \quad \text{for all } i.$$

We next provide a rate estimate for the projected-consensus algorithm (10.32). It is difficult to access the convergence rate in the absence of any specific structure on the constraint sets X_i. To deal with this, we consider a special case when the weights are time-invariant and equal, that is, $a_{ij}(k) = 1/m$ for all i, j, and k, and the intersection of the sets X_i has a nonempty interior. In particular, we have the following rate result [33].

THEOREM 10.6   *Let the constraint sets $X_1, \ldots, X_m$ be closed convex subsets of $\mathbb{R}^n$. Let $X = \cap_{i=1}^{m} X_i$, and assume that there is a vector $\bar{x} \in X$ and a scalar $\delta > 0$ such that*

$$\{z \mid \|z - \bar{x}\| \leq \delta\} \subset X.$$

Also, let Assumption 4 hold. Let the sequences $\{x_i(k)\}$, $i = 1\ldots, m$ be generated by the algorithm (10.32), where the weights are uniform, that is, $a_{ij}(k) = 1/m$ for all i, j, and k.

We then have

$$\sum_{i=1}^{m} \|x_i(k) - \tilde{x}\|^2 \leq \left(1 - \frac{1}{4R^2}\right)^k \sum_{i=1}^{m} \|x_i(0) - \tilde{x}\|^2 \qquad \textit{for all } k \geq 0,$$

where $\tilde{x} \in X$ is the common limit of the sequences $\{x_i(k)\}$, $i = 1\ldots, m$, and $R = \frac{1}{\delta} \sum_{i=1}^{m} \|x_i(0) - \tilde{x}\|$.

The result shows that the projected-consensus algorithm converges with a geometric rate under the interior-point and uniform-weights assumptions.

10.5   Future work

The models presented so far highlight a number of fruitful areas for future research. These include, but are not limited to the following topics.

10.5.1   Optimization with delays

The distributed-subgradient algorithm we presented in Section 10.3 [cf. (10.19)] assumes that at any time $k \geq 0$, agent i has access to estimates $x_j(k)$ of its neighbors. This may not be possible in communication networks where there are delays associated with transmission of agent estimates over a communication channel. A natural extension, therefore, is to study an asynchronous operation of the algorithm (10.19) using delayed agent values, in other words, agent i at time k has access to outdated values of agent j.

More formally, we consider the following update rule for agent i: suppose agent j sends its estimate $x_j(s)$ to agent i. If agent i receives the estimate $x_j(s)$ at time k, then the delay is $t_{ij}(k) = k - s$ and agent i assigns a weight $a_{ij}(k) > 0$ to the estimate $x_j(s)$. Otherwise, agent i uses $a_{ij}(k) = 0$. Hence, each agent i updates its estimate according to the following relation:

$$x_i(k+1) = \sum_{j=1}^{m} a_{ij}(k)x_j(k - t_{ij}(k)) - \alpha d_i(k) \qquad \text{for } k = 0, 1, 2, \ldots, \tag{10.36}$$

where the vector $x_i(0) \in \mathbb{R}^n$ is the initial estimate of agent i, the scalar $t_{ij}(k)$ is non-negative and it represents the delay of a message from agent j to agent i, while the scalar $a_{ij}(k)$ is a non-negative weight that agent i assigns to a delayed estimate $x_j(s)$ arriving from agent j at time k.

Establishing the convergence and rate properties of the update rule (10.36) is essential in understanding the robustness of the optimization algorithm to delays and dynamics associated with information exchange over finite bandwidth-communication channels. Under the assumption that all delay values are bounded [i.e., there exists a scalar $B > 0$ such that $t_{ij}(k) \leq B$ for all $i, j \in \mathcal{N}$ and all $k \geq 0$], the update rule (10.36) can be analyzed by considering an "augmented model", where we introduce "artificial" agents

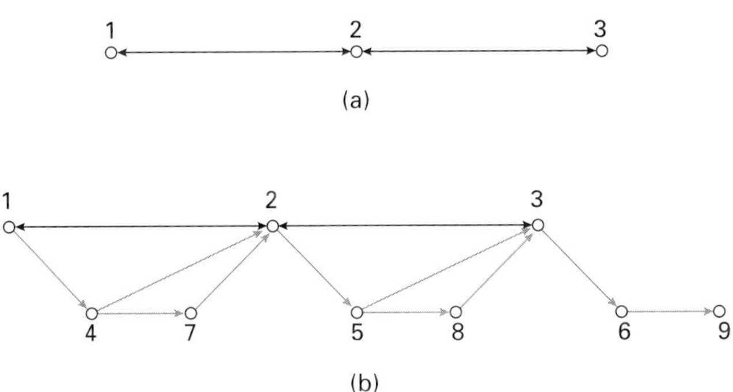

Figure 10.9   Plot (a) illustrates an agent network with 3 agents, where agents 1 and 2, and agents 2 and 3 communicate directly. Plot (b) illustrates the augmented network associated with the original network of part (a), when the delay value between agents is bounded by 3. The artificial agents introduced in the system are 4,...,9. Agents 4, 5, and 6 model the delay of 1, while agents 7, 8, and 9 model the delay of 2 for the original nodes 1, 2, and 3, respectively.

for handling the delayed information only. In particular, with each agent i of the original model, we associate a new agent for each of the possible values of the delay that a message originating from agent i may experience. In view of the bounded delay-values assumption, it suffices to add finitely many new agents handling the delays. This augmentation reduces the delayed multiagent model into a model without delays (see Figure 10.9).

The augmented-agent model is used in [37] to study the consensus problem in the presence of delays. In particular, this paper shows that agents reach consensus on a common decision even with delayed information and establishes convergence-rate results. Future work includes analyzing the optimization algorithm of (10.19) in the presence of delays. The analysis of the optimization algorithm is more challenging in view of the fact that due to delays, an agent may receive a different amount of information from different agents. This difference in the update frequencies results in the overall consensus value to be influenced more by some of the agents information (about their local-objective function). Thus, the value that agents reach a consensus on need not be the optimal solution of the problem of minimizing the sum of the local-objective functions of the agents. To address this issue, the optimization algorithm should be modified to include the update-frequency information in the agent-exchange model.

10.5.2   Optimization with constraints

Section 10.3 presents a distributed-subgradient method for solving the unconstrained-optimization problem (10.18). An important extension is to develop optimization methods for solving multiagent-optimization problems in which each agent i has a local, convex, closed constraint set $X_i \subseteq \mathbb{R}^n$ know by agent i only. Note that the case when there is a global constraint $C_g \subseteq \mathbb{R}^n$ is a special case of this problem with $X_i = C_g$ for all $i \in \mathcal{N}$ (see Introduction).

An immediate solution for this problem is to combine the subgradient algorithm (10.19) with the projected-consensus algorithm studied in Section 10.4.2. More specifically, we denote by $x_i(k)$ the estimate maintained by agent i at time slot k. Agent i updates this estimate by forming a convex combination of this estimate with the estimates received from its neighbors at time k, taking a step (with stepsize α) in the direction of the subgradient of its local convex-objective function f_i at $x_i(k)$, and taking the projection of this vector on its constraint set X_i, that is, agent i at time k generates its new estimate according to

$$ x_i(k+1) = P_{X_i}\left[\sum_{j=1}^{m} a_{ij}(k)x_j(k) - \alpha d_i(k) \right] \qquad \text{for } k = 0, 1, 2, \ldots . \qquad (10.37) $$

The convergence analysis of this algorithm involves combining the ideas and methods of Sections 10.3 and 10.4.2, in other words, understanding the behavior of the transition matrices, the approximate-subgradient method, and the projection errors. It is more challenging due to the dependencies of the error terms involved in the analysis.

When the global constraint set C_g has more structure, for example, when it can be expressed as finitely many equality and inequality constraints, it may be possible to develop "primal-dual algorithms", which combine the primal step (10.19) with a dual step for updating the dual solutions (or multipliers), as in Section 10.2. Primal-dual subgradient methods have been analyzed in recent work [40] for a model in which each agent has the same information set, in other words, at each time slot, each agent has access to the same estimate. It would be of great interest to combine this model with the multiagent model of Section 10.3 that incorporates different local-information structures.

10.5.3　Nonconvex local-objective functions

The distributed-optimization framework presented in Section 10.3 is very general in that it encompasses local-information structures, operates with time-varying connectivity, and optimizes general convex local-objective functions subject to convex constraints. However, there are many applications such as inelastic rate control for voice communication [18,49] and rendezvous problems with constraints [28] in which the local-objective functions and constraints are not convex. Under smoothness assumptions on the objective functions, the methods presented in this chapter can still be used and guarantee convergence to stationary points. An important future direction is the design of algorithms that can guarantee convergence to global optimal in the presence of nonconvexities and in decentralized environments.

10.6　Conclusions

This chapter presents a general framework for distributed optimization of a multiagent-networked system. In particular, we consider multiple agents, each with its own private local-objective function and local constraint, which exchange information over a network with time-varying connectivity. The goal is to design algorithms that the agents can use to cooperatively optimize a global-objective function, which is a function of the local-objective functions, subject to local and global constraints. A key characteristic of these algorithms is their operation within the informational constraints of the model, specified by the local-information structures and the underlying connectivity of the agents.

Our development focuses on two key approaches. The first approach uses Lagrangian-duality and dual-subgradient methods to design algorithms. These algorithms yield distributed methods for problems with separable structure (i.e., problems where local-objective functions and constraints decompose over the components of the decision vector), as illustrated in Section 10.2.3. Since these methods operate in the dual space, a particular interest is in producing primal near-feasible and near-optimal solutions using the information generated by the dual-subgradient algorithm. The analysis presented in Section 10.2 is from our recent work [36, 41], which discusses approximate primal-solution recovery and provides convergence-rate estimates on the generated solutions.

The second approach combines subgradient methods with consensus algorithms to optimize general, convex local-objective functions in decentralized settings. Even though the nonseparable structure of the local-objective functions does not immediately lead to decomposition schemes, the consensus part included in the algorithm serves as a mechanism to distribute the computations among the agents. The material presented in Sections 10.3 and 10.4 combines results from a series of recent papers [31–33,35,37,42].

This chapter illustrates the challenges associated with optimization-algorithm design for multiagent-networked systems. Optimization methodologies have played a key role in providing a systematic framework for the design of architectures and new protocols in many different networks. For such applications, it is clear that many of the assumptions we take for granted in the development and analysis of algorithms for solving constrained-optimization problems are not valid. These include assumptions such as global access to input data and ability to exchange real-valued variables instantly between different processors. Moreover, practical considerations divert our attention from complicated stepsize rules that can guarantee convergence to an optimal solution to simple stepsize rules (such as a constant stepsize rule), which may not guarantee convergence, but nevertheless can provide an "approximate solution" within reasonable time constraints. The importance of optimization methods for such practical applications motivate development of new frameworks and methods that can operate within the constraints imposed by the underlying networked system.

10.7   Problems

Exercise 10.1 Let $\{\mu_k\}$ be a dual sequence generated by the subgradient method with a constant stepsize α [cf. (10.8)]. Assume that the subgradients $\{g_k\}$ are uniformly bounded, that is, $\|g_k\| \leq L$ for all k. Assume also that the dual optimal solution set M^* is nonempty.

At each iteration k, consider an approximate dual solution generated by averaging the vectors $\mu_0, \ldots, \mu_{k-1}$, i.e.,

$$\hat{\mu}_k = \frac{1}{k} \sum_{i=0}^{k-1} \mu_i \qquad \text{for all } k \geq 1.$$

Show that

$$q(\hat{\mu}_k) \geq q^* - \frac{\text{dist}^2(\mu_0, M^*)}{2\alpha k} - \frac{\alpha L^2}{2} \qquad \text{for all } k \geq 1.$$

Exercise 10.2 (Minimum-cost network-flow problem)

Consider a directed connected graph $\mathcal{G} = (\mathcal{N}, \mathcal{E})$, where $\mathcal{N}$ is the node set and $\mathcal{E}$ is the edge set. At each node i, there is a given external-source flow b_i that enters (if $b_i > 0$) or leaves (if $b_i < 0$) node i. Let b be the vector $b = [b_i]_{i \in \mathcal{N}}$. We define the node-edge incidence matrix A as the $|\mathcal{N}| \times |\mathcal{E}|$ matrix given as follows: the (i,j)th entry $[A]_{ij}$ is given by $+1$ if edge j leaves node i; by -1 if edge j enters node i; and 0 otherwise.

Each edge has a convex cost function $f_i(x_i)$, where x_i denotes the flow on edge i. We are interested in finding a flow vector $x = [x_i]_{i \in \mathcal{E}}$ that minimizes the sum of the edge-cost functions. This problem can be formulated as an optimization problem as follows:

$$\begin{aligned}
\text{minimize} \quad & \sum_{i \in \mathcal{E}} f_i(x_i) \\
\text{subject to} \quad & Ax = b \\
& x \geq 0,
\end{aligned} \tag{10.38}$$

where the constraint $Ax = b$ captures the conservation of flow constraints.

Use Lagrangian-decomposition and the dual-subgradient algorithm to solve this problem. Show that this approach leads to a distributed-optimization method for solving problem (10.38).

Exercises 10.3–10.5 are the steps involved in proving the result of Theorem 10.4.

Exercise 10.3 Consider the quantized method given in (10.30) of Section 10.4.1. Define the transition matrices $\Phi(k, s)$ from time s to time k, as follows

$$\Phi(k, s) = A(k)A(k-1) \cdots A(s) \qquad \text{for all } s \text{ and } k \text{ with } k \geq s,$$

(see Section 10.3.1).

(a) Using the transition matrices and the decomposition of the estimate evolution of (10.34)–(10.35), show that the relation between $x_i(k+1)$ and the estimates

$x_1(0), \ldots, x_m(0)$ is given by

$$x_i^Q(k+1) = \sum_{j=1}^{m} [\Phi(k,0)]_{ij} x_j^Q(0) - \alpha \sum_{s=1}^{k} \sum_{j=1}^{m} [\Phi(k,s)]_{ij} \tilde{d}_j(s-1)$$

$$- \sum_{s=1}^{k} \sum_{j=1}^{m} [\Phi(k,s)]_{ij} \epsilon_j(s) - \alpha \tilde{d}_i(k) - \epsilon_i(k+1).$$

(b) Consider an auxiliary sequence $\{y(k)\}$ defined by

$$y(k) = \frac{1}{m} \sum_{j=1}^{m} x_i^Q(k).$$

Show that for all k,

$$y(k) = \frac{1}{m} \sum_{j=1}^{m} x_j^Q(0) - \frac{\alpha}{m} \sum_{s=1}^{k} \sum_{j=1}^{m} \tilde{d}_j(s-1) - \frac{1}{m} \sum_{s=1}^{k} \sum_{j=1}^{m} \epsilon_j(s).$$

(c) Suppose that Assumptions 3 and 4 hold. Using the relations in parts (a) and (b), and Theorem 10.2, show that

$$\|x_i^Q(k) - y(k)\| \le \beta^{\lceil \frac{k}{B} \rceil - 2} \sum_{j=1}^{m} \|x_j^Q(0)\| + \left(\alpha L + \frac{\sqrt{n}}{Q} \right) \left(2 + \frac{mB}{\beta(1-\beta)} \right).$$

Exercise 10.4 Let $y(k)$ be the sequence defined in Exercise 10.3. Consider the running averages $\hat{y}(k)$ of the vectors $y(k)$, given by

$$\hat{y}(k) = \frac{1}{k} \sum_{h=1}^{k} y(h) \qquad \text{for all } k \ge 1.$$

Let Assumptions 3 and 4 hold, and assume that the set X^* of optimal solutions of problem (10.18) is nonempty. Also, assume that the subgradients are uniformly bounded as in (10.23). Then, the average vectors $\hat{y}(k)$ satisfy for all $k \ge 1$,

$$f(\hat{y}(k)) \le f^* + \frac{\alpha L^2 \tilde{C}}{2} + \frac{2mLB}{k\beta(1-\beta)} \sum_{j=1}^{m} \|x_j^Q(0)\|$$

$$+ \frac{m}{2\alpha k} \left(\operatorname{dist}(y(0), X^*) + \alpha L + \frac{\sqrt{n}}{Q} \right)^2,$$

where $y(0) = \frac{1}{m} \sum_{j=1}^{m} x_j^Q(0)$, $\beta = 1 - \frac{\eta}{4m^2}$, and

$$\tilde{C} = 1 + 4m \left(1 + \frac{\sqrt{n}}{\alpha L Q} \right) \left(2 + \frac{mB}{\beta(1-\beta)} \right).$$

Exercise 10.5 Prove Theorem 10.4 using the results of Exercises 10.3 and 10.4.

References

[1] N. Aronszajn, "Theory of reproducing kernels," *Transactions of the American Mathematical Society*, vol. 68, no. 3, pp. 337–404, 1950.

[2] D.P. Bertsekas, A. Nedić, and A.E. Ozdaglar, *Convex Analysis and Optimization*. Belmont, MA: Athena Scientific, 2003.

[3] D.P. Bertsekas and J.N. Tsitsiklis, *Parallel and Distributed Computation: Numerical Methods*. Belmont, MA: Athena Scientific, 1997.

[4] D.P. Bertsekas, *Nonlinear Programming*. Belmont, MA: Athena Scientific, 1999.

[5] D. P. Bertsekas, A. Nedić, and A. Ozdaglar, "Min common/max crossing duality: A simple geometric framework for convex optimization and minimax theory," Massachusetts Institute of Technology, Tech. Report LIDS 2536, 2002.

[6] P.-A. Bliman and G. Ferrari-Trecate, "Average consensus problems in networks of agents with delayed communications," *Automatica*, vol. 44, no. 8, pp. 1985–95, 2008.

[7] V.D. Blondel, J.M. Hendrickx, A. Olshevsky, and J.N. Tsitsiklis, "Convergence in multiagent coordination, consensus, and flocking," *Proceedings of the IEEE CDC*, 2005, pp. 2996–3000.

[8] S. Boyd, A. Ghosh, B. Prabhakar, and D. Shah, "Gossip algorithms: Design, analysis, and applications," in *Proceedings of the IEEE INFOCOM*, vol. 3, 2005, pp. 1653–64.

[9] M. Cao, A.S. Morse, and B.D.O. Anderson, "Reaching a consensus in a dynamically changing environment: A graphical approach," *SIAM Journal on Control and Optimization*, vol. 47, no. 2, pp. 575–600, 2008.

[10] ——, "Reaching a consensus in a dynamically changing environment: Convergence rates, measurement delays, and asynchronous events," *SIAM Journal on Control and Optimization*, vol. 47, no. 2, pp. 601–23, 2008.

[11] M. Cao, D.A. Spielman, and A.S. Morse, "A lower bound on convergence of a distributed network consensus algorithm," *Proceedings of the IEEE CDC*, 2005, pp. 2356–61.

[12] R. Carli, F. Fagnani, P. Frasca, T. Taylor, and S. Zampieri, "Average consensus on networks with transmission noise or quantization," in *Proceedings of the European Control Conference*, 2007.

[13] R. Carli, F. Fagnani, A. Speranzon, and S. Zampieri, "Communication constraints in coordinated consensus problem," in *Proceedings of the IEEE American Control Conference*, 2006, pp. 4189–94.

[14] M. Chiang, S.H. Low, A.R. Calderbank, and J.C. Doyle, "Layering as optimization decomposition: a mathematical theory of network architectures," *Proceedings of the IEEE*, vol. 95, no. 1, pp. 255–312, 2007.

[15] F. Deutsch and H. Hundal, "The rate of convergence for the cyclic projections algorithm i: angles between convex sets," *Journal of Approximation Theory*, vol. 142, pp. 36–55, 2006.

[16] F. Deutsch, "Rate of convergence of the method of alternating projections," in *Parametric Optimization and Approximation*, vol. 76, B. Brosowski and F. Deutsch, eds. Basel: Birkhäuser, 1983, pp. 96–107.

[17] L.G. Gubin, B.T. Polyak, and E.V. Raik, "The method of projections for finding the common point of convex sets," *U.S.S.R Computational Mathematics and Mathematical Physics*, vol. 7, no. 6, pp. 1211–28, 1967.

[18] P. Hande, S. Zhang, and M. Chiang, "Distributed rate allocation for inelastic flows," *IEEE/ACM Transactions on Networking*, vol. 15, no. 6, pp. 1240–53, 2007.

[19] J.-B. Hiriart-Urruty and C. Lemaréchal, *Convex Analysis and Minimization Algorithms*. Berlin: Springer-Verlag, 1996.

[20] A. Jadbabaie, J. Lin, and S. Morse, "Coordination of groups of mobile autonomous agents using nearest neighbor rules," *IEEE Transactions on Automatic Control*, vol. 48, no. 6, pp. 988–1001, 2003.

[21] S. Kar and J. Moura. (2007). "Distributed consensus algorithms in sensor networks: Link and channel noise." Available: http://arxiv.org/abs/0711.3915

[22] A. Kashyap, T. Basar, and R. Srikant, "Quantized consensus," *Automatica*, vol. 43, no. 7, pp. 1192–203, 2007.

[23] F.P. Kelly, A.K. Maulloo, and D.K. Tan, "Rate control for communication networks: shadow prices, proportional fairness, and stability," *Journal of the Operational Research Society*, vol. 49, pp. 237–52, 1998.

[24] T. Larsson, M. Patriksson, and A. Strömberg, "Ergodic results and bounds on the optimal value in subgradient optimization," in *Operations Research Proceedings*, P. Kelinschmidt *et al.*, eds. Berlin: Springer, 1995, pp. 30–5.

[25] T. Larsson, M. Patriksson, and A. Strömberg, "Ergodic convergence in subgradient optimization," *Optimization Methods and Software*, vol. 9, pp. 93–120, 1998.

[26] ——, "Ergodic primal convergence in dual subgradient schemes for convex programming," *Mathematical Programming*, vol. 86, pp. 283–312, 1999.

[27] S. Low and D.E. Lapsley, "Optimization flow control, I: basic algorithm and convergence," *IEEE/ACM Transactions on Networking*, vol. 7, no. 6, pp. 861–74, 1999.

[28] J.R. Marden, G. Arslan, and J.S. Shamma, "Connections between cooperative control and potential games illustrated on the consensus problem," preprint, 2008.

[29] L. Moreau, "Stability of multiagent systems with time-dependent communication links," *IEEE Transactions on Automatic Control*, vol. 50, no. 2, pp. 169–82, 2005.

[30] A. Mutapcic, S. Boyd, S. Murali, D. Atienza, G. De Micheli, and R. Gupta, "Processor speed control with thermal constraints," submitted for publication, 2007.

[31] A. Nedić, A. Olshevsky, A. Ozdaglar, and J.N. Tsitsiklis, "Distributed subgradient methods and quantization effects," presented at the Proceedings of IEEE CDC, 2008.

[32] A. Nedić, A. Olshevsky, A. Ozdaglar, and J. N. Tsitsiklis, "On distributed averaging algorithms and quantization effects," *IEEE Transactions on Automatic Control*, submitted for publication, 2009. Available: http://arxiv.org/abs/0803.1202

[33] A. Nedić, A. Ozdaglar, and P.A. Parrilo, "Constrained consensus and optimization in multi-agent networks," LIDS Technical Report 2779. Available: http://arxiv.org/abs/0802.3922

[34] A. Nedić, A. Ozdaglar, and A. Rubinov, "Abstract convexity for non-convex optimization duality," *Optimization*, vol. 56, PP. 655–74, 2007.

[35] A. Nedić and A. Ozdaglar, "On the rate of convergence of distributed subgradient methods for multi-agent optimization," in *Proceedings of the IEEE CDC*, 2007, pp. 4711–6.

[36] ——, "Approximate primal solutions and rate analysis for dual subgradient methods," *SIAM Journal on Optimization*, submitted for publication.

[37] ——, "Convergence rate for consensus with delays," *Journal of Global Optimization*, submitted for publication.

[38] ——, "A geometric framework for nonconvex optimization duality using augmented Lagrangian functions," *Journal of Global Optimization*, vol. 40, no. 4, pp. 545–73, 2008.

[39] ——, "Separation of nonconvex sets with general augmenting functions," *Mathematics of Operations Research*, vol. 33, no. 3, pp. 587–605, 2008.

[40] ——, "Subgradient methods for saddle-point problems," *Journal of Optimization Theory and Applications*, submitted for publication.

[41] ——, "Subgradient methods in network resource allocation: rate analysis," *Proceedings of CISS*, 2008.

[42] ——, "Distributed subgradient method for multi-agent optimization," *IEEE Transactions on Automatic Control*, submitted for publication.

[43] J. Von Neumann, *Functional Operators*. Princeton: Princeton University Press, 1950.

[44] R. Olfati-Saber and R.M. Murray, "Consensus problems in networks of agents with switching topology and time-delays", *IEEE Transactions on Automatic Control*, vol. 49, no. 9, pp. 1520–33, 2004.

[45] A. Olshevsky and J.N. Tsitsiklis, *Convergence rates in distributed consensus averaging*, Proceedings of IEEE CDC, 2006, pp. 3387–3392.

[46] ——, "Convergence speed in distributed consensus and averaging", *SIAM Journal on Control and Optimization*, forthcoming.

[47] R. T. Rockafellar, *Convex analysis*. Princeton, NJ: Princeton University Press, 1970.

[48] S. Shakkottai and R. Srikant, "Network optimization and control", *Foundations and Trends in Networking*, vol. 2, no. 3, pp. 271–379.

[49] S. Shenker, "Fundamental design issues for the future internet", *IEEE Journal on Selected Areas in Communication*, vol. 13, no. 7, pp. 1176–1188, 1995.

[50] R. Srikant, *Mathematics of Internet congestion control*. Basel: Birkhauser, 2004.

[51] J.N. Tsitsiklis, D.P. Bertsekas, and M. Athans, "Distributed asynchronous deterministic and stochastic gradient optimization algorithms", *IEEE Transactions on Automatic Control*, vol. 31, no. 9, pp. 803–812, 1986.

[52] J.N. Tsitsiklis, *Problems in decentralized decision making and computation*, Ph.D. thesis, Dept. of Electrical Engineering and Computer Science, Massachusetts Institute of Technology, 1984.

[53] H. Uzawa, *Iterative methods in concave programming*, Studies in Linear and Nonlinear Programming, K. Arrow, L. Hurwicz, and H. Uzawa, eds., Palo Alto, CA: Stanford University Press, 1958, pp. 154–165.

[54] T. Vicsek, A. Czirok, E. Ben-Jacob, I. Cohen, and O. Schochet, "Novel type of phase transitions in a system of self-driven particles", *Physical Review Letters*, vol. 75, no. 6, pp. 1226–1229, 1995.

[55] L. Xiao, S. Boyd, and S.-J. Kim, "Distributed average consensus with least mean square deviation", *Journal of Parallel and Distributed Computing*, vol. 67, no. 1, pp. 33–46, 2007.

11 Competitive optimization of cognitive radio MIMO systems via game theory

Gesualso Scutari, Daniel P. Palomar, and Sergio Barbarossa

Game theory is a field of applied mathematics that describes and analyzes scenarios with interactive decisions. In recent years, there has been a growing interest in adopting cooperative and non-cooperative game-theoretic approaches to model many communications and networking problems, such as power control and resource sharing in wireless/wired and peer-to-peer networks. In this chapter we show how many challenging unsolved resource-allocation problems in the emerging field of *cognitive radio* (CR) networks fit naturally in the game-theoretical paradigm. This provides us with the mathematical tools necessary to analyze the proposed equilibrium problems for CR systems (e.g., existence and uniqueness of the solution) and to devise distributed algorithms along with their convergence properties. The proposed algorithms differ in performance, level of protection of the primary users, computational effort and signaling among primary and secondary users, convergence analysis, and convergence speed; which makes them suitable for many different CR systems. We also propose a more general framework suitable for investigating and solving more sophisticated equilibrium problems in CR systems when classical game theory may fail, based on *variation inequality* (VI) that constitutes a very general class of problems in nonlinear analysis.

11.1 Introduction and motivation

In recent years, increasing demand of wireless services has made the radio spectrum a very scarce and precious resource. Moreover, most current wireless networks characterized by fixed-spectrum assignment policies are known to be very inefficient considering that licensed bandwidth demands are highly varying along the time or space dimensions (according to the Federal Communications Commission [FCC], only 15% to 85% of the licensed spectrum is utilized on average [1]). Many recent works [2–4] have recognized that the most appropriate approach to tackle the great spectrum variability in time and space calls for *dynamic* access strategies that adapt transmission parameters (e.g., operating spectrum, modulation, transmission, power and communication technology) based on knowledge of the electromagnetic environment.

Cognitive radio originated as a possible solution to this problem [5] obtained by endowing the radio nodes with "cognitive capabilities", for example, the ability to sense the electromagnetic environment, make short-term predictions, and react intelligently

Convex Optimization in Signal Processing and Communication, eds. Daniel P. Palomar and Yonina C. Eldar. Published by Cambridge University Press © Cambridge University Press, 2010.

in order to optimize the usage of the available resources. Multiple debated positions have been proposed for implementing the CR idea [2–4], depending on the policy to be followed with respect to the *licensed* users, that is, the users who have acquired the right to transmit over specific portions of the spectrum buying the corresponding license. The most common strategies adopt a hierarchical access structure, distinguishing between *primary* users, or legacy spectrum holders, and *secondary* users, who access the licensed spectrum dynamically, under the constraint of not inducing any significant quality of service (QoS) degradations to the primary users.

Within this context, adopting a general *multiple input–multiple output* (MIMO) channel is natural to model the system of cognitive secondary users as a vector-interference channel, where the transmission over the generic qth MIMO channel with n_{T_q} transmit and n_{R_q} receive dimensions is given by the following baseband complex-valued signal model:

$$\mathbf{y}_q = \mathbf{H}_{qq}\mathbf{x}_q + \sum_{r \neq q} \mathbf{H}_{rq}\mathbf{x}_r + \mathbf{n}_q, \tag{11.1}$$

where $\mathbf{x}_q \in \mathbb{C}^{n_{T_q}}$ is the signal transmitted by source q, $\mathbf{y}_q \in \mathbb{C}^{n_{R_q}}$ is the received signal by destination q, $\mathbf{H}_{qq} \in \mathbb{C}^{n_{R_q} \times n_{T_q}}$ is the channel matrix between the qth transmitter and the intended receiver, $\mathbf{H}_{rq} \in \mathbb{C}^{n_{R_q} \times n_{T_r}}$ is the cross-channel matrix between source r and destination q, and $\mathbf{n}_q \in \mathbb{C}^{n_{R_q}}$ is a zero-mean, circularly symmetric, complex Gaussian noise vector with arbitrary (nonsingular) covariance matrix $\mathbf{R}_{n_q}$, collecting the effect of both thermal noise and interference generated by the primary users. The first term on the right-hand side of (11.1) is the useful signal for link q, the second and third terms represent the "multi-user interference" (MUI) received by secondary user q and generated by the other secondary users and the primary users, respectively. The power constraint for each transmitter is

$$\mathcal{E}\left\{\|\mathbf{x}_q\|_2^2\right\} = \mathsf{Tr}\left(\mathbf{Q}_q\right) \leq P_q, \tag{11.2}$$

where $\mathcal{E}\{\cdot\}$ denotes the expectation value, $\mathsf{Tr}(\cdot)$ is the trace operator, $\mathbf{Q}_q$ is the covariance matrix of the transmitted signal by user q, and P_q is the transmit power in units of energy per transmission.

The model in (11.1) represents a fairly general MIMO setup, describing multiuser transmissions (e.g., peer-to-peer links, multiple access, or broadcast channels) over multiple channels, which may represent frequency channels (as in OFDM systems) [6–9], time slots (as in TDMA systems) [6, 7, 9], or spatial channels (as in transmit/receive beamforming systems) [10].

Due to the distributed nature of the CR system, with neither a centralized control nor coordination among the secondary users, we focus on transmission techniques where no interference cancellation is performed and the MUI is treated as additive colored noise at each receiver. Each channel is assumed to change sufficiently slowly to be considered fixed during the whole transmission. Moreover, perfect channel state information at both transmitter and receiver sides of each link is assumed. This includes the direct channel

$\mathbf{H}_{qq}$ (but not the cross-channels $\{\mathbf{H}_{rq}\}_{r\neq q}$ from the other users), as well as the covariance matrix of noise plus MUI

$$\mathbf{R}_{-q}(\mathbf{Q}_{-q}) \triangleq \mathbf{R}_{n_q} + \sum_{r\neq q} \mathbf{H}_{rq}\mathbf{Q}_r\mathbf{H}_{rq}^H. \tag{11.3}$$

Within the assumptions made above, the maximum information rate on link q for a given set of user covariance matrices $\mathbf{Q}_1, \ldots, \mathbf{Q}_Q$, is [11]

$$R_q(\mathbf{Q}_q, \mathbf{Q}_{-q}) = \log \det \left(\mathbf{I} + \mathbf{H}_{qq}^H \mathbf{R}_{-q}^{-1}(\mathbf{Q}_{-q})\mathbf{H}_{qq}\mathbf{Q}_q \right), \tag{11.4}$$

where $\mathbf{Q}_{-q} \triangleq (\mathbf{Q}_r)_{r\neq q}$ is the set of all the users' covariance matrices, except the qth one.

In this chapter, we focus on opportunistic resource-allocation techniques in hierarchical CR systems as given in (11.1). In particular, our interest is in devising the most appropriate form of concurrent communications of cognitive users competing over the physical resources that primary users make available, under the constraint that the degradation induced on the primary users' performance is null or tolerable [2, 3]. While the definition of degradation may be formulated mathematically in a number of ways, one common definition involves the imposition of some form of interference constraints on the secondary users, whose choice and implementation are a complex and open regulatory issue. Both deterministic and probabilistic interference constraints have been suggested in the literature [2, 3]. In this chapter, we will consider in detail deterministic interference constraints, as described next.

11.1.1 Interference constraints: individual and conservative versus global and flexible

We envisage two classes of interference constraints termed "individual conservative" constraints and "global flexible" constraints.

Individual conservative constraints. These constraints are defined individually for each secondary user (with the disadvantage that sometimes this may result in them being too conservative) to control the overall interference caused on the primary receivers. Specifically, we have

- *Null-shaping constraints*:

$$\mathbf{U}_q^H \mathbf{Q}_q = \mathbf{0}, \tag{11.5}$$

where $\mathbf{U}_q \in \mathbb{C}^{n_{T_q} \times r_{U_q}}$ is a tall matrix whose columns represent the spatial and/or the frequency/time "directions" along which user q is not allowed to transmit.

- *Soft and peak power-shaping constraints*:

$$\mathsf{Tr}\left(\mathbf{G}_q^H \mathbf{Q}_q \mathbf{G}_q \right) \leq P_{\mathrm{SU},q}^{\mathrm{ave}} \quad \text{and} \quad \lambda_{\max}\left(\mathbf{G}_q^H \mathbf{Q}_q \mathbf{G}_q \right) \leq P_{\mathrm{SU},q}^{\mathrm{peak}}, \tag{11.6}$$

which represent a relaxed version of the null constraints with a constraint on the total-average and peak-average power radiated along the range space of matrix $\mathbf{G}_q \in \mathbb{C}^{n_{T_q} \times n_{G_q}}$, where $P_{\mathrm{SU},q}^{\mathrm{ave}}$ and $P_{\mathrm{SU},q}^{\mathrm{peak}}$ are the maximum average and average peak power, respectively, that can be transmitted along the spatial and/or the frequency directions spanned by $\mathbf{G}_q$.

The null constraints are motivated in practice by the interference-avoiding paradigm in CR communications (also called the *white-space filling approach*) [4, 12]: CR nodes sense the spatial, temporal, or spectral voids and adjust their transmission strategy to fill in the sensed white spaces. This white-space filling strategy is often considered to be the key motivation for the introduction and development of the CR idea, and has already been adopted as a core platform in emerging wireless access standards such as the IEEE 802.22-Wireless Regional Area Networks (WRANs standard) [13]. Observe that the structure of the null constraints in (11.5) has a very general form and includes, as particular cases, the imposition of nulls over: (a) frequency bands occupied by the primary users (the range space of $\mathbf{U}_q$ coincides with the subspace spanned by a set of IDFT vectors); (b) the time slots used by the primary users (the set of canonical vectors); (c) angular directions identifying the primary receivers as observed from the secondary transmitters (the set of steering vectors representing the directions of the primary receivers as observed from the secondary transmitters).

Opportunistic communications allow simultaneous transmissions between primary and secondary users, provided that the required QoS of the primary users is preserved (these are also called *interference-temperature controlled transmissions* [2, 12, 14]). This can be done using the individual "soft-shaping" constraints expressed in (11.6) that represent a constraint on the "total-average" and peak-average power allowed to be radiated (projected) along the directions spanned by the column space of matrix $\mathbf{G}_q$. For example, in a MIMO setup, the matrix $\mathbf{G}_q$ in (11.6) may contain, in its columns, the steering vectors identifying the directions of the primary receivers. By using these constraints, we assume that the power thresholds $P_{\mathrm{SU},q}^{\mathrm{ave}}$ and $P_{\mathrm{SU},q}^{\mathrm{peak}}$ at each secondary transmitter have been fixed in advance (imposed, e.g., by the network service provider, or legacy systems, or the spectrum body agency) so that the interference temperature-limit constraints at the primary receivers are met. For example, a possible (but conservative) choice for $P_{\mathrm{SU},q}^{\mathrm{ave}}$'s and $P_{\mathrm{SU},q}^{\mathrm{peak}}$'s is $P_{\mathrm{SU},q}^{\mathrm{ave}} = P_{\mathrm{PU}}^{\mathrm{ave}}/Q$ and $P_{\mathrm{SU},q}^{\mathrm{peak}} = P_{\mathrm{PU}}^{\mathrm{peak}}/Q$ for all q, where Q is the number of active secondary users, and $P_{\mathrm{PU}}^{\mathrm{ave}}$ and $P_{\mathrm{PU}}^{\mathrm{peak}}$ are the overall maximum-average and peak-average interference tolerable by the primary user. The assumption made above is motivated by all the practical CR scenarios where primary terminals are oblivious to the presence of secondary users, thus behaving as if no secondary activity was present (also called the *commons model*).

The imposition of the individual interference constraints requires an opportunity-identification phase, through a proper sensing mechanism: secondary users need to reliably detect weak primary signals of possibly different types over a targeted region and wide frequency band in order to identify white-space halls. Examples of solutions to this problem have recently been proposed in [3, 14–16]. The study of sensing in CR networks goes beyond the scope of this chapter. Thus, hereafter, we assume perfect sensing from the secondary users.

Individual interference constraints (possibly in addition to the null constraints) lead to totally distributed algorithms with no coordination between the primary and the secondary users, as we will show in the forthcoming sections. However, sometimes, they may be too restrictive and thus marginalize the potential gains offered by the dynamic resource-assignment mechanism. Since the interference temperature limit [2] is given by the *aggregate* interference induced by *all* the active secondary users to the primary users receivers, it seems natural to limit instead such an aggregate interference, rather than the individual soft power and peak power constraints. This motivates the following global interference constraints.

Global flexible constraints. These constraints, as opposed to the individual ones, are defined globally over all the secondary users:

$$\sum_{q=1}^{Q} \mathrm{Tr}\left(\mathbf{G}_{q,p}^{H}\mathbf{Q}_q\mathbf{G}_{q,p}\right) \le P_{\mathrm{PU},p}^{\mathrm{ave}} \quad \text{and} \quad \sum_{q=1}^{Q} \lambda_{\max}\left(\mathbf{G}_{q,p}^{H}\mathbf{Q}_q\mathbf{G}_{q,p}\right) \le P_{\mathrm{PU},p}^{\mathrm{peak}}, \quad (11.7)$$

where $P_{\mathrm{PU},p}^{\mathrm{ave}}$ and $P_{\mathrm{PU},p}^{\mathrm{peak}}$ are the maximum-average and peak-average interference tolerable by the pth primary user. As we will show in the forthcoming sections, these constraints, in general, lead to better performance for secondary users than imposing the conservative individual constraints. However, this gain comes at a price: the resulting algorithms require some signaling (albeit very reduced) from the primary to the secondary users. They can be employed in all CR networks where an interaction between the primary and the secondary users is allowed, as, for example, in the so-called "property-right" CR model (or "spectrum leasing"), where primary users own the spectral resource and possibly decide to lease part of it to secondary users in exchange for appropriate remuneration.

11.1.2 System design: a game-theoretical approach

Given the CR model in (11.1), the system design consists in finding out the set of covariance matrices of the secondary users satisfying a prescribed optimality criterion, under power and interference constraints in (11.2) and (11.5)–(11.7). One approach would be to design the transmission strategies of the secondary users using global-optimization techniques. However, this has some practical issues that are insurmountable in the CR context. First of all, it requires the presence of a central node having full knowledge of all the channels and interference structure at every receiver. But this poses a serious implementation problem in terms of scalability and amount of signaling to be exchanged among the nodes. The required extra signaling could, in the end, jeopardize the promise for higher efficiency. On top of that, recent results in [17] have shown that the network-utility maximization based on the rate functions is an NP-hard problem, under different choices of the system utility function; which means that there is no hope to obtain an algorithm, even centralized, that can efficiently compute a globally-optimal solution. Consequently, suboptimal algorithms have been proposed (see, e.g., [18, 19]), but they are centralized and may converge to poor spectrum-sharing strategies, due to the nonconvexity of the optimization problem. Thus, it seems natural to concentrate on decentralized strategies, where the cognitive users are able to self-enforce the negotiated agreements on the usage of the available resources (time, frequency, and space) without the intervention of a centralized authority. The philosophy underlying this approach is a "competitive-optimality" criterion, as every user aims for the transmission strategy that unilaterally maximizes its own payoff function. This form of equilibrium is, in fact, the well-known concept of *Nash equilibrium* (NE) in game theory.

Because of the inherently competitive nature of multi-user systems, it is not surprising indeed that game theory has already been adopted to solve distributively many resource-allocation problems in communications. An early application of game theory in a communication system is [20], where the information rates of the users were maximized with respect to the power allocation in a DSL system modeled as a frequency-selective (in practice, multicarrier) Gaussian interference channel. Extension of the basic problem to ad-hoc frequency-selective and MIMO networks were given in [6–9, 21] and [10, 22–24], respectively. However, results in the cited papers have been recognized not to be applicable to CR systems because they do not provide any mechanism to control the amount of interference generated by the secondary users on the primary users [2].

11.1.3   Outline

Within the CR context introduced so far, we formulate in the next sections the optimization problem for the transmission strategies of the secondary users under different combinations of power and individual/global interference constraints. Using the game-theoretic concept of NE as a competitive-optimality criterion, we propose various equilibrium problems that differ in the achievable trade-off between performance and amount of signaling among primary and secondary users. Using results from game theory and VI theory, we study, for each equilibrium problem, properties of the solution (e.g., existence and uniqueness) and propose many iterative, possibly asynchronous, distributed algorithms along with their convergence properties.

The rest of the chapter is organized as follows. Section 11.2 introduces some basic concepts and results on non-cooperative strategic form games that will be used extensively through the whole chapter. Section 11.3 deals with transmissions over unlicensed bands, where there are no constraints on the interference generated by the secondary users on the primary users. Section 11.4 considers CR systems under different individual interference constraints and proposes various NE problems. Section 11.5 focuses on the more challenging design of CR systems under global interference constraints and studies the NE problem using VI theory. Finally, Section 11.6 draws some conclusions.

11.1.4   Notation

The following notation is used in the chapter. Uppercase and lowercase boldface denote matrices and vectors, respectively. The operators $(\cdot)^*$, $(\cdot)^H$, $(\cdot)^\sharp$, $\mathcal{E}\{\cdot\}$, and $\mathrm{Tr}(\cdot)$ are conjugate, Hermitian, Moore-Penrose pseudo-inverse [25], expectation, and trace operators, respectively. The range space and null space are denoted by $\mathcal{R}(\cdot)$ and $\mathcal{N}(\cdot)$, respectively. The set of eigenvalues of an $n \times n$ Hermitian matrix $\mathbf{A}$ is denoted by $\{\lambda_i(\mathbf{A})\}_{i=1}^n$, whereas the maximum and the minimum eigenvalue are denoted by $\lambda_{\max}(\mathbf{A})$ and $\lambda_{\min}(\mathbf{A})$, respectively. The operators $\leq$ and $\geq$ for vectors and matrices are defined component wise, while $\mathbf{A} \succeq \mathbf{B}$ (or $\mathbf{A} \preceq \mathbf{B}$) means that $\mathbf{A} - \mathbf{B}$ is positive (or negative) semidefinite. The operator $\mathrm{Diag}(\cdot)$ is the diagonal matrix with the same diagonal elements as the matrix (or vector) argument; $\mathrm{bdiag}(\mathbf{A}, \mathbf{B}, \ldots)$ is the block diagonal matrix, whose diagonal blocks are the matrices $\mathbf{A}, \mathbf{B}, \ldots$; the operator $\perp$ for vector means that two vectors $\mathbf{x}$ and $\mathbf{y}$ are orthogonal, that is, $\mathbf{x} \perp \mathbf{y} \Leftrightarrow \mathbf{x}^H \mathbf{y} = 0$. The operators $(\cdot)^+$ and $[\,\cdot\,]_a^b$, with $0 \leq a \leq b$, are defined as $(x)^+ \triangleq \max(0, x)$ and $[\,\cdot\,]_a^b \triangleq \min(b, \max(x, a))$, respectively; when the argument of the operators is a vector or a matrix, then they are assumed to be applied component wise. The spectral radius of a matrix $\mathbf{A}$ is denoted by $\rho(\mathbf{A})$, and is defined as $\rho(\mathbf{A}) \triangleq \max\{|\lambda| : \lambda \in \sigma(\mathbf{A})\}$, with $\sigma(\mathbf{A})$ denoting the spectrum (set of eigenvalues) of $\mathbf{A}$ [26]. The operator $\mathbf{P}_{\mathcal{N}(\mathbf{A})}$ (or $\mathbf{P}_{\mathcal{R}(\mathbf{A})}$) denotes the orthogonal projection onto the null space (or the range space) of matrix $\mathbf{A}$ and it is given by $\mathbf{P}_{\mathcal{N}(\mathbf{A})} = \mathbf{N}_A(\mathbf{N}_A^H \mathbf{N}_A)^{-1} \mathbf{N}_A^H$ (or $\mathbf{P}_{\mathcal{R}(\mathbf{A})} = \mathbf{R}_A(\mathbf{R}_A^H \mathbf{R}_A)^{-1} \mathbf{R}_A^H$), where $\mathbf{N}_A$ (or $\mathbf{R}_A$) is any matrix whose columns are linear-independent vectors spanning $\mathcal{N}(\mathbf{A})$ (or $\mathcal{R}(\mathbf{A})$) [26]. The operator $[\mathbf{X}]_Q = \mathrm{argmin}_{\mathbf{Z} \in Q} \|\mathbf{Z} - \mathbf{X}\|_F$ denotes the matrix projection with respect to the Frobenius norm of matrix $\mathbf{X}$ onto the (convex) set Q, where $\|\mathbf{X}\|_F$ is defined as $\|\mathbf{X}\|_F \triangleq \left(\mathrm{Tr}(\mathbf{X}^H \mathbf{X}) \right)^{1/2}$ [26]. We denote by $\mathbf{I}_n$ the $n \times n$ identity matrix and by $r_X \triangleq \mathrm{rank}(\mathbf{X})$ the rank of matrix $\mathbf{X}$. The sets $\mathbb{C}$, $\mathbb{R}$, $\mathbb{R}_+$, $\mathbb{R}_-$, $\mathbb{R}_{++}$, $\mathbb{N}_+$, $\mathbb{S}^n$, and $\mathbb{S}_+^n$ (or $\mathbb{S}_{++}^n$) stand for the set of complex, real, non-negative real, non-positive real, positive real, non-negative integer numbers, and $n \times n$ complex Hermitian, and positive semidefinite (or definite) matrices, respectively.

11.2   Strategic non-cooperative games: basic solution concepts and algorithms

In this section we introduce non-cooperative strategic form games and provide some basic results dealing with the solution concept of the Nash equilibrium (NE). We do not attempt to cover such topics in encyclopedic depth. We have restricted our exposition only to those results (not necessarily the most general ones in the literature of game theory) that will be used in the forthcoming sections to solve the proposed CR problems and make this chapter self-contained. The literature on the pure Nash-equilibrium problem is enormous; we refer the interested reader to [27–32] as entry points. A more recent survey on current state-of-the-art results on non-cooperative games is [33].

A "non-cooperative strategic form" game models a scenario where all players act independently and simultaneously according to their own self-interests and with no a-priori knowledge of other players' strategies. Stated in mathematical terms, we have the following:

DEFINITION 11.1   *A strategic form game is a triplet* $\mathcal{G} = \left\langle \Omega, (Q_i)_{i \in \Omega}, (u_i)_{i \in \Omega} \right\rangle$, *where:*

- $\Omega = \{1, 2, \ldots, Q\}$ *is the (finite) set of players;*

- Q_i *is a nonempty set of the available (pure) strategies (actions) for player i, also called the admissible strategy set of player i (assumed here to be independent of the other players' strategies);*[1]
- $u_i : Q_1 \times \cdots \times Q_Q \to \mathbb{R}$ *is the payoff (utility) function of player i that depends, in general, on the strategies of all players.*

We denote by $\mathbf{x}_i \in Q_i$ *a feasible strategy profile of player i, by* $\mathbf{x}_{-i} = (\mathbf{x}_j)_{j \neq i}$ *a tuple of strategies of all players except the ith, and by* $Q = Q_1 \times \cdots \times Q_Q$ *the set of feasible-strategy profiles of all players. We use the notation* $Q_{-i} = Q_1 \times Q_{i-1}, Q_{i+1}, \cdots \times Q_Q$ *to define the set of feasible-strategy profiles of all players except the ith. If all the strategy sets* Q_i *are finite, the game is called* finite; *otherwise* infinite.

The non-cooperative paradigm postulates the rationality of players' behaviors: each player i competes against the others by choosing a strategy profile $\mathbf{x}_i \in Q_i$ that maximizes its own payoff function $u_i(\mathbf{x}_i, \mathbf{x}_{-i})$, given the actions $\mathbf{x}_{-i} \in Q_{-i}$ of the other players. A non-cooperative strategic form game can be then represented as a set of *coupled* optimization problems

$$(\mathcal{G}) : \quad \begin{array}{l} \underset{\mathbf{x}_i}{\text{maximize}} \quad u_i(\mathbf{x}_i, \mathbf{x}_{-i}) \\[2mm] \text{subject to} \quad \mathbf{x}_i \in Q_i, \end{array} \qquad \forall i \in \Omega. \tag{11.8}$$

The problem of the ith player in (11.8) is to determine, for each fixed but arbitrary tuple $\mathbf{x}_{-i}$ of the other players' strategies, an optimal strategy $\mathbf{x}_i^\star$ that solves the maximization problem in the variable $\mathbf{x}_i \in Q_i$.

A desirable solution to (11.8) is one in which every (rational) player acts in accordance with its incentives, maximizing its own payoff function. This idea is best captured by the notion of the Nash equilibrium, formally defined next.

DEFINITION 11.2 *Given a strategic form game* $\mathcal{G} = \langle \Omega, (Q_i)_{i \in \Omega}, (u_i)_{i \in \Omega} \rangle$, *an action profile* $\mathbf{x}^\star \in Q$ *is a pure-strategy Nash equilibrium of* $\mathcal{G}$ *if the following condition holds for all* $i \in \Omega$:

$$u_i(\mathbf{x}_i^\star, \mathbf{x}_{-i}^\star) \geq u_i(\mathbf{x}_i, \mathbf{x}_{-i}^\star), \quad \forall \mathbf{x}_i \in Q_i. \tag{11.9}$$

In words, a Nash equilibrium is a (self-enforcing) strategy profile with the property that no *single* player can unilaterally benefit from a deviation from it, given that all the other players act according to it. It is useful to restate the definition of NE in terms of a fixed-point solution to the best-response multifunction (i.e., "point-to-set map").

DEFINITION 11.3 *Let* $\mathcal{G} = \langle \Omega, (Q_i)_{i \in \Omega}, (u_i)_{i \in \Omega} \rangle$ *be a strategic form game. For any given* $\mathbf{x}_{-i} \in Q_{-i}$, *define the best-response multifunction* $\mathcal{B}_i(\mathbf{x}_{-i})$ *of player i as*

$$\mathcal{B}_i(\mathbf{x}_{-i}) \triangleq \{\mathbf{x}_i \in Q_i \mid u_i(\mathbf{x}_i, \mathbf{x}_{-i}) \geq u_i(\mathbf{y}_i, \mathbf{x}_{-i}), \ \forall \mathbf{y}_i \in Q_i\}, \tag{11.10}$$

---

[1] The focus on more general games where the strategy set of the players may depend on the other players' actions (usually termed as the generalized Nash-equilibrium problem) goes beyond the scope of this section. We refer the interested reader to [33] and references therein.

that is, the set of the optimal solutions to the ith optimization problem in (11.8), given $\mathbf{x}_{-i} \in \mathcal{Q}_{-i}$ *(assuming that the maximum in (11.10) exists). We also introduce the multi-function mapping* $\mathcal{B} : \mathcal{Q} \rightrightarrows \mathcal{Q}$ *defined as* $\mathcal{B}(\mathbf{x}) : \mathcal{Q} \ni \mathbf{x} \rightrightarrows \mathcal{B}_1(\mathbf{x}_{-1}) \times \mathcal{B}_2(\mathbf{x}_{-2}) \times \cdots \times \mathcal{B}_Q(\mathbf{x}_{-Q})$. *A strategy profile* $\mathbf{x}^\star \in \mathcal{Q}$ *is a pure-strategy NE of* $\mathcal{G}$ *if, and only if*

$$\mathbf{x}^\star \in \mathcal{B}(\mathbf{x}^\star). \tag{11.11}$$

If $\mathcal{B}(\mathbf{x})$ *is a single-valued function (denoted, in such a case, as* $\mathsf{B}(\mathbf{x})$*), then* $\mathbf{x}^\star \in \mathcal{Q}$ *is a pure, strategy NE if, and only if* $\mathbf{x}^\star = \mathsf{B}(\mathbf{x}^\star)$.

This alternative formulation of the equilibrium solution may be useful to address some essential issues of the equilibrium problems, such as the existence and uniqueness of solutions, stability of equilibria, design of effective algorithms for finding equilibrium solutions, thus paving the way to the application of the fixed-point machinery. In fact, in general, the uniqueness or even the existence of a pure-strategy Nash equilibrium is not guaranteed; neither is convergence to an equilibrium (of best-response based algorithms) when one exists (some basic existence and uniqueness results in the form useful for our purposes will be discussed in Section 11.2.1). Sometimes, however, the structure of a game is such that one is able to establish one or more of these desirable properties, as for example happens in potential games [34] or supermodular games [35], which have recently received some attention in the signal-processing and communication communities as a useful tool to solve various power-control problems in wireless communications [36–38].

Finally, it is important to remark that, even when the NE is unique, it need not be "Pareto efficient".

DEFINITION 11.4 *Given a strategic form game* $\mathcal{G} = \langle \Omega, (\mathcal{Q}_i)_{i \in \Omega}, (u_i)_{i \in \Omega} \rangle$, *and two action profiles* $\mathbf{x}^{(1)}, \mathbf{x}^{(2)} \in \mathcal{Q}$, $\mathbf{x}^{(1)}$ *is said to be Pareto-dominant on* $\mathbf{x}^{(2)}$ *if* $u_i(\mathbf{x}^{(1)}) \geq u_i(\mathbf{x}^{(2)})$ *for all* $i \in \Omega$, *and* $u_j(\mathbf{x}^{(1)}) > u_j(\mathbf{x}^{(2)})$ *for at least one* $j \in \Omega$. *A strategy profile* $\mathbf{x} \in \mathcal{Q}$ *is Pareto efficient (optimal) if there exists no other feasible strategy that dominates* $\mathbf{x}$.

This means that there might exist proper coalitions among the players yielding an outcome of the game with the property that there is always (at least) one player who cannot profit by deviating by that action profile. In other words, an NE may be vulnerable to deviations by coalitions of players, even if it is not vulnerable to unilateral deviation by a single player. However, Pareto optimality in general comes at the price of a centralized optimization, which requires the full knowledge of the strategy sets and the payoff functions of all players. Such a centralized approach is not applicable in many practical applications in signal processing and communications, for example, in emerging wireless networks, such as sensor networks, ad-hoc networks, cognitive radio systems, and pervasive computing systems. The NE solutions, instead, are more suitable to be computed using a decentralized approach that requires no exchange of information among the players. Different refinements of the NE concept have also been proposed in the literature to overcome some shortcomings of the NE solution (see, e.g., [29, 39]).

The definition of NE as given in Definition 11.2 covers only pure strategies. One can restate the NE concept to contain mixed strategies, in other words, the possibility of choosing a randomization over a set of pure strategies. A mixed-strategy NE of a strategic game is then defined as an NE of its mixed extension (see, e.g., [27, 40] for details). An interesting result dealing with Nash equilibria in mixed-strategy is that every *finite* strategic game has a mixed-strategy NE [41], which in general does not hold for pure-strategies. In this chapter, we focus only on pure strategy Nash equilibria of non-cooperative strategic form games with infinite strategy sets.

11.2.1   Existence and uniqueness of the NE

Several different approaches have been proposed in the literature to study properties of the Nash solutions, such as existence, (local/global) uniqueness, and to devise numerical algorithms to solve the NE problem. The three most frequent methods are: (a) interpreting the Nash equilibria as fixed-point solutions, (b) reducing the NE problem to a variational-inequality problem, and (c) transforming the equilibrium problem into an optimization problem. Each of these methods leads to alternative conditions and algorithms. We focus next only on the former approach and refer the interested reader to [32,33,42] and [27,43] as examples of the application of the other techniques.

Existence of a Nash solution

The study of the existence of equilibria under weaker and weaker assumptions has been investigated extensively in the literature (see, e.g., [41, 44–47]). A good overview of the relevant literature is [33]. For the purpose of this chapter, it is enough to recall an existence result that is one of the simplest of the genre, based on the interpretation of the NE as a fixed point of the best-response multifunction (cf. Definition 11.3) and the existence result from the "Kakutani fixed-point theorem".

THEOREM 11.1 Kakutani's fixed-point theorem   *Given $\mathcal{X} \subseteq \mathbb{R}^n$, let $\mathcal{S}(\mathbf{x}) : \mathcal{X} \ni \mathbf{x} \rightrightarrows \mathcal{S}(\mathbf{x}) \subseteq \mathcal{X}$ be a multifunction. Suppose that the following hold:*

(a) $\mathcal{X}$ is a nonempty, compact, and convex set;
(b) $\mathcal{S}(\mathbf{x})$ is a convex-valued correspondence (i.e., $\mathcal{S}(\mathbf{x})$ is a convex set for all $\mathbf{x} \in \mathcal{X}$) and has a closed graph (i.e., if $\{\mathbf{x}^{(n)}, \mathbf{y}^{(n)}\} \to \{\mathbf{x}, \mathbf{y}\}$ with $\mathbf{y}^{(n)} \in \mathcal{S}(\mathbf{x}^{(n)})$, then $\mathbf{y} \in \mathcal{S}(\mathbf{x})$).

Then, there exists a fixed point of $\mathcal{S}(\mathbf{x})$.

Given $\mathcal{G} = \langle \Omega, (\mathcal{Q}_i)_{i \in \Omega}, (u_i)_{i \in \Omega} \rangle$ with best-response $\mathcal{B}(\mathbf{x})$, it follows from Definition 11.3 and Theorem 11.1 that conditions (a) and (b) applied to $\mathcal{B}(\mathbf{x})$ are sufficient to guarantee the existence of an NE. To make condition (b) less abstract, we use Theorem 11.1 in a simplified form, which provides a set of sufficient conditions for assumption (b) that represent classical existence results in the game-theory literature [44–47].

THEOREM 11.2 Existence of an NE   *Consider a strategic form game $\mathcal{G} = \langle \Omega, (\mathcal{Q}_i)_{i \in \Omega}, (u_i)_{i \in \Omega} \rangle$, where Ω is a finite set. Suppose that*

(a) Each $\mathcal{Q}_i$ is a non empty, compact, and convex subset of a finite-dimensional Euclidean space;
(b) One of the two following conditions holds:

(1) Each payoff function $u_i(\mathbf{x}_i, \mathbf{x}_{-i})$ is continuous on $\mathcal{Q}$, and, for any given $\mathbf{x}_{-i} \in \mathcal{Q}_{-i}$, it is quasi-concave on $\mathcal{Q}_i$;

(2) Each payoff function $u_i(\mathbf{x}_i, \mathbf{x}_{-i})$ is continuous on $\mathcal{Q}$, and, for any given $\mathbf{x}_{-i} \in \mathcal{Q}_{-i}$, the following optimization problem

$$\max_{\mathbf{x}_i \in \mathcal{Q}_i} u_i(\mathbf{x}_i, \mathbf{x}_{-i}) \tag{11.12}$$

admits a unique (globally) optimal solution.

Then, game $\mathcal{G}$ admits a pure-strategy NE.

The assumptions in Theorem 11.2 are only sufficient for the existence of a fixed point. However, this does not mean that some of them can be relaxed. For example, the convexity assumption in the existence condition (Theorem 11.1(a) and Theorem 11.2(a)) cannot, in general, be removed, as the simple one-dimensional example $f(x) = -x$ and $\mathcal{X} = \{-c, c\}$, with $c \in \mathbb{R}$, shows. Furthermore, a pure-strategy NE may fail to exist if the quasi-concavity assumption (Theorem 11.2(b.1)) is relaxed, as shown in the following example. Consider a two-player game, where the payers pick points $\mathbf{x}_1$ and $\mathbf{x}_2$ on the unit circle, and the payoff functions of the two players are $u_1(\mathbf{x}_1, \mathbf{x}_2) = \|\mathbf{x}_1 - \mathbf{x}_2\|$ and $u_2(\mathbf{x}_1, \mathbf{x}_2) = -\|\mathbf{x}_1 - \mathbf{x}_2\|$, where $\|\cdot\|$ denotes the Euclidean norm. In this game there is no pure-strategy NE. In fact, if both players pick the same location, player 1 has an incentive to deviate; whereas if they pick different locations, player 2 has an incentive to deviate.

The relaxation of the assumptions in Theorem 11.2 has been the subject of a fairly intense study. Relaxations of the (a) continuity assumptions, (b) compactness assumptions, and (c) quasi-concavity assumption have all been considered in the literature. The relevant literature is discussed in detail in [33]. More recent advanced results can be found in [42].

Uniqueness of a Nash solution

The study of uniqueness of a solution for the Nash problem is more involved and available results are scarce. Some classical works on the subject are [46, 48, 49] and more recently [43, 50, 51], where different uniqueness conditions have been derived, most of them valid for games having special structure. Since the games considered in this chapter satisfy Theorem 11.2(b.2), in the following we focus on this special class of games and provide some basic results, based on the uniqueness of fixed points of single-valued functions. A simple uniqueness result is given in the following (see, e.g., [52, 53]).

THEOREM 11.3 Uniqueness of the NE   *Let* $\mathsf{B}(\mathbf{x}) : \mathcal{X} \ni \mathbf{x} \to \mathsf{B}(\mathbf{x}) \in \mathcal{X}$ *be a function, mapping* $\mathcal{X} \subseteq \mathbb{R}^n$ *into itself. Suppose that* B *is a contraction in some vector norm* $\|\cdot\|$, *with modulus* $\alpha \in [0, 1)$:

$$\left\| \mathsf{B}(\mathbf{x}^{(1)}) - \mathsf{B}(\mathbf{x}^{(2)}) \right\| \le \alpha \left\| \mathbf{x}^{(1)} - \mathbf{x}^{(2)} \right\|, \quad \forall \mathbf{x}^{(1)}, \mathbf{x}^{(2)} \in \mathcal{X}. \tag{11.13}$$

Then, there exists at most one fixed point of B. *If, in addition,* $\mathcal{X}$ *is closed, then there exists a unique fixed point.*   □

Alternative sufficient conditions still requiring some properties on the best-response function B can be obtained, by observing that the fixed point of B is unique if the function $\mathsf{T}(\mathbf{x}) \triangleq \mathbf{x} - \mathsf{B}(\mathbf{x})$ is one-to-one. Invoking results from mathematical analysis, many conditions can be obtained guaranteeing that T is one-to-one. For example, assuming that T is continuously differentiable and denoting by $\mathbf{J}(\mathbf{x})$ the Jacobian matrix of T at $\mathbf{x}$, some frequently applied conditions are the following: (a) all leading principal minors of $\mathbf{J}(\mathbf{x})$ are positive (i.e., $\mathbf{J}(\mathbf{x})$ is a P-matrix [54]); (b) all leading principal minors of $\mathbf{J}(\mathbf{x})$ are negative (i.e., $\mathbf{J}(\mathbf{x})$ is an N-matrix [54]); (c) matrix $\mathbf{J}(\mathbf{x}) + \mathbf{J}(\mathbf{x})^T$ is positive (or negative) semidefinite, and between any pair of points $\mathbf{x}^{(1)} \neq \mathbf{x}^{(2)}$ there is a point $\mathbf{x}^{(0)}$ such that $\mathbf{J}(\mathbf{x}^{(0)}) + \mathbf{J}(\mathbf{x}^{(0)})^T$ is positive (or negative) definite [53].

11.2.2   Convergence to a fixed point

We focus on asynchronous-iterative algorithms, since they are particularly suitable for CR applications. More specifically, we consider a general fixed-point problem – the NE problem in (11.11) – and describe a fairly general class of totally asynchronous algorithms following [52], along with a convergence theorem of broad applicability. According to the "totally asynchronous scheme", all the players of game $\mathcal{G} = \left(\Omega, (\mathcal{Q}_i)_{i \in \Omega}, (u_i)_{i \in \Omega}\right)$ maximize their own payoff function in a *totally asynchronous* way, meaning that some players are allowed to update their strategy more frequently than the others, and they might perform their updates using *outdated* information about the strategy profile used by the others. To provide a formal description of the algorithm, we need to introduce some preliminary definitions, as given next.

We assume w.l.o.g. that the set of times at which one or more players update their strategies is the discrete set $\mathcal{T} = \mathbb{N}_+ = \{0, 1, 2, \ldots\}$. Let $\mathbf{x}_i^{(n)}$ denote the strategy profile of user i at the nth iteration, and let $\mathcal{T}_i \subseteq \mathcal{T}$ be the set of times at which player i updates its own strategy $\mathbf{x}_i^{(n)}$ (thus, implying that, at time $n \notin \mathcal{T}_i$, $\mathbf{x}_i^{(n)}$ is left unchanged). Let $\tau_j^i(n)$ denote the most recent time at which the strategy profile from player j is perceived by player i at the nth iteration (observe that $\tau_j^i(n)$ satisfies $0 \leq \tau_j^i(n) \leq n$). Hence, if player i updates its strategy at the nth iteration, then it maximizes its payoff function using the following outdated strategy profile of the other players:

$$\mathbf{x}_{-i}^{(\tau^i(n))} \triangleq \left(\mathbf{x}_1^{(\tau_1^i(n))}, \ldots, \mathbf{x}_{i-1}^{(\tau_{i-1}^i(n))}, \mathbf{x}_{i+1}^{(\tau_{i+1}^i(n))}, \ldots, \mathbf{x}_Q^{(\tau_Q^i(n))}\right). \tag{11.14}$$

The overall system is said to be totally asynchronous if the following assumptions are satisfied for each i: (A1) $0 \leq \tau_j^i(n) \leq n$; (A2) $\lim_{k \to \infty} \tau_j^i(n_k) = +\infty$; and (A3) $|\mathcal{T}_i| = \infty$; where $\{n_k\}$ is a sequence of elements in $\mathcal{T}_i$ that tends to infinity. Assumptions (A1)–(A3) are standard in asynchronous-convergence theory [52], and they are fulfilled in any practical implementation. In fact, (A1) simply indicates that, at any given iteration n, each player i can use only the strategy profile $\mathbf{x}_{-i}^{(\tau^i(n))}$ adopted by the other players in the previous iterations (to preserve causality). Assumption (A2) states that, for any given iteration index n_k, the values of the components of $\mathbf{x}_{-i}^{(\tau^i(n))}$ in (11.14) generated prior to n_k, are not used in the updates of $\mathbf{x}_i^{(n)}$, when n becomes sufficiently larger than n_k; which guarantees that old information is eventually purged from the system. Finally, assumption (A3) indicates that no player fails to update its own strategy as time n goes on.

Using the above definitions, the totally asynchronous algorithm based on the multifunction $\mathcal{B}(\mathbf{x})$ is described in Algorithm 1. Observe that Algorithm 1 contains as special cases a plethora of algorithms, each one obtained by a possible choice of the scheduling of the users in the updating procedure (i.e., the parameters $\{\tau_r^q(n)\}$ and $\{\mathcal{T}_q\}$). Examples are the "sequential" (Gauss–Seidel scheme) and the "simultaneous" (Jacobi scheme) updates, where the players update their own strategies *sequentially* and *simultaneously*, respectively. Moreover, variations of such a totally asynchronous scheme, for example, including constraints on the maximum tolerable delay in the updating and on the use of the outdated information (which leads to the so-called partially asynchronous algorithms), can also be considered [52]. A fairly general convergence theorem for Algorithm 1 is given in Theorem 11.4, whose proof is based on [52].

Algorithm 1 Totally asynchronous algorithm

1 : Set $n = 0$ and choose any feasible $\mathbf{x}_i^{(0)}$, $\forall i \in \Omega$;

2 : repeat

$$
3: \qquad \mathbf{x}_i^{(n+1)} = \begin{cases} \mathbf{x}_i^\star \in \mathcal{B}_i\left(\mathbf{x}_{-i}^{(\tau^i(n))}\right), & \text{if } n \in \mathcal{T}_i, \\ \mathbf{x}_i^{(n)}, & \text{otherwise,} \end{cases} \qquad \forall i \in \Omega; \qquad (11.15)
$$

4 : until the prescribed convergence criterion is satisfied

THEOREM 11.4 Asynchronous-convergence theorem   *Given Algorithm 1 based on a multifunction $\mathcal{B}(\mathbf{x}) : \mathcal{X} \ni \mathbf{x} \rightrightarrows \mathcal{B}_1(\mathbf{x}_{-1}) \times \mathcal{B}_2(\mathbf{x}_{-2}) \times \cdots \times \mathcal{B}_Q(\mathbf{x}_{-Q}) \subseteq \mathcal{X}$, suppose that assumptions (A1)–(A3) hold true and that there exists a sequence of nonempty sets $\{\mathcal{X}(n)\}$ with*

$$
\ldots \subset \mathcal{X}(n+1) \subset \mathcal{X}(n) \subset \ldots \subset \mathcal{X}, \qquad (11.16)
$$

satisfying the next two conditions:

(a) (Synchronous-convergence condition) *For all $\mathbf{x} \in \mathcal{X}(n)$ and n,*

$$
\mathcal{B}(\mathbf{x}) \subseteq \mathcal{X}(n+1). \qquad (11.17)
$$

Furthermore, if $\{\mathbf{x}^{(n)}\}$ is a sequence such that $\mathbf{x}^{(n)} \in \mathcal{X}(n)$, for every n, then every limit point of $\{\mathbf{x}^{(n)}\}$ is a fixed point of $\mathcal{B}(\cdot)$.

(b) (Box condition) *For every n, there exist sets $\mathcal{X}_i(n) \subset \mathcal{X}_i$ such that $\mathcal{X}(n)$ can be written as a Cartesian product*

$$
\mathcal{X}(n) = \mathcal{X}_1(n) \times \ldots \times \mathcal{X}_Q(n). \qquad (11.18)
$$

Then, every limit point of $\{\mathbf{x}^{(n)}\}$ generated by Algorithm 1 and starting from $\mathbf{x}^{(0)} \in \mathcal{X}(0)$ is a fixed point of $\mathcal{B}(\cdot)$.

The challenge in applying the asynchronous-convergence theorem is to identify a suitable sequence of sets $\{\mathcal{X}(n)\}$. This is reminiscent of the process of identifying a Lyapunov function in the stability analysis of nonlinear dynamic systems (the sets $\mathcal{X}(k)$ play conceptually the role of the level set of a Lyapunov function). For the purpose of this chapter, it is enough to restrict our focus to single-value, best-response functions and consider sufficient conditions for (11.16)–(11.18) in Theorem 11.4, as detailed next.

Given the game $\mathcal{G} = \langle \Omega, (\mathcal{Q}_i)_{i \in \Omega}, (u_i)_{i \in \Omega} \rangle$ with the best-response function $\mathsf{B}(\mathbf{x}) = (\mathsf{B}_i(\mathbf{x}_{-i}))_{i \in \Omega}$, where each $\mathsf{B}_i(\mathbf{x}_{-i}) : \mathcal{Q}_{-i} \ni \mathbf{x}_{-i} \to \mathsf{B}_i(\mathbf{x}_{-i}) \in \mathcal{Q}_i$, let us introduce the following block-maximum vector norm $\|\cdot\|_{\mathrm{block}}$ on $\mathbb{R}^n$, defined as

$$\|\mathsf{B}\|_{\mathrm{block}} = \max_{i \in \Omega} \|\mathsf{B}_i\|_i, \tag{11.19}$$

where $\|\cdot\|_i$ is any vector norm on $\mathbb{R}^{n_i}$. Suppose that each $\mathcal{Q}_i$ is a closed subset of $\mathbb{R}^{n_i}$ and that $\mathsf{B}(\mathbf{x})$ is a contraction with respect to the block-maximum norm, in other words,

$$\left\| \mathsf{B}(\mathbf{x}^{(1)}) - \mathsf{B}(\mathbf{x}^{(2)}) \right\|_{\mathrm{block}} \leq \alpha \left\| \mathbf{x}^{(1)} - \mathbf{x}^{(2)} \right\|_{\mathrm{block}}, \quad \forall \mathbf{x}^{(1)}, \mathbf{x}^{(2)} \in \mathcal{Q}, \tag{11.20}$$

with $\alpha \in [0, 1)$. Then, there exists a unique fixed point $\mathbf{x}^\star$ of $\mathsf{B}(\mathbf{x})$ (cf. Theorem 11.3) – the NE of $\mathcal{G}$ – and the asynchronous-convergence theorem holds. In fact, it is not difficult to show that, under (11.20), conditions (11.16)–(11.18) in Theorem 11.4 are satisfied with the following choice for the sets $\mathcal{X}_i(k)$:

$$\mathcal{X}_i(k) = \left\{ \mathbf{x} \in \mathcal{Q} \mid \left\| \mathbf{x}_i - \mathbf{x}_i^\star \right\|_2 \leq \alpha^k w_i \left\| \mathbf{x} - \mathbf{x}^{(0)} \right\|_{\mathrm{block}} \right\} \subset \mathcal{Q}_i, \qquad k \geq 1, \tag{11.21}$$

where $\mathbf{x}^{(0)} \in \mathcal{Q}$ is the initial point of the algorithm. Note that, because of the uniqueness of the fixed point of $\mathsf{B}(\mathbf{x})$ under (11.20), the statement on convergence in Theorem 11.4 can be made stronger: for any initial vector $\mathbf{x}^{(0)} \in \mathcal{Q}$, the sequence $\{\mathbf{x}^{(n)}\}$ generated by Algorithm 1 converges to the fixed point of $\mathsf{B}(\mathbf{x})$.

11.3   Opportunistic communications over unlicensed bands

We start considering the CR system in (11.1), under the transmit power constraints (11.2) only. This models transmissions over unlicensed bands, where multiple systems coexist, thus interfering with each other, and there are no constraints on the maximum amount of interference that each transmitter can generate. The results obtained in this case provide the building blocks that are instrumental in studying the equilibrium problems including interference constraints, as described in the next sections.

The rate-maximization game among the secondary users in the presence of the power constraints (11.2) is formally defined as

$$(\mathcal{G}_{\mathrm{pow}}) : \quad \begin{array}{c} \underset{\mathbf{Q}_q}{\text{maximize}} \quad R_q(\mathbf{Q}_q, \mathbf{Q}_{-q}) \\ \text{subject to} \quad \mathbf{Q}_q \in \mathcal{Q}_q, \end{array} \quad \forall q \in \Omega, \qquad (11.22)$$

where $\Omega \triangleq \{1, 2, \cdots, Q\}$ is the set of players (the secondary users), $R_q(\mathbf{Q}_q, \mathbf{Q}_{-q})$ is the payoff function of player q, defined in (11.4), and $\mathcal{Q}_q$ is the set of admissible strategies (the covariance matrices) of player q, defined as

$$\mathcal{Q}_q \triangleq \left\{ \mathbf{Q} \in \mathbb{S}_+^{n_{T_q}} \,|\, \mathrm{Tr}\{\mathbf{Q}\} = P_q \right\}. \qquad (11.23)$$

Observe that there is no loss of generality in considering in (11.23) the power constraint with equality rather than inequality as stated in (11.2), since at the optimum to each problem in (11.22), the power constraint must be satisfied with equality. To write the Nash equilibria of game $\mathcal{G}_{\mathrm{pow}}$ in a convenient form, we introduce the MIMO waterfilling operator. Given $q \in \Omega$, $n_q \in \{1, 2, \ldots, n_{T_q}\}$, and some $\mathbf{X} \in \mathbb{S}_+^{n_q}$, the MIMO waterfilling function $\mathsf{WF}_q : \mathbb{S}_+^{n_q} \ni \mathbf{X} \to \mathbb{S}_+^{n_q}$ is defined as

$$\mathsf{WF}_q(\mathbf{X}) \triangleq \mathbf{U}_X \left(\mu_{q,X} \mathbf{I}_{r_X} - \mathbf{D}_X^{-1} \right)^+ \mathbf{U}_X^H, \qquad (11.24)$$

where $\mathbf{U}_X \in \mathbb{C}^{n_q \times r_X}$ and $\mathbf{D}_X > \mathbf{0}$ are the (semi-)unitary matrix of the eigenvectors and the diagonal matrix of the $r_X \triangleq \mathrm{rank}(\mathbf{X}) \le n_q$ (positive) eigenvalues of $\mathbf{X}$, respectively, and $\mu_{q,X} > 0$ is the water-level chosen to satisfy $\mathrm{Tr}\left\{ (\mu_{q,X} \mathbf{I}_{r_X} - \mathbf{D}_X^{-1})^+ \right\} = P_q$. Using the above definitions, the solution to the single-user optimization problem in (11.22) – the best-response of player q for any given $\mathbf{Q}_{-q} \succeq \mathbf{0}$ – is the well-known waterfilling solution (e.g., [11])

$$\mathbf{Q}_q^\star = \mathsf{WF}_q(\mathbf{H}_{qq}^H \mathbf{R}_{-q}^{-1}(\mathbf{Q}_{-q}) \mathbf{H}_{qq}), \qquad (11.25)$$

implying that the Nash equilibria of game $\mathcal{G}_{\mathrm{pow}}$ are the solutions of the following fixed-point matrix equation (cf. Definition 11.3):

$$\mathbf{Q}_q^\star = \mathsf{WF}_q(\mathbf{H}_{qq}^H \mathbf{R}_{-q}^{-1}(\mathbf{Q}_{-q}^\star) \mathbf{H}_{qq}) \,, \quad \forall q \in \Omega. \qquad (11.26)$$

REMARK 11.1 *On the Nash equilibria.* The main difficulty in the analysis of the solutions to (11.26) comes from the fact that the optimal eigenvector matrix $\mathbf{U}_q^\star = \mathbf{U}_q(\mathbf{Q}_{-q}^\star)$ of each user q (see (11.24)) depends, in general, on the strategies $\mathbf{Q}_{-q}^\star$ of all the other users, through a very complicated implicit relationship – the eigendecomposition of the equivalent channel matrix $\mathbf{H}_{qq}^H \mathbf{R}_{-q}^{-1}(\mathbf{Q}_q^\star) \mathbf{H}_{qq}$. To overcome this issue, we provide next an equivalent expression of the waterfilling solution enabling us to express the Nash equilibria in (11.26) as a fixed point of a more tractable mapping. This alternative expression is based on the recent interpretation of the MIMO waterfilling mapping as a proper projector operator [7, 10, 55]. Based on this result, we can then derive sufficient conditions for the uniqueness of the NE and convergence of asynchronous-distributed algorithms, as detailed in Sections 11.3.4 and 11.3.5, respectively.

11.3.1   Properties of the multiuser waterfilling mapping

In this section we derive some interesting properties of the multiuser MIMO waterfilling mapping. These results will be instrumental in studying the games we propose in this chapter. The main result of the section is a contraction theorem for the multiuser MIMO waterfilling mapping, valid for arbitrary channel matrices. Results in this section are based on recent works [10, 24].

For the sake of notation, through the whole section we refer to the best-response $\mathsf{WF}_q(\mathbf{H}_{qq}^H \mathbf{R}_{-q}^{-1}(\mathbf{Q}_{-q})\mathbf{H}_{qq})$ of each user q in (11.25) as $\mathsf{WF}_q(\mathbf{Q}_{-q})$, making explicit only the dependence on the strategy profile $\mathbf{Q}_{-q}$ of the other players.

11.3.2   MIMO waterfilling as a projector

The interpretation of the MIMO waterfilling solution as a matrix projection is based on the following result:

LEMMA 11.1   *Let* $\mathbb{S}^n \ni \mathbf{X}_0 = \mathbf{U}_0 \mathbf{D}_0 \mathbf{U}_0^H$, *where* $\mathbf{U}_0 \in \mathbb{C}^{n \times n}$ *is unitary and* $\mathbf{D}_0 = \mathrm{Diag}(\{d_{0,k}\}_{k=1}^n)$, *and let* $\mathcal{Q}$ *be the convex set defined as*

$$\mathcal{Q} \triangleq \left\{ \mathbf{Q} \in \mathbb{S}_+^n \mid \mathrm{Tr}\{\mathbf{Q}\} = P_T \right\}. \tag{11.27}$$

The matrix projection $[\mathbf{X}_0]_\mathcal{Q}$ *of* $\mathbf{X}_0$ *onto* $\mathcal{Q}$ *with respect to the Frobenius norm, defined as*

$$[\mathbf{X}_0]_\mathcal{Q} = \operatorname*{argmin}_{\mathbf{X} \in \mathcal{Q}} \|\mathbf{X} - \mathbf{X}_0\|_F^2 \tag{11.28}$$

takes the following form:

$$[\mathbf{X}_0]_\mathcal{Q} = \mathbf{U}_0 \left(\mathbf{D}_0 - \mu_0 \mathbf{I} \right)^+ \mathbf{U}_0^H, \tag{11.29}$$

where μ_0 *satisfies the constraint* $\mathrm{Tr}\{(\mathbf{D}_0 - \mu_0 \mathbf{I})^+\} = P_T$.

Proof   Using $\mathbf{X}_0 = \mathbf{U}_0 \mathbf{D}_0 \mathbf{U}_0^H$, the objective function in (11.28) becomes

$$\|\mathbf{X} - \mathbf{X}_0\|_F^2 = \left\|\widetilde{\mathbf{X}} - \mathbf{D}_0\right\|_F^2, \tag{11.30}$$

where $\widetilde{\mathbf{X}}$ is defined as $\widetilde{\mathbf{X}} \triangleq \mathbf{U}_0^H \mathbf{X} \mathbf{U}_0$ and we used the unitary invariance of the Frobenius norm [26]. Since

$$\left\|\widetilde{\mathbf{X}} - \mathbf{D}_0\right\|_F^2 \geq \left\|\mathrm{Diag}(\widetilde{\mathbf{X}}) - \mathbf{D}_0\right\|_F^2, \tag{11.31}$$

with equality if and only if $\widetilde{\mathbf{X}}$ is diagonal, and the power constraint $\mathrm{Tr}\{\mathbf{X}\} = \mathrm{Tr}\{\widetilde{\mathbf{X}}\} = P_T$ depends only on the diagonal elements of $\widetilde{\mathbf{X}}$, it follows that the optimal $\widetilde{\mathbf{X}}$ must be diagonal, that is, $\widetilde{\mathbf{X}} = \mathrm{Diag}(\{d_k\}_{k=1}^n)$. The matrix-valued problem in (11.28) reduces then to the following vector (strictly) convex-optimization problem

$$
\begin{aligned}
\underset{\mathbf{d} \geq \mathbf{0}}{\text{minimize}} \quad & \sum_{k=1}^{n} \left(d_k - d_{0,k}\right)^2 \\
\text{subject to} \quad & \sum_{k=1}^{n} d_k = P_T,
\end{aligned}
\tag{11.32}
$$

whose unique solution $\{d_k^\star\}$ is given by $d_k^\star = (d_{0,k} - \mu_0)^+$, with $k = 1, \cdots, n$, where μ_0 is chosen to satisfy $\sum_{k=1}^{n}(d_{0,k} - \mu_0)^+ = P_T$. ∎

Using the above result we can obtain the alternative expression of the waterfilling solution $\mathrm{WF}_q(\mathbf{Q})$ in (11.25) as given next.

LEMMA 11.2 MIMO waterfilling as a projector *The MIMO waterfilling operator* $\mathrm{WF}_q(\mathbf{Q}_{-q})$ *in (11.25) can be equivalently written as*

$$
\mathrm{WF}_q(\mathbf{Q}_{-q}) = \left[-\left(\left(\mathbf{H}_{qq}^H \mathbf{R}_{-q}^{-1}(\mathbf{Q}_{-q}) \mathbf{H}_{qq} \right)^\sharp + c_q \mathbf{P}_{\mathcal{N}(\mathbf{H}_{qq})} \right) \right]_{\mathcal{Q}_q},
\tag{11.33}
$$

where c_q is a positive constant that can be chosen independent of $\mathbf{Q}_{-q}$ (cf. [24]), and $\mathcal{Q}_q$ is defined in (11.23).

Proof Given $q \in \Omega$ and $\mathbf{Q}_{-q} \in \mathcal{Q}_{-q}$, using the eigendecomposition $\mathbf{H}_{qq}^H \mathbf{R}_{-q}^{-1}(\mathbf{Q}_{-q}) \mathbf{H}_{qq} = \mathbf{U}_{q,1} \mathbf{D}_{q,1} \mathbf{U}_{q,1}^H$, where $\mathbf{U}_{q,1} = \mathbf{U}_{q,1}(\mathbf{Q}_{-q}) \in \mathbb{C}^{n_{Tq} \times r_{\mathbf{H}_{qq}}}$ is semi-unitary and $\mathbf{D}_{q,1} = \mathbf{D}_{q,1}(\mathbf{Q}_{-q}) = \mathrm{diag}(\{\lambda_i\}_{i=1}^{r_{\mathbf{H}_{qq}}}) > \mathbf{0}$ (we omit in the following the dependence of $\mathbf{Q}_{-q}$ for the sake of notation), and introducing the unitary matrix $\mathbf{U}_q \triangleq (\mathbf{U}_{q,1}, \mathbf{U}_{q,2}) \in \mathbb{C}^{n_{Tq} \times n_{Tq}}$ (note that $\mathcal{R}(\mathbf{U}_{q,2}) = \mathcal{N}(\mathbf{H}_{qq})$), we have, for any given $c_q \in \mathbb{R}$,

$$
\left(\mathbf{H}_{qq}^H \mathbf{R}_{-q}^{-1} \mathbf{H}_{qq} \right)^\sharp + c_q \mathbf{P}_{\mathcal{N}(\mathbf{H}_{qq})} = \mathbf{U}_q \begin{pmatrix} \mathbf{D}_{q,1}^{-1} & \mathbf{0} \\ \mathbf{0} & c_q \mathbf{I}_{n_{Tq} - r_{\mathbf{H}_{qq}}} \end{pmatrix} \mathbf{U}_q^H \triangleq \mathbf{U}_q \widetilde{\mathbf{D}}_q^{-1} \mathbf{U}_q^H,
\tag{11.34}
$$

where $\widetilde{\mathbf{D}}_q^{-1} \triangleq \mathrm{bdiag}(\mathbf{D}_q^{-1}, c_q \mathbf{I}_{n_{Tq} - r_{\mathbf{H}_{qq}}})$. It follows from Lemma 11.1 that, for any given $c_q \in \mathbb{R}_{++}$,

$$
\left[-\left(\left(\mathbf{H}_{qq}^H \mathbf{R}_{-q}^{-1} \mathbf{H}_{qq} \right)^\sharp + c_q \mathbf{P}_{\mathcal{N}(\mathbf{H}_{qq})} \right) \right]_{\mathcal{Q}_q} = \mathbf{U}_q \left(\mu_q \mathbf{I}_{n_{Tq}} - \widetilde{\mathbf{D}}_q^{-1} \right)^+ \mathbf{U}_q^H,
\tag{11.35}
$$

where μ_q is chosen to satisfy the constraint $\mathrm{Tr}((\mu_q \mathbf{I}_{n_{Tq}} - \widetilde{\mathbf{D}}_q^{-1})^+) = P_q$. Since each $P_q < \infty$, there exists a (sufficiently large) constant $0 < c_q < \infty$, such that $(\mu_q - c_q)^+ = 0$, and thus the right-hand side (RHS) of (11.35) becomes

$$\left[-\left(\left(\mathbf{H}_{qq}^H\mathbf{R}_{-q}^{-1}\mathbf{H}_{qq}\right)^{\sharp} + c_q\mathbf{P}_{\mathcal{N}(\mathbf{H}_{qq})}\right)\right]_{\mathcal{Q}_q} = \mathbf{U}_{q,1}\left(\mu_q\mathbf{I}_{r_{H_{qq}}} - \mathbf{D}_{q,1}^{-1}\right)^+\mathbf{U}_{q,1}^H, \quad (11.36)$$

which coincides with the desired solution in (11.25). ∎

Observe that, for each $q \in \Omega$, $\mathbf{P}_{\mathcal{N}(\mathbf{H}_{qq})}$ in (11.33) depends only on the channel matrix $\mathbf{H}_{qq}$ (through the right singular vectors of $\mathbf{H}_{qq}$ corresponding to the zero singular values) and not on the strategies of the other users, since $\mathbf{R}_{-q}(\mathbf{Q}_{-q})$ is positive definite for all $\mathbf{Q}_{-q} \in \mathcal{Q}_{-q}$.

Lemma 11.2 can be further simplified if the (direct) channels $\mathbf{H}_{qq}$'s are full column-rank matrices: given the nonsingular matrix $\mathbf{H}_{qq}^H\mathbf{R}_{-q}^{-1}(\mathbf{Q}_{-q})\mathbf{H}_{qq}$, the MIMO waterfilling operator $\mathsf{WF}_q(\mathbf{Q}_{-q})$ in (11.25) can be equivalently written as

$$\mathsf{WF}_q(\mathbf{Q}_{-q}) = \left[-\left(\mathbf{H}_{qq}^H\mathbf{R}_{-q}^{-1}(\mathbf{Q}_{-q})\mathbf{H}_{qq}\right)^{-1}\right]_{\mathcal{Q}_q}. \quad (11.37)$$

Non-expansive property of the waterfilling operator. Thanks to the interpretation of the MIMO waterfilling in (11.25) as a projector, building on [52, properly 3.2], one can easily obtain the following non-expansive property of the waterfilling function.

LEMMA 11.3   *The matrix projection $[\cdot]_{\mathcal{Q}_q}$ onto the convex set $\mathcal{Q}_q$ defined in (11.23) satisfies the following non-expansive property:*

$$\left\|[\mathbf{X}]_{\mathcal{Q}_q} - [\mathbf{Y}]_{\mathcal{Q}_q}\right\|_F \leq \|\mathbf{X} - \mathbf{Y}\|_F, \quad \forall\, \mathbf{X}, \mathbf{Y} \in \mathbb{C}^{n_{T_q} \times n_{T_q}}. \quad (11.38)$$

11.3.3   Contraction properties of the multiuser MIMO waterfilling mapping

Building on the interpretation of the waterfilling operator as a projector, we can now focus on the contraction properties of the multiuser MIMO waterfilling operator. We will consider w.l.o.g. only the case where all the direct channel matrices are either full row-rank or full column-rank. The rank-deficient case, in fact, can be cast into the full column-rank case by a proper transformation of the original rank-deficient channel matrices into lower-dimensional, full column-rank matrices, as shown in Section 11.3.4.

Intermediate definitions

To derive the contraction properties of the MIMO waterfilling mapping we need the following intermediate definitions. Given the multiuser waterfilling mapping

$$\mathsf{WF}(\mathbf{Q}) = (\mathsf{WF}_q(\mathbf{Q}_{-q}))_{q\in\Omega} : \mathcal{Q} \mapsto \mathcal{Q}, \quad (11.39)$$

where $\mathsf{WF}_q(\mathbf{Q}_{-q})$ is defined in (11.25), we introduce the following block-maximum norm on $\mathbb{C}^{n\times n}$, with $n = n_{T_1} + \ldots + n_{T_Q}$, defined as

$$\|\mathsf{WF}(\mathbf{Q})\|_{F,\text{block}}^{\mathbf{w}} \triangleq \max_{q\in\Omega} \frac{\|\mathsf{WF}_q(\mathbf{Q}_{-q})\|_F}{w_q}, \quad (11.40)$$

where $\mathbf{w} \triangleq [w_1, \ldots, w_Q]^T > \mathbf{0}$ is any given positive weight vector. Let $\|\cdot\|_{\infty,\text{vec}}^{\mathbf{w}}$ be the *vector*-weighted maximum norm, defined as

$$\|\mathbf{x}\|_{\infty,\text{vec}}^{\mathbf{w}} \triangleq \max_{q \in \Omega} \frac{|x_q|}{w_q}, \quad \text{for} \quad \mathbf{w} > \mathbf{0}, \quad \mathbf{x} \in \mathbb{R}^Q, \tag{11.41}$$

and let $\|\cdot\|_{\infty,\text{mat}}^{\mathbf{w}}$ denote the *matrix* norm induced by $\|\cdot\|_{\infty,\text{vec}}^{\mathbf{w}}$, given by [26]

$$\|\mathbf{A}\|_{\infty,\text{mat}}^{\mathbf{w}} \triangleq \max_q \frac{1}{w_q} \sum_{r=1}^{Q} \left|[\mathbf{A}]_{qr}\right| w_r, \quad \text{for} \quad \mathbf{A} \in \mathbb{R}^{Q \times Q}. \tag{11.42}$$

Finally, we introduce the non-negative matrices $\mathbf{S}_{\text{pow}}, \mathbf{S}_{\text{pow}}^{\text{up}}, \widetilde{\mathbf{S}}_{\text{pow}}^{\text{up}} \in \mathbb{R}_+^{Q \times Q}$ defined as

$$\left[\mathbf{S}_{\text{pow}}\right]_{qr} \triangleq \begin{cases} \rho\left(\mathbf{H}_{rq}^H \mathbf{H}_{qq}^{\sharp H} \mathbf{H}_{qq}^{\sharp} \mathbf{H}_{rq}\right), & \text{if } r \neq q, \\ 0, & \text{otherwise,} \end{cases} \tag{11.43}$$

$$\left[\mathbf{S}_{\text{pow}}^{\text{up}}\right]_{qr} \triangleq \begin{cases} \text{innr}_q \cdot \rho\left(\mathbf{H}_{rq}^H \mathbf{H}_{rq}\right) \rho\left(\mathbf{H}_{qq}^{\sharp H} \mathbf{H}_{qq}^{\sharp}\right), & \text{if } r \neq q, \\ 0, & \text{otherwise} \end{cases} \tag{11.44}$$

$$\left[\widetilde{\mathbf{S}}_{\text{pow}}^{\text{up}}\right]_{qr} \triangleq \begin{cases} \left[\mathbf{S}_{\text{pow}}\right]_{qr}, & \text{if } \text{rank}(\mathbf{H}_{qq}) = n_{R_q}, \\ \left[\mathbf{S}_{\text{pow}}^{\text{up}}\right]_{qr}, & \text{otherwise,} \end{cases} \tag{11.45}$$

where the "interference-plus-noise to noise ratio" (innr_q) is given by

$$\text{innr}_q \triangleq \frac{\rho\left(\mathbf{R}_{n_q} + \sum_{r \neq q} P_r \mathbf{H}_{rq} \mathbf{H}_{rq}^H\right)}{\lambda_{\min}\left(\mathbf{R}_{n_q}\right)} \geq 1, \quad q \in \Omega. \tag{11.46}$$

Note that $\mathbf{S}_{\text{pow}} \leqslant \mathbf{S}_{\text{pow}}^{\text{up}} \leqslant \widetilde{\mathbf{S}}_{\text{pow}}$ implying $\|\mathbf{S}_{\text{pow}}\|_{\infty,\text{mat}}^{\mathbf{w}} < \|\mathbf{S}_{\text{pow}}^{\text{up}}\|_{\infty,\text{mat}}^{\mathbf{w}} < \|\widetilde{\mathbf{S}}_{\text{pow}}\|_{\infty,\text{mat}}^{\mathbf{w}}$, for all $\mathbf{w} > \mathbf{0}$.

Case of full row-rank (fat/square) channel matrices

We start by assuming that the channel matrices $\{\mathbf{H}_{qq}\}_{q \in \Omega}$ are full row-rank. The contraction property of the waterfilling mapping is given in the following.

THEOREM 11.5 Contraction property of WF mapping *Suppose that* $\text{rank}(\mathbf{H}_{qq}) = n_{R_q}$, $\forall q \in \Omega$. *Then, for any given* $\mathbf{w} \triangleq [w_1, \ldots, w_Q]^T > \mathbf{0}$, *the* WF *mapping defined in* (11.39) *is Lipschitz continuous on* $\mathcal{Q}$:

$$\left\|\mathsf{WF}(\mathbf{Q}^{(1)}) - \mathsf{WF}(\mathbf{Q}^{(2)})\right\|_{F,\text{block}}^{\mathbf{w}} \leq \|\mathbf{S}_{\text{pow}}\|_{\infty,\text{mat}}^{\mathbf{w}} \left\|\mathbf{Q}^{(1)} - \mathbf{Q}^{(2)}\right\|_{F,\text{block}}^{\mathbf{w}}, \tag{11.47}$$

for all $\mathbf{Q}^{(1)}, \mathbf{Q}^{(2)} \in \mathcal{Q}$, *where* $\|\cdot\|_{F,\text{block}}^{\mathbf{w}}$, $\|\cdot\|_{\infty,\text{mat}}^{\mathbf{w}}$ *and* $\mathbf{S}_{\text{pow}}$ *are defined in* (11.40), (11.42), *and* (11.43), *respectively. Furthermore, if the following condition is satisfied*

$$\|\mathbf{S}_{\text{pow}}\|_{\infty,\text{mat}}^{\mathbf{w}} < 1, \quad \text{for some } \mathbf{w} > \mathbf{0}, \tag{11.48}$$

then, the WF *mapping is a* block-contraction *with modulus* $\beta = \|\mathbf{S}_{\text{pow}}\|_{\infty,\text{mat}}^{\mathbf{w}}$.

Proof   Given $\mathbf{Q}^{(1)} = \left(\mathbf{Q}_q^{(1)}, \ldots, \mathbf{Q}_Q^{(1)}\right) \in \mathcal{Q}$ and $\mathbf{Q}^{(2)} = \left(\mathbf{Q}_1^{(2)}, \ldots, \mathbf{Q}_Q^{(2)}\right) \in \mathcal{Q}$, let us define, for each $q \in \Omega$,

$$e_{\text{WF}_q} \triangleq \left\| \text{WF}_q\left(\mathbf{Q}_{-q}^{(1)}\right) - \text{WF}_q\left(\mathbf{Q}_{-q}^{(2)}\right) \right\|_F \quad \text{and} \quad e_q \triangleq \left\| \mathbf{Q}_q^{(1)} - \mathbf{Q}_q^{(2)} \right\|_F, \tag{11.49}$$

where, according to Lemma 11.2, each component $\text{WF}_q(\mathbf{Q}_{-q})$ can be rewritten as in (11.33). Then, we have:

$$e_{\text{WF}_q} = \left\| \left[-\left(\mathbf{H}_{qq}^H \mathbf{R}_q^{-1}(\mathbf{Q}_{-q}^{(1)})\mathbf{H}_{qq}\right)^\sharp - c_q \mathbf{P}_{\mathcal{N}(\mathbf{H}_{qq})} \right]_{\mathcal{Q}_q} \right.$$
$$\left. - \left[-\left(\mathbf{H}_{qq}^H \mathbf{R}_q^{-1}(\mathbf{Q}_{-q}^{(2)})\mathbf{H}_{qq}\right)^\sharp - c_q \mathbf{P}_{\mathcal{N}(\mathbf{H}_{qq})} \right]_{\mathcal{Q}_q} \right\|_F \tag{11.50}$$

$$\leq \left\| \left(\mathbf{H}_{qq}^H \mathbf{R}_q^{-1}(\mathbf{Q}_{-q}^{(1)})\mathbf{H}_{qq}\right)^\sharp - \left(\mathbf{H}_{qq}^H \mathbf{R}_q^{-1}(\mathbf{Q}_{-q}^{(2)})\mathbf{H}_{qq}\right)^\sharp \right\|_F \tag{11.51}$$

$$= \left\| \mathbf{H}_{qq}^\sharp \left(\sum_{r \neq q} \mathbf{H}_{rq} \left(\mathbf{Q}_r^{(1)} - \mathbf{Q}_r^{(2)}\right) \mathbf{H}_{rq}^H \right) \mathbf{H}_{qq}^{\sharp H} \right\|_F \tag{11.52}$$

$$\leq \sum_{r \neq q} \rho\left(\mathbf{H}_{rq}^H \mathbf{H}_{qq}^{\sharp H} \mathbf{H}_{qq}^\sharp \mathbf{H}_{rq} \right) \left\| \mathbf{Q}_r^{(1)} - \mathbf{Q}_r^{(2)} \right\|_F \tag{11.53}$$

$$= \sum_{r \neq q} [\mathbf{S}_{\text{pow}}]_{qr} \left\| \mathbf{Q}_r^{(1)} - \mathbf{Q}_r^{(2)} \right\|_F = \sum_{r \neq q} [\mathbf{S}_{\text{pow}}]_{qr} e_r, \quad \forall \mathbf{Q}^{(1)}, \mathbf{Q}^{(2)} \in \mathcal{Q}, \tag{11.54}$$

where (11.50) follows from (11.33) (Lemma 11.2); (11.51) follows from the non-expansive property of the projector with respect to the Frobenius norm as given in (11.38) (Lemma 11.3); (11.52) follows from the reverse-order law for Moore–Penrose pseudo-inverses (see, e.g., [56]), valid under the assumption $\text{rank}(\mathbf{H}_{qq}) = n_{R_q}, \forall q \in \Omega$;[2] (11.53) follows from the triangle inequality [26] and

$$\left\| \mathbf{A}\mathbf{X}\mathbf{A}^H \right\|_F \leq \rho\left(\mathbf{A}^H \mathbf{A} \right) \|\mathbf{X}\|_F, \tag{11.55}$$

and in (11.54) we have used the definition of $\mathbf{S}_{\text{pow}}$ given in (11.43).

Introducing the vectors

$$\mathbf{e}_{\text{WF}} \triangleq [e_{\text{WF}_1}, \ldots, e_{\text{WF}_Q}]^T, \quad \text{and} \quad \mathbf{e} \triangleq [e_1, \ldots, e_Q]^T, \tag{11.56}$$

---

[2] Note that in the case of (strictly) *full column-rank* matrix $\mathbf{H}_{qq}$, the reverse-order law for $(\mathbf{H}_{qq}^H \mathbf{R}_q^{-1} \mathbf{H}_{qq})^\sharp$ does not hold true (the necessary and sufficient conditions given in [56, Theorem 2.2] are not satisfied). In fact, in such a case, it follows from (the matrix version of) the Kantorovich inequality [57, Chapter 11] that $\left(\mathbf{H}^H \mathbf{R}\mathbf{H}\right)^\sharp \preceq \mathbf{H}^\sharp \mathbf{R}\mathbf{H}^{\sharp H}$.

with e_{WF_q} and e_q defined in (11.49), the set of inequalities in (11.54) can be rewritten in vector form as

$$\mathbf{0} \leq \mathbf{e}_{\mathsf{WF}} \leq \mathbf{S}_{\mathrm{pow}}\, \mathbf{e}, \quad \forall \mathbf{Q}^{(1)}, \mathbf{Q}^{(2)} \in \mathcal{Q}. \tag{11.57}$$

Using the weighted maximum norm $\|\cdot\|_{\infty,\mathrm{vec}}^{\mathbf{w}}$ defined in (11.41) in combination with (11.57), we have, for any given $\mathbf{w} > \mathbf{0}$ (recall that $\|\cdot\|_{\infty,\mathrm{vec}}^{\mathbf{w}}$ is a monotonic norm),[3]

$$\|\mathbf{e}_{\mathsf{WF}}\|_{\infty,\mathrm{vec}}^{\mathbf{w}} \leq \left\|\mathbf{S}_{\mathrm{pow}}\mathbf{e}\right\|_{\infty,\mathrm{vec}}^{\mathbf{w}} \leq \left\|\mathbf{S}_{\mathrm{pow}}\right\|_{\infty,\mathrm{mat}}^{\mathbf{w}} \|\mathbf{e}\|_{\infty,\mathrm{vec}}^{\mathbf{w}}, \tag{11.58}$$

$\forall \mathbf{Q}^{(1)}, \mathbf{Q}^{(2)} \in \mathcal{Q}$, where $\|\cdot\|_{\infty,\mathrm{mat}}^{\mathbf{w}}$ is the matrix norm induced by the vector norm $\|\cdot\|_{\infty,\mathrm{vec}}^{\mathbf{w}}$ in (11.41) and defined in (11.42) [26]. Finally, introducing (11.40) in (11.58), we obtain the desired result as stated in (11.47). $\qquad\blacksquare$

Negative result. As stated in Theorem 11.5, the waterfilling mapping WF satisfies the Lipschitz property in (11.47) if the channel matrices $\{\mathbf{H}_{qq}\}_{q\in\Omega}$ are full row-rank. Surprisingly, if the channels are not full row-rank matrices, the property in (11.47) *does not* hold for *every* given set of matrices $\{\mathbf{H}_{qq}\}_{q\in\Omega}$, implying that the WF mapping is not a contraction under (11.48) for all $\{\mathbf{H}_{qq}\}_{q\in\Omega}$ and stronger conditions are needed, as given in the next section. A simple counter-example is given in [24].

Case of full column-rank (tall) channel matrices

The main difficulty in deriving contraction properties of the MIMO multiuser waterfilling mapping in the case of (strictly) tall channel matrices $\{\mathbf{H}_{qq}\}_{q\in\Omega}$ is that one cannot use the reverse-order law of generalized inverses, as was done in the proof of Theorem 11.5 (see (11.51)–(11.52)). To overcome this issue, we develop a different approach based on the mean-value theorem for complex matrix-valued functions, as detailed next.

Mean-value theorem for complex matrix-valued functions. The mean-value theorem for scalar real functions is one of the most important and basic theorems in functional analysis (see, e.g., [57, Chapter 5-Theorem 10], [58, Theorem 5.10]). The generalization of the (differential version of the) theorem to vector-valued real functions that one would expect does not hold, meaning that for real vector-valued functions $\mathbf{f} : \mathcal{D} \subseteq \mathbb{R}^m \mapsto \mathbb{R}^n$ in general

$$\nexists t \in (0,1) \mid \mathbf{f}(\mathbf{y}) - \mathbf{f}(\mathbf{x}) = \mathbf{D}_{\mathbf{x}}\mathbf{f}(t\,\mathbf{y} + (1-t)\,\mathbf{x})(\mathbf{y} - \mathbf{x}), \tag{11.59}$$

for any $\mathbf{x}, \mathbf{y} \in \mathcal{D}$ and $\mathbf{x} \neq \mathbf{y}$, where $\mathbf{D}_{\mathbf{x}}\mathbf{f}$ denotes the Jacobian matrix of $\mathbf{f}$. One of the simplest examples to illustrate (11.59) is the following. Consider the real vector-valued function $\mathbf{f}(x) = [x^{\alpha}, x^{\beta}]^T$, with $x \in \mathbb{R}$ and, for example, $\alpha = 2$, $\beta = 3$. There exists no value of $t \in (0,1)$ such that $\mathbf{f}(1) = \mathbf{f}(0) + \mathbf{D}_t\mathbf{f}(t)$.

Many extensions and variations of the main value theorem exist in the literature, either for (real/complex) scalar or real vector-valued functions (see, e.g., [59, 60], [53, Chapter 3.2]). Here, we focus on the following.

---

[3] A vector norm $\|\cdot\|$ is monotonic if $\mathbf{x} \geq \mathbf{y}$ implies $\|\mathbf{x}\| \geq \|\mathbf{y}\|$.

LEMMA 11.4 [24]   *Let* $\mathbf{F}(\mathbf{X}) : \mathcal{D} \subseteq \mathbb{C}^{m \times n} \mapsto \mathbb{C}^{p \times q}$ *be a complex matrix-valued function defined on a convex set* $\mathcal{D}$, *assumed to be continuous on* $\mathcal{D}$ *and differentiable on the interior of* $\mathcal{D}$, *with Jacobian matrix* $\mathbf{D_X F}(\mathbf{X})$.[4] *Then, for any given* $\mathbf{X}, \mathbf{Y} \in \mathcal{D}$, *there exists some* $t \in (0, 1)$ *such that*

$$\|\mathbf{F}(\mathbf{Y}) - \mathbf{F}(\mathbf{X})\|_F \leq \|\mathbf{D_X F}((t\,\mathbf{Y} + (1 - t)\,\mathbf{X}))\,\mathrm{vec}(\mathbf{Y} - \mathbf{X})\|_2 \tag{11.60}$$

$$\leq \|\mathbf{D_X F}\,((t\,\mathbf{Y} + (1 - t)\,\mathbf{X}))\|_{2,\mathrm{mat}}\,\|\mathbf{Y} - \mathbf{X}\|_F, \tag{11.61}$$

where $\|\mathbf{A}\|_{2,\mathrm{mat}} \triangleq \sqrt{\rho(\mathbf{A}^H \mathbf{A})}$ *denotes the spectral norm of* $\mathbf{A}$.

We can now provide the contraction theorem for the WF mapping valid also for the case in which the channels $\{\mathbf{H}_{qq}\}_{q \in \Omega}$ are full column-rank matrices.

THEOREM 11.6 Contraction property of WF mapping   *Suppose that* $\mathrm{rank}(\mathbf{H}_{qq}) = n_{T_q}$, $\forall q \in \Omega$. *Then, for any given* $\mathbf{w} \triangleq [w_1, \ldots, w_Q]^T > \mathbf{0}$, *the mapping* WF *defined in* (11.39) *is Lipschitz continuous on* $\mathcal{Q}$:

$$\left\|\mathsf{WF}(\mathbf{Q}^{(1)}) - \mathsf{WF}(\mathbf{Q}^{(2)})\right\|_{F,\mathrm{block}}^{\mathbf{w}} \leq \left\|\mathbf{S}_{\mathrm{pow}}^{\mathrm{up}}\right\|_{\infty,\mathrm{mat}}^{\mathbf{w}} \left\|\mathbf{Q}^{(1)} - \mathbf{Q}^{(2)}\right\|_{F,\mathrm{block}}^{\mathbf{w}}, \tag{11.62}$$

for all $\mathbf{Q}^{(1)}, \mathbf{Q}^{(2)} \in \mathcal{Q}$, *where* $\|\cdot\|_{F,\mathrm{block}}^{\mathbf{w}}$, $\|\cdot\|_{\infty,\mathrm{mat}}^{\mathbf{w}}$ *and* $\mathbf{S}_{\mathrm{pow}}^{\mathrm{up}}$ *are defined in* (11.40), (11.42), *and* (11.44), *respectively. Furthermore, if the following condition is satisfied*[5]

$$\left\|\mathbf{S}_{\mathrm{pow}}^{\mathrm{up}}\right\|_{\infty,\mathrm{mat}}^{\mathbf{w}} < 1, \qquad \textit{for some } \mathbf{w} > \mathbf{0}, \tag{11.63}$$

then, the mapping WF *is a block-contraction with modulus* $\beta = \left\|\mathbf{S}_{\mathrm{pow}}^{\mathrm{up}}\right\|_{\infty,\mathrm{mat}}^{\mathbf{w}}$.

Proof   The proof follows the same guidelines as that of Theorem 11.5, with the key difference that, in the case of (strictly) full column-rank direct channel matrices, we cannot use the reverse-order law of pseudo-inverses as was done to obtain (11.51)–(11.52) in the proof of Theorem 11.5. We apply instead the mean-value theorem in Lemma 11.4, as detailed next. For technical reasons, we introduce first a proper, complex matrix-valued function $\mathbf{F}_q(\mathbf{Q}_{-q})$ related to the MIMO multiuser waterfilling mapping $\mathsf{WF}_q(\mathbf{Q}_{-q})$ in (11.24) and, using Lemma 11.4, we study the Lipschitz properties of the function on $\mathcal{Q}_{-q}$. Then, building on this result, we show that the WF mapping satisfies (11.62).

Given $q \in \Omega$, let us introduce the following complex matrix-valued function $\mathbf{F}_q : \mathcal{Q}_{-q} \ni \mathbf{Q}_{-q} \mapsto \mathbb{S}_{++}^{n_{T_q}}$, defined as:

$$\mathbf{F}_q(\mathbf{Q}_{-q}) \triangleq \left(\mathbf{H}_{qq}^H \mathbf{R}_{-q}^{-1}(\mathbf{Q}_{-q})\mathbf{H}_{qq}\right)^{-1}. \tag{11.64}$$

---

[4] We define the Jacobian matrix of a differentiable, complex matrix-valued function following the approach in [61], meaning that we treat the complex differential of the complex variable and its complex conjugate as independent. This approach simplifies the derivation of many complex-derivative expressions. We refer the interested reader to [24, 61] for details.

[5] Milder conditions than (11.63) are given in [24], whose proof is much more involved and thus omitted here because of the space limitation.

Observe that the function $\mathbf{F}_q(\mathbf{Q}_{-q})$ is continuous on $\mathcal{Q}_{-q}$ (implied from the continuity of $\mathbf{R}_{-q}^{-1}(\mathbf{Q}_{-q})$ at any $\mathbf{Q}_{-q} \succeq \mathbf{0}$ [25, Theorem 10.7.1]) and differentiable on the interior of $\mathcal{Q}_{-q}$. The Jacobian matrix of $\mathbf{F}_q(\mathbf{Q}_{-q})$ is [24]:

$$\mathbf{D}_{\mathbf{Q}_{-q}}\mathbf{F}(\mathbf{Q}_{-q}) = \left[\mathbf{G}_{1\,q}^*(\mathbf{Q}_{-q}) \otimes \mathbf{G}_{1\,q}(\mathbf{Q}_{-q}),\dots,\mathbf{G}_{q-1\,q}^*(\mathbf{Q}_{-q}) \otimes \mathbf{G}_{q-1\,q}(\mathbf{Q}_{-q}),\dots,\right.$$
$$\left.\mathbf{G}_{q+1\,q}^*(\mathbf{Q}_{-q}) \otimes \mathbf{G}_{q+1\,q}(\mathbf{Q}_{-q}),\dots,\mathbf{G}_{Q\,q}^*(\mathbf{Q}_{-q}) \otimes \mathbf{G}_{Q\,q}(\mathbf{Q}_{-q})\right],$$
$$(11.65)$$

where

$$\mathbf{G}_{rq}(\mathbf{Q}_{-q}) \triangleq \left(\mathbf{H}_{qq}^H\mathbf{R}_{-q}^{-1}(\mathbf{Q}_{-q})\mathbf{H}_{qq}\right)^{-1}\mathbf{H}_{qq}^H\mathbf{R}_{-q}^{-1}(\mathbf{Q}_{-q})\mathbf{H}_{rq}. \qquad (11.66)$$

It follows from Lemma 11.4 that, for any two different points $\mathbf{Q}_{-q}^{(1)}, \mathbf{Q}_{-q}^{(2)} \in \mathcal{Q}_{-q}$, with $\mathbf{Q}_{-q}^{(i)} = [\mathbf{Q}_1^{(i)},\dots,\mathbf{Q}_{q-1}^{(i)},\mathbf{Q}_{q+1}^{(i)},\dots,\mathbf{Q}_Q^{(i)}]$ for $i = 1, 2$, there exists some $t \in (0, 1)$ such that, introducing

$$\boldsymbol{\Delta} \triangleq t\mathbf{Q}_{-q}^{(1)} + (1 - t)\mathbf{Q}_{-q}^{(2)}, \qquad (11.67)$$

we have:

$$\left\|\mathbf{F}_q\left(\mathbf{Q}_{-q}^{(1)}\right) - \mathbf{F}_q\left(\mathbf{Q}_{-q}^{(2)}\right)\right\|_F \leq \left\|\mathbf{D}_{\mathbf{Q}_{-q}}\mathbf{F}_q(\boldsymbol{\Delta})\,\mathrm{vec}\left(\mathbf{Q}_{-q}^{(1)} - \mathbf{Q}_{-q}^{(2)}\right)\right\|_2 \qquad (11.68)$$

$$\leq \sum_{r \neq q}\left\|\mathbf{G}_{rq}^*(\boldsymbol{\Delta}) \otimes \mathbf{G}_{rq}(\boldsymbol{\Delta})\right\|_{2,\mathrm{mat}}\left\|\mathbf{Q}_r^{(1)} - \mathbf{Q}_r^{(2)}\right\|_F \qquad (11.69)$$

$$= \sum_{r \neq q}\rho\left(\mathbf{G}_{rq}^H(\boldsymbol{\Delta})\mathbf{G}_{rq}(\boldsymbol{\Delta})\right)\left\|\mathbf{Q}_r^{(1)} - \mathbf{Q}_r^{(2)}\right\|_F, \qquad (11.70)$$

where (11.68) follows from (11.60) (Lemma 11.4); (11.69) follows from the structure of $\mathbf{D}_{\mathbf{Q}_{-q}}\mathbf{F}_q$ (see (11.65)) and the triangle inequality [26]; and in (11.70) we used

$$\rho\left[\left(\mathbf{G}_{rq}^T \otimes \mathbf{G}_{rq}^H\right)\left(\mathbf{G}_{rq}^* \otimes \mathbf{G}_{rq}\right)\right] = \left(\rho\left[\mathbf{G}_{rq}^H\mathbf{G}_{rq}\right]\right)^2. \qquad (11.71)$$

Observe that, differently from (11.53)–(11.54), the factor $\alpha_{rq}(\boldsymbol{\Delta}) \triangleq \rho\left[\mathbf{G}_{rq}^H(\boldsymbol{\Delta})\mathbf{G}_{rq}(\boldsymbol{\Delta})\right]$ in (11.70) depends, in general, on both $t \in (0, 1)$ and the covariance matrices $\mathbf{Q}_{-q}^{(1)}$ and $\mathbf{Q}_{-q}^{(2)}$ through $\boldsymbol{\Delta}$ (see (11.67)):

$$\alpha_{rq}(\boldsymbol{\Delta}) = \rho\left[\mathbf{H}_{rq}^H\mathbf{R}_{-q}^{-1}(\boldsymbol{\Delta})\mathbf{H}_{qq}\left(\mathbf{H}_{qq}^H\mathbf{R}_{-q}^{-1}(\boldsymbol{\Delta})\mathbf{H}_{qq}\right)^{-1}\right.$$
$$\left.\times \left(\mathbf{H}_{qq}^H\mathbf{R}_{-q}^{-1}(\boldsymbol{\Delta})\mathbf{H}_{qq}\right)^{-1}\mathbf{H}_{qq}^H\mathbf{R}_{-q}^{-1}(\boldsymbol{\Delta})\mathbf{H}_{rq}\right], \qquad (11.72)$$

where in (11.72) we used (11.66). Interestingly, in the case of square (nonsingular) channel matrices $\mathbf{H}_{qq}$, (11.72) reduces to $\alpha_{rq}(\boldsymbol{\Delta}) = \rho\left[\mathbf{H}_{rq}^H\mathbf{H}_{qq}^{\sharp H}\mathbf{H}_{qq}^{\sharp}\mathbf{H}_{rq}\right] = [\mathbf{S}_{\mathrm{pow}}]_{qr}$, where $\mathbf{S}_{\mathrm{pow}}$ is defined in (11.43), thus recovering the same contraction factor for the WF mapping as in Theorem 11.5. In the case of (strictly) full column-rank matrices $\mathbf{H}_{qq}$, an upper bound of $\alpha_{rq}(\boldsymbol{\Delta})$, independent of $\boldsymbol{\Delta}$ is [24]

$$\alpha_{rq}(\boldsymbol{\Delta}) < \mathrm{innr}_q \cdot \rho\left(\mathbf{H}_{rq}^H \mathbf{H}_{rq}\right) \rho\left(\mathbf{H}_{qq}^{\sharp H} \mathbf{H}_{qq}^{\sharp}\right), \tag{11.73}$$

where innr_q is defined in (11.46). The Lipschitz property of the WF mapping as given in (11.62) comes from (11.70) and (11.73), using the same steps as in the proof of Theorem 11.5. ■

Comparing Theorems 11.5 and 11.6, one infers that conditions for the multiuser MIMO waterfilling mapping to be a block-contraction in the case of (strictly) full column-rank channel matrices are stronger than those required when the channels are full row-rank matrices.

Case of full-rank channel matrices

In the case in which the (direct) channel matrices are either full row-rank or full column-rank, we have the following contraction theorem for the WF mapping.

THEOREM 11.7 Contraction property of WF mapping *Suppose that, for each $q \in \Omega$, either $\mathrm{rank}(\mathbf{H}_{qq}) = n_{R_q}$ or $\mathrm{rank}(\mathbf{H}_{qq}) = n_{T_q}$. Then, for any given $\mathbf{w} \triangleq [w_1, \ldots, w_Q]^T > \mathbf{0}$, the WF mapping defined in (11.39) is Lipschitz continuous on $\mathcal{Q}$:*

$$\left\|\mathsf{WF}(\mathbf{Q}^{(1)}) - \mathsf{WF}(\mathbf{Q}^{(2)})\right\|_{F,\mathrm{block}}^{\mathbf{w}} \leq \left\|\widetilde{\mathbf{S}}_{\mathrm{pow}}^{\mathrm{up}}\right\|_{\infty,\mathrm{mat}}^{\mathbf{w}} \left\|\mathbf{Q}^{(1)} - \mathbf{Q}^{(2)}\right\|_{F,\mathrm{block}}^{\mathbf{w}}, \tag{11.74}$$

for all $\mathbf{Q}^{(1)}, \mathbf{Q}^{(2)} \in \mathcal{Q}$, where $\|\cdot\|_{F,\mathrm{block}}^{\mathbf{w}}$, $\|\cdot\|_{\infty,\mathrm{mat}}^{\mathbf{w}}$ and $\widetilde{\mathbf{S}}_{\mathrm{pow}}^{\mathrm{up}}$ are defined in (11.40), (11.42), and (11.45), respectively. Furthermore, if the following condition is satisfied

$$\left\|\widetilde{\mathbf{S}}_{\mathrm{pow}}^{\mathrm{up}}\right\|_{\infty,\mathrm{mat}}^{\mathbf{w}} < 1, \quad \textit{for some } \mathbf{w} > \mathbf{0}, \tag{11.75}$$

then, the WF mapping is a block-contraction with modulus $\beta = \|\widetilde{\mathbf{S}}_{\mathrm{pow}}\|_{\infty,\mathrm{mat}}^{\mathbf{w}}$.

11.3.4 Existence and uniqueness of the Nash equilibrium

We can now study game $\mathcal{G}_{\mathrm{pow}}$ and derive conditions for the uniqueness of the NE, as given next.

THEOREM 11.8 *Game $\mathcal{G}_{\mathrm{pow}}$ always admits an NE, for any set of channel matrices and transmit power of the users. Furthermore, the NE is unique if[6]*

$$\rho(\widetilde{\mathbf{S}}_{\mathrm{pow}}^{\mathrm{up}}) < 1, \tag{C1}$$

where $\widetilde{\mathbf{S}}_{\mathrm{pow}}^{\mathrm{up}}$ is defined in (11.45).

---

[6] Milder conditions are given in [24].

Proof The existence of an NE of game $\mathcal{G}_{\text{pow}}$ for any set of channel matrices and power budget follows directly from Theorem 11.2 (i.e., compact convex-strategy sets and continuous, quasi-concave payoff functions). As far as the uniqueness of the NE is concerned, a sufficient condition is that the waterfilling mapping in (11.24) be a contraction with respect to some norm (Theorem 11.3). Hence, the sufficiency of (C1) in the case of full (column/row) rank channel matrices $\{\mathbf{H}_{qq}\}_{q\in\Omega}$ comes from Theorem 11.7 and the equivalence of the following two statements [52, Corollary 6.1]: (a) there exists some $\mathbf{w} > \mathbf{0}$ such that $\|\widetilde{\mathbf{S}}_{\text{pow}}^{\text{up}}\|_{\infty,\text{mat}}^{\mathbf{w}} < 1$; and (b) $\rho(\widetilde{\mathbf{S}}_{\text{pow}}^{\text{up}}) < 1$.

We focus now on the more general case in which the channel matrices $\mathbf{H}_{qq}$ may be rank deficient and prove that condition (C1) is still sufficient to guarantee the uniqueness of the NE. For any $q \in \overline{\Omega} \triangleq \{q \in \Omega \mid r_{H_{qq}} \triangleq \text{rank}(\mathbf{H}_{qq}) < \min(n_{T_q}, n_{R_q})\}$ and $\mathbf{Q}_{-q} \succeq \mathbf{0}$, the best-response $\mathbf{Q}_q^\star = \mathsf{WF}_q(\mathbf{H}_{qq}^H \mathbf{R}_{-q}^{-1}(\mathbf{Q}_{-q})\mathbf{H}_{qq})$ – the solution to the rate-maximization problem in (11.22) for a given $\mathbf{Q}_{-q} \succeq \mathbf{0}$ – will be orthogonal to the null space of $\mathbf{H}_{qq}$, implying $\mathbf{Q}_q^\star = \mathbf{V}_{q,1}\overline{\mathbf{Q}}_q^\star\mathbf{V}_{q,1}^H$ for some $\overline{\mathbf{Q}}_q^\star \in \mathbb{S}_+^{r_{H_{qq}}}$ such that $\text{Tr}(\overline{\mathbf{Q}}_q^\star) = P_q$, where $\mathbf{V}_{q,1} \in \mathbb{C}^{n_{T_q} \times r_{H_{qq}}}$ is a semiunitary matrix such that $\mathcal{R}(\mathbf{V}_{q,1}) = \mathcal{N}(\mathbf{H}_{qq})^\perp$. Thus, the best-response of each user $q \in \overline{\Omega}$ belongs to the following class of matrices:

$$\mathbf{Q}_q = \mathbf{V}_{q,1}\overline{\mathbf{Q}}_q\mathbf{V}_{q,1}^H, \qquad \text{with} \qquad \overline{\mathbf{Q}}_q \in \overline{\mathcal{Q}}_q \triangleq \left\{\mathbf{X} \in \mathbb{S}_+^{r_{H_{qq}}} \mid \text{Tr}(\mathbf{X}) = P_q\right\}. \qquad (11.76)$$

Using (11.76) and introducing the (possibly) lower-dimensional covariance matrices $\overline{\mathbf{Q}}_q$'s and the modified channel matrices $\widetilde{\mathbf{H}}_{rq}$'s, defined respectively as

$$\widetilde{\mathbf{Q}}_q \triangleq \begin{cases} \overline{\mathbf{Q}}_q \in \mathbb{S}_+^{r_{H_{qq}}}, & \text{if } q \in \overline{\Omega}, \\ \mathbf{Q}_q \in \mathbb{S}_+^{n_{T_q}}, & \text{otherwise}, \end{cases} \qquad \text{and} \qquad \widetilde{\mathbf{H}}_{rq} \triangleq \begin{cases} \mathbf{H}_{rq}\mathbf{V}_{r,1}, & \text{if } r \in \overline{\Omega}, \\ \mathbf{H}_{rq}, & \text{otherwise}, \end{cases}$$

$$(11.77)$$

game $\mathcal{G}_{\text{pow}}$ can be recast in the following lower-dimensional game $\widetilde{\mathcal{G}}_{\text{pow}}$, defined as

$$(\widetilde{\mathcal{G}}_{\text{pow}}): \quad \begin{array}{ll} \underset{\widetilde{\mathbf{Q}}_q \succeq \mathbf{0}}{\text{maximize}} & \log\det\left(\mathbf{I} + \widetilde{\mathbf{H}}_{qq}^H \widetilde{\mathbf{R}}_{-q}^{-1}(\widetilde{\mathbf{Q}}_{-q})\widetilde{\mathbf{H}}_{qq}\widetilde{\mathbf{Q}}_q\right) \\ \text{subject to} & \text{Tr}(\widetilde{\mathbf{Q}}_q) = P_q, \end{array} \qquad \forall q \in \Omega,$$

$$(11.78)$$

where $\widetilde{\mathbf{R}}_{-q}(\widetilde{\mathbf{Q}}_{-q}) \triangleq \mathbf{R}_{n_{R_q}} + \sum_{r \neq q} \widetilde{\mathbf{H}}_{rq}\widetilde{\mathbf{Q}}_r\widetilde{\mathbf{H}}_{rq}^H$. It turns out that conditions guaranteeing the uniqueness of the NE of game $\widetilde{\mathcal{G}}_{\text{pow}}$ are sufficient also for the uniqueness of the NE of $\mathcal{G}_{\text{pow}}$.

Observe that, in the game $\widetilde{\mathcal{G}}_{\text{pow}}$, all channel matrices $\widetilde{\mathbf{H}}_{qq}$ are full-rank matrices. We can thus use Theorem 11.7 and obtain the following sufficient condition for the uniqueness of the NE of both games $\widetilde{\mathcal{G}}_{\text{pow}}$ and $\mathcal{G}_{\text{pow}}$:

$$\rho(\widetilde{\mathbf{S}}_{\text{pow}}) < 1, \qquad (11.79)$$

with

$$
\left[\widetilde{\mathbf{S}}_{\mathrm{pow}}\right]_{qr} \triangleq
\begin{cases}
\rho\left(\widetilde{\mathbf{H}}_{rq}^{H}\mathbf{H}_{qq}^{\sharp H}\mathbf{H}_{qq}^{\sharp}\widetilde{\mathbf{H}}_{rq}\right), & \text{if } r \neq q,\ r_{H_{qq}} = n_{R_q}, \\[2mm]
\widetilde{\mathrm{innr}}_q \cdot \rho\left(\widetilde{\mathbf{H}}_{rq}^{H}\widetilde{\mathbf{H}}_{rq}\right)\rho\left(\widetilde{\mathbf{H}}_{qq}^{\sharp H}\widetilde{\mathbf{H}}_{qq}^{\sharp}\right), & \text{if } r \neq q,\ r_{H_{qq}} < n_{R_q}, \\[2mm]
0 & \text{if } r = q,
\end{cases}
\tag{11.80}
$$

and $\widetilde{\mathrm{innr}}_q$ is defined as in (11.46), where each $\mathbf{H}_{rq}$ is replaced by $\widetilde{\mathbf{H}}_{rq}$. The sufficiency of (11.79) for (C1) follows from $\mathbf{0} \leq \widetilde{\mathbf{S}}_{\mathrm{pow}} \leq \widetilde{\mathbf{S}}_{\mathrm{pow}}^{\mathrm{up}} \implies \rho(\widetilde{\mathbf{S}}_{\mathrm{pow}}) \leq \rho(\widetilde{\mathbf{S}}_{\mathrm{pow}}^{\mathrm{up}})$ [54, Corollary 2.2.22]; which completes the proof. ∎

To give additional insight into the physical interpretation of sufficient conditions for the uniqueness of the NE, we provide the following.

COROLLARY 11.1 *If* $\mathrm{rank}(\mathbf{H}_{qq}) = n_{R_q}$ *for all* $q \in \Omega$, *then a sufficient condition for (C1) in Theorem 11.8 is given by one of the two following sets of conditions:*

Low received MUI:
$$
\frac{1}{w_q}\sum_{r \neq q}\rho\left(\mathbf{H}_{rq}^{H}\mathbf{H}_{qq}^{\sharp H}\mathbf{H}_{qq}^{\sharp}\mathbf{H}_{rq}\right)w_r < 1, \quad \forall q \in \Omega, \tag{C2}
$$

Low generated MUI:
$$
\frac{1}{w_r}\sum_{q \neq r}\rho\left(\mathbf{H}_{rq}^{H}\mathbf{H}_{qq}^{\sharp H}\mathbf{H}_{qq}^{\sharp}\mathbf{H}_{rq}\right)w_q < 1, \quad \forall r \in \Omega, \tag{C3}
$$

where $\mathbf{w} \triangleq [w_1,\ldots,w_Q]^{T}$ *is any positive vector.*

If $\mathrm{rank}(\mathbf{H}_{qq}) \leq n_{T_q}$, *for all* $q \in \Omega$, *then a sufficient condition for (C1) is given by one of the two following sets of conditions:*[7]

Low received MUI:
$$
\frac{1}{w_q}\sum_{r \neq q}\mathrm{innr}_q \cdot \rho\left(\mathbf{H}_{rq}^{H}\mathbf{H}_{rq}\right)\rho\left(\mathbf{H}_{qq}^{\sharp H}\mathbf{H}_{qq}^{\sharp}\right)w_r < 1,\ \forall q \in \Omega,
$$
$$
\tag{C4}
$$

Low generated MUI:
$$
\frac{1}{w_r}\sum_{q \neq r}\mathrm{innr}_q \cdot \rho\left(\mathbf{H}_{rq}^{H}\mathbf{H}_{rq}\right)\rho\left(\mathbf{H}_{qq}^{\sharp H}\mathbf{H}_{qq}^{\sharp}\right)w_q < 1,\ \forall r \in \Omega,
$$
$$
\tag{C5}
$$

where the innr_q*'s are defined in (11.46).* □

REMARK 11.2 *On the uniqueness conditions.* Conditions (C2)–(C3) and (C4)–(C5) provide a physical interpretation of the uniqueness of the NE: as expected, the uniqueness of the NE is ensured if the interference among the links is sufficiently small. The importance of (C2)–(C3) and (C4)–(C5) is that they quantify how small the interference must be to guarantee that the equilibrium is indeed unique. Specifically, conditions (C2) and (C4) can be interpreted as a constraint on the maximum amount of interference that each receiver can tolerate, whereas (C3) and (C5) introduce an upper bound on the

---

[7] The case in which some channel matrices $\mathbf{H}_{qq}$ are (strictly) tall and some others are fat or there are rank deficient channel matrices can be similarly addressed (cf. [24]).

maximum level of interference that each transmitter is allowed to generate. Surprisingly, the above conditions differ if the channel matrices $\{\mathbf{H}_{qq}\}_{q\in\Omega}$ are (strictly) tall or fat.

11.3.5 Distributed algorithms

In this section we focus on distributed algorithms that converge to the NE of game $\mathcal{G}_{\mathrm{pow}}$. We consider totally asynchronous distributed algorithms, as described in Section 11.2.2. Using the same notation as introduced in Section 11.2.2, the asynchronous MIMO IWFA is formally described in Algorithm 2, where $\mathbf{Q}_q^{(n)}$ denotes the covariance matrix of the vector signal transmitted by user q at the nth iteration, and $\mathsf{T}_q(\mathbf{Q}_{-q})$ in (11.82) is the best-response function of user q:

$$\mathsf{T}_q(\mathbf{Q}_{-q}) \triangleq \mathsf{WF}_q\left(\mathbf{H}_{qq}^H\mathbf{R}_{-q}^{-1}(\mathbf{Q}_{-q})\mathbf{H}_{qq}\right), \tag{11.81}$$

with $\mathsf{WF}_q(\cdot)$ defined in (11.24). The algorithm is totally asynchronous, meaning that one can use any arbitrary schedule $\{\tau_r^q(n)\}$ and $\{\mathcal{T}_q\}$ satisfying the standard assumptions (A1)–(A3), given in Section 11.2.2.

Algorithm 2 MIMO asynchronous IWFA

1 : Set $n = 0$ and choose any feasible $\mathbf{Q}_q^{(0)}$;

2 : repeat

$$3: \qquad \mathbf{Q}_q^{(n+1)} = \begin{cases} \mathsf{T}_q\left(\mathbf{Q}_{-q}^{(\tau^q(n))}\right), & \text{if } n \in \mathcal{T}_q, \\ \mathbf{Q}_q^{(n)}, & \text{otherwise;} \end{cases} \qquad \forall q \in \Omega \qquad (11.82)$$

4 : until the prescribed convergence criterion is satisfied

Sufficient conditions guaranteeing the global convergence of the algorithm are given in Theorem 11.9, whose proof follows from results given in Section 11.2.2.

THEOREM 11.9 *Suppose that condition (C1) of Theorem 11.8 is satisfied. Then, any sequence $\{\mathbf{Q}^{(n)}\}_{n=1}^{\infty}$ generated by the asynchronous MIMO IWFA, described in Algorithm 2, converges to the unique NE of game $\mathcal{G}_{\mathrm{pow}}$, for any set of feasible initial conditions and updating schedule satisfying (A1)–(A3).*

REMARK 11.3 *Features of Algorithm 2.* Algorithm 2 contains as special cases a plethora of algorithms, each one obtained by a possible choice of the scheduling of the users in the updating procedure (i.e., the parameters $\{\tau_r^q(n)\}$ and $\{\mathcal{T}_q\}$). Two well-known special cases are the *sequential* and the *simultaneous* MIMO IWFA, where the users update their own strategies *sequentially* [7, 8, 10, 21] and *simultaneously* [7, 8, 10, 55, 62], respectively. Interestingly, since condition (C1) does not depend on the particular choice of $\{\mathcal{T}_q\}$ and $\{\tau_r^q(n)\}$, the important result coming from the convergence analysis is that all the algorithms resulting as special cases of the asynchronous MIMO IWFA are guaranteed to globally converge to the unique NE of the game, under the same set of convergence conditions. Moreover they have the following desired properties:

- *Low complexity and distributed nature.* The algorithm can be implemented in a distributed way, since each user, to compute his best response $\mathsf{T}_q(\cdot)$ in (11.81), only needs to measure the overall interference-plus-noise covariance matrix $\mathbf{R}_{-q}(\mathbf{Q}_{-q})$ and waterfill over the equivalent channel $\mathbf{H}_{qq}^H \mathbf{R}_{-q}^{-1}(\mathbf{Q}_{-q})\mathbf{H}_{qq}$.
- *Robustness.* Algorithm 2 is robust against missing or outdated updates of secondary users. This feature strongly relaxes the constraints on the synchronization of the users' updates with respect to those imposed, for example, by the simultaneous- or sequential-updating schemes.

- *Fast convergence behavior.* The simultaneous version of the proposed algorithm converges in a very few iterations, even in networks with many active secondary users. As expected, the sequential IWFA is slower than the simultaneous IWFA, especially if the number of active secondary users is large, since each user is forced to wait for all the users scheduled ahead, before updating his own covariance matrix. As an example, in Figure 11.1 we compare the performance of the sequential and simultaneous IWFA, in terms of convergence speed, for a given set of MIMO channel realizations. We consider a cellular network composed of 7 (regular) hexagonal cells, sharing the same spectrum. For the sake of simplicity, we assume that in each cell there is only one active link, corresponding to the transmission from the BS (placed at the center of the cell) to an MT placed in a corner of the cell. The overall network is thus modeled as eight 4×4 MIMO interference wideband channels, according to (11.1). In Figure 11.1, we show the rate evolution of the links of three cells corresponding to the sequential IWFA and simultaneous IWFA as a function of the iteration index n. To make the figure not excessively overcrowded, we plot only the curves of 3 out of 8 links.

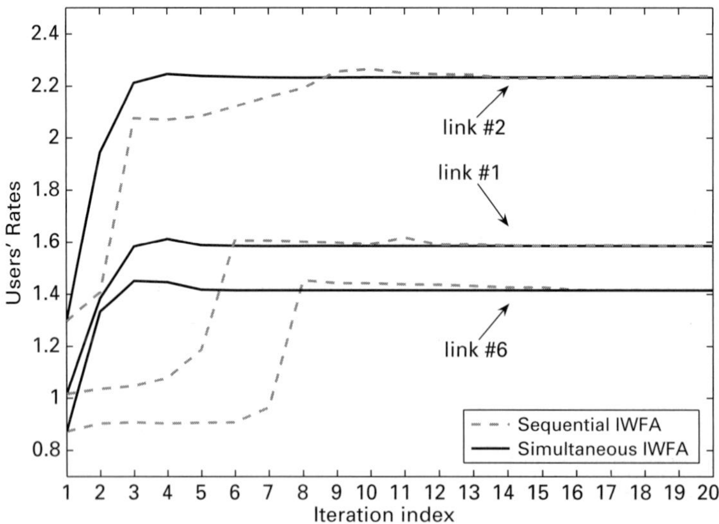

Figure 11.1   Rates of the MIMO links versus iterations: sequential IWFA (dashed-line curves) and simultaneous IWFA (solid-line curves).

11.4 Opportunistic communications under individual interference constraints

In this section, we focus now on the more general resource-allocation problem under interference constraints as given in (11.6). We start considering power constraints (11.2) and individual null constraints (11.5), since they are suitable to model the white-space filling paradigm. More specifically, the problem formulated leads directly to what we call game $\mathcal{G}_{\text{null}}$. We also consider an alternative game formulation, $\mathcal{G}_{\infty}$, with improved convergence properties; however, it does not correspond to any physical scenario so it is a rather artificial formulation. The missing ingredient is provided by another game formulation, $\mathcal{G}_{\alpha}$, that does have a good physical interpretation and asymptotically is equivalent to $\mathcal{G}_{\infty}$ (in the sense specified next); thus inheriting the improved convergence properties as well as the physical interpretation. After that, we consider more general opportunistic communications by also allowing soft-shaping interference constraints (11.6) through the game $\mathcal{G}_{\text{soft}}$.

11.4.1 Game with null constraints

Given the rate functions in (11.4), the rate-maximization game among the secondary users in the presence of the power constraints (11.2) and the null constraints (11.5) is formally defined as:

$$(\mathcal{G}_{\text{null}}): \quad \begin{array}{ll} \underset{\mathbf{Q}_q \succeq \mathbf{0}}{\text{maximize}} & R_q(\mathbf{Q}_q, \mathbf{Q}_{-q}) \\ \text{subject to} & \mathrm{Tr}\left(\mathbf{Q}_q\right) \leq P_q, \quad \mathbf{U}_q^H \mathbf{Q}_q = \mathbf{0} \end{array} \quad \forall q \in \Omega, \qquad (11.83)$$

where $R_q(\mathbf{Q}_q, \mathbf{Q}_{-q})$ is defined in (11.4). Without the null constraints, the solution of each optimization problem in (11.83) would lead to the MIMO waterfilling solution, as studied in Section 11.3. The presence of the null constraints modifies the problem, and the solution for each user is not necessarily a waterfilling anymore. Nevertheless, we show now that introducing a proper projection matrix enables the solutions of (11.83) to still be efficiently computed via a waterfilling-like expression. To this end, we rewrite game $\mathcal{G}_{\text{null}}$ in the form of game $\mathcal{G}_{\text{pow}}$ in (11.22), as detailed next.

We need the following intermediate definitions. For any $q \in \Omega$, given $r_{H_{qq}} \triangleq \mathrm{rank}(\mathbf{H}_{qq})$ and $r_{U_q} \triangleq \mathrm{rank}(\mathbf{U}_q)$, with $r_{U_q} < n_{T_q}$ w.l.o.g., let $\mathbf{U}_q^{\perp} \in \mathbb{C}^{n_{T_q} \times r_{U_q^{\perp}}}$ be the semiunitary matrix orthogonal to $\mathbf{U}_q$ (note that $\mathcal{R}(\mathbf{U}_q^{\perp}) = \mathcal{R}(\mathbf{U}_q)^{\perp}$), with $r_{U_q^{\perp}} \triangleq \mathrm{rank}(\mathbf{U}_{qq}^{\perp}) = n_{T_q} - r_{U_q}$ and $\mathbf{P}_{\mathcal{R}(\mathbf{U}_q^{\perp})} = \mathbf{U}_q^{\perp} \mathbf{U}_q^{\perp H}$ be the orthogonal projection onto $\mathcal{R}(\mathbf{U}_q^{\perp})$. We can then rewrite the null constraint $\mathbf{U}_q^H \mathbf{Q}_q = \mathbf{0}$ in (11.83) as

$$\mathbf{Q}_q = \mathbf{P}_{\mathcal{R}(\mathbf{U}_q^{\perp})} \mathbf{Q}_q \mathbf{P}_{\mathcal{R}(\mathbf{U}_q^{\perp})}. \qquad (11.84)$$

At this point, the problem can be further simplified by noting that the constraint $\mathbf{Q}_q = \mathbf{P}_{\mathcal{R}(\mathbf{U}_q^{\perp})} \mathbf{Q}_q \mathbf{P}_{\mathcal{R}(\mathbf{U}_q^{\perp})}$ in (11.83) is redundant, provided that the original channels $\mathbf{H}_{rq}$ are replaced with the modified channels $\mathbf{H}_{rq}\mathbf{P}_{\mathcal{R}(\mathbf{U}_r^\perp)}$. The final formulation then becomes

$$
\begin{aligned}
&\underset{\mathbf{Q}_q \succeq \mathbf{0}}{\text{maximize}} && \log \det \left(\mathbf{I} + \mathbf{P}_{\mathcal{R}(\mathbf{U}_q^\perp)} \mathbf{H}_{qq}^H \tilde{\mathbf{R}}_{-q}^{-1}(\mathbf{Q}_{-q}) \mathbf{H}_{qq} \mathbf{P}_{\mathcal{R}(\mathbf{U}_q^\perp)} \mathbf{Q}_q \right) && \forall q \in \Omega, \\
&\text{subject to} && \mathsf{Tr}(\mathbf{Q}_q) \leq P_q
\end{aligned}
$$

(11.85)

where

$$
\tilde{\mathbf{R}}_{-q}(\mathbf{Q}_{-q}) \triangleq \mathbf{R}_{n_q} + \sum_{r \neq q} \mathbf{H}_{rq} \mathbf{P}_{\mathcal{R}(\mathbf{U}_r^\perp)} \mathbf{Q}_r \mathbf{P}_{\mathcal{R}(\mathbf{U}_r^\perp)} \mathbf{H}_{rq}^H \succ \mathbf{0}.
$$

(11.86)

Indeed, for any user q, any optimal solution $\mathbf{Q}_q^\star$ in (11.85) – the MIMO waterfilling solution – will be orthogonal to the null space of $\mathbf{H}_{qq}\mathbf{P}_{\mathcal{R}(\mathbf{U}_q^\perp)}$, whatever $\tilde{\mathbf{R}}_{-q}(\mathbf{Q}_{-q})$ is (recall that $\tilde{\mathbf{R}}_{-q}(\mathbf{Q}_{-q}) \succ \mathbf{0}$ for all feasible $\mathbf{Q}_{-q}$), implying $\mathcal{R}(\mathbf{Q}_q^\star) \subseteq \mathcal{R}(\mathbf{U}_q^\perp)$.

Building on the equivalence of (11.83) and (11.85), we can focus on the game in (11.85) and apply the framework developed in Section 11.3.1 to fully characterize game $\mathcal{G}_{\text{null}}$, by deriving the structure of the Nash equilibria and the conditions guaranteeing the existence/uniqueness of the equilibrium and the global convergence of the proposed distributed algorithms. We address these issues in the next sections.

Nash equilibria: existence and uniqueness

To write the Nash equilibria of game $\mathcal{G}_{\text{null}}$ in a convenient form, we need the following notations and definitions. Given the game in (11.85), we introduce the set $\tilde{\Omega}$ of user indexes associated with the rank-deficient matrices $\mathbf{H}_{qq}\mathbf{U}_q^\perp$, defined as

$$
\tilde{\Omega} \triangleq \left\{ q \in \Omega : r_{H_{qq}U_q^\perp} \triangleq \text{rank}\left(\mathbf{H}_{qq}\mathbf{U}_q^\perp \right) < \min\left(n_{R_q}, r_{U_q^\perp} \right) \right\},
$$

(11.87)

and the semiunitary matrices $\mathbf{V}_{q,1} \in \mathbb{C}^{r_{U_q^\perp} \times r_{H_{qq}U_q^\perp}}$ such that $\mathcal{R}(\mathbf{V}_{q,1}) = \mathcal{N}\left(\mathbf{H}_{qq}\mathbf{U}_q^\perp \right)^\perp$. To obtain weak conditions guaranteeing the uniqueness of the NE and convergence of the proposed algorithms, it is useful to also introduce: the modified channel matrices $\tilde{\mathbf{H}}_{rq} \in \mathbb{C}^{n_{R_q} \times r_{H_{rr}U_r^\perp}}$, defined as

$$
\tilde{\mathbf{H}}_{rq} = \begin{cases} \mathbf{H}_{rq}\mathbf{U}_r^\perp \mathbf{V}_{r,1}, & \text{if } r \in \tilde{\Omega}, \\[2mm] \mathbf{H}_{rq}\mathbf{U}_r^\perp, & \text{otherwise,} \end{cases} \qquad \forall r, q \in \Omega,
$$

(11.88)

the interference-plus-noise to noise ratios $\widetilde{\text{innr}}_q$s, defined as

$$
\widetilde{\text{innr}}_q \triangleq \frac{\rho\left(\mathbf{R}_{n_q} + \sum_{r \neq q} P_r \tilde{\mathbf{H}}_{rq} \tilde{\mathbf{H}}_{rq}^H \right)}{\lambda_{\min}\left(\mathbf{R}_{n_q} \right)} \geq 1,
$$

(11.89)

and the non-negative matrices $\mathbf{S}_{\mathrm{null}} \in \mathbb{R}_{+}^{Q \times Q}$ defined as

$$
[\mathbf{S}_{\mathrm{null}}]_{qr} \triangleq
\begin{cases}
\widetilde{\mathrm{innr}}_q \cdot \rho\left(\tilde{\mathbf{H}}_{rq}^H \tilde{\mathbf{H}}_{rq}\right) \rho\left(\tilde{\mathbf{H}}_{qq}^{\sharp H} \tilde{\mathbf{H}}_{qq}^{\sharp}\right), & \text{if } r \neq q, \\
0, & \text{otherwise.}
\end{cases}
\tag{11.90}
$$

Using the above definitions, the full characterization of the Nash equilibria of $\mathcal{G}_{\mathrm{null}}$ is stated in the following theorem, whose proof follows similar steps to that of Theorem 11.8 and thus is omitted.

THEOREM 11.10 Existence and uniqueness of the NE of $\mathcal{G}_{\mathrm{null}}$ *Consider the game $\mathcal{G}_{\mathrm{null}}$ in (11.83) and suppose w.l.o.g. that $r_{U_q} < n_{T_q}$, for all $q \in \Omega$. Then, the following hold:*

(a) *There always exists an NE, for any set of channel matrices, power constraints for the users, and null-shaping constraints;*
(b) *All the Nash equilibria are the solutions to the following set of nonlinear, matrix-value, fixed-point equations:*

$$
\mathbf{Q}_q^{\star} = \mathbf{U}_q^{\perp} \mathsf{WF}_q \left(\mathbf{U}_q^{\perp H} \mathbf{H}_{qq}^H \mathbf{R}_{-q}^{-1}(\mathbf{Q}_{-q}^{\star}) \mathbf{H}_{qq} \mathbf{U}_q^{\perp} \right) \mathbf{U}_q^{\perp H}, \quad \forall q \in \Omega,
\tag{11.91}
$$

with $\mathsf{WF}_q(\cdot)$ and $\mathbf{R}_{-q}(\mathbf{Q}_{-q})$ defined in (11.24) and (11.3), respectively;
(c) *The NE is unique if[8]*

$$
\rho(\mathbf{S}_{\mathrm{null}}) < 1,
\tag{C6}
$$

with $\mathbf{S}_{\mathrm{null}}$ defined in (11.90).

REMARK 11.4   *Structure of the Nash equilibria.* The structure of the Nash equilibria as given in (11.91) shows that the null constraints in the transmissions of secondary users can be handled without affecting the computational complexity: given the strategies $\mathbf{Q}_{-q}$ of the others, the optimal covariance matrix $\mathbf{Q}_q^{\star}$ of each user q can be efficiently computed via a MIMO waterfilling solution, provided that the original channel matrix $\mathbf{H}_{qq}$ is replaced by $\mathbf{H}_{qq} \mathbf{U}_q^{\perp}$. Observe that the structure of $\mathbf{Q}_q^{\star}$ in (11.91) has an intuitive interpretation: to guarantee that each user q does not transmit over a given subspace (spanned by the columns of $\mathbf{U}_q$), *whatever* the strategies of the other users are, while maximizing its information rate, it is enough to induce in the original channel matrix $\mathbf{H}_{qq}$ a null space that (at least) coincides with the subspace where the transmission is not allowed. This is precisely what is done in the payoff functions in (11.85) by replacing $\mathbf{H}_{qq}$ with $\mathbf{H}_{qq} \mathbf{P}_{\mathcal{R}(\mathbf{U}_q^{\perp})}$.

REMARK 11.5   *Physical interpretation of uniqueness conditions.* Similarly to (C1), condition (C6) quantifies how small the interference among secondary users must be to guarantee the uniqueness of the NE of the game. What affects the uniqueness of the equilibrium is only the MUI generated by secondary users in the subspaces orthogonal to

---

[8] Milder (but less easy to check) uniqueness conditions than (C6) are given in [24].

$\mathcal{R}(\mathbf{U}_q)$'s, that is, the subspaces where secondary users are allowed to transmit (note that all the Nash equilibria $\{\mathbf{Q}_q^\star\}_{q\in\Omega}$ satisfy $\mathcal{R}(\mathbf{Q}_q^\star) \subseteq \mathcal{R}(\mathbf{U}_q^\perp)$, for all $q \in \Omega$). Interestingly, one can also obtain uniqueness conditions that are independent of the null constraints $\{\mathbf{U}_q\}_{q\in\Omega}$: it is sufficient to replace in (C6) the modified channels $\widetilde{\mathbf{H}}_{rq}$ with the original channel matrices $\mathbf{H}_{rq}$ [63]. This means that if the NE is unique in a game without null constraints, then it is also unique with null constraints, which is not a trivial statement.

Observe that all the conditions above depend, among all, on the interference generated by the primary users and the power budgets of the secondary users through the $\widetilde{\mathrm{innr}}_q$; which is an undesired result. We overcome this issue in Section 11.4.2.

Distributed algorithms

To reach the Nash equilibria of game $\mathcal{G}_{\mathrm{null}}$ while satisfying the null constraints (11.5), one can use the asynchronous IWFA as given in Algorithm 2, where the best-response $\mathsf{T}_q(\mathbf{Q}_{-q})$ of each user q in (11.82) is replaced by the following:

$$\mathsf{T}_q(\mathbf{Q}_{-q}) \triangleq \mathbf{U}_q^\perp \mathsf{WF}_q \left(\mathbf{U}_q^{\perp H} \mathbf{H}_{qq}^H \mathbf{R}_{-q}^{-1}(\mathbf{Q}_{-q})\mathbf{H}_{qq}\mathbf{U}_q^\perp \right) \mathbf{U}_q^{\perp H}, \qquad (11.92)$$

where the MIMO waterfilling operator WF_q is defined in (11.24). Observe that such an algorithm has the same good properties of the algorithm proposed to reach the Nash equilibria of game $\mathcal{G}_{\mathrm{pow}}$ in (11.22) (see Remark 4 in Section 11.3.5). In particular, even in the presence of null constraints, the best-response of each player q can be efficiently and locally computed via a MIMO waterfilling-like solution, provided that each channel $\mathbf{H}_{qq}$ is replaced by the channel $\mathbf{H}_{qq}\mathbf{U}_q^\perp$. Furthermore, thanks to the inclusion of the null constraints in the game, the game-theoretical formulation, the proposed asynchronous IWFA based on the mapping $\mathsf{T}_q(\mathbf{Q}_{-q})$ in (11.92), does not suffer from the main drawback of the classical sequential IWFA [20, 62, 64], that is, the violation of the interference-temperature limits [2]. The convergence properties of the algorithm are given in the following theorem (the proof follows from results in Section 11.2.2).

THEOREM 11.11 *Suppose that condition (C6) of Theorem 11.10 is satisfied. Then, any sequence $\{\mathbf{Q}^{(n)}\}_{n=1}^\infty$ generated by the asynchronous MIMO IWFA, described in Algorithm 1 and based on mapping in (11.92), converges to the unique NE of game $\mathcal{G}_{\mathrm{null}}$, for any set of feasible initial conditions and updating schedule satisfying (A1)–(A3).*

11.4.2   Game with null constraints via virtual noise shaping

We have seen how to deal efficiently with null constraints in the rate-maximization game. However, condition (C6) guaranteeing the uniqueness of the NE as well as the convergence of the asynchronous IWFA depends, among all, on the interference generated by the primary users (through the innr_q's), which is an undesired result. In such a case, the NE might not be unique and there is no guarantee that the proposed algorithms converge to an equilibrium. To overcome this issue, we propose here an alternative approach to impose null constraints (11.5) on the transmissions of secondary users, based on the introduction of "virtual interferers." This leads to a new game with more relaxed uniqueness and convergence conditions. The solutions of this new game are "in general" different to the Nash equilibria of $\mathcal{G}_{\text{null}}$, but the two games are numerically shown to have almost the same performance in terms of sum-rate.

The idea behind this alternative approach can be easily understood if one considers the transmission over SISO frequency-selective channels, where all the channel matrices have the same eigenvectors (the DFT vectors): to avoid the use of a given subchannel, it is sufficient to introduce a "virtual" noise with sufficiently high power over that subchannel. The same idea cannot be directly applied to the MIMO case, as arbitrary MIMO channel matrices have different right/left singular vectors from each other. Nevertheless, we show how to bypass this difficulty to design the covariance matrix of the virtual noise (to be added to the noise covariance matrix of each secondary receiver), so that all the Nash equilibria of the game satisfy the null constraints along the specified directions. For the sake of notation simplicity and because of the space limitation, we focus here only on the case of square, nonsingular channel matrices $\mathbf{H}_{qq}$, in other words, $r_{H_{qq}} = n_{R_q} = n_{T_q}$ for all $q \in \Omega$. Let us consider the following strategic non-cooperative game:

$$(\mathcal{G}_\alpha): \quad \begin{array}{ll} \underset{\mathbf{Q}_q \succeq \mathbf{0}}{\text{maximize}} & \log \det \left(\mathbf{I} + \mathbf{H}_{qq}^H \mathbf{R}_{-q,\alpha}^{-1}(\mathbf{Q}_{-q}) \mathbf{H}_{qq} \mathbf{Q}_q \right) \\ \text{subject to} & \text{Tr}\left(\mathbf{Q}_q \right) \le P_q \end{array}, \qquad \forall q \in \Omega,$$

$$(11.93)$$

where

$$\mathbf{R}_{-q,\alpha}(\mathbf{Q}_{-q}) \triangleq \mathbf{R}_{n_q} + \sum_{r \neq q} \mathbf{H}_{rq} \mathbf{Q}_r \mathbf{H}_{rq}^H + \alpha \hat{\mathbf{U}}_q \hat{\mathbf{U}}_q^H \succ \mathbf{0}, \qquad (11.94)$$

denotes the MUI-plus-noise covariance matrix observed by secondary user q, plus the covariance matrix $\alpha \hat{\mathbf{U}}_q \hat{\mathbf{U}}_q^H$ of the virtual interference along $\mathcal{R}(\hat{\mathbf{U}}_q)$, where $\hat{\mathbf{U}}_q \in \mathbb{C}^{n_{R_q} \times r_{\hat{U}_q}}$ is a (strictly) tall matrix assumed to be full column-rank with $r_{\hat{U}_q} \triangleq \text{rank}(\hat{\mathbf{U}}_q) < r_{H_{qq}} (= n_{T_q} = n_{R_q})$ w.l.o.g., and α is a positive constant. Our interest is on deriving the asymptotic properties of the solutions of $\mathcal{G}_\alpha$, as $\alpha \to +\infty$, and the structure of $\hat{\mathbf{U}}_q$'s making the null constraints (11.5) satisfied.

To this end, we introduce the following intermediate definitions first. For each q, define the (strictly) tall full column-rank matrix $\hat{\mathbf{U}}_q^\perp \in \mathbb{C}^{n_{R_q} \times r_{\hat{U}_q^\perp}}$, with $r_{\hat{U}_q^\perp} = n_{R_q} - r_{\hat{U}_q} = \text{rank}(\hat{\mathbf{U}}_q^\perp)$ and such that $\mathcal{R}(\hat{\mathbf{U}}_q^\perp) = \mathcal{R}(\hat{\mathbf{U}}_q)^\perp$, and the modified (strictly fat) channel matrices $\hat{\mathbf{H}}_{rq} \in \mathbb{C}^{r_{\hat{U}_q^\perp} \times n_{T_r}}$:

$$\hat{\mathbf{H}}_{rq} = \hat{\mathbf{U}}_q^{\perp H} \mathbf{H}_{rq} \qquad \forall r, q \in \Omega. \qquad (11.95)$$

We then introduce the auxiliary game $\mathcal{G}_\infty$, defined as:

$$(\mathcal{G}_\infty): \quad \begin{array}{ll} \underset{\mathbf{Q}_q \succeq \mathbf{0}}{\text{maximize}} & \log \det \left(\mathbf{I} + \hat{\mathbf{H}}_{qq}^H \hat{\mathbf{R}}_{-q}^{-1}(\mathbf{Q}_{-q}) \hat{\mathbf{H}}_{qq} \mathbf{Q}_q \right) \\ \text{subject to} & \text{Tr}\left(\mathbf{Q}_q \right) \le P_q \end{array}, \qquad \forall q \in \Omega,$$

$$(11.96)$$

where

$$\hat{\mathbf{R}}_{-q}(\mathbf{Q}_{-q}) \triangleq \hat{\mathbf{U}}_q^{\perp H} \mathbf{R}_{n_q} \mathbf{U}_q^{\perp} + \sum_{r \neq q} \hat{\mathbf{H}}_{rq} \mathbf{Q}_r \hat{\mathbf{H}}_{rq}^H. \tag{11.97}$$

Building on the results obtained in Section 11.3.1, we study both games $\mathcal{G}_\alpha$ and $\mathcal{G}_\infty$, and derive the relationship between the Nash equilibria of $\mathcal{G}_\alpha$ and $\mathcal{G}_\infty$, showing that, under milder conditions, the two games are asymptotically equivalent (in the sense specified next), which will provide an alternative way to impose the null constraints (11.5).

Nash equilibria: existence and uniqueness

We introduce the non-negative matrices $\mathbf{S}_{\infty,1}, \mathbf{S}_{\infty,2} \in \mathbb{R}_+^{Q \times Q}$, defined as

$$\left[\mathbf{S}_{\infty,1}\right]_{qr} \triangleq \begin{cases} \rho\left(\hat{\mathbf{H}}_{rq}^H \hat{\mathbf{H}}_{qq}^{\sharp H} \hat{\mathbf{H}}_{qq}^{\sharp} \hat{\mathbf{H}}_{rq}\right), & \text{if } r \neq q, \\ 0, & \text{otherwise,} \end{cases} \tag{11.98}$$

$$\left[\mathbf{S}_{\infty,2}\right]_{qr} \triangleq \begin{cases} \rho\left(\mathbf{H}_{rq}^H \mathbf{H}_{qq}^{-H} \mathbf{P}_{\mathcal{R}(\mathbf{U}_q^{\perp})} \mathbf{H}_{qq}^{-1} \mathbf{H}_{rq}\right), & \text{if } r \neq q, \\ 0, & \text{otherwise.} \end{cases} \tag{11.99}$$

Game $\mathcal{G}_\alpha$. The full characterization of game $\mathcal{G}_\alpha$ is given in the following theorem, whose proof is based on existence and uniqueness results given in Section 11.2 and the contraction properties of the multiuser waterfilling mapping as derived in Section 11.3.1.

THEOREM 11.12 Existence and uniqueness of the NE of $\mathcal{G}_\alpha$ *Consider the game $\mathcal{G}_\alpha$ in (11.93), the following hold:*

(a) There always exists an NE, for any set of channel matrices, transmit power of the users, virtual interference matrices $\hat{\mathbf{U}}_q \hat{\mathbf{U}}_q^H$'s, and $\alpha \geq 0$;

(b) All the Nash equilibria are the solutions to the following set of nonlinear, matrix-value, fixed-point equations:

$$\mathbf{Q}_{q,\alpha}^\star = \mathsf{WF}_q\left(\mathbf{H}_{qq}^H \mathbf{R}_{-q,\alpha}^{-1}(\mathbf{Q}_{-q,\alpha}^\star)\mathbf{H}_{qq}\right), \quad \forall q \in \Omega, \tag{11.100}$$

with $\mathsf{WF}_q(\cdot)$ defined in (11.24);
(c) The NE is unique if

$$\rho(\mathbf{S}_{\mathrm{pow}}) < 1, \tag{C7}$$

with $\mathbf{S}_{\mathrm{pow}}$ defined in (11.43).

REMARK 11.6 *On the properties of game $\mathcal{G}_\alpha$.* Game $\mathcal{G}_\alpha$ has some interesting properties, namely: (a) the nash equilibria depend on α and the virtual interference covariance matrices $\hat{\mathbf{U}}_q \hat{\mathbf{U}}_q^H$'s, whereas uniqueness condition (C7) *does not*; and (b) as desired, the uniqueness of the NE is not affected by the presence of the primary users. Exploring this degree of freedom, one can thus choose, under condition (C7), the proper set of α and $\hat{\mathbf{U}}_q \hat{\mathbf{U}}_q^H$'s so that the (unique) NE of the game satisfies the null constraints (11.5), while keeping the uniqueness property of the equilibrium unaltered and independent of both $\hat{\mathbf{U}}_q \hat{\mathbf{U}}_q^H$'s and the interference level generated by the primary users. It is not difficult to realize that the optimal design of α and $\hat{\mathbf{U}}_q \hat{\mathbf{U}}_q^H$'s in $\mathcal{G}_\alpha$ passes through the properties of game $\mathcal{G}_\infty$, as detailed next.

Game $\mathcal{G}_\infty$. The properties of game $\mathcal{G}_\infty$ are given in the following.

THEOREM 11.13 Existence and uniqueness of the NE of $\mathcal{G}_\infty$ *Consider the game $\mathcal{G}_\infty$ in (11.96) and suppose w.l.o.g. that $r_{\hat{U}_q} < r_{H_{qq}} (= n_{R_q} = n_{T_q})$, for all $q \in \Omega$. Then, the following hold:*

(a) *There always exists an NE, for any set of channel matrices, transmit power of the users, and virtual interference matrices $\hat{\mathbf{U}}_q$'s;*

(b) *All the Nash equilibria are the solutions to the following set of nonlinear, matrix-value, fixed-point equations:*

$$\mathbf{Q}_{q,\infty}^\star = \mathsf{WF}_q\left(\hat{\mathbf{H}}_{qq}^H \hat{\mathbf{R}}_{-q}^{-1}(\mathbf{Q}_{-q,\infty}^\star)\hat{\mathbf{H}}_{qq}\right), \quad \forall q \in \Omega, \tag{11.101}$$

with $\mathsf{WF}_q(\cdot)$ defined in (11.24), and satisfy

$$\mathcal{R}(\mathbf{Q}_{q,\infty}^\star) \perp \mathcal{R}(\mathbf{H}_{qq}^{-1}\hat{\mathbf{U}}_q), \quad \forall q \in \Omega; \tag{11.102}$$

(c) *The NE is unique if*

$$\rho(\mathbf{S}_{\infty,1}) < 1, \tag{C8}$$

with $\mathbf{S}_{\infty,1}$ defined in (11.98).

REMARK 11.7 *Null constraints and virtual noise directions.* Condition (11.102) provides the desired relationship between the directions of the virtual noise to be introduced in the noise covariance matrix of the user (see (11.97)) – the matrix $\hat{\mathbf{U}}_q$ – and the real directions along which user q will not allocate any power, in other words, the matrix $\mathbf{U}_q$. It turns out that if user q is not allowed to allocate power along $\mathbf{U}_q$, it is sufficient to choose in (11.97) $\hat{\mathbf{U}}_q \triangleq \mathbf{H}_{qq}\mathbf{U}_q$. Exploring this choice, the structure of the Nash equilibria of game $\mathcal{G}_\infty$ can be further simplified, as given next.

COROLLARY 11.2 *Consider the game $\mathcal{G}_\infty$ and the null constraints (11.5) with $r_{H_{qq}} = n_{R_q} = n_{T_q}$ and $\hat{\mathbf{U}}_q = \mathbf{H}_{qq}\mathbf{U}_q$, for all $q \in \Omega$. Then, the following hold:*

(a) *All the Nash equilibria are the solutions to the following set of nonlinear, matrix-value, fixed-point equations:*

$$\mathbf{Q}_{q,\infty}^\star = \mathbf{U}_q^\perp \mathsf{WF}_q\left(\left(\mathbf{U}_q^{\perp H}\mathbf{H}_{qq}^{-1}\mathbf{R}_{-q}(\mathbf{Q}_{-q,\infty}^\star)\mathbf{H}_{qq}^{-H}\mathbf{U}_q^\perp\right)^{-1}\right)\mathbf{U}_q^{\perp H}, \quad \forall q \in \Omega, \tag{11.103}$$

with $\mathsf{WF}_q(\cdot)$ defined in (11.24);

(b) The NE is unique if

$$\rho(\mathbf{S}_{\infty,2}) < 1, \tag{C9}$$

with $\mathbf{S}_{\infty,2}$ defined in (11.99). $\square$

Observe that, since $\mathcal{R}(\mathbf{Q}^\star_{q,\infty}) \subseteq \mathcal{R}(\mathbf{U}^\perp_q)$, any solution $\mathbf{Q}^\star_{q,\infty}$ to (11.103) will be orthogonal to $\mathbf{U}_q$, whatever the strategies $\mathbf{Q}^\star_{-q,\infty}$ of the other secondary users are. Thus, all the Nash equilibria in (11.103) satisfy the null constraints (11.5).

At this point, however, one may ask: what is the physical meaning of a solution to (11.103)? Does it still correspond to a waterfilling over a real MIMO channel and thus to the maximization of mutual information? The interpretation of game $\mathcal{G}_\infty$ and its solutions passes through game $\mathcal{G}_\alpha$: we indeed prove next that the solutions to (11.103) can be reached as Nash equilibria of game $\mathcal{G}_\alpha$ for sufficiently large $\alpha > 0$.

Relationship between game $\mathcal{G}_\alpha$ and $\mathcal{G}_\infty$. The asymptotic behavior of the Nash equilibria of $\mathcal{G}_\alpha$ as $\alpha \to +\infty$, is given in the following (the proof can be found in [63]).

THEOREM 11.14 *Consider games $\mathcal{G}_\alpha$ and $\mathcal{G}_\infty$, with $r_{\hat{\mathbf{U}}_q} < r_{\mathbf{H}_{qq}}(= n_{T_q} = n_{R_q})$ for all $q \in \Omega$, and suppose that condition (C7) in Theorem 11.12 is satisfied. Then, the following hold:*

(a) $\mathcal{G}_\alpha$ and $\mathcal{G}_\infty$ admit a unique NE, denoted by $\mathbf{Q}^\star_\alpha$ and $\mathbf{Q}^\star_\infty$, respectively;
(b) The two games are asymptotically equivalent, in the sense that

$$\lim_{\alpha \to +\infty} \mathbf{Q}^\star_\alpha = \mathbf{Q}^\star_\infty. \tag{11.104}$$

Invoking Theorem 11.14 and Corollary 11.2 we obtained the following desired property of game $\mathcal{G}_\alpha$: under condition (C7) of Theorem 11.12, the (unique) NE of $\mathcal{G}_\alpha$ tends to satisfy the null constraints (11.5) for sufficiently large α (see (11.103) and (11.104)), provided that the virtual interference matrices $\{\hat{\mathbf{U}}_q\}_{q \in \Omega}$ in (11.94) are chosen according to Corollary 11.2. This approach provides an alternative way to impose the null constraints (11.5).

Distributed algorithms

To reach the Nash equilibria of game $\mathcal{G}_\alpha$ while satisfying the null constraints (11.5) (for sufficiently large α), one can use the asynchronous IWFA as given in Algorithm 2, where the best-response $\mathsf{T}_q(\mathbf{Q}_{-q})$ in (11.82) is replaced by

$$\mathsf{T}_{q,\alpha}(\mathbf{Q}_{-q}) \triangleq \mathsf{WF}_q\left(\mathbf{H}^H_{qq}\mathbf{R}^{-1}_{-q,\alpha}(\mathbf{Q}_{-q,})\mathbf{H}_{qq}\right), \tag{11.105}$$

where the MIMO waterfilling operator WF_q is defined in (11.24). Observe that such an algorithm has the same good properties of the algorithm proposed to reach the Nash equilibria of game $\mathcal{G}_{\mathrm{null}}$ in (11.83). In particular, the best-response of each player q can be still efficiently and locally computed via a MIMO waterfilling-like solution, provided that the virtual interference covariance matrix $\alpha\mathbf{U}_q\mathbf{U}^H_q$ is added to the MUI covariance matrix

$\mathbf{R}_{-q}(\mathbf{Q}_{-q})$ measured at the q-th receiver. The convergence properties of the algorithm are given in the following.

THEOREM 11.15   *Consider games $\mathcal{G}_\alpha$ and $\mathcal{G}_\infty$, with $r_{\hat{U}_q} < r_{H_{qq}} (= n_{T_q} = n_{R_q})$ for all $q \in \Omega$, and suppose that condition (C7) of Theorem 11.12 is satisfied. Then, the following hold:*

(a) As $n \to \infty$, the asynchronous MIMO IWFA, described in Algorithm 2 and based on mapping in (11.105), converges uniformly with respect to $\alpha \in \mathbb{R}_+$ to the unique NE of game $\mathcal{G}_\alpha$, for any set of feasible initial conditions, and updating schedule satisfying (A1)–(A3);

(b) The sequence $\mathbf{Q}_\alpha^{(n)} = \left(\mathbf{Q}_{q,\alpha}^{(n)} \right)_{q \in \Omega}$ generated by the algorithm satisfies:

$$\lim_{n \to +\infty} \lim_{\alpha \to +\infty} \mathbf{Q}_\alpha^{(n)} = \lim_{\alpha \to +\infty} \lim_{n \to +\infty} \mathbf{Q}_\alpha^{(n)} = \mathbf{Q}_\infty^\star, \qquad (11.106)$$

where $\mathbf{Q}_\infty^\star$ is the (unique) NE of game $\mathcal{G}_\infty$.

REMARK 11.8   *On the convergence/uniqueness conditions.* Condition (C7) guaranteeing the global convergence of the asynchronous IWFA to the unique NE of $\mathcal{G}_\alpha$ (for *any* $\alpha > 0$) has the desired property of being independent of both the interference generated by the primary users and the power budgets of the secondary users, which is the main difference with the uniqueness and convergence condition (C6) associated with game $\mathcal{G}_{\text{null}}$ in (11.83).

Example 11.1 Comparison of uniqueness/convergence conditions   Since the uniqueness/convergence conditions given so far depend on the channel matrices $\{\mathbf{H}_{rq}\}_{r,q \in \Omega}$, there is a nonzero probability that they will not be satisfied for a given channel realization drawn from a given probability space. To quantify the adequacy of our conditions, we tested them over a set of random channel matrices whose elements are generated as circularly symmetric, complex Gaussian random variables with variance equal to the inverse of the square distance between the associated transmitter-receiver links ("flat-fading channel model"). We consider a hierarchical CR network as depicted in Figure 11.2(a), composed of 3 secondary-user MIMO links and one primary user (the base station BS), sharing the same band. To preserve the QoS of the primary users, null constraints are imposed on the secondary users in the direction of the receiver of the primary user. In Figure 11.2(b), we plot the probability that conditions (C6) and (C7) are satisfied versus the intra-pair distance $d \in (0;1)$ (normalized by the cell's side) (see Figure 11.2(a)) between each secondary transmitter and the corresponding receiver (assumed for the simplicity of representation to be equal for all the secondary links), for different values of the transmit/receive antennas. Since condition (C6) depends on the interference generated by the primary user and the power budgets of the secondary users, we considered two different values of the SNR at the receivers of the secondary users, namely $\mathrm{snr}_q \triangleq P_q/\sigma_{q,\text{tot}}^2 = 0\,\mathrm{dB}$ and $\mathrm{snr}_q = 8\,\mathrm{dB}$, for all $q \in \Omega$, where $\sigma_{q,\text{tot}}^2$ is

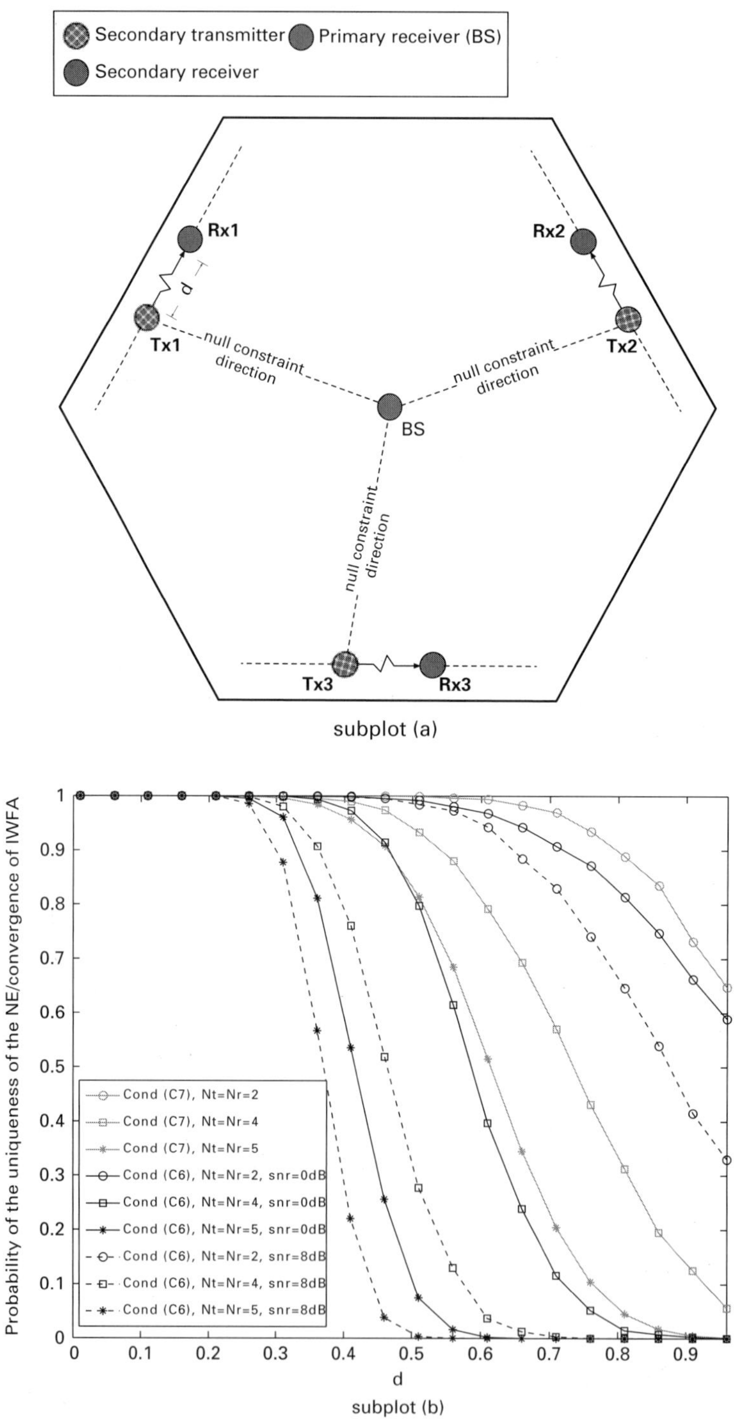

Figure 11.2   (a) CR MIMO system; (b) Probability of the uniqueness of the NE of games $\mathcal{G}_{\text{null}}$ and $\mathcal{G}_\alpha$ and convergence of the asynchronous IWFA as a function of the normalized intra-pair distance $d \in (0, 1)$.

the variance of thermal noise plus the interference generated by the primary user over all the substreams.

As expected, the probability of the uniqueness of the NE of both games $\mathcal{G}_{\text{null}}$ and $\mathcal{G}_{\alpha}$ and convergence of the IWFAs increases as each secondary transmitter approaches its receiver, corresponding to a decrease of the overall MUI. Moreover, condition (C6) is confirmed to be stronger than (C7) whatever the number of transmit/receive antennas, the intra-pair distance d, and the SNR value are, implying that game $\mathcal{G}_{\alpha}$ admits weaker (more desirable) uniqueness/convergence conditions than those of the original game $\mathcal{G}_{\text{null}}$.

Example 11.2 Performance of $\mathcal{G}_{\text{null}}$ and $\mathcal{G}_{\infty}$ As an example, in Figure 11.3, we compare games $\mathcal{G}_{\text{null}}$ and $\mathcal{G}_{\infty}$ in terms of sum-rate. All the Nash equilibria are computed using Algorithm 2 with mapping in (11.81) for game $\mathcal{G}_{\text{null}}$ and (11.103) for game $\mathcal{G}_{\infty}$. Specifically, in Figure 11.3(a), we plot the sum-rate at the (unique) NE of the games $\mathcal{G}_{\text{null}}$ and $\mathcal{G}_{\infty}$ for the CR network depicted in Figure 11.2(a) as a function of the intra-pair distance $d \in (0, 1)$ among the links, for different numbers of transmit/receive antennas. In Figure 11.3(b), we plot the outage sum-rate for the same systems as in Figure 11.3(a) and $d = 0.5$. For each secondary user, a null constraint in the direction of the receiver of the primary user is imposed. From the figures one infers that games $\mathcal{G}_{\text{null}}$ and $\mathcal{G}_{\infty}$ have almost the same performance in terms of sum-rate at the NE; even if in the game $\mathcal{G}_{\infty}$, given the strategies of the others, each player does not maximize its own rate, as happens in the game $\mathcal{G}_{\text{null}}$. This is due to the fact that the Nash equilibria of game $\mathcal{G}_{\text{null}}$ are in general not Pareto efficient.

In conclusion, the above results indicate that game $\mathcal{G}_{\alpha}$, with sufficiently large α, may be a valid alternative to game $\mathcal{G}_{\text{null}}$ to impose the null constraints (11.5), with more relaxed conditions for convergence.

11.4.3 Game with null and soft constraints

We focus now on the rate maximization in the presence of both null- and soft-shaping constraints. The resulting game can be formulated as follows:

$$
(\mathcal{G}_{\text{soft}}) : \quad
\begin{aligned}
&\underset{\mathbf{Q}_q \succeq \mathbf{0}}{\text{maximize}} && R_q(\mathbf{Q}_q, \mathbf{Q}_{-q}) \\
&\text{subject to} && \mathrm{Tr}\left(\mathbf{G}_q^H \mathbf{Q}_q \mathbf{G}_q\right) \leq P_{\text{SU},q}^{\text{ave}} \\
& && \lambda_{\max}\left(\mathbf{G}_q^H \mathbf{Q}_q \mathbf{G}_q\right) \leq P_{\text{SU},q}^{\text{peak}} && \forall q \in \Omega, \\
& && \mathbf{U}_q^H \mathbf{Q}_q = \mathbf{0}
\end{aligned}
\tag{11.107}
$$

where we have included both types of individual soft-shaping constraints as well as null-shaping constraints, and the transmit power constraint (11.2) has been absorbed

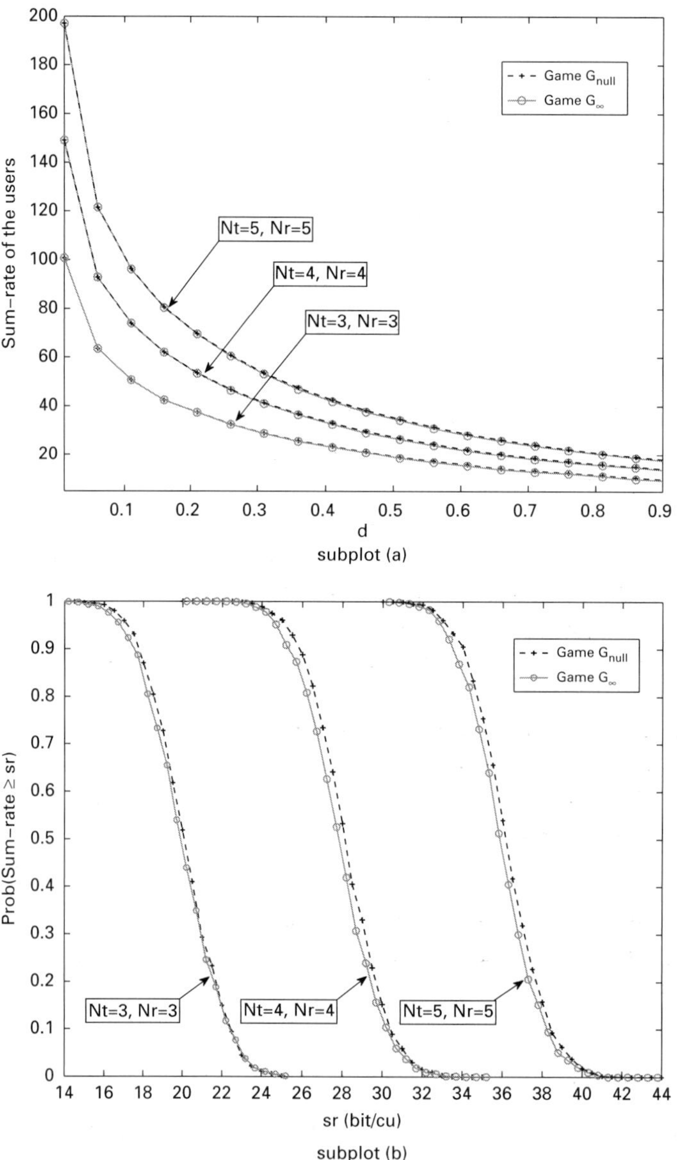

Figure 11.3 Performance of games $\mathcal{G}_{\text{null}}$ and $\mathcal{G}_\infty$ in terms of Nash equilibria for the CR MIMO system given in Figure 11.2(a): (a) Average sum-rate at the NE versus the normalized intra-pair distance $d \in (0, 1)$ for $d = 0.5$; (b) "Cumulative distribution function" (CDF) of the sum-rate for the games $\mathcal{G}_{\text{null}}$ (plus-mark, dashed-dot, line curves) and $\mathcal{G}_\infty$ (circle-mark, solid line curves).

into the trace soft constraint for convenience. For this, it is necessary that each $r_{G_q} \triangleq \text{rank}\left(\mathbf{G}_q\right) = n_{T_q}$; otherwise there would be no power constraint along $\mathcal{N}(\mathbf{G}_q^H)$ (if user q is allowed to transmit along $\mathcal{N}(\mathbf{G}_q^H)$, in other words, $\mathcal{N}(\mathbf{G}_q^H) \cap \mathcal{R}(\mathbf{U}_q)^\perp \neq \emptyset$). It is worth pointing out that, in practice, a transmit power constraint (11.2) in (11.107) will be dominated by the trace-shaping constraint, which motivates the absence in (11.107) of an explicit power constraint as in (11.2). More specifically, constraint (11.2) becomes redundant whenever $P_{\mathrm{SU},q}^{\mathrm{ave}} \leq P_q \lambda_{\min}(\mathbf{G}_q\mathbf{G}_q^H)$. On the other hand, if $P_{\mathrm{SU},q}^{\mathrm{ave}} \geq P_q \lambda_{\max}(\mathbf{G}_q\mathbf{G}_q^H)$, then constraint (11.6) can be removed without loss of optimality, and game $\mathcal{G}_{\mathrm{soft}}$ reduces in the form of game $\mathcal{G}_{\mathrm{null}}$. In the following, we then focus on the former case only.

Nash equilibria: existence and uniqueness

Before studying game $\mathcal{G}_{\mathrm{soft}}$, we need the following intermediate definitions. For any $q \in \Omega$, define the tall matrix $\overline{\mathbf{U}}_q \in \mathbb{C}^{n_{G_q} \times r_{U_q}}$ as $\overline{\mathbf{U}}_q \triangleq \mathbf{G}_q^{\sharp}\mathbf{U}_q$ (recall that $n_{G_q} \geq n_{T_q} > r_{U_q}$), and introduce: the semiunitary matrix $\overline{\mathbf{U}}_q^{\perp} \in \mathbb{C}^{n_{G_q} \times r_{\overline{U}_q^{\perp}}}$ orthogonal to $\overline{\mathbf{U}}_q$, with $r_{\overline{U}_q^{\perp}} = n_{G_q} - r_{U_q} = \mathrm{rank}(\overline{\mathbf{U}}_q^{\perp})$, the set of modified channels $\overline{\mathbf{H}}_{rq} \in \mathbb{C}^{n_{R_q} \times r_{\overline{U}_r^{\perp}}}$, defined as

$$\overline{\mathbf{H}}_{rq} = \mathbf{H}_{rq}\mathbf{G}_r^{\sharp H}\overline{\mathbf{U}}_r^{\perp}, \quad \forall r, q \in \Omega, \tag{11.108}$$

the interference-plus-noise to noise ratios $\overline{\mathrm{innr}}_q$'s, defined as

$$\overline{\mathrm{innr}}_q \triangleq \frac{\rho\left(\mathbf{R}_{n_q} + \sum_{r \neq q} P_r \overline{\mathbf{H}}_{rq}\overline{\mathbf{H}}_{rq}^H\right)}{\lambda_{\min}(\mathbf{R}_{n_q})} \geq 1 \quad q \in \Omega, \tag{11.109}$$

and the non-negative matrix $\mathbf{S}_{\mathrm{soft}} \in \mathbb{R}_+^{Q \times Q}$:

$$[\mathbf{S}_{\mathrm{soft}}]_{qr} \triangleq \begin{cases} \overline{\mathrm{innr}}_q \cdot \rho\left(\overline{\mathbf{H}}_{rq}^H\overline{\mathbf{H}}_{rq}\right) \rho\left(\overline{\mathbf{H}}_{qq}^{\sharp H}\overline{\mathbf{H}}_{qq}^{\sharp}\right), & \text{if } r \neq q, \\ 0, & \text{otherwise.} \end{cases} \tag{11.110}$$

These definitions are useful to obtain sufficient conditions for the uniqueness of the NE of $\mathcal{G}_{\mathrm{soft}}$. Finally, we introduce for any $q \in \Omega$ and given $n_q \in \{1, 2, \ldots, n_{T_q}\}$, the *modified* MIMO waterfilling operator $\overline{\mathsf{WF}}_q : \mathbb{S}_+^{n_q \times n_q} \ni \mathbf{X} \rightarrow \mathbb{S}_+^{n_q \times n_q}$, defined as

$$\overline{\mathsf{WF}}_q(\mathbf{X}) \triangleq \mathbf{U}_X \left[\mu_{q,X}\mathbf{I}_{r_X} - \mathbf{D}_X^{-1}\right]_0^{P_q^{\mathrm{peak}}} \mathbf{U}_X^H, \tag{11.111}$$

where $\mathbf{U}_X \in \mathbb{C}^{n_q \times r_X}$ and $\mathbf{D}_q \in \mathbb{R}_{++}^{r_X \times r_X}$ are defined as in (11.24) and $\mu_{q,X} > 0$ is the water-level chosen to satisfy $\mathrm{Tr}\left\{\left[\mu_{q,X}\mathbf{I}_{r_X} - \mathbf{D}_X^{-1}\right]_0^{P_q^{\mathrm{peak}}}\right\} = \min(P_q, r_X P_q^{\mathrm{peak}})$ (see, e.g., [65] for practical algorithms to compute the water-level $\mu_{q,X}$ in (11.111)). Using the above definitions, we can now characterize the Nash equilibria of game $\mathcal{G}_{\mathrm{soft}}$, as shown next.

THEOREM 11.16 Existence and structure of the NE of $\mathcal{G}_{\mathrm{soft}}$ *Consider the game $\mathcal{G}_{\mathrm{soft}}$ in (11.107), and suppose w.l.o.g. that $r_{G_q} = n_{T_q}$, for all $q \in \Omega$ (all matrices $\mathbf{G}_q$ are full row-rank). Then, the following hold:*

(a) *There always exists an NE, for any set of channel matrices and null/soft-shaping constraints;*

(b) *If, in addition, $r_{U_q} < r_{H_{qq}}$ and $\operatorname{rank}(\mathbf{H}_{qq}\mathbf{G}_q^{\sharp H}\overline{\mathbf{U}}_q^\perp) = r_{\overline{U}_q^\perp}$ for all $q \in \Omega$, all the Nash equilibria are the solutions to the following set of nonlinear, matrix-value, fixed-point equations:*

$$\mathbf{Q}_q^\star = \mathbf{G}_q^{\sharp H}\overline{\mathbf{U}}_q^\perp \overline{\mathsf{WF}}_q\left(\overline{\mathbf{H}}_{qq}^H \mathbf{R}_{-q}^{-1}(\mathbf{Q}_{-q}^\star)\overline{\mathbf{H}}_{qq}\right)\overline{\mathbf{U}}_q^{\perp H}\mathbf{G}_q^\sharp, \quad \forall q \in \Omega, \tag{11.112}$$

with $\overline{\mathsf{WF}}_q(\cdot)$ and $\mathbf{R}_{-q}(\mathbf{Q}_{-q})$ defined in (11.111) and (11.3), respectively.

Proof   The proof of theorem is based on the following intermediate result.

LEMMA 11.5   *Given $\mathbb{S}_+^{n_T} \ni \mathbf{R}_H = \mathbf{V}_H\mathbf{\Lambda}_H\mathbf{V}_H^H$, with $r_{R_H} = \operatorname{rank}(\mathbf{R}_H)$, the solution to the following optimization problem*

$$\begin{array}{ll}
\underset{\mathbf{Q}\succeq 0}{\text{maximize}} & \log\det(\mathbf{I} + \mathbf{R}_H\mathbf{Q}) \\[1em]
\text{subject to} & \operatorname{Tr}(\mathbf{Q}) \le P_T, \\[1em]
& \lambda_{\max}(\mathbf{Q}) \le P^{\text{peak}},
\end{array} \tag{11.113}$$

with $P_T \le P^{\text{peak}} r_{R_H}$, is unique and it is given by

$$\mathbf{Q}^\star = \mathbf{V}_{H,1}\left[\mu\mathbf{I}_{r_{R_H}} - \mathbf{\Lambda}_{H,1}^{-1}\right]_0^{P^{\text{peak}}} \mathbf{V}_{H,1}^H, \tag{11.114}$$

where $\mathbf{V}_{H,1} \in \mathbb{C}^{n_T \times r_{R_H}}$ is the semiunitary matrix of the eigenvectors of matrix $\mathbf{R}_H$ corresponding to the r_{R_H} positive eigenvalues in the diagonal matrix $\mathbf{\Lambda}_{H,1}$, and $\mu > 0$ satisfies $\operatorname{Tr}\left([\mu\mathbf{I}_{r_{R_H}} - \mathbf{\Lambda}_{H,1}^{-1}]_0^{P^{\text{peak}}}\right) = P_T.$

Under $r_{G_q} = n_{T_q}$, for all $q \in \Omega$, game $\mathcal{G}_{\text{soft}}$ admits at least an NE, since it satisfies Theorem 11.2.

We prove now (11.112). To this end, we rewrite $\mathcal{G}_{\text{soft}}$ in (11.107) in a more convenient form. Introducing the transformation:

$$\overline{\mathbf{Q}}_q \triangleq \mathbf{G}_q^H\mathbf{Q}_q\mathbf{G}_q, \qquad \forall q \in \Omega \tag{11.115}$$

one can rewrite $\mathcal{G}_{\text{soft}}$ in terms of $\overline{\mathbf{Q}}_q$ as

$$\begin{array}{ll}
\underset{\overline{\mathbf{Q}}_q\succeq 0}{\text{maximize}} & \log\det\left(\mathbf{I} + \mathbf{P}_{\mathcal{R}(\overline{\mathbf{U}}_q^\perp)}\mathbf{G}_q^\sharp\mathbf{H}_{qq}^H\overline{\mathbf{R}}_{-q}^{-1}(\overline{\mathbf{Q}}_{-q})\mathbf{H}_{qq}\mathbf{G}_q^{\sharp H}\mathbf{P}_{\mathcal{R}(\overline{\mathbf{U}}_q^\perp)}\overline{\mathbf{Q}}_q\right) \\[1em]
\text{subject to} & \operatorname{Tr}(\overline{\mathbf{Q}}_q) \le P_{\text{SU},q}^{\text{ave}} \\[1em]
& \lambda_{\max}(\overline{\mathbf{Q}}_q) \le P_{\text{SU},q}^{\text{peak}} \\[1em]
& \overline{\mathbf{Q}}_q = \mathbf{P}_{\mathcal{R}(\overline{\mathbf{U}}_q^\perp)}\overline{\mathbf{Q}}_q\mathbf{P}_{\mathcal{R}(\overline{\mathbf{U}}_q^\perp)}
\end{array} \quad \forall q \in \Omega, \tag{11.116}$$

where $\overline{\mathbf{R}}_{-q}(\overline{\mathbf{Q}}_{-q}) \triangleq \mathbf{R}_{n_q} + \sum_{r \neq q} \mathbf{H}_{rq} \mathbf{G}_r^{\sharp H} \mathbf{P}_{\mathcal{R}(\overline{\mathbf{U}}_r^\perp)} \overline{\mathbf{Q}}_r \mathbf{H}_{rq}^H \mathbf{G}_r^{\sharp} \mathbf{P}_{\mathcal{R}(\overline{\mathbf{U}}_r^\perp)}$. Observe now that the

power constraint $\mathrm{Tr}(\overline{\mathbf{Q}}_q) \leq P_{\mathrm{SU},q}^{\mathrm{ave}}$ in (11.116) can be replaced with $\mathrm{Tr}(\overline{\mathbf{Q}}_q) \leq \overline{P}_{\mathrm{SU},q}^{\mathrm{ave}}$

w.l.o.g., where $\overline{P}_{\mathrm{SU},q}^{\mathrm{ave}} \triangleq \min(P_{\mathrm{SU},q}^{\mathrm{ave}}, r_{\overline{U}_q^\perp} P_{\mathrm{SU},q}^{\mathrm{peak}})$. Indeed, because of the null constraint,

any solution $\overline{\mathbf{Q}}_q^\star$ to (11.116) will satisfy $\mathrm{rank}(\overline{\mathbf{Q}}_q^\star) \leq r_{\overline{U}_q^\perp}$, *whatever* the strategies $\overline{\mathbf{Q}}_{-q}$ of

the others are, implying $\mathrm{Tr}(\overline{\mathbf{Q}}_q) = \sum_{k=1}^{r_{\overline{U}_q^\perp}} \lambda_k(\overline{\mathbf{Q}}_q) \leq P_{\mathrm{SU},q}^{\mathrm{ave}}$ (the eigenvalues $\lambda_k(\overline{\mathbf{Q}}_q)$ are

assumed to be arranged in decreasing order); which, together with $\lambda_{\max}(\overline{\mathbf{Q}}_q) \leq P_{\mathrm{SU},q}^{\mathrm{peak}}$,

leads to the desired equivalence. Using $\mathrm{rank}(\mathbf{H}_{qq} \mathbf{G}_q^{\sharp H} \overline{\mathbf{U}}_q^\perp) = r_{\overline{U}_q^\perp}$ and invoking Lemma

11.5, the game in (11.116) can be further simplified to

$$
\begin{aligned}
&\underset{\overline{\mathbf{Q}}_q \succeq \mathbf{0}}{\text{maximize}} && \log \det \left(\mathbf{I} + \mathbf{P}_{\mathcal{R}(\overline{\mathbf{U}}_q^\perp)} \mathbf{G}_q^{\sharp} \mathbf{H}_{qq}^H \overline{\mathbf{R}}_{-q}^{-1}(\overline{\mathbf{Q}}_{-q}) \mathbf{H}_{qq} \mathbf{G}_q^{\sharp H} \mathbf{P}_{\mathcal{R}(\overline{\mathbf{U}}_q^\perp)} \overline{\mathbf{Q}}_q \right) \\
&\text{subject to} && \mathrm{Tr}(\overline{\mathbf{Q}}_q) \leq P_{\mathrm{SU},q}^{\mathrm{ave}} && \forall q \in \Omega. \\
& && \lambda_{\max}(\overline{\mathbf{Q}}_q) \leq P_{\mathrm{SU},q}^{\mathrm{peak}}
\end{aligned}
$$

$$(11.117)$$

Indeed, according to (11.114) in Lemma 11.5, any optimal solution $\overline{\mathbf{Q}}_q^\star$ to (11.117) will

satisfy $\mathcal{R}(\overline{\mathbf{Q}}_q^\star) \subseteq \mathcal{R}(\overline{\mathbf{U}}_q^\perp)$, implying that the null constraint $\overline{\mathbf{Q}}_q = \mathbf{P}_{\mathcal{R}(\overline{\mathbf{U}}_q^\perp)} \overline{\mathbf{Q}}_q \mathbf{P}_{\mathcal{R}(\overline{\mathbf{U}}_q^\perp)}$ in

(11.116) is redundant.

Given the game in (11.117), all the Nash equilibria satisfy the following MIMO waterfilling-like equation (Lemma 11.5):

$$
\overline{\mathbf{Q}}_q^\star = \overline{\mathsf{WF}}_q \left(\mathbf{P}_{\mathcal{R}(\overline{\mathbf{U}}_q^\perp)} \mathbf{G}_q^{\sharp} \mathbf{H}_{qq}^H \overline{\mathbf{R}}_{-q}^{-1}(\overline{\mathbf{Q}}_{-q}^\star) \mathbf{H}_{qq} \mathbf{G}_q^{\sharp H} \mathbf{P}_{\mathcal{R}(\overline{\mathbf{U}}_q^\perp)} \right) \tag{11.118}
$$

$$
= \overline{\mathbf{U}}_q^\perp \overline{\mathsf{WF}}_q \left(\overline{\mathbf{U}}_q^{\perp H} \mathbf{G}_q^{\sharp} \mathbf{H}_{qq}^H \overline{\mathbf{R}}_{-q}^{-1}(\overline{\mathbf{Q}}_{-q}^\star) \mathbf{H}_{qq} \mathbf{G}_q^{\sharp H} \overline{\mathbf{U}}_q^\perp \right) \overline{\mathbf{U}}_q^{\perp H}, \quad \forall q \in \Omega. \tag{11.119}
$$

The structure of the Nash equilibria of game $\mathcal{G}_{\mathrm{soft}}$ in (11.107) as given in (11.112) follows directly from (11.115) and (11.119). $\blacksquare$

REMARK 11.9 *On the structure of the Nash equilibria.* The structure of the Nash equilibria in (11.112) states that the optimal transmission strategy of each user leads to a diagonalizing transmission with a proper power allocation, after pre/post-multiplication by matrix $\mathbf{G}_q^{\sharp H} \overline{\mathbf{U}}_q^\perp$. Thus, even in the presence of soft constraints, the optimal transmission strategy of each user q, given the strategies $\mathbf{Q}_{-q}$ of the others, can be efficiently computed via a MIMO waterfilling-like solution. Note that the Nash equilibria in (11.112) satisfy the null constrains in (11.5), since $\mathcal{R}(\overline{\mathbf{U}}_q^\perp)^\perp = \mathcal{R}\left(\mathbf{G}_q^{\sharp} \mathbf{U}_q\right)$, implying $\mathbf{U}_q^H \mathbf{G}_q^{\sharp H} \overline{\mathbf{U}}_q^\perp = \mathbf{0}$ and thus $\mathcal{R}(\mathbf{Q}_q^\star) \perp \mathcal{R}\left(\mathbf{U}_q\right)$, for all $\mathbf{Q}_{-q} \succeq \mathbf{0}$ and $q \in \Omega$.

We provide now a more convenient expression for the Nash equilibria given in (11.112), that will be instrumental in deriving conditions for the uniqueness of the equilibrium and the convergence of the distributed algorithms. Introducing the convex closed sets $\overline{\mathcal{Q}}_q$ defined as

$$\overline{\mathcal{Q}}_q \triangleq \left\{ \mathbf{X} \in \mathbb{S}_+^{n_{T_q}} \mid \mathrm{Tr}\{\mathbf{X}\} = \overline{P}_{\mathrm{SU},q}^{\mathrm{ave}}, \quad \lambda_{\max}(\mathbf{X}) \leq P_{\mathrm{SU},q}^{\mathrm{peak}} \right\}, \tag{11.120}$$

where $\overline{P}_{\mathrm{SU},q}^{\mathrm{ave}} \triangleq \min(P_{\mathrm{SU},q}^{\mathrm{ave}}, r_{\overline{U}_q^{\perp}} P_{\mathrm{SU},q}^{\mathrm{peak}})$, we have the following equivalent expression for the MIMO waterfilling solutions in (11.112), whose proof is similar to that of Lemma 11.2 and thus is omitted.

LEMMA 11.6 NE as a projection *The set of nonlinear, matrix-value, fixed-point equations in (11.112) can be equivalently rewritten as*

$$\mathbf{Q}_q^{\star} = \mathbf{G}_q^{\sharp H} \overline{\mathbf{U}}_q^{\perp} \left[-\left(\left(\overline{\mathbf{H}}_{qq}^H \mathbf{R}_{-q}^{-1}(\mathbf{Q}_{-q}^{\star}) \overline{\mathbf{H}}_{qq} \right)^{\sharp} + c_q \mathbf{P}_{\mathcal{N}(\overline{\mathbf{H}}_{qq})} \right) \right]_{\overline{\mathcal{Q}}_q} \overline{\mathbf{U}}_q^{\perp H} \mathbf{G}_q^{\sharp}, \qquad \forall q \in \Omega, \tag{11.121}$$

where c_q is a positive constant that can be chosen independent of $\mathbf{Q}_{-q}$ (cf. [63]) and $\overline{\mathcal{Q}}_q$ is defined in (11.120).

Using Lemma 11.6, we can study contraction properties of the multiuser MIMO waterfilling mapping $\overline{\mathsf{WF}}$ in (11.112) via (11.121) (following the same approach as in Theorem 11.7) and obtain sufficient conditions guaranteeing the uniqueness of the NE of game $\mathcal{G}_{\mathrm{soft}}$, as given next.

THEOREM 11.17 Uniqueness of the NE *The solution to (11.121) is unique if*

$$\rho(\mathbf{S}_{\mathrm{soft}}) < 1, \tag{C10}$$

where $\mathbf{S}_{\mathrm{soft}}$ is defined in (11.110). $\qquad\qquad\square$

Condition (C10) is also sufficient for the convergence of the distributed algorithms to the unique NE of $\mathcal{G}_{\mathrm{soft}}$, as detailed in the next section.

Distributed algorithms

Similarly to games $\mathcal{G}_{\mathrm{null}}$ and $\mathcal{G}_\alpha$, the Nash equilibria of game $\mathcal{G}_{\mathrm{soft}}$ can be reached using the asynchronous IWFA algorithm given in Algorithm 2, based on the mapping

$$\overline{\mathsf{T}}_q(\mathbf{Q}_{-q}) \triangleq \mathbf{G}_q^{\sharp H} \overline{\mathbf{U}}_q^{\perp} \overline{\mathsf{WF}}_q \left(\overline{\mathbf{H}}_{qq}^H \mathbf{R}_{-q}^{-1}(\mathbf{Q}_{-q}) \overline{\mathbf{H}}_{qq} \right) \overline{\mathbf{U}}_q^{\perp H} \mathbf{G}_q^{\sharp}, \qquad q \in \Omega, \tag{11.122}$$

where the MIMO waterfilling operator is defined in (11.111) and the modified channels $\overline{\mathbf{H}}_{qq}$'s are defined in (11.108). Observe that such an algorithm has the same good properties of the algorithm proposed to reach the Nash equilibria of game $\mathcal{G}_{\mathrm{null}}$ in (11.83) (see Remark 4 in Section 11.3.5), such as: low-complexity, distributed and asynchronous nature, and fast convergence behavior. Moreover, thanks to our game-theoretical formulation including null and/or soft shaping constraints, the algorithm does not suffer from the main drawback of the classical sequential IWFA [20, 62, 64], that is, the violation of the interference-temperature limits [2]. The convergence properties of the algorithm are given in the following.

THEOREM 11.18 *Suppose that condition (C10) in Theorem 11.17 is satisfied. Then, any sequence $\{\mathbf{Q}^{(n)}\}_{n=1}^{\infty}$ generated by the asynchronous MIMO IWFA, described in Algorithm 2 and based on the mapping in (11.122), converges to the unique solution to (11.121), for any set of feasible initial conditions, and updating schedule satisfying* (A1)–(A3).

11.5 Opportunistic communications under global interference constraints

We focus now on the design of the CR system in (11.1), including the global interference constraints in (11.7), instead of the conservative individual constraints considered so far. This problem has been formulated and studied in [66]. Because of the space limitation, here we provide only some basic results without proofs. For the sake of simplicity, we focus only on block transmissions over SISO frequency-selective channels. It is well known that, in such a case, multicarrier transmission is capacity-achieving for large block-length [11]. This allows the simplification of the system model in (11.1), since each channel matrix $\mathbf{H}_{rq}$ becomes an $N \times N$ Toeplitz circulant matrix with eigendecomposition $\mathbf{H}_{rq} = \mathbf{F}\mathbf{D}_{rq}\mathbf{F}^{H}$, where $\mathbf{F}$ is the normalized IFFT matrix, i.e., $[\mathbf{F}]_{ij} \triangleq e^{j2\pi(i-1)(j-1)/N}/\sqrt{N}$ for $i,j = 1,\ldots,N$, N is the length of the transmitted block, $\mathbf{D}_{rq} = \mathrm{diag}(\{H_{rq}(k)\}_{k=1}^{N})$ is the diagonal matrix whose kth diagonal entry is the frequency-response of the channel between source r and destination q at carrier k, and $\mathbf{R}_{n_q} = \mathrm{diag}(\{\sigma_q^2(k)\}_{k=1}^{N})$.

Under this setup, the strategy of each secondary user q becomes the power allocation $\mathbf{p}_q = \{p_q(k)\}_{k=1}^{N}$ over the N carriers and the payoff function in (11.4) reduces to the information rate over the N parallel channels

$$r_q(\mathbf{p}_q, \mathbf{p}_{-q}) = \sum_{k=1}^{N} \log \left(1 + \frac{\left|H_{qq}(k)\right|^2 p_q(k)}{\sigma_q^2(k) + \sum_{r \neq q} \left|H_{rq}(k)\right|^2 p_r(k)} \right). \qquad (11.123)$$

Local power constraints and global interference constraints are imposed on the secondary users. The admissible strategy set of each player q associated with local power constraints is then

$$\mathcal{P}_q \triangleq \left\{ \mathbf{p} : \sum_{k=1}^{N} p(k) \leq P_q, \quad \mathbf{0} \leq \mathbf{p} \leq \mathbf{p}_q^{\max} \right\}, \qquad (11.124)$$

where we also included possibly (local) spectral-mask constraints $\mathbf{p}_q^{\max} = (p_q^{\max}(k))_{k=1}^{N}$. In the case of transmissions over frequency-selective channels, the global interference constraints in (11.7) impose an upper bound on the value of the per-carrier and total interference (the interference-temperature limit [2]) that can be tolerated by each primary user $p = 1, \cdots, P$, and reduce to [66]

$$\text{(total interference)}: \quad \sum_{q=1}^{Q}\sum_{k=1}^{N}\left|H_{q,p}(k)\right|^2 p_q(k) \le P_{p,\text{tot}}^{\text{ave}}$$

$$\text{(per-carrier interference)}: \quad \sum_{q=1}^{Q}\left|H_{q,p}(k)\right|^2 p_q(k) \le P_{p,k}^{\text{peak}}, \quad \forall k = 1, \cdots, N,$$

$$(11.125)$$

where $H_{q,p}(k)$ is the channel-transfer function between the transmitter of the qth secondary user and the receiver of the pth primary user, and $P_{p,\text{tot}}^{\text{ave}}$ and $P_{p,k}^{\text{peak}}$ are the interference-temperature limit and the maximum interference over subcarrier k tolerable by the p-th primary user, respectively. These limits are chosen by each primary user, according to its QoS requirements.

The aim of each secondary user is to maximize its own rate $r_q(\mathbf{p}_q, \mathbf{p}_{-q})$ under the local power constraints in (11.124) and the global interference constraints in (11.125). Note that the interference constraints introduce a global coupling among the admissible power allocations of all the players. This means that now the secondary users are not allowed to choose their power allocations individually, since this would lead to an infeasible strategy profile, where the global interference constraints, in general, are not satisfied. To keep the resource power allocation as decentralized as possible while imposing global interference constraints, the basic idea proposed in [66] is to introduce a proper pricing mechanism, controlled by the primary users, through a penalty in the payoff function of each player, so that the interference generated by all the secondary users will depend on these prices. The challenging goal is then to find the proper decentralized-pricing mechanism that guarantees the global interference constraints are satisfied while the secondary users reach an equilibrium. Stated in mathematical terms, we have the following NE problem [66]:

$$\underset{\mathbf{p}_q \ge \mathbf{0}}{\text{maximize}} \quad r_q(\mathbf{p}_q, \mathbf{p}_{-q}) - \sum_{p=1}^{P}\sum_{k=1}^{N}\lambda_{p,k}^{\text{peak}}\left|H_{q,p}(k)\right|^2 p_q(k) - \sum_{p=1}^{P}\lambda_{p,\text{tot}}\sum_{k=1}^{N}\left|H_{q,p}(k)\right|^2 p_q(k)$$

$$\text{subject to} \quad \mathbf{p}_q \in \mathcal{P}_q$$

$$(11.126)$$

for all $q \in \Omega$, where the prices $\lambda_{p,\text{tot}}$ and $\lambda_p^{\text{peak}} = \{\lambda_{p,k}^{\text{peak}}\}_{k=1}^{N}$ are chosen such that the following complementarity conditions are satisfied:

$$0 \le \lambda_{p,\text{tot}} \perp P_{p,\text{tot}}^{\text{ave}} - \sum_{q=1}^{Q}\sum_{k=1}^{N}\left|H_{q,p}(k)\right|^2 p_q(k) \ge 0, \quad \forall p,$$

$$0 \le \lambda_{p,k}^{\text{peak}} \perp P_{p,k}^{\text{peak}} - \sum_{q=1}^{Q}\left|H_{q,p}(k)\right|^2 p_q(k) \ge 0, \quad \forall p,k,$$

$$(11.127)$$

These constraints state that the per-carrier/global interference constraints must be satisfied together with nonnegative pricing; in addition, they imply that if one constraint is trivially satisfied with strict inequality then the corresponding price should be zero (no punishment is needed in that case). With a slight abuse of terminology, we will refer in the following to the problem in (11.126) with the complementarity constraints (11.127) as game $\mathcal{G}_{\mathrm{VI}}$.

11.5.1 Equilibrium solutions: existence and uniqueness

The coupling among the strategies of the players of $\mathcal{G}_{\mathrm{VI}}$ due to the global interference constraints presents a new challenge for the analysis of this class of Nash games that cannot be addressed using results from game-theory or game-theoretical models proposed in the literature [6–9, 21, 62]. For this purpose, we need the framework given by the more advanced theory of finite-dimensional VIs [32, 67] that provides a satisfactory resolution to the game $\mathcal{G}_{\mathrm{VI}}$, as detailed next. We first introduce the following definitions. Define the joint admissible strategy set of game $\mathcal{G}_{\mathrm{VI}}$ as

$$
\mathcal{K} \triangleq \mathcal{P} \cap \left\{
\begin{array}{ll}
\mathbf{p}: \ \displaystyle\sum_{q=1}^{Q}\sum_{k=1}^{N} \left|H_{q,p}(k)\right|^2 p_q(k) \leq P_{p,\mathrm{tot}}^{\mathrm{ave}}, & \forall p = 1, \cdots, P \\[4mm]
\displaystyle\sum_{q=1}^{Q} \left|H_{q,p}(k)\right|^2 p_q(k) \leq P_{p,k}^{\mathrm{peak}}, & \forall p = 1, \cdots, P, \ k = 1, \cdots, N
\end{array}
\right\},
$$

$$(11.128)$$

with $\mathcal{P} = \mathcal{P}_1 \times \cdots \times \mathcal{P}_Q$, and the vector function $\mathbf{F} : \mathcal{K} \ni \mathbf{p} \mapsto \mathbf{F}(\mathbf{p}) \in \mathbb{R}_{-}^{QN}$

$$
\mathbf{F}(\mathbf{p}) \triangleq \begin{pmatrix} \mathbf{F}_1(\mathbf{p}) \\ \vdots \\ \mathbf{F}_Q(\mathbf{p}) \end{pmatrix}, \ \text{where each} \ \mathbf{F}_q(\mathbf{p}) \triangleq \left(-\frac{\left|H_{qq}(k)\right|^2}{\sigma_q^2(k) + \sum_r \left|H_{rq}(k)\right|^2 p_r(k)} \right)_{k=1}^{N}.
$$

$$(11.129)$$

Finally, to rewrite the solutions to $\mathcal{G}_{\mathrm{VI}}$ in a convenient form, we introduce the interference-plus-noise to noise ratios $\mathrm{innr}_{rq}(k)$s, defined as

$$
\mathrm{innr}_{rq}(k) \triangleq \frac{\sigma_r^2(k) + \sum_t \left|H_{tr}(k)\right|^2 p_t^{\mathrm{max}}(k)}{\sigma_q^2(k)},
$$

$$(11.130)$$

and, for each q and given $\mathbf{p}_{-q} \geq \mathbf{0}$ and $\boldsymbol{\lambda} \geq \mathbf{0}$, define the waterfilling-like mapping wf_q as

$$
\left[\mathsf{wf}_q\left(\mathbf{p}_{-q}; \boldsymbol{\lambda}\right)\right]_k \triangleq \left[\frac{1}{\mu_q + \gamma_q(k; \boldsymbol{\lambda})} - \frac{\sigma_q^2(k) + \sum_{r \neq q} |H_{rq}(k)|^2 p_r(k)}{|H_{qq}(k)|^2} \right]_0^{p_q^{\mathrm{max}}(k)},
$$

$$(11.131)$$

with $k = 1, \cdots, N$, where $\gamma_q(k; \lambda) = \sum_{p=1}^{P} \left| H_{q,p}(k) \right|^2 (\lambda_{p,k}^{\mathrm{peak}} + \lambda_{p,\mathrm{tot}})$ and $\mu_q \geq 0$ is chosen to satisfy the power constraint $\sum_{k=1}^{N} \left[\mathsf{wf}_q \left(\mathbf{p}_{-q}; \lambda \right) \right]_k \leq P_q$ ($\mu_q = 0$ if the inequality is strictly satisfied).

THEOREM 11.19 [66]   *Consider the NE problem $\mathcal{G}_{\mathrm{VI}}$ in (11.126), the following hold:*

(a) *$\mathcal{G}_{\mathrm{VI}}$ is equivalent to the VI problem defined by the pair $(\mathcal{K}, \mathbf{F})$, which is to find a vector $\mathbf{p}^\star \in \mathcal{K}$ such that*

$$\left(\mathbf{p} - \mathbf{p}^\star \right)^T \mathbf{F}(\mathbf{p}^\star) \geq 0, \quad \forall \mathbf{p} \in \mathcal{K}, \tag{11.132}$$

with $\mathcal{K}$ and $\mathbf{F}(\mathbf{p})$ defined in (11.128) and (11.129), respectively;

(b) *There always exists a solution to the VI problem in (11.132), for any given set of channels, power budgets, and interference constraints;*

(c) *Given the set of the optimal prices $\hat{\lambda} = \{\hat{\lambda}_p^{\mathrm{peak}}, \hat{\lambda}_{p,\mathrm{tot}}\}_{p=1}^{P}$, the optimal power-allocation vector $\mathbf{p}^\star(\hat{\lambda}) = (\mathbf{p}_q^\star(\hat{\lambda}))_{q=1}^{Q}$ of the secondary users at an NE of game $\mathcal{G}_{\mathrm{VI}}$ is the solution to the following vector, waterfilling-like, fixed-point equation:*

$$\mathbf{p}_q^\star(\hat{\lambda}) = \mathsf{wf}_q \left(\mathbf{p}_{-q}^\star(\hat{\lambda}); \hat{\lambda} \right), \quad \forall q \in \Omega, \tag{11.133}$$

with wf_q defined in (11.131);

(d) *The optimal power-allocation vector $\mathbf{p}^\star$ of game $\mathcal{G}_{\mathrm{VI}}$ is unique if the two following sets of conditions are satisfied:[9]*

$$
\begin{aligned}
\text{Low received MUI:} \quad & \sum_{r \neq q} \max_k \left\{ \frac{\left| H_{rq}(k) \right|^2}{\left| H_{qq}(k) \right|^2} \cdot \mathrm{innr}_{rq}(k) \right\} < 1, \quad \forall q \in \Omega, \\
\text{Low generated MUI:} \quad & \sum_{q \neq r} \max_k \left\{ \frac{\left| H_{rq}(k) \right|^2}{\left| H_{qq}(k) \right|^2} \cdot \mathrm{innr}_{rq}(k) \right\} < 1, \quad \forall r \in \Omega,
\end{aligned}
\tag{C11}
$$

with $\mathrm{innr}_{rq}(k)$ defined in (11.130).

The equivalence between the game $\mathcal{G}_{\mathrm{VI}}$ in (11.126) and the VI problem in (11.132), as stated in Theorem 11.19(a) is in the following sense: if $\mathbf{p}^\star$ is a solution of the VI$(\mathcal{K}, \mathbf{F})$, then there exists a set of prices $\lambda^\star = (\lambda_p^\star, \lambda_{p,\mathrm{tot}}^\star)_{p=1}^{P} \geq \mathbf{0}$ such that $(\mathbf{p}^\star, \lambda^\star)$ is an equilibrium pair of $\mathcal{G}_{\mathrm{VI}}$; conversely if $(\mathbf{p}^\star, \lambda^\star)$ is an equilibrium of $\mathcal{G}_{\mathrm{VI}}$, then $\mathbf{p}^\star$ is a solution of the VI$(\mathcal{K}, \mathbf{F})$. Finally, observe that condition (C11) has the same good interpretations as those obtained for the games introduced so far: the uniqueness of the NE of $\mathcal{G}_{\mathrm{VI}}$ is guaranteed if the interference among the secondary users is not too high, in the sense specified by (C11).

---

[9] Milder conditions are given in [66].

11.5.2 Distributed algorithms

To obtain efficient algorithms that distributively compute both the optimal power allocations of the secondary users and prices, we can borrow from the wide literature of solutions methods for VIs [32, 67]. Many alternative algorithms have been proposed in [66] to solve game $\mathcal{G}_{\mathrm{VI}}$ that differ in: (a) the signaling among primary and secondary users needed to be implemented, (b) the computational effort, (c) the convergence speed, and (d) the convergence analysis. Because of the space limitation, here we focus only on one of them, based on the projection algorithm (with constant stepsize) [67, Algorithm 12.1.4] and formally described in Algorithm 3, where the waterfilling mapping wf_q is defined in (11.131).

Algorithm 3 Projection algorithm with constant stepsize

1 : Set $n = 0$, initialize $\boldsymbol{\lambda} = \boldsymbol{\lambda}^{(0)} \geq \mathbf{0}$, and choose the step size $\tau > 0$

2 : repeat

3 : Given $\boldsymbol{\lambda}^{(n)}$, compute $\mathbf{p}^{\star}(\boldsymbol{\lambda}^{(n)})$ as the solution to the fixed-point equation

4 :

$$\mathbf{p}_q^{\star}(\boldsymbol{\lambda}^{(n)}) = \mathsf{wf}_q\left(\mathbf{p}_{-q}^{\star}(\boldsymbol{\lambda}^{(n)}); \boldsymbol{\lambda}^{(n)}\right), \quad \forall q \in \Omega \tag{11.134}$$

5 : Update the price vectors: for all $p = 1, \cdots, P$, compute

6 :

$$\lambda_{p,\mathrm{tot}}^{(n+1)} = \left[\lambda_{p,\mathrm{tot}}^{(n)} - \tau\left(P_{p,\mathrm{tot}}^{\mathrm{ave}} - \sum_{q=1}^{Q}\sum_{k=1}^{N}\left|H_{q,p}(k)\right|^2 p_q^{\star}(k; \boldsymbol{\lambda}^{(n)})\right)\right]^+ \tag{11.135}$$

7 :

$$\lambda_{p,k}^{(n+1)} = \left[\lambda_{p,k}^{(n)} - \tau\left(P_{p,k}^{\mathrm{peak}} - \sum_{q=1}^{Q}\left|H_{q,p}(k)\right|^2 p_q^{\star}(k; \boldsymbol{\lambda}^{(n)})\right)\right]^+, \quad \forall k = 1, \cdots, N \tag{11.136}$$

8 : until the prescribed convergence criterion is satisfied

The algorithm can be interpreted as follows. In the main loop, at the nth iteration, each primary user p measures the received interference generated by the secondary users and, locally and independently from the other primary users, adjusts its own set of prices $\boldsymbol{\lambda}_p^{(n)}$ accordingly, via a simple projection scheme (see (11.135) and (11.136)). The primary users broadcast their own prices $\boldsymbol{\lambda}_p^{(n)}$'s to the secondary users, who then play the game in (11.126) keeping fixed the prices to the value $\boldsymbol{\lambda}^{(n)}$. The Nash equilibria of such a game are the fixed points of mapping $\mathsf{wf} = (\mathsf{wf}_q)_{q \in \Omega}$ as given in (11.133), with $\hat{\boldsymbol{\lambda}} = \boldsymbol{\lambda}^{(n)}$. Interestingly, the secondary users can reach these solutions using any algorithm falling within the class of asynchronous IWFAs as described in Algorithm 2 (e.g., simultaneous or sequential) and based on mapping $\mathsf{wf} = (\mathsf{wf}_q)_{q \in \Omega}$ in (11.133). Convergence properties of Algorithm 3 are given in the following.

THEOREM 11.20 [66]   *Suppose that condition (C11) in Theorem 11.19 is satisfied. Then, there exists some $\tau_0 > 0^{10}$ such that Algorithm 3 asymptotically converges to a solution to $\mathcal{G}_{\mathrm{VI}}$ in (11.126), for any set of feasible initial conditions and $\tau \in (0, \tau_0)$.*

REMARK 11.10   *Features of Algorithm 3.* Even though the per-carrier and global-interference constraints impose a coupling among the feasible power-allocation strategies of the secondary users, the equilibrium of game $\mathcal{G}_{\mathrm{VI}}$ can be reached using iterative algorithms that are fairly distributed with a minimum signaling from the primary to the secondary users. In fact, in Algorithm 3, the primary users, to update their prices, only need to measure the interference generated by the secondary users, which can be performed locally and independently from the other primary users. Regarding the secondary users (see (11.131)), once $\gamma_q(k; \boldsymbol{\lambda})$s are given, the optimal power allocation can be computed locally by each secondary user, since only the measure of the received MUI over the N subcarriers is needed. However, the computation of $\gamma_q(k; \boldsymbol{\lambda})$s requires a signaling among the primary and secondary users: at each iteration, the primary users have to broadcast the new values of the prices and the secondary users estimate the $\gamma_q(k; \boldsymbol{\lambda})$s. Note that, under the assumption of channel reciprocity, the computation of each term $\gamma_q(k; \boldsymbol{\lambda})$ does not requires the estimate from each secondary user of the (cross-)channel transfer functions between his transmitter and the primary receivers.

Example 11.3 Comparison of proposed algorithms   As a numerical example, in Figure 11.4, we compare some of the algorithms proposed in this chapter in terms of interference generated against the primary users. We consider a CR system composed of 6 secondary links randomly distributed within a hexagonal cell and one primary user (the BS at the center of the cell). In Figure 11.4(a) we plot the "power spectral density" (PSD) of the interference due to the secondary users at the receiver of the primary user, generated using the classical IWFA [20, 62, 64], the IWFA with individual interference constraints (i.e., a special case of Algorithm 2 applied to game $\mathcal{G}_{\mathrm{soft}}$) that we call "conservative" IWFA, and the IWFA with global-interference constraints (based on Algorithm 3) that we call "flexible" IWFA. For the sake of simplicity, we consider only a constant interference threshold over the whole spectrum occupied by the primary user, in other words, $P_{p,k}^{\mathrm{peak}} = 0.01$ for all $k = 1, \cdots, N$. We clearly see from the picture that while classical IWFA violates the interference constraints, both conservative and flexible IWFAs satisfy them, but the global interference constraints impose less stringent conditions on the transmit power of the secondary users than those imposed by the individual interference constraints. However, this comes at the price of more signaling from the primary to the secondary users. Interestingly, for the example considered in the figure, Algorithm 3 converges quite fast, as shown in Figure 11.4(b), where we plot the worst-case violation

---

[10] An expression for τ_0 is given in [66].

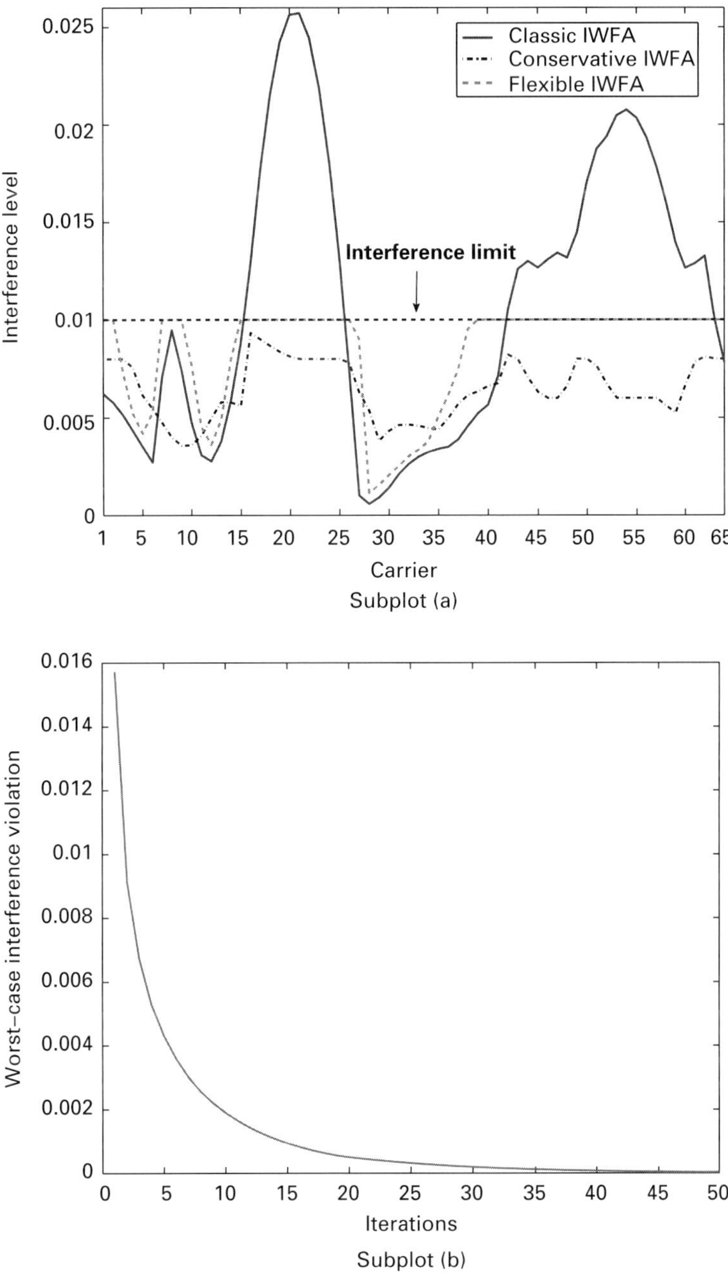

Figure 11.4 Comparison of different algorithms: (a) power spectral density of the interference profile at the primary user's receiver generated by the secondary users; (b) worst-case violation of the interference constraint achieved by Algorithm 3 (flexible IWFA).

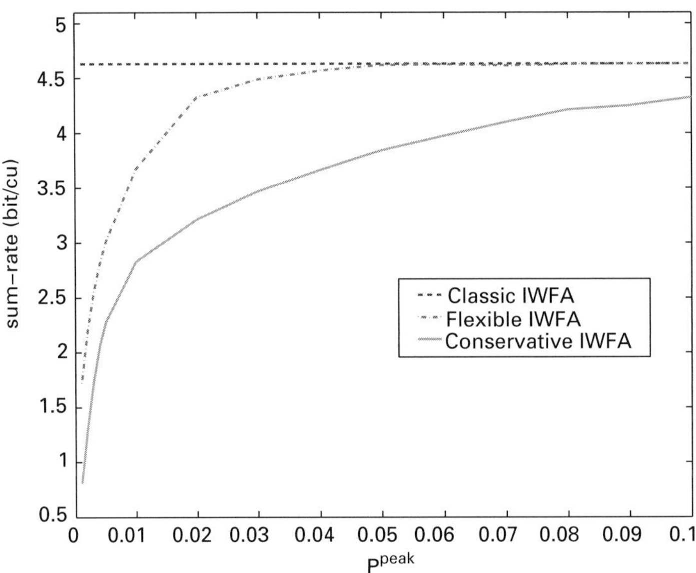

Figure 11.5   Conservative IWFA versus flexible IWFA: achievable sum-rate as a function of the maximum-tolerable interference at the primary receiver.

of the interference constraint achieved by the algorithm versus the number of iterations of the outer loop.

Finally, in Figure 11.5, we compare the conservative IWFA and the flexible IWFA in terms of achievable sum-rate as a function of the maximum tolerable interference at the primary receiver, within the same setup described above (we considered the same interference threshold P^{peak} for all the subcarriers). As expected, the flexible IWFA exhibits a much better performance, thanks to less stringent constraints on the transmit powers of the secondary users.

11.6   Conclusions

In this chapter we have proposed different equilibrium models to formulate and solve resource allocation problems in CR systems, using a competitive optimality principle based on the NE concept. We have seen how game theory and the more general VI theory provide the natural framework to address and solve some of the challenging issues in CR, namely: (a) the establishment of conditions guaranteeing that the dynamical interaction among cognitive nodes, under different constraints on the transmit spectral mask and on interference induced to primary users, admits a (possibly unique) equilibrium; and (b) the design of decentralized algorithms able to reach the equilibrium points, with minimal coordination among the nodes. The proposed algorithms differ in the trade-off between performance (in terms of information rate) achievable by the secondary users and the amount of information to be exchanged between the primary and the secondary users. Thus the algorithms are valid candidates to be applied to both main paradigms having emerged for CR systems, namely the common model and the spectral leasing approach. Results proposed in this chapter are based on recent works [6–8, 10, 24, 63, 66, 68].

Acknowledgments

We are grateful to Francisco Facchinei and Jong-Shi Pang for very helpful discussions. The research reported here was supported in part by the NSFC/RGC N_HKUST604/08 research grant, and in part by the SURFACE project funded by the European Community under Contract IST-4-027187-STP-SURFACE.

References

[1] FCC Spectrum Policy Task Force. (2002). Report of the spectrum efficiency working group. Available: www.fcc.gov/sptf/files/SEWGFinalReport1.pdf

[2] S. Haykin, "Cognitive radio: brain-empowered wireless communications," *IEEE Journal on Selected Areas in Communications*, vol. 23, no. 2, pp. 201–20, 2005.

[3] Q. Zhao and B. Sadler, "A survey of dynamic spectrum access," *IEEE Communications Magazine*, vol. 24, no. 3, pp. 79–89, 2009.

[4] A. Goldsmith, S. A. Jafar, I. Maric, and S. Srinivasa, "Breaking spectrum gridlock with cognitive radios: an information theoretic perspective," presented at the *Proceedings of the IEEE*, 2009.

[5] J. Mitola, "Cognitive radio for flexible mobile multimedia communication," in *IEEE International Workshop on Mobile Multimedia IEEE 1999 International Workshop on Mobile Multimedia Communications (MoMuC 1999)*, San Diego, CA, Nov. 15–17, 1999, pp. 3–10.

[6] G. Scutari, D. P. Palomar, and S. Barbarossa, "Optimal linear precoding strategies for wideband noncooperative systems based on game theory—part I: Nash equilibria," *IEEE Transactions on Signal Processing*, vol. 56, no. 3, pp. 1230–49, 2008.

[7] ——, "Optimal linear precoding strategies for wideband noncooperative systems based on game theory—part II: Algorithms," *IEEE Transactions on Signal Processing*, vol. 56, no. 3, pp. 1250–67, 2008.

[8] ——, "Asynchronous iterative water-filling for Gaussian frequency-selective interference channels," *IEEE Transactions on Information Theory*, vol. 54, no. 7, pp. 2868–78, 2008.

[9] J.-S. Pang, G. Scutari, F. Facchinei, and C. Wang, "Distributed power allocation with rate constraints in Gaussian parallel interference channels," *IEEE Transactions on Information Theory*, vol. 54, no. 8, pp. 3471–89, 2008.

[10] G. Scutari, D. P. Palomar, and S. Barbarossa, "Competitive design of multiuser MIMO systems based on game theory: a unified view," *IEEE Journal on Selected Areas in Communications*, vol. 26, no. 7, pp. 1089–103, 2008.

[11] T. M. Cover and J. A. Thomas, *Elements of Information Theory*. New York: John Wiley & Sons, 1991.

[12] N. Devroye, P. Mitran, and V. Tarokh, "Limits on communications in a cognitive radio channel," *IEEE Communications Magazine*, vol. 44, no. 6, pp. 44–9, 2006.

[13] *Working Group on Wireless Regional Area Networks*. IEEE Std. 802.22. Available: http://www.ieee802.org/22/

[14] N. Devroye, M. Vu, and V. Tarokh, "Cognitive radio networks: highlights of information theoretic limits, models, and design," *IEEE Signal Processing Magazine*, vol. 25, no. 6, pp. 12–23, 2008.

[15] B. Wild and K. Ramchandran, "Detecting primary receivers for cognitive radio applications," in *Proceedings of the IEEE 2005 Symposium on New Frontiers Dynamic Spectrum Access Networks (DYSPAN 2005)*, Baltimore, MA, 2005, pp. 124–30.

[16] Z. Quan, S. Cui, H. V. Poor, and A. H. Sayed, "Collaborative wideband sensing for cognitive radios: an overview of challenges and solutions," *IEEE Signal Processing Magazine*, vol. 25, no. 6, pp. 60–73, 2008.

[17] Z.-Q. Luo and S. Zhang, "Spectrum management: complexity and duality," *Journal of Selected Topics in Signal Processing*, vol. 2, no. 1, pp. 57–72, 2008.

[18] W. Yu and R. Lui, "Dual methods for nonconvex spectrum optimization of multicarrier systems," *IEEE Transactions on Communications*, vol. 54, no. 7, pp. 1310–22, 2006.

[19] S. Ye and R. S. Blum, "Optimized signaling for MIMO interference systems with feedback," *IEEE Transactions on Signal Processing*, vol. 51, no. 11, pp. 2839–48, 2003.

[20] W. Yu, G. Ginis, and J. M. Cioffi, "Distributed multiuser power control for digital subscriber lines," *IEEE Journal on Selected Areas in Communications*, vol. 20, no. 5, pp. 1105–15, 2002.

[21] Z.-Q. Luo and J.-S. Pang, "Analysis of iterative waterfilling algorithm for multiuser power control in digital subscriber lines," *EURASIP Journal on Applied Signal Processing*, vol. 2006, pp. 1–10, 2006.

[22] G. Arslan, M. F. Demirkol, and Y. Song, "Equilibrium efficiency improvement in MIMO interference systems: a decentralized stream control approach," *IEEE Transactions on Wireless Communications*, vol. 6, no. 8, pp. 2984–93, 2007.

[23] E. Larsson and E. Jorswieck, "Competition versus collaboration on the MISO interference channel," *IEEE Journal on Selected Areas in Communications*, vol. 26, no. 7, pp. 1059–69, 2008.

[24] G. Scutari, D. P. Palomar, and S. Barbarossa, "The MIMO iterative waterfilling algorithm," *IEEE Transactions on Signal Processing*, vol. 57, no. 5, pp. 1917–35, 2009.

[25] S. L. Campbell and C. D. Meyer, *Generalized Inverses of Linear Transformations*. New York: Dover Publications, 1991.

[26] R. A. Horn and C. R. Johnson, *Matrix Analysis*. New York: Cambridge University Press, 1985.

[27] J.-P. Aubin, *Mathematical Method for Game and Economic Theory*. New York: Dover Publications, 2007.

[28] T. Basar and G. J. Olsder, *Dynamic Noncooperative Game Theory*, 2nd ed. London: Academic Press, 1989.

[29] E. van Damme, *Stability and Perfection of Nash Equilibria*, 2nd ed. Berlin: Springer-Verlag, 1996.

[30] D. Fudenberg and J. Tirole, *Game Theory*. Cambridge: MIT Press, 1991.

[31] C. B. Garcia and W. I. Zangwill, *Pathways to Solutions, Fixed Points, and Equilibria*. New Jersey: Prentice-Hall, 1981.

[32] I. V. Konnov, *Equilibrium Models and Variational Inequalities*. Amsterdam: Elsevier B.V., 2007.

[33] F. Facchinei and C. Kanzow, "Generalized Nash equilibrium problems," *A Quarterly Journal of Operations Research (4OR)*, vol. 5, no. 3, pp. 173–210, 2007.

[34] D. Monderer and L. S. Shapley, "Potential games," *Games and Economic Behavior*, vol. 14, no. 1, pp. 124–43, 1996.

[35] D. Topkis, *Supermodularity and Complementarity*. Princeton, NJ: Princeton University Press, 1998.

[36] E. Altman and L. Wynter, "Equilibrium, games, and pricing in transportation and telecommunications networks," *Networks and Spacial Economics*, vol. 4, no. 1, pp. 7–21, 2004.

[37] E. Altman, T. Boulogne, R. E. Azouzi, T. Jimenez, and L. Wynter, "A survey on networking games," *Computers and Operations Research*, vol. 33, pp. 286–311, 2006.

[38] A. B. MacKenzie and S. B. Wicker, "Game theory and the design of self-configuring, adaptive wireless networks," *IEEE Communication Magazine*, vol. 39, no. 11, pp. 126–31, 2001.

[39] A. Chinchuluun, P. M. Pardalos, A. Migdalas, and L. Pitsoulis, *Pareto Optimality, Game Theory and Equilibria*, Series: Springer Optimization and Its Applications, vol. 17. New York: Springer-Verlag, 2008.

[40] M. J. Osborne and A. Rubinstein, *A Course in Game Theory*. Cambridge, MA: MIT Press, July 2004.

[41] J. Nash, "Equilibrium points in n-person game," in *National Academy of Science*, vol. 36, 1950, pp. 48–9.

[42] F. Facchinei and J.-S. Pang, "Nash equilibria: The variational approach," in *Convex Optimization in Signal Processing and Communications*, D. P. Palomar and Y. C. Eldar, Eds. London: Cambridge University Press, 2009.

[43] F. Szidarovszky, "Nonlinear games," in *Pareto Optimality, Game Theory And Equilibria*, A. Chinchuluun, P. M. Pardalos, A. Migdalas, and L. Pitsoulis, Eds. New York: Springer, 2008, vol. 17, ch. 16, pp. 95–117.

[44] I. L. Glicksburg, "A social equilibrium existence theorem," in *American Mathematical Society*, vol. 38, 1952, pp. 170–4.

[45] G. Debreu, "A social equilibrium existence theorem," in *National Academy of Science*, vol. 38, no. 2, Oct. 1952, pp. 886–93.

[46] J. Rosen, "Existence and uniqueness of equilibrium points for concave n-person games," *Econometrica*, vol. 33, no. 3, pp. 520–34, 1965.

[47] K. Fan, "A generalization of Tychonoff's fixed point theorem," *Mathematische Annalen*, vol. 42, pp. 305–10, 1961.

[48] S. Uryasev and R. Y. Rubinstein, "On relaxation algorithms in computation of noncooperative equilibria," *IEEE Transactions on Automatic Control*, vol. 39, no. 6, pp. 1263–7, 1994.

[49] D. Gabay and H. Moulin, "On the uniqueness and stability of Nash-equilibria in noncooperative games," in *Applied Stochastic Control in Econometrics and Management Science*. A. Bensoussan et al., eds. Amsterdam: North-Holland Publishing Company, 1980, pp. 271–293.

[50] A. Simsek, A. Ozdaglar, and D. Acemoglu, "Uniqueness of generalized equilibrium for box constrained problems and applications," paper presented at the *43rd Allerton Conference on Communications, Control, and Computing*, Allerton House, Monticello, IL, 28–30, 2005.

[51] ——, "Generalized Poincare-Hopf theorem for compact nonsmooth regions," *Mathematics of Operations Research*, vol. 32, no. 1, pp. 193–214, 2007.

[52] D. P. Bertsekas and J. N. Tsitsiklis, *Parallel and Distributed Computation: Numerical Methods*, 2nd ed. Nashua, NH: Athena Scientific Press, 1989.

[53] J. M. Ortega and W. C. Rheinboldt, *Iterative Solution of Nonlinear Equations in Several Variables*. Philadelphia, PA: Society for Industrial Mathematics (SIAM), 1987.

[54] R. W. Cottle, J.-S. Pang, and R. E. Stone, *The Linear Complementarity Problem*. Boston, MA: Cambridge Academic Press, 1992.

[55] G. Scutari, "Competition and cooperation in wireless communication networks," Ph.D. dissertation, INFOCOM Dept., University of Rome, "La Sapienza", 2004.

[56] B. Zheng and Z. Xiong, "A new equivalent condition of the reverse order law for g-inverses of multiple matrix products," *Electronic Journal of Linear Algebra*, vol. 17, no. 3, pp. 1–8, 2008.

[57] J. R. Magnus and H. Neudecker, *Matrix Differential Calculus, with Applications in Statistics and Econometrics*. New York: John Wiley & Sons, 1999.

[58] W. Rudin, *Principles of Mathematical Analysis*, 3rd ed. McGraw-Hill Publishing Co, 1976.

[59] R. M. Mcleod, "Mean value theorems for vector valued functions," *Edinburgh Mathematical Society*, vol. 14, pp. 197–209, Series II 1965.

[60] M. Furi and M. Martelli, "On the mean value theorem, inequality, and inclusion," *The American Mathematical Monthly*, vol. 9, pp. 840–6, 1991.

[61] A. Hjørungnes and D. Gesbert, "Complex-valued matrix differentiation: techniques and key results," *IEEE Transactions on Signal Processing*, vol. 55, no. 4, pp. 2740–6, 2007.

[62] J. Huang, R. Cendrillon, M. Chiang, and M. Moonen, "Autonomous spectrum balancing for digital subscriber lines," *IEEE Transactions on Signal Processing*, vol. 55, no. 8, pp. 4241–57, 2007.

[63] G. Scutari and D. P. Palomar, "MIMO cognitive radio: A game theoretical approach," *IEEE Transactions on Signal Processing,* accepted for publication.

[64] R. Etkin, A. Parekh, and D. Tse, "Spectrum sharing for unlicensed bands," *IEEE Journal on Selected Areas in Communications*, vol. 25, no. 3, pp. 517–28, 2007.

[65] D. P. Palomar and J. Fonollosa, "Practical algorithms for a family of waterfilling solutions," *IEEE Transactions on Signal Processing*, vol. 53, no. 2, pp. 686–95, 2005.

[66] J.-S. Pang, G. Scutari, D. P. Palomar, and F. Facchinei, "Cognitive radio wireless systems: flexible designs via variational inequality theory," *IEEE Transactions on Signal Processing,* accepted for publication.

[67] F. Facchinei and J.-S. Pang, *Finite-Dimensional Variational Inequalities and Complementarity Problem*. New York: Springer-Verlag, 2003.

[68] G. Scutari, D. P. Palomar, and S. Barbarossa, "Cognitive MIMO radio: competitive optimality design based on subspace projections," *IEEE Signal Processing Magazine*, vol. 25, no. 6, pp. 46–59, 2008.

12 Nash equilibria: the variational approach

Francisco Facchinei and Jong-Shi Pang

Non-cooperative game theory is a branch of game theory for the resolution of conflicts among players (or economic agents), each behaving selfishly to optimize their own well-being subject to resource limitations and other constraints that may depend on the rivals' actions. While many telecommunication problems have traditionally been approached by using optimization, game models are being increasingly used; they seem to provide meaningful models for many applications where the interaction among several agents is by no means negligible, for example, the choice of power allocations, routing strategies, and prices. Furthermore, the deregulation of telecommunication markets and the explosive growth of the Internet pose many new problems that can be effectively tackled with game-theoretic tools. In this chapter, we present a comprehensive treatment of Nash equilibria based on the variational inequality and complementarity approach, covering the topics of existence of equilibria using degree theory, global uniqueness of an equilibrium using the P-property, local-sensitivity analysis using degree theory, iterative algorithms using fixed-point iterations, and a descent approach for computing variational equilibria based on the regularized Nikaido–Isoda function. We illustrate the existence theory using a communication game with QoS constraints. The results can be used for the further study of conflict resolution of selfish agents in telecommunication.

12.1 Introduction

The literature on non-cooperative games is vast. Rather than reviewing this extensive literature, we refer the readers to the recent survey [20], which we will use as the starting point of this chapter. Indeed, our goal herein is to go beyond the cited survey and present a detailed mathematical treatment of the Nash-equilibrium problem as a class of (possibly multi-valued) variational inequalities, following the path presented in the two-volume monograph [21] on the latter subject. Our point of view of the Nash-equilibrium problem is that it is an extension of an optimization problem; therefore, the variational approach, supported by its deep and broad foundation, offers a constructive and powerful platform for fruitful developments.

This chapter is organized in seven sections. The next section presents the setting of the problem and introduces the variational approach. An existence theory is the main topic of Section 12.3 which is complemented by Section 12.4 dealing with the uniqueness

Convex Optimization in Signal Processing and Communication, eds. Daniel P. Palomar and Yonina C. Eldar. Published by Cambridge University Press © Cambridge University Press, 2010.

of equilibria. Sensitivity to data variations is the subject of Section 12.5, while two families of iterative algorithms are analyzed in Section 12.6. Finally, in Section 12.7, we illustrate some of the general results by applying them to a communication game with quality-of-service constraints. The chapter is concluded with a brief summary.

12.2   The Nash-equilibrium problem

The mathematical definition of the basic Nash-equilibrium problem is as follows. There are N players each of whom has a strategy set $X^\nu \subseteq \mathbb{R}^{n_\nu}$ and an objective function $\theta_\nu : \mathbb{R}^n \to \mathbb{R}$, where $n \triangleq \sum_{\nu=1}^{N} n_\nu$. By definition, a tuple $x^* \equiv (x^{*,\nu})_{\nu=1}^{N} \in \widehat{X} \triangleq \prod_{\nu=1}^{N} X^\nu$ is a "Nash equilibrium" (NE) of the game $\mathcal{G}$ defined by the pair $(\widehat{X}, \Theta)$, where Θ denotes the tuple $(\theta_\nu)_{\nu=1}^{N}$, if, for every $\nu = 1, \ldots, N$,

$$\theta_\nu(x^*) \le \theta_\nu(x^\nu, x^{*,-\nu}), \quad \forall x^\nu \in X^\nu.$$

In words, a feasible tuple x^*, in other words, a tuple x^* satisfying $x^{*,\nu} \in X^\nu$ for all ν, is an NE if no player can improve his objective by unilaterally deviating from the chosen strategy $x^{*,\nu}$. A key word underlying this definition is the adverb "unilaterally." In particular, an NE does not rule out the possibility for the players to each lower their respective objectives by colluding. First proved by Nash [40,41] for the mixed extension of a finite game, a fundamental fact of the above game is that if each set X^ν is compact and convex, each θ_ν is continuous, and for each $x^{-\nu} \in X^{-\nu} \triangleq \prod_{\nu'\neq\nu} X^{\nu'}$, the function $\theta_\nu(\bullet, x^{-\nu}) : \mathbb{R}^{n_\nu} \to \mathbb{R}$ is convex, then an NE of the game exists.

Extending the above basic setting, we consider a game in which each player's strategy set X^ν is no longer a fixed set; instead, it becomes a moving set $\Xi^\nu(x^{-\nu}) \subseteq \mathbb{R}^{n_\nu}$ that varies with the rival's strategy $x^{-\nu}$. Abstractly, each such set $\Xi^\nu(x^{-\nu})$ is the image of a multifunction Ξ^ν that maps the strategy space $\widehat{\mathbb{R}}^\nu \triangleq \prod_{\nu'\neq\nu} \mathbb{R}^{n_{\nu'}}$ of player ν's rivals into player ν's strategy space $\mathbb{R}^{n_\nu}$. Denoted alternatively by $\mathcal{G}(\Xi, \Theta)$ and $\mathcal{G}(\Xi, \partial\Theta)$ (see Subsection 12.2.2 for an explanation of the latter notation), the resulting game is of the generalized kind, in which each player's strategy set is dependent on its rivals' strategies; this dependence is in addition to that in each player's objective function θ_ν. In the generalized game $\mathcal{G}(\Xi, \Theta)$, a tuple $x^* \equiv (x^{*,\nu})_{\nu=1}^{N}$ is, by definition, a Nash equilibrium if, for every $\nu = 1, \ldots, N$, $x^{*,\nu}$ is a global minimizer of player ν's optimization problem:

$$\underset{x^\nu \in \Xi^\nu(x^{*,-\nu})}{\text{minimize}} \ \theta_\nu(x^\nu, x^{*,-\nu}), \tag{12.1}$$

or equivalently,

$$\theta_\nu(x^*) \le \theta_\nu(x^\nu, x^{*,-\nu}), \quad \forall x^\nu \in \Xi^\nu(x^{*,-\nu}).$$

Throughout the chapter, we assume that each $\Xi^\nu(x^{-\nu})$ is a closed-convex, albeit not necessarily bounded, subset of $\mathbb{R}^{n_\nu}$. Despite the convexity of each set $\Xi^\nu(x^{-\nu})$, the graph of the multifunction Ξ^ν, in other words, the set $\text{gph}(\Xi^\nu) \triangleq \{x : x^\nu \in \Xi^\nu(x^{-\nu})\}$, is not necessarily a convex subset of $\mathbb{R}^n$. An example is given by $\Xi^\nu(x^{-\nu}) = \{x^\nu : g_\nu(x^\nu, x^{-\nu}) \leq 0\}$ with $g_\nu(\bullet, x^{-\nu})$ being a convex function for fixed $x^{-\nu}$ but with g_ν not convex in $(x^\nu, x^{-\nu})$ jointly. In this case, $\text{gph}(\Xi^\nu) = \{x : g_\nu(x) \leq 0\}$ is the level set of a nonconvex function g_ν.

An important special case of the generalized game $\mathcal{G}(\Xi, \Theta)$ arises when $\text{gph}(\Xi^\nu)$ is convex for all $\nu = 1, \cdots, N$. A further special case is when all these graphs are equal to a common (convex) set $\mathbf{C}$ in $\mathbb{R}^n$. Thus there are at least these two aspects of the game $\mathcal{G}(\Xi, \partial\Theta)$ that one needs to deal with: (a) the possible nonconvexity of the graphs of the Ξ^ν, whose intersection defines the feasible set of the game, and (b) the distinctiveness of these player-dependent strategy graphs. Another important generality to be concerned with is the boundedness of these graphs and that of the sets $\Xi^\nu(x^{-\nu})$ for given $x^{-\nu}$. Finally, the fact that the players' objective functions θ_ν are allowed to be non-differentiable makes the analysis of the game more challenging. For simplicity, we make the following standing assumption throughout this chapter: the functions θ_ν are defined and continuous on all of $\mathbb{R}^n$, with $\theta_\nu(\bullet, x^{-\nu})$ being convex for all $x^{-\nu} \in \widehat{\mathbb{R}}^\nu$.

In Section 12.7, we will discuss an interesting game in signal processing. Here, we consider two simple games to illustrate the setting introduced so far. The first one is a standard Nash problem, while the latter one is of the generalized kind.

Example 12.1 Consider a game with 2 players each with one decision variable; that is, $n_1 = n_2 = 1$ and thus $n = 2$. For simplicity, we let $x \in \mathbb{R}$ and $y \in \mathbb{R}$ denote these 2 players' strategies, respectively. Let the players' problems be

$$\begin{array}{ll} \underset{x}{\text{minimize}} \quad (x - y)^2 & \text{and} \quad \underset{y}{\text{minimize}} \quad xy + y^2 \\ \text{subject to} \quad 0 \leq x \leq 1, & \text{subject to} \quad -1 \leq y \leq 1. \end{array}$$

The optimal solutions are given by

$$\mathcal{S}_1(y) = \begin{cases} 0 & \text{if } y < 0, \\ y & \text{if } 0 \leq y \leq 1, \\ 1 & \text{if } y > 1 \end{cases} \quad \text{and} \quad \mathcal{S}_2(x) = \begin{cases} 1 & \text{if } x < -2, \\ -x/2 & \text{if } -2 \leq x \leq 2, \\ -1 & \text{if } x > 2. \end{cases}$$

It is easy to check that the unique fixed point of the map: $\mathcal{S}_1 \times \mathcal{S}_2$, in other words, a pair (x, y) such that $x = \mathcal{S}_1(y)$ and $y = \mathcal{S}_2(x)$, is $(0, 0)$, which is the unique NE of this game.

Example 12.2 Consider the game of Example 12.1 except that the two players' optimization problems are, respectively,

$$\begin{array}{ll} \underset{x}{\text{minimize}} \quad (x - 1)^2 & \text{and} \quad \underset{y}{\text{minimize}} \quad (y - \tfrac{1}{2})^2 \\ \text{subject to} \quad x + y \leq 1, & \text{subject to} \quad x + y \leq 1. \end{array}$$

The optimal-solution sets are given by

$$
\mathcal{S}_1(y) = \begin{cases} 1 & \text{if } y \le 0, \\ 1 - y & \text{if } y \ge 0, \end{cases} \qquad \text{and} \qquad \mathcal{S}_2(x) = \begin{cases} \frac{1}{2} & \text{if } x \le \frac{1}{2}, \\ 1 - x & \text{if } x \ge \frac{1}{2}. \end{cases}
$$

It is easy to check that the set of fixed points of the map $\mathcal{S}_1 \times \mathcal{S}_2$ consists of all pairs $(\alpha, 1 - \alpha)$ for every $\alpha \in [1/2, 1]$; any such pair is an NE of the game. Thus this game has infinitely many equilibria.

It should be remarked on that the general setting of the game $\mathcal{G}(\Xi, \partial\Theta)$ is quite broad and that we are able to derive several interesting results for it. However, when it comes to the practical implementation and convergence analysis of algorithms, the differentiability of the functions θ_ν invariably plays an essential role.

12.2.1   Review of multifunctions

In order to deal with the aforementioned generalities of the game $\mathcal{G}(\Xi, \partial\Theta)$, we introduce some notations and review some terminology and concepts of set-valued maps, which we call multifunctions. Details for this review can be found in [21, Subsection 2.1.3] and in many standard texts such as the classic [5] and the more contemporary monograph [4].

For a given set S in a Euclidean space, we let $\overline{S}$ and ∂S denote, respectively, the closure and boundary of S. The domain of a multifunction $\mathbf{\Phi}$ from the Euclidean space $\mathbb{R}^n$ into $\mathbb{R}^m$ is the subset $\mathrm{dom}(\mathbf{\Phi}) \triangleq \{x \in \mathbb{R}^n : \mathbf{\Phi}(x) \ne \emptyset\}$. The graph of $\mathbf{\Phi}$ is the set $\mathrm{gph}(\mathbf{\Phi}) \triangleq \{(x, y) \in \mathbb{R}^{n+m} : y \in \mathbf{\Phi}(x)\}$. A multifunction $\mathbf{\Phi}$ is lower semicontinuous at a point $\bar{x} \in \mathrm{dom}(\mathbf{\Phi})$ if for every open set $\mathcal{U}$ such that $\mathbf{\Phi}(\bar{x}) \cap \mathcal{U} \ne \emptyset$, there exists an open neighborhood $\mathcal{N}$ of $\bar{x}$ such that $\mathbf{\Phi}(x) \cap \mathcal{U} \ne \emptyset$ for every $x \in \mathcal{N}$; thus in particular, $\mathcal{N} \subset \mathrm{dom}(\mathbf{\Phi})$. If $\mathbf{\Phi}$ is lower semicontinuous at every point in its domain, then $\mathrm{dom}(\mathbf{\Phi})$ must be an open set. The multifunction $\mathbf{\Phi}$ is upper semicontinuous at a point $\bar{x} \in \mathrm{dom}(\mathbf{\Phi})$ if for every open set $\mathcal{V}$ containing $\mathbf{\Phi}(\bar{x})$, there exists an open neighborhood $\mathcal{N}$ of $\bar{x}$ such that $\mathbf{\Phi}(x) \subset \mathcal{V}$ for every $x \in \mathcal{N}$. If $\mathbf{\Phi}$ is both lower and upper semicontinuous at $\bar{x}$, then $\mathbf{\Phi}$ is said to be continuous there. These continuity concepts have some sequential consequences. Specifically, lower semicontinuity of $\mathbf{\Phi}$ at $\bar{x}$ implies that for every $\bar{y} \in \mathbf{\Phi}(\bar{x})$ and every sequence $\{x^k\} \subset \mathrm{dom}(\mathbf{\Phi})$ converging to $\bar{x}$, there exists $y^k \in \mathbf{\Phi}(x^k)$ for all k such that the sequence $\{y^k\}$ converges to $\bar{y}$. Upper semicontinuity of $\mathbf{\Phi}$ at $\bar{x}$ implies that $\mathbf{\Phi}$ is closed at $\bar{x}$; that is, if $\{x^k\}$ is a sequence converging to $\bar{x}$, and $\{y^k\}$ is a sequence converging to a vector $\bar{y}$ such that $y^k \in \mathbf{\Phi}(x^k)$ for all k sufficiently large, then $\bar{y} \in \mathbf{\Phi}(\bar{x})$. In turn, closedness of $\mathbf{\Phi}$ at $\bar{x}$ implies that $\mathbf{\Phi}(\bar{x})$ is a closed set. The renowned Kakutani fixed-point theorem states that if $\mathbf{\Phi} : K \subset \mathrm{dom}(\mathbf{\Phi}) \to K$, with K compact and convex, then it is an upper-semicontinuous and convex-valued multifunction (meaning that $\mathbf{\Phi}(x)$ is a convex subset of K for all $x \in K$), and $\mathbf{\Phi}$ then has a fixed point in K, that is, $\bar{x} \in K$ exists such that $\bar{x} \in \mathbf{\Phi}(\bar{x})$.

A prominent example of a multifunction in convex analysis is the subdifferential of a convex function $\theta : \mathbb{R}^n \to \bar{\mathbb{R}}$ [50], which is the set:

$$\partial\theta(x) \triangleq \{ u \in \mathbb{R}^n : \theta(y) \geq \theta(x) + (y - x)^T u \ \text{ for all } y \in \mathbb{R}^n \}.$$

In addition to being upper semicontinuous, this multifunction has the important property of being monotone, that is, for any two pairs (x, x') and (u, u') with $u \in \partial\theta(x)$ and $u' \in \partial\theta(x')$, it holds that $(x - x')^T (u - u') \geq 0$.

An important consequence of the continuity of a multifunction is the closedness of the optimal-solution map of an optimization problem under perturbation. Let $\mathbb{R}^s$ be the space of parameters p, and suppose that a function $\theta : \mathbb{R}^{n+s} \to \mathbb{R}$ is given. Let Φ be a multifunction from $\mathbb{R}^s$ to $\mathbb{R}^n$. We consider, for each value of the parameter $p \in \mathbb{R}^s$, an optimization problem and the corresponding optimal-solution set:

$$S^{\text{opt}}(p) \triangleq \operatorname*{argmin}_{x \in \Phi(p)} \ \theta(x, p), \tag{12.2}$$

of which the players' optimization problems (12.1) are examples. Note that S^{opt} is itself a multifunction mapping of the space of the parameter p into the space of the primary decision variable x.

PROPOSITION 12.1 *If Φ is continuous at p^* and θ is a continuous function, then the optimal solution map S^{opt} is closed at p^*. In particular, if S^{opt} is single-valued near p^*, then it is continuous at p^*.*

Proof Let $\{p^k\}$ be a sequence converging to p^* and $\{x^k\}$ be a sequence converging to x^* such that $x^k \in S^{\text{opt}}(p^k)$ for all k. By the upper semicontinuity of Φ, it follows that $x^* \in S^{\text{opt}}(p^*)$. It remains to show that $\theta(x^*, p^*) \leq \theta(x, p^*)$ for all $x \in \Phi(p^*)$. By the lower semicontinuity of Φ, there exists a sequence $\{v^k\}$ converging to x such that $v^k \in \Phi(p^k)$ for all k. Hence, $\theta(x^k, p^k) \leq \theta(v^k, p^k)$. Passing to the limit $k \to \infty$ yields the desired conclusion readily. $\blacksquare$

As we will see in Section 12.3, the continuity of the multifunction Ξ is essential to the derivation of existence results of an equilibrium. In many practical applications, each player's strategy set $\Xi^\nu(x^{-\nu})$ is defined by finitely many inequalities:

$$\Xi^\nu(x^{-\nu}) \triangleq \left\{ x^\nu : A^\nu x^\nu = b^\nu(x^{-\nu}), \text{ and } g^\nu(x^\nu, x^{-\nu}) \leq 0, \right\}, \tag{12.3}$$

where $A^\nu \in \mathbb{R}^{\ell_\nu \times n_\nu}$ is a given matrix, and $b^\nu : \widehat{\mathbb{R}}^\nu \to \mathbb{R}^{\ell_\nu}$ and $g^\nu : \mathbb{R}^n \to \mathbb{R}^{m_\nu}$ are given vector functions with each $g_i^\nu(\bullet, x^{-\nu})$ being convex for all $i = 1, \cdots, m_\nu$. Note that there are no nonlinear equality constraints in $\Xi^\nu(x^{-\nu})$; the presence of such constraints will introduce nonconvexity in this constraint set.

The continuity of the multifunction Ξ^ν with $\Xi^\nu(x^{-\nu})$ given by (12.3) is ensured by a well-known "constraint qualification" (CQ) in nonlinear programming known as the "Mangasarian–Fromovitz constraint qualification" [38] or, in short, MFCQ. Specifically, this CQ holds at a point $\bar{x}$ satisfying the general inequality system: [$g(x) \leq 0$ and

$h(x) = 0]$, where $g : \mathbb{R}^n \to \mathbb{R}^m$ and $h : \mathbb{R}^n \to \mathbb{R}^\ell$ are differentiable vector functions, if (a) the gradients of the equality constraints at $\bar{x}$ are linearly independent, that is, the family of vectors:

$$\left\{ \nabla h_j(\bar{x}) \right\}_{j=1}^{\ell}$$

is linearly independent, and (b) a vector $\bar{v} \in \mathbb{R}^n$ exists such that

$$\begin{aligned} \bar{v}^T \nabla h_j(\bar{x}) &= 0, \quad j = 1, \cdots, \ell \\ \bar{v}^T \nabla g_i(\bar{x}) &< 0, \quad i \in \mathcal{I}(\bar{x}). \end{aligned}$$

where

$$\mathcal{I}(\bar{x}) \triangleq \{ i : g_i(\bar{x}) = 0 \}$$

is the index set of active (inequality) constraints at $\bar{x}$.

The proof of the following result can be found in [21, Proposition 4.7.1 and Corollary 4.7.3]; a much more general version of this result is due originally to Robinson [48] and serves as the basis of modern sensitivity theory of nonlinear programs under data perturbation [10].

PROPOSITION 12.2   *Let $\Xi^\nu(x^{-\nu})$ be given by (12.3). Suppose that each $g_i^\nu(\bullet, x^{-\nu})$ is convex for every $x^{-\nu}$ near $x^{*,-\nu}$, $b^\nu(x^{-\nu})$ is continuous near $x^{*,-\nu}$, and g^ν is continuously differentiable at $(x^{*,\nu}, x^{*,-\nu})$, where $x^{*,\nu} \in \Xi^\nu(x^{*,-\nu})$, and the MFCQ holds at $x^{*,\nu}$ for the system in $\Xi^\nu(x^{*,-\nu})$, then Ξ^ν is continuous at $x^{*,-\nu}$.*

As the following simple example shows, the MFCQ is an essential condition that cannot be easily dropped, even in the polyhedral case.

Example 12.3   Let $K(y) \triangleq \{x \geq 0 : x = y\} \subset \mathbb{R}$. This multifunction clearly is not continuous at $\bar{y} = 0$ because $K(y)$ is empty for any $y < 0$.

12.2.2   Connection to variational inequalities

The notation Ξ in the game $\mathcal{G}(\Xi, \partial\Theta)$ refers to the multifunction mapping of the players' strategy space $\mathbb{R}^n$ into itself with $\Xi(x) \triangleq \prod_{\nu=1}^{N} \Xi^\nu(x^{-\nu})$, where $x \triangleq (x^\nu)_{\nu=1}^{N}$. The notation $\partial\Theta$ in the game $\mathcal{G}(\Xi, \partial\Theta)$ refers to the multifunction of (partial) subgradients of the players' objective functions with respect to their own variables; specifically,

$$\partial\Theta(x) \triangleq \prod_{\nu=1}^{N} \partial_{x^\nu}\theta_\nu(x^\nu, x^{-\nu}),$$

where each $\partial_{x^\nu}\theta_\nu(x^\nu, x^{-\nu})$ is the subdifferential of the convex function $\theta_\nu(\bullet, x^{-\nu})$ at x^ν; that is, $\partial_{x^\nu}\theta_\nu(x^\nu, x^{-\nu})$ is the set of vectors $u^\nu \in \mathbb{R}^{n_\nu}$ such that

$$\theta_\nu(\widehat{x}^\nu, x^{-\nu}) - \theta_\nu(x^\nu, x^{-\nu}) \geq (\widehat{x}^\nu - x^\nu)^T u^\nu, \quad \forall \widehat{x}^\nu.$$

If $\theta_\nu(\bullet, x^{-\nu})$ is C^1, then $\partial_{x^\nu}\theta_\nu(x^\nu, x^{-\nu})$ is the singleton $\{\nabla_{x^\nu}\theta_\nu(x^\nu, x^{-\nu})\}$ consisting of the single partial gradient of $\theta_\nu(\bullet, x^{-\nu})$ at x^ν. In this differentiable case, we write $\mathcal{G}(\Xi, \nabla\Theta)$ instead of $\mathcal{G}(\Xi, \partial\Theta)$. Both notations refer to the game defined by the same pair (Ξ, Θ); an advantage of the different notations is that they easily distinguish the differentiable case ($\nabla\Theta$) from the non-differential case ($\partial\Theta$).

Note that a necessary condition for $\bar{x}$ to be an NE of the game $\mathcal{G}(\Xi, \Theta)$ is that $\bar{x}$ must belong to the set of fixed points of Ξ, in other words, $\bar{x} \in \mathcal{F}_\Xi \triangleq \{x : x \in \Xi(x)\}$; thus this set of fixed points has a lot to do with the Nash game; for one thing, $\mathcal{F}_\Xi$ must be a nonempty set for an equilibrium to exist. In the sequel we will call this set also the "feasible set" of the game $\mathcal{G}(\Xi, \Theta)$. It is not difficult to see that $\mathcal{F}_\Xi = \bigcap_{\nu=1}^N \mathrm{gph}(\Xi^\nu)$. If subsets $\mathbf{C} \subset \mathbb{R}^n$ and $X^\nu \subset \mathbb{R}^{n_\nu}$ exist such that $\mathrm{gph}(\Xi^\nu) = \{x \in \mathbf{C} : x^\nu \in X^\nu\}$ for all $\nu = 1, \cdots, N$, then $\mathcal{F}_\Xi = \mathbf{C} \cap \widehat{X}$. This is the case where each player ν of the game has some private constraints described by the individual set X^ν that is defined by player ν's variables only, and some common constraints defined by the set $\mathbf{C}$ that couples all players' variables; the latter "coupled" constraints defined by $\mathbf{C}$ are in contrast to the "separable" constraints defined by the Cartesian set $\widehat{X}$. An example of such a strategy set is:

$$\Xi^{-\nu}(x^{-\nu}) = \{x^\nu \in X^\nu : g(x^\nu, x^{-\nu}) \leq 0\},$$

where $g : \mathbb{R}^n \to \mathbb{R}^m$ is a given vector function involving all players' variables. Note that in this case, the coupled constraints of each player (defined by the vector function g) are the same as those of its rivals.

In general, given a pair of multifunctions $(\mathbf{K}, \mathbf{\Phi})$, the "generalized quasi-variational inequality", denoted GQVI $(\mathbf{K}, \mathbf{\Phi})$ is the problem of finding a pair of vectors $(\bar{x}, \bar{u})$, with $\bar{x} \triangleq (\bar{x}^\nu)_{\nu=1}^N$ and $\bar{u} \triangleq (\bar{u}^\nu)_{\nu=1}^N$, such that $\bar{x} \in \mathbf{K}(\bar{x})$, $\bar{u} \in \mathbf{\Phi}(\bar{x})$ and

$$(x - \bar{x})^T \bar{u} \geq 0, \quad \forall x \in \mathbf{\Phi}(\bar{x}).$$

If $\mathbf{K}$ is single-valued with $\mathbf{K}(x) = K$ for all x and $\mathbf{\Phi}$ remains multi-valued, then the GQVI $(\mathbf{K}, \mathbf{\Phi})$ becomes the generalized variational inequality GVI $(K, \mathbf{\Phi})$; if $\mathbf{K}$ is multi-valued and $\mathbf{\Phi}$ is equal to the single-valued map Φ, then the GQVI $(\mathbf{K}, \mathbf{\Phi})$ becomes the quasi-variational inequality QVI $(\mathbf{K}, \Phi)$; finally, if $\mathbf{K} = K$ for all x and $\mathbf{\Phi} = \Phi$ is single-valued, then the GQVI $(\mathbf{K}, \mathbf{\Phi})$ becomes the variational inequality VI (K, Φ). An early study of the GQVI can be found in [12]; see also [21, Section 2.8] where a more contemporary treatment of the QVI is presented.

By the chosen notation for the game $\mathcal{G}(\Xi, \partial\Theta)$, the following result that connects the generalized Nash with the GQVI is not surprising, and is easy to prove by using the optimality conditions for convex problems.

PROPOSITION 12.3 *Suppose that for every v, $\Xi^v(x^{-v})$ is a convex set and $\theta_v(\bullet, x^{-v})$ is a convex function for every x^{-v}. A tuple x^* is an NE of the game $\mathcal{G}(\Xi, \partial\Theta)$ if, and only if x^*, along with a suitable $u^* \in \partial\Theta(x^*)$, is a solution of the GQVI $(\Xi, \partial\Theta)$.*

It follows from Proposition 12.3 that if $\Xi^v(x^{-v}) = X^v$ for all x^{-v}, then the game $\mathcal{G}(\Xi, \partial\Theta)$ is equivalent to the GVI $(\widehat{X}, \partial\Theta)$; if $\Xi^v(x^{-v})$ is a moving set and $\theta_v(\bullet, x^{-v})$ is differentiable, then the game $\mathcal{G}(\Xi, \nabla\Theta)$ is equivalent to the QVI $(\Xi, \nabla\Theta)$, where $\nabla\Theta(x) \triangleq \left(\nabla_{x_v}\theta_v(x)\right)_{v=1}^N$; finally, if $\Xi^v(x^{-v}) = X^v$ for all x^{-v} and $\theta_v(\bullet, x^{-v})$ is differentiable, then the game $\mathcal{G}(\Xi, \nabla\Theta)$ is equivalent to the VI $(\widehat{X}, \nabla\Theta)$. Thus, depending on the level of generality, the variational formulation of a Nash game ranges from the simplest (the basic version of the game) to the most general (where the objective functions are not differentiable and there are distinct coupled constraints among the players).

The following result deals with the special case of the game where the players share the same coupled constraints.

PROPOSITION 12.4 *Assume that subsets $\mathbf{C} \subset \mathbb{R}^n$ and $X^v \subset \mathbb{R}^{n_v}$ exist such that $\mathrm{gph}(\Xi^v) = \{x \in \mathbf{C} : x^v \in X^v\}$ for all v. Suppose further that each $\theta_v(\bullet, x^{-v})$ is a convex function for every x^{-v}. Every solution of GVI $(\mathbf{C} \cap \widehat{X}, \partial\Theta)$ is an NE of the game $\mathcal{G}(\Xi, \partial\Theta)$. In particular, if each $\theta_v(\bullet, x^{-v})$ is differentiable, then every solution of the VI $(\mathbf{C} \cap \widehat{X}, \nabla\Theta)$ is an NE.*

Proof Let x^* be a solution of GVI $(\mathbf{C} \cap \widehat{X}, \partial\Theta)$. Then $x^* \in \Xi(x^*)$. Moreover, for every $x^v \in \Xi^v(x^{*,-v})$, we have $(x^v, x^{*,-v}) \in \mathbf{C} \cap \widehat{X}$; hence $u^v \in \partial_{x_v}\theta_v(x^*)$ exists such that $0 \leq (x^v - x^{*,v})^T u^v$. Since the above holds for all v, it follows that x^* is a NE. ∎

As illustrated by the example below, in the setting of Proposition 12.4, it is possible for the game $\mathcal{G}(\Xi, \partial\Theta)$ to have an equilibrium that is not a solution of the GVI $(\mathbf{C} \cap \widehat{X}, \partial\Theta)$; thus, while the latter GVI may be helpful in analyzing the former game, these two problems are not completely equivalent in that the GVI may omit some equilibria. What is worst is that the game may have an equilibrium while the GVI has no solution. We will return to discuss more about this issue when we introduce the Karush–Kuhn–Tucker formulation of the variational problem in the next subsection. For ease of reference, we follow the terminology in [20, Definition 3.10] and call the solutions of the GVI $(\mathbf{C} \cap \widehat{X}, \partial\Theta)$ variational equilibria.

Example 12.4 Consider a game with 2 players whose strategies are labeled $x \in \mathbb{R}$ and $y \in \mathbb{R}$, respectively, and whose optimization problems are as follows:

$$\left\{\begin{array}{ll} \underset{x}{\text{minimize}} & \tfrac{1}{2}(x-y)^2 \\ \text{subject to} & x - y = 0 \\ \text{and} & 0 \leq x \leq 1 \end{array}\right\} \quad \text{and} \quad \left\{\begin{array}{ll} \underset{y}{\text{minimize}} & -2y \\ \text{subject to} & x - y = 0 \\ \text{and} & 0 \leq y \leq 1 \end{array}\right\}.$$

For this problem, we have $\mathbf{C} \triangleq \{(x, y) \in \mathbb{R}^2 : x = y\}$ and $X^1 = X^2 = [0, 1]$. It is easy to see that every pair $(x, y) \in \mathbf{C} \times [0, 1]^2$ is an NE. Yet, the VI $(\mathbf{C} \cap \widehat{X}, F)$, which is to find a pair $(x, y) \in \mathbf{C} \times [0, 1]^2$ such that

$$
\begin{pmatrix} x' - x \\ y' - y \end{pmatrix}^T \begin{pmatrix} x - y \\ -2 \end{pmatrix} \geq 0, \quad \forall\, (x', y') \in \mathbf{C} \times [0, 1]^2,
$$

has $(1, 1)$ as the only solution. Thus this game has only one variational equilibrium but has a continuum of Nash equilibria.

Consider next a variant of the game with $X^1 = X^2 = \mathbb{R}$. Thus the players' optimization problems are, respectively:

$$
\begin{cases} \underset{x}{\text{minimize}} & \frac{1}{2}(x - y)^2 \\ \text{subject to} & x - y = 0 \end{cases} \quad \text{and} \quad \begin{cases} \underset{y}{\text{minimize}} & -2y \\ \text{subject to} & x - y = 0 \end{cases}.
$$

For this modified game, all pairs in $\mathbf{C}$ remain Nash equilibria; yet the VI $(\mathbf{C}, F)$ has no solution. Thus for this game, Nash equilibria exist but variational equilibria do not.

12.2.3 The Karush–Kuhn–Tucker conditions

Let $\Xi^\nu(x^\nu)$ be given by (12.3) with the constraint function g^ν being continuously differentiable on $\mathbb{R}^n$. By well-known nonlinear-programming theory, it follows that if x^* is an NE of the game $\mathcal{G}(\Xi, \partial\Theta)$, and if the MFCQ holds at $x^{*,\nu}$ for the set $\Xi^\nu(x^{*,-\nu})$, then a subgradient $u^\nu \in \partial_{x^\nu} \theta_\nu(x^*)$ and multipliers $\lambda^\nu \in \mathbb{R}^{m_\nu}$ and $\mu^\nu \in \mathbb{R}^{\ell_\nu}$ exist such that

$$
\begin{aligned}
& u^\nu + (A^\nu)^T \mu^\nu + \sum_{i=1}^{m_\nu} \nabla_{x^\nu} g_i^\nu(x^*)\, \lambda_i^\nu = 0 \\
& 0 \leq \lambda^\nu \perp g^\nu(x^*) \leq 0 \\
& 0 = b^\nu(x^{*,-\nu}) - A^\nu x^{*,\nu}, \text{ and } \mu^\nu \text{ free in sign,}
\end{aligned}
\tag{12.4}
$$

where the $\perp$ notation is the usual perpendicularity sign, which in this context is an equivalent expression of the complementary-slackness condition between the constraints and the associated multipliers. Conversely, if each $g_i^\nu(\bullet, x^{*,-\nu})$ is convex and the above "Karush–Kuhn–Tucker" (KKT) system holds for all ν, then x^* is an NE of the game $\mathcal{G}(\Xi, \partial\Theta)$. The concatenation of the system (12.4) for all ν constitutes the KKT system for this game. The concatenated system defines a multi-valued (mixed) complementarity problem in the tuple of variables: (x, λ, μ), where $\lambda \triangleq (\lambda^\nu)_{\nu=1}^N$ and $\mu \triangleq (\mu^\nu)_{\nu=1}^N$.

We discuss a special case of the game $\mathcal{G}(\Xi, \nabla\Theta)$ where, for every ν,

$$
\Xi^\nu(x^{-\nu}) = \{ x^\nu \in \mathbb{R}_+^{n_\nu} : g^\nu(x^\nu, x^{-\nu}) \leq 0 \}
\tag{12.5}
$$

and $\theta_\nu(\bullet, x^{-\nu})$ is differentiable. In this case, (12.4) becomes

$$0 \leq x^\nu \perp \nabla_{x^\nu} \theta_\nu(x) + \sum_{i=1}^{m_\nu} \nabla_{x^\nu} g_i^\nu(x^*) \lambda_i^\nu \geq 0$$

$$0 \leq \lambda^\nu \perp g^\nu(x^*) \leq 0,$$

which, when concatenated for all ν, leads to the "nonlinear-complementarity problem" (NCP) formulation of the game:

$$0 \leq \begin{pmatrix} x \\ \lambda \end{pmatrix} \triangleq \mathbf{z} \perp \mathbf{F}(\mathbf{z}) \triangleq \begin{pmatrix} \left(\nabla_{x^\nu} \theta_\nu(x) + \displaystyle\sum_{i=1}^{m_\nu} \nabla_{x^\nu} g_i^\nu(x^*) \lambda_i^\nu \right)_{\nu=1}^N \\ -\left(g^\nu(x) \right)_{\nu=1}^N \end{pmatrix} \geq 0. \quad (12.6)$$

The further special case where g^ν are all equal to the same function $g : \mathbb{R}^n \to \mathbb{R}^m$ is worth some additional discussion. In this case, the function $\mathbf{F}$ becomes

$$\mathbf{F}(\mathbf{z}) = \begin{pmatrix} \left(\nabla_{x^\nu} \theta_\nu(x) + \displaystyle\sum_{i=1}^{m} \nabla_{x^\nu} g_i(x^*) \lambda_i^\nu \right)_{\nu=1}^N \\ \left. \begin{bmatrix} -g(x) \\ \vdots \\ -g(x) \end{bmatrix} \right\} N \text{ times} \end{pmatrix},$$

which is clearly different from the function

$$\widehat{\mathbf{F}}(x, \widehat{\lambda}) \triangleq \begin{pmatrix} \left(\nabla_{x^\nu} \theta_\nu(x) + \displaystyle\sum_{i=1}^{m} \nabla_{x^\nu} g_i(x^*) \lambda_i \right)_{\nu=1}^N \\ -g(x) \end{pmatrix}, \quad \text{where } \widehat{\lambda} \triangleq (\lambda_i)_{i=1}^m.$$

If the set $\mathbf{C} \triangleq \{x \geq 0 : g(x) \leq 0\}$ is convex, which holds if each $g_i(x)$ is convex in all its variables jointly, then, under the MFCQ, it follows that the NCP:

$$0 \leq \begin{pmatrix} x \\ \widehat{\lambda} \end{pmatrix} \triangleq \widehat{\mathbf{z}} \perp \widehat{\mathbf{F}}(\widehat{\mathbf{z}}) \geq 0$$

is equivalent to the VI $(\mathbf{C}, \nabla\Theta)$. In turn, the NCP $(\widehat{\mathbf{F}})$ is related to the NCP defined by the function $\mathbf{F}$ in the following way. A pair $(x, \widehat{\lambda})$ is a solution of the NCP $(\widehat{\mathbf{F}})$ if, and only if the pair (x, λ), where $\lambda^\nu = \widehat{\lambda}$ for all ν, is a solution of the NCP $(\mathbf{F})$. Thus unless there is a common vector of multipliers for all players associated with an NE of the game $\mathcal{G}(\Xi, \nabla\Theta)$, such an NE will not be a solution of the VI $(\mathbf{C}, \nabla\Theta)$. This provides an explanation for the non-equivalence between the latter VI and the game in general, as illustrated by Example 12.4.

The classic paper [51] contains an extensive treatment of the game with common-coupled-convex constraints defined by differentiable functions and bounded player strategy sets via the KKT system of the game. The recent survey [20] terms this the jointly convex case. For such a specialized game, the paper [51] introduced the concept of a normalized Nash equilibrium. While this is at first glance a generalization of the case of common multipliers corresponding to the (common) coupled constraints, a closer look reveals that via a suitable scaling, such a normalized equilibrium is essentially the same as the standard equilibrium of a game with common multipliers. There are two main inadequacies in Rosen's treatment: one is the absence of player-dependent coupled constraints; and the second one is the assumed boundedness of the strategy sets. In contrast, the existence theory presented in Section 12.3 is aimed at alleviating both inadequacies.

12.2.4 Regularization and single-valued formulations

The variational formulations presented so far for the game $\mathcal{G}(\Xi, \partial\Theta)$ involve multifunctions. It is possible to obtain some single-valued formulations by regularizing the players' optimization problems. The main benefits of such regularization and the resulting single-valued formulations are twofold: (a) the players' objective functions $\theta_\nu(\bullet, x^{-\nu})$ are not required to be differentiable nor strictly (or strongly) convex; more importantly, (b) we can use a fixed-point theory, or the more powerful degree theory of continuous functions to provide an existence theory for Nash equilibria.

Specifically, for any tuple $y \equiv (y^\nu)_{\nu=1}^N$, let $\widehat{x}^\nu(y)$ denote the unique optimal solution of the following optimization problem:

$$\underset{x^\nu \in \Xi^\nu(y^{-\nu})}{\text{minimize}} \quad \theta_\nu(x^\nu, y^{-\nu}) + \tfrac{1}{2} \| x^\nu - y^\nu \|^2. \tag{12.7}$$

The existence and uniqueness of $\widehat{x}^\nu(y)$ can easily be proved if the convexity of the function $\theta_\nu(\bullet, y^{-\nu})$ and the convexity and nonemptiness of the set $\Xi^\nu(y^{-\nu})$ hold. This optimal solution is characterized by the multi-valued variational inequality: a subgradient $u^\nu \in \partial_{x^\nu}\theta_\nu(\widehat{x}^\nu(y), y^{-\nu})$ exists such that

$$(x^\nu - \widehat{x}^\nu(y))^T \left[u^\nu + \widehat{x}^\nu(y) - y^\nu \right] \geq 0, \quad \forall x^\nu \in \Xi^\nu(y^{-\nu}). \tag{12.8}$$

We call $\widehat{x}^\nu(y)$ player ν's proximal response to the rivals' strategy $y^{-\nu}$ that is regularized by strategy y^ν. We also call $\widehat{x}(y) \triangleq (\widehat{x}^\nu(y))_{\nu=1}^N$ the game's proximal-response map. The following proposition formally connects the proximal-response map to an NE.

PROPOSITION 12.5 *Suppose that for every ν, $\Xi^\nu(x^{-\nu})$ is a closed-convex set and $\theta_\nu(\bullet, x^{-\nu})$ is an convex function for every $x^{-\nu}$. A tuple x^* is an NE of the game $\mathcal{G}(\Xi, \partial\Theta)$ if, and only if x^* is a fixed point of the map $\widehat{x}$; that is, if, and only if $x^{*,\nu} = \widehat{x}^\nu(x^*)$ for all ν.*

Proof If $x^* = \widehat{x}(x^*)$, then (12.8) with $y = x^*$ implies

$$(x^\nu - x^{*,\nu})^T u^\nu \geq 0, \quad \forall x^\nu \in \Xi^\nu(x^{*,-\nu}),$$

which is a sufficient condition for x^* to be an optimal solution of player v's optimization problem: $\underset{x \in \Xi^v(x^{*,-v})}{\text{minimize}}\ \theta_v(x^v, x^{*,-v})$. The converse holds because (a) $\widehat{x}^v(x^*)$ is the unique optimal solution of (12.7) with $y = x^*$, and (b) $x^{*,v}$ is also an optimal solution to the latter problem by the variational characterization of its optimal solution. $\blacksquare$

The proximal-response map becomes a familiar map in variational-inequality theory when the players' objective functions $\theta_v(\bullet, y^{-v})$ are differentiable. In this case, the inequality (12.8) becomes:

$$(x^v - \widehat{x}^v(y))^T \left[\nabla_{x^v} \theta_v(\widehat{x}^v(y), y^{-v}) + \widehat{x}^v(y) - y^v \right] \geq 0, \quad \forall x^v \in \Xi^v(y^{-v}).$$

Thus $\widehat{x}^v(y) = \Pi_{\Xi^v(y^{-v})}(y^v - \nabla_{x^v}\theta_v(\widehat{x}^v(y), y^{-v}))$, where $\Pi_S(\bullet)$ denotes the Euclidean projector [21, Chapter 4] onto the closed-convex set S. Hence, x^* is a fixed point of the proximal-response map $\widehat{x}$ if, and only if x^* is a zero of the map $y \mapsto y - \Pi_{\Xi(y)}(y - \nabla\Theta(y))$, which is precisely the natural map of the QVI $(\Xi, \nabla\Theta)$ [21, Subsection 2.8]. Adopting this terminology for the general case, we call the map $\mathbf{N}^{\mathcal{G}} : y \mapsto y - \widehat{x}(y)$ the natural map of the game $\mathcal{G}(\Xi, \Theta)$. Note that

$$\mathbf{N}^{\mathcal{G}}(y) = y - \underset{x \in \Xi(y)}{\text{argmin}} \left[\sum_{v=1}^{N} \theta_v(x^v, y^{-v}) + \tfrac{1}{2}\|x - y\|^2 \right]. \tag{12.9}$$

Moreover, the domain of the proximal-response map $\widehat{x}$, and thus of the natural map $\mathbf{N}^{\mathcal{G}}$, coincides with the domain of the multifunction Ξ, and both are equal to

$$\text{dom}(\widehat{x}) = \text{dom}(\Xi) = \prod_{v=1}^{N} \text{dom}(\Xi^v) = \left\{ (y^v)_{v=1}^{N} : y^{-v} \in \text{dom}(\Xi^v) \quad \forall v \right\}.$$

By Proposition 12.1, the proximal-response map $\widehat{x}$ is continuous on its domain, provided that each Ξ^v is continuous (as a multifunction). The regularization of the players' optimization problems can be used to define a minimax formulation for the Nash game. This formulation is based on the (regularized) Nikaido–Isoda (scalar-valued) function [42] defined, for $y \in \text{dom}(\Xi)$, by:

$$\varphi_{\text{NI}}(y) \triangleq \sum_{v=1}^{N} \left\{ \theta_v(y) - \underset{x^v \in \Xi^v(y^{-v})}{\min} \left[\theta_v(x^v, y^{-v}) + \tfrac{1}{2}\|x^v - y^v\|^2 \right] \right\}$$

$$= \sum_{v=1}^{N} \left[\theta_v(y^v, y^{-v}) - \theta_v(\widehat{x}^v(y), y^{-v}) - \tfrac{1}{2}\|\widehat{x}^v(y) - y^v\|^2 \right].$$

Some important properties of the function $\varphi_{\text{NI}}(y)$ are summarized below.

PROPOSITION 12.6 *Suppose that for every v, $\Xi^v(x^{-v})$ is a closed-convex set and $\theta_v(\bullet, x^{-v})$ is a convex function for every x^{-v}. The following statements are valid.*

(a) $\varphi_{\text{NI}}(y)$ is a well-defined, continuous function on $\text{dom}(\Xi)$, and non-negative on $\mathcal{F}_{\Xi}$.

(b) A tuple x^* is an NE of the game $\mathcal{G}(\Xi, \partial\Theta)$ if, and only if x^* minimizes $\varphi_{\mathrm{NI}}(y)$ on $\mathcal{F}_{\Xi}$ and $\varphi_{\mathrm{NI}}(x^*) = 0$.

(c) If each θ_ν is differentiable and $\Xi^\nu(x^{-\nu}) = X^\nu$ for all $x^{-\nu}$, then φ_{NI} is differentiable with

$$\nabla_{y^{\nu'}} \varphi_{\mathrm{NI}}(y) = \sum_{\nu=1}^{N} \nabla_{y^{\nu'}} \theta_\nu(y) - \sum_{\nu \neq \nu'} \nabla_{y^{\nu'}} \theta_\nu(\widehat{x}^\nu(y), y^{-\nu}) - y^{\nu'} + \widehat{x}^{\nu'}(y).$$

Proof For part (a), only the nonnegativity of $\varphi_{\mathrm{NI}}(y)$ requires a proof. This is easy because for $y \in \mathcal{F}_{\Xi}$, we have $y^\nu \in \Xi^\nu(y^{-\nu})$ for every ν; the nonnegativity of each summand in $\varphi_{\mathrm{NI}}(y)$ is then obvious. For part (b), note that $\varphi_{\mathrm{NI}}(x^*) = 0$ with $x^* \in \mathcal{F}_{\Xi}$ if and only if for every ν, $x^{*,\nu} \in \Xi^\nu(x^{*,-\nu})$ and $\theta_\nu(x^*) = \min\limits_{x^\nu \in \Xi^\nu(x^{*,-\nu})} \left[\theta_\nu(x^\nu, x^{*,-\nu}) + \frac{1}{2} \| x^\nu - x^{*,\nu} \|^2 \right]$. Since the right-hand minimization problem has a unique solution, it follows that the equality is equivalent to $x^{*,\nu} = \widehat{x}^\nu(x^{*,-\nu})$ for every ν; or equivalently, to x^* being an NE. Part (c) is an immediate consequence of the well-known Danskin theorem regarding the differentiability of the optimal objective function of a parametric program, see [17] and [21, Theorem 10.2.1]. $\blacksquare$

The un-regularized Nikaido–Isoda function has been used extensively by several authors in the design of "relaxation algorithms" for computing Nash equilibrium with bounded-strategy sets; see the papers [7, 31, 32, 54] which also contain applications to environmental problems. The idea of using regularization in the treatment of Nash equilibria is not new; see, for example, [23] where this idea was employed so that Brouwer's fixed-point theorem, instead of Kakutani's fixed-point theorem, could be used for dealing with standard Nash games under the boundedness assumption of the players' strategy sets. Nevertheless, using regularization to deal with the existence issue associated with unbounded-strategy sets for generalized Nash-equilibrium problems is believed to be novel.

12.3   Existence theory

We begin our presentation of an existence theory with a slight extension of a classic result of Ichiishi [27] that pertains to the case of bounded coupled-strategy sets. (In the notation of the proposition below, Ichiishi's result assumes that $\emptyset \neq \Xi(y) \subseteq \widehat{K}$ for all $y \in \widehat{K}$ and $\widehat{K}$ is the Cartesian product $\prod_{\nu=1}^{N} K^\nu$ where each K^ν is compact convex.) The proof of the proposition is based on a simple application of Brouwer's fixed-point theorem to the proximal-response map.

PROPOSITION 12.7   *Let each $\Xi^\nu : \widehat{\mathbb{R}}^\nu \to \mathbb{R}^{n_\nu}$ be a continuous convex-valued multifunction on $\mathrm{dom}(\Xi^\nu)$; let each $\theta_\nu : \mathbb{R}^n \to \mathbb{R}$ be continuous. Assume that for each $x^{-\nu} \in \mathrm{dom}(\Xi^\nu)$, the function $\theta_\nu(\bullet, x^{-\nu})$ is convex. Suppose further that a nonempty* *compact-convex set $\widehat{K} \subseteq \mathbb{R}^n$ exists such that for every $y \in \widehat{K}$, $\Xi(y)$ is nonempty and $\widehat{x}(y) \in \widehat{K}$. Then the Nash game $\mathcal{G}(\Xi, \partial\Theta)$ has an equilibrium solution in $\widehat{K}$.*

Proof Consider the self-map $\widehat{x} : \widehat{K} \to \widehat{K}$, which is well defined and continuous. By Brouwer's fixed-point theorem, $\widehat{x}$ has a fixed point, which is necessarily an NE of the game. ∎

Our next task is to relax the rather restrictive assumption requiring the existence of the special set $\widehat{K}$. In particular, we are most interested in the case where the players' strategy sets are not bounded. The tool we employ to accomplish this relaxation is degree theory. Two highly recommended texts on degree theory are [22, 33]; for a brief summary that is sufficient for our purpose, see [43, Chapter 6] and [21, Subsection 2.2.1]. Here is a snapshot of the recipe in applying degree theory. By this theory, it suffices to identify a bounded open set Ω with $\overline{\Omega} \subseteq \mathrm{dom}(\Xi)$ such that the degree of the natural map $\mathbf{N}^{\mathcal{G}}$ with respect to Ω, which we denote $\deg(\mathbf{N}^{\mathcal{G}}, \Omega)$, is nonzero. When this holds, it follows that $\mathbf{N}^{\mathcal{G}}$ has a zero in Ω. Hence the existence of a Nash equilibrium of the game $\mathcal{G}(\Xi, \Theta)$ in $\overline{\Omega}$ follows readily. In turn, to show that $\deg(\mathbf{N}^{\mathcal{G}}, \Omega)$ is nonzero, we define a homotopy that continuously deforms the nature map $\mathbf{N}^{\mathcal{G}}$ of the game to a certain map that is known to have a nonzero degree with respect to Ω. Once such a homotopy is constructed, and if certain additional "boundary conditions" are met, we can apply the invariance property of the degree of a continuous mapping to complete the proof. A distinct advantage of using a degree-theoretic proof is that one obtains not only an existence result for the nominal game, but also the existence of Nash equilibria for all "nearby games", in other words, games where the players' objective functions are slight perturbations of those of the nominal game. This is due to the "nearness" property of the degree. See Proposition 12.8.

In what follows, we present an existence theorem of an NE for the game $\mathcal{G}(\Xi, \partial\Theta)$ that is applicable to a non-differentiable Θ and to unbounded strategy sets. The main postulate is the existence of a continuous selection of the strategy map Ξ with a certain non-ascent property. Using the topological concept of a retract, one can easily recover Ichiishi's result that requires the boundedness of the players' strategy sets. We first state and prove the theorem before addressing its postulates.

THEOREM 12.1 *Let each $\Xi^\nu : \widehat{\mathbb{R}}^\nu \to \mathbb{R}^{n_\nu}$ be a continuous convex-valued multifunction on its domain $\mathrm{dom}(\Xi^\nu)$; let each $\theta_\nu : \mathbb{R}^n \to \mathbb{R}$ be continuous. Assume that for each $x^{-\nu} \in \mathrm{dom}(\Xi^\nu)$, the function $\theta_\nu(\bullet, x^{-\nu})$ is convex. Suppose further that there exist a bounded open set Ω with $\overline{\Omega} \subseteq \mathrm{dom}(\Xi)$, a vector $x^{\mathrm{ref}} \triangleq \left(x^{\mathrm{ref},\nu}\right)_{\nu=1}^{N} \in \Omega$, and a continuous function $s : \overline{\Omega} \to \Omega$ such that*

(a) $s(y) \triangleq \left(s^\nu(y)\right)_{\nu=1}^{N} \in \Xi(y)$ for all $y \in \overline{\Omega}$;
(b) The open line segment joining x^{ref} and $s(y)$ is contained in Ω for all $y \in \overline{\Omega}$, and
(c) $L_< \cap \partial\Omega = \emptyset$, where

$$L_< \triangleq \left\{ (y^\nu)_{\nu=1}^{N} \in \mathcal{F}_\Xi : \begin{array}{l} \text{for each } \nu \text{ such that } y^\nu \neq s^\nu(y) \\ (y^\nu - s^\nu(y))^T u^\nu < 0 \text{ for some } u^\nu \in \partial_{x^\nu}\theta_\nu(y) \end{array} \right\}.$$

Then the Nash game $\mathcal{G}(\Xi, \partial\Theta)$ has an equilibrium solution in $\overline{\Omega}$.

Proof It suffices to show that the natural map $\mathbf{N}^{\mathcal{G}}$ has a zero in $\overline{\Omega}$. Assume for contradiction that this is not the case. We construct two homotopies: the first homotopy connecting the natural map $\mathbf{N}^{\mathcal{G}}(y)$ to the continuous map $y \mapsto y - s(y)$, and the second homotopy connecting the latter map to the identity map translated by x^{ref}. For $(t, y) \in [0, 1] \times \overline{\Omega}$, let

$$H(t, y) \triangleq y - \underset{x \in \Xi(y)}{\operatorname{argmin}} \left[t \sum_{v=1}^{N} \theta_v(x^v, y^{-v}) + \tfrac{1}{2} \| x - t y - (1 - t) s(y) \|^2 \right].$$

The function H is continuous with $H(1, y) = \mathbf{N}^{\mathcal{G}}(y)$ and

$$H(0, y) = y - \underset{x \in \Xi(y)}{\operatorname{argmin}} \tfrac{1}{2} \| x - s(y) \|^2 = y - \Pi_{\Xi(y)}(s(y)) = y - s(y)$$

by condition (a). Moreover, since $s(y) \in \Omega$ for all $y \in \Omega$, $H(0, \bullet)$ has no zeros on $\partial \Omega$. The same is true for $H(1, \bullet)$ by assumption. Assume for contradiction that $H(t, y) = 0$ for some $(t, y) \in (0, 1) \times \partial \Omega$. We then have for all v,

$$y^v = \underset{x^v \in \Xi^v(y^{-v})}{\operatorname{argmin}} \left[t \theta_v(x^v, y^{-v}) + \tfrac{1}{2} \| x^v - t y^v - (1 - t) s^v(y) \| \right]$$

so that $y \in \mathcal{F}_{\Xi}$. Moreover, by the variational principle of the right-hand minimization, a subgradient $u^v \in \partial_{x^v} \theta_v(y)$ exists such that

$$(x^v - y^v) \left[t u^v + (1 - t) (y^v - s^v(y)) \right] \geq 0, \qquad \forall x^v \in \Xi^v(y^{-v}).$$

In particular, for $x^v = s^v(y)$, we deduce

$$0 \leq (s^v(y) - y^v) \left[t u^v + (1 - t) (y^v - s^v(y)) \right]$$
$$< t (s^v(y) - y^v)^T u^v, \qquad \text{if} \quad y^v \neq s^v(y).$$

Hence the vector y belongs to the set $L_<$. But since $y \in \partial \Omega$ by its choice, we have obtained a contradiction to condition (c). This contradiction shows that the map H has no zeros on the boundary of its domain of definition: $[0, 1] \times \overline{\Omega}$. Thus, by the invariance property of the degree, we deduce that $\deg(H(1, \bullet), \Omega) = \deg(H(0, \bullet), \Omega)$. It remains to show that this common degree is nonzero. For this purpose, we construct the second homotopy:

$$\widehat{H}(t, y) \triangleq t (y - s(y)) + (1 - t)(y - x^{\text{ref}}), \qquad \forall (t, y) \in [0, 1] \times \overline{\Omega}.$$

We have $\widehat{H}(0, y) = y - x^{\text{ref}}$ whose unique zero is in Ω. Moreover, $\widehat{H}(1, y) = y - s(y)$ has no zero on $\partial \Omega$. If $\widehat{H}(t, y) = 0$ for some $t \in (0, 1)$, then

$$y = t s(y) + (1 - t) x^{\text{ref}},$$

showing that $y \in \Omega$ by (b). Hence $\widehat{H}$ has no zeros on the boundary of its domain of definition. Consequently,

$$\deg(H(1,\bullet),\Omega) = \deg(H(0,\bullet),\Omega) = \deg(\widehat{H}(1,\bullet),\Omega) = \deg(\widehat{H}(0,\bullet),\Omega) = 1,$$

where the last identity is a basic property of the degree. This string of degree-theoretic equalities shows that the natural map of the game must have a zero in $\overline{\Omega}$. $\blacksquare$

It should be pointed out that the natural map $\mathbf{N}^{\mathcal{G}}$ is employed only as a tool to establish the existence of an NE; the assumptions (a), (b), and (c) do not involve this map. Obviously, condition (b) holds if Ω is convex. We now discuss how Ichiishi's result can be derived from Theorem 12.1. The idea is to embed the set $\widehat{K}$ in a bounded open ball and consider the retract (i.e., a continuous mapping) of this ball into $\widehat{K}$. Formally, the argument is as follows. For simplicity, take $\widehat{K}$ to be the Cartesian product $\prod_{\nu=1}^{N} K^{\nu}$ where each K^{ν} is compact convex. For each ν, let $\mathbb{B}^{\nu}$ be an open ball, with closure $\overline{\mathbb{B}}^{\nu}$, such that $K^{\nu} \subset \mathbb{B}^{\nu}$. Let $\Omega \triangleq \prod_{\nu=1}^{N} \mathbb{B}^{\nu}$. Let $r^{\nu} : \overline{\mathbb{B}}^{\nu} \to K^{\nu}$ be a retract of $\overline{\mathbb{B}}^{\nu}$ onto K^{ν}, in other words, r^{ν} is continuous and r^{ν} is an identity map restricted to K^{ν}. For every $y \in \overline{\Omega}$, let $\widehat{\Xi}(y) \triangleq \Xi(r(y))$ and $s(y) \triangleq \widehat{x}(r(y)) \in \widehat{K}$, where $r(y) \triangleq \left(r^{\nu}(y^{\nu})\right)_{\nu=1}^{N}$. Clearly, $s(y) \in \widehat{\Xi}(y)$. Thus s is a continuous map from $\overline{\Omega}$ into Ω. The composite map $\widehat{\Xi}$ is continuous on its domain. Let x^{ref} be an arbitrary element of Ω. For each $y \in \mathcal{F}_{\widehat{\Xi}}$, $y \in \Xi(r(y))$; thus $y \in \widehat{K}$. Hence $y = r(y)$. For every ν, there exists $w^{\nu} \in \partial_{x^{\nu}}\theta_{\nu}(s^{\nu}(y), r^{-\nu}(y^{-\nu})) = \partial_{x^{\nu}}\theta_{\nu}(s^{\nu}(y), y^{-\nu})$ such that

$$0 \le (y^{\nu} - s^{\nu}(y))^{T}\left[w^{\nu} + s^{\nu}(y) - r^{\nu}(y^{\nu})\right] = (y^{\nu} - s^{\nu}(y))^{T}\left[w^{\nu} + s^{\nu}(y) - y^{\nu}\right].$$

By the monotonicity of the subdifferential $\partial_{x^{\nu}}\theta_{\nu}(\bullet, y^{-\nu})$ it follows that for every $u^{\nu} \in \partial_{x^{\nu}}\theta_{\nu}(y)$, we have

$$(y^{\nu} - s^{\nu}(y))^{T}u^{\nu} \ge (y^{\nu} - s^{\nu}(y))^{T}w^{\nu} \ge \| y^{\nu} - s^{\nu}(y) \|^{2}.$$

Hence, for every element y in the set

$$L_{<} \triangleq \left\{ \begin{array}{l} (y^{\nu})_{\nu=1}^{N} \in \mathcal{F}_{\widehat{\Xi}} \; : \quad \text{for each } \nu \text{ such that } y^{\nu} \neq s^{\nu}(y) \\ (y^{\nu} - s^{\nu}(y))^{T}u^{\nu} < 0 \text{ for some } u^{\nu} \in \partial_{x^{\nu}}\theta_{\nu}(y) \end{array} \right\},$$

we must have $y = s(y) \in \Omega$. Consequently, $L_{<} \cap \partial\Omega = \emptyset$. We have now verified all the conditions in Theorem 12.1 for the game $\mathcal{G}(\widehat{\Xi}, \Theta)$. Therefore, this game has an NE $\widehat{y}$ satisfying $\widehat{y} \in \mathcal{F}_{\widehat{\Xi}}$, which implies that $r(\widehat{y}) = \widehat{y}$. Thus y is an NE for the game $\mathcal{G}(\Xi, \Theta)$.

To further explain conditions (a) and (c) of Theorem 12.1, we say that a vector d^{ν} is a direction of non-ascent of $\theta_{\nu}(\bullet, y^{-\nu})$ at y^{ν} if $(d^{\nu})^{T}u^{\nu} \le 0$ for some $u^{\nu} \in \partial_{x^{\nu}}\theta_{\nu}(y)$.

Roughly speaking, conditions (a) and (c) assert that the strategy map Ξ has a continuous selection s on $\overline{\Omega}$ with the property that for every $y \in \mathcal{F}_\Xi \cap \partial\Omega$, the vector $s^\nu(y) - y^\nu$, which is necessarily a feasible direction for the set $\Xi^\nu(y^{-\nu})$ at y^ν, is a direction of non-ascent of $\theta_\nu(\bullet, y^{-\nu})$ at y^ν. This condition allows us to restrict the search of an NE in $\overline{\Omega}$; the continuity of the selection s is needed for technical purposes.

To illustrate the usefulness of the degree-theoretic argument, let us consider a family of games defined by the pair $(\Xi, \Theta(\bullet; p))$, where $\Theta(x; p) \triangleq (\theta_\nu(x; p))_{\nu=1}^N$ is a parameterized tuple of objective functions with the vector p being the parameter. The next result shows that if for some p^*, the game $(\Xi, \Theta(\bullet; p^*))$ satisfies the assumptions of Theorem 12.1, then for all p sufficiently close to p^*, the perturbed game $(\Xi, \Theta(\bullet; p))$ also has an NE, provided that the parameterized objective functions θ_ν, under small perturbations of p, satisfy appropriate convexity and continuity assumptions.

PROPOSITION 12.8 *Suppose that the game $(\Xi, \Theta(\bullet; p^*))$ satisfies the assumptions of Theorem 12.1. Suppose further that there exist a neighborhood $\mathcal{P}$ of p^* and a constant $c > 0$ such that for every $p \in \mathcal{P}$, $\theta_\nu(\bullet; p)$ is continuous and $\theta_\nu(\bullet, x^{-\nu}; p)$ is convex for every ν and $x^{-\nu}$, and that for every $y \in \overline{\Omega}$ and every $u \triangleq (u^\nu)_{\nu=1}^N$ and $u^* \triangleq (u^{*,\nu})_{\nu=1}^N$ such that for every ν, $u^\nu \in \partial_{x^\nu}\theta_\nu(x^\nu, y^{-\nu}; p)$ and $u^{*,\nu} \in \partial_{x^\nu}\theta_\nu(x^\nu, y^{-\nu}; p^*)$ for some $x^\nu \in \Xi^\nu(y^{-\nu})$,*

$$\| u - u^* \| \le c \| p - p^* \|. \tag{12.10}$$

Then for every p sufficiently close to p^*, the game $(\Xi, \Theta(\bullet; p))$ has an NE.

Proof Let $\widehat{x}(\bullet; p)$ denote the proximal-response map of the game $(\Xi, \Theta(\bullet; p))$ and $\mathbf{N}(y; p) \triangleq y - \widehat{x}(y; p)$ be the game's natural map. We derive an upper bound for the difference

$$\| \mathbf{N}(y; p) - \mathbf{N}(y; p^*) \| = \| \widehat{x}(y; p) - \widehat{x}(y; p^*) \|$$

for $y \in \overline{\Omega}$. By the variational principle for the players' proximal-optimal responses, there exist $u \triangleq (u^\nu)_{\nu=1}^N$ and $u^* \triangleq (u^{*,\nu})_{\nu=1}^N$ such that for every ν, $u^\nu \in \partial_{x^\nu}\theta_\nu(\widehat{x}^\nu(y; p), y^{-\nu}; p)$, $u^{*,\nu} \in \partial_{x^\nu}\theta_\nu(\widehat{x}^\nu(y; p^*), y^{-\nu}; p^*)$,

$$\sum_{\nu=1}^N (\widehat{x}^\nu(y; p) - \widehat{x}^\nu(y; p^*))^T \left[u^{*,\nu} + \widehat{x}^\nu(y; p^*) - y^\nu \right] \ge 0$$

and

$$\sum_{\nu=1}^N (\widehat{x}^\nu(y; p^*) - \widehat{x}^\nu(y; p))^T \left[u^\nu + \widehat{x}^\nu(y; p) - y^\nu \right] \ge 0.$$

Adding the two inequalities, rearranging terms, and using the Cauchy–Schwartz inequality and (12.10), we deduce, by taking any $\widehat{u}^{*,\nu} \in \partial_{x^\nu}\theta_\nu(\widehat{x}^\nu(y;p), y^{-\nu}; p^*)$,

$$\| \widehat{x}(y;p) - \widehat{x}(y;p^*) \|^2 \leq \sum_{\nu=1}^{N} (\widehat{x}^\nu(y;p) - \widehat{x}^\nu(y;p^*))^T (u^{*,\nu} - \widehat{u}^{*,\nu})$$

$$+ \sum_{\nu=1}^{N} (\widehat{x}^\nu(y;p^*) - \widehat{x}^\nu(y;p))^T (u^\nu - \widehat{u}^{*,\nu})$$

$$\leq \| \widehat{x}(y;p^*) - \widehat{x}(y;p) \| \, \| u^* - \widehat{u}^* \|,$$

where the second inequality is due to the monotonicity of the subdifferential $\partial_{x^\nu}\theta_\nu(\bullet, y^{-\nu}; p^*)$. Hence, for all $y \in \overline{\Omega}$,

$$\| \widehat{x}(y;p) - \widehat{x}(y;p^*) \| \leq c \, \| p - p^* \|.$$

Thus, by restricting the neighborhood $\mathcal{P}$ if necessary, we can bound $\| \mathcal{N}(y;p) - \mathcal{N}(y;p^*) \|$ uniformly on $\overline{\Omega}$ as small as needed. This is enough for the application of the nearness property of the degree to conclude that $\deg(\mathbf{N}(\bullet;p), \Omega)$ is well defined and equal to $\deg(\mathbf{N}(\bullet;p^*), \Omega)$ for all p sufficiently close to p^*. Hence for each such p, the perturbed game $(\Xi, \Theta(\bullet;p))$ has an NE.   ∎

12.3.1   Special cases

The existence Theorem 12.1 is fairly broad, yet abstract. In what follows, we present two special cases, the first of which pertains to the case in which $\Xi^\nu(x^{-\nu})$ is independent of $x^{-\nu}$. The following corollary gives a sufficient condition for the existence of an equilibrium solution to the basic Nash game with fixed, but possibly unbounded, strategy sets; this condition turns out to be necessary when there is a single player; that is, when the game reduces to a standard convex-optimization problem. Moreover, a slightly modified condition is sufficient for the set of Nash equilibria to be bounded. Among other things, this result shows that it is, in general, not easy to derive a broader set of conditions for the existence of a Nash equilibrium for a game with separable strategy sets, because any such conditions must be applicable to the case of a single player, which then must imply the condition obtained here. Subsequently, we will extend the necessity result to a class of games with multiple players under a certain "P_0-property"; see Proposition 12.10. We recall the notation: $\widehat{X} \triangleq \prod_{\nu=1}^{N} X^\nu$.

COROLLARY 12.1 *Let X^ν be a closed-convex subset of $\mathbb{R}^{n_\nu}$; let $\theta_\nu : \mathbb{R}^n \to \mathbb{R}$ be continuous. Let $\theta_\nu(\bullet, x^{-\nu})$ be convex for every $x^{-\nu}$. Consider the following two statements:*

(a) A vector $x^{\mathrm{ref},\nu} \in X^\nu$ exists for every ν such that the set

$$\widehat{L}_< \triangleq \left\{ \begin{array}{l} (y^\nu)_{\nu=1}^N \in \widehat{X} \;:\; \text{for each } \nu \text{ such that } y^\nu \neq x^{\mathrm{ref},\nu}, \\ (y^\nu - x^{\mathrm{ref},\nu})^T u^\nu \;<\; 0 \text{ for some } u^\nu \in \partial_{x^\nu} \theta_\nu(y) \end{array} \right\}$$

is bounded.
(b) The Nash game $\mathcal{G}(\widehat{X}, \partial\Theta)$ has an equilibrium solution.

It holds that $(a) \Rightarrow (b)$; conversely, $(b) \Rightarrow (a)$ when $N = 1$. Finally, if the larger set

$$\widehat{L}_\leq \triangleq \left\{ \begin{array}{l} (y^\nu)_{\nu=1}^N \in \widehat{X} \;:\; \text{for each } \nu, \\ (y^\nu - x^{\mathrm{ref},\nu})^T u^\nu \leq 0 \text{ for some } u^\nu \in \partial_{x^\nu} \theta_\nu(y) \end{array} \right\}$$

is bounded, then the set of Nash equilibria is bounded. $\qquad\square$

Proof $(a) \Rightarrow (b)$. Let Ω be a bounded open set containing $\widehat{L}_<$ and x^{ref}. It follows that $\widehat{L}_< \cap \partial\Omega = \emptyset$. With $\Xi^\nu(x^{-\nu}) = X^\nu$ for all $x^{-\nu}$ and all ν, it follows that the assumptions of Theorem 12.1 are satisfied. Thus an NE exists.

$(b) \Rightarrow (a)$ if $N = 1$. In this case, the Nash problem becomes the optimization problem (omitting the respective superscript and subscript labels on the set X^1 and θ_1): minimize$_{x \in X}$ $\theta(x)$ and the condition (a) reads as: a vector $x^{\mathrm{ref}} \in X$ exists such that the set

$$\widehat{L}_< \triangleq \left\{ y \in X \;:\; y \neq x^{\mathrm{ref}} \;\Rightarrow\; (y - x^{\mathrm{ref}})^T u \;<\; 0 \text{ for some } u \in \partial\theta(y) \right\}$$

is bounded. We claim that $\widehat{L}_< = \{x^{\mathrm{opt}}\}$ if x^{ref} is taken to be a minimizer, denoted x^{opt}, of θ on X, which exists by assumption. Clearly, x^{opt} is an element of $\widehat{L}_<$. Conversely, let $y \in \widehat{L}_<$ be distinct from x^{ref}. We then have $(y - x^{\mathrm{opt}})^T u < 0$ for some $u \in \partial\theta(y)$. Since θ is convex, its subdifferential $\partial\theta$ is monotone, meaning that for any subgradient $s \in \partial\theta(x^{\mathrm{opt}})$, we have

$$0 \;\leq\; (y - x^{\mathrm{opt}})^T (u - s) \;<\; -(y - x^{\mathrm{opt}})^T s.$$

By the optimality of x^{opt}, a subgradient $\bar{s} \in \partial\theta(x^{\mathrm{opt}})$ exists such that $\bar{s}^T(y - x^{\mathrm{opt}}) \geq 0$. This contradiction establishes the equality $\widehat{L}_< = \{x^{\mathrm{opt}}\}$, and thus statement (a).

To show the last assertion, it suffices to show that the set of Nash equilibria is a subset of $\widehat{L}_\leq$. Let $x^* \triangleq (x^{*,\nu})_{\nu=1}^N$ be an NE. Then for every ν, there exists $u^\nu \in \partial_{x^\nu} \theta_\nu(x^*)$ such that $(x^{\mathrm{ref},\nu} - x^{*,\nu})^T u \geq 0$. Hence x^* satisfies the definition of an element in the set $\widehat{L}_\leq$. $\qquad\blacksquare$

In the next existence result, we impose a growth property, known as coercivity, on the players' objective functions; namely, we postulate that for some x^{ref},

$$\lim_{\substack{\|y\| \to \infty \\ y \in \mathcal{F}_\Xi}} \frac{\displaystyle\max_{1 \leq \nu \leq N} \left[\min_{u^\nu \in \partial_{x^\nu} \theta_\nu(y)} (y^\nu - x^{\mathrm{ref},\nu})^T u^\nu \right]}{\|y\|} = \infty. \qquad (12.11)$$

COROLLARY 12.2 *Let each* $\Xi^\nu : \widehat{\mathbb{R}}^\nu \to \mathbb{R}^{n_\nu}$ *be a continuous convex-valued multi-function with* $\mathrm{dom}(\Xi^\nu) = \widehat{\mathbb{R}}^\nu$; *let* $\theta_\nu : \mathbb{R}^n \to \mathbb{R}$ *be continuous. Assume that* $\theta_\nu(\bullet, x^{-\nu})$ *is convex for every* $x^{-\nu}$. *Suppose that a vector* $x^{\mathrm{ref}} \triangleq \left(x^{\mathrm{ref},\nu}\right)_{\nu=1}^N \in \bigcap_{y \in \mathbb{R}^n} \Xi(y)$ *exists satisfying* (12.11). *Then the Nash game* $\mathcal{G}(\Xi, \partial\Theta)$ *has an equilibrium solution.* $\square$

Proof The coercivity condition (12.11) implies that a scalar $R > \|x^{\mathrm{ref}}\|$ exists such that for all $y \in \mathcal{F}_\Xi$ with $\|y\| \geq R$,

$$\max_{1 \leq \nu \leq N} \left[\min_{u^\nu \in \partial_{x^\nu} \theta_\nu(y)} (\widehat{x}^\nu(y) - x^{\mathrm{ref},\nu})^T u^\nu \right] > 0.$$

With Ω taken to be the open Euclidean ball with radius R, it is not difficult to see that the assumptions in Theorem 12.1 are satisfied. Thus this theorem readily yields the desired existence of an NE. ∎

Whereas coercivity is not an uncommon property in practice, the requirement on the reference vector x^{ref} is the restrictive assumption of Corollary 12.2.

12.3.2   A game with prices

In what follows, we consider a special instance of the game $\mathcal{G}(\Xi, \partial\Theta)$ that has interesting applications, also in telecommunications, as we shall see later on. In this game there are P distinguished players, each called a primary player and labeled by $p = 1, \cdots, P$, who are imposing fees on N players, each called a secondary player, through an exogenous price vector $\rho \in \mathbb{R}^P$ as a way to control the secondary players' activities. In turn, the primary players' goal is to set the prices to ensure that the secondary players' activities satisfy certain prescribed restrictions. Mathematically, the interesting feature of this game is that unboundedness is prominent in the primary players' optimization problem as we see below. This extended game with prices is reminiscent of the classical Arrow–Debreu general equilibrium model formulated as a generalized game; see [3] for the original model and [20, Subsection 2.1] for the latter game formulation; a major difference is that prices in the Arrow–Debreu model are normalized to be unity, whereas prices in this model below are not so restricted; indeed, the admissible set of prices may be unbounded.

Specifically, the players' optimization problems are as follows. For secondary player ν, the optimization problem is parameterized by $(x^{-\nu}, \rho)$:

$$\underset{x^\nu \in \Xi^\nu(x^{-\nu})}{\mathrm{minimize}} \quad \widehat{\theta}_\nu(x, \rho) \triangleq \left[\theta_\nu(x^\nu, x^{-\nu}) + \rho^T B^\nu x^\nu \right]. \tag{12.12}$$

For some given vectors $\alpha^\nu > 0$, the primary users wish to set the prices so that the complementarity condition is satisfied:

$$0 \leq \rho \perp \sum_{\nu=1}^N (\alpha^\nu - B^\nu x^\nu) \geq 0. \tag{12.13}$$

Note that these complementarity conditions are the optimality condition for the simple non-negatively constrained-optimization problem in the price vector ρ, parameterized by $x \triangleq \left(x^\nu\right)_{\nu=1}^N$:

$$\underset{\rho \geq 0}{\text{minimize}} \quad \widehat{\theta}_{N+1}(x,\rho) \triangleq \rho^T \sum_{\nu=1}^N (\alpha^\nu - B^\nu x^\nu). \tag{12.14}$$

The constraint of (12.14) is the unbounded non-negative orthant. The price-optimization problem (12.14) is akin to the "market clearing mechanism" in the Arrow–Debreu general equilibrium problem.

Thus, solving the problem (12.12)-(12.13) is equivalent to solving the game $\mathcal{G}(\widehat{\Xi}, \partial\widehat{\Theta})$, where $\widehat{\Xi}(x,\rho) \triangleq \Xi(x) \times \mathbb{R}_+^P$ and

$$\widehat{\Theta}(x,\rho) \triangleq \begin{pmatrix} \left(\widehat{\theta}_\nu(x,\rho)\right)_{\nu=1}^N \\[2mm] \rho^T \sum_{\nu=1}^N (\alpha^\nu - B^\nu x^\nu) \end{pmatrix}.$$

For this game the players' proximal-response maps are respectively: for $(y,\eta) \in \text{dom}(\Xi) \times \mathbb{R}_+^P$,

$$\widehat{x}^\nu(y,\eta) \triangleq \underset{x^\nu \in \Xi^\nu(y^{-\nu})}{\text{argmin}} \quad \theta_\nu(x^\nu, y^{-\nu}) + \eta^T B^\nu x^\nu + \tfrac{1}{2} \| x^\nu - y^\nu \|^2,$$

$$\widehat{\rho}(y,\eta) \triangleq \underset{\rho \geq 0}{\text{argmin}} \quad \rho^T \sum_{\nu=1}^N (\alpha^\nu - B^\nu y^\nu) + \tfrac{1}{2} \| \rho - \eta \|^2.$$

Rather than treating this game as one with $(N+1)$ players, we consider it as one with N players who are subject to some additional constraint coming from (12.13). Specifically, inspired by the aggregated definition of the natural map (12.9), we will consider the following regularized-optimization problem with a coupled constraint:

$$\underset{x \in \Xi(y)}{\text{minimize}} \quad \left[\sum_{\nu=1}^N \theta_\nu(x^\nu, y^{-\nu}) + \tfrac{1}{2} \| x - y \|^2 \right]$$
$$\text{subject to} \quad \sum_{\nu=1}^N (\alpha^\nu - B^\nu x^\nu) \geq 0. \tag{12.15}$$

This problem has an optimal solution, which must necessarily be unique, provided that the set

$$\Xi^E(y) \triangleq \Xi(y) \cap \left\{ x : \sum_{\nu=1}^N (\alpha^\nu - B^\nu x^\nu) \geq 0 \right\}$$

is nonempty. In this case, we let $\widehat{x}^{E}(y) \triangleq \left(\widehat{x}^{E,\nu}(y)\right)_{\nu=1}^{N}$ denote the unique-optimal solution. To connect this map to the game $\mathcal{G}(\widehat{\Xi}, \partial\widehat{\Theta})$, we need to impose a constraint qualification at $\widehat{x}^{E}(y)$ so that we can introduce multipliers of the constraint $\sum_{\nu=1}^{N}(\alpha^{\nu} - B^{\nu}x^{\nu}) \geq 0$. For instance, if each $\Xi(y)$ is a finitely representable set defined by differentiable inequalities, it suffices to assume that the MFCQ holds at $\widehat{x}^{E}(y)$. Note that this CQ is not needed if $\Xi^{\nu}(y^{-\nu})$ is polyhedral. Under this assumption, it follows that $\widehat{x}^{E}(y)$ is characterized by the following variational condition: $\widehat{x}^{E}(y) \in \Xi(y)$ and there exist λ and $u \triangleq \left(u^{\nu}\right)_{\nu=1}^{N}$ such that $u^{\nu} \in \partial_{x^{\nu}}\theta_{\nu}(\widehat{x}^{E,\nu}(y), y^{-\nu})$ for all ν,

$$0 \leq \lambda \perp \sum_{\nu=1}^{N}(\alpha^{\nu} - B^{\nu}\widehat{x}^{E,\nu}(y)) \geq 0,$$

and for every ν,

$$(z^{\nu} - \widehat{x}^{E,\nu}(y))^{T}\left[u^{\nu} + (B^{\nu})^{T}\lambda + \widehat{x}^{E,\nu}(y) - y^{\nu}\right] \geq 0, \quad \forall z^{\nu} \in \Xi^{\nu}(y^{-\nu}).$$

Similar to Proposition 12.5, we can show that every fixed point of $\widehat{x}^{E}$, if it exists, is an NE of the extended game. Based on this observation, we obtain the following existence result for this extended game; the result is an extension of Proposition 12.7 and can be proved by applying Brouwer's fixed-point theorem to the map $\widehat{x}^{E}$.

THEOREM 12.2   *Let each* $\Xi^{\nu} : \widehat{\mathbb{R}}^{\nu} \to \mathbb{R}^{n_{\nu}}$ *be a convex-valued multifunction such that* Ξ^{E} *is continuous on its domain. Assume that for each* $x^{-\nu} \in \mathrm{dom}(\Xi^{\nu})$, *the function* $\theta_{\nu}(\bullet, x^{-\nu})$ *is convex. Suppose further that a compact-convex set* $\widehat{K}$ *exists such that for every* $y \in \widehat{K}$, $\Xi^{E}(y)$ *is nonempty,* $\widehat{x}^{E}(y) \in \widehat{K}$, *and the MFCQ holds at* $\widehat{x}^{E}(y)$. *Then the Nash game* $\mathcal{G}(\widehat{\Xi}, \partial\widehat{\Theta})$ *has an equilibrium solution in* $\widehat{K}$.

12.3.3   Jointly convex constraints

We next discuss the existence of a variational equilibrium for the game $\mathcal{G}(\Xi, \partial\Theta)$ under the setting of Proposition 12.4. Relying on the GVI $(\widehat{C}, \partial\Theta)$ formulation for such an equilibrium, where $\widehat{C} \triangleq C \cap \widehat{X}$, we let, for a given $y \triangleq (y^{\nu})_{\nu=1}^{N} \in \widehat{C}$, $\widehat{x}^{J}(y) \triangleq \left(\widehat{x}^{J,\nu}(y)\right)_{\nu=1}^{N}$ be the unique solution of the following strongly convex-optimization problem with a separable objective function:

$$\underset{x \in \widehat{C}}{\mathrm{minimize}} \quad \sum_{\nu=1}^{N}\left[\theta_{\nu}(x^{\nu}, y^{-\nu}) + \tfrac{1}{2}\|x^{\nu} - y^{\nu}\|^{2}\right]. \tag{12.16}$$

By the variational principle of the latter program, it follows that $\widehat{x}^{J}(y)$ is characterized by the conditions: $\widehat{x}^{J}(y) \in \widehat{C}$ and there exists $\widehat{u} \triangleq (\widehat{u}^{\nu})_{\nu=1}^{N}$ with $u^{\nu} \in \partial_{x^{\nu}}\theta_{\nu}(\widehat{x}^{J,\nu}(y), y^{-\nu})$

for every ν such that

$$\sum_{\nu=1}^{N} (x^\nu - \widehat{x}^{\,\mathrm{J},\nu}(y))^T \left[u^\nu + \widehat{x}^{\,\mathrm{J},\nu}(y) - y^\nu \right] \geq 0, \quad \forall x \in \widehat{C}. \tag{12.17}$$

Note that when $\widehat{C} = \prod_{\nu=1}^{N} \Xi^\nu(y^{-\nu})$, then (12.17) is equivalent to the aggregation of (12.8) over all ν. Nevertheless, when $\widehat{C}$ is not such a Cartesian product, the equivalence breaks down. From the inequality (12.17), it is not difficult to verify that a tuple x^* is a solution of the GVI $(\widehat{C}, \partial\Theta)$ if, and only if, $\widehat{x}^{\,\mathrm{J}}(x^*) = x^*$. Based on this observation, we obtain the following result that is the analog of Theorem 12.1 for the game described in Proposition 12.4. Not relying on the multi-valued continuity of the strategy sets, this result provides sufficient conditions for the existence of a variational equilibrium to a Nash game in the presence of possibly unbounded-coupled constraints shared by all players.

THEOREM 12.3 *Let $\theta_\nu : \mathbb{R}^n \to \mathbb{R}$ be continuous such that $\theta_\nu(\bullet, x^{-\nu})$ is convex for every $x^{-\nu}$. Suppose that there exist closed-convex subsets $C \subset \mathbb{R}^n$ and $X^\nu \subset \mathbb{R}^{n_\nu}$ such that $\mathrm{gph}(\Xi^\nu) = \{x \in C : x^\nu \in X^\nu\}$ for all ν. If there exist a bounded open set Ω and a vector $x^{\mathrm{ref}} \in \Omega \cap \widehat{C}$ such that $L_< \cap \partial\Omega = \emptyset$, where*

$$L_< \triangleq \left\{ y \in \widehat{C} : \exists\, u \in \partial\Theta(y) \text{ such that } (y - x^{\mathrm{ref}})^T u < 0 \right\},$$

then the Nash game $\mathcal{G}(\Xi, \partial\Theta)$ has a variational equilibrium in Ω.

Proof It suffices to define the homotopy:

$$H(t, y) \triangleq y - \underset{x \in \widehat{C}}{\mathrm{argmin}} \left[t \sum_{\nu=1}^{N} \theta_\nu(x^\nu, y^{-\nu}) + \tfrac{1}{2} \| x - t\, y^\nu - (1 - t)\, x^{\mathrm{ref}} \|^2 \right]$$

for $(t, y) \in [0, 1] \times \overline{\Omega}$ and follow the same degree-theoretic proof as before. The details are omitted. ∎

12.3.4 The NCP approach

When the sets $\Xi^\nu(x^{-\nu})$ are defined by differentiable inequalities and the functions θ_ν are differentiable, an NCP-based approach may be used to establish the existence of a solution to the game. For this purpose, let $\Xi^\nu(x^{-\nu})$ be given by (12.5) and consider the NCP (12.6) formulation of the game $\mathcal{G}(\Xi, \partial\Theta)$. A degree-theoretic proof of the following theorem can be found in [21, Theorem 2.6.1].

THEOREM 12.4 *Suppose that for every ν and $x^{-\nu}$, $\theta_\nu(\bullet, x^{-\nu})$ is convex and differentiable and $\Xi^\nu(x^{-\nu})$ is given by (12.5) with $g_i^\nu(\bullet, x^{-\nu})$ being convex and differentiable for all $i = 1, \cdots, m_\nu$. Suppose that the solutions (if they exist) of the NCP: $0 \leq \mathbf{z} \perp$*

$\mathbf{F}(\mathbf{z}) + \tau \mathbf{z} \geq 0$ *over all scalars* $\tau > 0$ *are bounded, where* $\mathbf{F}$ *is defined in (12.6). Then the NCP (12.6), and thus the game* $\mathcal{G}(\Xi, \partial\Theta)$ *has a solution.*

It would be interesting to contrast the NCP approach in Theorem 12.4 with the approach based on the proximal-response map, especially since both sets of results are proved by degree theory. The assumption of Theorem 12.4 postulates that solutions of the complementarity conditions:

$$0 \leq \begin{pmatrix} x^{\nu} \\ \lambda^{\nu} \end{pmatrix} \perp \begin{pmatrix} \nabla_{x^{\nu}} \theta_{\nu}(x) + \sum_{i=1}^{m_{\nu}} \nabla_{x^{\nu}} g_i^{\nu}(x) \lambda_i^{\nu} + \tau\, x^{\nu} \\ -g^{\nu}(x) + \tau\, \lambda^{\nu} \end{pmatrix} \geq 0, \quad \text{all } \nu$$

over all scalars $\tau > 0$ are bounded. In contrast, the proximal-response map-based approach provides sufficient conditions for the map $\widehat{x}(y)$ to have a fixed point, where $\widehat{x}(y)$ is characterized, in the present context, by the complementarity conditions: for all ν,

$$0 \leq \begin{pmatrix} \widehat{x}^{\nu}(y) \\ \lambda^{\nu} \end{pmatrix} \perp \begin{pmatrix} \nabla_{x^{\nu}} \theta_{\nu}(\widehat{x}^{\nu}(y), y^{-\nu}) + \sum_{i=1}^{m_{\nu}} \nabla_{x^{\nu}} g_i^{\nu}(\widehat{x}^{\nu}(y), y^{-\nu}) \lambda_i^{\nu} + \widehat{x}^{\nu}(y) - y^{\nu} \\ -g^{\nu}(\widehat{x}^{\nu}(y), y^{-\nu}) \end{pmatrix} \geq 0.$$

A main difference between these two sets of complementarity conditions is that in the former set, the constraint $-g^{\nu}(x) \geq 0$ is augmented by the term $\tau\lambda^{\nu}$ that involves the multiplier λ^{ν}, whereas there is no such augmentation in the latter set of complementarity conditions. This difference translates into a game-theoretic interpretation of the regularization approach in terms of players' responses, whereas there is no such interpretation in the NCP approach derived from the formulation (12.6).

12.4   Uniqueness theory

Global uniqueness results for Nash equilibria are scarce; typically, the only way to derive such a result is when there is an equivalent formulation of the game as a GVI or as a (possibly multi-valued) NCP in terms of the constraint multipliers. In particular, uniqueness of Nash equilibria is generally not expected to hold when there are player-dependent coupled constraints (see e.g., [19]); invariably, the analysis is very much problem related. Nevertheless, the P-property or one of its variants is typically what one would need to derive a uniqueness result. As we will see at the end of Subsection 12.6.1, this property is also sufficient for the convergence of a fixed-point iteration for computing the unique NE of the partitioned game $\mathcal{G}(\widehat{X}, \nabla\Theta)$. In this section, we begin our general discussion with the game $\mathcal{G}(\widehat{X}, \partial\Theta)$, where Θ is not necessarily differentiable; we then specialize the results to the differentiable case and obtain sharper conclusions.

Specifically, let $\widehat{K} \triangleq \prod_{i=1}^{N} K^i$ be a Cartesian product of sets $K^i \subset \mathbb{R}^{n_i}$. A multifunction $\Phi : \widehat{K} \to \mathbb{R}^n$ is said to have the P-property on $\widehat{K}$ if for every pair of distinct vectors

$$\widehat{x} \triangleq \left(\widehat{x}^{\,i}\right)_{i=1}^{N} \neq \widetilde{x} \triangleq \left(\widetilde{x}^{\,i}\right)_{i=1}^{N} \ \text{ in } \ \widehat{K} \ \text{ and for every pair } \ \widehat{y} \triangleq \left(\widehat{y}^{\,i}\right)_{i=1}^{N} \in \Phi(\widehat{x}) \ \text{ and}$$
$$\widetilde{y} \triangleq \left(\widetilde{y}^{\,i}\right)_{i=1}^{N} \in \Phi(\widetilde{x}),$$

$$\max_{1 \le i \le N} \ (\widehat{x}^{\,i} - \widetilde{x}^{\,i})^{T}(\widehat{y}^{\,i} - \widetilde{y}^{\,i}) > 0.$$

Originally introduced in the paper [39] for single-valued functions and with each K^{i} being an interval on the real line, the above definition is quite broad and includes a number of important special cases. Foremost is the case where $N = 1$; in this case the definition reduces to that of a strictly monotone-multifunction which is well known in monotone-operator theory. When $N > 1$ and Φ is a single-valued mapping, the definition becomes that of a P-function which is well known in the theory of partitioned-variational inequalities [21, Definition 3.5.8]. In general, it is easy to show that if Φ has the P-property on the partitioned set $\widehat{K}$, then the GVI $(\widehat{K}, \Phi)$ has at most one solution, an observation that can be proved in the same way as in the single-valued case [21, Proposition 3.5.10(a)]. Specializing this fact to the Nash game, we have the following result.

PROPOSITION 12.9 *Suppose that for every v and x^{-v}, $\theta_{v}(\bullet, x^{-v})$ is convex. The following two statements hold.*

(a) *Suppose that for every v and all x^{-v}, $\Xi^{v}(x^{-v}) = X^{v}$ is closed and convex and $\partial\Theta$ has the P-property on $\widehat{X}$. Then the game $\mathcal{G}(\widehat{X}, \partial\Theta)$ has at most one Nash equilibrium;*
(b) *Suppose that closed-convex subsets $\mathbf{C} \subset \mathbb{R}^{n}$ and $X^{v} \subset \mathbb{R}^{n_{v}}$ exist such that* $\mathrm{gph}(\Xi^{v}) = \{x \in \mathbf{C} : x^{v} \in X^{v}\}$ *for all v. Suppose further that $\partial\Theta$ is strictly monotone on $\mathbf{C} \cap \widehat{X}$. Then the game $\mathcal{G}(\Xi, \partial\Theta)$ has at most one variational equilibrium.*

Weakening the P-property, we say that the multifunction $\Phi : \widehat{K} \to \mathbb{R}^{n}$ has the P_{0}-property on $\widehat{K}$ if for every pair of distinct vectors $\widehat{x} \triangleq \left(\widehat{x}^{\,i}\right)_{i=1}^{N} \neq \widetilde{x} \triangleq \left(\widetilde{x}^{\,i}\right)_{i=1}^{N}$ in $\widehat{K}$ and for every pair $\widehat{y} \triangleq \left(\widehat{y}^{\,i}\right)_{i=1}^{N} \in \Phi(\widehat{x})$ and $\widetilde{y} \triangleq \left(\widetilde{y}^{\,i}\right)_{i=1}^{N} \in \Phi(\widetilde{x})$, there exists an index i such that $\widehat{x}^{\,i} \neq \widetilde{x}^{\,i}$ and

$$(\widehat{x}^{\,i} - \widetilde{x}^{\,i})^{T}(\widehat{y}^{\,i} - \widetilde{y}^{\,i}) \ge 0.$$

When $N = 1$, this definition reduces to that of a monotone multifunction Φ. It turns out that if the multi-valued subdifferential $\partial\Theta$ has the P_{0} property, then the sufficient condition in part (a) of Corollary 12.1 for the existence of an NE is also necessary.

PROPOSITION 12.10 *Let X^{v} be a closed-convex subset of $\mathbb{R}^{n_{v}}$ and let $\theta_{v}(\bullet, x^{-v})$ be a convex function for every x^{-v}.*

(a) *Suppose that $\partial\Theta$ has the P_{0}-property on $\widehat{X}$. The game $\mathcal{G}(\widehat{X}, \partial\Theta)$ has a Nash equilibrium if, and only if, a vector $x^{\mathrm{ref}} \triangleq \left(x^{\mathrm{ref},v}\right)_{v=1}^{N} \in \widehat{X}$ exists such that the set $\widehat{L}_{<}$ is* *bounded, where*

$$
\widehat{L}_< \triangleq \left\{ \begin{array}{l} (y^\nu)_{\nu=1}^N \in \widehat{X} : \ \textit{for each } \nu \textit{ such that } y^\nu \neq x^{\mathrm{ref},\nu}, \\ (y^\nu - x^{\mathrm{ref},\nu})^T u^\nu \ < \ 0 \textit{ for some } u^\nu \in \partial_{x^\nu} \theta_\nu(y) \end{array} \right\}.
$$

(b) Suppose that $\partial\Theta$ has the P-property on $\widehat{X}$. The game $\mathcal{G}(\widehat{X}, \partial\Theta)$ has a unique Nash equilibrium if, and only if a vector $x^{\mathrm{ref}} \in \widehat{X}$ exists such that the set $\widehat{L}_<$ is bounded.

Proof For (a), it suffices to show the "only if" statement. Let x^* be an NE. We claim that with $x^{\mathrm{ref}} = x^*$, the set $L_<$ in Theorem 12.1 consists of the vector x^* only. Indeed, let $y \neq x^*$ be an element of $L_<$. Let $u \in \partial\Theta(y)$ be such that for every ν such that $y^\nu \neq x^{*,\nu}$, $(y^\nu - x^{*,\nu})^T u^\nu \ < \ 0$. Since x^* is an NE, there exists $v \in \partial\Theta(x^*)$ such that $(y^\nu - x^{*,\nu})^T v^\nu \geq 0$. By the P_0-property of $\partial\Theta$ on $\widehat{X}$, an index ν exists such that $y^\nu \neq x^{*,\nu}$ and $(y^\nu - x^\nu)^T (u^\nu - v^\nu) \geq 0$. But this is a contradiction. Thus the claim is proved. Part (b) is immediate from part (a) and Proposition 12.9. $\blacksquare$

Similar to part (b) of Proposition 12.9, we can derive an analog of Proposition 12.10 for variational equilibria. Specifically, recalling Theorem 12.3, we present a necessary and sufficient condition for a "monotone" game with jointly convex constraints to have an NE. The significance of this result is that it establishes the converse of the previous existence theorem in the monotone case.

PROPOSITION 12.11 *Let $\theta_\nu(\bullet, x^{-\nu})$ be convex for every $x^{-\nu}$. Suppose that there exist closed-convex subsets $\mathbf{C} \subset \mathbb{R}^n$ and $X^\nu \subset \mathbb{R}^{n_\nu}$ such that $\mathrm{gph}(\Xi^\nu) = \{x \in \mathbf{C} : x^\nu \in X^\nu\}$ for all ν. If $\partial\Theta$ is (strictly) monotone on $\widehat{\mathbf{C}}$, then the game $\mathcal{G}(\Xi, \partial\Theta)$ has a (unique) variational equilibrium if, and only if there exists a vector $x^{\mathrm{ref}} \in \widehat{\mathbf{C}}$ such that the set below is bounded:*

$$
\left\{ y \in \widehat{\mathbf{C}} : \ \exists u \in \partial\Theta(y) \textit{ such that } (y - x^{\mathrm{ref}})^T u \ < \ 0 \right\}.
$$

Proof The proof is similar to that of part (a) of Proposition 12.10. $\blacksquare$

Variational equilibria of monotone games, in other words, games satisfying the assumptions of Proposition 12.11, are solutions to monotone GVIs, for which there is a rich theory. In particular, necessary and sufficient conditions for such equilibria to be bounded can be found in [16].

12.4.1   A matrix-theoretic criterion

In what follows, we consider the game $\mathcal{G}(\widehat{X}, \nabla\Theta)$ and provide a sufficient condition for the game to have a unique solution under the assumption that each θ_ν is twice continuously differentiable with bounded second derivatives; see Theorem 12.5. The condition is a certain matrix-theoretic, quasi-diagonal dominance property that ensures a uniform P-property of $\nabla\Theta$ on $\widehat{X}$. A key in the proof is the following consequence of a mean-value theorem applied to the univariate, differentiable, scalar-valued function:

$$\tau \in [0,1] \mapsto (\widehat{x}^{\,\nu} - \widetilde{x}^{\,\nu})^T \nabla_{x^\nu} \theta_\nu (\tau \widehat{x} + (1-\tau)\widetilde{x});$$

namely, for any two tuples $\widehat{x} \triangleq (\widehat{x}^{\,\nu})_{\nu=1}^N$ and $\widetilde{x} \triangleq (\widetilde{x}^{\,\nu})_{\nu=1}^N$ in $\widehat{X}$, there exists $\tau_\nu \in (0,1)$ such that with $z^\nu \triangleq \tau_\nu \widehat{x} + (1-\tau_\nu)\widetilde{x} \in \widehat{X}$,

$$
\begin{aligned}
(\widehat{x}^{\,\nu} - \widetilde{x}^{\,\nu})^T &[\nabla_{x^\nu}\theta_\nu(\widehat{x}) - \nabla_{x^\nu}\theta_\nu(\widetilde{x})] \\
&= (\widehat{x}^{\,\nu} - \widetilde{x}^{\,\nu})^T \sum_{\nu'=1}^N \nabla^2_{x^\nu x^{\nu'}}\theta_\nu(z^\nu)\,(\widehat{x}^{\,\nu'} - \widetilde{x}^{\,\nu'}),
\end{aligned}
\tag{12.18}
$$

where $\nabla^2_{x^\nu x^{\nu'}}\theta_\nu(z)$ is the Jacobian matrix of $\nabla_{x^\nu}\theta_\nu(z)$ with respect to the $x^{\nu'}$-variables; that is,

$$\left(\nabla^2_{x^\nu x^{\nu'}}\theta_\nu(z)\right)_{ij} \triangleq \frac{\partial}{\partial x_j^{\nu'}} \left(\frac{\partial \theta_\nu(z)}{\partial x_i^\nu}\right) \quad \forall\, i = 1, \cdots, m_\nu \text{ and } j = 1, \cdots, m_{\nu'}.$$

The expression (12.18) motivates the definition of the block-partitioned matrix:

$$\mathbf{M}(\mathbf{z}) \triangleq \left[M_{\nu\nu'}(z^\nu)\right]_{\nu,\nu'=1}^N, \quad \text{where} \quad M_{\nu\nu'}(z^\nu) \triangleq \nabla^2_{x^\nu x^{\nu'}}\theta_\nu(z^\nu),$$

for an N-tuple $\mathbf{z} \triangleq (z^\nu)_{\nu=1}^N$ with $z^\nu \in \widehat{X}$ for all $\nu = 1, \cdots, N$. Note that the off-diagonal blocks $M_{\nu\nu'}(z^\nu)$ and $M_{\nu'\nu}(z^{\nu'})$ for $\nu \neq \nu'$ are evaluated at two tuples z^ν and $z^{\nu'}$, both in $\widehat{X}$ but possibly distinct. Under the convexity of the functions $\theta_\nu(\bullet, x^{-\nu})$ on X^ν for all $x^{-\nu} \in X^{-\nu}$, the diagonal blocks $M_{\nu\nu}(z^\nu)$ are positive semidefinite matrices; but the entire matrix $\mathbf{M}(\mathbf{z})$ need not be so. Nevertheless, if $\mathbf{M}(\mathbf{z})$ is positive (semi)definite, then $\mathbf{M}(\mathbf{z})$ must be a P (P_0)-matrix on the Cartesian product $\widehat{X}$; in other words, the linear map $x \in \mathbb{R}^n \mapsto \mathbf{M}(\mathbf{z})x \in \mathbb{R}^n$ is a P (P_0)-function on $\widehat{X}$; this means that for every pair of distinct tuples $\widehat{x} \triangleq (\widehat{x}^{\,\nu})_{\nu=1}^N \in \widehat{X}$ and $\widetilde{x} \triangleq (\widetilde{x}^{\,\nu})_{\nu=1}^N \in \widehat{X}$:

- for the P-case:

$$\max_{1 \leq \nu \leq N} (\widehat{x}^{\,\nu} - \widetilde{x}^{\,\nu})^T \left[\sum_{\nu'=1}^N M_{\nu\nu'}(z^\nu)\left(\widehat{x}^{\,\nu'} - \widetilde{x}^{\,\nu'}\right)\right] > 0;$$

- for the P_0-case: there exists ν such that $\widehat{x}^{\,\nu} \neq \widetilde{x}^{\,\nu}$ and

$$(\widehat{x}^{\,\nu} - \widetilde{x}^{\,\nu})^T \left[\sum_{\nu'=1}^N M_{\nu\nu'}(z^\nu)\left(\widehat{x}^{\,\nu'} - \widetilde{x}^{\,\nu'}\right)\right] \geq 0.$$

The claim that positive (semi)definiteness implies the P (P_0) property is clear because of the inequality:

$$\sum_{v=1}^{N} \left(\widehat{x}^v - \widetilde{x}^v\right)^T \left[\sum_{v'=1}^{N} M_{vv'}(z^v)\left(\widehat{x}^{v'} - \widetilde{x}^{v'}\right)\right]$$

$$\leq N \max_{1 \leq v \leq N} \left(\widehat{x}^v - \widetilde{x}^v\right)^T \left[\sum_{v'=1}^{N} M_{vv'}(z^v)\left(\widehat{x}^{v'} - \widetilde{x}^{v'}\right)\right].$$

The proposition below gives a sufficient condition for the pair $(\widehat{X}, \nabla\Theta)$ to have the P_0 (P)-property in terms of the matrices $\mathbf{M}(\mathbf{z})$. Its proof follows immediately from the equality (12.18) and is omitted.

PROPOSITION 12.12 *Let each θ_v be twice continuously differentiable. If for every tuple $\mathbf{z} \triangleq \left(z^v\right)_{v=1}^{N}$ with each $z^v \in \widehat{X}$, the matrix $\mathbf{M}(\mathbf{z})$ is P_0 (P) on $\widehat{X}$, then $\nabla\Theta$ is a P_0 (P)-function on $\widehat{X}$.*

Whereas the class of real square P_0 (P)-matrices are well studied (see e.g., [15]), checking the P_0 (P)-property of a partitioned matrix on a Cartesian product of lower-dimensional sets is not easy; this is particularly hard in the context of Proposition 12.12 which involves an assumption for all tuples $\mathbf{z}$. In what follows, we will define a single matrix that will do the job. As an intermediate step, we derive a sufficient condition for the matrix $\mathbf{M}(\mathbf{z})$ to have the P_0 (P)-property on $\widehat{X}$ in terms of an $N \times N$ matrix. (Recall that $\mathbf{M}(\mathbf{z})$ is of order $n \triangleq \sum_{v=1}^{N} n_v$.) For this purpose, define the key scalars: for each $z \in \widehat{X}$,

$$\zeta_v(z) \triangleq \text{ smallest eigenvalue of } \nabla^2_{x^v}\theta_v(z) \tag{12.19}$$

and

$$\xi_{vv'}(z) \triangleq \left\| \nabla^2_{x^v x^{v'}}\theta_v(z) \right\|, \qquad \forall v \neq v'. \tag{12.20}$$

Note that in general, $\xi_{vv'}(z) \neq \xi_{v'v}(z)$ for $v \neq v'$. Under the convexity of the functions $\theta_v(\bullet, x^{-v})$ on X^v for all $x^{-v} \in X^{-v}$, it follows that $\zeta_v(z)$ is a non-negative scalar for every v for every $z \in \widehat{X}$. For an N-tuple $\mathbf{z} \triangleq \left(z^v\right)_{v=1}^{N}$ with $z^v \in \widehat{X}$ for all $v = 1, \cdots, N$, define the $N \times N$ (asymmetric) matrix $\Upsilon(\mathbf{z})$:

$$\begin{bmatrix}
\zeta_1(z^1) & -\xi_{12}(z^1) & -\xi_{13}(z^1) & \cdots & -\xi_{1N}(z^1) \\
-\xi_{21}(z^2) & \zeta_2(z^2) & -\xi_{23}(z^2) & \cdots & -\xi_{2N}(z^2) \\
\vdots & \vdots & \ddots & \vdots & \vdots \\
-\xi_{(N-1)1}(z^{N-1}) & -\xi_{(N-1)2}(z^{N-1}) & \cdots & \zeta_{N-1}(z^{N-1}) & -\xi_{(N-1)N}(z^{N-1}) \\
-\xi_{N1}(z^N) & -\xi_{N2}(z^N) & \cdots & -\xi_{NN-1}(z^N) & \zeta_N(z^N)
\end{bmatrix}$$

whose off-diagonal entries are all nonpositive; thus $\Upsilon(\mathbf{z})$ is a Z-matrix. There is an extensive theory of Z-matrices [6]; in particular, checking the P (P_0)-property of a Z-matrix is very easy. (We will be using properties of Z-matrices freely in what follows.) Note that each row of $\Upsilon(\mathbf{z})$ involves a possibly distinct $z^\nu \in \widehat{X}$. To unify these matrices, let

$$\zeta^\nu_{\min} \triangleq \inf_{z \in \widehat{X}} \zeta_\nu(z) \quad \text{and} \quad \xi^{\nu\nu'}_{\max} \triangleq \sup_{z \in \widehat{X}} \xi_{\nu\nu'}(z),$$

assume that all the sup above are finite, and define

$$\Upsilon \triangleq \begin{bmatrix} \zeta^1_{\min} & -\xi^{12}_{\max} & -\xi^{13}_{\max} & \cdots & -\xi^{1N}_{\max} \\ -\xi^{21}_{\max} & \zeta^2_{\min} & -\xi^{23}_{\max} & \cdots & -\xi^{2N}_{\max} \\ \vdots & \vdots & \ddots & \vdots & \vdots \\ -\xi^{(N-1)1}_{\max} & -\xi^{(N-1)2}_{\max} & \cdots & \zeta^{N-1}_{\min} & -\xi^{(N-1)N}_{\max} \\ -\xi^{N1}_{\max} & -\xi^{N2}_{\max} & \cdots & -\xi^{N(N-1)}_{\max} & \zeta^N_{\min} \end{bmatrix}. \tag{12.21}$$

Note that $\Upsilon(\mathbf{z}) \geq \Upsilon$ for all such N-tuples $\mathbf{z}$. This matrix componentwise inequality is key to the following proposition.

PROPOSITION 12.13 *If the matrix $\Upsilon(\mathbf{z})$ is P_0 (P), then the partitioned matrix $\mathbf{M}(\mathbf{z})$ is P_0 (P) on $\widehat{X}$. If Υ is P_0 (P), then $\Upsilon(\mathbf{z})$ is P_0 (P) for every N-tuple $\mathbf{z} \triangleq \left(z^\nu\right)^N_{\nu=1}$ with $z^\nu \in \widehat{X}$ for all $\nu = 1, \cdots, N$.*

Proof For the first assertion, it suffices to note the inequality:

$$(x^\nu)^T \sum_{\nu'=1}^N \left[M_{\nu\nu'}(z^\nu)x^{\nu'} \right] \geq e_\nu \left(\Upsilon(\mathbf{z})e \right)_\nu \tag{12.22}$$

where e is the N-vector with components $e_\nu \triangleq \| x^\nu \|$. The second assertion of the proposition follows from the fact that $\Upsilon(\mathbf{z}) \geq \Upsilon$ for all such N-tuples $\mathbf{z}$; since both are Z-matrices, the P (P_0)-property of Υ therefore implies that of $\Upsilon(\mathbf{z})$; see [6]. $\blacksquare$

Note that if Υ is a P-matrix, then $\zeta^\nu_{\min} > 0$ for all ν. Thus an implicit consequence of the P-assumption of the matrix Υ is the uniform positive definiteness of the matrices $\nabla^2_\nu \theta_\nu(x)$ on $\widehat{X}$, which implies the uniformly strong convexity of $\theta_\nu(\bullet, x^{-\nu})$ for all $x^{-\nu} \in X^{-\nu}$. It turns out that if the single matrix Υ is P, then the game $\mathcal{G}(\widehat{X}, \nabla\Theta)$ has a unique NE. This existence and uniqueness result does not require $\widehat{X}$ to be bounded. A similar conclusion holds for the variational equilibria of the game $\mathcal{G}(\widehat{C}, \nabla\Theta)$.

THEOREM 12.5 *Let X^ν be a closed-convex subset of $\mathbb{R}^{n_\nu}$ and let $\theta_\nu(\bullet, x^{-\nu})$ be a convex function for every $x^{-\nu}$. Suppose further that θ_ν is twice continuously differentiable with bounded second derivatives on $\widehat{X}$.*

(a) If Υ is a P-matrix, then the game $\mathcal{G}(\widehat{X}, \nabla\Theta)$ has a unique NE.

(b) Let $\mathbf{C}$ be a closed-convex subset of $\mathbb{R}^n$. If Υ is positive definite, then the game $\mathcal{G}(\widehat{\mathbf{C}}, \nabla\Theta)$ has a unique variational equilibrium.

Proof We prove only statement (a). By Proposition 12.10, it suffices to verify that the set

$$
\left\{ (y^\nu)_{\nu=1}^N \in \widehat{X} \ : \ \begin{array}{l} \text{for each } \nu \text{ such that } y^\nu \neq x^{\text{ref},\nu}, \\ (y^\nu - x^{\text{ref},\nu})^T \nabla_{x^\nu}\theta_\nu(y) < 0 \end{array} \right\}, \tag{12.23}
$$

for a fixed $x^{\text{ref},\nu} \in \widehat{X}$, is bounded. Let $y \triangleq (y^\nu)_{\nu=1}^N$ be an arbitrary vector in the above set. By (12.18) and (12.22), we have

$$
\max_{1 \leq \nu \leq N} (y^\nu - x^{\text{ref},\nu})^T \left[\nabla_{x^\nu}\theta_\nu(y) - \nabla_{x^\nu}\theta_\nu(x^{\text{ref}}) \right] \geq \max_{1 \leq \nu \leq N} e_\nu \, (\Upsilon e)_\nu ,
$$

where $e_\nu \triangleq \| y^\nu - x^{\text{ref},\nu} \|$ for all ν. By the P-property of the matrix Υ, it follows that a constant $c > 0$ such that for all vectors $v \in \mathbb{R}^N$,

$$
\max_{1 \leq \nu \leq N} v_\nu \, (\Upsilon v)_\nu \geq c \, \| v \|^2.
$$

Hence,

$$
\max_{1 \leq \nu \leq N} (y^\nu - x^{\text{ref},\nu})^T \left[\nabla_{x^\nu}\theta_\nu(y) - \nabla_{x^\nu}\theta_\nu(x^{\text{ref}}) \right] \geq c \, \| y - x^{\text{ref}} \|^2.
$$

This inequality is sufficient to establish the boundedness of the set (12.23). ∎

It turns out that the P-property of the matrix Υ is sufficient not only for the existence and uniqueness of an NE of the game $\mathcal{G}(\widehat{X}, \nabla\Theta)$, but this property is also sufficient for the convergence of a fixed-point iteration for computing the unique NE. This topic is discussed in Subsection 12.6.1.

12.5 Sensitivity analysis

In this section, we discuss the sensitivity of an NE of the game $\mathcal{G}(\widehat{X}, \nabla\Theta)$ under data perturbations, via the sensitivity theory of the VI $(\widehat{X}, \nabla\Theta)$; see [21, Chapter 5]; thus we assume throughout this section the differentiability of the players' objective functions. Due to the Cartesian structure of the set $\widehat{X}$, results in the cited reference can be sharpened and interpreted in the context of the game. In general, both the functions $\theta_\nu(\bullet, x^{-\nu})$ and the sets X^ν can be subject to data perturbations; for simplicity, we restrict our discussion to the case where only the functions $\theta_\nu(\bullet, x^{-\nu})$ are dependent on a certain parameter while the strategy sets X^ν remain constants. Thus, we are interested in the family of games

$$
\left\{ \mathcal{G}(\widehat{X}, \Theta(\bullet; p)) \ : \ p \in \mathcal{P} \right\} \tag{12.24}
$$

where p is a parameter varying in the set $\mathcal{P}$. Extensions to the case where each set $X^\nu(p)$ is finitely representable by differentiable inequalities parameterized by p can be carried out via the KKT system of the game; see [21, Section 5.4].

Focusing on the family of games (12.24), the setting is that an pair (x^*, p^*) is given, where x^* is an NE of the game $\mathcal{G}(\widehat{X}, \Theta(\bullet; p^*))$ corresponding to $p = p^*$. We are interested in the existence of equilibria near x^* of the perturbed games $\mathcal{G}(\widehat{X}, \Theta(\bullet; p))$ for p sufficiently near p^*. This is a "stability" property of x^*. In order for this to happen, it is desirable for x^* to be a "locally unique", or "isolated" solution of the nominal game at p^*. Therefore, our first order of business is to study the topic of local uniqueness of an equilibrium. Besides the significance in sensitivity analysis, the isolatedness property is central to the convergence analysis of a "Newton method" for the computation of the equilibrium. Due to space limitation, we will not discuss the latter topic and refer the reader to [21, Chapter 7] for details.

We recall the definition of the tangent cone of a set $K \subset \mathbb{R}^n$ at a point $x^* \in K$, which we denote $\mathcal{T}(K; x^*)$. Specifically, elements in $\mathcal{T}(K; x^*)$ are vectors v for which a sequence of positive scalars $\{\tau_k\} \downarrow 0$ and a sequence of vectors $\{x^k\} \subset K$ converging to x^* exist such that

$$v = \lim_{k \to \infty} \frac{x^k - x^*}{\tau_k}.$$

Given a differentiable objective function θ, the critical cone [49] of the pair (K, θ) at x^* is by definition the intersection $\mathcal{T}(K; x^*) \cap \nabla\theta(x^*)^\perp$, where $\nabla\theta(x^*)^\perp$ denotes the orthogonal complement of the gradient $\nabla\theta(x^*)$. The fundamental importance of the critical cone is well recognized in the sensitivity theory [10] of the optimization problem: $\min_{x \in K} \theta(x)$; for properties and a brief historical account about this cone, see [21, Subsection 3.3.1 and Section 3.8, respectively].

12.5.1 Local uniqueness

The fundamental object in the local properties of the NE x^* of the game $\mathcal{G}(\widehat{X}, \nabla\Theta)$, where each θ_v is twice continuously differentiable, is the pair $(\mathbf{M}(x^*), \mathcal{C}(x^*))$, where

$$\mathbf{M}(x^*) \triangleq \left[\nabla^2_{x^v x^{v'}} \theta_v(x^*) \right]_{v,v'=1}^N$$

is a block-partitioned matrix of the second derivatives of Θ at x^* and

$$\mathcal{C}(x^*) \triangleq \prod_{v=1}^N \left[\mathcal{T}(X^v; x^{*,v}) \cap \nabla_{x^v}\theta_v(x^*)^\perp \right]$$

is the product of the critical cones of the players' optimization problems: $\min_{x^v \in X^v} \theta_v(x^v, x^{*,-v})$ at $x^{*,v}$. We call the elements of $\mathcal{C}(x^*)$ "critical directions" of the game $\mathcal{G}(\widehat{X}, \Theta)$. We say that a block-partitioned matrix $M \triangleq [M_{vv'}]_{v,v'=1}^N$ is strictly semicopositive on the Cartesian product of cones: $C \triangleq \prod_{v=1}^N C^v$ if for every nonzero vector $x \triangleq (x^\nu)_{\nu=1}^N \in C$,

$$\max_{1 \leq \nu \leq N} (x^\nu)^T \sum_{\nu'=1}^N M_{\nu\nu'} x_{\nu'} > 0.$$

If M has the partitioned P-property on C, then M is strictly semicopositive on C, but not conversely.

PROPOSITION 12.14 *Let X^ν be a closed-convex subset of $\mathbb{R}^{n_\nu}$ and let $\theta_\nu(\bullet, x^{-\nu})$ be a convex function for every $x^{-\nu}$. Suppose that θ_ν is twice continuously differentiable. Let x^* be an NE of the game $\mathcal{G}(\widehat{X}, \Theta)$. If $\mathbf{M}(x^*)$ is strictly semicopositive on $\mathcal{C}(x^*)$, then x^* is an isolated NE.*

Proof Assume for contradiction that $\left\{ x^k \triangleq \left(x^{k,\nu} \right)_{\nu=1}^N \right\}$ is a sequence of Nash equilibria converging to x^* and such that $x^k \neq x^*$ for all k. The sequence of normalized vectors $\left\{ \dfrac{x^k - x^*}{\|x^k - x^*\|} \right\}$ must have at least one accumulation point. Without loss of generality, we may assume that this sequence has a limit, which we denote $\widehat{v} \triangleq (\widehat{v}^\nu)_{\nu=1}^N$. Clearly $\widehat{v}$ is nonzero and belongs to $\prod_{\nu=1}^N \mathcal{T}(X^\nu; x^{*,\nu})$. We claim that $\widehat{v}$ belongs to $\mathcal{C}(x^*)$; for this, it suffices to show that $(\widehat{v}^\nu)^T \nabla_{x^\nu}\theta_\nu(x^*) = 0$ for every ν. We have, for all k and ν

$$(x^{k,\nu} - x^{*,\nu})^T \nabla_{x^\nu}\theta_\nu(x^*) \geq 0 \quad \text{and} \quad (x^{*,\nu} - x^{k,\nu})^T \nabla_{x^\nu}\theta_\nu(x^k) \geq 0.$$

Dividing these two inequalities by $\|x^k - x^*\|$ and passing to the limit $k \to \infty$, we easily deduce that $(\widehat{v}^\nu)^T \nabla_{x^\nu}\theta_\nu(x^*) = 0$ for every ν. Adding the same two inequalities, we deduce, for some z^ν on the line segment joining x^k and x^*,

$$0 \geq (x^{k,\nu} - x^{*,\nu})^T \left[\nabla_{x^\nu}\theta_\nu(x^k) - \nabla_{x^\nu}\theta_\nu(x^*) \right]$$

$$= \left(x^{k,\nu} - x^{*,\nu} \right)^T \sum_{\nu'=1}^N \nabla^2_{x^\nu x^{\nu'}}\theta_\nu(z^\nu) \left(x^{k,\nu'} - x^{*,\nu'} \right).$$

Dividing by $\|x^k - x^*\|^2$ and passing to the limit $k \to \infty$, we deduce that

$$(\widehat{v}^\nu)^T \sum_{\nu'=1}^N \nabla^2_{\nu\nu'}\theta_\nu(z^*)\widehat{v}^{\nu'} \leq 0,$$

which holds for all ν. By the strict semicopositivity of $\mathbf{M}(x^*)$, it follows that $\widehat{v} = 0$, which is a contradiction. $\blacksquare$

When each X^ν is a polyhedral set, it is possible to establish the local uniqueness of an NE under a sufficient condition weaker than strict semicopositivity. Quoting [21,

Proposition 3.3.7], we remark that in this polyhedral case, it is sufficient to postulate that $(\mathcal{C}(x^*), \mathbf{M}(x^*))$ has the R_0-property, that is, the implication below holds:

$$\left[\mathcal{C}(x^*) \ni v \perp \mathbf{M}(x^*)v \in \mathcal{C}(x^*)^* \right] \Rightarrow v = 0, \tag{12.25}$$

where

$$\mathcal{C}(x^*)^* \triangleq \{ d \ : \ d^T v \geq 0, \text{ for all } v \in \mathcal{C}(x^*) \}$$

is the dual cone of $\mathcal{C}(x^*)$. In turn, the latter R_0-condition can be interpreted in terms of the equilibrium to a certain "linearized game", which we denote $\mathcal{LG}(x^*)$. This linearized game is a natural approximation of the original game $\mathcal{G}(\widehat{X}, \Theta)$ wherein we linearize the constraints and introduce a second-order approximation of the objectives. Specifically, in the game $\mathcal{LG}(x^*)$, with y^{-v} as the parameter, player v's optimization problem is

$$\underset{y^v}{\text{minimize}} \quad \nabla_{x^v}\theta_v(x^*)^T y^v + \tfrac{1}{2}(y^v)^T \nabla^2_{x^v}\theta_v(x^*)y^v + (y^v)^T \sum_{v \neq v'=1}^{N} \nabla^2_{x^v x^{v'}}\theta_v(x^*)y^{v'}$$

$$\text{subject to} \quad y^v \in \mathcal{T}(X^v; x^{*,v}).$$

The linearized game $\mathcal{LG}(x^*)$ is a homogeneous instance of a "polymatrix game" [18]. Including coupled constraints in general, the latter game is one where the players' optimization problems are convex-quadratic programs parameterized by their rivals' variables. The cited reference studies the application of Lemke's method [15] in linear-complementarity theory for the computation of an equilibrium of the polymatrix game with common multipliers of the coupled constraints.

Since $\nabla_{x^v}\theta_v(x^*) \in \mathcal{T}(X^v; x^{*,v})^*$, it follows that zero is always an NE to the game $\mathcal{LG}(x^*)$. The following corollary shows that the R_0-property of the pair $(\mathcal{C}(x^*), \mathbf{M}(x^*))$ is equivalent to this game having no nonzero equilibrium that is also a critical direction of the original game $\mathcal{G}(\widehat{X}, \Theta)$. This corollary is a specialization of a result first proved by Reinoza [47] for an affine-variational inequality; see also [21, Proposition 3.3.7]. The interpretation in the context of the Nash equilibrium problem is new.

COROLLARY 12.3 *Let each X^v be a polyhedron and let $\theta_v(\bullet, x^{-v})$ be a convex function for every x^{-v}. Suppose that θ_v is twice continuously differentiable. Let x^* be an NE of the game $\mathcal{G}(\widehat{X}, \Theta)$. The following two statements are equivalent.*

(a) *The implication (12.25) holds.*
(b) *The only NE of the linearized game $\mathcal{LG}(x^*)$ that is a critical direction of the game $\mathcal{G}(\widehat{X}, \Theta)$ is the zero equilibrium.*

Either condition implies that x^* is an isolated NE of the game $\mathcal{G}(\widehat{X}, \Theta)$. $\square$

Proof We prove only the equivalence of (a) and (b). A tuple y is an equilibrium of the game $\mathcal{LG}(x^*)$ if and only if for each v,

$$\mathcal{T}(X^v; x^{*,v}) \ni y^v \perp \nabla_{x^v}\theta_v(x^*) + \sum_{v'=1}^{N} \nabla^2_{x^v x^{v'}}\theta_v(x^*)y^{v'} \in \mathcal{T}(X^v; x^{*,v})^*. \tag{12.26}$$

By polyhedrality,

$$\left[\mathcal{T}(X^{\nu};x^{*,\nu}) \cap \nabla_{x^{\nu}}\theta_{\nu}(x^{*}) \right]^{*} = \mathcal{T}(X^{\nu};x^{*,\nu})^{*} + \mathbb{R}\nabla_{x^{\nu}}\theta_{\nu}(x^{*}).$$

Hence if y satisfies (12.26) for all ν, then $\mathbf{M}(x^{*})y \in \mathcal{C}(x^{*})^{*}$. If y is also a critical direction of the game $\mathcal{G}(\widehat{X},\Theta)$, then $\nabla_{x^{\nu}}\theta_{\nu}(x^{*})^{T}y^{\nu} = 0$ for all ν. Hence $y^{T}\mathbf{M}(x^{*})y = 0$. Consequently (a) implies (b).

Conversely, suppose v satisfies the right-hand conditions of (12.25). For every ν, v^{ν} belongs to $\mathcal{T}(X^{\nu};x^{*,\nu})$ and $\displaystyle\sum_{\nu'=1}^{N} \nabla^{2}_{x^{\nu}x^{\nu'}}\theta_{\nu}(x^{*})v^{\nu'} \in \mathcal{T}(X^{\nu};x^{*,\nu})^{*}+\mathbb{R}\nabla_{x^{\nu}}\theta_{\nu}(x^{*})$. Thus, for some scalar α_{ν} (possibly negative), $\alpha_{\nu}\nabla_{x^{\nu}}\theta_{\nu}(x^{*})+\displaystyle\sum_{\nu'=1}^{N} \nabla^{2}_{x^{\nu}x^{\nu'}}\theta_{\nu}(x^{*})v^{\nu'} \in \mathcal{T}(X^{\nu};x^{*,\nu})^{*}$. Since $\nabla_{x^{\nu}}\theta_{\nu}(x^{*})$ belongs to $\mathcal{T}(X^{\nu};x^{*,\nu})^{*}$, by a simple scaling, it follows that

$$\mathcal{T}(X^{\nu};x^{*,\nu}) \ni y^{\nu} \perp \nabla_{x^{\nu}}\theta_{\nu}(x^{*}) + \sum_{\nu'=1}^{N} \nabla^{2}_{x^{\nu}x^{\nu'}}\theta_{\nu}(x^{*})y^{\nu'} \in \mathcal{T}(X^{\nu};x^{*,\nu})^{*},$$

where $y^{\nu} \triangleq \dfrac{1}{1+|\alpha_{\nu}|+\alpha_{\nu}} v^{\nu}$ is a positive multiple v^{ν} and is thus a critical direction too. Consequently, (b) implies (a). ∎

12.5.2 Stability of an equilibrium

Returning to the non-polyhedral case, we consider the family of parameterized games (12.24). with the goal of showing that if the nominal game $\mathcal{G}(\widehat{X},\Theta(\bullet;p^{*}))$ has an NE x^{*} satisfying the assumption of Proposition 12.14, and if the perturbed function $\Theta(\bullet;p)$ satisfies the assumptions of Proposition 12.8, then for all p sufficiently near p^{*}, the perturbed game $\mathcal{G}(\widehat{X},\Theta(\bullet;p))$ has an NE that is close to x^{*}. It should be pointed out that this result is different from Proposition 12.8 in two aspects: one, the assumption here is on a given NE of the base game at p^{*}; two, the conclusion not only asserts existence of equilibria of the perturbed games, but also nearness of at least one perturbed equilibrium to x^{*}. In spite of these differences, the same proof technique is applicable here. For this purpose, we first establish a "continuation" of Proposition 12.14.

PROPOSITION 12.15 *In the setting of Proposition 12.14, an open neighborhood Ω of x^{*} exists such that* $\deg(\mathbf{N}^{\mathcal{G}};\Omega) = 1$.

Proof Consider the homotopy:

$$H(t,y) \triangleq y - \operatorname*{argmin}_{x\in\widehat{X}}\left[t\sum_{\nu=1}^{N}\theta_{\nu}(x^{\nu},y^{-\nu}) + \tfrac{1}{2}\left\| x - ty - (1-t)x^{*} \right\|^{2} \right].$$

Since $H(0,y) = y - x^{*}$, it suffices to show that an open neighborhood Ω of x^{*} exists such that $H(t,y) \neq 0$ for all $(t,y) \in [0,1] \times \partial\Omega$. Assume for contradiction that this is false. Then there exists a sequence of scalars $\{t_k\} \subset [0, 1]$ and a sequence of tuples $\{y^k\}$ with $y^k \neq x^*$ for all k such that $H(t_k, y^k) = 0$ for all k and $\{y^k\}$ converges to x^*. Clearly $t_k > 0$ for all k. Moreover, since x^* is an isolated NE, it follows that $t_k < 1$ for all k sufficiently large. Without loss of generality, we may assume that $t_k \in (0, 1)$ for all k. We have

$$y^k = \underset{x \in \widehat{X}}{\operatorname{argmin}} \left[t_k \sum_{\nu=1}^{N} \theta_\nu(x^\nu, y^{k, -\nu}) + \tfrac{1}{2} \left\| x - t_k\, y^k - (1 - t_k)\, y^k \right\|^2 \right],$$

which implies that for all $x \in \widehat{X}$,

$$(x^\nu - y^{k,\nu})^T \left[t_k\, \nabla_{x^\nu} \theta_\nu(y^k) + (1 - t_k)\,(y^{k,\nu} - x^{*,\nu}) \right] \geq 0,$$

which yields

$$(x^{*,\nu} - y^{k,\nu})^T \nabla_{x^\nu} \theta_\nu(y^k) \geq 0.$$

We also have

$$(y^{k,\nu} - x^{*,\nu})^T \nabla_{x^\nu} \theta_\nu(x^*) \geq 0.$$

By an argument similar to that of the proof of Proposition 12.14, we can derive a contradiction to the strict semicopositivity assumption. ∎

Based on Proposition 12.15, we can now establish the desired stability of the NE in the setting of Proposition 12.14.

PROPOSITION 12.16 *Let X^ν be a closed-convex subset of $\mathbb{R}^{n_\nu}$. Suppose that a neighborhood $\mathcal{P}$ of p^*, an open neighborhood $\mathcal{N}$ of x^*, and a scalar $c > 0$ exist such that for every $p \in \mathcal{P}$, $\theta_\nu(\bullet; p)$ is twice continuously differentiable and $\theta_\nu(\bullet, x^{-\nu}; p)$ is convex for every ν and $x^{-\nu}$, and that for every $x \in \widehat{X} \cap \overline{\mathcal{N}}$,*

$$\| \nabla\Theta(x; p) - \nabla\Theta(x; p^*) \| \leq c \| p - p^* \|.$$

Let x^* be an NE of the game $\mathcal{G}(\widehat{X}, \theta_\nu(\bullet; p^*))$ such that $\mathbf{M}(x^*)$ is strictly semicopositive on $C(x^*)$. Then open neighborhoods $\widehat{\mathcal{P}}$ of p^* and $\widehat{\mathcal{N}}$ of x^* exist such that for every $p \in \widehat{\mathcal{P}}$, the game $(\Xi, \Theta(\bullet; p))$ has an NE in $\widehat{\mathcal{N}}$. Moreover,

$$\lim_{p \to p^*} \sup \left\{ \| x(p) - x^* \| : x(p) \in \widehat{\mathcal{N}} \text{ is an NE of the game } (\Xi, \Theta(\bullet; p)) \right\} = 0.$$

Proof The first assertion of the proposition follows from Proposition 12.15 and the nearest property of the degree; cf. the proof of Proposition 12.8. The second assertion of the proposition follows from the isolatedness of x^*. ∎

12.6   Iterative algorithms

In this section, we discuss some iterative algorithms for computing an NE, with an emphasis on algorithms amenable to decomposition. Since a Nash equilibrium is characterized as a fixed point of the proximal-response map, a natural approach is to consider, starting at a given iterate y^0, the fixed-point iteration $y^{k+1} \triangleq \widehat{x}(y^k)$ for $k = 0, 1, 2, \cdots$. There are also averaging schemes of this basic fixed-point iteration; see [36, 37] for an extensive study of such schemes, which also discuss applications to the VI. Prior to the latter two references, the papers [44, 45] analyzed the convergence of many iterative methods for solving variational inequalities; particularly noteworthy is the latter paper [44] that specialized in partitioned VIs on Cartesian product sets. The "relaxation algorithms" presented in [7, 31, 32, 54] are averaging schemes for the unregularized proximal map of the Nash game. The text [9] is a good source of parallel and distributed algorithms and their convergence for optimization problems and variational inequalities.

12.6.1   Non-expansiveness of the proximal responses

In this subsection, we analyze the proximal-response map $\widehat{x}$ for the game $\mathcal{G}(\widehat{X}, \nabla\Theta)$ under the assumption that each objective function θ_ν is twice continuously differentiable. For ease of reference, we repeat the definition of $\widehat{x}$:

$$\widehat{x}(y) \triangleq \operatorname*{argmin}_{x \in \widehat{X}} \left[\sum_{\nu=1}^{N} \theta_\nu(x^\nu, y^{-\nu}) + \tfrac{1}{2}\|x - y\|^2 \right], \quad \text{for } y \in \widehat{X},$$

which decomposes into N subproblems, one for each player:

$$\widehat{x}^\nu(y) = \operatorname*{argmin}_{x^\nu \in X^\nu} \left[\theta_\nu(x^\nu, y^{-\nu}) + \tfrac{1}{2}\|x^\nu - y^\nu\|^2 \right], \quad \forall\, \nu = 1, \cdots, N. \tag{12.27}$$

Thus, the implementation of a fixed-point scheme based on $\widehat{x}$ can be carried out in a distributed and independent manner. At the end of each iteration, each player's update is made known to all players before a new iteration begins. The assumed differentiability of the objection functions facilitates the convergence analysis of such a scheme. Note that player ν's regularized-optimization problem (12.27) is strongly convex; therefore each $\widehat{x}^\nu(y)$ can be computed by a host of efficient algorithms from convex programming [8, 11].

Our goal is to derive conditions under which the proximal-response map is either a contraction or a non-expansive map. In the former case, the fixed-point iteration $y^{k+1} \triangleq \widehat{x}(y^k)$ converges to the unique NE of the game $\mathcal{G}(\widehat{X}, \nabla\Theta)$; in the latter case, an averaging scheme such as: $z^{k+1} = z^k + \tau_k(\widehat{x}(z^k) - z^k)$, where the sequence of positive scalars $\{\tau_k\} \downarrow 0$ satisfies $\sum_{k=0}^{\infty} \tau_k = \infty$, will achieve the same result, provided that an NE exists.

For any two vectors y and y' we have, by the variational principle,

$$(z^\nu - \widehat{x}^\nu(y))^T \left[\nabla_{x^\nu} \theta_\nu(\widehat{x}^\nu(y), y^{-\nu}) + \widehat{x}^\nu(y) - y^\nu \right] \geq 0, \quad \forall z^\nu \in X^\nu,$$

$$(z^\nu - \widehat{x}^\nu(y'))^T \left[\nabla_{x^\nu} \theta_\nu(\widehat{x}^\nu(y'), y'^{,-\nu}) + \widehat{x}^\nu(y') - y'^{,\nu} \right] \geq 0, \quad \forall z^\nu \in X^\nu.$$

Substituting $z^\nu = \widehat{x}^\nu(y')$ into the former inequality and $z^\nu = \widehat{x}^\nu(y)$ into the latter, adding the two resulting inequalities, we obtain, with $z^\nu \triangleq \tau_\nu(\widehat{x}^\nu(y), y^{-\nu}) + (1 - \tau_\nu)(\widehat{x}^\nu(y'), y'^{,-\nu})$ for some $\tau_\nu \in (0, 1)$,

$$0 \leq (\widehat{x}^\nu(y') - \widehat{x}^\nu(y))^T \left[\nabla_{x^\nu} \theta_\nu(\widehat{x}^\nu(y), y^{-\nu}) + \widehat{x}^\nu(y) - y^\nu \right]$$

$$+ (\widehat{x}^\nu(y) - \widehat{x}^\nu(y'))^T \left[\nabla_{x^\nu} \theta_\nu(\widehat{x}^\nu(y'), y'^{,-\nu}) + \widehat{x}^\nu(y') - y'^{,\nu} \right]$$

$$= (\widehat{x}^\nu(y') - \widehat{x}^\nu(y))^T \left[\nabla_{x^\nu} \theta_\nu(\widehat{x}^\nu(y), y^{-\nu}) - \nabla_{x^\nu} \theta_\nu(\widehat{x}^\nu(y'), y'^{,-\nu}) \right]$$

$$- \| \widehat{x}^\nu(y') - \widehat{x}^\nu(y) \|^2 + (\widehat{x}^\nu(y') - \widehat{x}^\nu(y))^T (y'^{,\nu} - y^\nu)$$

$$= (\widehat{x}^\nu(y') - \widehat{x}^\nu(y))^T \left[\nabla_{x^\nu}^2 \theta_\nu(z^\nu) \right] (\widehat{x}^\nu(y) - \widehat{x}^\nu(y'))$$

$$+ (\widehat{x}^\nu(y') - \widehat{x}^\nu(y))^T \sum_{\nu' \neq \nu} \left[\nabla_{x^\nu x^{\nu'}}^2 \theta_\nu(z^\nu) \right] (y^{\nu'} - y'^{,\nu'})$$

$$- \| \widehat{x}^\nu(y') - \widehat{x}^\nu(y) \|^2 + (\widehat{x}^\nu(y') - \widehat{x}^\nu(y))^T (y'^{,\nu} - y^\nu).$$

Recalling the scalars $\zeta_\nu(z)$ and $\xi_{\nu\nu'}(z)$ defined in (12.19) and (12.20), and letting $\xi_{\nu\nu}(z) \triangleq 1$, we deduce from the above inequalities:

$$(1 + \zeta_\nu(z^\nu)) \, \| \widehat{x}^\nu(y') - \widehat{x}^\nu(y) \| \leq \sum_{\nu'=1}^{N} \xi_{\nu\nu'}(z^\nu) \, \| y^{\nu'} - y'^{,\nu'} \|.$$

For an N-tuple $\mathbf{z} \triangleq (z^\nu)_{\nu=1}^{N}$ with $z^\nu \in \widehat{X}$ for all $\nu = 1, \cdots, N$, defining the $N \times N$ (asymmetric) matrix $\Gamma(\mathbf{z})$:

$$\begin{bmatrix}
\dfrac{1}{1 + \zeta_1(z^1)} & \dfrac{\xi_{12}(z^1)}{1 + \zeta_1(z^1)} & \dfrac{\xi_{13}(z^1)}{1 + \zeta_1(z^1)} & \cdots & \dfrac{\xi_{1N}(z^1)}{1 + \zeta_1(z^1)} \\[2em]
\dfrac{\xi_{21}(z^2)}{1 + \zeta_2(z^2)} & \dfrac{1}{1 + \zeta_2(z^2)} & \dfrac{\xi_{23}(z^2)}{1 + \zeta_2(z^2)} & \cdots & \dfrac{\xi_{2N}(z^2)}{1 + \zeta_2(z^2)} \\[2em]
\vdots & \vdots & \ddots & \vdots & \vdots \\[2em]
\dfrac{\xi_{(N-1)1}(z^{N-1})}{1 + \zeta_{N-1}(z^{N-1})} & \dfrac{\xi_{(N-1)2}(z^{N-1})}{1 + \zeta_{N-1}(z^{N-1})} & \cdots & \dfrac{1}{1 + \zeta_{N-1}(z^{N-1})} & \dfrac{\xi_{(N-1)N}(z^{N-1})}{1 + \zeta_{N-1}(z^{N-1})} \\[2em]
\dfrac{\xi_{N1}(z^N)}{1 + \zeta_N(z^N)} & \dfrac{\xi_{N2}(z^N)}{1 + \zeta_N(z^N)} & \cdots & \dfrac{\xi_{N(N-1)}(z^N)}{1 + \zeta_N(z^N)} & \dfrac{1}{1 + \zeta_N(z^N)}
\end{bmatrix}$$

we obtain

$$
\begin{pmatrix} \| \widehat{x}^1(y') - \widehat{x}^1(y) \| \\ \vdots \\ \| \widehat{x}^N(y') - \widehat{x}^N(y) \| \end{pmatrix} \le \Gamma(\mathbf{z}) \begin{pmatrix} \| y^1 - y'^{,1} \| \\ \vdots \\ \| y^N - y'^{,N} \| \end{pmatrix}. \tag{12.28}
$$

In what follows, we let $\rho(A)$ denote the spectral radius of a matrix A. By the well-known Perron–Frobenius theory of non-negative matrices, it follows that $\rho(A) \ge \rho(B)$ if $A \ge B \ge 0$. We recall that a vector norm $\||\bullet\||$ is monotonic if $x \ge y \ge 0$ implies $\||x\|| \ge \||y\||$. Such a vector norm induces a matrix norm via the standard definition $\||A\|| \triangleq \max_{\||x\||=1} \||Ax\||$. For such a matrix norm, it holds that if $A \ge B \ge 0$ (componentwise), then $\||A\|| \ge \||B\||$.

Now, assuming that the second-order derivatives of the objective functions are bounded on $\widehat{X}$, we can define, similarly to the matrix $\mathbf{\Upsilon}$ in (12.21), the following matrix:

$$
\mathbf{\Gamma} \triangleq \begin{bmatrix}
\dfrac{1}{1+\zeta^1_{\min}} & \dfrac{\xi^{12}_{\max}}{1+\zeta^1_{\min}} & \dfrac{\xi^{13}_{\max}}{1+\zeta^1_{\min}} & \cdots & \dfrac{\xi^{1N}_{\max}}{1+\zeta^1_{\min}} \\[2ex]
\dfrac{\xi^{21}_{\max}}{1+\zeta^2_{\min}} & \dfrac{1}{1+\zeta^2_{\min}} & \dfrac{\xi^{23}_{\max}}{1+\zeta^2_{\min}} & \cdots & \dfrac{\xi^{2N}_{\max}}{1+\zeta^2_{\min}} \\[2ex]
\vdots & \vdots & \ddots & \vdots & \vdots \\[2ex]
\dfrac{\xi^{(N-1)1}_{\max}}{1+\zeta^{N-1}_{\min}} & \dfrac{\xi^{(N-1)2}_{\max}}{1+\zeta^{N-1}_{\min}} & \cdots & \dfrac{1}{1+\zeta^{N-1}_{\min}} & \dfrac{\xi^{(N-1)N}_{\max}}{1+\zeta^{N-1}_{\min}} \\[2ex]
\dfrac{\xi^{N1}_{\max}}{1+\zeta^N_{\min}} & \dfrac{\xi^{N2}_{\max}}{1+\zeta^N_{\min}} & \cdots & \dfrac{\xi^{N(N-1)}_{\max}}{1+\zeta^N_{\min}} & \dfrac{1}{1+\zeta^N_{\min}}
\end{bmatrix}.
$$

The following result follows easily from the basic inequality (12.28), the fact that $\mathbf{\Gamma}(\mathbf{z}) \le \mathbf{\Gamma}$ for all tuples $\mathbf{z} \triangleq \left(z^i\right)_{i=1}^N$ with $z^i \in \widehat{X}$ for all $i = 1, \cdots, N$, and the remarks above.

PROPOSITION 12.17 *Assume that the second-order derivatives of the objective functions are bounded on $\widehat{X}$. If $\||\mathbf{\Gamma}\|| \le 1$ for some monotonic norm, then the proximal-response map is nonexpansive in the sense that for any two tuples y and y' in $\widehat{X}$:*

$$
\left\| \begin{pmatrix} \| \widehat{x}^1(y') - \widehat{x}^1(y) \| \\ \vdots \\ \| \widehat{x}^N(y') - \widehat{x}^N(y) \| \end{pmatrix} \right\| \le \left\| \begin{pmatrix} \| y^1 - y'^{,1} \| \\ \vdots \\ \| y^N - y'^{,N} \| \end{pmatrix} \right\|.
$$

If $\rho(\mathbf{\Gamma}) < 1$, then a positive scalar $a < 1$ and a monotonic norm $\vert\,\Vert \bullet \Vert\,\vert$ exists such that

$$\left\Vert \left(\begin{array}{c} \Vert \widehat{x}^1(y') - \widehat{x}^1(y) \Vert \\ \vdots \\ \Vert \widehat{x}^N(y') - \widehat{x}^N(y) \Vert \end{array} \right) \right\Vert \leq a \left\Vert \left(\begin{array}{c} \Vert y^1 - y'^{,1} \Vert \\ \vdots \\ \Vert y^N - y'^{,N} \Vert \end{array} \right) \right\Vert ;$$

in other words, the proximal-response map is a contraction.

If $\rho(\mathbf{\Gamma})$ is less than unity, then the matrix $\mathbb{I}_N - \mathbf{\Gamma}$ must be a P-matrix, where $\mathbb{I}_N$ is the identity matrix of order N. Since the P-property is invariant under multiplication by positive diagonal matrices, it follows that the matrix $\mathbf{\Upsilon}$ in (12.21) is also a P-matrix. Conversely, if $\mathbf{\Upsilon}$ in (12.21) is a P-matrix, then reversing the argument, it follows that $\rho(\mathbf{\Gamma}) < 1$. Thus, the matrix-theoretic condition in Theorem 12.5 that is sufficient for the existence and uniqueness of the NE of the game $\mathcal{G}(\widehat{X}, \nabla\Theta)$ also ensures the contraction of the fixed-point iteration $y^{k+1} = \widehat{x}(y^k)$, whose limit is the claimed NE.

12.6.2 Descent methods for variational equilibria

A main advantage of the proximal-response-based, fixed-point iterative method for computing an NE on partitioned player-strategy spaces is that the resulting method can be very easily implemented in a distributed computing environment. Yet this approach is handicapped by a drawback of its convergence theory; namely, it requires a bounded assumption of the second derivatives of the players' objective functions, which for unbounded strategy sets could be somewhat restrictive; the P-property of the matrix $\mathbf{\Upsilon}$ is potentially another restriction in certain applications.

In what follows, we discuss the adoption of a standard-descent method for optimization problems to the computation of a variational equilibrium of a game with jointly convex constraints, via the solution of the VI $(\widehat{C}, \nabla\Theta)$. Related discussion can be found in the three recent papers [25, 26]. Throughout this subsection, we assume that each function θ_ν is twice continuously differentiable on an open set containing $\widehat{C}$ but make no assumption about the boundedness of the second derivatives on $\widehat{C}$. Specifically, we consider the minimization of the Nikaido–Isoda function on the set $\widehat{C}$:

$$\underset{y \in \widehat{C}}{\text{minimize}}\ \varphi_{\mathrm{NI}}(y) \triangleq \sum_{\nu=1}^{N} \left[\theta_\nu(y^\nu, y^{-\nu}) - \theta_\nu(\widehat{x}^{\mathrm{J},\nu}(y), y^{-\nu}) - \tfrac{1}{2} \Vert \widehat{x}^{\mathrm{J},\nu}(y) - y^\nu \Vert^2 \right]$$

where

$$\widehat{x}^{\mathrm{J}}(y) \triangleq \underset{x \in \widehat{C}}{\text{argmin}}\ \sum_{\nu=1}^{N} \left[\theta_\nu(x^\nu, y^{-\nu}) + \tfrac{1}{2} \Vert x^\nu - y^\nu \Vert^2 \right].$$

By Proposition 12.6, we have, for all $v' = 1, \cdots, N$,

$$\nabla_{y^{v'}} \varphi_{\mathrm{NI}}(y) = \sum_{v=1}^{N} \nabla_{y^{v'}} \theta_v(y) - \sum_{v \neq v'} \nabla_{y^{v'}} \theta_v(\widehat{x}^{\mathrm{J},v}(y), y^{-v}) - y^{v'} + \widehat{x}^{\mathrm{J},v'}(y).$$

Thus,

$$(y^{v'} - \widehat{x}^{\mathrm{J},v'}(y))^T \nabla_{y^{v'}} \varphi_{\mathrm{NI}}(y) = (y^{v'} - \widehat{x}^{\mathrm{J},v'}(y))^T \nabla_{y^{v'}} \theta_{v'}(y) - \| y^{v'} - \widehat{x}^{\mathrm{J},v'}(y) \|^2$$

$$+ (y^{v'} - \widehat{x}^{\mathrm{J},v'}(y))^T \sum_{v \neq v'} \left[\nabla_{y^{v'}} \theta_v(y) - \nabla_{y^{v'}} \theta_v(\widehat{x}^{\mathrm{J},v}(y), y^{-v}) \right]$$

$$= (y^{v'} - \widehat{x}^{\mathrm{J},v'}(y))^T \sum_{v=1}^{N} \left[\nabla_{y^{v'}} \theta_v(y) - \nabla_{y^{v'}} \theta_v(\widehat{x}^{\mathrm{J},v}(y), y^{-v}) \right]$$

$$+ (y^{v'} - \widehat{x}^{\mathrm{J},v'}(y))^T \left[\nabla_{y^{v'}} \theta_{v'}(\widehat{x}^{\mathrm{J},v'}(y), y^{-v'}) + \widehat{x}^{\mathrm{J},v'}(y) - y^{v'} \right].$$

Summing up the above inequalities for $v' = 1, \cdots, N$, we deduce, by the variational characterization (12.17) of $\widehat{x}^{\mathrm{J}}(y)$,

$$(y - \widehat{x}^{\mathrm{J}}(y))^T \nabla \varphi_{\mathrm{NI}}(y)$$

$$\geq \sum_{v'=1}^{N} (y^{v'} - \widehat{x}^{\mathrm{J},v'}(y))^T \sum_{v=1}^{N} \left[\nabla_{y^{v'}} \theta_v(y) - \nabla_{y^{v'}} \theta_v(\widehat{x}^{\mathrm{J},v}(y), y^{-v}) \right].$$

Applying the mean-value theorem to the univariate, differentiable, scalar-valued function:

$$\tau \in [0,1] \mapsto \sum_{v'=1}^{N} (y^{v'} - \widehat{x}^{\mathrm{J},v'}(y))^T \sum_{v=1}^{N} \nabla_{y^{v'}} \theta_v(y^v(\tau), y^{-v}),$$

where $y^v(\tau) \triangleq \tau y^v + (1 - \tau)\widehat{x}^{\mathrm{J},v}(y)$, we deduce the existence of a scalar $\widehat{\tau} \in (0, 1)$ such that

$$(y - \widehat{x}^{\mathrm{J}}(y))^T \nabla \varphi_{\mathrm{NI}}(y)$$

$$\geq \sum_{v'=1}^{N} (y^{v'} - \widehat{x}^{\mathrm{J},v'}(y))^T \sum_{v=1}^{N} \nabla^2_{y^{v'} y^v} \theta_v(y^v(\widehat{\tau}), y^{-v}) (y^v - \widehat{x}^{\mathrm{J},v}(y)).$$

Note that the tuple $(y^v(\widehat{\tau}), y^{-v})$ is a convex combination of y and $(\widehat{x}^{\mathrm{J},v}(y), y^{-v})$, both belonging to $\widehat{X}$. Similar to the matrix $\mathbf{M}(z)$, we define, for an N-tuple $z \triangleq (z^v)_{v=1}^{N} \in \widehat{X}$, the matrix

$$\widehat{\mathbf{M}}(z) \triangleq \left[\widehat{M}_{v'v}(z^v) \right]_{v,v'=1}^{N}, \quad \text{where} \quad M_{v'v}(z^v) \triangleq \nabla^2_{x^{v'} x^v} \theta_v(z^v).$$

It follows that for each $y \in \widehat{\mathbf{C}}$, there exist N tuples z^ν each being a convex combination of y and $(\widehat{x}^{J,\nu}(y), y^{-\nu})$, such that

$$(y - \widehat{x}^{J}(y))^{T} \nabla \varphi_{\mathrm{NI}}(y) \geq (y - \widehat{x}^{J}(y))^{T} \widehat{\mathbf{M}}(\mathbf{z})(y - \widehat{x}^{J}(y)). \tag{12.29}$$

Thus, assuming that the matrix $\widehat{\mathbf{M}}(\mathbf{z})$ is positive definite, we have that either $(\widehat{x}^{J}(y) - y)^{T} \nabla \varphi_{\mathrm{NI}}(y) < 0$ or $\widehat{x}^{J}(y) = y$. In the former case, the vector $d(y) \triangleq \widehat{x}^{J}(y) - y$ is a descent direction of the function φ_{NI} at y. In the latter case, it can be shown (see the proof of Proposition 12.18 below) that $\widehat{x}^{J}(y) = y$; thus y is a variational equilibrium of the game $\mathcal{G}(\widehat{\mathbf{C}}, \nabla\Theta)$. These facts pave the way to the use of standard line search methods for constrained optimization [8] for the minimization of φ_{NI} on the feasible set $\widehat{\mathbf{C}}$; this will typically require some additional assumption like the uniform positive definiteness of the matrix $\widehat{\mathbf{M}}(\mathbf{z})$. We leave the details to the interested reader.

When a descent method is applied to minimize φ_{NI} on $\widehat{\mathbf{C}}$, it typically will terminate with only a constrained stationary point, i.e., a point $y^* \in \widehat{\mathbf{C}}$ such that $(y - y^*)^{T} \varphi_{\mathrm{NI}}(y^*) \geq 0$ for all $y \in \widehat{\mathbf{C}}$. Since φ_{NI} is not a convex function, such a point is in general not even a local minimum. Thus, it is important to know when a constrained stationary point of φ_{NI} is a variational equilibrium of the game $\mathcal{G}(\widehat{\mathbf{C}}, \nabla\Theta)$. The above analysis actually has provided an answer to this issue which we formalize in the result below.

PROPOSITION 12.18 *Suppose that y^* is a stationary point of φ_{NI} on $\widehat{\mathbf{C}}$. If the matrix $\mathbf{M}(\mathbf{z})$ is positive semidefinite for all tuples $\mathbf{z}$, where for every ν, z^ν is any vector lying on the line segment joining y^* and $(x^\nu, y^{*,-\nu})$ for arbitrary $x^\nu \in X^\nu$, then y^* is a variational equilibrium of the game $\mathcal{G}(\widehat{\mathbf{C}}, \nabla\Theta)$.*

Proof On one hand, we have, by (12.29),

$$(y - \widehat{x}^{J}(y))^{T} \nabla \varphi_{\mathrm{NI}}(y) \geq 0$$

by the positive semidefiniteness of $\mathbf{M}(\mathbf{z})$, on the other hand the right-hand side is non positive since y is a constrained stationary point of φ_{NI} on $\widehat{\mathbf{C}}$. Hence equality holds. Tracing back the derivation of (12.29), we easily deduce that $\widehat{x}^{J}(y) = y$. $\blacksquare$

12.7 A communication game

There are many Nash models in communication systems; a basic one was introduced in the thesis [13, 14, 56] and has since been studied extensively in several subsequent papers [34, 52, 55]. In this model, each of a finite number of users of a communication network selfishly chooses their power allocations to maximize a signal-to-noise objective function subject to a total-budget constraint. A variant of this model to include a jammer is formulated in [24]. There are also other models, such as the internet-switching problem [19, 30], and resource-allocation games. [28, 29]. The survey papers [1, 2] are also useful entry points to the literature on this topic. In this section, we consider a cognitive radio game with "quality-of-service" (QoS) constraints. Communicated to us by

Daniel P. Palomar and Gesualdo Scutari and described in detail in [53, 58], this model is an extension of the one studied in [46] that corresponds to the case of exogenous prices. We choose this model to illustrate the general existence theory because of several reasons: (a) it is a new model; (b) it combines several features of other models; (c) it involves unbounded player-dependent strategy sets; and (d) it illustrates both the NCP approach (Subsection 12.3.4) for solution existence in the case of exogenous prices, and the theory of extended games with endogenous prices (Subsection 12.3.2). Due to space limitations, we can only discuss the existence of equilibria.

12.7.1   A model with QoS constraints

Suppose that there are P primary users, Q secondary users, and N frequency bandwidths in a communication system. Let

$$\mathbf{x}_q \triangleq \begin{pmatrix} x_q(1) \\ \vdots \\ x_q(N) \end{pmatrix},$$

be user q's transmission on bandwidth k. Define the prices from the primary users as the pair $\widehat{\boldsymbol{\lambda}} \triangleq \left\{ \boldsymbol{\lambda}_{\text{tot}}, (\boldsymbol{\lambda}_p)_{p=1}^{P} \right\}$ where, for $p = 1, \cdots, P$,

$$\boldsymbol{\lambda}_{\text{tot}} \triangleq \begin{pmatrix} \lambda_{1,\text{tot}} \\ \vdots \\ \lambda_{P,\text{tot}} \end{pmatrix}, \quad \boldsymbol{\lambda}_p \triangleq \begin{pmatrix} \lambda_p(1) \\ \vdots \\ \lambda_p(N) \end{pmatrix}$$

with $\lambda_p(k)$ being the price imposed by primary user p over bandwidth k.

The secondary users minimize their total-transmit power plus the costs paid, subject to a minimum level of rate transmitted $\{L_q\}_{q=1}^{Q} > 0$, called a "quality of service" (QoS) constraint. Thus, parameterized by the rivals' allocations $\mathbf{x}_{-q} \triangleq (\mathbf{x}_r)_{r \neq q}$ and prices $\left\{ \boldsymbol{\lambda}_{\text{tot}}, (\boldsymbol{\lambda}_p)_{p=1}^{P} \right\}$, secondary user q solves the minimization problem:

$$\underset{\mathbf{x}_q}{\text{minimize}} \quad \sum_{k=1}^{N} x_q(k) + \sum_{p=1}^{P} \sum_{k=1}^{N} (\lambda_p(k) + \lambda_{p,\text{tot}}) \beta_{p,q}(k) x_q(k)$$

$$\text{subject to} \quad \mathbf{x}_q \geq 0$$

$$\text{and} \quad \sum_{k=1}^{N} \log\left(1 + \frac{x_q(k)}{\widehat{\sigma}_q^2(k) + \sum_{r \neq q} \widehat{h}_{qr}(k) x_r(k)} \right) \geq L_q, \tag{12.30}$$

where $\beta_{p,q}(k)$ represents the frequency-response of channel k between the secondary player q and the primary player p, $\widehat{h}_{qr}(k)$ is the normalized frequency-response of channel k between the two secondary players r and q, and $\widehat{\sigma}_q^2(k)$ is the variance of a Gaussian white noise on channel k for player q.

In contrast, the primary users choose the prices $\lambda_p(k)$ and $\lambda_{p,\mathrm{tot}}$ such that

$$0 \leq \lambda_p(k) \perp \alpha_p(k) - \sum_{q=1}^{Q} \beta_{p,q}(k)\, x_q(k) \geq 0 \tag{12.31}$$

$$0 \leq \lambda_{p,\mathrm{tot}} \perp \alpha_{p,\mathrm{tot}} - \sum_{k=1}^{N}\sum_{q=1}^{Q} \beta_{p,q}(k)\, x_q(k) \geq 0 \tag{12.32}$$

for all $p = 1,\cdots,P$ and $k = 1,\cdots,N$. In essence, while (12.31) is a per-tone interference constraint imposed by each primary user on the secondary users' transmissions, (12.32) is an aggregate of these interference constraints.

To ensure that both (12.31) and (12.32) are effective, we assume without loss of generality that $\alpha_p(k) < \alpha_{p,\mathrm{tot}} < \sum_{k'=1}^{N} \alpha_p(k')$ for all p and k. The two complementarity conditions (12.31) and (12.32) are the optimality conditions for the linear program in the prices $\left\{\lambda_{\mathrm{tot}}, \left(\lambda_p\right)_{p=1}^{P}\right\}$ parameterized by the secondary users' allocations $\mathbf{x}$:

$$\begin{aligned}
\underset{\lambda_{\mathrm{tot}}\geq 0,\ \lambda_p\geq 0}{\text{minimize}} \quad & \sum_{p=1}^{P} \lambda_{p,\mathrm{tot}}\left[\alpha_{p,\mathrm{tot}} - \sum_{k=1}^{N}\sum_{q=1}^{Q}\beta_{p,q}(k)\,x_q(k)\right] + \\
& \sum_{p=1}^{P}\sum_{k=1}^{N}\lambda_p(k)\left[\alpha_p(k) - \sum_{q=1}^{Q}\beta_{p,q}(k)\,x_q(k)\right].
\end{aligned} \tag{12.33}$$

In summary, this communication game is composed of the optimization problems (12.30) for the secondary users $q = 1,\cdots,Q$ and the single optimization problem (12.33) for the primary users. The model constants are as follows: $\beta_{p,q}(k) \geq 0$, $\widehat{h}_{qr}(k) \geq 0$ with $\widehat{h}_{qq}(k) = 1$, $L_q > 0$, $\alpha_p(k) > 0$, $\widehat{\sigma}_q(k) > 0$, and

$$\alpha_{p,\mathrm{tot}} < \sum_{k=1}^{N}\alpha_p(k), \quad \text{to avoid the redundancy of (12.32)}$$

for all (q,p,k). We refer to [46] for a more detailed interpretation of these constants.

12.7.2 Exogenous prices

Before analyzing the model where the prices are endogenous, we analyze the case, interesting on its own, where the prices are fixed and there are no primary users. Thus, we seek an equilibrium to a game where each (secondary) user's optimization problem is:

$$\underset{\mathbf{x}_q}{\text{minimize}} \quad \sum_{k=1}^{N} \widehat{\gamma}_q(k)\, x_q(k)$$

$$\text{subject to} \quad \mathbf{x}_q \geq 0$$

$$\text{and} \quad \sum_{k=1}^{N} \log\left(1 + \frac{x_q(k)}{\widehat{\sigma}_q^2(k) + \displaystyle\sum_{r \neq q} \widehat{h}_{qr}(k) x_r(k)}\right) \geq L_q, \tag{12.34}$$

where each $\widehat{\gamma}_q(k) > 0$ is a given constant that contains the exogenous prices. The model studied in [46] has $\gamma_q(k)$ all equal to unity. Since the Slater-constraint qualification, thus the MFCQ, holds for (12.34), this problem is equivalent to its KKT conditions, which are: with μ_q denoting the multiplier of the logarithmic constraint in (12.34),

$$0 \leq x_q(k) \perp \widehat{\gamma}_q(k) - \frac{\mu_q}{\widehat{\sigma}_q^2(k) + \displaystyle\sum_{r=1}^{Q} \widehat{h}_{qr}(k) x_r(k)} \geq 0$$

$$0 \leq \mu_q \perp \sum_{k=1}^{N} \log\left(1 + \frac{x_q(k)}{\widehat{\sigma}_q^2(k) + \displaystyle\sum_{r \neq q} \widehat{h}_{qr}(k) x_r(k)}\right) - L_q \geq 0. \tag{12.35}$$

To apply Theorem 12.4, suppose that there exists a sequence of positive scalars $\{\tau_q^\ell\}_{\ell=1}^{\infty}$, a sequence of vectors $\left\{\mathbf{x}^\ell \triangleq \left(\mathbf{x}_q^\ell\right)_{q=1}^{Q}\right\}_{\ell=1}^{\infty}$, and a sequence of multipliers $\left\{\boldsymbol{\mu} \triangleq \left(\mu_q^\ell\right)_{q=1}^{Q}\right\}_{\ell=1}^{\infty}$ such that $\lim_{\ell \to \infty} \|(\mathbf{x}^\ell, \boldsymbol{\mu}^\ell)\| = \infty$ and for each ℓ,

$$0 \leq x_q^\ell(k) \perp \widehat{\gamma}_q(k) - \frac{\mu_q^\ell}{\widehat{\sigma}_q^2(k) + x_q^\ell(k) + \displaystyle\sum_{r \neq q} \widehat{h}_{qr}(k) x_r^\ell(k)} + \tau_q^\ell x_q^\ell(k) \geq 0$$

$$0 \leq \mu_q^\ell \perp \sum_{k=1}^{N} \log\left(1 + \frac{x_q^\ell(k)}{\widehat{\sigma}_q^2(k) + \displaystyle\sum_{r \neq q} \widehat{h}_{qr}(k) x_r^\ell(k)}\right) - L_q + \tau_q^\ell \mu_q^\ell \geq 0.$$

We wish to derive a contradiction to the unboundedness of the sequence $\{\mathbf{x}^\ell, \boldsymbol{\mu}^\ell\}_{\ell=1}^{\infty}$. From the first complementarity condition, we deduce that $\mu_q^\ell > 0$, which implies, by the second complementarity condition,

$$\sum_{k=1}^{N} \log \left(1 + \frac{x_q^\ell(k)}{\widehat{\sigma}_q^2(k) + \sum_{r \neq q} \widehat{h}_{qr}(k)\, x_r^\ell(k)} \right) + \tau_\ell\, \mu_q^\ell = L_q.$$

Hence the sequence $\{\tau_\ell\, \mu_q^\ell\}$ is bounded. Moreover, we have, for each k,

$$\log \left(1 + \frac{x_q^\ell(k)}{\widehat{\sigma}_q^2(k) + \sum_{r \neq q} \widehat{h}_{qr}(k)\, x_r^\ell(k)} \right) \leq L_q,$$

or equivalently,

$$x_q^\ell(k) - \left(e^{L_q} - 1 \right) \sum_{r \neq q} \widehat{h}_{qr}(k)\, x_r^\ell(k) \leq \left(e^{L_q} - 1 \right) \widehat{\sigma}_q^2(k).$$

The above inequalities can be written in matrix form as:

$$H_k(\mathbf{L}) \begin{pmatrix} x_1^\ell(k) \\ \vdots \\ x_Q^\ell(k) \end{pmatrix} \leq \begin{pmatrix} \left(e^{L_1} - 1 \right) \widehat{\sigma}_1^2(k) \\ \vdots \\ \left(e^{L_Q} - 1 \right) \widehat{\sigma}_Q^2(k) \end{pmatrix},$$

where $\mathbf{L}$ is the vector of L_q for $q = 1, \cdots, Q$, and for all $k = 1, \cdots, N$,

$$H_k(\mathbf{L}) \triangleq \begin{bmatrix} 1 & -\left(e^{L_1} - 1 \right) \widehat{h}_{12}(k) & \cdots & -\left(e^{L_1} - 1 \right) \widehat{h}_{1Q}(k) \\ -\left(e^{L_1} - 1 \right) \widehat{h}_{21}(k) & 1 & \cdots & -\left(e^{L_2} - 1 \right) \widehat{h}_{2Q}(k) \\ \vdots & \vdots & \ddots & \vdots \\ -\left(e^{L_Q} - 1 \right) \widehat{h}_{Q1}(k) & -\left(e^{L_Q} - 1 \right) \widehat{h}_{Q2}(k) & \cdots & 1 \end{bmatrix}.$$

Note that $H_k(\mathbf{L})$ is a Z-matrix. If $H_k(\mathbf{L})$ is a P-matrix, then $H(k)^{-1}$ exists and is non-negative. Hence,

$$0 \leq \begin{pmatrix} x_1^\ell(k) \\ \vdots \\ x_Q^\ell(k) \end{pmatrix} \leq H_k(\mathbf{L})^{-1} \begin{pmatrix} \left(e^{L_1} - 1 \right) \widehat{\sigma}_1^2(k) \\ \vdots \\ \left(e^{L_Q} - 1 \right) \widehat{\sigma}_Q^2(k) \end{pmatrix},$$

showing that the sequence $\{\mathbf{x}^\ell\}$ is bounded. Thus $\{\boldsymbol{\mu}^\ell\}$ is unbounded, from which it follows that $\{\tau_\ell\}$ must converge to zero. But this contradicts the inequality:

$$\frac{\mu_q^\ell}{\widehat{\sigma}_q^2(k) + x_q^\ell(k) + \sum_{r \neq q} \widehat{h}_{qr}(k) x_r^\ell(k)} \leq \gamma_q(k) + \tau_q^\ell x_q^\ell(k), \qquad (12.36)$$

which implies that $\{\mu_q^\ell\}_{\ell=1}^\infty$ is bounded. This contradiction allows the application of Theorem 12.4 to establish the existence of an NE to this game.

PROPOSITION 12.19 *Suppose that each $H(k)$ is a P-matrix for $k = 1, \cdots, N$. The communication game with exogenous prices has a nonempty bounded set of Nash equilibria.*

Proof The boundedness of the Nash equilibria follows from the fact that any such equilibrium must satisfy the QoS constraints as equalities. The P-property of the matrices $H(k)$ then yields a bound for the equilibria similar to (12.36). ∎

Under appropriate assumptions, it is also possible to show uniqueness of the NE by using the P-function theory. However, note that this theory is not directly applicable to this game because the QoS constraints are coupled and the P-function theory requires the players' strategy sets to be separable. Therefore, the approach aims at first converting the QoS game into an equivalent partitioned VI on a Cartesian product of sets in order to apply the P-function theory. See [46] for details.

12.7.3   Endogenous prices

We return to the analysis of the model with both primary and secondary users, where the prices are endogenous. We follow the theory in Subsection 12.3.2. Define for each $\mathbf{x}_{-q}$, the set

$$\Xi_q(\mathbf{x}_{-q}) \triangleq \left\{ \mathbf{x}_q \geq 0 : \sum_{k=1}^N \log\left(1 + \frac{x_q(k)}{\widehat{\sigma}_q^2(k) + \sum_{r \neq q} \widehat{h}_{qr}(k) x_r(k)}\right) \geq L_q \right\},$$

which is nonempty, closed, and convex, has a nonempty interior, but is unbounded for all $\mathbf{x}_{-q} \geq 0$. The set $\mathcal{F}_\Xi$ of fixed points of the strategy map Ξ is equal to:

$$\mathcal{F}_\Xi = \left\{ \mathbf{x} \geq 0 : \sum_{k=1}^N \log\left(1 + \frac{x_q(k)}{\widehat{\sigma}_q^2(k) + \sum_{r \neq q} \widehat{h}_{qr}(k) x_r(k)}\right) \geq L_q, \quad \forall q \right\},$$

which is a nonconvex and unbounded set. The optimization problem (12.15) takes the form: for given $\mathbf{y} \triangleq \left\{ \left(y_q(k)\right)_{k=1}^{N} \right\}_{q=1}^{Q}$,

$$\underset{\mathbf{x}}{\text{minimize}} \quad \sum_{q=1}^{Q}\sum_{k=1}^{N} x_q(k) + \frac{1}{2}\sum_{q=1}^{Q}\sum_{k=1}^{N} \left(x_q(k) - y_q(k) \right)^2$$

$$\text{subject to} \quad x_q(k) \geq 0, \quad \forall q = 1,\cdots,Q,\, k = 1,\cdots,N$$

$$\sum_{k=1}^{N} \log\left(1 + \frac{x_q(k)}{\widehat{\sigma}_q^2(k) + \sum_{r\neq q}\widehat{h}_{qr}(k)y_r(k)} \right) \geq L_q, \quad \forall q = 1,\cdots,Q \tag{12.37}$$

$$\sum_{q=1}^{Q} \beta_{p,q}(k)\, x_q(k) \leq \alpha_p(k), \quad \forall p = 1,\cdots,P,\, k = 1,\cdots,N$$

$$\text{and} \quad \sum_{k=1}^{N}\sum_{q=1}^{Q} \beta_{p,q}(k)\, x_q(k) \leq \alpha_{p,\text{tot}}, \quad \forall k = 1,\cdots,N.$$

Our approach is to apply Theorem 12.2 to complete the existence proof. For this purpose, define the bounded polyhedron:

$$\mathcal{P} \triangleq \left\{ \begin{array}{l} \mathbf{x} \geq 0 : \ \displaystyle\sum_{q=1}^{Q} \beta_{p,q}(k)\, x_q(k) \leq \alpha_p(k), \quad \forall p = 1,\cdots,P,\, k = 1,\cdots,N \\[2em] \displaystyle\sum_{k=1}^{N}\sum_{q=1}^{Q} \beta_{p,q}(k)\, x_q(k) \leq \alpha_{p,\text{tot}}, \quad \forall p = 1,\cdots,P \end{array} \right\}$$

to serve as the set $\widehat{K}$ in the cited theorem. It suffices to guarantee that for each $\mathbf{y} \in \mathcal{P}$, there exists $\mathbf{x} \in \mathcal{P}$ satisfying

$$\sum_{k=1}^{N} \log\left(1 + \frac{x_q(k)}{\widehat{\sigma}_q^2(k) + \sum_{r\neq q}\widehat{h}_{qr}(k)y_r(k)} \right) > L_q, \quad \forall q = 1,\cdots,Q. \tag{12.38}$$

If this is true, then by joining such an $\mathbf{x}$ with any interior point of $\mathcal{P}$, it follows that any such combination that is sufficiently close to $\mathbf{x}$ must be an interior point of the set

$$\Xi^{\mathrm{E}}(\mathbf{y}) \triangleq \left\{ \mathbf{x} \in \mathcal{P} : \sum_{k=1}^{N} \log\left(1 + \frac{x_q(k)}{\widehat{\sigma}_q^2(k) + \sum_{r\neq q}\widehat{h}_{qr}(k)y_r(k)} \right) \geq L_q, \quad \forall q \right\}.$$

Thus the MFCQ holds at all feasible points of (12.37), and Theorem 12.2 applies. In turn, we can derive a sufficient condition for the existence of a vector $\mathbf{x} \in \mathcal{P}$ satisfying (12.38) for all $\mathbf{y} \in \mathcal{P}$. Note that with

$$R_{p,q}(k) \triangleq \max_{r \neq q} \frac{\widehat{h}_{qr}(k)}{\beta_{r,p}(k)} \quad \text{and} \quad \widehat{R}_q(k) \triangleq \max_{1 \leq p \leq P} R_{p,q}(k)\,\alpha_p(k),$$

we have

$$\sum_{k=1}^{N} \log\left(1 + \frac{x_q(k)}{\widehat{\sigma}_q^2(k) + \sum_{r \neq q} \widehat{h}_{qr}(k) y_r(k)}\right)$$

$$= \sum_{k=1}^{N} \log\left(1 + \frac{x_q(k)}{\widehat{\sigma}_q^2(k) + \sum_{r \neq q} \dfrac{\widehat{h}_{qr}(k)}{\beta_{r,p}(k)}\,\beta_{r,p}(k)\,y_r(k)}\right)$$

$$\geq \sum_{k=1}^{N} \log\left(1 + \frac{x_q(k)}{\widehat{\sigma}_q^2(k) + R_{p,q}(k) \sum_{r \neq q} \beta_{r,p}(k)\,y_r(k)}\right)$$

$$\geq \sum_{k=1}^{N} \log\left(1 + \frac{x_q(k)}{\widehat{\sigma}_q^2(k) + R_{p,q}(k)\,\alpha_p(k)}\right) \geq \sum_{k=1}^{N} \log\left(1 + \frac{x_q(k)}{\widehat{\sigma}_q^2(k) + \widehat{R}_q(k)}\right).$$

Summarizing the above analysis, the following result gives a sufficient condition for the existence of NE of the QoS game with endogenous prices. At this time, the uniqueness issue of such an equilibrium is not yet resolved.

PROPOSITION 12.20   *Suppose that there exists* $\mathbf{x}^{\mathrm{ref}} \in \mathcal{P}$ *such that*

$$\sum_{k=1}^{N} \log\left(1 + \frac{x_q^{\mathrm{ref}}(k)}{\widehat{\sigma}_q^2(k) + \widehat{R}_q(k)}\right) > L_q, \quad \forall q = 1, \cdots, Q.$$

Then the QoS game with endogenous prices has an NE.

Acknowledgments

We are grateful to Daniel P. Palomar and Gesualdo Scutari for suggesting the model in Subsection 12.7.1 and for stimulating discussions and to Christian Kanzow for useful observations on an earlier version of this chapter. The work of Facchinei was partially supported by MIUR, Research program PRIN 2007-9PLLN7 *Nonlinear Optimization,*

Variational Inequalities and Equilibrium Problems, Italy. The work of Pang is based on research supported by the U.S. National Science Foundation under grant CMMI-0802022.

References

[1] E. Altman, T. Boulogne, R. El-Azouzi, T. Jiménez, and L. Wynter, "A survey on networking games in telecommunications," *Computers & Operations Research*, vol. 33, pp.286–311, 2006.

[2] E. Altman and L. Wynter, "Equilibrium, games, and pricing in transportation and telecommunication networks," *Networks and Spatial Economics*, vol. 4, pp. 7–21, 2004.

[3] K.J. Arrow and G. Debreu, "Existence of an equilibrium for a competitive economy," *Econometrica*, vol. 22, pp. 265–90, 1954.

[4] J.P. Aubin and H. Frankowska, *Set-Valued Analysis*. Boston: Birkäuser, 1990.

[5] C. Berge, *Topological Spaces*. Edinburgh: Oliver and Boyd, 1963.

[6] A. Berman and R.J. Plemmons, *Nonnegative Matrices in the Mathematical Sciences*. Philadelphia: SIAM Classics in Applied Mathematics, no. 9, 1994.

[7] S. Berridge and J. Krawczyk (1997). Relaxation algorithms in finding Nash equilibria. Economic Working Papers Archive. Available: http://econwpa.wustl.edu/eprints/comp/papers/9707/9707002.abs

[8] D.P. Bertsekas, *Constrained Optimization and Lagrange Multiplier Methods*. Cambridge: Athena Scientific, 1996.

[9] D.P. Bertsekas and J.N. Tsitsiklis, *Parallel and Distributed Computation: Numerical Methods*. Cambridge: Athena Scientific, 1997.

[10] J.F. Bonnans and A. Shapiro, *Perturbation Analysis of Optimization Problems*. New York: Springer, 1998.

[11] S. Boyd, L. El Ghaoui, E. Feron, and V. Balakrishnan, *Linear Matrix Inequalities in System and Control Theory*. Philadelphia: Volume 15 of Studies in Applied Mathematics Society for Industrial and Applied Mathematics, 1994.

[12] D. Chan and J.S. Pang, "The generalized quasi-variational inequality," *Mathematics of Operations Research*, vol. 7, pp. 211–24, 1982.

[13] S.T. Chung, "*Transmission schemes for frequency selective gaussian interference channels*," Doctoral dissertation, Department of Electrical Engineering, Stanford University, 2003.

[14] S.T. Chung, S.J. Kim, J. Lee, and J.M. Cioffi, "A game-theoretic approach to power allocation in frequency-selective Gaussian interference channels," *Proceeding of the 2003 IEEE International Symposium on Information Theory*, Yokohama, Japan, 2003.

[15] R.W. Cottle, J.S. Pang, and R.E. Stone, *The Linear Complementarity Problem*. Cambridge: Academic Press, 1992.

[16] J.P. Crouzeix, "Pseudomonotone variational inequality problems: existence of solutions," *Mathematical Programming*, vol. 78 pp. 305–14, 1997.

[17] J.M. Danskin, "The theory of min-max with applications," *SIAM Journal on Applied Mathematics*, vol. 14, pp. 641–64, 1966.

[18] B.C. Eaves, "Polymatrix games with joint constraints," *SIAM Journal on Applied Mathematics*, vol. 24, pp. 418–23, 1973.

[19] F. Facchinei, A. Fischer, and V. Piccialli, "Generalized Nash equilibrium problems and Newton methods," *Mathematical Programming, Series B*, vol. 117, pp. 163–94, 2009.

[20] F. Facchinei and C. Kanzow, "Generalized Nash equilibrium problems," *4OR*, vol. 5, pp. 173–210, 2007.

[21] F. Facchinei and J.S. Pang, *Finite-Dimensional Variational Inequalities and Complementarity Problems*. New York: Springer, 2003.

[22] I. Fonseca and W. Gangbo, *Degree Theory in Analysis and Applications*. Oxford: Oxford University Press, 1995.

[23] J. Geanakoplos, "Nash and Walras equilibrium via Brouwer," *Economic Theory*, vol. 21, pp. 585–603, 2003.

[24] R.H. Gohary, Y. Huang, Z.Q. Luo, and J.S. Pang, "A generalized iterative water-filling algorithm for distributed power control in the presence of a jammer," *IEEE Transactions on Signal Processing*, to be published.

[25] A. von Heusinger and Ch. Kanzow, "SC^1 optimization reformulations of the generalized Nash equilibrium problem," *Optimization Methods and Software*, vol. 33, pp. 953–73, 2008.

[26] A. von Heusinger and Ch. Kanzow, "Optimization reformulations of the generalized Nash equilibrium problem using Nikaido-Isoda-type functions," *Computational Optimization and Applications*, to be published.

[27] T. Ichiishi, *Game Theory for Economic Analysis*. New York: Academic Press, 1983.

[28] R. Johari, S. Mannor, and J.N. Tsitsiklis, "Efficiency loss in a network resource allocation game: the case of elastic supply," *IEEE Transactions on Automatic Control*, vol. 50, pp. 1712–24, 2005.

[29] R. Johari and J.N. Tsitsiklis, "Efficiency loss in a network resource allocation game," *Mathematics of Operations Research*, vol. 29, pp. 407–35, 2004.

[30] A. Kesselman, S. Leonardi, and V. Bonifaci, "Game-theoretic analysis of internet switching with selfish users," in *Proceedings of the First International Workshop on Internet and Network Economics, WINE 2005*. Springer Lectures Notes in Computer Science # 3828, 2005, pp. 236–45; 2005.

[31] J.B. Krawczyk, "Coupled constraint Nash equilibria in environmental games," *Resource and Energy Economics*, vol. 27, pp. 157–81, 2005.

[32] J.B. Krawczyk and S. Uryasev, "Relaxation algorithms to find Nash equilibria with economic applications," *Environmental Modeling and Assessment*, vol. 5, pp. 63–73, 2000.

[33] N.G. Lloyd, *Degree Theory*. Cambridge: Cambridge University Press, 1978.

[34] Z.Q. Luo and J.S. Pang, "Analysis of iterative waterfilling algorithm for multiuser power control in digital subscriber lines," in *EURASIP Journal on Applied Signal Processing*, Article ID 24012. 10 pages, 2006.

[35] Z.Q. Luo, J.S. Pang, and D. Ralph, *Mathematical Programs With Equilibrium Constraints*. Cambridge, England: Cambridge University Press, 1996.

[36] T.L. Magnanti and G. Perakis, "Averaging schemes for variational inequalities and systems of equations," *Mathematics of Operations Research*, vol. 22, pp. 568–587, 1997.

[37] T.L. Magnanti and G. Perakis, "Computing fixed points by averaging," in *Transportation and Network Analysis – Current Trends*, P. Marcotte and M. Gendreau, eds. New York: Springer, 2001.

[38] O.L. Mangasarian and S. Fromovitz, "The Fritz John necessary optimality conditions in the presence of equality constraints," *Journal of Mathematical Analysis and Applications*, vol. 17 pp. 34–47, 1967.

[39] J.J. Moré and W.C. Rheinboldt, "On P- and S-functions and related class of n-dimensional nonlinear mappings," *Linear Algebra and its Applications*, vol. 6, pp. 45–68, 1973.

[40] J.F. Nash, "Equilibrium points in n-person games," *Proceedings of the National Academy of Sciences*, vol. 36, pp. 48–9, 1950.

[41] J.F. Nash, "Non-cooperative games," *Annals of Mathematics*, vol. 54, pp. 286—95, 1951.

[42] H. Nikaido and K. Isoda, "Note on noncooperative convex games," *Pacific Journal of Mathematics*, vol. 5, Supplement 1, pp. 807–15, 1955.

[43] J.M. Ortega and W.C. Wheinboldt, *Iterative Solution of Nonlinear Equations in Several Variables*. Philadelphia: SIAM Classics in Applied Mathematics, vol. 30, 2000.

[44] J.S. Pang, "Asymmetric variational inequality problems over product sets: applications and iterative methods," *Mathematical Programming*, vol. 31, pp. 206–19, 1985.

[45] J.S. Pang and D. Chan, "Iterative methods for variational and complementarity problems," *Mathematical Programming*, vol. 24, pp. 284–313, 1982.

[46] J.S. Pang, G. Scutari, F. Facchinei, and C. Wang, "Distributed power allocation with rate constraints in Gaussian frequency-selective channels," *IEEE Transactions on Information Theory*, vol. 54, pp. 3471–89, 2008.

[47] A. Reinoza, "The strong positivity conditions," *Mathematics of Operations Research*, vol. 10, pp. 54–62, 1985.

[48] S.M. Robinson, "Stability theory for systems of inequalities II: differentiable nonlinear systems," *SIAM Journal on Numerical Analysis*, vol. 13, pp. 496–513, 1976.

[49] S.M. Robinson, "Local structure of feasible sets in nonlinear programming. III. Stability and sensitivity," *Mathematical Programming Study*, vol. 30, pp. 45–66, 1987.

[50] R.T. Rockafellar, *Convex Analysis*. Princeton: Princeton University Press, 1970.

[51] J. Rosen, "Existence and uniqueness of equilibrium points for concave n-person games," *Econometrica*, vol. 33, pp. 520–34, 1965.

[52] G. Scutari, D.P. Palomar, and S. Barbarossa, "Asynchronous iterative water-filling for Gaussian frequency-selective interference channels," *IEEE Transactions on Information Theory*, vol. 54, pp. 2868–78, 2008.

[53] G. Scutari, D.P. Palomar, J.S. Pang, and F. Facchinei, "Flexible design for cognitive wireless systems: from game theory to variational inequality theory," *IEEE Signal Processing Magazine*, vol. 26, no. 5, pp. 107–23, 2009.

[54] S. Uryasev and R.Y. Rubinstein, "On relaxation algorithms in computation of noncooperative equilibria," *IEEE Transactions on Automatic Control*, vol. 39, pp. 1263–67, 1994.

[55] N. Yamashita and Z.-Q. Luo, "A nonlinear complementarity approach to multiuser power control for digital subscriber lines," *Optimization Methods and Software*, vol. 19, pp. 633–52, 2004.

[56] W. Yu, G. Ginis, and J.M. Cioffi, "Distributed multiuser power control for digital subscriber lines," *IEEE Journal on Selected Areas in Communications*, vol. 20, pp. 1105–15, 2002.

[57] A. von Heusinger and Ch. Kanzow, "Relaxation methods for generalized Nash equilibrium problems with inexact line search," *Journal of Optimization Theory & Applications*, submitted for publication.

[58] J.S. Jong, A. Scutari, D. Palomar, and F. Facchinei, "Design of cognitive radio systems under temperature–interference constraints: a variational inequality approach," *IEEE Transactions on Signal Processing*, submitted for publication.

Afterword

The past two decades have witnessed the onset of a surge of research in optimization. This includes theoretical aspects, algorithmic developments such as generalizations of interior-point methods to a rich class of convex-optimization problems, and many new engineering applications. The development of general-purpose software tools as well as the insight generated by the underlying theory have contributed to the emergence of convex optimization as a major signal-processing tool; this has made a significant impact on numerous problems previously considered intractable. Given this success of convex optimization, many new applications are continuously flourishing. This book aims at providing the reader with a series of tutorials on a wide variety of convex-optimization applications in signal processing and communications, written by worldwide leading experts, and contributing to the diffusion of these new developments within the signal-processing community. The topics included are automatic code generation for real-time solvers, graphical models for autoregressive processes, gradient-based algorithms for signal-recovery applications, semidefinite programming (SDP) relaxation with worst-case approximation performance, radar waveform design via SDP, blind non-negative source separation for image processing, modern sampling theory, robust broadband beamforming techniques, distributed multiagent optimization for networked systems, cognitive radio systems via game theory, and the variational-inequality approach for Nash-equilibrium solutions.

Index